T0258880

Wavelet Transforms

This textbook serves as an introduction to contemporary aspects of time-frequency analysis encompassing the theories of Fourier transforms, wavelet transforms and their respective offshoots.

This book is the first of its kind totally devoted to the treatment of continuous signals, and it systematically encompasses the theory of Fourier transforms, wavelet transforms, geometrical wavelet transforms and their ramifications. The authors intend to motivate and stimulate interest among mathematicians, computer scientists, engineers and physical, chemical and biological scientists.

This text is written from the ground up with target readers being senior undergraduate and first-year graduate students, and it can serve as a reference for professionals in mathematics, engineering and applied sciences.

Features

- Flexibility in the book's organization enables instructors to select chapters appropriate to courses of different lengths, emphasis and levels of difficulty

- Self-contained, the text provides an impetus to contemporary developments in the signal processing aspects of wavelet theory at the forefront of research

- A large number of worked-out examples is included

- Every major concept is presented with explanations, limitations and subsequent developments, with emphasis on applications in science and engineering

- A wide range of exercises in varying levels from elementary to challenging are incorporated so readers may develop both manipulative skills in the theory of wavelets and deeper insight

- Hints for selected exercises appear at the end

The origin of the theory of wavelet transforms dates back to the 1980s as an outcome of the intriguing efforts of mathematicians, physicists and engineers. Owing to the lucid mathematical framework and versatile applicability, the theory of wavelet transforms is now a nucleus of shared aspirations and ideas.

Firdous A. Shah earned his post-graduate degree in pure mathematics from the University of Kashmir and a PhD in applied mathematics from the Central University, Jamia Millia Islamia, New Delhi, India. He is an associate professor of mathematics at the University of Kashmir, South Campus, India. His research interests include the theory of wavelets, time-frequency analysis, abstract harmonic analysis, sampling theory and applications of wavelets to signal processing and mathematical biology. He is the author/co-author of over 150 journal articles and has co-authored two books with Prof. Lokenath Debnath (University of Texas) on wavelet transforms and their applications published by Birkhäuser, Springer. He has received numerous research grants from NBHM and SERB-DST, Govt. of India. He serves as an editorial board member of several reputed journals and is a lifetime member of several academic societies.

Azhar Y. Tantary is a native of Jammu and Kashmir, India. He completed his BSc and MSc degrees in mathematics from the University of Kashmir during the academic sessions 2010-2012 and 2013-2014, respectively. Presently, he is a researcher in the Department of Mathematics, South Campus, University of Kashmir. His broad field of research is wavelet theory and his research interests include integral transforms, sampling theory and mathematical signal processing. He has authored/co-authored more than 15 research articles in different journals of international repute. He has also presented his research at various national and international forums and has received several awards for excellence in research.

Textbooks in Mathematics

Series editors:
Al Boggess, Kenneth H. Rosen

Wavelet Transforms
Kith and Kin

Firdous A. Shah and Azhar Y. Tantary

CRC Press
Taylor & Francis Group
Boca Raton London New York

CRC Press is an imprint of the
Taylor & Francis Group, an **informa** business

A CHAPMAN & HALL BOOK

First edition published 2023
by CRC Press
6000 Broken Sound Parkway NW, Suite 300, Boca Raton, FL 33487-2742

and by CRC Press
4 Park Square, Milton Park, Abingdon, Oxon, OX14 4RN

CRC Press is an imprint of Taylor & Francis Group, LLC

LCCN: 2022014289

ISBN: 978-1-032-00796-0 (hbk)
ISBN: 978-1-032-34495-9 (pbk)
ISBN: 978-1-003-17576-6 (ebk)

DOI: 10.1201/9781003175766

Typeset in Nimbus
by KnowledgeWorks Global Ltd.

Dedicated to my parents, wife and kids,
To whom I owe my care and time.

-Firdous A. Shah

Dedicated foremost to my grandparents and parents,
Then to all whom I owe.

-Azhar Y. Tantary

Contents

Foreword by Prof. Srivastava

This monumental book presents a rather comprehensive treatment of the state-of-the-art developments on the theory and applications of the wavelet transforms and the various families of transforms that are associated with them. It is based essentially on more than 300 journal articles and other related publications on the subject.

The book begins with comprehensive and lucid coverage of the Fourier transform and its generalizations including the fractional Fourier transform, the families of linear canonical and special affine Fourier transforms, and the recently investigated quadratic-phase Fourier transform. The authors have successfully extended all of these concepts to their windowed counterparts and have paid special attention to the analysis of different classes of window functions, which serve as an antecedent to the concept of wavelets. In fact, a complete section is devoted to a remarkably healthy discussion on the recently introduced notion of the directional windowed Fourier transform. The potentially useful feature of the book is exclusive literature on the fundamental concept of the wavelet transform, which is then developed in order to incorporate the notions of the Stockwell transform, the two-dimensional wavelet transform, the ridgelet transform, the curvelet transform, the ripplet transform, the shearlet transform and the newly adopted bendlet transform. Moreover, with recent research trends both the classical wavelets as well as the other mentioned directional representation systems are elegantly extended to the fractional Fourier and the linear canonical domains. Remarkably, this monograph takes a smooth transition to the notion of wavelet transforms based upon different classes of orthogonal polynomials and special functions. This feature is conspicuously missing in all of the currently-available books and monographs on wavelet transforms.

The entire book is written in an enthusiastic, engaging and reader-friendly manner while being quite resolute in being as self-contained as possible. Moreover, each major concept is elucidated with illustrative examples supported with vibrant graphics, which shall significantly improve the understandability of the concepts.

In conclusion, the work *Wavelet Transforms: Kith and Kin* is undoubtedly a highly recommended state-of-the-art monograph which deals extensively with the notion of continuous wavelet transforms, together with their related recent developments of the Stockwell, ridgelet, curvelet, ripplet, shearlet and bendlet transforms, as well as the other widely scattered developments on the subject.

H. M. Srivastava
Department of Mathematics and Statistics
University of Victoria
Victoria, British Columbia
Canada
harimsri@math.uvic.ca

Foreword by Prof. Jorgensen

What is now referred to as wavelet ideas includes an ever-expanding list of powerful tools. Equally, it involves the active and rich theory of wavelets and a diverse and quickly expanding list of applications. Indeed, the subject has seen recent and exciting advances that continued to profoundly influence new areas in both pure and applied mathematics. While existing books do explore some of these trends, the purpose of the present book is to explore other such new areas, and in keeping with a student-accessible presentation. In particular, this volume aims to serve both as a reference and textbook. Indeed, the target audience is wide, including both senior graduate students and researchers in the field of wavelet theory. The presentation is further aimed at readers from such neighbouring fields as physics and engineering with an emphasis on transform theory as it relates to signal processing.

At the outset, comprehensive coverage of the family of Fourier transforms sets a firm base for the reader. On the road to time-frequency analysis the pedestal of windowed Fourier transform is elegantly developed to the windowed counterparts of the generalized Fourier transforms. Ever interesting is the inclusion of the directionally sensitive variant of the windowed Fourier transform together with an impetus to new directions of research. In the sequel, a detailed account of the concept of wavelet transforms together with the chronological developments adds to its charm. What excites me more is the fact that the book takes the reader from the group-theoretic setting en route to the conventional perspective on the fundamental concepts of wavelet and shearlet transforms. Another distinguishing feature is that each chapter concludes with a broad problem set. For classroom use this is important as the problems serve to help students gain deeper insight to select classroom projects and to develop better problem-solving skills. The overall presentation of all major concepts and the elucidation of minor details are highly applaudable. In addition, all of the important concepts are supported with illustrative examples and wonderful graphics, which pushes up the standard of the book to the next level.

To make a long story short, this book is a futuristic text on the theory of continuous wavelet transforms, together with their erstwhile and subsequent developments. The book is a very complete and in-depth treatment of the subject both from a firm theoretical, as well as intuitive, conceptual point of view. Nevertheless, the book provides skillful insight into a vividly growing research area and opens the reader's mind to a world yet completely unexplored. It encourages the reader to participate in the exciting quest for new horizons of knowledge. The title of the book, *Wavelet Transforms: Kith and Kin*, is quite appropriate as well as catchy patently describing the subject matter.

Palle E.T. Jorgensen
Department of Mathematics
University of Iowa
Iowa City, IA
USA
palle-jorgensen@uiowa.edu

Preface

The book before you is a result of an unwavering effort put into extracting the small pieces of information from the vast assemblage of literature. Writing the book shall be worthwhile if it infuses a love of the subject into young minds, motivates budding researchers and explores new directions of research in the subject. What else could offer a better start to the journey we are about to take than the following stanza from Charles Mackay ("The Old and the New," Voices from the Crowd, and Town Lyrics, 1857):

> The smallest effort is not lost,
> Each wavelet on the ocean tost,
> Aids in the ebb-tide or the flow,
> Each rain-drop makes some floweret blow,
> Each struggle lessens human woe.

The inception of wavelets about three decennaries ago not only opened up a new vista of mathematical analysis, but also revolutionized applied mathematics, computer science and engineering by serving as a potent tool for analyzing and processing univariate functions/signals containing singularities. Indeed, the wavelet theory has been formalized extensively due to the efforts of mathematicians, physicists and engineers in such a way that the subject is now considered a nucleus of shared aspirations and ideas. Wavelet theory has attained a highly distinguished position in harmonic analysis as the application potential is continuously piercing across the boundaries of different subjects. As of now, various directions of research have been established in the theory of wavelets and the subject has paved its way into diverse branches of science and engineering, with the most notable ones being signal and image processing, approximation theory, geophysics, astrophysics, quantum mechanics, computer science, statistics, economics and finance, control theory, differential and integral equations, numerical analysis, physiology, medicine, neural networks, chemistry, nano-technology and more. In the last couple of decades, wavelet theory has witnessed many ramifications in pursuit of geometrically efficient waveforms that could perform equally well while analyzing and processing multivariate functions/signals endowed with edges and other distributed singularities. Indeed, such geometric features are essential in the multivariate setting, since multivariate problems are typically governed by anisotropic phenomena such as singularities concentrated on lower-dimensional embedded manifolds. This led to the development of a new class of waveforms such as steerable wavelets, ridgelets, curvelets, ripplets, shearlets and bendlets. These geometric wavelets are collectively called "X-lets" and are primarily characterized by their anisotropic, multi-scale nature. The anisotropy of the aforementioned waveforms corresponds to their better directional selectivity and hence, are also called directional wavelets.

Owing to the prolificacy of the wavelet concept, a vast collection of lucubrations have been witnessed over the past few decades in the forms of books, journals, special issues and research articles on the subject. Among the multitude of books, many excellent monographs have appeared over the course of time, which treat the subject of wavelets from a wide range of perspectives and a large spectrum of possible applications. Therefore, the fundamental question that arises can be put as, "Why another book on wavelets?" The answer to the posed question and our main motivation for writing this book can be ascertained from the following points:

- Surprisingly, not a single book encapsulating all the aforementioned developments on the subject of wavelets has appeared hitherto. As such, to cater to the needs of students, academicians, practitioners and researchers, it is the need of the hour to have an exclusive treatise concerned with the theory of wavelets from their outset to more recent developments on the subject.

- According to Yevs Meyer, "Continuous wavelets are like a film of reality, where nothing is lost." However, none of the books in the open literature has treated the theory of continuous wavelet transforms from an undivided and wider perspective. As such the present book is the first monograph of its kind, exclusively devoted to exploring the theory of continuous wavelet transforms: kith and kin, both from the mathematical and signal processing perspectives.

- The distinguishing feature of the book is that it takes care of both the Fourier transforms as well as their generalizations, ranging from the fractional Fourier transform to the more recent quadratic-phase Fourier transform proceeding to the formulation of novel wavelet transforms underlying the generalized Fourier transforms.

- The book makes a smooth transition from elementary to advanced developments in the context of wavelet transforms, and the reader shall certainly enjoy the tour from the pedestal of Fourier transforms to windowed Fourier transforms to wavelet transforms, their friends and relatives. Nonetheless, this book differs from the rest in the way that it allows the reader to gain deeper insights regarding the notion of window functions, which are in a sense a precursor to the concept of wavelets. The meticulous analysis of window functions and their analyzing qualities instill a sense of curiosity and prepares the reader for gaining a firm ground prior to the exploration of the fundamental tools of time-frequency analysis.

- Yet another promising feature of this book is that the major concepts in the context of wavelet theory are elegantly described both from the group-theoretic as well as the conventional analytical perspectives. The group-theoretic approach is primarily intended to meet the needs of students and researchers who wish to have deeper insights into the theory of wavelets from the perspective of coherent states, which play a pivotal role in the development of quantum mechanics.

- As the saying goes, "understanding the basics is easy; appreciating the nuances takes years," and therefore, all the minute details in these fundamentally correlated concepts have been carefully identified and stressed upon at their respective places. Moreover, each major concept is elucidated by illustrative examples supported by vibrant graphics, which significantly improve the understanding of the readers.

- The book finds its way to the notion of wavelet transforms based on different classes of orthogonal polynomials and special functions, which is completely missing in all books on wavelet transforms available in the open literature. As such, the subject matter is quite interesting and certainly adds flavour to the treatise.

- On the flip side, each chapter concludes with a stack of problems extracted from the main content, which shall give the reader both a better understanding of the concepts as well as problemsolving skills.

- From the signal processing perspective, a fair exposition to certain applications is also provided in the book. To mention a few, the signal reconstruction is demonstrated via the celebrated sampling theorems, whereas the filtering procedure is unveiled via the notion of multiplicative filtering. Besides for accessing the time-frequency localization characteristics of various integral transforms, the respective Heisenberg's uncertainty principles and the resolutions in the spatial and spectral domains are examined in great detail. Occasionally, applications to certain prominent differential equations, such as the Laplace equation, wave equation and heat equation, are also incorporated into the text.

- The entire book is written in an enthusiastic, engaging and reader-friendly manner while being quite resolute in making the text self-contained.

Because of the above-cited facts, *Wavelet Transforms: Kith and Kin* is certainly a state-of-the-art book dealing with the notion of continuous wavelet transforms, together with their erstwhile and subsequent developments. The title of the book has been carefully selected in consonance to the theme, with the idiom "kith and kin" lucidly conveying that the subject matter includes both the proximate ramifications of the wavelet transforms, such as the Stockwell, ridgelet, curvelet, ripplet, shearlet and bendlet transforms, as well as the other remote developments on the subject. A brief outline of each chapter is given below:

- Chapter 1 deals with the theory of Fourier transforms and the subsequent ramifications, including the notion of fractional Fourier, linear canonical, special affine Fourier, quadratic-phase Fourier and the two-dimensional Fourier transforms. Besides the study of fundamental properties of the Fourier and generalized Fourier transforms, the formulation of the convolution and correlation theories are also carried out in the chapter. Nevertheless, several variants of the well-known Heisenberg's uncertainty principle pertaining to the classical and generalized Fourier transforms are also studied in detail. Special attention is given to illustrating the important concepts via lucid examples. The chapter concludes with an exclusive analysis of the two-dimensional Fourier transform, which is of great significance for the subsequent developments on the subject.

- Chapter 2 is exclusively devoted to an in-depth study of the windowed Fourier transforms. Most importantly, the chapter is reinforced with a detailed survey on the analysis of diverse types of window functions frequently employed in the literature and applications. Besides the fundamental notion of windowed Fourier transform, recent ramifications, including the windowed fractional Fourier transform, windowed linear canonical transform, windowed special affine Fourier transform and windowed quadratic-phase Fourier transform, are also discussed. Nevertheless, the chapter concludes with the more recent notion of directional windowed Fourier transform, which opens up a new research prospect in the context of windowed Fourier transforms.

- The sole aim of Chapter 3 is to provide comprehensive coverage of the theory of wavelet transforms and their kin, including the Stockwell transform, two-dimensional wavelet transform, ridgelet transform, curvelet transform, ripplet transform, shearlet transform and bendlet transform. The fundamental ideas and results involved in the formulation of the respective integral transforms are discussed in detail, with special attention given to the applications from a signal processing perspective. All the major concepts are illustrated with suitable examples followed by illustrative depictions.

- Based on the notion of generalized Fourier transforms, the theory of wavelets has also witnessed many significant lucubrations resulting in the birth of fractional wavelets, linear canonical wavelets and their modifications. Nevertheless, several directional representation systems have also been extended to the generalized Fourier domains. In

Chapter 4, an extensive study of the theory of generalized wavelet transforms is carried out, which includes the formulation of the fractional wavelet transform, fractional Stockwell transform, linear canonical wavelet transform, linear canonical Stockwell transform, linear canonical ridgelet transform, linear canonical curvelet and ripplet transforms, and the linear canonical shearlet transform.

- Chapter 5 is entirely concerned with a detailed study of several new classes of wavelet transforms based on certain well-known orthogonal polynomials and special functions. Among the class of orthogonal polynomials, the chapter focuses on the Laguerre and Legendre polynomials, leading to the notions of the Laguerre and Legendre wavelet transforms. This is followed by the formulation of another couple of wavelet transforms by using the fundamental notions of the Bessel transform and the Dunkl transform, which are based on the well-known class of special functions known as the Bessel and Dunkl functions. Nevertheless, complementary to these developments, another hybrid wavelet transform called the Mehler-Fock wavelet transform is also formulated in the sequel, which relies on a special class of the Legendre functions. The chapter concludes with the investigation of an interesting interface between the classical wavelet transform and the Hartley transform.

Despite the best efforts of everyone involved, some typographical errors doubtlessly remain. We do hope that these are both few and obvious and will cause minimal confusion. Finally, it is important to emphasize that the work presented in this book would not have been possible without the deliberations and interactions with many eminent personalities during the preparation of the treatise. It is our pleasure to express our sincere gratitude to Hari M. Srivastava, Palle E. Jorgensen, Ahmed I. Zayed, Lokenath Debnath, Charles K. Chui, Gitta Kutyniok, Wolfgang Sprößig, Hatem Mejjaoli, Shiv K. Kaushik, Lalit K. Vashisht, Santosh K. Upadhyay, Khalil Ahmad, and Abdullah and Pammy Manchanda. The insightful comments and valuable suggestions received from the above-cited individuals have immensely helped in grooming the text to its present form. We appreciate the comments received from the researchers in our lab including Waseem Zahoor, Aajaz Ahmad, Mohd. Irfan and Naveed Ahmad. Last but not least, our special thanks go to Bob Ross, acquiring editor, Beth Hawkins, editorial assistant, and the staff of CRC Press for their help and cooperation. The patience they exhibited throughout this process is incredible. The editors' faith proved absolutely unprecedented during multiple lengthy stretches when the authors themselves doubted whether the book would ever be published. The first author is deeply indebted to his wife, Dr. Saima Habeeb, for her understanding and tolerance while the book was being written, especially for taking care of the lovely kids, Muhammad Moris and Haya Hazel. The second author is highly obliged for being fortunate enough to have wonderfully supportive parents (Dr. M.Y. Tantary and Parveena) who left no stone unturned in the wondrous upbringing of their children. In addition, the second author would like to extend sincere thanks to Dr. M.Y. Tantary, who not only proved to be a great father but also acted like a good friend and a wonderful mentor. The second author is also delighted to take this opportunity to express his immense love for his grandparents, parents, brothers and endearing sisters, especially the youngest, Noorain Fatima. Lastly, the second author feels immensely indebted to his esteemed PhD supervisor Dr. Firdous A. Shah, who proved to be an inspiration and a reliable shoulder to lean on.

Firdous A. Shah
Azhar Y. Tantary   **Dated: September 01, 2022**

List of Figures

.

1

The Fourier Transforms

"The deep study of nature is the most fruitful source of mathematical discoveries. By offering to research a definite end, this study has the advantage of excluding vague questions and useless calculations; besides it is a sure means of forming analysis itself and of discovering the elements which it most concerns us to know, and which natural science ought always to conserve."

-Joseph Fourier

1.1 Introduction

The French scientist Joseph Fourier (1768–1830) was the first to introduce the remarkable idea of expansion of a function in terms of trigonometric series [112]. However, Fourier had not paid much heed to the underlying rigorous mathematical analysis. The reason being elegantly delineated by the American mathematician Edward V. Vleck (1863–1943) as "Fourier uses his mathematics with the delightful freedom and naivete of the physicist or astronomer who trusts in a mathematical providence." The integral formulas for the coefficients of the Fourier expansion were already known to Leonardo Euler (1707–1783) and others. In fact, Fourier developed his new idea for finding the solution of heat (or Fourier) equation in terms of Fourier series so that the Fourier series can be used as a practical tool for determining the Fourier series solution of partial differential equations under prescribed boundary conditions. The Fourier series of a function $f(t)$ defined on the interval $(-L, L)$ is given by

$$f(t) = \sum_{n \in \mathbb{Z}} c_n\, e^{in\pi t/L}, \tag{1.1}$$

where the Fourier coefficients are

$$c_n = \frac{1}{2L} \int_{-L}^{L} f(t)\, e^{-in\pi t/L} dt. \tag{1.2}$$

In order to obtain a representation for a general, non-periodic function defined for all real t, it seems desirable to take limit as $L \to \infty$, that leads to the formulation of the famous Fourier integral formula:

$$f(t) = \frac{1}{\sqrt{2\pi}} \int_{\mathbb{R}} \left\{ \frac{1}{\sqrt{2\pi}} \int_{\mathbb{R}} f(z)\, e^{-i\omega z}\, dz \right\} e^{i\omega t} d\omega. \tag{1.3}$$

The integral expression in the brackets of (1.3) is often referred as the Fourier integral or the Fourier transform of $f(t)$ and is usually denoted by $\hat{f}(\omega)$. That is,

$$\hat{f}(\omega) = \frac{1}{\sqrt{2\pi}} \int_{\mathbb{R}} f(z)\, e^{-i\omega z} dz. \tag{1.4}$$

DOI: 10.1201/9781003175766-1

The integral transformation (1.4) is also alternatively abbreviated as $\mathscr{F}[f]$, where $\mathscr{F}$ is the Fourier transform operator. Physically, the Fourier transform (1.4) can be interpreted as an integral superposition of an infinite number of sinusoidal oscillations with different wave-number ω (or different wavelengths $\lambda = 2\pi/\omega$). In lieu to this, the Fourier series (1.1) represents the resolution of a given function into an infinite but discrete set of harmonic components. As of now, the theory of Fourier transforms has evolved into a widely recognized discipline of harmonic analysis and has been successfully applied in diverse scientific and engineering pursuits [48, 51, 57, 151, 217, 219].

Despite the success and profound impact of Fourier transforms, the theory is being continuously revisited to cater to the needs of scientific and engineering communities, which has resulted in the advent of fractional Fourier transform, linear canonical transform, special affine Fourier transform and quadratic-phase Fourier transform. One of the remarkable features of these phase-space transforms is that they all generalize the classical Fourier transform and thus all the mathematical developments are compatible with the Fourier analysis. Unlike the classical Fourier transform, the generalized phase-space transforms are endowed with certain extra degrees of freedom, which are of substantial importance in handling diverse problems arising in different branches of science and engineering. For instance, in the context of signal processing, the generalized Fourier transforms are best matched for the class of chirp-like signals, which reveal their characteristics along non-orthogonal directions in the time-frequency plane. The prolificacy of the generalized Fourier transforms has given birth to many new theories beyond the Fourier domain, including convolution theory, sampling theory, filtering theory, time-frequency representations and so on.

This chapter deals with the theory of Fourier transforms and the subsequent ramifications within the realm of the well-known spaces of square integrable functions, formally called as the "L^2-spaces." The *raison dêtere* for choosing the dominion as the spaces of square integrable functions is that the Fourier and generalized Fourier operators enjoy several pleasant L^2-properties. Besides studying all fundamental properties of the Fourier and generalized Fourier transforms, we also formulate the important results such as the Parseval's relations, inversion formulae and also elucidate the notions of convolution, correlation and so on. A sound discourse on the Heisenberg's uncertainty principles associated with the classical and generalized Fourier transforms is also presented. Special attention is given to illustrate the important concepts via lucid examples supported with interpretive graphics. Nevertheless, the salient feature of the chapter is the critical analysis on the fundamental signal processing aspects of sampling and multiplicative filtering. The chapter is culminated with a thorough discussion on the outgrowth of the classical Fourier transform to the two-dimensional setting, encapsulating the notion of polar Fourier transform, which is to be invoked in the subsequent developments on the subject.

1.2   The Fourier Transform

"The sine and cosine series, by which one can represent an arbitrary function in a given interval, enjoy among other remarkable properties that of being convergent. This property did not escape the great geometer [Fourier] who began, through the introduction of the representation of functions just mentioned, a new class for the applications of analysis."

-Peter G. Dirichlet

We begin by introducing some notation that will be used throughout this monograph. The set of natural numbers (positive integers) is denoted by $\mathbb{N}$ and the set of integers by $\mathbb{Z}$. The fields of real and complex numbers are denoted by $\mathbb{R}$ and $\mathbb{C}$, respectively. Elements of the fields $\mathbb{R}$ and $\mathbb{C}$ are called as "scalars." The function spaces that we shall be dealing with are the well-known L^p-spaces, $1 \leq p < \infty$ consisting of all the complex-valued Lebesgue integrable functions f defined on $\mathbb{R}^n$, $n \in \mathbb{N}$ and satisfying

$$\left\| f \right\|_p = \left\{ \int_{\mathbb{R}^n} |f(\mathbf{t})|^p d\mathbf{t} \right\}^{1/p} < \infty, \quad \mathbf{t} \in \mathbb{R}^n. \tag{1.5}$$

For the one-dimensional Euclidean space $\mathbb{R}$, the corresponding variable is simply denoted as t. In order to emphasize the domain $\mathbb{R}^n$, these function spaces are denoted as $L^p(\mathbb{R}^n)$ and the number $\left\| f \right\|_p$ is called as the "L^p-norm." These classes of functions turn out to be Banach spaces, since the Cauchy sequences of functions in L^p converge to a limit function belonging to L^p. Although, we do not require any rigorous knowledge of the Banach spaces for an understanding of the theory of wavelets in this introductory book, however, the reader ought to know some elementary properties of the L^p-spaces. The L^p-spaces for the cases $p = 1$, $p = 2$, $0 < p < 1$, and $1 < p < \infty$ are different in structure, importance and technique, and these spaces play a very special role in many mathematical investigations. The case $p = 2$ is very special, because the L^2-space constitutes a Hilbert space; that is, there is an inner product defined on the square integrable functions, which extends the idea of the usual vector space dot product in a lucid and insightful fashion. Physically, the L^2-space is particularly useful for analyzing signals, as the L^2-norm of any signal is often interpreted as its energy. A signal, for instance a sound signal, can be viewed as a one-dimensional function $f(t)$ that indicates the intensity of the sound at time t. Typical examples of higher-dimensional signals include the image and video signals.

In the rest of this section, we shall study the notion of Fourier transform in the context of the Hilbert space $L^2(\mathbb{R})$. To begin with, we note that for $1 < p < \infty$, the L^p-spaces do not descend as p increases. For example, if $U(t)$ denotes the unit step function on $\mathbb{R}$, then $f(t) = \frac{1}{t} U(t - 1) \in L^2(\mathbb{R})$ but does not belong to $L^1(\mathbb{R})$. Hence, even though a function $f \in L^2(\mathbb{R})$, the Fourier integral (1.4) need not exist; therefore, it cannot be used directly for defining the Fourier transform in $L^2(\mathbb{R})$. To circumvent this difficulty, we adopt a limiting approach for defining the Fourier transform in $L^2(\mathbb{R})$. We choose the dense subspace $L^1(\mathbb{R}) \cap L^2(\mathbb{R})$ of $L^2(\mathbb{R})$ containing all the continuous functions with compact support and then invoke the density arguments to extend the Fourier transform to the Hilbert space $L^2(\mathbb{R})$. Therefore, for any $f \in L^2(\mathbb{R})$, the Fourier integral (1.4) makes sense as

$$\hat{f}(\omega) = \lim_{n \to \infty} f_n := \frac{1}{\sqrt{2\pi}} \left\{ \lim_{n \to \infty} \int_{-n}^{n} f(t) \, e^{-i\omega t} dt \right\}, \tag{1.6}$$

where the limit is taken with respect to the L^2-norm and $\{ f_n \}_{n=1}^{\infty}$ is a sequence of continuous functions with compact support, such that $f_n \to f$ as $n \to \infty$. Moreover, it can be easily verified that the above limit exists and is independent of a particular sequence approximating f. For further useful details regarding the extension of the Fourier integral (1.4) from $L^1(\mathbb{R})$ to $L^2(\mathbb{R})$, we refer the interested reader to the monographs [94, 95, 51]. Here, it is imperative to mention that, in the rest of this monograph, we shall not be concerned with the aforementioned nuances and explore the subject in the context of L^2-spaces.

To begin with, we take a tour of the notion of the Fourier transform and study the fundamental properties such as the Parseval's relation and inversion formula. Several physical aspects of the Fourier transform, particularly with reference to signal processing, are illustrated via a bunch of examples. Moreover, the concepts of convolution and correlation in the Fourier domain are also examined in great detail. Nevertheless, the celebrated Shannon's sampling theorem and the well-known Heisenberg's uncertainty principle are also incorporated to add flavour to the text.

1.2.1   Definition and examples

This subsection is exclusively devoted for an in-depth exposition of the notion of Fourier transform. We begin with the formal definition and then proceed to a cluster of illustrative examples.

Definition 1.2.1. The Fourier transform of any function $f \in L^2(\mathbb{R})$ is denoted by $\mathscr{F}[f]$ or $\hat{f}$ and is defined as

$$\mathscr{F}[f](\omega) = \hat{f}(\omega) := \frac{1}{\sqrt{2\pi}} \int_{\mathbb{R}} f(t) \, e^{-i\omega t} \, dt. \tag{1.7}$$

Definition 1.2.1 allows us to make the following comments regarding the notion of the Fourier transform:

(i) The Fourier transform $\hat{f}(\omega)$ measures the frequency component of a signal $f(t)$ that vibrates with frequency ω and $\hat{f}(\omega)$ is called the "frequency spectrum" of the signal.

(ii) In literature, certain tangible differences in the definition of Fourier transform can also be witnessed. For instance, the normalization factor of $1/\sqrt{2\pi}$ in Definition 1.2.1 gives a symmetric appearance to the Fourier transform and the corresponding inversion formula; however, it can be omitted by attaching a factor of $1/2\pi$ with the inversion formula. Nevertheless, some authors make use of the substitution $\sigma = \omega/2\pi$ in (1.7) so that the integral is free from any external factors. That is,

$$\hat{f}(\sigma) = \int_{\mathbb{R}} f(t) \, e^{-2\pi i \sigma t} \, dt. \tag{1.8}$$

The expression (1.8) does not use the angular frequency ω, but rather the frequency σ. This effectively makes the formulas look symmetrical, though it saddles us with many 2π-factors in the exponent.

(iii) The Fourier transform (1.7) corresponding to any function $f \in L^2(\mathbb{R})$ is defined almost everywhere on $\mathbb{R}$ and $\hat{f} \in L^2(\mathbb{R})$.

(iv) Another form of the Fourier transform (1.7) used frequently in probability theory replaces the kernel $e^{-i\omega t}$ with its conjugate $e^{i\omega t}$. In this case, if f is the probability density function of the random variable x, then

$$h(x) = \frac{1}{\sqrt{2\pi}} \int_{\mathbb{R}} f(t) \, e^{ixt} \, dt, \tag{1.9}$$

is called the "characteristic function" of f.

(v) The Fourier transform (1.7) is complex-valued, in general, therefore, it can be expressed in the polar form as

$$\hat{f}(\omega) = \mathrm{Real}\big(\hat{f}(\omega)\big) + i \, \mathrm{Imaginary}\big(\hat{f}(\omega)\big) = A(\omega) \, e^{i\theta(\omega)}, \tag{1.10}$$

where $A(\omega) = \big|\hat{f}(\omega)\big|$ is called the "amplitude spectrum" of the signal $f(t)$, and $\theta(\omega) = \arg\big\{\hat{f}(\omega)\big\}$ is called the "phase spectrum" of $f(t)$.

(vi) It is customary to define the time-frequency plane such that time and frequency are mutually orthogonal coordinates. Representations of a signal along these orthogonal domains are related by the Fourier transform.

"A good stack of examples, as large as possible, is indispensable for a thorough understanding of any concept, and when I want to learn something new, I make it my first job to build one."

-Paul Halmos

The following stack of examples leads to elucidation of the concept of Fourier transform given in (1.7).

Example 1.2.1. (Rectangular Pulse): The rectangular pulse is an even function and is defined by

$$f(t) = \begin{cases} 1, & -\frac{T}{2} \leq t \leq \frac{T}{2}, \quad T > 0 \\ 0, & \text{otherwise.} \end{cases} \tag{1.11}$$

In order to compute the Fourier transform of the rectangular pulse (1.11), we proceed as follows:

$$\hat{f}(\omega) = \frac{1}{\sqrt{2\pi}} \int_{-T/2}^{T/2} e^{-i\omega t} \, dt$$

$$= \frac{1}{\sqrt{2\pi}} \left(\frac{e^{-i\omega t}}{-i\omega} \right)_{-T/2}^{T/2}$$

$$= \frac{1}{\sqrt{2\pi}} \left(\frac{1}{-i\omega} \right) \left(e^{-\frac{i\omega T}{2}} - e^{-\frac{i\omega(-T)}{2}} \right)$$

$$= \frac{1}{\sqrt{2\pi}} \left(\frac{2}{2i\omega} \right) \left(e^{\frac{i\omega T}{2}} - e^{-\frac{i\omega T}{2}} \right)$$

$$= \frac{1}{\sqrt{2\pi}} \left(\frac{2}{\omega} \right) \left(\frac{e^{\frac{i\omega T}{2}} - e^{-\frac{i\omega T}{2}}}{2i} \right)$$

$$= \frac{1}{\sqrt{2\pi}} \left(\frac{2}{\omega} \right) \sin \left(\frac{\omega T}{2} \right)$$

$$= \frac{T}{\sqrt{2\pi}} \frac{\sin \left(\frac{\omega T}{2} \right)}{\left(\frac{\omega T}{2} \right)}$$

$$= \frac{T}{\sqrt{2\pi}} \operatorname{sinc} \left(\frac{\omega T}{2} \right). \tag{1.12}$$

For $T = 1$, the rectangular pulse f and the corresponding Fourier transform $\hat{f}$ are plotted in Figure 1.1. As already pointed out that the Fourier transform $\hat{f}(\omega)$ measures the frequency content of f that vibrates at ω. Since the function given by (1.11) is constant over its domain of influence, therefore the Fourier transform is expected to be concentrated about zero, because a constant function vibrates at zero frequency. This attributive property of the Fourier transform is intuitively clear in Figure 1.1, wherein the larger values of $\hat{f}(\omega)$ occur when ω is near zero.

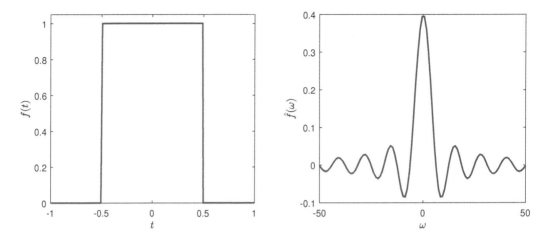

FIGURE 1.1: The rectangular pulse and its Fourier transform.

Nevertheless, the Fourier transform of the rectangular function (1.11) could also be computed in a relatively compact form by using the fact that the rectangular pulse is an even function. Note that

$$\hat{f}(\omega) = \frac{1}{\sqrt{2\pi}} \int_{\mathbb{R}} \left(f(t) \cos \omega t + i \, f(t) \sin \omega t \right) dt. \tag{1.13}$$

The second integrand in (1.13) being an odd function has integral equal to zero over the whole real line. Thus, we have

$$\begin{aligned}
\hat{f}(\omega) &= \frac{1}{\sqrt{2\pi}} \int_{\mathbb{R}} f(t) \cos \omega t \, dt \\
&= \frac{2}{\sqrt{2\pi}} \int_{0}^{T/2} f(t) \cos \omega t \, dt \\
&= \frac{2}{\sqrt{2\pi}} \left(\frac{\sin \left(\frac{\omega T}{2} \right)}{\omega} \right) \\
&= \frac{T}{\sqrt{2\pi}} \left(\frac{\sin \left(\frac{\omega T}{2} \right)}{\left(\frac{\omega T}{2} \right)} \right) \\
&= \frac{T}{\sqrt{2\pi}} \, \text{sinc} \left(\frac{\omega T}{2} \right).
\end{aligned}$$

Example 1.2.2. (Triangular Function): Consider the triangular function

$$f(t) = \begin{cases} t + \dfrac{T}{2}, & -\dfrac{T}{2} \le t \le 0 \\ \dfrac{T}{2} - t, & 0 \le t \le \dfrac{T}{2} \\ 0, & \text{otherwise,} \end{cases} \qquad T > 0. \tag{1.14}$$

The triangular function is clearly an even function over $[-T/2, T/2]$. Therefore, the Fourier transform is given by

$$\hat{f}(\omega) = \frac{1}{\sqrt{2\pi}} \int_{\mathbb{R}} f(t) \cos \omega t \, dt$$

$$= \frac{2}{\sqrt{2\pi}} \int_0^{T/2} \left(\frac{T}{2} - t \right) \cos \omega t \, dt$$

$$= \frac{2}{\sqrt{2\pi}} \left\{ \int_0^{T/2} \frac{T}{2} \cos \omega t \, dt - \int_0^{T/2} t \cos \omega t \, dt \right\}$$

$$= \frac{2}{\sqrt{2\pi}} \left\{ \frac{T}{2} \left(\frac{\sin \omega t}{\omega} \right)_0^{T/2} - \left(\frac{t \sin \omega t}{\omega} \right)_0^{T/2} + \int_0^{T/2} \frac{\sin \omega t}{\omega} \, dt \right\}$$

$$= \frac{2}{\sqrt{2\pi}} \left\{ \frac{T}{2} \left(\frac{\sin \omega T/2}{T/2} \right) - \frac{T}{2} \left(\frac{\sin \omega T/2}{\omega} \right) + \frac{1}{\omega} \left(\frac{\cos \omega t}{\omega} \right)_0^{T/2} \right\}$$

$$= \frac{2}{\sqrt{2\pi}} \left\{ \frac{T^2}{4} \operatorname{sinc} \left(\frac{\omega T}{2} \right) - \frac{T^2}{4} \operatorname{sinc} \left(\frac{\omega T}{2} \right) - \frac{1}{\omega^2} \left(\cos \left(\frac{\omega T}{2} \right) - 1 \right) \right\}$$

$$= \frac{2}{\sqrt{2\pi}} \left(\frac{1 - \cos \left(\frac{\omega T}{2} \right)}{\omega^2} \right)$$

$$= \frac{2}{\sqrt{2\pi}} \frac{2 \sin^2 \frac{\omega T}{4}}{\omega^2}$$

$$= \frac{2}{\sqrt{2\pi}} \frac{2 \sin^2 \frac{\omega T}{4}}{\left(\frac{\omega^2 T^2}{4^2} \right) \left(\frac{4^2}{T^2} \right)}$$

$$= \frac{4}{\sqrt{2\pi}} \left(\frac{\sin \frac{\omega T}{4}}{\frac{\omega T}{4}} \right)^2 \frac{T^2}{16}$$

$$= \frac{1}{\sqrt{2\pi}} \left(\frac{T}{2} \right)^2 \left(\operatorname{sinc} \frac{\omega T}{4} \right)^2. \tag{1.15}$$

Note that the Fourier transform of (1.14) decays at the rate $1/\omega^2$ as $\omega \to \infty$, which is faster than the decay of the Fourier transform of rectangle pulse (1.11). This faster decay is due to the continuity of the triangular function. For $T = 1$, the triangular function (1.14) and the Fourier transform (1.15) are plotted in Figure 1.2.

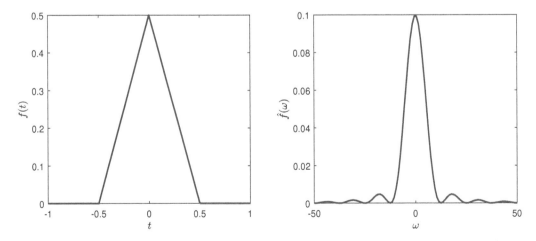

FIGURE 1.2: The triangular function and its Fourier transform.

Example 1.2.3. Consider the following function:

$$f(t) = \begin{cases} \cos(3t), & -\pi \le t \le \pi \\ 0, & \text{otherwise.} \end{cases} \tag{1.16}$$

Clearly, f is an even function; therefore, only the cosine part contributes to the Fourier transform. Thus,

$$\begin{aligned} \hat{f}(\omega) &= \frac{1}{\sqrt{2\pi}} \int_{\mathbb{R}} f(t) \cos(\omega t)\, dt \\ &= \frac{1}{\sqrt{2\pi}} \int_{\mathbb{R}} \cos(3t) \cos(\omega t)\, dt \\ &= \frac{1}{\sqrt{2\pi}} \int_{\mathbb{R}} \left\{ \frac{\cos\left((3+\omega)t\right) - \cos\left((3-\omega)t\right)}{2} \right\} dt \\ &= \frac{1}{\sqrt{2\pi}} \left\{ \frac{\sin\left((3+\omega)\pi\right)}{(3+\omega)} - \frac{\sin\left((3-\omega)\pi\right)}{(3-\omega)} \right\} dt \\ &= \frac{2\omega \sin(\omega\pi)}{\sqrt{2\pi}\,(9-\omega^2)}. \end{aligned} \tag{1.17}$$

The function (1.16) and the corresponding Fourier transform (1.17) are plotted in Figure 1.3. It is quite evident that the Fourier transform has spikes at $\omega = \pm 3$, which is due to the fact that the function (1.16) vibrates with a frequency of 3 over the interval $[-\pi, \pi]$.

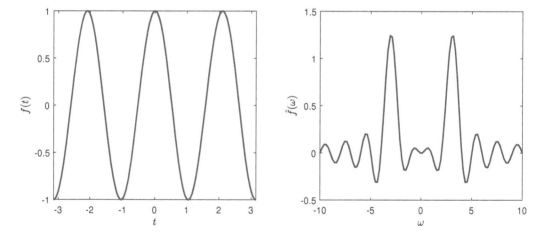

FIGURE 1.3: The truncated cosine-wave and its Fourier transform.

Example 1.2.4. Consider the following function:

$$f(t) = \begin{cases} \sin(3t), & -\pi \le t \le \pi \\ 0, & \text{otherwise.} \end{cases} \tag{1.18}$$

Clearly, f is an odd function; therefore, only the sine part contributes to the Fourier transform. Thus, we have

$$\hat{f}(\omega) = \frac{i}{\sqrt{2\pi}} \int_{\mathbb{R}} f(t) \sin(\omega t)\, dt = -\frac{3\sqrt{2}\,i\,\sin(\omega\pi)}{\sqrt{\pi}\,(9-\omega^2)}. \tag{1.19}$$

The function (1.18) and the imaginary part of the corresponding Fourier transform (1.19) are plotted in Figure 1.4.

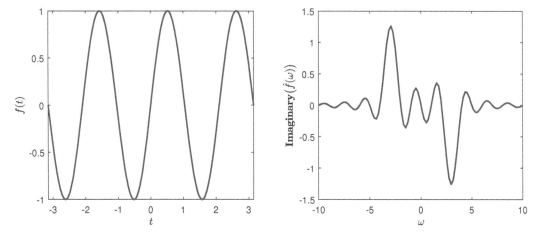

FIGURE 1.4: The truncated sine-wave and its Fourier transform.

Remark 1.2.1. It is worth noticing that the Fourier transform of the waveforms (1.14), (1.16) and (1.18) decays faster than that of the rectangular pulse (1.11). This faster decay is due to the continuity of triangular, sine and cosine waveforms.

Example 1.2.5. (The Gaussian Function): The Gaussian function is one of the most important functions in probability theory and signal analysis. In fact, the Gaussian function is the pedestal for the well-known Gabor and Stockwell transforms. Consider the following Gaussian function:

$$f(t) = \frac{1}{\sigma\sqrt{2\pi}} e^{-t^2/2\sigma^2}, \quad \sigma > 0, \tag{1.20}$$

which is normalized with respect to L^1-norm. The Fourier transform of (1.20) is computed as

$$
\begin{aligned}
\hat{f}(\omega) &= \frac{1}{\sqrt{2\pi}} \left\{ \frac{1}{\sigma\sqrt{2\pi}} \int_{\mathbb{R}} e^{-t^2/2\sigma^2} e^{-i\omega t}\, dt \right\} \\
&= \frac{1}{\sqrt{2\pi}} \left\{ \frac{1}{\sigma\sqrt{2\pi}} \int_{\mathbb{R}} e^{(-1/2\sigma^2)t^2 + (-i\omega)t}\, dt \right\}.
\end{aligned}
\tag{1.21}
$$

Application of the standard Gaussian integral

$$\int_{\mathbb{R}} e^{-at^2 + bt}\, dt = \sqrt{\frac{\pi}{a}}\, e^{b^2/4a}$$

simplifies the task of computation of the Fourier transform to a significant extent as

$$
\begin{aligned}
\hat{f}(\omega) &= \frac{1}{\sqrt{2\pi}} \left\{ \frac{1}{\sigma\sqrt{2\pi}} \left(\frac{\sqrt{\pi}\, e^{(-i\omega)^2/(4(1/2\sigma^2))}}{\sqrt{1/2\sigma^2}} \right) \right\} \\
&= \frac{1}{\sigma\sqrt{2\pi}} \left(\sigma\sqrt{2\pi}\, e^{2\sigma^2(-\omega^2/4)} \right) \frac{1}{\sqrt{2\pi}} \\
&= \frac{1}{\sqrt{2\pi}} e^{-\sigma^2\omega^2/2}.
\end{aligned}
\tag{1.22}
$$

It is worth noticing that the Fourier transform of the Gaussian function is again a Gaussian function; however, (1.22) is not a normalized Gaussian function under L^1-norm. For $\sigma = 1$, the Gaussian function (1.20) and its Fourier transform (1.22) are plotted in Figure 1.5.

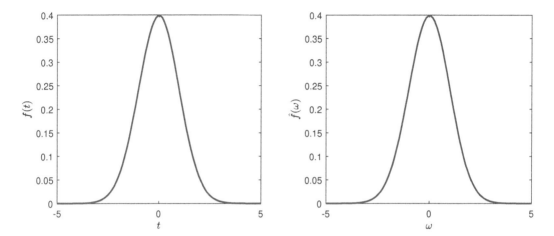

FIGURE 1.5: The L^1-normalized Gaussian function and its Fourier transform.

On the other hand, if we consider the L^2-normalized Gaussian function given by the following expression:

$$f(t) = \frac{1}{\pi^{1/4}\sqrt{\sigma}}\, e^{-t^2/2\sigma^2}, \tag{1.23}$$

then the corresponding Fourier transform can be obtained as

$$
\begin{aligned}
\hat{f}(\omega) &= \frac{1}{\pi^{1/4}\sqrt{\sigma}} \left\{ \frac{1}{\sqrt{2\pi}} \int_{\mathbb{R}} e^{(-1/2\sigma^2)t^2 + (-i\omega)t}\, dt \right\} \\
&= \frac{1}{\pi^{1/4}\sqrt{\sigma}} \left\{ \frac{\pi}{\sqrt{1/2\sigma^2}}\, e^{-\omega^2\sigma^2/2}\, \frac{1}{\sqrt{2\pi}} \right\} \\
&= \frac{1}{\pi^{1/4}\sqrt{\sigma}} \left(\frac{\sigma\sqrt{2\pi}}{\sqrt{2\pi}}\, e^{-\omega^2\sigma^2/2} \right) = \frac{\sqrt{\sigma}}{\pi^{1/4}}\, e^{-\omega^2\sigma^2/2},
\end{aligned}
\tag{1.24}
$$

which is again a Gaussian function. Moreover, it is interesting to note that the Fourier transform (1.24) is indeed normalized with respect to the L^2-norm as

$$\left\| \hat{f} \right\|_2^2 = \int_{\mathbb{R}} \left| \frac{\sqrt{\sigma}}{\pi^{1/4}}\, e^{-\omega^2\sigma^2/2} \right|^2 d\omega = \frac{\sigma}{\sqrt{\pi}} \int_{\mathbb{R}} e^{-\omega^2\sigma^2}\, d\omega = \frac{\sigma}{\sqrt{\pi}} \left(\frac{\sqrt{\pi}}{\sqrt{\sigma^2}} \right) = 1. \tag{1.25}$$

Hence, we conclude that in any case the Fourier transform of a Gaussian function is again a Gaussian function. Also, if the Gaussian function $f(t)$ is normalized with respect to L^2-norm, then $\hat{f}$ is also normalized. Moreover, we note that in both the cases, the parameter σ can be used to control the width of the Gaussian pulse and the corresponding Fourier transform. In fact, the slimmer the Gaussian, the wider the corresponding Fourier transform. For $\sigma = 1$, the L^2-normalized Gaussian function (1.23) and the Fourier transform (1.24) are plotted in Figure 1.6.

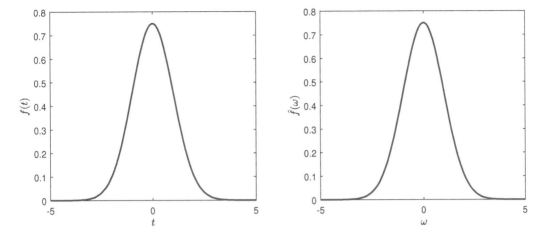

FIGURE 1.6: The L^2-normalized Gaussian function and its Fourier transform.

Remark 1.2.2. The factor $1/2$ appearing in the exponent of (1.20) and (1.23) is handy in applications as the following is true for this representation:

$$\sigma = \sqrt{2 \ln 2} \times \text{HWHM (Half width at half maximum)} = 1.177 \times \text{HWHM}$$

Theoretically, the factor $1/2$ appearing in the exponent could have been merged with σ.

Example 1.2.6. (Unilateral Exponential Function): Consider the unilateral exponential function

$$f(t) = \begin{cases} e^{-\lambda t}, & t \geq 0, \ \lambda > 0 \\ 0, & \text{otherwise.} \end{cases} \tag{1.26}$$

The Fourier transform of (1.26) is given by

$$\begin{aligned} \hat{f}(\omega) &= \frac{1}{\sqrt{2\pi}} \int_0^\infty e^{-\lambda t} e^{-i\omega t}\, dt \\ &= \frac{1}{\sqrt{2\pi}} \int_0^\infty e^{-\lambda t} \big(\cos(\omega t) - i\, \sin(\omega t)\big)\, dt \\ &= \frac{1}{\sqrt{2\pi}} \left\{ \int_0^\infty e^{-\lambda t} \cos(\omega t)\, dt - i \int_0^\infty e^{-\lambda t} \sin(\omega t)\, dt \right\} \\ &= \frac{1}{\sqrt{2\pi}} \left(\frac{\lambda}{\lambda^2 + \omega^2} + \frac{-i\omega}{\lambda^2 + \omega^2} \right). \end{aligned} \tag{1.27}$$

Evidently, the Fourier transform (1.27) is complex-valued, as $f(t)$ is neither even nor odd. For $\lambda = 1$, the function (1.26), its amplitude spectrum and the real and imaginary parts of the Fourier transform (1.27) are jointly plotted in Figure 1.7. From Figure 1.7, we infer that the real part has a Lorentzain shape whereas the imaginary part has a dispersion shape. Moreover, the importance of the unilateral exponential function can be visualized from the fact that the damped oscillation used to describe the emission of a particle, (for example, a proton) from an excited nuclear state with a life time of k (meaning, that the excited state depopulates according to $e^{-t/k}$) results in a Lorentzian-shaped emission line. The exponential relaxation process results in Lorentzian-shaped spectral lines as is the case with the nuclear magnetic resonance.

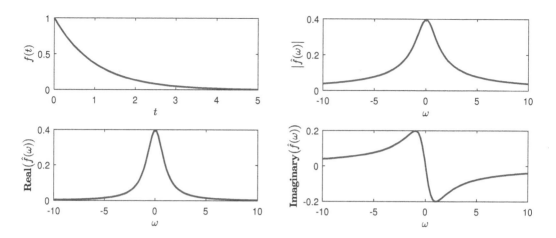

FIGURE 1.7: The unilateral exponential function and its Fourier transform.

Example 1.2.7. (Bilateral Exponential Function): Consider the bilateral exponential function

$$f(t) = e^{-|t|/\lambda}, \quad \lambda > 0. \tag{1.28}$$

As $f(t)$ is an even function, therefore, the corresponding Fourier transform is given by

$$\hat{f}(\omega) = \frac{1}{\sqrt{2\pi}} \int_{\mathbb{R}} e^{-|t|/\lambda} e^{-i\omega t} \, dt$$

$$= \frac{2}{\sqrt{2\pi}} \int_0^{\infty} e^{-t/\lambda} \cos(\omega t) \, dt = \frac{2}{\sqrt{2\pi}} \left(\frac{\lambda}{1 + \omega^2 \lambda^2} \right). \tag{1.29}$$

Note that, the bilateral exponential is even; therefore, the corresponding Fourier transform is real-valued. For $\lambda = 1$, the function (1.28) and its Fourier transform (1.29) are plotted in Figure 1.8. From Figure 1.8, we observe that the Fourier transform is of Lorentzian shape.

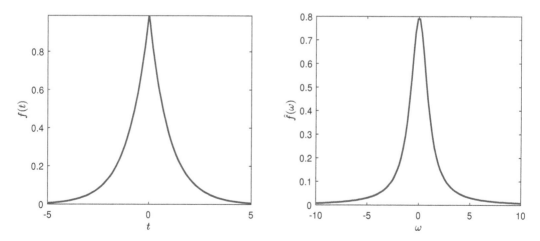

FIGURE 1.8: The bilateral exponential function and its Fourier transform.

Example 1.2.8. (Cosine Function): The cosine function is defined by

$$f(t) = \begin{cases} \cos\left(\dfrac{\pi t}{T}\right), & -\dfrac{T}{2} \le t \le \dfrac{T}{2}, \quad T > 0 \\ 0, & \text{otherwise.} \end{cases} \tag{1.30}$$

The Fourier transform of (1.30) can be computed as follows:

$$\hat{f}(\omega) = \frac{1}{\sqrt{2\pi}} \int_{-T/2}^{T/2} \cos\left(\frac{\pi t}{T}\right) e^{-i\omega t}\, dt$$

$$= \frac{1}{\sqrt{2\pi}} \int_{-T/2}^{T/2} \left(\frac{e^{i\pi t/T} + e^{-i\pi t/T}}{2}\right) e^{-i\omega t}\, dt$$

$$= \frac{1}{2\sqrt{2\pi}} \int_{-T/2}^{T/2} \left(e^{i\left(\frac{\pi}{T}-\omega\right)t} + e^{-i\left(\frac{\pi}{T}+\omega\right)t}\right) dt$$

$$= \frac{1}{2\sqrt{2\pi}} \left\{ \left(\frac{e^{i\left(\frac{\pi}{T}-\omega\right)t}}{i\left(\frac{\pi}{T}-\omega\right)}\right)_{-T/2}^{T/2} + \left(\frac{e^{-i\left(\frac{\pi}{T}+\omega\right)t}}{-i\left(\frac{\pi}{T}+\omega\right)}\right)_{-T/2}^{T/2} \right\}$$

$$= \frac{1}{2\sqrt{2\pi}} \left\{ \left(\frac{e^{i\left(\frac{\pi}{T}-\omega\right)\left(\frac{T}{2}\right)} - e^{-i\left(\frac{\pi}{T}-\omega\right)\left(\frac{T}{2}\right)}}{i\left(\frac{\pi}{T}-\omega\right)}\right) + \left(\frac{e^{-i\left(\frac{\pi}{T}+\omega\right)\left(\frac{T}{2}\right)} - e^{i\left(\frac{\pi}{T}+\omega\right)\left(\frac{T}{2}\right)}}{i\left(\frac{\pi}{T}+\omega\right)}\right) \right\}$$

$$= \frac{1}{\sqrt{2\pi}} \left\{ \frac{\sin\left(\frac{\pi}{2} - \frac{\omega T}{2}\right)}{\left(\frac{\pi}{T}-\omega\right)} + \frac{\sin\left(\frac{\pi}{2} + \frac{\omega T}{2}\right)}{\left(\frac{\pi}{T}+\omega\right)} \right\}$$

$$= \frac{1}{\sqrt{2\pi}} \left\{ \frac{\cos\left(\frac{\omega T}{2}\right)}{\left(\frac{\pi}{T}-\omega\right)} + \frac{\cos\left(\frac{\omega T}{2}\right)}{\left(\frac{\pi}{T}+\omega\right)} \right\}$$

$$= \frac{T}{\sqrt{2\pi}} \cos\left(\frac{\omega T}{2}\right) \left\{ \frac{1}{(\pi - \omega T)} + \frac{1}{(\pi + \omega T)} \right\}. \tag{1.31}$$

For $T = 1$, the function (1.30) and the corresponding Fourier transform (1.31) are plotted in Figure 1.9.

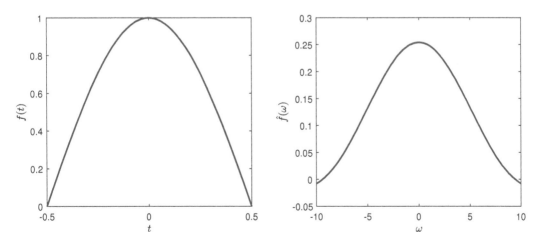

FIGURE 1.9: The cosine function and its Fourier transform.

Example 1.2.9. (The Hanning Function): The Hanning function is defined by

$$f(t) = \begin{cases} \frac{1}{2}\left(1 + \cos\left(\frac{2\pi t}{T}\right)\right) = \cos^2\left(\frac{\pi t}{T}\right), & -\frac{T}{2} \le t \le \frac{T}{2}, \quad T > 0 \\ 0, & \text{otherwise.} \end{cases} \tag{1.32}$$

We shall calculate the Fourier transform of (1.32) as

$$\hat{f}(\omega) = \frac{1}{\sqrt{2\pi}} \int_{-T/2}^{T/2} \cos^2\left(\frac{\pi t}{T}\right) e^{-i\omega t}\, dt$$

$$= \frac{1}{\sqrt{2\pi}} \int_{-T/2}^{T/2} \left(\frac{e^{i\pi t/T} + e^{-i\pi t/T}}{2}\right)^2 e^{-i\omega t}\, dt$$

$$= \frac{1}{4\sqrt{2\pi}} \int_{-T/2}^{T/2} \left(e^{i\left(\frac{2\pi}{T}-\omega\right)t} + e^{-i\left(\frac{2\pi}{T}+\omega\right)t} + 2\, e^{-i\omega t}\right) dt$$

$$= \frac{1}{4\sqrt{2\pi}} \left\{ \left(\frac{e^{i\left(\frac{2\pi}{T}-\omega\right)t}}{i\left(\frac{2\pi}{T}-\omega\right)}\right)_{-T/2}^{T/2} + \left(\frac{e^{-i\left(\frac{2\pi}{T}+\omega\right)t}}{-i\left(\frac{2\pi}{T}+\omega\right)}\right)_{-T/2}^{T/2} + 2\left(\frac{e^{-i\omega t}}{-i\omega}\right)_{-T/2}^{T/2} \right\}$$

$$= \frac{1}{\sqrt{2\pi}} \left\{ \frac{\sin\left(\frac{2\pi}{T}-\omega\right)\left(\frac{T}{2}\right)}{2\left(\frac{2\pi}{T}-\omega\right)} + \frac{\sin\left(\frac{2\pi}{T}+\omega\right)\left(\frac{T}{2}\right)}{2\left(\frac{2\pi}{T}+\omega\right)} + \frac{\sin\left(\frac{\omega T}{2}\right)}{\omega} \right\}$$

$$= \frac{1}{\sqrt{2\pi}} \left\{ \frac{\sin\left(\pi - \frac{\omega T}{2}\right)}{\left(\frac{4\pi}{T} - 2\omega\right)} + \frac{\sin\left(\pi + \frac{\omega T}{2}\right)}{\left(\frac{4\pi}{T} + 2\omega\right)} + \frac{\sin\left(\frac{\omega T}{2}\right)}{\omega} \right\}$$

$$= \frac{1}{\sqrt{2\pi}} \left\{ \frac{T\sin\left(\frac{\omega T}{2}\right)}{4\left(\pi - \frac{\omega T}{2}\right)} - \frac{T\sin\left(\frac{\omega T}{2}\right)}{4\left(\pi + \frac{\omega T}{2}\right)} + \frac{T\sin\left(\frac{\omega T}{2}\right)}{2\left(\frac{\omega T}{2}\right)} \right\}$$

$$= \frac{T}{4\sqrt{2\pi}} \sin\left(\frac{\omega T}{2}\right) \left\{ \frac{1}{\left(\pi - \frac{\omega T}{2}\right)} - \frac{1}{\left(\pi + \frac{\omega T}{2}\right)} + \frac{2}{\left(\frac{\omega T}{2}\right)} \right\}. \tag{1.33}$$

For $T = 1$, the Hanning function (1.32) and the corresponding Fourier transform (1.33) are plotted in Figure 1.10.

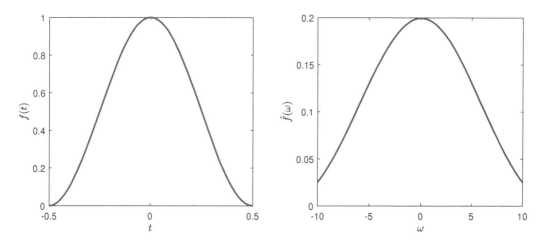

FIGURE 1.10: The Hanning function and its Fourier transform.

Example 1.2.10. (The Hamming Function): The Hamming function is defined by

$$f(t) = \begin{cases} \alpha + (1 - \alpha) \cos^2 \left(\dfrac{\pi t}{T} \right), & -\dfrac{T}{2} \leq t \leq \dfrac{T}{2}, \quad T > 0 \\ 0, & \text{otherwise}, \end{cases} \tag{1.34}$$

where α is a real parameter. Then, it can be easily verified that the Fourier transform corresponding to (1.34) is given by

$$\hat{f}(\omega) = \frac{T}{4\sqrt{2\pi}} \sin \left(\frac{\omega T}{2} \right) \left\{ \frac{1 - \alpha}{\left(\pi - \frac{\omega T}{2} \right)} - \frac{1 - \alpha}{\left(\pi + \frac{\omega T}{2} \right)} + \frac{2(1 + \alpha)}{\left(\frac{\omega T}{2} \right)} \right\}. \tag{1.35}$$

For $\alpha = 0.5$ and $T = 1$, the Hamming function (1.34) and the corresponding Fourier transform (1.35) are plotted in Figure 1.11.

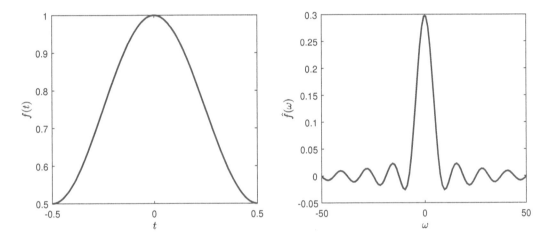

FIGURE 1.11: The Hamming function and its Fourier transform.

Example 1.2.11. (The Blackman-Harris Function): The Blackman-Harris function is defined by

$$f(t) = \begin{cases} \displaystyle\sum_{n=0}^{3} a_n \cos \left(\dfrac{2n\pi t}{T} \right), & -\dfrac{T}{2} \leq t \leq \dfrac{T}{2}, \quad T > 0 \\ 0, & \text{otherwise}, \end{cases} \tag{1.36}$$

where a_n, $n = 0, 1, 2, 3$ are real coefficients. The Fourier transform corresponding to (1.36) is computed below:

$$\hat{f}(\omega) = \frac{1}{\sqrt{2\pi}} \int_{-T/2}^{T/2} \left(\sum_{n=0}^{3} a_n \cos \left(\frac{2n\pi t}{T} \right) \right) e^{-i\omega t} \, dt$$

$$= \frac{1}{\sqrt{2\pi}} \int_{-T/2}^{T/2} \sum_{n=0}^{3} a_n \cos \left(\frac{2n\pi t}{T} \right) \cos(\omega t) \, dt$$

$$= \frac{2}{\sqrt{2\pi}} \sum_{n=0}^{3} a_n \int_{0}^{T/2} \cos \left(\frac{2n\pi t}{T} \right) \cos(\omega t) \, dt$$

$$= \frac{1}{\sqrt{2\pi}} \sum_{n=0}^{3} a_n \int_0^{T/2} \left\{ \cos\left(\frac{2n\pi}{T} + \omega\right) t + \cos\left(\frac{2n\pi}{T} + \omega\right) t \right\} dt$$

$$= \frac{1}{\sqrt{2\pi}} \sum_{n=0}^{3} a_n \left\{ \frac{\sin\left(\frac{2n\pi}{T} + \omega\right)\frac{T}{2}}{\left(\frac{2n\pi}{T} + \omega\right)} + \frac{\sin\left(\frac{2n\pi}{T} - \omega\right)\frac{T}{2}}{\left(\frac{2n\pi}{T} - \omega\right)} \right\}$$

$$= \frac{1}{\sqrt{2\pi}} \sum_{n=0}^{3} a_n \left\{ \frac{\sin\left(n\pi + \frac{\omega T}{2}\right)}{\left(\frac{2n\pi}{T} + \omega\right)} + \frac{\sin\left(n\pi - \frac{\omega T}{2}\right)}{\left(\frac{2n\pi}{T} - \omega\right)} \right\}$$

$$= \frac{1}{\sqrt{2\pi}} \sum_{n=0}^{3} a_n \left\{ \frac{(-1)^n \sin\left(\frac{\omega T}{2}\right)}{\left(\frac{2n\pi}{T} + \omega\right)} - \frac{(-1)^n \sin\left(\frac{\omega T}{2}\right)}{\left(\frac{2n\pi}{T} - \omega\right)} \right\}$$

$$= \frac{1}{\sqrt{2\pi}} \sum_{n=0}^{3} a_n \left(\frac{T}{2n\pi + \omega T} - \frac{T}{2n\pi - \omega T} \right) \sin\left(\frac{\omega T}{2}\right)$$

$$= \frac{T}{\sqrt{2\pi}} \sin\left(\frac{\omega T}{2}\right) \sum_{n=0}^{3} a_n \left(\frac{1}{2n\pi + \omega T} - \frac{1}{2n\pi - \omega T} \right). \tag{1.37}$$

For $a_n = 1$, $n = 0, 1, 2, 3$ and $T = 1$, the Blackman-Harris function (1.36) and the corresponding Fourier transform (1.37) are plotted in Figure 1.12.

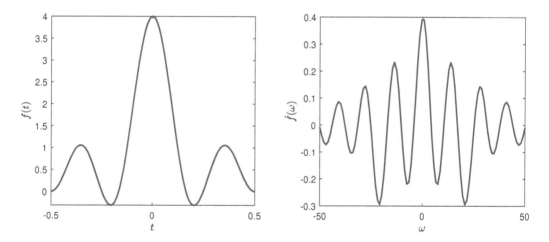

FIGURE 1.12: The Blackman-Harris function and its Fourier transform.

Example 1.2.12. (The Kaiser-Bessel Function): Developed by James Kaiser $(1929 - 2020)$ at Bell Laboratories, the Kaiser-Bessel function is one of the vital functions in digital signal processing with many pleasant properties and is defined by

$$f(t) = \begin{cases} \dfrac{I_0\left(\beta\sqrt{1 - \left(\frac{2t}{T}\right)^2}\right)}{I_0(\beta)}, & -\dfrac{T}{2} \le t \le \dfrac{T}{2}, \quad T > 0 \\ 0, & \text{otherwise,} \end{cases} \tag{1.38}$$

where

$$I_0(t) = \sum_{m=0}^{\infty} \frac{1}{(m!)^2} \left(\frac{t}{2}\right)^{2m}, \tag{1.39}$$

is the zero-order modified Bessel function of first kind and β is a parameter which may be chosen at will.

In order to compute the Fourier transform of the Kaiser-Bessel function (1.38), we observe that

$$\sqrt{2\pi}\,\hat{f}(\omega) = \frac{2}{I_0(\beta)} \int_0^{T/2} I_0\left(\beta\sqrt{1 - \left(\frac{2t}{T}\right)^2}\right) \cos(\omega t)\, dt. \tag{1.40}$$

Using (1.39) and interchanging the order of integration and summation, we can express (1.40) as

$$
\begin{aligned}
\sqrt{2\pi}\,\hat{f}(\omega) &= \frac{2}{I_0(\beta)} \sum_{m=0}^{\infty} \frac{1}{(m!)^2} \int_0^{T/2} \left(\frac{\beta\sqrt{1 - \left(\frac{2t}{T}\right)^2}}{2}\right) \cos(\omega t)\, dt \\
&= \frac{2}{I_0(\beta)} \sum_{m=0}^{\infty} \frac{1}{(m!)^2} \left(\frac{\beta}{2}\right)^{2m} \int_0^{T/2} \left(1 - \left(\frac{2t}{T}\right)^2\right)^m \cos(\omega t)\, dt \\
&= \frac{T}{I_0(\beta)} \sum_{m=0}^{\infty} \frac{1}{(m!)^2} \left(\frac{\beta}{2}\right)^{2m} \int_0^1 (1 - t^2)^m \cos\left(\frac{\omega T t}{2}\right) dt. \tag{1.41}
\end{aligned}
$$

Invoking the integral representation of the Bessel functions $J_k(\cdot)$, we have

$$J_k(z) = \frac{2\left(\frac{z}{2}\right)^k}{\pi^{1/2}\,\Gamma\left(k + \frac{1}{2}\right)} \int_0^1 (1 - t^2)^{k-1/2} \cos(zt)\, dt. \tag{1.42}$$

Plugging $k - 1/2 = m$ into the expression (1.42), we obtain

$$J_{m+1/2}\left(\frac{\omega T}{2}\right) = \frac{2\left(\frac{\omega T}{4}\right)^{m+1/2}}{\pi^{1/2}\,\Gamma(m+1)} \int_0^1 (1 - t^2)^{k-1/2} \cos\left(\frac{\omega T t}{2}\right) dt. \tag{1.43}$$

Using (1.43) in (1.41), we get

$$
\begin{aligned}
\sqrt{2\pi}\,\hat{f}(\omega) &= \frac{T}{I_0(\beta)} \sum_{m=0}^{\infty} \frac{1}{(m!)^2} \left(\frac{\beta}{2}\right)^{2m} \frac{\pi^{1/2}\,\Gamma(m+1)}{2\left(\frac{\omega T}{4}\right)^{m+1/2}} J_{m+1/2}\left(\frac{\omega T}{2}\right) \\
&= \frac{T\pi^{1/2}}{2I_0(\beta)} \sum_{m=0}^{\infty} \frac{1}{m!} \left(\frac{\beta^{2m}}{2^{2m}}\right) \frac{4^{m+1/2}}{(\omega T)^{m+1/2}} J_{m+1/2}\left(\frac{\omega T}{2}\right) \\
&= \frac{T\pi^{1/2}}{2I_0(\beta)} \sum_{m=0}^{\infty} \frac{2^{-2m}}{m!} \beta^{2m} \frac{4^m\, 2}{(\omega T)^{m+1/2}} J_{m+1/2}\left(\frac{\omega T}{2}\right) \\
&= \frac{T\pi^{1/2}}{I_0(\beta)} \sum_{m=0}^{\infty} \frac{\beta^{2m}}{m!} \frac{1}{(\omega T)^{m+1/2}} J_{m+1/2}\left(\frac{\omega T}{2}\right)
\end{aligned}
$$

$$= \frac{T\pi^{1/2}}{I_0(\beta)\,(\omega T)^{m+1/2}} \sum_{m=0}^{\infty} \frac{\beta^{2m}}{m!} \frac{1}{(\omega T)^m} \, J_{m+1/2}\left(\frac{\omega T}{2}\right)$$

$$= \frac{T\pi^{1/2}}{I_0(\beta)\,(\omega T)^{m+1/2}} \sum_{m=0}^{\infty} \frac{\beta^{2m}\,2^{-m}}{m!\left(\frac{\omega T}{2}\right)^m} \, J_{m+1/2}\left(\frac{\omega T}{2}\right). \tag{1.44}$$

Also, note that the multiplication theorem for Bessel functions reads:

$$J_k(\sigma z) = \sigma^k \sum_{m=0}^{\infty} \frac{(-1)^m \left(\sigma^2 - 1\right)^m \left(\frac{z}{2}\right)^m}{m!} \, J_{k+m}(z), \tag{1.45}$$

which is valid for any complex value σ. Therefore, for $k = 1/2$ and $z = \omega T/2$, we can rewrite (1.44) as

$$\sqrt{2\pi}\,\hat{f}(\omega) = \frac{T\pi^{1/2}}{I_0(\beta)\,(\omega T)^{1/2}} \sum_{m=0}^{\infty} \frac{1}{m!} \left(\frac{\beta^2 4^{-1}\omega T}{(\omega T/2)^2}\right)^m J_{m+1/2}\left(\frac{\omega T}{2}\right)$$

$$= \frac{T\pi^{1/2}}{I_0(\beta)\,(\omega T)^{1/2}} \sum_{m=0}^{\infty} \frac{1}{m!} \left(\frac{\beta^2}{(\omega T/2)^2}\right)^m \left(\frac{\omega T}{4}\right)^m J_{m+1/2}\left(\frac{\omega T}{2}\right). \tag{1.46}$$

Upon comparing (1.45) with (1.46), we set

$$(-1)\left(\sigma^2 - 1\right) = \frac{\beta^2}{(\omega T/2)^2} \Rightarrow \sigma = \sqrt{1 - \frac{\beta^2}{(\omega T/2)^2}}$$

and

$$m + 1/2 = m + k \Rightarrow k = 1/2.$$

Then, expression (1.44) becomes

$$\sqrt{2\pi}\,\hat{f}(\omega) = \frac{T\pi^{1/2}}{I_0(\beta)\,(\omega T)^{1/2}} \left\{ \frac{1}{\left(1 - \frac{\beta^2}{(\omega T/2)^2}\right)^{1/4}} \, J_{1/2}\left(\left(1 - \frac{\beta^2}{(\omega T/2)^2}\right)^2 \frac{\omega T}{2}\right) \right\}$$

$$= \frac{T\pi^{1/2}}{I_0(\beta)\,(\omega T)^{1/2}} \left\{ \frac{\left(\frac{\omega T}{2}\right)^{1/2}}{\left(\frac{\omega^2 T^2}{4} - \beta^2\right)^{1/4}} \, J_{1/2}\left(\left(\frac{\omega^2 T^2}{4} - \beta^2\right)^{1/2}\right) \right\}$$

$$= \frac{T\pi^{1/2}}{\sqrt{2}\,I_0(\beta)} \left\{ \frac{1}{\left(\frac{\omega^2 T^2}{4} - \beta^2\right)^{1/4}} \, J_{1/2}\left(\left(\frac{\omega^2 T^2}{4} - \beta^2\right)^{1/2}\right) \right\},$$

which further yields

$$\hat{f}(\omega) = \frac{T}{2\,I_0(\beta)} \left\{ \frac{1}{\left(\frac{\omega^2 T^2}{4} - \beta^2\right)^{1/4}} \, J_{1/2}\left(\left(\frac{\omega^2 T^2}{4} - \beta^2\right)^{1/2}\right) \right\}. \tag{1.47}$$

Since the explicit expression for $J_{1/2}(z)$ is given by

$$J_{1/2}(z) = \left(\frac{2}{\pi z}\right)^{1/2} \sin z,$$

therefore, it follows that

$$\hat{f}(\omega) = \frac{T}{2\,I_0(\beta)} \left\{ \frac{1}{\left(\frac{\omega^2 T^2}{4} - \beta^2\right)^{1/4}} \frac{\sqrt{2}}{\sqrt{\pi} \left(\frac{\omega^2 T^2}{4} - \beta^2\right)^{1/4}} \sin\left(\sqrt{\frac{\omega^2 T^2}{4} - \beta^2}\right) \right\}$$

$$= \left\{ \frac{T}{2\,I_0(\beta)} \frac{1}{\left(\frac{\omega^2 T^2}{4} - \beta^2\right)^{1/2}} \frac{\sqrt{2}}{\sqrt{\pi}} \right\} \sin\left(\sqrt{\frac{\omega^2 T^2}{4} - \beta^2}\right)$$

$$= \frac{T}{I_0(\beta)\sqrt{2\pi}} \left\{ \frac{\sin\left(\sqrt{\frac{\omega^2 T^2}{4} - \beta^2}\right)}{\sqrt{\frac{\omega^2 T^2}{4} - \beta^2}} \right\}. \tag{1.48}$$

It is pertinent to mention that the expression (1.48) is valid for all values of ω. In particular, for the case

$$\left(\frac{\omega T}{2}\right)^2 < \beta^2,$$

we have

$$\hat{f}(\omega) = \frac{T}{I_0(\beta)\sqrt{2\pi}} \left\{ \frac{\sinh\left(\sqrt{\beta^2 - \frac{\omega^2 T^2}{4}}\right)}{\sqrt{\beta^2 - \frac{\omega^2 T^2}{4}}} \right\}. \tag{1.49}$$

Thus, we conclude that the Fourier transform of the Kaiser-Bessel function (1.38) is given by

$$\hat{f}(\omega) = \begin{cases} \dfrac{T}{I_0(\beta)\sqrt{2\pi}} \left\{ \dfrac{\sinh\left(\sqrt{\beta^2 - \frac{\omega^2 T^2}{4}}\right)}{\sqrt{\beta^2 - \frac{\omega^2 T^2}{4}}} \right\}, & \beta > \left|\dfrac{\omega T}{2}\right| \\[4ex] \dfrac{T}{I_0(\beta)\sqrt{2\pi}} \left\{ \dfrac{\sin\left(\sqrt{\frac{\omega^2 T^2}{4} - \beta^2}\right)}{\sqrt{\frac{\omega^2 T^2}{4} - \beta^2}} \right\}, & \beta \le \left|\dfrac{\omega T}{2}\right|. \end{cases} \tag{1.50}$$

For $T = 1$ and different values of β, the Kaiser-Bessel function (1.38) and the corresponding Fourier transform (1.50) are plotted in Figures 1.13–1.16.

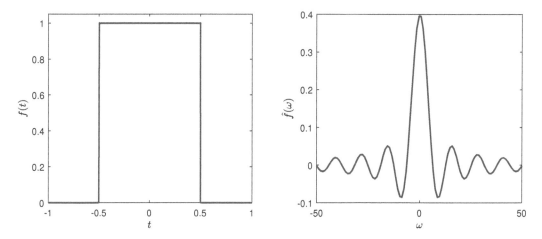

FIGURE 1.13: The Kaiser-Bessel function and its Fourier transform for $\beta = 0$.

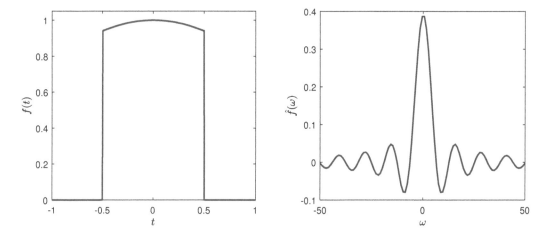

FIGURE 1.14: The Kaiser-Bessel function and its Fourier transform for $\beta = 0.5$.

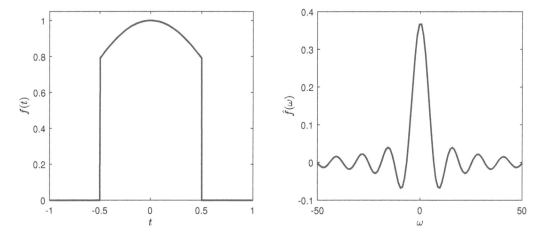

FIGURE 1.15: The Kaiser-Bessel function and its Fourier transform for $\beta = 1$.

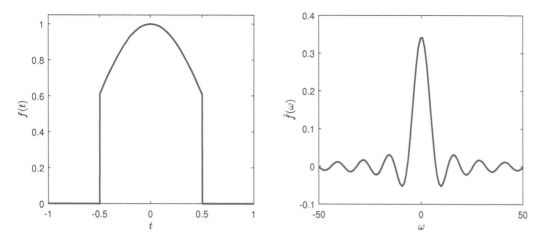

FIGURE 1.16: The Kaiser-Bessel function and its Fourier transform for $\beta = 1.5$.

1.2.2 Basic properties of the Fourier transform

"Mathematics compares the most diverse phenomena and discovers the secret analogies that unite them."

-Joseph Fourier

In this subsection, we study some basic properties of the Fourier transform. To facilitate the narrative, we define the fundamental operators, viz; translation , dilation and modulation operators acting on any $f \in L^2(\mathbb{R})$ as

- Translation: $\mathcal{T}_a f(t) = f(t - a)$,

- Modulation: $\mathcal{M}_b f(t) = e^{ibt} f(t)$,

- Dilation: $\mathcal{D}_c f(t) = \dfrac{1}{\sqrt{|c|}} f\left(\dfrac{t}{c}\right)$,

where $a, b, c \in \mathbb{R}$ and $c \neq 0$. Each of the above defined operators is a unitary operator from $L^2(\mathbb{R})$ onto itself. The following results can be easily verified:

(i) $\mathcal{T}_a \mathcal{M}_b f(t) = e^{ib(t-a)} f(t-a)$,

(ii) $\mathcal{M}_b \mathcal{T}_a f(t) = e^{ibt} f(t-a)$,

(iii) $\mathcal{D}_c \mathcal{T}_a f(t) = \dfrac{1}{\sqrt{|c|}} f\left(\dfrac{t-a}{c}\right)$,

(iv) $\mathcal{T}_a \mathcal{D}_c f(t) = \dfrac{1}{\sqrt{|c|}} f\left(\dfrac{t-a}{c}\right)$,

(v) $\mathcal{M}_b \mathcal{D}_c f(t) = \dfrac{1}{\sqrt{|c|}} \exp\left\{\dfrac{ibt}{c}\right\} f\left(\dfrac{t}{c}\right)$,

(vi) $\mathcal{D}_c \mathcal{M}_b f(t) = \dfrac{1}{\sqrt{|c|}} \exp\left\{\dfrac{ibt}{c}\right\} f\left(\dfrac{t}{c}\right)$.

In the following theorem, we assemble some basic properties of the Fourier transform defined in (1.7).

Theorem 1.2.1. *Let $f, g \in L^2(\mathbb{R})$ and $\alpha, \beta \in \mathbb{C}$, $k, \omega_0 \in \mathbb{R}$, $\lambda \in \mathbb{R} \setminus \{0\}$ are scalars. Then, the Fourier transform defined by (1.7) satisfies the following properties:*

(i) *Linearity:* $\mathscr{F}\big[\alpha f + \beta g\big](\omega) = \alpha \, \mathscr{F}\big[f\big](\omega) + \beta \, \mathscr{F}\big[g\big](\omega)$,

(ii) *Translation:* $\mathscr{F}\big[\mathcal{T}_k f\big](\omega) = e^{-i\omega k} \, \mathscr{F}\big[f\big](\omega)$,

(iii) *Modulation:* $\mathscr{F}\big[\mathcal{M}_{\omega_0} f\big](\omega) = \mathscr{F}\big[f\big](\omega - \omega_0)$,

(iv) *Scaling:* $\mathscr{F}\big[f(\lambda t)\big](\omega) = \dfrac{1}{|\lambda|} \, \mathscr{F}\big[f\big]\left(\dfrac{\omega}{\lambda}\right)$,

(v) *Translation and scaling:* $\mathscr{F}\big[f(\lambda t - k)\big](\omega) = \dfrac{e^{-i\omega k/\lambda}}{|\lambda|} \, \mathscr{F}\big[f\big]\left(\dfrac{\omega}{\lambda}\right)$,

(vi) *Conjugation:* $\mathscr{F}\big[\overline{f}\big](\omega) = \overline{\mathscr{F}\big[f\big](-\omega)}$.

Proof. (i) Using Definition 1.2.1, we have

$$\mathscr{F}\big[\alpha f + \beta g\big](\omega) = \frac{1}{\sqrt{2\pi}} \int_{\mathbb{R}} (\alpha f + \beta g)(t) \, e^{-i\omega t} \, dt$$

$$= \frac{1}{\sqrt{2\pi}} \left\{ \alpha \int_{\mathbb{R}} f(t) \, e^{-i\omega t} \, dt + \beta \int_{\mathbb{R}} g(t) \, e^{-i\omega t} \, dt \right\}$$

$$= \alpha \, \mathscr{F}\big[f\big](\omega) + \beta \, \mathscr{F}\big[g\big](\omega).$$

(ii) For the translation operator $\mathcal{T}_k$, $k \in \mathbb{R}$, we observe that

$$\mathscr{F}\big[\mathcal{T}_k f\big](\omega) = \frac{1}{\sqrt{2\pi}} \int_{\mathbb{R}} \mathcal{T}_k f(t) \, e^{-i\omega t} \, dt$$

$$= \frac{1}{\sqrt{2\pi}} \int_{\mathbb{R}} f(t - k) \, e^{-i\omega t} \, dt$$

$$= \frac{1}{\sqrt{2\pi}} \int_{\mathbb{R}} f(z) \, e^{-i\omega(k+z)} \, dz$$

$$= \frac{1}{\sqrt{2\pi}} \int_{\mathbb{R}} f(z) \, e^{-i\omega k} \, e^{-i\omega z} \, dz$$

$$= e^{-i\omega k} \, \mathscr{F}\big[f\big](\omega).$$

(iii) For the modulation operator $\mathcal{M}_{\omega_0}$, $\omega_0 \in \mathbb{R}$, we have

$$\mathscr{F}\big[\mathcal{M}_{\omega_0} f\big](\omega) = \frac{1}{\sqrt{2\pi}} \int_{\mathbb{R}} \mathcal{M}_{\omega_0} f(t) \, e^{-i\omega t} \, dt$$

$$= \frac{1}{\sqrt{2\pi}} \int_{\mathbb{R}} e^{i\omega_0 t} f(t) \, e^{-i\omega t} \, dt$$

$$= \frac{1}{\sqrt{2\pi}} \int_{\mathbb{R}} f(t) \, e^{-i(\omega - \omega_0)t} \, dt$$

$$= \mathscr{F}\big[f\big](\omega - \omega_0).$$

(iv) Upon scaling the argument of the given signal f by $\lambda \in \mathbb{R} \setminus \{0\}$, we have

$$
\begin{aligned}
\mathscr{F}\left[f(\lambda t)\right](\omega) &= \frac{1}{\sqrt{2\pi}} \int_{\mathbb{R}} f(\lambda t)\, e^{-i\omega t}\, dt \\
&= \frac{1}{\sqrt{2\pi}} \int_{\mathbb{R}} f(z)\, e^{-i\omega z/\lambda}\, \frac{dz}{|\lambda|} \\
&= \frac{1}{|\lambda|}\, \mathscr{F}[f]\left(\frac{\omega}{\lambda}\right).
\end{aligned}
$$

(v) Upon the joint influence of the translation and scaling operations, we observe that

$$
\begin{aligned}
\mathscr{F}\left[f(\lambda t - k)\right](\omega) &= \frac{1}{\sqrt{2\pi}} \int_{\mathbb{R}} f(\lambda t - k)\, e^{-i\omega t}\, dt \\
&= \frac{1}{\sqrt{2\pi}} \int_{\mathbb{R}} f(z)\, e^{-i\omega(k+z)/\lambda}\, \frac{dz}{|\lambda|} \\
&= \frac{e^{-i\omega k/\lambda}}{\sqrt{2\pi}} \int_{\mathbb{R}} f(z)\, e^{-i\omega z/\lambda}\, \frac{dz}{|\lambda|} \\
&= \frac{e^{-i\omega k/\lambda}}{|\lambda|}\, \mathscr{F}[f]\left(\frac{\omega}{\lambda}\right).
\end{aligned}
$$

(vi) Finally, under the conjugation of the input signal, the Fourier transform behaves as

$$
\begin{aligned}
\mathscr{F}\left[\bar{f}\right](\omega) &= \frac{1}{\sqrt{2\pi}} \int_{\mathbb{R}} \overline{f(t)}\, e^{-i\omega t}\, dt \\
&= \overline{\frac{1}{\sqrt{2\pi}} \int_{\mathbb{R}} f(t)\, e^{-i(-\omega)t}\, dt} \\
&= \overline{\mathscr{F}[f](-\omega)}.
\end{aligned}
$$

This completes the proof of Theorem 1.2.1. $\qquad\square$

Having furnished the proof of Theorem 1.2.1, lets ponder on the illustration of the above cited properties. Firstly, we note that the translation property of the Fourier transform implies that shifting in the spatial domain leads to modulation in the frequency domain. To demonstrate this property explicitly, we choose the function f as the rectangular pulse (1.11), whose Fourier transform is given by (1.12). Upon applying the translation operator $\mathcal{T}_k$ with $k = T/2$ and $T > 0$ to the function f, we obtain a new function g defined by $g(t) = f(t - T/2)$. Thus, we have

$$
g(t) = \begin{cases} 1, & -\frac{T}{2} \le t - \frac{T}{2} \le \frac{T}{2} \\ 0, & \text{otherwise} \end{cases} = \begin{cases} 1, & 0 \le t \le T \\ 0, & \text{otherwise}. \end{cases} \tag{1.51}
$$

Evidently, the Fourier transform of (1.51) is given by

$$
\begin{aligned}
\hat{g}(\omega) &= \frac{T}{\sqrt{2\pi}}\, \text{sinc}\left(\frac{\omega T}{2}\right) e^{-i\omega T/2} \\
&= \frac{T}{\sqrt{2\pi}}\, \text{sinc}\left(\frac{\omega T}{2}\right) \left\{ \cos\left(\frac{\omega T}{2}\right) - i \sin\left(\frac{\omega T}{2}\right) \right\}. \tag{1.52}
\end{aligned}
$$

From (1.52), we observe that the real part of $\hat{g}(\omega)$ gets modulated by $\cos\left(\omega T/2\right)$ and besides this, there is also an imaginary part associated with the Fourier transform, unlike the case with $\hat{f}(\omega)$. Moreover, it is pertinent to mention that, if we are concerned only about the magnitude of Fourier transforms of f and g; that is $\left|\hat{f}(\omega)\right|$ and $\left|\hat{g}(\omega)\right|$, then no change can be seen in the respective plots. This means that, we can shift the function $f(t)$ along the t-axis as much as we desire without any influence on the power-spectrum. However, in the polar representation, we can observe the shift in the phase as

$$\tan\phi = \frac{\text{Imaginary part}}{\text{Real part}} = -\frac{\sin\left(\omega T/2\right)}{\cos\left(\omega T/2\right)} = -\tan\left(\frac{\omega T}{2}\right),$$

which implies that $\phi = -\left(\dfrac{\omega T}{2}\right)$. For $T=1$, the function $g(t)$, its amplitude spectrum and the real, imaginary parts of the Fourier transform (1.52) are plotted in Figure 1.17.

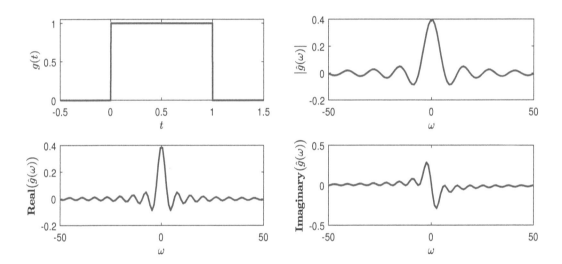

FIGURE 1.17: The translated function $g(t)$ and its Fourier transform.

As a consequence of the modulation property, we infer that complex modulation in the input signal corresponds to a shift in the Fourier spectrum. On the other hand, if the given signal f is subjected to real modulations, then we have

$$\mathscr{F}\left[\cos(\omega_0 t)\, f(t)\right](\omega)$$

$$= \frac{1}{\sqrt{2\pi}} \int_{\mathbb{R}} \cos(\omega_0 t)\, f(t)\, e^{-i\omega t}\, dt$$

$$= \frac{1}{\sqrt{2\pi}} \int_{\mathbb{R}} f(t)\left(\cos(\omega_0 t)\cos(\omega t) - i\cos(\omega_0 t)\sin(\omega t)\right) dt$$

$$= \frac{1}{\sqrt{2\pi}} \int_{\mathbb{R}} f(t)\left(\frac{\cos(\omega+\omega_0)+\cos(\omega-\omega_0)}{2} - i\,\frac{\sin(\omega+\omega_0)+\sin(\omega-\omega_0)}{2}\right) dt$$

$$= \frac{1}{\sqrt{2\pi}} \int_{\mathbb{R}} f(t)\left(\frac{\cos(\omega+\omega_0)-i\sin(\omega+\omega_0)}{2} + \frac{\cos(\omega-\omega_0)-\sin(\omega-\omega_0)}{2}\right) dt$$

$$= \frac{1}{\sqrt{2\pi}}\left\{\frac{1}{2}\int_{\mathbb{R}} f(t)\, e^{-i(\omega+\omega_0)t}\, dt + \frac{1}{2}\int_{\mathbb{R}} f(t)\, e^{-i(\omega-\omega_0)t}\, dt\right\}$$

$$= \frac{\hat{f}(\omega+\omega_0)+\hat{f}(\omega-\omega_0)}{2}. \tag{1.53}$$

In a similar fashion, we can show that

$$\mathscr{F}\left[\sin(\omega_0 t)\, f(t)\right](\omega) = i\left(\frac{\hat{f}(\omega + \omega_0) - \hat{f}(\omega - \omega_0)}{2}\right). \tag{1.54}$$

Now, we consider the rectangular pulse $f(t)$ defined in (1.11), which is subjected to the real modulation to yield a new function $h(t) = \cos(\omega_0 t)\, f(t)$. Then, the Fourier transform corresponding to $h(t)$ is given by

$$\hat{h}(\omega) = \frac{T}{2\sqrt{2\pi}}\left\{ \text{sinc}\left(\frac{(\omega + \omega_0)T}{2}\right) + \text{sinc}\left(\frac{(\omega - \omega_0)T}{2}\right)\right\}. \tag{1.55}$$

It is quite evident from (1.55) that $\hat{h}(\omega)$ consists of two frequency peaks at $\omega = \pm\omega_0$ and the amplitude naturally gets split evenly. Clearly, for $\omega_0 = 0$, we only have the central peak at $\omega = 0$. Therefore, increasing ω_0 splits this peak into two peaks, moving to the left and the right. Moreover, if we do not want to see the negative frequencies, we may flip the negative half-plane to obtain a single peak at $\omega = \omega_0$ with twice the intensity of the original. For $\omega_0 = 10$ and $T = 1$, the function $h(t)$ and its Fourier transform (1.55) are depicted in Figure 1.18.

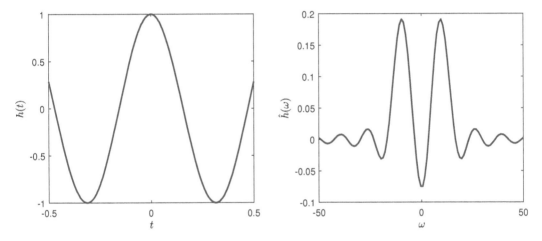

FIGURE 1.18: The modulated function $h(t)$ and its Fourier transform.

Finally, we turn our attention to the scaling property of the Fourier transform. For $\lambda > 1$, the graph of $f(\lambda t)$ is a compressed version of the graph of $f(t)$. Moreover, the dominant frequencies of $f(\lambda t)$ are larger than those of f by a factor of λ. Such a behaviour of the Fourier transform is explicitly illustrated by choosing following function:

$$f(t) = \begin{cases} \sin(3t), & -\pi \le t \le \pi \\ 0, & \text{otherwise}, \end{cases} \tag{1.56}$$

whose Fourier transform is computed in (1.19) and is given by

$$\hat{f}(\omega) = -\frac{3\sqrt{2}\, i \, \sin(\omega\pi)}{\sqrt{\pi}\,(9 - \omega^2)}. \tag{1.57}$$

Consequently, the dyadic scaled version of the waveform (1.56) takes the following form:

$$s(t) := f(2t) = \begin{cases} \sin(6t), & -\frac{\pi}{2} \le t \le \frac{\pi}{2} \\ 0, & \text{otherwise}, \end{cases} \tag{1.58}$$

with the corresponding Fourier transform as

$$\hat{s}(\omega) = -\frac{3i \sin\left(\omega\pi/2\right)}{\sqrt{2\pi}\left(9 - \omega^2/4\right)}. \tag{1.59}$$

The scaled function $s(t)$ and the imaginary part of the corresponding Fourier transform (1.59) are plotted in Figure 1.19. Moreover, from Figure 1.19, we observe that increasing the frequency of a signal has the effect of stretching the graph of the Fourier transform. The dominant frequency of $f(t)$ is 3, whereas the dominant frequency of $s(t)$ is 6. Thus, the maximum value of $\left|\hat{f}(\omega)\right|$ occurs at $\omega = 3$, whereas the maximum value of $\left|\hat{s}(\omega)\right|$ occurs at $\omega = 6$. Hence, the graph of the later is obtained by stretching the graph of the former by a factor of 2.

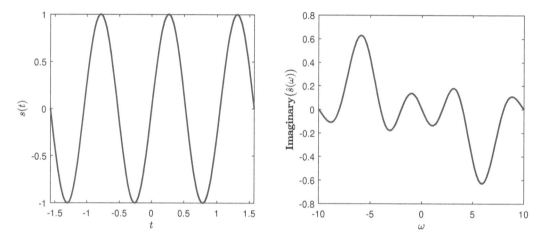

FIGURE 1.19: The scaled function $s(t)$ and its Fourier transform.

From the above discussion, we conclude that for $\lambda > 1$, the graph of $s(t) = f(\lambda t)$ is compressed, which in turn speeds up the frequency and hence, the graph of $\hat{s}(\omega)$ is a stretched version of the graph of $\hat{f}(\omega)$. In case $0 < \lambda < 1$, then the graph of $s(t)$ is a stretched copy of the graph of $f(t)$, which slows the frequency so that the graph of $\hat{s}(\omega)$ is a compressed version of the graph of $\hat{f}(\omega)$. Finally, for the special case $\lambda = -1$, we have $\hat{s}(\omega) = \hat{f}(-\omega)$; that is, turning around the time axis (looking into the past) results in turning around the frequency axis.

"The introduction of the delta function into our analysis will not be in itself a source of lack of rigour in the theory.... The delta function is thus merely a convenient notation. The only lack of rigour in the theory arises from the fact that we perform operations on the abstract symbols, such as differentiation and integration with respect to parameters occurring in them, which are not rigorously defined. When these operations are permissible, the delta function may be used freely for dealing with the representatives of the abstract symbols, as though it were a continuous function, without leading to incorrect results."

-Paul A. Dirac

Before proceeding to further developments on the Fourier transform, we ought to deal with one of the intriguing mathematical objects, namely the Dirac's delta, which is often informally referred as the Dirac delta function or simply as delta function. Indeed the Dirac's delta is the best known and widely studied among a class of mathematical entities, called as "generalized functions;" which describe many abstract notions occurring in physical sciences in a simple and insightful way. For instance, the impulsive force, point mass, point charge, point dipole and the frequency response of a harmonic oscillator in a non-dissipating medium are all aptly described via the notion of generalized functions. The generalized functions are of critical significance in the theory of Fourier transforms, as they can resolve certain intricacies that occur in the classical mathematical analysis. For example, every locally integrable function can be considered as the integral of some generalized function, and thus becomes infinitely differentiable in a new sense. Nevertheless, several function sequences arise in mathematical analysis which do not converge to a limit function in ordinary sense can be found to converge to a generalized function. Therefore, in many ways the notion of generalized functions not only simplifies the rules of mathematical analysis, but also facilitates the applications in physical sciences.

The notion of Dirac delta function $\delta(t)$ first appeared in 1920 in the work of the theoretical physicist Paul Dirac $(1902 - 1984)$ on quantum mechanics, by defining its value as zero anywhere, except when its argument is equal to zero; in this case the value is $+\infty$. Technically, the Dirac delta function is expressed via the characteristics $\delta(t) = 0, t \neq 0$ and $\int_{\mathbb{R}} \delta(t) \, dt = 1$. However, the aforementioned properties cannot be satisfied by any function in the classical mathematics, hence the name "Dirac delta function" is somewhat misleading as $\delta(t)$ is not essentially a function in the classical sense. But, $\delta(t)$ can be regarded as the limit of a sequence of ordinary functions. An elegant example of such a sequence of functions is the sequence of Gaussian functions:

$$G_n(t) = \sqrt{\frac{n}{\pi}} \, e^{-nt^2}, \quad n \in \mathbb{N}. \tag{1.60}$$

Since, the exponential term in (1.60) grows faster than the algebraic term, so it follows that $G_n(t) \to 0, t \neq 0$ as $n \to \infty$. Also, using the standard Gaussian integral, we have $\int_{\mathbb{R}} G_n(t) \, dt = 1$. Furthermore, we observe that

$$1 = \lim_{n \to \infty} \int_{\mathbb{R}} G_n(t) \, dt = \int_{\mathbb{R}} \lim_{n \to \infty} G_n(t) \, dt. \tag{1.61}$$

Hence, we infer that $\delta(t)$ can be regarded as the limit of the sequence of Gaussian functions (1.60). The first few members of the sequence of the Gaussian functions (1.60) are plotted in Figure 1.20 and it is quite intuitively clear that the limit function must be $\delta(t)$.

On the other hand, if we consider a pulse of height $k > 0$ as

$$H_k(t) = \begin{cases} k, & -\frac{1}{2k} \leq t \leq \frac{1}{2k} \\ 0, & \text{otherwise,} \end{cases} \tag{1.62}$$

then we observe that the height of the pulse (1.62) increases as the width goes on diminishing, which keeps the area unchanged (normalized to 1). In fact, we have $\lim_{k \to \infty} H_k(t) = \delta(t)$ and for this reason $\delta(t)$ is also called as an "impulse."

Nevertheless, in the late 1940s, Laurent Schwartz $(1915-2002)$ presented an alternative approach leading to the notion of delta function. This approach is based on the idea that if a function f is continuous at $t = a$, then $\delta(t)$ is defined via its fundamental shift property:

$$\int_{\mathbb{R}} f(t) \, \delta(t - a) \, dt = f(a). \tag{1.63}$$

Or equivalently,

$$\int_{\mathbb{R}} f(t)\,\delta(t)\,dt = f(0). \tag{1.64}$$

As of now, the concept of the delta function is clear and simple in the context of modern mathematics and has become very useful in science and engineering. Physically, the delta function represents a point mass, that is, a particle of unit mass located at the origin. In view of the aforementioned details, it follows that a point particle can be regarded as the limit of a sequence of continuous mass distribution. The Dirac delta function is also interpreted as a probability measure in terms of the formula (1.63).

Yet another representation of the Dirac's delta, which shall be frequently used in the subsequent developments in this book, is given by the integral expression

$$\delta(\omega) = \frac{1}{2\pi}\int_{\mathbb{R}} e^{i\omega t}\,dt. \tag{1.65}$$

In fact, some authors often multiply the integrand in (1.65) by a damping factor $e^{-\alpha|x|}$ and then take the limit as $\alpha \to 0$. This, manipulation will not affect the essence of the Dirac's delta that everything gets oscillated or averaged away for all the frequencies $\omega \neq 0$, whereas for $\omega = 0$, the integration will be over the integrand 1 from $-\infty$ to ∞; that is, the result will have to be $+\infty$. Moreover, as a consequence of (1.65), we have

$$\delta(\omega) = \frac{1}{\sqrt{2\pi}}\left\{\frac{1}{\sqrt{2\pi}}\int_{\mathbb{R}} e^{i\omega t}\,dt\right\} = \frac{1}{\sqrt{2\pi}}\,\mathscr{F}^{-1}\big[1\big](\omega), \tag{1.66}$$

so that $\mathscr{F}\big[1\big](\omega) = \sqrt{2\pi}\,\delta(\omega)$. Hence, the constant 1 can be represented by a single spectral component $\omega = 0$ and no other component occurs in the frequency spectrum.

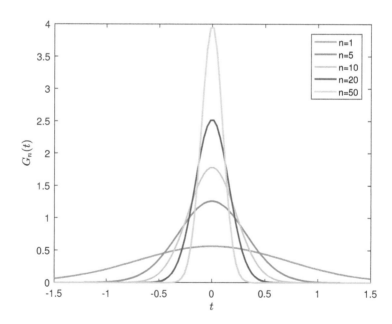

FIGURE 1.20: Few members of the sequence of Gaussian functions.

Next, we present the Plancherel theorem which is an assemblage of some of the vital properties of the Fourier transform defined in (1.7).

Theorem 1.2.2. (Plancherel Theorem): *For any pair of functions* $f, g \in L^2(\mathbb{R})$, *the following assertions are true:*

(i) *(Inversion Formula): If* $\mathscr{F}[f](\omega)$ *is the Fourier transform corresponding to* $f \in L^2(\mathbb{R})$ *and* $\mathscr{F}^{-1} : L^2(\mathbb{R}) \to L^2(\mathbb{R})$ *denotes the inverse Fourier operator, given by*

$$\mathscr{F}^{-1}\left(\mathscr{F}[f](\omega)\right)(t) = \frac{1}{\sqrt{2\pi}} \int_{\mathbb{R}} \mathscr{F}[f](\omega) \, e^{i\omega t} \, d\omega. \tag{1.67}$$

Then, we have

$$\mathscr{F}^{-1}\left(\mathscr{F}[f](\omega)\right)(t) = f(t). \tag{1.68}$$

That is, any function $f \in L^2(\mathbb{R})$ *can be retracted from the corresponding Fourier transform via the inversion formula (1.67).*

(ii) *(Parseval's Formula): The following orthogonality relation holds:*

$$\left\langle \mathscr{F}[f], \mathscr{F}[g] \right\rangle_2 = \left\langle f, g \right\rangle_2. \tag{1.69}$$

Thus, whenever the functions f *and* g *are orthogonal in the natural domain, then the corresponding Fourier transforms are also orthogonal in the frequency domain. In particular, the following norm equality holds:*

$$\left\| \mathscr{F}[f] \right\|_2 = \left\| f \right\|_2. \tag{1.70}$$

That is, the "information content" of the function f-*defined as integral over the square of the magnitude of* $f(t)$-*is just as large as the "information content" of the corresponding Fourier transform* $\mathscr{F}[f]$-*defined as integral over the square of the magnitude of* $\mathscr{F}[f](\omega)$.

(iii) *(Isomorphism): The mapping* $f \to \mathscr{F}[f]$ *is a Hilbert space isomorphism from* $L^2(\mathbb{R})$ *onto itself.*

Proof. **(i)** We observe that,

$$\begin{aligned}
\mathscr{F}^{-1}\left(\mathscr{F}[f](\omega)\right)(t) &= \frac{1}{\sqrt{2\pi}} \int_{\mathbb{R}} \mathscr{F}[f](\omega) \, e^{i\omega t} \, d\omega \\
&= \frac{1}{\sqrt{2\pi}} \int_{\mathbb{R}} \left\{ \frac{1}{\sqrt{2\pi}} \int_{\mathbb{R}} f(x) \, e^{-i\omega x} \, dx \right\} e^{i\omega t} \, d\omega \\
&= \int_{\mathbb{R}} f(x) \left\{ \frac{1}{2\pi} \int_{\mathbb{R}} e^{i\omega(t-x)} \, d\omega \right\} dx \\
&= \int_{\mathbb{R}} f(x) \, \delta(t-x) \, dx \\
&= f(t),
\end{aligned}$$

where the last equality is due to the fundamental property of the well-known Dirac delta function.

(ii) Note that,

$$\left\langle \mathscr{F}[f], \mathscr{F}[g] \right\rangle_2 = \int_{\mathbb{R}} \mathscr{F}[f](\omega) \, \overline{\mathscr{F}[g](\omega)} \, d\omega$$

$$= \int_{\mathbb{R}} \left\{ \frac{1}{\sqrt{2\pi}} \int_{\mathbb{R}} f(x) \, e^{-i\omega x} \, dx \right\} \left\{ \frac{1}{\sqrt{2\pi}} \int_{\mathbb{R}} \overline{g(y)} \, e^{i\omega y} \, dy \right\} d\omega$$

$$= \int_{\mathbb{R}} \int_{\mathbb{R}} f(x) \, \overline{g(y)} \left\{ \frac{1}{2\pi} \int_{\mathbb{R}} e^{i\omega(y-x)} \, d\omega \right\} dx \, dy$$

$$= \int_{\mathbb{R}} \int_{\mathbb{R}} f(x) \, \overline{g(y)} \, \delta(y - x) \, dx \, dy$$

$$= \int_{\mathbb{R}} f(x) \, \overline{g(x)} \, dx$$

$$= \left\langle f, g \right\rangle_2.$$

Moreover, plugging $f = g$ into the Parseval's formula (1.69), yields the desired norm equality.

(iii) Evidently, the map $\mathscr{F} : L^2(\mathbb{R}) \to L^2(\mathbb{R})$ is both one-to-one and onto. Therefore, we conclude that $f \to \mathscr{F}[f]$ is a Hilbert space isomorphism from $L^2(\mathbb{R})$ onto itself.

This completes the proof of Theorem 1.2.2.   □

Remark 1.2.3. The assertions made in Theorem 1.2.2 follow as a consequence of shift property of the Dirac delta function. Therefore, we infer that the Plancherel theorem holds in "almost everywhere" sense. Having stated that, such nuances may not be highlighted repeatedly in the subsequent developments on the subject.

We shall conclude this subsection by examining the derivatives of the Fourier transform of a given function f. As we know that smoother the function f, the more rapidly $\hat{f}$ will decay at infinity and conversely. Thus, more rapidly f decays at infinity, the smoother $\hat{f}$ will be. There are various ways to measure the smoothness of a given function f, but here we will measure the smoothness of f by counting the number of derivatives it has.

Theorem 1.2.3. *Assume that $f \in L^2(\mathbb{R})$ is such that $tf(t)$ is both integrable as well as square integrable in the sense of Lebesgue. Then, we have*

$$\frac{d\hat{f}(\omega)}{d\omega} = (-i) \, \mathscr{F}\left[tf(t)\right](\omega). \tag{1.71}$$

Proof. From the elementary results on calculus, we note that

$$\frac{d\hat{f}}{d\omega} = \lim_{h \to 0} \left(\frac{\hat{f}(\omega + h) - \hat{f}(\omega)}{h} \right) = \lim_{h \to 0} \left\{ \frac{1}{\sqrt{2\pi}} \int_{\mathbb{R}} e^{-i\omega t} f(t) \left(\frac{e^{-iht} - 1}{h} \right) dt \right\}. \tag{1.72}$$

Observe that

$$\left| e^{-i\omega t} f(t) \left(\frac{e^{-iht} - 1}{h} \right) \right| = |f(t)| \left| \frac{e^{-\frac{iht}{2}} \left(e^{\frac{iht}{2}} - e^{-\frac{iht}{2}} \right)}{h} \right| = |tf(t)| \left| \frac{\sin\left(\frac{ht}{2}\right)}{\left(\frac{ht}{2}\right)} \right| \leq |tf(t)|.$$

Also,

$$\lim_{h \to 0} \left(\frac{e^{-iht} - 1}{h} \right) = -it.$$

Hence, by virtue of the well-known Lebesgue Dominated Convergence Theorem, we obtain

$$\frac{d\hat{f}(\omega)}{d\omega} = \frac{1}{\sqrt{2\pi}} \int_{\mathbb{R}} \lim_{h \to 0} \left\{ e^{-i\omega t} f(t) \left(\frac{e^{-iht} - 1}{h} \right) \right\} dt$$

$$= -\frac{i}{\sqrt{2\pi}} \int_{\mathbb{R}} t f(t) e^{-i\omega t} \, dt$$

$$= (-i) \mathscr{F}\left[t f(t)\right](\omega).$$

This completes the proof of Theorem 1.2.3. □

Corollary 1.2.4. *If $f \in L^2(\mathbb{R})$ such that $t^n f(t)$ is both Lebesgue integrable and square integrable for finite $n \in \mathbb{N}$, then the nth derivative of $\hat{f}$ exists and is given by*

$$\frac{d^n \hat{f}(\omega)}{d\omega^n} = (-i)^n \mathscr{F}\left[t^n f(t)\right](\omega). \tag{1.73}$$

Proof. The proof follows from Theorem 1.2.3 together with the principle of mathematical induction. □

Remark 1.2.4. For $\omega = 0$, the expression (1.73) yields

$$\left[\frac{d^n \hat{f}(\omega)}{d\omega^n}\right]_{\omega=0} = (-i)^n \int_{\mathbb{R}} t^n f(t) \, dt = (-i)^n m_n, \tag{1.74}$$

where m_n represents the n-th moment of $f(t)$. Thus, the moments $m_1, m_2, \ldots, m_n$ can be calculated from (1.73).

1.2.3 Convolution and correlation

The notion of convolution is one of the widely applied concepts of mathematics with application areas ranging from functional analysis to different fields of signal and image processing. Mathematically, the convolution can be viewed as an operation convolving two functions f and g into a third function h, which gives the area overlap between the two functions as a function of the amount by which one of the original functions is translated. The convolution theorem associated with a given convolution can be viewed as a manifestation of the behaviour of the convolution in the transformed domain. In the realm of integral transformations, it is highly desirable that the transformation of two convoluted functions yields a pointwise product of the transformations of the individual functions. The notion of convolution is one of the core research areas in the theory of Fourier transforms and has been widely applied in different areas of science and engineering including quantum mechanics, operator theory and many diverse aspects of signal and image processing [51, 146, 217]. Some prime characteristics of the convolution in the Fourier dominion are mentioned below:

(i) The convolution exhibits commutative and associative properties which play a pivotal role in its applications to signal processing, as any given cascade of systems can be handled in a convenient and lucid manner.

(ii) The convolution in the time domain involves a single integral expression and is as such befitting for the design of computationally efficient filters.

(iii) The Fourier transform of two convoluted signals corresponds to a simple multiplication of the respective Fourier transforms and hence, the multiplicative filters can be easily achieved in the Fourier domain.

Below, we present the formal definition of the convolution operation in the Fourier domain.

Definition 1.2.2. For any $f, g \in L^2(\mathbb{R})$, the convolution operation is denoted by $*$ and is defined as

$$(f * g)(z) = \int_{\mathbb{R}} f(t)\, g(z - t)\, dt. \tag{1.75}$$

As a consequence of Definition 1.2.2, we can infer that the convolution of g against f is a moving average, which is centered at time t. Thus, apart from (1.75), the convolution can also be expressed as

$$(f * g)(z) = \int_{\mathbb{R}} f(z - t)\, g(t)\, dt. \tag{1.76}$$

Some fundamental properties of the convolution operation (1.75) are assembled in the following theorem:

Theorem 1.2.5. *For any trio $f, g, h \in L^2(\mathbb{R})$ and $k \in \mathbb{R}$, $\lambda \in \mathbb{R} \setminus \{0\}$, the convolution operation $*$ defined in (1.75) satisfies the following properties:*

(i) *Commutativity:* $(f * g)(z) = (g * f)(z)$,

(ii) *Associativity:* $\big(f * (g * h)\big)(z) = \big((f * g) * h\big)(z)$,

(iii) *Distributivity:* $\big(f * (g + h)\big)(z) = (f * g)(z) + (f * h)(z)$,

(iv) *Translation:* $(f * g)(z - k) = \big(f(t - k) * g(t)\big)(z)$,

(v) *Scaling:* $(f * g)(\lambda z) = |\lambda|\big(f(\lambda t) * g(\lambda t)\big)(z)$,

(vi) *Reflection:* $(f * g)(-z) = \big(f(-t) * g(-t)\big)(z)$.

Proof. The proof of the theorem can be obtained easily and the same is omitted. $\qquad\square$

Next, our aim is to illustrate the notion of convolution via some technical examples.

Example 1.2.13. Consider the following pair of rectangular functions:

$$f(t) = \begin{cases} 1, & -\frac{T}{2} \le t \le \frac{T}{2}, \quad T > 0 \\ 0, & \text{otherwise}, \end{cases} \quad \text{and} \quad g(t) = \begin{cases} 1, & 0 \le t \le T, \quad T > 0 \\ 0, & \text{otherwise}. \end{cases} \tag{1.77}$$

Then, we observe that

$$g(z - t) = \begin{cases} 1, & 0 \le z - t \le T \\ 0, & \text{otherwise}. \end{cases} = \begin{cases} 1, & z - T \le t \le z, \quad T > 0 \\ 0, & \text{otherwise}. \end{cases} \tag{1.78}$$

To find the convolution function $(f * g)(z)$, we need first to understand the nature of overlapping of the functions $f(t)$ and $g(z - t)$. For this, we depict the overlapping process in Figure 1.21. According to Definition 1.2.2, we need to compute the mirror of the function g (that is, invert the domain of g) to obtain $g(-\cdot)$ and then translate the function along the spatial axis to observe the nature of overlap of the two functions. We note that the first overlap occurs at $z = -T/2$ and the last overlap occurs at $z = 3T/2$ (see Figure 1.21). At the limits $z = -T/2$ and $z = 3T/2$, we start and finish with an overlap of zero (note that the two functions are just adjoined to each other), whereas at $z = T/2$ both the functions completely overlap each other. At $z = T/2$, the value of the convolution integral is exactly T, whereas the integral rises (falls) at a linear rate to the left (right) of $z = T/2$.

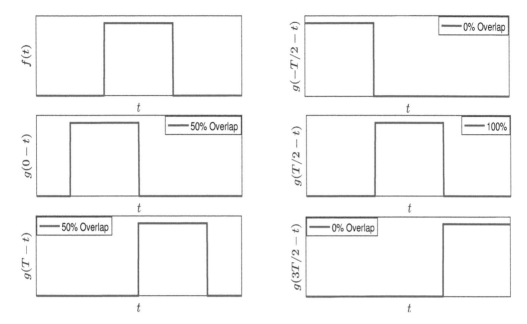

FIGURE 1.21: The process of overlap in the convolution of f and g.

From the graphical illustration, we observe that the convolution of the pair of functions defined via (1.77) is given by

$$
(f * g)(z) = \begin{cases} z + \dfrac{T}{2}, & -\dfrac{T}{2} \le z < \dfrac{T}{2} \\ T, & z = \dfrac{T}{2} \\ \dfrac{3T}{2} - z, & \dfrac{T}{2} < z \le \dfrac{3T}{2}. \end{cases}
\tag{1.79}
$$

The convolution function (1.79) is plotted in Figure 1.22. From Figure 1.22, we observe that the function $(f * g)(z)$ is non-zero over the interval $[-T/2, 3T/2]$ with length $2T$, which is twice as big as that of the individual functions f and g defined in (1.77).

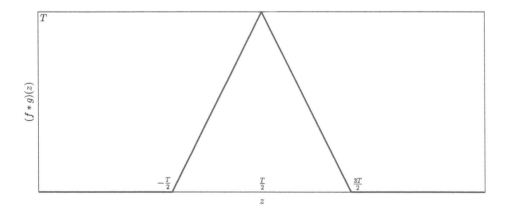

FIGURE 1.22: The convolution function $(f * g)(z)$ given in (1.79).

Example 1.2.14. Choosing f as the unilateral exponential function (1.26) and g as the Gaussian function (1.20). Then, by virtue of Definition 1.2.2, we have

$$
\begin{aligned}
\left(f * g\right)(z) &= \frac{1}{\sigma\sqrt{2\pi}} \int_0^\infty e^{-t/\lambda}\, e^{-(z-t)^2/2\sigma^2}\, dt \\
&= \frac{1}{\sigma\sqrt{2\pi}} \int_0^\infty e^{-t/\lambda}\, e^{-(z^2+t^2-2zt)/2\sigma^2}\, dt \\
&= \frac{e^{-z^2/2\sigma^2}}{\sigma\sqrt{2\pi}} \int_0^\infty e^{-t/\lambda}\, e^{-t^2/2\sigma^2}\, e^{zt/\sigma^2}\, dt \\
&= \frac{1}{\sigma\sqrt{2\pi}} \exp\left\{\frac{\sigma^2}{2\lambda^2} - \frac{z}{\lambda}\right\} \int_0^\infty \exp\left\{-\frac{\left(t-\left(z-\frac{\sigma^2}{\lambda}\right)\right)^2}{2\sigma^2}\right\} dt.
\end{aligned}
\tag{1.80}
$$

Making the substitution $y = t - (z - \sigma^2/\lambda)$, we can simplify the expression (1.80) as follows:

$$
\begin{aligned}
\left(f * g\right)(z) &= \frac{1}{\sigma\sqrt{2\pi}} \exp\left\{\frac{\sigma^2}{2\lambda^2} - \frac{z}{\lambda}\right\} \int_{\frac{\sigma^2}{\lambda}-z}^\infty \exp\left\{-\frac{y^2}{2\sigma^2}\right\} dy \\
&= \frac{1}{2} \exp\left\{\frac{\sigma^2}{2\lambda^2} - \frac{z}{\lambda}\right\} \operatorname{erfc}\left(\frac{\sigma}{\sqrt{2}\lambda} - \frac{z}{\sigma\sqrt{2}}\right),
\end{aligned}
\tag{1.81}
$$

where $\operatorname{erf}(\cdot)$ denotes the error function and $\operatorname{erfc}(\cdot) = 1 - \operatorname{erf}(\cdot)$ is the complimentary error function with the defining equation:

$$
\operatorname{erf}(t) = \frac{2}{\sqrt{\pi}} \int_0^t e^{-x^2}\, dx.
\tag{1.82}
$$

The integral (1.82) is a special sigmoid function that occurs often in probability, statistics, and partial differential equations. Another intimately intertwined function is the imaginary error function $\operatorname{erfi}(\cdot)$, which is defined as $\operatorname{erfi}(t) = -i\operatorname{erf}(it)$. For a pictorial illustration, the error function $\operatorname{erf}(t)$ and the complimentary error function $\operatorname{erfc}(t)$ are shown in Figure 1.23. Moreover, for different choices of the parameters σ and λ, the convolution functions given by (1.81) are plotted in Figure 1.24.

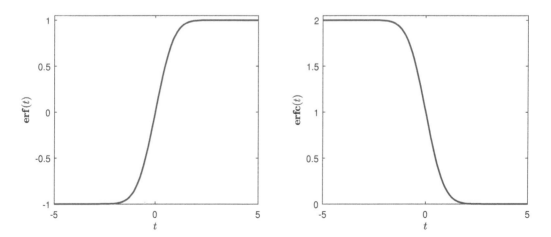

FIGURE 1.23: The error function $\operatorname{erf}(t)$ and the complimentary error function $\operatorname{erfc}(t)$.

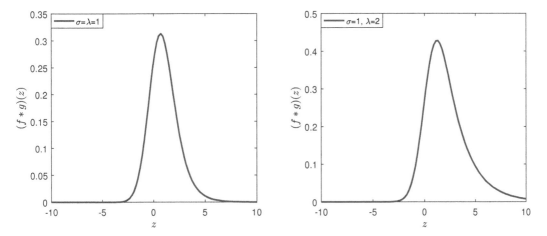

FIGURE 1.24: The convolution function $(f * g)(z)$ given in (1.81).

Example 1.2.15. Consider the following pair of Gaussian functions:

$$f(t) = \frac{1}{\sigma_1 \sqrt{2\pi}} e^{-t^2/2\sigma_1^2} \quad \text{and} \quad g(t) = \frac{1}{\sigma_2 \sqrt{2\pi}} e^{-t^2/2\sigma_2^2}, \quad \sigma_1, \sigma_2 > 0. \tag{1.83}$$

Using Definition 1.2.2 and invoking the standard Gaussian integral, we can compute the convolution of the pair of functions defined in (1.83) as

$$
\begin{aligned}
(f * g)(z) &= \frac{1}{2\pi \, \sigma_1 \sigma_2} \int_{\mathbb{R}} e^{-t^2/2\sigma_1^2} \, e^{-(z-t)^2/2\sigma_2^2} \, dt \\
&= \frac{1}{2\pi \, \sigma_1 \sigma_2} \int_{\mathbb{R}} e^{-t^2/2\sigma_1^2} \, e^{-(z^2+t^2-2zt)^2/2\sigma_2^2} \, dt \\
&= \frac{1}{2\pi \, \sigma_1 \sigma_2} \int_{\mathbb{R}} e^{-z^2/2\sigma_2^2} \, e^{-(1/2\sigma_1^2 + 1/2\sigma_2^2)t^2} \, e^{zt/\sigma_2^2} \, dt \\
&= \frac{e^{-z^2/2\sigma_2^2}}{2\pi \, \sigma_1 \sigma_2} \int_{\mathbb{R}} \exp\left\{ -\left(\frac{1}{2\sigma_1^2} + \frac{1}{2\sigma_2^2} \right) t^2 + \left(\frac{z}{\sigma_2^2} \right) t \right\} \, dt \\
&= \frac{e^{-z^2/2\sigma_2^2}}{2\pi \, \sigma_1 \sigma_2} \left[\frac{\sqrt{\pi}}{\sqrt{\left(\frac{1}{2\sigma_1^2} + \frac{1}{2\sigma_2^2} \right)}} \exp\left\{ \frac{\left(\frac{z}{\sigma_2^2} \right)^2}{4\left(\frac{1}{2\sigma_1^2} + \frac{1}{2\sigma_2^2} \right)} \right\} \right] \\
&= \frac{1}{\sqrt{2\pi(\sigma_1^2 + \sigma_2^2)}} \exp\left\{ -\frac{z^2}{2}\left(\frac{1}{\sigma_2^2} - \frac{\sigma_1^2}{\sigma_2^2(\sigma_1^2 + \sigma_2^2)} \right) \right\} \\
&= \frac{1}{\sqrt{2\pi(\sigma_1^2 + \sigma_2^2)}} \exp\left\{ -\frac{z^2}{2(\sigma_1^2 + \sigma_2^2)} \right\}. \tag{1.84}
\end{aligned}
$$

From the expression (1.84), it is quite evident that the convolution of any pair of Gaussian functions dictated by the controlling parameters $\sigma_1, \sigma_2 > 0$ is again a Gaussian function with the controlling parameter $\sqrt{\sigma_1^2 + \sigma_2^2}$. For different values of σ_1 and σ_2, the convolution functions given by (1.84) are plotted in Figure 1.25.

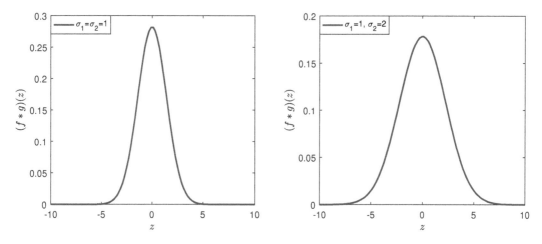

FIGURE 1.25: The convolution function $(f * g)(z)$ given in (1.84).

Following is the convolution theorem associated with the convolution operation defined in (1.75):

Theorem 1.2.6. *For any pair of functions* $f, g \in L^2(\mathbb{R})$, *we have*

$$\mathscr{F}\Big[(f * g)\Big](\omega) = \sqrt{2\pi}\, \mathscr{F}\Big[f\Big](\omega)\, \mathscr{F}\Big[g\Big](\omega). \tag{1.85}$$

Proof. In view of Definition 1.2.2, we obtain

$$
\begin{aligned}
\mathscr{F}\Big[(f * g)\Big](\omega) &= \frac{1}{\sqrt{2\pi}} \int_{\mathbb{R}} (f * g)(z)\, e^{-i\omega z}\, dz \\
&= \frac{1}{\sqrt{2\pi}} \int_{\mathbb{R}} \left\{ \int_{\mathbb{R}} f(t)\, g(z - t)\, dt \right\} e^{-i\omega z}\, dz \\
&= \frac{1}{\sqrt{2\pi}} \int_{\mathbb{R}} f(t) \left\{ \int_{\mathbb{R}} g(z - t)\, e^{-i\omega z}\, dz \right\} dt.
\end{aligned}
\tag{1.86}
$$

Making the substitution $z - t = y$ inside the second integral appearing on the R.H.S of (1.86), we get

$$
\begin{aligned}
\mathscr{F}\Big[(f * g)\Big](\omega) &= \frac{1}{\sqrt{2\pi}} \int_{\mathbb{R}} f(t) \left\{ \int_{\mathbb{R}} g(y)\, e^{-i\omega(t+y)}\, dy \right\} dt \\
&= \int_{\mathbb{R}} f(t) \left\{ \frac{1}{\sqrt{2\pi}} \int_{\mathbb{R}} g(y)\, e^{-i\omega y}\, dy \right\} e^{-i\omega t}\, dt \\
&= \mathscr{F}\Big[g\Big](\omega) \left\{ \int_{\mathbb{R}} f(t)\, e^{-i\omega t}\, dt \right\} \\
&= \sqrt{2\pi}\, \mathscr{F}\Big[g\Big](\omega) \left\{ \frac{1}{\sqrt{2\pi}} \int_{\mathbb{R}} f(t)\, e^{-i\omega t}\, dt \right\} \\
&= \sqrt{2\pi}\, \mathscr{F}\Big[f\Big](\omega)\, \mathscr{F}\Big[g\Big](\omega).
\end{aligned}
$$

This completes the proof of Theorem 1.2.6.   □

Remark 1.2.5. Theorem 1.2.6 asserts that the convolution of two functions in the spatial domain leads to a simple multiplication of their respective spectra in the Fourier domain. This behaviour of the convolution (1.75) plays a crucial role in diverse aspects of signal processing, particularly in the sampling and filtering theories.

Theorem 1.2.7. *Given a pair of functions $f, g \in L^2(\mathbb{R})$, with their respective Fourier transforms as $\hat{f}$ and $\hat{g}$. If $h \in L^2(\mathbb{R})$ is another function defined as $h(t) = f(t) \cdot g(t)$, then we have*

$$\mathscr{F}\big[h\big](\omega) = \frac{1}{\sqrt{2\pi}} \left(\mathscr{F}\big[f\big] * \mathscr{F}\big[g\big]\right)(\omega). \tag{1.87}$$

Proof. Using Definition 1.2.1 together with the definition of inverse Fourier transform (1.67), we get

$$
\begin{aligned}
\mathscr{F}\big[h\big](\omega) &= \frac{1}{\sqrt{2\pi}} \int_{\mathbb{R}} h(t)\, e^{-i\omega t}\, dt \\
&= \frac{1}{\sqrt{2\pi}} \int_{\mathbb{R}} \big(f(t) \cdot g(t)\big)\, e^{-i\omega t}\, dt \\
&= \frac{1}{\sqrt{2\pi}} \int_{\mathbb{R}} \left\{ \frac{1}{\sqrt{2\pi}} \int_{\mathbb{R}} \mathscr{F}\big[f\big](\xi)\, e^{i\xi t}\, d\xi \right\} \left\{ \frac{1}{\sqrt{2\pi}} \int_{\mathbb{R}} \mathscr{F}\big[g\big](\eta)\, e^{i\eta t}\, d\eta \right\} e^{-i\omega t}\, dt \\
&= \frac{1}{\sqrt{2\pi}} \int_{\mathbb{R}} \int_{\mathbb{R}} \mathscr{F}\big[f\big](\xi)\, \mathscr{F}\big[g\big](\eta) \left\{ \frac{1}{2\pi} \int_{\mathbb{R}} e^{i(\xi+\eta-\omega)t}\, dt \right\} d\xi\, d\eta \\
&= \frac{1}{\sqrt{2\pi}} \int_{\mathbb{R}} \int_{\mathbb{R}} \mathscr{F}\big[f\big](\xi)\, \mathscr{F}\big[g\big](\eta)\, \delta(\xi + \eta - \omega)\, d\xi\, d\eta \\
&= \frac{1}{\sqrt{2\pi}} \int_{\mathbb{R}} \mathscr{F}\big[f\big](\xi)\, \mathscr{F}\big[g\big](\omega - \xi)\, d\xi \\
&= \frac{1}{\sqrt{2\pi}} \left(\mathscr{F}\big[f\big] * \mathscr{F}\big[g\big]\right)(\omega).
\end{aligned}
$$

This completes the proof of Theorem 1.2.7. □

Remark 1.2.6. By virtue of Theorem 1.2.7, we conclude that the product of a pair of functions in the spatial domain leads to convolution in the Fourier domain. For this reason, it is sometimes called as "inverse convolution theorem."

Next, our aim is to study the notion of correlation and formulate the corresponding correlation theorem. The correlation is ideally suited for examining that whether there is anything in common between a given pair of functions. Technically, the correlation can be viewed as a convolution operation when one function is axis reversed and complex conjugated.

Definition 1.2.3. For any pair of functions $f, g \in L^2(\mathbb{R})$, the cross-correlation or simply correlation is defined by

$$\big(f \otimes g\big)(z) = \int_{\mathbb{R}} f(t)\, \overline{g(t - z)}\, dt. \tag{1.88}$$

In the next theorem, we assemble some fundamental properties of the correlation operation defined in (1.88).

Theorem 1.2.8. *For any $f, g, h \in L^2(\mathbb{R})$ and $k \in \mathbb{R}$, $\lambda \in \mathbb{R} \setminus \{0\}$, the correlation operation $\otimes$ satisfies the following properties:*

(i) *Non-commutativity:* $(f \otimes g)(z) = \overline{(g \otimes f)(-z)}$,

(ii) *Distributivity:* $(f \otimes (g + h))(z) = (f \otimes g)(z) + (f \otimes h)(z)$,

(iii) *Translation:* $(f \otimes g)(z - k) = (f(t - k) \otimes g(t))(z)$,

(iv) *Scaling:* $(f \otimes g)(\lambda z) = |\lambda|(f(\lambda t) \otimes g(\lambda t))(z)$,

(v) *Reflection:* $(f \otimes g)(-z) = (f(-t) \otimes g(-t))(z)$.

Proof. The proof of Theorem 1.2.8 can be obtained quite straightforwardly and is, therefore, omitted. □

Below, we have the correlation theorem corresponding to the correlation operation $\otimes$ given in Definition 1.2.3.

Theorem 1.2.9. *For any $f, g \in L^2(\mathbb{R})$, we have*

$$\mathscr{F}\Big[(f \otimes g)\Big](\omega) = \sqrt{2\pi}\, \mathscr{F}\Big[f\Big](\omega)\, \overline{\mathscr{F}\Big[g\Big](\omega)}. \tag{1.89}$$

Proof. Applying Definition 1.2.3, we obtain

$$\begin{aligned}
\mathscr{F}\Big[(f \otimes g)\Big](\omega) &= \frac{1}{\sqrt{2\pi}} \int_{\mathbb{R}} (f \otimes g)(z)\, e^{-i\omega z}\, dz \\
&= \frac{1}{\sqrt{2\pi}} \int_{\mathbb{R}} \left\{ \int_{\mathbb{R}} f(t)\, \overline{g(t - z)}\, dt \right\} e^{-i\omega z}\, dz \\
&= \int_{\mathbb{R}} f(t) \left\{ \frac{1}{\sqrt{2\pi}} \int_{\mathbb{R}} \overline{g(t - z)}\, e^{-i\omega z}\, dz \right\} dt \\
&= \int_{\mathbb{R}} f(t) \left\{ \frac{1}{\sqrt{2\pi}} \int_{\mathbb{R}} \overline{g(-(z - t))}\, e^{-i\omega z}\, dz \right\} dt. \tag{1.90}
\end{aligned}$$

Using the scaling, translation and conjugation properties of the Fourier transform studied in Theorem 1.2.1, the expression (1.90) can be simplified as

$$\begin{aligned}
\mathscr{F}\Big[(f \otimes g)\Big](\omega) &= \int_{\mathbb{R}} f(t)\, \mathscr{F}\Big[\overline{g}\Big](-\omega)\, e^{-i\omega t}\, dt \\
&= \int_{\mathbb{R}} f(t)\, \overline{\mathscr{F}\Big[g\Big](\omega)}\, e^{-i\omega t}\, dt \\
&= \sqrt{2\pi}\, \overline{\mathscr{F}\Big[g\Big](\omega)} \left\{ \frac{1}{\sqrt{2\pi}} \int_{\mathbb{R}} f(t)\, e^{-i\omega t}\, dt \right\} \\
&= \sqrt{2\pi}\, \mathscr{F}\Big[f\Big](\omega)\, \overline{\mathscr{F}\Big[g\Big](\omega)}.
\end{aligned}$$

This completes the proof of Theorem 1.2.9. □

A particular and interesting case of Definition 1.2.3 arises when the function g is replaced with f, in this case (1.88) is called the "auto-correlation" of f. That is, the auto-correlation of a function f is the cross-correlation of f with itself. One may wonder, for what purpose we would want to check for what $f(t)$ has in common with $f(t)$. Auto-correlation, however, seems to go way beyond than that. For instance, in signal processing, the auto-correlation is heavily used in improving the signal-to-noise ratio (SNR) abreast of many other advantages. If $h(z)$ denotes the auto-correlation of the function $f(t)$ with itself, then

$$h(z) = \int_{\mathbb{R}} f(t)\, \overline{f(t - z)}\, dt. \tag{1.91}$$

As a consequence of Theorem 1.2.9, we obtain the Fourier transform of (1.91) as

$$\mathscr{F}\Big[h\Big](\omega) = \sqrt{2\pi}\,\mathscr{F}\Big[f\Big](\omega)\,\overline{\mathscr{F}\Big[f\Big](\omega)} = \sqrt{2\pi}\,\Big|\mathscr{F}\Big[f\Big](\omega)\Big|^2. \tag{1.92}$$

That is the Fourier transform of the auto-correlated function is a $\sqrt{2\pi}$-multiple of the squared magnitude of the original function. As it appears, the squared magnitude always pleases the eye if we want to do cosmetics to a noisy spectrum. The big spectral components will grow even bigger when squared, whereas, the smaller ones will get diminished.

The auto-correlation is also used in the femtosecond measuring devices. A femtosecond is one part in a thousand trillion (US) or a thousand billion (British) of a second. Although the duration of time is very short, however, modern devices can easily produce such short duration laser pulses. In order to measure such shorter durations, we use a system known as the "auto-correlator." Initially, the laser pulse is split into two pulses, which are allowed to travel a slightly different optical length using mirrors and are then combined afterwards. The detector is an "optical coincidence" which yields an output only if both the pulses overlap. By tuning the optical path (using the nanometer screw), one can shift one pulse over the other, that is perform a cross-correlation of the pulse with itself. This entire system is called auto-correlation.

1.2.4 Shannon's sampling theorem

"The fundamental problem of communication is that of reproducing at one point either exactly or approximately a message selected at another point."

-Claude E. Shannon

The celebrated Shannon's sampling theorem in the Fourier domain is one of the remarkable, profound and elegant concepts of digital signal processing which serves as a bridge between the analog and digital signals. The theorem asserts that a band-limited signal can be completely reconstructed from its values at regularly spaced times. More explicitly, the theorem says that if a given signal has a range of n-frequencies (measured in cycles/sec) and if $2n$ evenly spaced samples per second are taken, then the signal can be perfectly reconstructed. If we take any additional samples, it will turn out to be redundant, while as if we take fewer samples, we will lose some quality. This is a fundamental result in the field of information theory, in particular, telecommunications and signal processing [266, 292, 323]. It is commonly called the "Shannon's sampling theorem" and is also known as "Shannon-Whittaker sampling theorem." We begin with a definition which is central to this theorem.

Definition 1.2.4. A function $f \in L^2(\mathbb{R})$ is said to be "Ω-band-limited" if its Fourier transform has a compact support in the sense that there exists a constant $\Omega > 0$ such that

$$\mathscr{F}\Big[f\Big](\omega) = 0, \quad \forall\, |\omega| > \Omega. \tag{1.93}$$

If Ω is the smallest value for which (1.93) holds, then it is called as "bandwidth" of the signal. In that case, the natural frequency $\nu = \Omega/2\pi$ is called the "Nyquist frequency" and $2\nu = \Omega/\pi$ is the "Nyquist rate."

As an illustration to Definition 1.2.4, we note that the audibility of humans is band-limited in the sense that the human ear can only detect sounds in the frequency range 20 Hz to 20 kHz (1 Hz = 1000 cycles/sec). Even though, some species make sounds below 20

Hz and others above 20 kHz, humans cannot hear such sounds due to constraints on the audible spectrum.

In the following theorem, we demonstrate that a band-limited signal can be reconstructed from its values at regularly spaces times.

Theorem 1.2.10. *If $f \in L^2(\mathbb{R})$ is continuous and band-limited with bandwidth Ω. Then, $f(t)$ can be reconstructed via*

$$f(t) = \sum_{n \in \mathbb{Z}} f\left(\frac{n\pi}{\Omega}\right) \frac{\sin\left(\Omega t - n\pi\right)}{\left(\Omega t - n\pi\right)}. \tag{1.94}$$

Proof. Since f is band-limited with bandwidth Ω; that is, $\hat{f}(\omega) = 0$, for $|\omega| > \Omega$, therefore, using the inversion formula (1.67) pertaining to the Fourier transform, we have

$$f(t) = \frac{1}{\sqrt{2\pi}} \int_{-\Omega}^{\Omega} \hat{f}(\omega)\, e^{i\omega t}\, d\omega. \tag{1.95}$$

Equation (1.95) holds for all t as the integral $\int_{-\Omega}^{\Omega} \hat{f}(\omega)\, e^{i\omega t}\, d\omega$ is a continuous function of t and f is given to be continuous. Since $\hat{f} \in L^2[-\Omega, \Omega]$, so we can expand $\hat{f}(\omega)$ into a Fourier series over the interval $[-\Omega, \Omega]$ as

$$\hat{f}(\omega) = \sum_{n \in \mathbb{Z}} c_n\, e^{-i\pi n\omega/\Omega}, \tag{1.96}$$

where

$$c_n = \frac{1}{2\Omega} \int_{-\Omega}^{\Omega} \hat{f}(\omega)\, e^{i\pi n\omega/\Omega}\, d\omega. \tag{1.97}$$

Since $\hat{f}(\omega) = 0$, for $|\omega| \geq \Omega$, the limits of integration in (1.97) can be changed to $-\infty$ to ∞, so that

$$
\begin{aligned}
c_n &= \frac{1}{2\Omega} \int_{\mathbb{R}} \hat{f}(\omega)\, e^{i(\pi n/\Omega)\omega}\, d\omega \\
&= \frac{\sqrt{2\pi}}{2\Omega} \left\{ \frac{1}{\sqrt{2\pi}} \int_{\mathbb{R}} \hat{f}(\omega)\, e^{i(\pi n/\Omega)\omega}\, d\omega \right\} \\
&= \frac{\sqrt{2\pi}}{2\Omega}\, f\left(\frac{n\pi}{\Omega}\right).
\end{aligned}
\tag{1.98}
$$

After substituting (1.98) in (1.96), we obtain

$$\hat{f}(\omega) = \frac{\sqrt{2\pi}}{2\Omega} \sum_{n \in \mathbb{Z}} f\left(\frac{n\pi}{\Omega}\right) e^{-in\pi\omega/\Omega}. \tag{1.99}$$

Implementing (1.99) in the relation (1.96), we obtain

$$
\begin{aligned}
f(t) &= \frac{\sqrt{2\pi}}{2\Omega} \sum_{n \in \mathbb{Z}} \frac{1}{\sqrt{2\pi}} \int_{-\Omega}^{\Omega} f\left(\frac{n\pi}{\Omega}\right) e^{-in\pi\omega/\Omega}\, e^{i\omega t}\, dt \\
&= \frac{1}{2\Omega} \sum_{n \in \mathbb{Z}} f\left(\frac{n\pi}{\Omega}\right) \int_{-\Omega}^{\Omega} e^{\left(it - \frac{in\pi}{\Omega}\right)\omega}\, d\omega
\end{aligned}
$$

$$= \frac{1}{2\Omega} \sum_{n\in\mathbb{Z}} f\left(\frac{n\pi}{\Omega}\right) \left[\frac{e^{\left(it - \frac{in\pi}{\Omega}\right)\omega}}{\left(it - \frac{in\pi}{\Omega}\right)} \right]_{-\Omega}^{\Omega}$$

$$= \sum_{n\in\mathbb{Z}} f\left(\frac{n\pi}{\Omega}\right) \left[\frac{e^{i(\Omega t - n\pi)} - e^{-i(\Omega t - n\pi)}}{2i(\Omega t - n\pi)} \right]$$

$$= \sum_{n\in\mathbb{Z}} f\left(\frac{n\pi}{\Omega}\right) \frac{\sin\left(\Omega t - n\pi\right)}{(\Omega t - n\pi)}.$$

This completes the proof of Theorem 1.2.10. □

Remark 1.2.7. The convergence in (1.94) is somewhat slow as the coefficients (in absolute value) decay as $1/n$. The convergence rate can be increased so that the terms behave like $1/n^2$ or better, by a technique called as "over-sampling." On the other hand, if a signal is sampled below the Nyquist rate, then the reconstructed signal will not only be missing high frequency components, but it will also have the energy in those components transferred to low frequencies that may not have been in the original signal at all. This phenomenon is called as "aliasing."

Example 1.2.16. Consider the function f defined via the Fourier transform

$$\hat{f}(\omega) = \begin{cases} \sqrt{2\pi}\,(1 - \omega^2), & |\omega| \le 1 \\ 0, & |\omega| > 1. \end{cases} \tag{1.100}$$

Then, with the aid of Fourier inversion formula (1.67), we may write

$$f(t) = \frac{1}{\sqrt{2\pi}} \int_{\mathbb{R}} \hat{f}(\omega) e^{i\omega t}\, d\omega$$

$$= \frac{1}{\sqrt{2\pi}} \int_{-1}^{1} \sqrt{2\pi}\,(1 - \omega^2) e^{i\omega t}\, d\omega$$

$$= \int_{-1}^{1} e^{i\omega t}\, d\omega - \int_{-1}^{1} \omega^2 e^{i\omega t}\, d\omega$$

$$= \left[\frac{e^{i\omega t}}{it} \right]_{-1}^{1} - \left[\omega^2 \int e^{i\omega t}\, d\omega \right]_{-1}^{1} + \int_{-1}^{1} 2\omega \left(\frac{e^{i\omega t}}{it} \right) d\omega$$

$$= \left[\frac{e^{i\omega t}}{it} \right]_{-1}^{1} - \left[\omega^2 \left(\frac{e^{i\omega t}}{it} \right) \right]_{-1}^{1} + \frac{2}{it} \left\{ \omega \left[\frac{e^{i\omega t}}{it} \right]_{-1}^{1} - \int_{-1}^{1} \frac{e^{i\omega t}}{it}\, d\omega \right\}$$

$$= \frac{\left(e^{it} - e^{-it}\right)}{it} - \frac{\left(e^{it} - e^{-it}\right)}{it} - \frac{2\left(e^{it} + e^{-it}\right)}{t^2} + \frac{2\left(e^{it} - e^{-it}\right)}{it^3}$$

$$= \frac{4\sin t - 4t\cos t}{t^3}. \tag{1.101}$$

The function f given by (1.101) is plotted in Figure 1.26. Since, $\hat{f}(\omega) = 0$ for $|\omega| > 1$, therefore, choosing $\Omega = 1$ in (1.94), we obtain a series representation of the signal (1.101) as follows:

$$f(t) = \sum_{n\in\mathbb{Z}} f(n\pi) \frac{\sin(t - n\pi)}{(t - n\pi)}. \tag{1.102}$$

The partial sum of the first 30 terms in the series (1.102) is plotted in Figure 1.27, which has a striking resemblance to Figure 1.26.

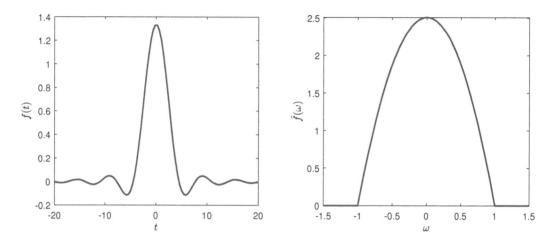

FIGURE 1.26: The signal $f(t)$ and its Fourier transform (band-limited signal).

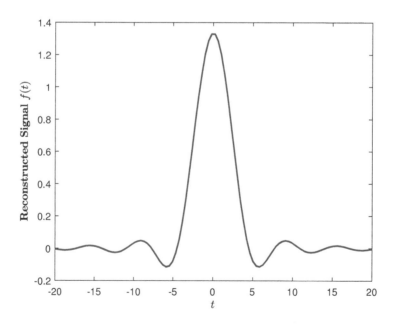

FIGURE 1.27: The reconstructed signal.

"The American mathematician and computer scientist [Shannon] who conceived and laid the foundations for information theory. His theories laid the groundwork for the electronic communications networks that now lace the earth."

-IEEE Information Theory Society (2017)

Before concluding the subsection, we would like to emphasize on the remarkability of Theorem 1.2.10. Without the fear of exaggeration, the Shannon's sampling theorem is the pedestal of modern communication systems. For a simple explanation, lets take a look at how we might use it to transmit several phone conversations simultaneously over a single channel. Since, the maximum possible (natural) frequency that we can hear is about 20 kHz, so a phone conversation $f(t)$ is effectively a band-limited signal. In fact, the dominant frequencies in most phone conversations are below 1 kHz, which we will take as the Nyquist frequency. In that case, the Nyquist rate is clearly 2 kHz, so we need to sample the signal every half-milisecond. How many phone conversations can we send in this manner? Transmission lines typically send about 56 thousand bits of information per second. If each sample of information can be represented by 7 bits, then we can transmit 8 thousand samples per second, or 8 every milisecond, or 4 every half-milisecond. By tagging and interlacing signals, we can transmit the samples from four conversations. At the receiving end, we can use the series expansion (1.94) to reconstruct the signal from the samples given by $f(n/2)$, with n being an integer and time in miliseconds.

1.2.5 Uncertainty principle for the Fourier transform

"The first gulp from the glass of natural sciences will turn you into an atheist, but at the bottom of the glass God is waiting for you."

-Werner K. Heisenberg

One of the fundamental cornerstones of quantum mechanics is the uncertainty principle established by Werner Heisenberg $(1901 - 1976)$, which states that the position and the momentum of a particle cannot be both determined explicitly but only in probabilistic sense with a certain uncertainty [111]. That is, increasing the knowledge of the position, decreases the knowledge of the momentum of the particle and vice versa. On the flip side, the uncertainty principle also has a useful interpretation in the context of harmonic analysis. This aspect of the uncertainty principle was expounded by Norbert Wiener $(1894 - 1964)$ in a lecture in Göttingen in 1925. Unfortunately, no written record of this lecture seems to have survived, apart from the non-technical account in Wiener's autobiography [312]. The formal account of the uncertainty principle, which synced with the minds of signal analysts can be traced back to the Cabot's fundamental work in 1946. Since then, the uncertainty principle has attained a central place in harmonic analysis as it expresses ones limitations in dealing with the spatial and frequency domains simultaneously [36, 137]. Here, we study the uncertainty principle in the context of the Fourier transform, which leads to the assertion that a function cannot be simultaneously well localized in both the time and frequency domains. That is, there is always a trade-off of localization in the time and frequency domains. To facilitate the intent, we have the following definition:

Definition 1.2.5. Given any square integrable function f on $\mathbb{R}$, the dispersion about a point t_0 is defined as

$$\Delta_{t_0}^2 f = \frac{\displaystyle\int_{\mathbb{R}} (t - t_0)^2 \left| f(t) \right|^2 dt}{\displaystyle\int_{\mathbb{R}} \left| f(t) \right|^2 dt} = \frac{1}{\left\| f \right\|_2^2} \left\{ \int_{\mathbb{R}} (t - t_0)^2 \left| f(t) \right|^2 dt \right\}. \qquad (1.103)$$

The dispersion of a function f about the point t_0 measures the deviation or spread of the graph of f with reference to t_0. A small value of the dispersion implies that the graph is concentrated about the reference point, whereas a large value of the dispersion means that the graph is spread out with regard to the reference point. If we set

$$h(t) = \frac{|f(t)|^2}{\displaystyle\int_{\mathbb{R}} |f(t)|^2 dt} = \frac{|f(t)|^2}{\|f\|_2^2}, \tag{1.104}$$

then we observe that $h(t)$ is non-negative and

$$\int_{\mathbb{R}} h(t)\, dt = \frac{1}{\|f\|_2^2} \int_{\mathbb{R}} |f(t)|^2 dt = \frac{\|f\|_2^2}{\|f\|_2^2} = 1. \tag{1.105}$$

Thus, $h(t)$ can be thought of as a probability density function. If t_0 is the mean of this density, then $\Delta_{t_0}^2 f$ is simply the variance. Moreover, in analogy to (1.103), the dispersion of the Fourier transform $\hat{f}$ about the point ω_0 is defined as

$$\Delta_{\omega_0}^2 \hat{f} = \frac{\displaystyle\int_{\mathbb{R}} (\omega - \omega_0)^2 |\hat{f}(\omega)|^2 d\omega}{\displaystyle\int_{\mathbb{R}} |\hat{f}(\omega)|^2 d\omega} = \frac{1}{\|\hat{f}\|_2^2} \left\{ \int_{\mathbb{R}} (\omega - \omega_0)^2 |\hat{f}(\omega)|^2 d\omega \right\}. \tag{1.106}$$

By virtue of the Parseval's formula (1.70), we can express (1.106) in the following fashion:

$$\Delta_{\omega_0}^2 \hat{f} = \frac{1}{\|f\|_2^2} \left\{ \int_{\mathbb{R}} (\omega - \omega_0)^2 |\hat{f}(\omega)|^2 d\omega \right\}. \tag{1.107}$$

In case the value of $\Delta_{\omega_0}^2 \hat{f}$ is small, the frequency range of f must be concentrated about ω_0, whereas the frequency range would be spread out in case $\Delta_{\omega_0}^2 \hat{f}$ is large.

Following is the Heisenberg's uncertainty principle associated with the Fourier transform:

Theorem 1.2.11. *For any $f \in L^2(\mathbb{R})$ that vanishes at $\pm\infty$, the following inequality holds:*

$$\Delta_{t_0}^2 f \cdot \Delta_{\omega_0}^2 \hat{f} \geq \frac{1}{4}, \qquad t_0, \omega_0 \in \mathbb{R}. \tag{1.108}$$

Proof. To begin with, we note that any function $f \in L^2(\mathbb{R})$ can be expressed in the following fashion:

$$f(t) = \left(\left(\frac{d}{dt} - i\omega_0 \right)(t - t_0) \right) f(t) - \left((t - t_0)\left(\frac{d}{dt} - i\omega_0 \right) \right) f(t). \tag{1.109}$$

The assertion (1.109) follows immediately from the fact that

$$\frac{d}{dt}\left((t - t_0) f(t) \right) - i\omega_0 (t - t_0) f(t) - (t - t_0)\left(\frac{df}{dt} - i\omega_0 f(t) \right)$$
$$= (t - t_0) f'(t) + f(t) - i\omega_0 (t - t_0) f(t) - (t - t_0) f'(t) + i\omega_0 (t - t_0) f(t)$$
$$= f(t).$$

By taking the L^2-inner product on both sides of (1.109) with respect to f, we obtain

$$\left\langle \left(\frac{d}{dt} - i\omega_0\right)(t - t_0)f(t), f(t)\right\rangle_2 - \left\langle (t - t_0)\left(\frac{df}{dt} - i\omega_0\right)f(t), f(t)\right\rangle_2 = \langle f, f\rangle_2$$

$$= \|f\|_2^2. \quad (1.110)$$

In order to simplify the first expression on the L.H.S of (1.110), we proceed in the following manner:

$$\left\langle \left(\frac{d}{dt} - i\omega_0\right)(t - t_0)f(t), f(t)\right\rangle_2 = \int_{\mathbb{R}} \left(\frac{d}{dt} - i\omega_0\right)(t - t_0)f(t)\,\overline{f(t)}\,dt$$

$$= \int_{\mathbb{R}} \left\{(t - t_0)f'(t) + f(t) - i\omega_0(t - t_0)f(t)\right\}\overline{f(t)}\,dt$$

$$= \int_{\mathbb{R}} (t - t_0)f'(t)\overline{f(t)}\,dt + \int_{\mathbb{R}} |f(t)|^2\,dt$$

$$- i\omega_0 \int_{\mathbb{R}} (t - t_0)f(t)\,\overline{f(t)}\,dt. \quad (1.111)$$

Applying the product rule of integration to simplify the first integral appearing on the R.H.S of (1.111), we obtain

$$\left\langle \left(\frac{d}{dt} - i\omega_0\right)(t - t_0)f(t), f(t)\right\rangle_2 = \left[(t - t_0)f(t)\,\overline{f(t)}\right]_{-\infty}^{\infty} - \int_{\mathbb{R}} \frac{d}{dt}\left((t - t_0)\overline{f(t)}\right)f(t)\,dt$$

$$+ \int_{\mathbb{R}} |f(t)|^2\,dt - i\omega_0 \int_{\mathbb{R}} (t - t_0)f(t)\,\overline{f(t)}\,dt. \quad (1.112)$$

Since f vanishes at $\pm\infty$, the expression (1.112) boils down to

$$\left\langle \left(\frac{d}{dt} - i\omega_0\right)(t - t_0)f(t), f(t)\right\rangle_2$$

$$= -\int_{\mathbb{R}} \frac{d}{dt}\left((t - t_0)\overline{f(t)}\right)f(t)\,dt + \int_{\mathbb{R}} |f(t)|^2\,dt - i\omega_0 \int_{\mathbb{R}} (t - t_0)f(t)\,\overline{f(t)}\,dt$$

$$= -\int_{\mathbb{R}} (t - t_0)\left(\frac{d}{dt}\,\overline{f(t)}\right)f(t)\,dt - \int_{\mathbb{R}} |f(t)|^2\,dt + \int_{\mathbb{R}} |f(t)|^2\,dt$$

$$- \int_{\mathbb{R}} i\omega_0(t - t_0)f(t)\,\overline{f(t)}\,dt$$

$$= \int_{\mathbb{R}} (t - t_0)\left(-\frac{d}{dt}\,\overline{f(t)}\right)f(t)\,dt - \int_{\mathbb{R}} i\omega_0(t - t_0)f(t)\,\overline{f(t)}\,dt$$

$$= \int_{\mathbb{R}} (t - t_0)f(t)\left(-\frac{d}{dt}\,\overline{f(t)} - i\omega_0\,\overline{f(t)}\right)dt$$

$$= \int_{\mathbb{R}} (t - t_0)f(t)\overline{\left\{\left(-\frac{d}{dt} + i\omega_0\right)f(t)\right\}}\,dt$$

$$= \left\langle (t - t_0)f(t), \left(-\frac{d}{dt} + i\omega_0\right)f(t)\right\rangle_2. \quad (1.113)$$

Similarly, we can show that

$$\left\langle (t - t_0)\left\{\left(\frac{d}{dt} - i\omega_0\right)f(t)\right\}, f(t)\right\rangle_2 = \left\langle \left(\frac{d}{dt} - i\omega_0\right)f(t), (t - t_0)f(t)\right\rangle_2. \quad (1.114)$$

Substituting (1.113) and (1.114) in (1.112), we obtain

$$\left\langle (t - t_0)\, f(t), \left(-\frac{d}{dt} + i\omega_0 \right) f(t) \right\rangle_2 - \left\langle \left(\frac{d}{dt} - i\omega_0 \right) f(t), (t - t_0)\, f(t) \right\rangle_2 = \left\| f \right\|_2^2.$$

(1.115)

Implementation of the triangle inequality in (1.115) yields

$$\left\| f \right\|_2^2 \leq \left| \left\langle (t - t_0)\, f(t), \left(-\frac{d}{dt} + i\omega_0 \right) f(t) \right\rangle_2 \right| + \left| \left\langle \left(\frac{d}{dt} - i\omega_0 \right) f(t), (t - t_0)\, f(t) \right\rangle_2 \right|.$$

(1.116)

Applying the Cauchy-Schwarz inequality to (1.116), we obtain the following norm inequality:

$$\left\| f \right\|_2^2 \leq 2 \left\| \left(\frac{d}{dt} - i\omega_0 \right) f(t) \right\|_2 \left\| (t - t_0)\, f(t) \right\|_2.$$

(1.117)

Moreover, we observe that

$$\begin{aligned}
\mathscr{F}\left[\frac{d}{dt} f(t) \right](\omega) &= \frac{1}{\sqrt{2\pi}} \int_{\mathbb{R}} \left(\frac{d}{dt} f(t) \right) e^{-i\omega t}\, dt \\
&= \frac{1}{\sqrt{2\pi}} \left\{ \left[f(t)\, e^{-i\omega t} \right]_{-\infty}^{\infty} - (-i\omega) \int_{\mathbb{R}} f(t)\, e^{-i\omega t}\, dt \right\} \\
&= i\omega \left\{ \frac{1}{\sqrt{2\pi}} \int_{\mathbb{R}} f(t)\, e^{-i\omega t}\, dt \right\} \\
&= i\omega\, \mathscr{F}\left[f \right](\omega).
\end{aligned}$$

(1.118)

Therefore, we have

$$\begin{aligned}
\mathscr{F}\left[\left(\frac{d}{dt} - i\omega_0 \right) f(t) \right](\omega) &= i\omega\, \mathscr{F}\left[\frac{d}{dt} f(t) \right](\omega) - i\omega_0\, \mathscr{F}\left[f \right](\omega) \\
&= i(\omega - \omega_0)\, \mathscr{F}\left[f \right](\omega).
\end{aligned}$$

(1.119)

Using the Parseval's formula (1.70) together with (1.119), we can express the pre-factor on the R.H.S of (1.117) as

$$\left\| \left(\frac{d}{dt} - i\omega_0 \right) f(t) \right\|_2 = \left\| (\omega - \omega_0)\, \hat{f}(\omega) \right\|_2.$$

(1.120)

Hence, inequality (1.117) can be recast as

$$\left\| (t - t_0)\, f(t) \right\|_2 \left\| (\omega - \omega_0)\, \hat{f}(\omega) \right\|_2 \geq \frac{1}{2} \left\| f \right\|_2^2.$$

(1.121)

On the other hand, we note that

$$\Delta_{t_0}^2 f = \frac{\left\| (t - t_0)\, f(t) \right\|_2^2}{\left\| f \right\|_2^2} \quad \text{and} \quad \Delta_{\omega_0}^2 \hat{f} = \frac{\left\| (\omega - \omega_0)\, \hat{f}(\omega) \right\|_2^2}{\left\| f \right\|_2^2}.$$

(1.122)

Finally, upon squaring (1.121) on both sides and then using (1.122), we obtain the desired Heisenberg's uncertainty inequality:

$$\Delta_{t_0}^2 f \cdot \Delta_{\omega_0}^2 \hat{f} \geq \frac{1}{4}.$$

This completes the proof of Theorem 1.2.11.   □

Remark 1.2.8. In case the reference points t_0 and ω_0 in Theorem 1.2.11 are chosen to be $t_0 = \omega_0 = 0$, then the Heisenberg's uncertainty inequality (1.108) becomes

$$\Delta^2 f \cdot \Delta^2 \hat{f} \geq \frac{1}{4}. \tag{1.123}$$

It is pertinent to mention that equality holds in (1.123) if and only if equality holds in (1.116) and (1.117) with $t_0 = 0 = \omega_0$. Indeed, this is the case when

$$\frac{d}{dt} f(t) = kt\, f(t), \tag{1.124}$$

where k is a scalar. This leads to the Gaussian signal $f(t) = C\, e^{-bt^2}$, where C is a constant and $b = -k/2 > 0$. Note that the positivity of b ensures that the function is indeed square integrable.

Remark 1.2.9. From a probabilistic sense, it would suffice to prove Theorem 1.2.11 around the zero mean values $t_0 = \omega_0 = 0$. This is because any square integrable function f can be translated by t_0 and modulated by ω_0 to obtain a function having zero mean values in the spatial and Fourier domains, and having the respective variances same as that of f and $\hat{f}$. This is due to the fact that modulation causes a simple shift in the Fourier domain that clearly does not affect the variance of the Fourier transform. From similar arguments, and using the duality between modulation and shifting in the Fourier pairs f and $\hat{f}$, we conclude that shifting also does not change the variance. Hence, it can be easily verified that equality holds in (1.108) if and only if f is a translated and modulated Gaussian: $f(t) = C\, e^{i\omega_0 t} e^{-b(t-t_0)^2}$, where $C \in \mathbb{C}$ and $b \in \mathbb{R}^+$.

Remark 1.2.10. Most often in literature, the statement of Theorem 1.2.11 is devoid of the condition that the given square integrable function f vanishes at $\pm\infty$. In fact, the realistic assumption regarding the finiteness of the quantities $\Delta_{t_0}^2 f$ and $\Delta_{\omega_0}^2 \hat{f}$, where $t_0, \omega_0 \in \mathbb{R}$, suffices to conclude the argument. An elegant proof of the uncertainty inequality (1.108) on such lines is briefly sketched in [111].

Before concluding the subsection, we present an intuitive explanation to the Heisenberg's uncertainty principle. One consequence of the uncertainty inequality (1.108) is that the dispersion of a given signal f about any point $t_0 \in \mathbb{R}$; that is, $\Delta_{t_0}^2 f$ and the dispersion of the corresponding Fourier transform about $\omega_0 \in \mathbb{R}$; that is, $\Delta_{\omega_0}^2 \hat{f}$ cannot be simultaneously small. Consider the function

$$f_s(t) = \sqrt{s}\, e^{-st^2}. \tag{1.125}$$

For $s = 1$ and $s = 5$, the functions given by (1.125) are plotted in Figure 1.28. Note that as s increases, the exponent becomes more and more negative and therefore, the dispersion of (1.125) decreases, hence, the graph becomes more and more concentrated about origin. The Fourier transform corresponding to (1.125) can be computed by invoking the standard Gaussian integral as

$$\begin{aligned}
\hat{f}_s(\omega) &= \frac{1}{\sqrt{2\pi}} \int_{\mathbb{R}} f_s(t)\, e^{-i\omega t}\, dt \\
&= \frac{1}{\sqrt{2\pi}} \int_{\mathbb{R}} \sqrt{s}\, e^{-st^2}\, e^{-i\omega t}\, dt \\
&= \frac{\sqrt{s}}{\sqrt{2\pi}} \int_{\mathbb{R}} e^{-st^2 + (-i\omega)t}\, dt
\end{aligned}$$

$$= \frac{\sqrt{s}}{\sqrt{2\pi}} \left\{ \sqrt{\frac{\pi}{s}}\, e^{-\omega^2/4s} \right\}$$

$$= \frac{1}{\sqrt{2}}\, e^{-\omega^2/4s}. \tag{1.126}$$

Therefore, except for the factor $1/\sqrt{2}$, the Fourier transform (1.126) has the same general negative exponential form as the signal (1.125). Thus, the graph of $\hat{f}_s(\omega)$ has the same general shape as that of $f_s(t)$. However, there is one notable difference: The factor s appears in the denominator of the exponent in (1.126) in lieu to (1.125), where the factor s appears in the numerator of the exponent. In conclusion, if s increases the dispersion of $\hat{f}_s(\omega)$ also increases, whereas the dispersion of $f_s(t)$ decreases. That is, it is not possible to choose a value of the parameter s that the dispersions of both $\hat{f}_s(\omega)$ and $f_s(t)$ are simultaneously small. For $s = 1$ and $s = 5$, the respective Fourier transforms given by (1.126) are plotted in Figure 1.29.

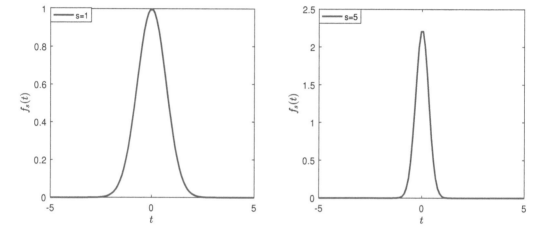

FIGURE 1.28: The function $f_s(t)$.

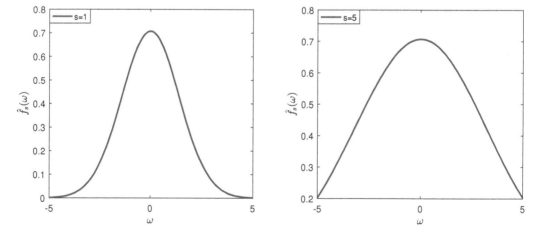

FIGURE 1.29: The Fourier transform $\hat{f}_s(\omega)$.

1.3   The Fractional Fourier Transform

"In his discovery of calculus, Leibniz first introduced the idea of a symbolic method and used the symbol $D^n y$ for the nth derivative, where n is a non-negative integer. L̇. Hŏspital asked Leibniz about the possibility that n be a fraction; *what if n= 1/2?* Leibniz replied, *it will lead to a paradox.* But he added prophetically, *from this apparent paradox, one day useful consequences will be drawn.*"

-Gottfried W. Leibniz

While solving some deep problems in quantum mechanics arising from classical quadratic Hamiltonians, Victor Namias [214] proposed the fractional Fourier transform (FrFT) in 1980. As a generalization of the celebrated Fourier transform, the fractional Fourier transform gained its ground intermittently and profoundly influenced several branches of science and engineering including signal and image processing, quantum mechanics, neural networks, differential equations, optics, pattern recognition, radar, sonar and communication systems [15, 102, 169, 220, 221, 250, 287, 288].

In time-frequency analysis, one normally uses a plane with two orthogonal axes corresponding to time and frequency, respectively, as is portrayed in Figure 1.30. If we consider

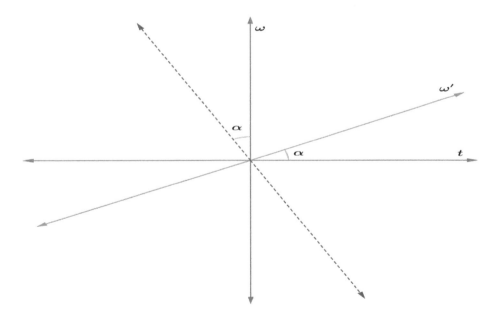

FIGURE 1.30: Time-frequency and time-fractional-frequency planes.

a signal $f(t)$ represented along time axis t and its Fourier transform $\hat{f}(\omega)$ represented along the frequency axis ω, then we can view the Fourier transform operator as a change in the representation of the signal $f(t)$ corresponding to a counter-clockwise axis rotation of $\pi/2$ radians. This interpretation of the Fourier transform operator originates from the fact that the repeated application of the Fourier transform corresponds to an axis-reversal of the signal; that is, $\mathscr{F}\big(\mathscr{F}\big[f(t)\big](\omega)\big) = f(-t)$. Note that two successive rotations of $\pi/2$ of the t-axis result in an axis directed along $-t$. In this context, if one is interested in a linear operator that would correspond to a rotation by an angle α which is not a multiple of $\pi/2$, or equivalently, what would be a representation of the signal $f(t)$ along an axis ω' making an arbitrary angle α with the time axis? then the following properties of the operator are desired:

(i) *Zero rotation:* $R_0 = I$, the identity operator,

(ii) *Consistency with the Fourier transform:* $R_{\pi/2} = \mathscr{F}$, the Fourier transform operator,

(iii) *Additivity of rotation:* $R_\alpha R_\beta = R_{\alpha+\beta}$,

(iv) *2π-rotation:* $R_{2\pi} = I$.

It is worth noticing that the property (iv) is a consequence of the properties (ii) and (iii) and the fact that four successive applications of the Fourier transform correspond to the identity operator: $R_{2\pi} = R_{4(\pi/2)} = \mathscr{F}^4 = I$.

1.3.1 Definition and basic properties

The aforementioned properties are the underpinnings of the notion of fractional Fourier transform. Due to the fact that the fractional Fourier transform can be interpreted as a rotation of the classical Fourier transform, it is sometimes also referred as the angular Fourier transform. Formally, we have the following definition:

Definition 1.3.1. For any $f \in L^2(\mathbb{R})$, the α-angle fractional Fourier transform is denoted by $\mathscr{F}^\alpha[f]$ and is defined as

$$\mathscr{F}^\alpha[f](\omega) = \widehat{f}^\alpha(\omega) = \int_{\mathbb{R}} f(t)\, \mathcal{K}_\alpha(t,\omega)\, dt, \tag{1.127}$$

where $\mathcal{K}_\alpha(t,\omega)$ is called as the "kernel" of the fractional Fourier transform and is given by

$$\mathcal{K}_\alpha(t,\omega) = \begin{cases} \sqrt{\dfrac{1 - i\cot\alpha}{2\pi}}\, \exp\left\{ \dfrac{i(t^2 + \omega^2)\cot\alpha}{2} - i\omega t \csc\alpha \right\}, & \alpha \neq n\pi \\ \delta(t - \omega), & \alpha = 2n\pi \\ \delta(t + \omega), & \alpha = (2n \pm 1)\pi, \quad n \in \mathbb{Z}. \end{cases} \tag{1.128}$$

Definition 1.3.1 allows us to mark the following important points regarding the fractional Fourier transform:

(i) The square root factor which precedes the exponential in the kernel (1.128) can be written in a number of ways. A useful form of the this factor is $\sqrt{2i\, e^{i\alpha}/\pi \sin\alpha}$.

(ii) For $\alpha = 0$, the FrFT operator $\mathscr{F}^0$ is precisely the identity operator I, whereas for $\alpha = \pi/2$, the FrFT given by (1.127) boils down to the classical Fourier transform (1.7). In general, in case α is an integral multiple of $\pi/2$, then the FrFT corresponds to repeated applications of the classical Fourier transform as

$$\mathscr{F}^\alpha[f](\omega) = \mathscr{F}^{n\pi/2}[f](\omega) = (\mathscr{F})^n[f](\omega), \quad n \in \mathbb{Z}, \tag{1.129}$$

where $\mathscr{F}$ denotes the classical Fourier operator. Nevertheless, if we take $\alpha = a\pi/2$, $a \in \mathbb{R}$, then we observe that

$$\mathscr{F}^{\frac{a\pi}{2}}[f](\omega) = (\mathscr{F})^a[f](\omega). \tag{1.130}$$

Consequently, we have

$$\mathscr{F}^{\frac{a\pi}{2}+4n}\Big[f\Big](\omega) = (\mathscr{F})^a (\mathscr{F})^{4n}\Big[f\Big](\omega) = (\mathscr{F})^a \Big[f\Big](\omega). \qquad (1.131)$$

This is because the operator $(\mathscr{F})^2$ corresponds to axis reversal, whereas the operator $(\mathscr{F})^4$ corresponds to identity operator. Therefore, the range of a is usually restricted to the interval $[0, 4)$.

(iii) The class of all FrFT operators parameterized by α has a group structure called the "elliptic group."

(iv) In analogy to the Fourier decomposition of a signal, the FrFT can be regarded as the decomposition of a signal in terms of chirps. Here, it is imperative to mention that a chirp is a signal in which the frequency increases (up-chirp) or decreases (down-chirp) with time. This type of signal occurs due to a non-linear dependence between frequency and the propagation speed of the wave components. The name "chirp" is inspired by the biological phenomenon of bird vocalization and is a reference to the chirping sound made by birds.

Indeed, for any $f \in L^2(\mathbb{R})$, the fractional Fourier transform (1.127) can be alternatively expressed as follows:

$$\begin{aligned} \mathscr{F}^\alpha\Big[f\Big](\omega) &= \sqrt{1 - i\cot\alpha}\left\{\frac{1}{\sqrt{2\pi}}\int_{\mathbb{R}} \exp\left\{\frac{i\omega^2 \cot\alpha}{2}\right\} G(t)\, e^{-i\omega t\csc\alpha}\, dt\right\} \\ &= \sqrt{1 - i\cot\alpha}\left\{\exp\left\{\frac{i\omega^2 \cot\alpha}{2}\right\} \mathscr{F}\Big[G\Big](\omega\csc\alpha)\right\}, \quad \alpha \neq n\pi, \qquad (1.132) \end{aligned}$$

where

$$G(t) = \exp\left\{\frac{it^2 \cot\alpha}{2}\right\} f(t). \qquad (1.133)$$

From the expression (1.132), we infer that the fractional Fourier transform (1.127) corresponds to the following steps:

- A product by a chirp to the input signal $f(t) \to \exp\left\{\dfrac{it^2 \cot\alpha}{2}\right\} f(t)$.

- The computation of the Fourier transform with argument scaled by $\csc\alpha$.

- Another product by chirp $\mathscr{F}\Big[G\Big](\omega\csc\alpha) \to \exp\left\{\dfrac{i\omega^2 \cot\alpha}{2}\right\} \mathscr{F}\Big[G\Big](\omega\csc\alpha)$.

- A product by the complex amplitude factor $\sqrt{1 - i\cot\alpha}$.

Therefore, the fractional Fourier transform given in (1.127) can be viewed as a chirp-Fourier-chirp transformation. Since, chirps have constant magnitude of 1, therefore, if $f \in L^2(\mathbb{R})$, then its product by a chirp is again in $L^2(\mathbb{R})$. Thus, the fractional Fourier transform of a signal $f \in L^2(\mathbb{R})$ exists in the same context as the Fourier transform exists.

In the following theorem, we assemble some fundamental properties of the fractional Fourier kernel (1.128).

Theorem 1.3.1. *The fractional Fourier kernel defined in (1.128) satisfies the following properties:*

(i) $\mathcal{K}_\alpha(t,\omega) = \mathcal{K}_\alpha(\omega,t)$,

(ii) $\overline{\mathcal{K}_\alpha(t,\omega)} = \mathcal{K}_{-\alpha}(t,\omega)$,

(iii) $\mathcal{K}_\alpha(-t,\omega) = \mathcal{K}_\alpha(t,-\omega)$,

(iv) $\displaystyle\int_{\mathbb{R}} \mathcal{K}_\alpha(t,\omega)\,\mathcal{K}_\beta(\omega,z)\,d\omega = \mathcal{K}_{\alpha+\beta}(t,z)$,

(v) $\displaystyle\int_{\mathbb{R}} \mathcal{K}_\alpha(t,\omega)\,\overline{\mathcal{K}_\alpha(t,\xi)}\,dt = \delta(\omega - \xi)$.

Proof. For the sake of brevity, we omit the proof. □

Remark 1.3.1. The property (v) of the fractional Fourier kernel (1.128) indicates that the kernel functions $\mathcal{K}_\alpha(t,\omega)$ taken as functions of t with parameter ω, or equivalently, as functions of ω with parameter t, constitute an orthonormal set.

Next, our aim is to demonstrate that the fractional Fourier transform (1.127) is invertible in the sense that any square integrable function on $\mathbb{R}$ can be retrieved from the corresponding fractional Fourier transform. In this direction, we have the following theorem:

Theorem 1.3.2. *Any $f \in L^2(\mathbb{R})$ can be recovered from the fractional Fourier transform $\mathscr{F}^\alpha[f](\omega)$ via the inversion formula:*

$$f(t) = \mathscr{F}^{-\alpha}\Big(\mathscr{F}^\alpha[f](\omega)\Big)(t) := \int_{\mathbb{R}} \mathscr{F}^\alpha[f](\omega)\,\mathcal{K}_{-\alpha}(t,\omega)\,d\omega. \tag{1.134}$$

That is, the inverse of an FrFT with angle α is an FrFT with angle $-\alpha$.

Proof. Firstly, we show that the fractional Fourier transform (1.127) satisfies the index additive property. For any pair of fractional parameters α, β and any $f \in L^2(\mathbb{R})$, we have

$$\begin{aligned}
\mathscr{F}^\alpha\Big(\mathscr{F}^\beta[f](\xi)\Big)(\omega) &= \int_{\mathbb{R}} \mathscr{F}^\beta[f](\xi)\,\mathcal{K}_\alpha(\xi,\omega)\,d\xi \\
&= \int_{\mathbb{R}} \left\{ \int_{\mathbb{R}} f(t)\,\mathcal{K}_\beta(t,\xi)\,dt \right\} \mathcal{K}_\alpha(\xi,\omega)\,d\xi \\
&= \int_{\mathbb{R}} f(t) \left\{ \int_{\mathbb{R}} \mathcal{K}_\alpha(\xi,\omega)\,\mathcal{K}_\beta(t,\xi)\,d\xi \right\} dt \\
&= \int_{\mathbb{R}} f(t)\,\mathcal{K}_{\alpha+\beta}(t,\omega)\,dt \\
&= \mathscr{F}^{\alpha+\beta}[f](\omega). \tag{1.135}
\end{aligned}$$

As a consequence of the index additive property (1.135), we infer that the inverse of an FrFT with angle α is another FrFT with angle $-\alpha$. This can be alternatively verified by observing that

$$\begin{aligned}
\int_{\mathbb{R}} \mathscr{F}^\alpha[f](\omega)\,\mathcal{K}_{-\alpha}(t,\omega)\,d\omega &= \int_{\mathbb{R}} \left\{ \int_{\mathbb{R}} f(x)\,\mathcal{K}_\alpha(x,\omega)\,dx \right\} \mathcal{K}_{-\alpha}(t,\omega)\,d\omega \\
&= \int_{\mathbb{R}} f(x) \left\{ \int_{\mathbb{R}} \mathcal{K}_\alpha(x,\omega)\,\mathcal{K}_{-\alpha}(t,\omega)\,d\omega \right\} dx \\
&= \int_{\mathbb{R}} f(x) \left\{ \int_{\mathbb{R}} \mathcal{K}_\alpha(x,\omega)\,\mathcal{K}_{-\alpha}(\omega,t)\,d\omega \right\} dx
\end{aligned}$$

$$= \int_{\mathbb{R}} f(x)\,\delta(x-t)\,dx$$

$$= f(t).$$

This completes the proof of Theorem 1.3.2. □

Remark 1.3.2. It is quite evident from (1.134) that the fractional Fourier transform offers an expansion of a signal f on an orthonormal basis constituted by the set of functions $\mathcal{K}_{-\alpha}(t,\omega)$, with ω acting as a parameter for spanning the set of basis functions. In fact, the basis functions are chirps constituted by complex exponentials with linear frequency modulation. Moreover, for different values of ω, these basis only differ by a time-shift and a phase factor that depends on ω; that is

$$\mathcal{K}_{\alpha}(t,\omega) = \exp\left\{-\frac{i\omega^2 \tan\alpha}{2}\right\}\mathcal{K}_{\alpha}(t - \omega\sec\alpha, 0). \tag{1.136}$$

Theorem 1.3.3. (Parseval's Formula): *For any pair of functions $f, g \in L^2(\mathbb{R})$, we have*

$$\left\langle \mathscr{F}^{\alpha}\big[f\big],\, \mathscr{F}^{\alpha}\big[g\big] \right\rangle_2 = \left\langle f, g \right\rangle_2. \tag{1.137}$$

Particularly, if $f = g$, then the following energy preserving relation holds:

$$\left\| \mathscr{F}^{\alpha}\big[f\big] \right\|_2^2 = \left\| f \right\|_2^2. \tag{1.138}$$

Proof. For the given pair of functions $f, g \in^2 (\mathbb{R})$, we shall invoke Definition 1.3.1, so that

$$\left\langle \mathscr{F}^{\alpha}\big[f\big],\, \mathscr{F}^{\alpha}\big[g\big] \right\rangle_2 = \int_{\mathbb{R}} \mathscr{F}^{\alpha}\big[f\big](\omega)\,\overline{\mathscr{F}^{\alpha}\big[g\big]}(\omega)\,d\omega$$

$$= \int_{\mathbb{R}} \left\{ \int_{\mathbb{R}} f(t)\,\mathcal{K}_{\alpha}(t,\omega)\,dt \right\} \left\{ \int_{\mathbb{R}} \overline{g(x)}\,\overline{\mathcal{K}_{\alpha}(x,\omega)}\,dx \right\} d\omega$$

$$= \int_{\mathbb{R}} \int_{\mathbb{R}} f(t)\,\overline{g(x)} \left\{ \int_{\mathbb{R}} \mathcal{K}_{\alpha}(t,\omega)\,\overline{\mathcal{K}_{\alpha}(x,\omega)}\,d\omega \right\} dt\,dx$$

$$= \int_{\mathbb{R}} \int_{\mathbb{R}} f(t)\,\overline{g(x)} \left\{ \int_{\mathbb{R}} \mathcal{K}_{\alpha}(t,\omega)\,\mathcal{K}_{-\alpha}(x,\omega)\,d\omega \right\} dt\,dx. \tag{1.139}$$

Using the symmetry property of the fractional Fourier kernel (1.128), we can express the relation (1.139) as

$$\left\langle \mathscr{F}^{\alpha}\big[f\big],\, \mathscr{F}^{\alpha}\big[g\big] \right\rangle_2 = \int_{\mathbb{R}} \int_{\mathbb{R}} f(t)\,\overline{g(x)} \left\{ \int_{\mathbb{R}} \mathcal{K}_{\alpha}(t,\omega)\,\mathcal{K}_{-\alpha}(\omega,x)\,d\omega \right\} dt\,dx$$

$$= \int_{\mathbb{R}} \int_{\mathbb{R}} f(t)\,\overline{g(x)}\,\delta(t - x)\,dt\,dx$$

$$= \int_{\mathbb{R}} f(t)\,\overline{g(t)}\,dt$$

$$= \left\langle f, g \right\rangle_2, \tag{1.140}$$

which is the desired Parseval's relation for the fractional Fourier transform (1.127). Moreover, plugging $f = g$ in (1.140) yields the energy preserving relation:

$$\left\| \mathscr{F}^{\alpha}\big[f\big] \right\|_2^2 = \left\| f \right\|_2^2.$$

This completes the proof of Theorem 1.3.3. □

Remark 1.3.3. The Parseval's relation (1.137) and the energy preserving property (1.138) can be viewed as an implication of the fact that the fractional Fourier transform is based on a set of orthonormal functions. Moreover, note that the squared magnitude of the Fourier transform of a signal $f(t)$; that is, $\left|\mathscr{F}[f](\omega)\right|^2$ is often termed as the "energy spectrum" of the signal f and is interpreted as the distribution of the energy of the signal among different frequencies $e^{i\omega t}$. In analogy to this, the squared magnitude of the fractional Fourier transform of a signal f with angle α is called as the "fractional energy spectrum" of the signal and can be interpreted as the energy distribution of the signal into different chirps $\mathcal{K}_{-\alpha}(t, \omega)$.

In the following theorem, we study some fundamental properties of the fractional Fourier transform.

Theorem 1.3.4. *Let $f, g \in L^2(\mathbb{R})$ and $c_1, c_2 \in \mathbb{C}$, $k, \omega_0 \in \mathbb{R}$, $\lambda \in \mathbb{R} \setminus \{0\}$ are scalars. Then, the fractional Fourier transform (1.127) satisfies the following properties:*

(i) *Linearity:* $\mathscr{F}^\alpha\Big[c_1\, f + c_2\, g\Big](\omega) = c_1\, \mathscr{F}^\alpha\Big[f\Big](\omega) + c_2\, \mathscr{F}^\alpha\Big[g\Big](\omega),$

(ii) *Translation:* $\mathscr{F}^\alpha\Big[f(t-k)\Big](\omega) = \exp\left\{\dfrac{ik^2 \cos\alpha \sin\alpha - 2i\omega k \sin\alpha}{2}\right\} \mathscr{F}^\alpha\Big[f\Big](\omega)(\omega - k\cos\alpha),$

(iii) *Modulation:* $\mathscr{F}^\alpha\Big[e^{i\omega_0 t} f(t)\Big](\omega) = \exp\left\{\dfrac{2i\omega\omega_0 \cos\alpha - i\omega_0^2 \sin\alpha \cos\alpha}{2}\right\} \mathscr{F}^\alpha\Big[f\Big](\omega - \omega_0 \sin\alpha),$

(iv) *Scaling:* $\mathscr{F}^\alpha\Big[f(\lambda t)\Big](\omega) = \sqrt{\dfrac{1 - i\cot\alpha}{\lambda^2(1 - \cot\beta)}}\, \exp\left\{\dfrac{i\lambda^2\omega^2 \cot\beta\left(1 - \dfrac{\csc^2\alpha}{\lambda^4 \csc^2\alpha}\right)}{2}\right\}$

$$\times\, \mathscr{F}^\alpha\Big[f\Big]\left(\frac{\omega \sin\beta}{\lambda \sin\alpha}\right),\ \beta = \operatorname{arccot}\left(\cot\alpha/\lambda^2\right),$$

(v) *Parity:* $\mathscr{F}^\alpha\Big[f(-t)\Big](\omega) = \mathscr{F}^\alpha\Big[f\Big](-\omega),$

(vi) *Conjugation:* $\mathscr{F}^\alpha\Big[\overline{f}\Big](\omega) = \overline{\mathscr{F}^\alpha\Big[f\Big](-\omega)}.$

Proof. To be brief, we shall only prove the scaling property, the rest of the properties can be obtained quite straightforwardly.

(iv) Upon scaling the input signal $f(t)$ by $\lambda \in \mathbb{R} \setminus \{0\}$, we have

$$\mathscr{F}^\alpha\Big[f(\lambda t)\Big](\omega) = \int_{\mathbb{R}} f(\lambda t)\, \mathcal{K}_\alpha(t, \omega)\, dt$$

$$= \frac{1}{|\lambda|}\sqrt{\frac{1 - i\cot\alpha}{2\pi}} \int_{\mathbb{R}} f(z)$$

$$\times \exp\left\{\frac{i\big((z/\lambda)^2 + \omega^2\big)\cot\alpha - i\omega(z/\lambda)\csc\alpha}{2}\right\} dz$$

$$= \sqrt{\frac{1 - i\cot\alpha}{2\pi\lambda^2}} \int_{\mathbb{R}} f(z)\exp\left\{\frac{i\big(z^2 + (\lambda\omega)^2\big)\left(\dfrac{\cot\alpha}{\lambda^2}\right) - i\omega z\left(\dfrac{\csc\alpha}{\lambda}\right)}{2}\right\} dz.$$

For $\beta = \operatorname{arccot}\left(\cot\alpha/\lambda^2\right) = \arctan\left(\lambda^2\tan\alpha\right)$, the above expression can be re-expressed as

$$\mathscr{F}^\alpha\left[f(\lambda t)\right](\omega) = \sqrt{\frac{1-i\cot\alpha}{2\pi\lambda^2}}\int_{\mathbb{R}} f(z)\exp\left\{\frac{i(z^2+(\lambda\omega)^2)\cot\beta - i\omega z\left(\frac{\csc\alpha}{\lambda\csc\beta}\right)\csc\beta}{2}\right\}dz$$

$$= \sqrt{\frac{1-i\cot\alpha}{2\pi\lambda^2}}\int_{\mathbb{R}} f(z)\exp\left\{\frac{i\left(z^2+\left(\frac{\omega\csc\alpha}{\lambda\csc\beta}\right)^2\right)\cot\beta}{2}\right.$$

$$\left. -i\left(\frac{\omega\csc\alpha}{\lambda\csc\beta}\right)z\csc\beta - \frac{i\left(\frac{\omega\csc\alpha}{\lambda\csc\beta}\right)^2\cot\beta - i\lambda^2\omega^2\cot\beta}{2}\right\}dz$$

$$= \sqrt{\frac{1-i\cot\alpha}{2\pi\lambda^2}}\exp\left\{\frac{i\lambda^2\omega^2\cot\beta\left(1-\frac{\csc^2\alpha}{\lambda^4\csc^2\beta}\right)}{2}\right\}\int_{\mathbb{R}} f(z)$$

$$\times\exp\left\{\frac{i\left(z^2+\left(\frac{\omega\csc\alpha}{\lambda\csc\beta}\right)^2\right)\cot\beta - 2i\left(\frac{\omega\csc\alpha}{\lambda\csc\beta}\right)z\csc\beta}{2}\right\}dz$$

$$= \sqrt{\frac{1-i\cot\alpha}{\lambda^2(1-i\cot\beta)}}\exp\left\{\frac{i\lambda^2\omega^2\cot\beta\left(1-\frac{\csc^2\alpha}{\lambda^4\csc^2\beta}\right)}{2}\right\}$$

$$\times\int_{\mathbb{R}} f(z)\,\mathcal{K}_\beta\left(z,\frac{\omega\csc\alpha}{\lambda\csc\beta}\right)dz$$

$$= \sqrt{\frac{1-i\cot\alpha}{\lambda^2(1-i\cot\beta)}}\exp\left\{\frac{i\lambda^2\omega^2\cot\beta\left(1-\frac{\csc^2\alpha}{\lambda^4\csc^2\beta}\right)}{2}\right\}\mathscr{F}^\alpha[f]\left(\frac{\omega\sin\beta}{\lambda\sin\alpha}\right).$$

The scaling property is one of the most important properties of the fractional Fourier transform which shows the effect of the change of units (or scaling) of the independent variable t. Recall that in the classical Fourier domain, the scaling in the input signal simply corresponds to a scaling of Fourier transform and a scaling in amplitude. However, in case of fractional Fourier transform, the scaling in the input signal by λ corresponds to a scaling by $\frac{\sin\beta}{\lambda\sin\alpha}$ in the frequency variable, a (complex) amplitude scaling, a product by a chirp and most importantly, a change in the angle at which at which we compute the transform from α to β. This angle change is understood by taking into consideration the fact that in the classical Fourier domain, the contraction (scaling) of time axis in the time-frequency plane corresponds to expansion of the frequency axis. Thus, in case of fractional Fourier transform, the axis along which we are computing the transformation, which was initially at α with the usual time axis will move to a new position at angle β.

$\square$

1.3.2 Fractional convolution and correlation

Although many properties of the classical Fourier transform are inherited by the fractional Fourier transform, however, certain important characteristics of the Fourier transform do not extend directly to the fractional Fourier transform (1.127). For instance, the symmetry properties of the Fourier transform for even and odd real functions are not inherited by the fractional Fourier transform with an arbitrary angle α. This is because the fractional Fourier transform of a real function need not be Hermitian. Besides, the classical convolution and correlation operations in the Fourier domain do not retain their elegance in the fractional Fourier domain. Many important properties of the classical convolution operation, including commutativity, associativity, convolution theorem do not hold simultaneously for a given fractional convolution. For this reason, the convolution theory in the fractional Fourier domain is still an open field of research and numerous definitions of the fractional convolution have appeared in the literature [8, 16, 122, 220, 221, 291].

In this subsection, our motive is to explore the convolution theory in the fractional Fourier domain. Our aim is to study a duo of important convolution operations associated with the fractional Fourier transform (1.127) and also investigate upon the fundamental properties. Nevertheless, we shall also examine the correlation operations abreast to these fractional convolutions.

(A) Chirp-free weighted convolution

Here, we shall study the chirp-free convolution associated with the fractional Fourier transform. The name "chirp-free" is coined due to the fact that the associated convolution theorem does not contain any chirp multiplier. However, as we shall demonstrate in the sequel, such a convolution does not satisfy the commutative and associative properties. Nevertheless, the distributive property holds good. As such, this convolution plays a significant role for describing the operation of a parallel system with added outputs and allows to replace such systems by a single system. Therefore, such a convolution operation can be of substantial importance in the realm of multi-channel sampling and reconstruction [256, 292, 323].

Definition 1.3.2. Given any pair of functions $f, g \in L^2(\mathbb{R})$, the chirp-free fractional convolution is denoted by $\circledast_\alpha$ and is defined as

$$(f \circledast_\alpha g)(z) = \int_\mathbb{R} f(t)\, g(z-t) \exp\left\{ \frac{i(t^2 - z^2)\cot\alpha}{2} \right\} dt. \qquad (1.141)$$

In the following theorem, we assemble some elementary properties of the fractional convolution operation (1.141).

Theorem 1.3.5. *For any $f, g, h \in L^2(\mathbb{R})$ and the scalars $k \in \mathbb{R}$, $\lambda \in \mathbb{R} \setminus \{0\}$, the fractional convolution operation $\circledast_\alpha$ defined in (1.141) has the following properties:*

(i) *Non-commutativity:* $(f \circledast_\alpha g)(z) \neq (g \circledast_\alpha f)(z)$,

(ii) *Non-associativity:* $(f \circledast_\alpha (g \circledast_\alpha h))(z) \neq ((f \circledast_\alpha g) \circledast_\alpha h)(z)$,

(iii) *Distributivity:* $(f \circledast_\alpha (g + h))(z) = (f \circledast_\alpha g)(z) + (f \circledast_\alpha h)(z)$,

(iv) *Translation:* $(f \circledast_\alpha g)(z - k) = (f(y - k) \circledast_\alpha G_\alpha(y))(z), \quad G_\alpha(t) = e^{itk\cot\alpha} g(t)$,

(v) *Scaling:* $(f \circledast_\alpha g)(\lambda z) = |\lambda| (f(\lambda y) \circledast_\beta g(\lambda y))(z), \quad \beta = \operatorname{arccot}(\lambda^2 \cot\alpha)$,

(vi) *Parity:* $(f \circledast_\alpha g)(-z) = (f(-t) \circledast_\alpha g(-t))(z)$.

Proof. The proof of Theorem 1.3.5 follows directly from Definition 1.3.2. □

Next, our goal is to examine the nature of convolution theorem associated with the fractional convolution operation $\circledast_\alpha$ defined in (1.141).

Theorem 1.3.6. *For any $f, g \in L^2(\mathbb{R})$, we have*

$$\mathscr{F}^\alpha\Big[(f \circledast_\alpha g)\Big](\omega) = \sqrt{2\pi}\,\mathscr{F}^\alpha\Big[f\Big](\omega)\,\mathscr{F}\Big[g\Big](\omega \csc \alpha). \tag{1.142}$$

Proof. Using Definition 1.3.1, we can compute the fractional Fourier transform corresponding to (1.141) as follows:

$$\mathscr{F}^\alpha\Big[(f \circledast_\alpha g)\Big](\omega) = \int_{\mathbb{R}} (f \circledast_\alpha g)(z)\,\mathcal{K}_\alpha(z, \omega)\,dz$$

$$= \int_{\mathbb{R}} \left\{ \int_{\mathbb{R}} f(t)\,g(z - t)\,\exp\left\{\frac{i(t^2 - z^2)\cot \alpha}{2}\right\} dt \right\} \mathcal{K}_\alpha(z, \omega)\,dz$$

$$= \int_{\mathbb{R}} f(t) \left\{ \int_{\mathbb{R}} g(z - t)\,\exp\left\{\frac{i(t^2 - z^2)\cot \alpha}{2}\right\} \mathcal{K}_\alpha(z, \omega)\,dz \right\} dt. \tag{1.143}$$

Making the substitution $z - t = y$ into the second integral on the R.H.S of (1.143), we obtain

$$\mathscr{F}^\alpha\Big[(f \circledast_\alpha g)\Big](\omega)$$

$$= \int_{\mathbb{R}} f(t) \left\{ \int_{\mathbb{R}} g(y)\,\exp\left\{\frac{i(t^2 - (t + y)^2)\cot \alpha}{2}\right\} \mathcal{K}_\alpha(t + y, \omega)\,dy \right\} dt$$

$$= \sqrt{\frac{1 - i\cot \alpha}{2\pi}} \int_{\mathbb{R}} f(t) \left\{ \int_{\mathbb{R}} g(y)\,\exp\left\{\frac{i(t^2 - t^2 - y^2 - 2ty)\cot \alpha}{2}\right\} \right.$$

$$\left. \times \exp\left\{\frac{i(t^2 + y^2 + 2ty + \omega^2)\cot \alpha}{2} - i\omega t \csc \alpha - i\omega y \csc \alpha \right\} dy \right\} dt$$

$$= \sqrt{\frac{1 - i\cot \alpha}{2\pi}} \int_{\mathbb{R}} f(t)\,\exp\left\{\frac{i(t^2 + \omega^2)\cot \alpha}{2} - i\omega t \csc \alpha\right\} \left\{ \int_{\mathbb{R}} g(y)\,e^{-i\omega y \csc \alpha}\,dy \right\} dt$$

$$= \sqrt{2\pi} \int_{\mathbb{R}} f(t)\,\mathcal{K}_\alpha(t, \omega) \left\{ \frac{1}{\sqrt{2\pi}} \int_{\mathbb{R}} g(y)\,e^{-i\omega y \csc \alpha}\,dy \right\} dt$$

$$= \sqrt{2\pi}\,\mathscr{F}^\alpha\Big[f\Big](\omega)\,\mathscr{F}\Big[g\Big](\omega \csc \alpha).$$

This completes the proof of Theorem 1.3.6. □

Abreast to the notion of fractional convolution (1.141), we have the following definition of the fractional correlation operation:

Definition 1.3.3. *For any pair of functions $f, g \in L^2(\mathbb{R})$, the fractional correlation operation is denoted by $\otimes_\alpha$ and is defined as*

$$(f \otimes_\alpha g)(z) = \int_{\mathbb{R}} f(t)\,\overline{g(t - z)}\,\exp\left\{\frac{i(t^2 - z^2)\cot \alpha}{2}\right\} dt. \tag{1.144}$$

Following is the correlation theorem associated with the notion of fractional correlation operation $\otimes_\alpha$ defined in (1.144):

Theorem 1.3.7. *For any $f, g \in L^2(\mathbb{R})$, we have*

$$\mathscr{F}^\alpha \Big[\big(f \otimes_\alpha g \big) \Big](\omega) = \sqrt{2\pi}\, \mathscr{F}^\alpha \big[f \big](\omega)\, \overline{\mathscr{F} \big[g \big]}(\omega \csc \alpha). \tag{1.145}$$

Proof. The proof can be obtained in a manner analogous to that of Theorem 1.3.6. □

Remark 1.3.4. The fractional convolution (1.141) and fractional correlation (1.144) are both dictated by the fractional parameter α appearing in the exponential weight function. Evidently, for $\alpha = \pi/2$, Definition 1.3.2 as well as Definition 1.3.3 boil down to the notion of classical convolution and correlation operations given by (1.75) and (1.88), respectively.

(B) Another weighted convolution

Here, we present another variant of the fractional convolution operation which upholds the classical convolution theorem in the sense that, except for a chirp, the fractional Fourier transform of two convoluted signals corresponds to the product of their respective fractional spectra. Nevertheless, we also demonstrate that such a convolution structure satisfies the fundamental properties of commutativity, associativity and distributivity.

Definition 1.3.4. For any pair of functions $f, g \in L^2(\mathbb{R})$, the weighted convolution operation $\odot_\alpha$ is defined as

$$\big(f \odot_\alpha g \big)(z) = \int_\mathbb{R} f(t)\, g(z - t)\, \exp \Big\{ it(t - z) \cot \alpha \Big\}\, dt. \tag{1.146}$$

We now assemble some fundamental properties of the fractional convolution operation (1.146). Most importantly, we shall demonstrate that the weighted convolution $\odot_\alpha$ is indeed commutative, associative and distributive.

Theorem 1.3.8. *For any $f, g, h \in L^2(\mathbb{R})$ and the scalars $k \in \mathbb{R}$, $\lambda \in \mathbb{R} \setminus \{0\}$, the weighted convolution $\odot_\alpha$ defined in (1.146) has the following properties:*

(i) *Commutativity:* $\big(f \odot_\alpha g \big)(z) = \big(g \odot_\alpha f \big)(z),$

(ii) *Associativity:* $\big(f \odot_\alpha (g \odot_\alpha h) \big)(z) = \big((f \odot_\alpha g) \odot_\alpha h \big)(z),$

(iii) *Distributivity:* $\big(f \odot_\alpha (g + h) \big)(z) = \big(f \circledast_\alpha g \big)(z) + \big(f \circledast_\alpha h \big)(z),$

(iv) *Translation:* $\big(f \odot_\alpha g \big)(z - k) = \big(f(y - k) \odot_\alpha G_\alpha(y) \big)(z), \quad G_\alpha(t) = e^{ikt \cot \alpha}\, g(t),$

(v) *Scaling:* $\big(f \odot_\alpha g \big)(\lambda z) = \big| \lambda \big| \big(f(\lambda y) \odot_\beta g(\lambda y) \big)(z),$

(vi) *Parity:* $\big(f \odot_\alpha g \big)(-z) = \big(f(-t) \odot_\alpha g(-t) \big)(z).$

Proof. To keep the long story short, the proof is a straightaway consequence of Definition 1.3.4. □

Next, our aim is to obtain the convolution theorem associated with the weighted convolution (1.146). It is pertinent to mention that the convolution theorem is in consistence with the classical convolution theorem in the Fourier domain in the sense that, except for a chirp, the fractional Fourier transform of the convolution of two functions is the product of their respective factional Fourier transforms.

Theorem 1.3.9. *For any $f, g \in L^2(\mathbb{R})$, we have*

$$\mathscr{F}^\alpha \Big[\big(f \odot_\alpha g \big) \Big](\omega) = \sqrt{\frac{2\pi}{1 - i \cot \alpha}}\, \exp \left\{ -\frac{i\omega^2 \cot \alpha}{2} \right\}\, \mathscr{F}^\alpha \big[f \big](\omega)\, \mathscr{F}^\alpha \big[g \big](\omega). \tag{1.147}$$

Proof. Applying Definition 1.3.1, we have

$$\mathscr{F}^\alpha\Big[(f \odot_\alpha g)\Big](\omega) = \int_\mathbb{R} (f \odot_\alpha g)(z)\, \mathcal{K}_\alpha(z,\omega)\, dz$$

$$= \sqrt{\frac{1 - i\cot\alpha}{2\pi}} \int_\mathbb{R} \left\{ \int_\mathbb{R} f(t)\, g(z-t)\, \exp\left\{it(t-z)\cot\alpha\right\} dt \right\}$$

$$\times \exp\left\{ \frac{i(z^2 + \omega^2)\cot\alpha}{2} - i\omega z\csc\alpha \right\} dz$$

$$= \sqrt{\frac{1 - i\cot\alpha}{2\pi}} \int_\mathbb{R} f(t) \left\{ \int_\mathbb{R} g(z-t)\, \exp\left\{it(t-z)\cot\alpha\right\} \right.$$

$$\left. \times \exp\left\{ \frac{i(z^2 + \omega^2)\cot\alpha}{2} - i\omega z\csc\alpha \right\} dz \right\} dt. \qquad (1.148)$$

Making use of the substitution $z - t = y$ into the inner integral on the R.H.S of (1.148), we obtain

$$\mathscr{F}^\alpha\Big[(f \odot_\alpha g)\Big](\omega)$$

$$= \sqrt{\frac{1 - i\cot\alpha}{2\pi}} \int_\mathbb{R} f(t) \left\{ \int_\mathbb{R} g(y)\, \exp\left\{it(-y)\cot\alpha\right\} \right.$$

$$\left. \times \exp\left\{ \frac{i(t^2 + y^2 + 2ty + \omega^2)\cot\alpha}{2} - i\omega t\csc\alpha - i\omega y\csc\alpha \right\} dy \right\} dt$$

$$= \sqrt{\frac{1 - i\cot\alpha}{2\pi}} \int_\mathbb{R} f(t)\, \exp\left\{ \frac{i(t^2 + \omega^2)\cot\alpha}{2} - i\omega t\csc\alpha \right\} \exp\left\{ -\frac{i\omega^2\cot\alpha}{2} \right\}$$

$$\times \left\{ \int_\mathbb{R} g(y)\, \exp\left\{ \frac{i(y^2 + \omega^2)\cot\alpha}{2} - i\omega y\csc\alpha \right\} dy \right\} dt$$

$$= \sqrt{\frac{2\pi}{1 - i\cot\alpha}} \exp\left\{ -\frac{i\omega^2\cot\alpha}{2} \right\} \int_\mathbb{R} f(t)\, \mathcal{K}_\alpha(t,\omega) \left\{ \int_\mathbb{R} g(y)\, \mathcal{K}_\alpha(y,\omega)\, dy \right\} dt$$

$$= \sqrt{\frac{2\pi}{1 - i\cot\alpha}} \exp\left\{ -\frac{i\omega^2\cot\alpha}{2} \right\} \mathscr{F}^\alpha\Big[f\Big](\omega)\, \mathscr{F}^\alpha\Big[g\Big](\omega).$$

This completes the proof of Theorem 1.3.9. $\qquad\square$

In the sequel, we have the definition of weighted correlation associated with the convolution operation (1.146)

Definition 1.3.5. For any $f, g \in L^2(\mathbb{R})$, the weighted correlation operation is denoted by $\oslash_\alpha$ and is defined as

$$(f \oslash_\alpha g)(z) = \int_\mathbb{R} f(t)\, \overline{g(t-z)}\, \exp\left\{iz(t-z)\cot\alpha\right\} dt. \qquad (1.149)$$

Following is the correlation theorem associated with the weighted correlation operation given in (1.149):

Theorem 1.3.10. *For any $f, g \in L^2(\mathbb{R})$, we have*

$$\mathscr{F}^\alpha\Big[(f \oslash_\alpha g)\Big](\omega) = \sqrt{\frac{2\pi}{1 + i\cot\alpha}} \exp\left\{ \frac{i\omega^2\cot\alpha}{2} \right\} \mathscr{F}^\alpha\Big[f\Big](\omega)\, \overline{\mathscr{F}^\alpha\Big[g\Big](\omega)}. \qquad (1.150)$$

Proof. The proof can be accomplished by analogously following the strategy in the proof of Theorem 1.3.9. $\qquad\square$

Remark 1.3.5. For $\alpha = \pi/2$, Definition 1.3.4 as well as Definition 1.3.5 yield the classical convolution and correlation operations given by (1.75) and (1.88), respectively.

1.3.3 Uncertainty principle for the fractional Fourier transform

"The exact knowledge of the present allows the future to be calculated-not the conclusion but the hypothesis is false."

-Werner K. Heisenberg

In this subsection, we derive the Heisenberg's uncertainty principle for the fractional Fourier transform (1.127). It is imperative to mention that the underlying inequality shall be formulated about the zero mean values $t_0 = \omega_0 = 0$, as this will not change the essence of the uncertainty principle.

Theorem 1.3.11. *For any non-trivial function $f \in L^2(\mathbb{R})$, the following uncertainty inequality holds:*

$$\left\{ \int_{\mathbb{R}} |\omega|^2 \left| \mathscr{F}^\alpha \big[f \big] (\omega) \right|^2 d\omega \right\} \left\{ \int_{\mathbb{R}} |\eta|^2 \left| \mathscr{F}^\beta \big[f \big] (\eta) \right|^2 d\eta \right\} \geq \frac{|\sin(\alpha - \beta)|^2}{4} \|f\|_2^4, \qquad (1.151)$$

where α and β are chosen such that $\beta = \alpha - \gamma$ and $\sin\alpha, \sin\beta, \sin\gamma \neq 0$.

Proof. As a consequence of the definition of the fractional Fourier transform (1.127), we have

$$\mathscr{F}^\alpha \big[f \big] (\omega) = \sqrt{\frac{1 - i \cot\alpha}{2\pi}} \int_{\mathbb{R}} f(t) \exp\left\{ \frac{i(t^2 + \omega^2) \cot\alpha}{2} - i\omega t \csc\alpha \right\} dt$$

$$= \sqrt{\frac{1 - i \cot\alpha}{2\pi}} \exp\left\{ \frac{i\omega^2 \cot\alpha}{2} \right\} \int_{\mathbb{R}} f(t) \exp\left\{ \frac{it^2 \cot\alpha}{2} - i\omega t \csc\alpha \right\} dt,$$

which further yields

$$\exp\left\{ -\frac{i\omega^2 \cot\alpha}{2} \right\} \mathscr{F}^\alpha \big[f \big] (\omega) = \sqrt{\frac{1 - i \cot\alpha}{2\pi}} \int_{\mathbb{R}} f(t) \exp\left\{ \frac{it^2 \cot\alpha}{2} - i\omega t \csc\alpha \right\} dt$$

$$:= G(\omega). \qquad (1.152)$$

In view of $\left| \mathscr{F}^\alpha \big[f \big] (\omega) \right|^2 = |G(\omega)|^2$, we observe that the following integral equation holds:

$$\int_{\mathbb{R}} |\omega|^2 \left| \mathscr{F}^\alpha \big[f \big] (\omega) \right|^2 d\omega = \int_{\mathbb{R}} |\omega|^2 |G(\omega)|^2 d\omega. \qquad (1.153)$$

Therefore, for the function

$$g(t) = \frac{1}{\sqrt{2\pi}} \int_{\mathbb{R}} G(\omega) e^{i\omega t} d\omega, \qquad (1.154)$$

the Heisenberg's uncertainty inequality in the Fourier domain, given by (1.123), implies that

$$\left\{ \int_{\mathbb{R}} |t|^2 |g(t)|^2 dt \right\} \left\{ \int_{\mathbb{R}} |\omega|^2 |G(\omega)|^2 d\omega \right\} \geq \frac{1}{4} \|f\|_2^4. \qquad (1.155)$$

Let $\gamma \in \mathbb{R}$ be such that $\sin\gamma \neq 0$. Then, by virtue of scaling property of the Lebesgue integral, the first integral in the L.H.S of inequality (1.155) can be expressed as

$$\int_{\mathbb{R}} |t|^2 |g(t)|^2 dt = \int_{\mathbb{R}} \left| \frac{t}{\sin\gamma} \right|^2 \left| g\left(\frac{t}{\sin\gamma} \right) \right|^2 d\left(\frac{t}{\sin\gamma} \right)$$

$$= \frac{1}{|\sin\gamma|} \int_{\mathbb{R}} \left| \frac{t}{\sin\gamma} \right|^2 \left| g\left(\frac{t}{\sin\gamma} \right) \right|^2 dt. \qquad (1.156)$$

Using (1.154), we have

$$\left| g\left(\frac{t}{\sin\gamma} \right) \right|^2 = \left| \frac{1}{\sqrt{2\pi}} \int_{\mathbb{R}} G(\omega) \exp\left\{ i\omega \left(\frac{t}{\sin\gamma} \right) \right\} d\omega \right|^2. \qquad (1.157)$$

Invoking the relation (1.152), we can simplify the expression on the R.H.S of (1.157) in the following manner:

$$\left| g\left(\frac{t}{\sin\gamma} \right) \right|^2 = \frac{1}{\sqrt{2\pi}} \left| \int_{\mathbb{R}} \exp\left\{ -\frac{i\omega^2 \cot\alpha}{2} \right\} \mathscr{F}^{\alpha}[f](\omega) \exp\left\{ i\omega \left(\frac{t}{\sin\gamma} \right) \right\} d\omega \right|^2$$

$$= |\sin\gamma|^2 \left| \sqrt{\frac{1 - i\cot\alpha}{2\pi}} \exp\left\{ -\frac{it^2 \cot\gamma}{2} \right\} \right.$$

$$\left. \times \int_{\mathbb{R}} \mathscr{F}^{\alpha}[f](\omega) \exp\left\{ -\frac{i\omega^2 \cot\gamma}{2} \right\} \exp\left\{ \frac{i\omega t}{\sin\gamma} \right\} d\omega \right|^2$$

$$= |\sin\gamma|^2 \left| \int_{\mathbb{R}} \mathscr{F}^{\alpha}[f](\omega) \, \mathcal{K}_{-\gamma}(t,\omega) \, d\omega \right|^2$$

$$= |\sin\gamma|^2 \left| \mathscr{F}^{-\gamma}\left(\mathscr{F}^{\alpha}[f](\omega) \right)(t) \right|^2. \qquad (1.158)$$

Plugging (1.158) in (1.156), we obtain

$$\int_{\mathbb{R}} |t|^2 |g(t)|^2 dt = \int_{\mathbb{R}} \left| \frac{t}{\sin\gamma} \right|^2 \left| g\left(\frac{t}{\sin\gamma} \right) \right|^2 d\left(\frac{t}{\sin\gamma} \right)$$

$$= \int_{\mathbb{R}} \left| \frac{t}{\sin\gamma} \right|^2 \left| \mathscr{F}^{-\gamma}\left(\mathscr{F}^{\alpha}[f](\omega) \right)(t) \right|^2 dt$$

$$= \int_{\mathbb{R}} \left| \frac{t}{\sin\gamma} \right|^2 \left| \mathscr{F}^{\alpha-\gamma}[f](t) \right|^2 dt. \qquad (1.159)$$

Consequently, upon setting $\beta = \alpha - \gamma$, we get

$$\int_{\mathbb{R}} |t|^2 |g(t)|^2 dt = \int_{\mathbb{R}} \frac{|t|^2}{|\sin(\alpha - \beta)|^2} \left| \mathscr{F}^{\beta}[f](t) \right|^2 dt. \qquad (1.160)$$

For the sake of compliance with the formal definition of fractional Fourier transform, we shall set $t = \eta$ in the R.H.S of (1.160), so that the inequality (1.155) takes the following form:

$$\left\{ \int_{\mathbb{R}} |\omega|^2 \left| \mathscr{F}^{\alpha}[f](\omega) \right|^2 d\omega \right\} \left\{ \int_{\mathbb{R}} |\eta|^2 \left| \mathscr{F}^{\beta}[f](\eta) \right|^2 d\eta \right\} \geq \frac{|\sin(\alpha - \beta)|^2}{4} \|f\|_2^4.$$

This completes the proof of Theorem 1.3.11. $\qquad\qquad \square$

Before concluding the discourse on the Heisenberg-type uncertainty inequality associated with the fractional Fourier transform (1.127), we note that Theorem 1.3.11 allows us to make the following deductions:

(i) Inequality (1.151) demonstrate that the Heisenberg's uncertainty inequality in the fractional Fourier domain governs the localization of the function representations along two angles, viz; α and β. Particularly, if $\alpha \to 0$ and $\beta \to \pi/2$, then (1.151) yields the classical Heisenberg's uncertainty principle in the Fourier domain. Moreover, if we assume that $|\alpha - \beta| = 2n\pi + \pi/2$, $n \in \mathbb{N}$, then inequality (1.151) boils down to the following inequality:

$$\left\{ \int_{\mathbb{R}} |\omega|^2 \left| \mathscr{F}^{\alpha} [f](\omega) \right|^2 d\omega \right\} \left\{ \int_{\mathbb{R}} |\eta|^2 \left| \mathscr{F}^{\beta} [f](\eta) \right|^2 d\eta \right\} \geq \frac{1}{4} \left\| f \right\|_2^4. \qquad (1.161)$$

This suggests that the two variables ω and η lie in orthogonal domains. In case, $|\alpha - \beta| \neq 2n\pi + \pi/2$, $n \in \mathbb{N}$, then these variables lie in non-orthogonal domains.

(ii) Since the inequality (1.151) is deduced from the classical Heisenberg's uncertainty principle in the Fourier domain, therefore, the function that leads to equality in (1.151) can be easily deduced to be

$$f(t) = K \exp \left\{ -\frac{i(A \cos \beta - \cos \alpha)t^2}{(A \sin \beta - \sin \alpha)} \right\}, \qquad (1.162)$$

where K and A are constants.

(iii) For $\beta = 0$, inequality (1.151) yields a typical uncertainty inequality, governing the localization of an input signal with respect to the localization of the corresponding fractional Fourier transform as

$$\left\{ \int_{\mathbb{R}} |t|^2 \left| f(t) \right|^2 dt \right\} \left\{ \int_{\mathbb{R}} |\omega|^2 \left| \mathscr{F}^{\alpha} [f](\omega) \right|^2 d\omega \right\} \geq \frac{|\sin \alpha|^2}{4} \left\| f \right\|_2^4. \qquad (1.163)$$

1.4   The Linear Canonical Transform

"Natural science, does not simply describe and explain nature; it is part of the interplay between nature and ourselves."

-Werner K. Heisenberg

The origin of the theory of linear canonical transforms dates back to early 1970s with the independent seminal works of Collins Jr. [77] in paraxial optics, and Moshinsky and Quesne [212, 213] in quantum mechanics, to study the conservation of information and uncertainty under linear maps of phase space. It was only in 1990s that both these independent works begin to be referred jointly in the open literature [37, 54, 55, 320]. The linear canonical transform (LCT) is a three free-parameter class of linear integral transforms, which encompasses a number of well-known unitary transformations as well as signal processing and optics-related mathematical operations, for example, the Fourier transform, the fractional Fourier transform, the Fresnel transform and the scaling operations [157, 222, 320]. Due to the extra degrees of freedom and simple geometrical manifestation, the LCT is more flexible than other transforms and is as such suitable as well as powerful tool for investigating deep problems in science and engineering. The application areas for LCT have been growing from last two decades at a very rapid rate and it has been applied in a number of fields including optics, quantum physics, time-frequency analysis, filter design, phase reconstruction, pattern recognition, radar analysis, holographic three-dimensional television and many more. For more about linear canonical transformations and their applications, the reader is referred to the monograph [320].

In this section, our motive is to present a detailed analysis regarding the LCT and explore the theory from a mathematical and signal processing perspective. For notational convenience, we shall write a 2×2 real matrix $M = \begin{pmatrix} A & B \\ C & D \end{pmatrix}$ as $M = (A, B; C, D)$. We shall remain confined to the realm of real, unimodular matrices in which case the LCT turns to be a unitaray operator in the Hilbert space $L^2(\mathbb{R})$; the space of finite energy signals.

1.4.1 Definition and basic properties

This subsection is devoted for initiating the study of linear canonical transform abreast of its fundamental properties. Below we have the formal definition of LCT.

Definition 1.4.1. Given a real, unimodular matrix $M = (A, B; C, D)$, the linear canonical transform of any $f \in L^2(\mathbb{R})$ with respect to the matrix M is defined by

$$\mathscr{L}_M\Big[f\Big](\omega) = \int_{\mathbb{R}} f(t)\, \mathcal{K}_M(t, \omega)\, dt, \tag{1.164}$$

where $\mathcal{K}_M(t, \omega)$ is called as "kernel" of the LCT and is given by

$$\mathcal{K}_M(t, \omega) = \begin{cases} \dfrac{1}{\sqrt{2\pi i B}} \exp\left\{ \dfrac{i(At^2 - 2t\omega + D\omega^2)}{2B} \right\}, & B \neq 0 \\[3mm] \sqrt{D} \exp\left\{ \dfrac{iCD\omega^2}{2} \right\} f(D\omega), & B = 0. \end{cases} \tag{1.165}$$

Definition 1.4.1 allows us to make the following points regarding the notion of linear canonical transform:

(i) For the parameter matrix $M = (0, 1; -1, 0)$, the linear canonical transform (1.164) boils down to the classical Fourier transform (1.7), except for a factor of $1/\sqrt{i}$.

(ii) For $M = (\cos\alpha, \sin\alpha; -\sin\alpha, \cos\alpha)$, $\alpha \neq n\pi$, the linear canonical transform (1.164), when multiplied by the factor $2i\sqrt{e^{i\alpha}}$, yields the fractional Fourier transform (1.127).

(iii) In case $M = (1, B; 0, 1)$, $B \neq 0$, the linear canonical transform (1.164) reduces to the well-known Fresnel transform given by

$$\mathscr{L}_M\Big[f\Big](\omega) = \frac{1}{\sqrt{2\pi i B}} \int_{\mathbb{R}} f(t)\, \exp\left\{ \frac{i(t - \omega)^2}{2B} \right\} dt. \tag{1.166}$$

(iv) The phase-space transform (1.164) is lossless if and only if the matrix $M = (A, B; C, D)$ is unimodular; that is, $AD - BC = 1$.

(v) For the case $B = 0$, the linear canonical transform (1.164) corresponds to a scaling transformation coupled with amplitude and quadratic phase modulation. Moreover, multiplication by Gaussian or chirp function is obtained by switching the matrix to $M = (1, 0; C, 1)$ and the scaling operation can be regarded as a particular case when $M = (1/D, 0; 0, D)$.

(vi) It is worth noticing that only three free parameters, viz; A, B and D, are employed in Definition 1.4.1 and apart from the unimodular condition of the matrix $M = (A, B; C, D)$, there is no other constraint involving all the four parameters. In fact, the linear canonical transform can also be defined by using only three parameters; however, the four-parameter representation plays a significant role in the applications of LCT to optical systems. Thus, we infer that the linear canonical transform is 3-dimensional family of integral transforms and can be visualized as the action of the special linear group $SL(2, \mathbb{R})$ on the time-frequency plane. Apart from modelling many general optical systems, the geometrical manifestations of the transformation (1.164) demonstrate that the LCT maps any convex body into another convex body and the unimodular condition imposed in Definition 1.4.1 guarantees that the area of the body is preserved under the transformation.

(vii) The linear canonical transform given in (1.164) enjoys both the additive and associative properties; that is, $\mathscr{L}_M \mathscr{L}_N = \mathscr{L}_{MN}$ and $(\mathscr{L}_M \mathscr{L}_N) \mathscr{L}_P = \mathscr{L}_M (\mathscr{L}_N \mathscr{L}_P)$, for any trio of 2×2 real, unimodular matrices M, N and P. The additive operation implies that any cascading of linear canonical transformations can be obtained by a single linear canonical transformation, whereas the associative operation guarantees the fact that any cascade of systems can be arbitrarily regrouped. Such properties of the linear canonical transform make it befitting for evaluating the output of a cascade of such systems in a simple and insightful manner.

(viii) For any $f \in L^2(\mathbb{R})$, the linear canonical transform defined in (1.164) is expressible as

$$\mathscr{L}_M \big[f\big](\omega) = \frac{1}{\sqrt{2\pi i B}} \exp\left\{\frac{iD\omega^2}{2B}\right\} \int_{\mathbb{R}} f(t) \exp\left\{\frac{iAt^2}{2B}\right\} \exp\left\{-\frac{i\omega t}{B}\right\} dt$$

$$= \frac{1}{\sqrt{iB}} \exp\left\{\frac{iD\omega^2}{2B}\right\} \mathscr{F}\big[G\big]\left(\frac{\omega}{B}\right), \tag{1.167}$$

where

$$G(t) = f(t) \exp\left\{\frac{iAt^2}{2B}\right\}. \tag{1.168}$$

Hence, the computation of linear canonical transform essentially corresponds to:

- Multiplying the input with a chirp signal; that is, $f(t) \rightarrow \exp\left\{\frac{iAt^2}{2B}\right\} f(t)$.

- Computing the B^{-1}-scaled Fourier transform of the chirped signal $(iB)^{-1/2} G(t)$.

- Multiplying the output with another chirp signal: $\exp\left\{\frac{iD\omega^2}{2B}\right\}$.

From the above description, it is clear that the linear canonical transform (1.164) can be regarded as a chirp-Fourier-chirp transformation as depicted in Figure 1.31.

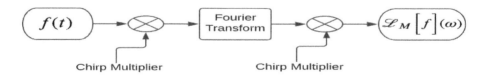

FIGURE 1.31: LCT as a chirp-Fourier-chirp transformation.

(ix) The applications of linear canonical transform are similar to those of the fractional Fourier transform; however, the former is much more flexible, especially for optical system analysis and synthesis. Moreover, in the context of time-frequency analysis, the fractional Fourier transform just corresponds to a rotation operation; however, the linear canonical transform corresponds to a twisting operation. This is due to the fact that, unlike the fractional Fourier transform, the chirping in Definition 1.4.1 is performed at different rates along the time and frequency axes. A visual description of the twisting operation of the linear canonical transform on the time-frequency plane is eloquently given in Figure 1.32.

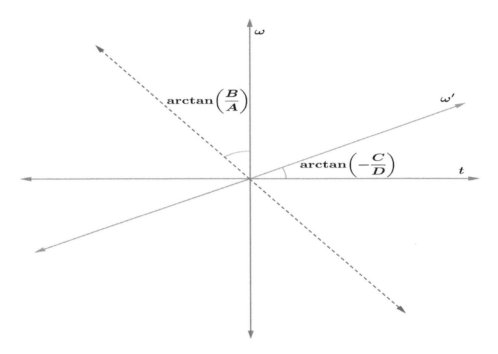

FIGURE 1.32: Twisting operation of LCT on the time-frequency plane.

(x) From a signal processing perspective, for a given matrix $M = (A, B; C, D)$ associated with the linear canonical transform, the quantities B/A and C/D are particularly interesting. This is because these ratios determine the twisting operations carried out via the linear canonical transformation. However, in certain practical aspects, for instance in the context of filter designing, the entity B/A is relatively more important than C/D as B/A determines the principle direction, whereas C/D determines the shearing of the canonical convolution operations. In such practical requirements, it always suffices to control the value B/A, regardless of the value of C/D.

In the following theorem, we assemble some elementary properties of the linear canonical kernel (1.165).

Theorem 1.4.1. *The kernel of the linear canonical transformation given by (1.165) satisfies the following properties:*

(i) $\mathcal{K}_{M^{-1}}(\omega, t) = \overline{\mathcal{K}_M(t, \omega)}$,

(ii) $\mathcal{K}_M(-t, \omega) = \mathcal{K}_M(t, -\omega)$,

(iii) $\int_{\mathbb{R}} \mathcal{K}_M(t, \omega)\, \mathcal{K}_{M^{-1}}(\omega', t)\, dt = \delta(\omega' - \omega),$

(iv) $\int_{\mathbb{R}} \mathcal{K}_M(t, \omega)\, \mathcal{K}_{M^{-1}}(\omega, t')\, d\omega = \delta(t' - t),$

(v) $\int_{\mathbb{R}} \mathcal{K}_M(t, \omega)\, \mathcal{K}_N(\omega', t)\, dt = \mathcal{K}_{MN}(\omega, \omega').$

Proof. For the sake of brevity, we omit the proof. □

Remark 1.4.1. The properties (iii) and (iv) in Theorem 1.4.1 imply that the kernel functions $\mathcal{K}_M(t, \omega)$ defined in (1.165) constitute an orthonormal set, when considered either as functions of t with parameter ω, or as functions of ω with parameter t. Moreover, it is imperative to mention that, sometimes in practice, the pre-factor $1/\sqrt{2\pi iB}$ appearing in the kernel defined in (1.165) is modified slightly, according to need. For instance, one of the frequently used modifications is $1/\sqrt{2\pi|B|}$, however, such modifications does not affect on the set of properties given in Theorem 1.4.1.

Below, we shall exhibit that any signal can be retracted from the corresponding linear canonical transform.

Theorem 1.4.2. *If $\mathscr{L}_M\big[f\big](\omega)$ is the linear canonical transform of any $f \in L^2(\mathbb{R})$, then the following inversion formula holds:*

$$f(t) = \mathscr{L}_{M^{-1}}\Big(\mathscr{L}_M\big[f\big](\omega)\Big)(t) := \int_{\mathbb{R}} \mathscr{L}_M\big[f\big](\omega)\, \mathcal{K}_{M^{-1}}(\omega, t)\, d\omega. \tag{1.169}$$

Proof. Using Definition 1.4.1 together with Theorem 1.4.1, we obtain

$$\int_{\mathbb{R}} \mathscr{L}_M\big[f\big](\omega)\, \mathcal{K}_{M^{-1}}(\omega, t)\, d\omega = \int_{\mathbb{R}} \left\{ \int_{\mathbb{R}} f(x)\, \mathcal{K}_M(x, \omega)\, dx \right\} \mathcal{K}_{M^{-1}}(\omega, t)\, d\omega$$

$$= \int_{\mathbb{R}} f(x) \left\{ \int_{\mathbb{R}} \mathcal{K}_M(x, \omega)\, \mathcal{K}_{M^{-1}}(\omega, t)\, d\omega \right\} dx$$

$$= \int_{\mathbb{R}} f(x)\, \delta(x - t)\, dx$$

$$= f(t).$$

Thus, we conclude that the inverse of linear canonical transform with respect to a real, unimodular matrix M is dictated by the matrix M^{-1}. This completes the proof of Theorem 1.4.2. □

Remark 1.4.2. In view of property (i) in Theorem 1.4.1, the inversion formula (1.169) can be alternatively expressed as

$$f(t) = \int_{\mathbb{R}} \mathscr{L}_M\big[f\big](\omega)\, \overline{\mathcal{K}_M(t, \omega)}\, d\omega. \tag{1.170}$$

It is worth noticing that the linear canonical transform with respect to a real, unimodular matrix $M = (A, B; C, D)$ offers an expansion of a signal f on an orthonormal basis constituted by the set of functions $\mathcal{K}_{M^{-1}}(\omega, t)$, with ω acting as a parameter for spanning the set of basis functions.

Theorem 1.4.3. (Parseval's Formula): *For any pair of functions* $f, g \in L^2(\mathbb{R})$, *we have*

$$\left\langle \mathscr{L}_M[f], \mathscr{L}_M[g] \right\rangle_2 = \left\langle f, g \right\rangle_2. \tag{1.171}$$

Particularly, if $f = g$, *then the following energy preserving relation holds:*

$$\left\| \mathscr{L}_M[f] \right\|_2^2 = \left\| f \right\|_2^2. \tag{1.172}$$

Proof. Using Definition 1.4.1 for the given pair of functions $f, g \in^2 (\mathbb{R})$ and invoking the well-known Fubini's theorem to change the order of integration, we have

$$\left\langle \mathscr{L}_M[f], \mathscr{L}_M[g] \right\rangle_2 = \int_{\mathbb{R}} \mathscr{L}_M[f](\omega)\, \overline{\mathscr{L}_M[g](\omega)}\, d\omega$$

$$= \int_{\mathbb{R}} \left\{ \int_{\mathbb{R}} f(t)\, \mathcal{K}_M(t,\omega)\, dt \right\} \left\{ \int_{\mathbb{R}} \overline{g(x)\, \mathcal{K}_M(x,\omega)}\, dx \right\} d\omega$$

$$= \int_{\mathbb{R}} \int_{\mathbb{R}} f(t)\, \overline{g(x)} \left\{ \int_{\mathbb{R}} \mathcal{K}_M(t,\omega)\, \overline{\mathcal{K}_M(x,\omega)}\, d\omega \right\} dt\, dx$$

$$= \int_{\mathbb{R}} \int_{\mathbb{R}} f(t)\, \overline{g(x)} \left\{ \int_{\mathbb{R}} \mathcal{K}_M(t,\omega)\, \mathcal{K}_{M^{-1}}(\omega,x)\, d\omega \right\} dt\, dx$$

$$= \int_{\mathbb{R}} \int_{\mathbb{R}} f(t)\, \overline{g(x)}\, \delta(x - t)\, dt\, dx$$

$$= \int_{\mathbb{R}} f(t)\, \overline{g(t)}\, dt$$

$$= \left\langle f, g \right\rangle_2, \tag{1.173}$$

which is the desired Parseval's relation for the linear canonical transform (1.164). Moreover, plugging $f = g$ in (1.173), we obtain the following energy preserving relation:

$$\left\| \mathscr{L}_M[f] \right\|_2^2 = \left\| f \right\|_2^2.$$

This completes the proof of Theorem 1.4.3. $\qquad\square$

Remark 1.4.3. The Parseval's formula (1.171) and the energy preserving relation (1.172) are indeed owed to the fact that the linear canonical transform (1.164) relies upon an orthonormal set of functions. Moreover, the squared magnitude of the linear canonical transform of a signal f with respect to a real, unimodular matrix $M = (A, B; C, D)$ is referred as the linear canonical energy spectrum of the signal and can be interpreted as the energy distribution of the signal into different chirps $\mathcal{K}_{M^{-1}}(\omega, t)$.

Next, we focus to investigate upon some of the elementary properties of linear canonical transform (1.164). In this direction, we have the following theorem:

Theorem 1.4.4. *Let* $\alpha, \beta \in \mathbb{C}$, $k, \omega_0 \in \mathbb{R}$ *and* $\lambda \in \mathbb{R} \setminus \{0\}$. *Then, for any* $f, g \in L^2(\mathbb{R})$, *the linear canonical transform defined in (1.164) satisfies the following properties:*

(i) *Linearity:* $\mathscr{L}_M[\alpha f + \beta g](\omega) = \alpha\, \mathscr{L}_M[f](\omega) + \beta\, \mathscr{L}_M[g](\omega),$

(ii) *Translation:* $\mathscr{L}_M[f(t - k)](\omega) = \exp\left\{ i\left(\omega k C - \dfrac{k^2 AC}{2} \right) \right\} \mathscr{L}_M[f](\omega - Ak),$

(iii) *Modulation:* $\mathscr{L}_M\left[e^{i\omega_0 t}f(t)\right](\omega) = \exp\left\{\dfrac{iD(2\omega\omega_0 - B\omega_0^2)}{2}\right\}\mathscr{L}_M\left[f\right](\omega - B\omega_0),$

(iv) *Scaling:* $\mathscr{L}_M\left[f\left(\dfrac{t}{\lambda}\right)\right](\omega) = |\lambda|\,\mathscr{L}_N\left[f\right](\lambda\omega), \quad N = \left(A\lambda^2, B; C, D/\lambda^2\right),$

(v) *Parity:* $\mathscr{L}_M\left[f(-t)\right](\omega) = \mathscr{L}_M\left[f\right](-\omega),$

(vi) *Conjugation:* $\mathscr{L}_M\left[\overline{f}\right](\omega) = \overline{\mathscr{L}_{M^{-1}}\left[f\right](\omega)}, \quad M^{-1} = (D, -B; -C, A), \text{ provided } A = D,$

(vii) *Translation and modulation:* $\mathscr{L}_M\left[e^{i\omega_0 t}f(t - k)\right](\omega)$

$$= \exp\left\{-i\left(\dfrac{ACk^2}{2} + \dfrac{DB\omega_0^2}{2} + BCk\omega_0 - Ck\omega - D\omega\omega_0\right)\right\}\mathscr{L}_M\left[f\right](\omega - B\omega_0 - Ak).$$

Proof. **(i)** Applying Definition 1.4.1 for the given pair of functions $f, g \in L^2(\mathbb{R})$, we have

$$\mathscr{L}_M\left[\alpha f + \beta g\right](\omega) = \int_{\mathbb{R}}\left(\alpha f(t) + \beta g(t)\right)\mathcal{K}_M(t, \omega)\,dt$$

$$= \alpha \int_{\mathbb{R}} f(t)\,\mathcal{K}_M(t, \omega)\,dt + \beta \int_{\mathbb{R}} g(t)\,\mathcal{K}_M(t, \omega)\,dt$$

$$= \alpha\,\mathscr{L}_M\left[f\right](\omega) + \beta\,\mathscr{L}_M\left[g\right](\omega).$$

(ii) For any $k \in \mathbb{R}$, we have

$$\mathscr{L}_M\left[f(t - k)\right](\omega) = \dfrac{1}{\sqrt{2\pi iB}}\int_{\mathbb{R}} f(t - k)\,\exp\left\{\dfrac{i(At^2 - 2t\omega + D\omega^2)}{2B}\right\}dt$$

$$= \dfrac{1}{\sqrt{2\pi iB}}\int_{\mathbb{R}} f(z)\,\exp\left\{\dfrac{i(A(z + k)^2 - 2(z + k)\omega + D\omega^2)}{2B}\right\}dz$$

$$= \dfrac{1}{\sqrt{2\pi iB}}\int_{\mathbb{R}} f(z)\,\exp\left\{\dfrac{i(Az^2 - 2z(\omega - Ak) + D(\omega - Ak)^2)}{2B}\right\}$$

$$\times \exp\left\{\dfrac{i(A(1 - AD)k^2 + 2(AD - 1)k\omega)}{2B}\right\}dz$$

$$= \exp\left\{i\left(\omega kC - \dfrac{k^2 AC}{2}\right)\right\}\mathscr{L}_M\left[f\right](\omega - Ak).$$

(iii) In case the square integrable function f is modulated by $\omega_0 \in \mathbb{R}$, then the corresponding linear canonical transform can be computed as

$$\mathscr{L}_M\left[e^{i\omega_0 t}f(t)\right](\omega) = \dfrac{1}{\sqrt{2\pi iB}}\int_{\mathbb{R}} e^{i\omega_0 t}f(t)\,\exp\left\{\dfrac{i(At^2 - 2t\omega + D\omega^2)}{2B}\right\}dt$$

$$= \dfrac{1}{\sqrt{2\pi iB}}\int_{\mathbb{R}} f(t)\,\exp\left\{\dfrac{i(At^2 - 2t(\omega - B\omega_0) + D\omega^2)}{2B}\right\}dt$$

$$= \dfrac{1}{\sqrt{2\pi iB}}\int_{\mathbb{R}} f(t)\,\exp\left\{\dfrac{i(At^2 - 2t(\omega - B\omega_0) + D(\omega - B\omega_0)^2)}{2B}\right\}$$

$$\times \exp\left\{\dfrac{iD(2\omega\omega_0 - B\omega_0^2)}{2}\right\}dt$$

$$= \exp\left\{\dfrac{iD(2\omega\omega_0 - B\omega_0^2)}{2}\right\}\mathscr{L}_M\left[f\right](\omega - B\omega_0).$$

(iv) For $\lambda \in \mathbb{R} \setminus \{0\}$, we have

$$
\begin{aligned}
\mathscr{L}_M\left[f\left(\frac{t}{\lambda}\right)\right](\omega) &= \frac{1}{\sqrt{2\pi i B}} \int_{\mathbb{R}} f(t/\lambda) \exp\left\{\frac{i(At^2 - 2t\omega + D\omega^2)}{2B}\right\} dt \\
&= \frac{|\lambda|}{\sqrt{2\pi i B}} \int_{\mathbb{R}} f(z) \exp\left\{\frac{i(A(z\lambda)^2 - 2z\omega\lambda + D\omega^2)}{2B}\right\} dz \\
&= \frac{|\lambda|}{\sqrt{2\pi i B}} \int_{\mathbb{R}} f(z) \exp\left\{\frac{i((A\lambda^2)z - 2z(\omega\lambda) + (D/\lambda^2)(\lambda\omega)^2)}{2B}\right\} dz \\
&= |\lambda|\,\mathscr{L}_N\left[f\right](\lambda\omega), \quad N = (A\lambda^2, B; C, D/\lambda^2).
\end{aligned}
$$

Here, it is imperative to mention that the scaling operation has a direct influence on the unimodular matrix underlying the linear canonical transform. Most importantly, the compression and relaxation inflicted by the scaling parameter λ on the linear canonical spectrum of the input signal are interchangeable by suitable modifications to the unimodular matrix. To observe this behaviour of the linear canonical transform, we proceed as

$$
\begin{aligned}
\mathscr{L}_M\left[f\left(\frac{t}{\lambda}\right)\right](\omega) &= \frac{1}{\sqrt{2\pi i B}} \int_{\mathbb{R}} f(t/\lambda) \exp\left\{\frac{i(At^2 - 2t\omega + D\omega^2)}{2B}\right\} dt \\
&= \frac{|\lambda|}{\sqrt{2\pi i B}} \int_{\mathbb{R}} f(z) \exp\left\{\frac{i(A(z\lambda)^2 - 2z\omega\lambda + D\omega^2)}{2B}\right\} dz \\
&= \frac{|\lambda|}{\sqrt{2\pi i B}} \int_{\mathbb{R}} f(z) \exp\left\{\frac{i\lambda^2(Az^2 - 2z(\omega/\lambda) + D(\omega/\lambda)^2)}{2B}\right\} dz \\
&= \frac{1}{\sqrt{2\pi i(B/\lambda^2)}} \int_{\mathbb{R}} f(z) \exp\left\{\frac{i(Az^2 - 2z(\omega/\lambda) + D(\omega/\lambda)^2)}{2(B/\lambda^2)}\right\} dz \\
&= \mathscr{L}_P\left[f\right]\left(\frac{\omega}{\lambda}\right), \quad P = (A, B/\lambda^2; \lambda^2 C, D).
\end{aligned}
$$

Note that the parameter λ^2 multiplied to the matrix element C has the effect of making the matrix $P = (A, B/\lambda^2, \lambda^2 C, D)$ unimodular.

(v) Using the Definition 1.4.1, we have

$$
\begin{aligned}
\mathscr{L}_M\left[f(-t)\right](\omega) &= \frac{1}{\sqrt{2\pi i B}} \int_{\mathbb{R}} f(-t) \exp\left\{\frac{i(At^2 - 2t\omega + D\omega^2)}{2B}\right\} dt \\
&= \frac{1}{\sqrt{2\pi i B}} \int_{\mathbb{R}} f(z) \exp\left\{\frac{i(Az^2 + 2z\omega + D\omega^2)}{2B}\right\} dz \\
&= \frac{1}{\sqrt{2\pi i B}} \int_{\mathbb{R}} f(z) \exp\left\{\frac{i(Az^2 - 2z(-\omega) + D\omega^2)}{2B}\right\} dz \\
&= \mathscr{L}_M\left[f\right](-\omega).
\end{aligned}
$$

Here, it is important to note that the parity property also follows directly from (iv) by plugging $\lambda = -1$.

(vi) In order to obtain LCT of the complex-conjugated input, we consider any real, unimodular matrix $M = (A, B; C, D)$ and denote $\widehat{M} = (D, B; C, A)$. Then, it can be easily verified that $\mathcal{K}_{\widehat{M}}(\omega, t) = \mathcal{K}_M(t, \omega)$. Thus, for any $f \in L^2(\mathbb{R})$, we have

$$\mathscr{L}_{\widehat{M}}\left(\overline{\mathscr{L}_M\left[f\right](\omega')}\right)(t) = \int_{\mathbb{R}} \overline{\mathscr{L}_M\left[f\right](\omega')}\, \mathcal{K}_{\widehat{M}}(\omega',t)\, d\omega'$$

$$= \int_{\mathbb{R}} \left\{ \int_{\mathbb{R}} \overline{f(x)}\, \overline{\mathcal{K}_M(x,\omega')}\, dx \right\} \mathcal{K}_{\widehat{M}}(\omega',t)\, d\omega'$$

$$= \int_{\mathbb{R}} \int_{\mathbb{R}} \overline{f(x)}\, \mathcal{K}_{M^{-1}}(\omega',x)\, \mathcal{K}_M(t,\omega')\, dx\, d\omega'$$

$$= \int_{\mathbb{R}} \overline{f(x)}\, \delta(x-t)\, dx$$

$$= \overline{f(t)},$$

so that

$$\mathscr{L}_M\left[\overline{f}\right](\omega) = \int_{\mathbb{R}} \overline{f(t)}\, \mathcal{K}_M(t,\omega)\, dt$$

$$= \int_{\mathbb{R}} \overline{\mathscr{L}_{\widehat{M^{-1}}}\left(\overline{\mathscr{L}_{M^{-1}}\left[f\right](\omega')}\right)(t)}\, \mathcal{K}_M(t,\omega)\, dt$$

$$= \int_{\mathbb{R}} \left\{ \int_{\mathbb{R}} \overline{\mathscr{L}_{M^{-1}}\left[f\right](\omega')}\, \mathcal{K}_{\widehat{M^{-1}}}(\omega',t)\, d\omega' \right\} \mathcal{K}_M(t,\omega)\, dt$$

$$= \int_{\mathbb{R}} \int_{\mathbb{R}} \overline{\mathscr{L}_{M^{-1}}\left[f\right](\omega')}\, \mathcal{K}_{M^{-1}}(t,\omega')\, \mathcal{K}_M(t,\omega)\, dt\, d\omega'$$

$$= \int_{\mathbb{R}} \overline{\mathscr{L}_{M^{-1}}\left[f\right](\omega')}\left\{ \int_{\mathbb{R}} \mathcal{K}_{M^{-1}}(t,\omega')\, \mathcal{K}_M(t,\omega)\, dt \right\} d\omega'$$

$$= \int_{\mathbb{R}} \overline{\mathscr{L}_{M^{-1}}\left[f\right](\omega')}\left\{ \int_{\mathbb{R}} \mathcal{K}_{M^{-1}}(t,\omega')\, \mathcal{K}_M(\omega,t)\, dt \right\} d\omega'$$

$$= \int_{\mathbb{R}} \overline{\mathscr{L}_{M^{-1}}\left[f\right](\omega')}\, \delta(\omega'-\omega)\, d\omega'$$

$$= \overline{\mathscr{L}_{M^{-1}}\left[f\right](\omega)},$$

where the relationship $\mathcal{K}_M(t,\omega) = \mathcal{K}_M(\omega,t)$ used in the last but third equality holds, provided $A = D$.

Indeed, the proof of property (vi) could also be obtained by directly making use of the assumption $A = D$, so that $\mathcal{K}_{M^{-1}}(\omega,t) = \mathcal{K}_{M^{-1}}(t,\omega)$, which implies that

$$\mathscr{L}_M\left[\overline{f}\right](\omega) = \int_{\mathbb{R}} \overline{f(t)}\, \mathcal{K}_M(t,\omega)\, dt$$

$$= \int_{\mathbb{R}} \overline{f(t)}\, \overline{\mathcal{K}_{M^{-1}}(\omega,t)}\, dt$$

$$= \int_{\mathbb{R}} \overline{f(t)}\, \overline{\mathcal{K}_{M^{-1}}(t,\omega)}\, dt$$

$$= \overline{\mathscr{L}_{M^{-1}}\left[f\right](\omega)}.$$

However, we followed a bit longer approach, with the intent to comment about the notion of the so-called "reverse system." It is worth to mention that the unimodular matrix $\widehat{M} = (D, B; C, A)$ corresponding to any $M = (A, B; C, D)$ plays an important role in the theory of linear canonical transforms, particularly in understanding the behaviour of the reverse system. Also, it is important to note that the reverse system is not the same as the inverse system $\mathscr{L}_{M^{-1}}$. In fact, if $\mathscr{L}_M$ is the forward system, then in the reverse system $\mathscr{L}_{\widehat{M}}$, the signal phases are reversed and the signals propagate in the opposite direction, which means that the frequency variable has to be reversed, and that $\mathscr{L}_M[f]$ now acts the input signal, while as $\overline{f} = \mathscr{L}_{\widehat{M}}(\mathscr{L}_M[f])$ is the output signal.

(vii) Finally, we compute LCT of the shifted and modulated version of the input signal. For $\omega_0, k \in \mathbb{R}$, we have

$$\mathscr{L}_M\left[e^{i\omega_0 t}f(t-k)\right](\omega) = \frac{1}{\sqrt{2\pi i B}}\int_{\mathbb{R}} e^{i\omega_0 t}f(t-k)\exp\left\{\frac{i(At^2 - 2t\omega + D\omega^2)}{2B}\right\}dt$$

$$= \frac{1}{\sqrt{2\pi i B}}\int_{\mathbb{R}} e^{i\omega_0(k+z)}f(z)$$

$$\times \exp\left\{\frac{i(A(k+z)^2 - 2(k+z)\omega + D\omega^2)}{2B}\right\}dz$$

$$= \frac{1}{\sqrt{2\pi i B}}\int_{\mathbb{R}} f(z)\exp\left\{i\omega_0 k + \frac{iAk^2}{2B} - \frac{ik\omega}{B}\right\}$$

$$\times \exp\left\{\frac{i(Az^2 - 2(\omega - B\omega_0 - Ak)z + D(\omega - B\omega_0 - Ak)^2)}{2B}\right\}$$

$$\times \exp\left\{-\frac{iD(A^2k^2 - 2Ak\omega + 2ABk\omega_0 + B^2\omega_0^2 - 2B\omega\omega_0)}{2B}\right\}dz$$

$$= \exp\left\{-\frac{iD(A^2k^2 - 2Ak\omega + 2ABk\omega_0 + B^2\omega_0^2 - 2B\omega\omega_0)}{2B}\right\}$$

$$\times \exp\left\{i\omega_0 k + \frac{iAk^2}{2B} - \frac{ik\omega}{B}\right\}\mathscr{L}_M[f](\omega - B\omega_0 - Ak)$$

$$= \exp\left\{\frac{iA(1-AD)k^2}{2B} + \frac{i(AD-1)k\omega}{B} - i(1-AD)k\omega_0\right.$$

$$\left. +iD\omega\omega_0 - \frac{iDB\omega_0^2}{2} - \frac{iD\omega_0}{2}\right\}\mathscr{L}_M[f](\omega - B\omega_0 - Ak)$$

$$= \exp\left\{-i\left(\frac{ACk^2}{2} + \frac{DB\omega_0^2}{2} + BCk\omega_0 - Ck\omega - D\omega\omega_0\right)\right\}$$

$$\times \mathscr{L}_M[f](\omega - B\omega_0 - Ak).$$

This completes the proof of Theorem 1.4.4. $\qquad\square$

1.4.2 Linear canonical convolution and correlation

In this subsection, we present the notion of convolution and correlation associated with the linear canonical transform. The convolution and correlation theory in the linear canonical domain plays a central role in the applications of LCT and has been paid considerable attention in recent years [96, 256, 264, 309]. Unfortunately, the usual convolution and correlation theorems do not hold in the linear canonical domain and for this reason, several different definitions of the linear canonical convolution and correlation operations have been proposed in the recent literature [332]. However, we shall focus our attention only on few notable approaches towards the convolution and correlation operations in linear canonical domain.

(A) Chirp-free weighted convolution

Firstly, we present the most simple and elegant linear canonical convolution, which is in compliance to both the classical convolution in the Fourier domain (1.75) as well the fractional convolution (1.141). Here, it is imperative to mention that such a convolution operation plays a pivotal role in many subsequent developments in the present book.

Definition 1.4.2. Given a unimodular matrix $M = (A, B; C, D)$ and any pair of functions $f, g \in L^2(\mathbb{R})$. The chirp-free weighted convolution associated with the linear canonical transform is denoted by $\otimes_M$ and is defined as

$$(f \otimes_M g)(z) = \int_{\mathbb{R}} f(t)\, g(z - t)\, \exp\left\{ \frac{iA(t^2 - z^2)}{2B} \right\} dt. \tag{1.174}$$

Some important characteristics of the linear canonical convolution (1.174) are described in the following theorem:

Theorem 1.4.5. *For any $f, g, h \in L^2(\mathbb{R})$ and the scalars $k \in \mathbb{R}$, $\lambda \in \mathbb{R} \setminus \{0\}$, the linear canonical convolution operation $\otimes_M$ defined in (1.174) satisfies the following properties:*

  (i) *Non-commutativity:* $(f \otimes_M g)(z) \neq (g \otimes_M f)(z)$,

 (ii) *Non-associativity:* $(f \otimes_M (g \otimes_M h))(z) \neq ((f \otimes_M g) \otimes_M h)(z)$,

(iii) *Distributivity:* $(f \otimes_M (g + h))(z) = (f \otimes_M g)(z) + (f \otimes_M h)(z)$,

 (iv) *Translation:* $(f \otimes_M g)(z - k) = (f(y - k) \otimes_M G_M(y))(z)$, $G_M(t) = e^{iAtk/B} g(t)$,

  (v) *Scaling:* $(f \otimes_M g)(\lambda z) = |\lambda| (f(\lambda y) \otimes_N g(\lambda y))(z)$, $N = (A, B/\lambda^2; \lambda^2 C, D)$,

 (vi) *Parity:* $(f \otimes_M g)(-z) = (f(-t) \otimes_M g(-t))(z)$.

Proof. All the assertions in Theorem 1.4.5 can be verified easily, so we omit the formal proof.   □

In the following theorem, we demonstrate that indeed the convolution theorem pertaining to the linear canonical convolution operation $\otimes_M$ defined in (1.174) is chirp-free.

Theorem 1.4.6. *For any $f, g \in L^2(\mathbb{R})$, we have*

$$\mathscr{L}_M\big[(f \otimes_M g)\big](\omega) = \sqrt{2\pi}\, \mathscr{L}_M\big[f\big](\omega)\, \mathscr{F}\big[g\big]\left(\frac{\omega}{B}\right). \tag{1.175}$$

Proof. Taking the linear canonical transform of Definition 1.4.2, we obtain

$$\mathscr{L}_M\big[(f \otimes_M g)\big](\omega) = \int_{\mathbb{R}} (f \otimes_M g)(z)\, \mathcal{K}_M(z, \omega)\, dz$$

$$= \int_{\mathbb{R}} \left\{ \int_{\mathbb{R}} f(t)\, g(z - t) \exp\left\{ \frac{iA(t^2 - z^2)}{2B} \right\} dt \right\} \mathcal{K}_M(z, \omega)\, dz$$

$$= \int_{\mathbb{R}} f(t) \left\{ \int_{\mathbb{R}} g(z - t) \exp\left\{ \frac{iA(t^2 - z^2)}{2B} \right\} \mathcal{K}_M(z, \omega)\, dz \right\} dt. \tag{1.176}$$

Substituting $z - t = y$ into the second integral on the R.H.S of (1.176) yields

$$\mathscr{L}_M\big[(f \otimes_M g)\big](\omega)$$

$$= \int_{\mathbb{R}} f(t) \left\{ \int_{\mathbb{R}} g(y) \exp\left\{ \frac{iA(t^2 - (t + y)^2)}{2B} \right\} \mathcal{K}_M(t + y, \omega)\, dy \right\} dt$$

$$= \frac{1}{\sqrt{2\pi i B}} \int_{\mathbb{R}} f(t) \left\{ \int_{\mathbb{R}} g(y) \exp\left\{ \frac{iA(t^2 - t^2 - y^2 - 2ty)}{2B} \right\} \right.$$

$$\times \exp\left\{\frac{i\left(A(t^2 + y^2 + 2ty) - 2(t+y)\omega + D\omega^2\right)}{2B}\right\} dy \right\} dt$$

$$= \frac{1}{\sqrt{2\pi i B}} \int_{\mathbb{R}} f(t) \exp\left\{\frac{i(At^2 - 2t\omega + D\omega^2)}{2B}\right\} \left\{\int_{\mathbb{R}} g(y) \exp\left\{-iy\left(\frac{\omega}{B}\right)\right\} dy\right\} dt$$

$$= \sqrt{2\pi} \int_{\mathbb{R}} f(t)\, \mathcal{K}_M(t,\omega) \left\{\frac{1}{\sqrt{2\pi}} \int_{\mathbb{R}} g(y) \exp\left\{-iy\left(\frac{\omega}{B}\right)\right\} dy\right\} dt$$

$$= \sqrt{2\pi}\, \mathscr{L}_M\big[f\big](\omega)\, \mathscr{F}\big[g\big]\left(\frac{\omega}{B}\right).$$

This completes the proof of Theorem 1.4.6. □

On the flip side to Definition 1.4.2, we have the following definition of the linear canonical correlation operation:

Definition 1.4.3. Given a unimodular matrix $M = (A, B; C, D)$, the chirp-free weighted linear canonical correlation of any pair of functions $f, g \in L^2(\mathbb{R})$ is denoted by $\star_M$ and is defined as

$$(f \star_M g)(z) = \int_{\mathbb{R}} f(t)\, \overline{g(t-z)} \exp\left\{\frac{iA(t^2 - z^2)}{2B}\right\} dt. \tag{1.177}$$

The correlation theorem associated with the linear canonical correlation given in Definition 1.4.3 is next.

Theorem 1.4.7. *For any $f, g \in L^2(\mathbb{R})$, we have*

$$\mathscr{L}_M\Big[(f \star_M g)\Big](\omega) = \sqrt{2\pi}\, \mathscr{L}_M\big[f\big](\omega)\, \overline{\mathscr{F}\big[g\big]\left(\frac{\omega}{B}\right)}. \tag{1.178}$$

Proof. The proof of Theorem 1.4.7 follows in a manner analogous to the proof of Theorem 1.4.6. □

Remark 1.4.4. The linear canonical convolution and correlation operations defined in (1.174) and (1.177) are governed by the unimodular matrix $M = (A, B; C, D)$. For $M = (0, 1; -1, 0)$, these convolution and correlation operations boil down to the classical convolution and correlation in the Fourier domain defined in (1.75) and (1.88), respectively. Moreover, the unimodular matrix $M = (\cos\alpha, \sin\alpha; -\sin\alpha, \cos\alpha)$, $\alpha \neq n\pi$, $n \in \mathbb{Z}$ yields the convolution and correlation operations for the fractional Fourier transform defined in (1.141) and (1.144), respectively. Hence, we infer that the above defined linear canonical convolution and correlation operations are prolific in nature and circumscribe both the classical as well as the fractional convolution operations.

(B) Another weighted convolution

In the following definition, we present another weighted convolution operation associated with the linear canonical transform (1.164).

Definition 1.4.4. Given a unimodular matrix $M = (A, B; C, D)$ and any pair of functions $f, g \in L^2(\mathbb{R})$. The weighted linear canonical convolution $\circledast_M$ is defined as

$$(f \circledast_M g)(z) = \int_{\mathbb{R}} f(t)\, g(z-t) \exp\left\{-\frac{iAt(z-t)}{B}\right\} dt. \tag{1.179}$$

In the following theorem, we examine the characteristic features of weighted linear canonical convolution given by Definition 1.4.4.

Theorem 1.4.8. *For any $f, g, h \in L^2(\mathbb{R})$ and the scalars $k \in \mathbb{R}$, $\lambda \in \mathbb{R} \setminus \{0\}$, the linear canonical convolution operation $\circledast_M$ defined in (1.179) satisfies:*

(i) *Commutativity:* $(f \circledast_M g)(z) = (g \circledast_M f)(z)$,

(ii) *Associativity:* $(f \circledast_M (g \circledast_M h))(z) = ((f \circledast_M g) \circledast_M h)(z)$,

(iii) *Distributivity:* $(f \circledast_M (g + h))(z) = (f \circledast_M g)(z) + (f \circledast_M h)(z)$,

(iv) *Translation:* $(f \circledast_M g)(z - k) = (f(\cdot - k) \circledast_M G_M(\cdot))(z)$, $G_M(t) = e^{iAtk/B} g(t)$,

(v) *Scaling:* $(f \circledast_M g)(\lambda z) = |\lambda| (f(\lambda y) \circledast_N g(\lambda y))(z)$, $N = (A, B/\lambda^2; \lambda^2 C, D)$,

(vi) *Parity:* $(f \circledast_M g)(-z) = (f(-t) \circledast_M g(-t))(z)$.

Proof. Being brief in the ongoing discourse, we ought to omit the proof. $\qquad\square$

Following is convolution theorem corresponding to the weighted linear canonical convolution given by Definition 1.4.4:

Theorem 1.4.9. *For any pair of functions $f, g \in L^2(\mathbb{R})$, we have*

$$\mathscr{L}_M \left[(f \circledast_M g) \right](\omega) = \sqrt{2\pi i B} \, \exp\left\{ -\frac{iD\omega^2}{2B} \right\} \mathscr{L}_M \left[f \right](\omega) \, \mathscr{L}_M \left[g \right](\omega). \qquad (1.180)$$

Proof. Using the Definition 1.4.1, we have

$$\mathscr{L}_M \left[(f \circledast_M g) \right](\omega) = \frac{1}{\sqrt{2\pi i B}} \int_{\mathbb{R}} (f \circledast_M g)(z) \exp\left\{ \frac{i(Az^2 - 2z\omega + D\omega^2)}{2B} \right\} dz$$

$$= \frac{1}{\sqrt{2\pi i B}} \int_{\mathbb{R}} \left\{ \int_{\mathbb{R}} f(t) \, g(z - t) \exp\left\{ -\frac{iAt(z - t)}{B} \right\} dt \right\}$$

$$\times \exp\left\{ \frac{i(Az^2 - 2z\omega + D\omega^2)}{2B} \right\} dz$$

$$= \frac{1}{\sqrt{2\pi i B}} \int_{\mathbb{R}} f(t) \left\{ \int_{\mathbb{R}} g(z - t) \exp\left\{ -\frac{iAt(z - t)}{B} \right\} \right.$$

$$\left. \times \exp\left\{ \frac{i(Az^2 - 2z\omega + D\omega^2)}{2B} \right\} dz \right\} dt. \qquad (1.181)$$

Making use of the substitution $z - t = y$ in the second integral on the R.H.S of (1.181), we get

$$\mathscr{L}_M \left[(f \circledast_M g) \right](\omega)$$

$$= \frac{1}{\sqrt{2\pi i B}} \int_{\mathbb{R}} f(t) \left\{ \int_{\mathbb{R}} g(y) \exp\left\{ -\frac{iAty}{B} \right\} \exp\left\{ \frac{i(A(t + y)^2 - 2(t + y)\omega + D\omega^2)}{2B} \right\} dy \right\} dt$$

$$= \int_{\mathbb{R}} f(t) \exp\left\{ \frac{i(At^2 - 2t\omega)}{2B} \right\} \left\{ \frac{1}{\sqrt{2\pi i B}} \int_{\mathbb{R}} g(y) \exp\left\{ \frac{i(Ay^2 - 2y\omega + D\omega^2)}{2B} \right\} dy \right\} dt$$

$$= \int_{\mathbb{R}} f(t) \exp\left\{ \frac{i(At^2 - 2t\omega)}{2B} \right\} \left\{ \int_{\mathbb{R}} g(y) \, \mathcal{K}_M(y, \omega) \, dy \right\} dt$$

$$= \sqrt{2\pi i B} \, \exp\left\{ -\frac{iD\omega^2}{2B} \right\} \mathscr{L}_M \left[g \right](\omega) \left\{ \frac{1}{\sqrt{2\pi i B}} \int_{\mathbb{R}} f(t) \exp\left\{ \frac{i(At^2 - 2t\omega + D\omega^2)}{2B} \right\} dt \right\}$$

$$= \sqrt{2\pi i B} \, \exp\left\{ -\frac{iD\omega^2}{2B} \right\} \mathscr{L}_M \left[g \right](\omega) \left\{ \int_{\mathbb{R}} f(t) \, \mathcal{K}_M(t, \omega) \, dt \right\}$$

$$= \sqrt{2\pi i B} \, \exp\left\{-\frac{iD\omega^2}{2B}\right\} \mathscr{L}_M\Big[f\Big](\omega) \, \mathscr{L}_M\Big[g\Big](\omega).$$

This completes the proof of Theorem 1.4.9. □

Remark 1.4.5. From Theorem 1.4.8, we observe that the weighted linear canonical convolution (1.179) satisfies several fundamental properties, particularly the commutativity and associativity. Moreover, Theorem 1.4.9 implies that, except for a chirp, the convolution theorem is in analogy to the convolution theorem in the Fourier domain, in the sense that the linear canonical transform of two convoluted signals corresponds to the product of their respective linear canonical spectra. Hence, the convolution operation $\circledast_M$ defined in (1.179) extends the classical convolution in a neat fashion and maintains the theoretical elegance. However, it is worth noticing that there exists an extra chirp multiplier in Theorem 1.4.9 which sometimes imposes difficulty in real applications as it is challenging to accurately generate a chirp signal in practical engineering.

Alongside the notion of weighted linear canonical convolution (1.179), we have the following definition of weighted linear canonical correlation:

Definition 1.4.5. Given a unimodular matrix $M = (A, B; C, D)$. Then, for any pair of functions $f, g \in L^2(\mathbb{R})$, the weighted linear canonical correlation $\oslash_M$ is defined as

$$(f \oslash_M g)(z) = \int_{\mathbb{R}} f(t) \, \overline{g(t-z)} \, \exp\left\{\frac{iAz(t-z)}{B}\right\} dt. \tag{1.182}$$

The correlation theorem corresponding to the weighted linear canonical correlation (1.182) is given below:

Theorem 1.4.10. *For any pair of functions $f, g \in L^2(\mathbb{R})$, we have*

$$\mathscr{L}_M\Big[(f \oslash_M g)\Big](\omega) = \sqrt{2\pi(-i)B} \, \exp\left\{\frac{iD\omega^2}{2B}\right\} \mathscr{L}_M\Big[f\Big](\omega) \, \overline{\mathscr{L}_M\Big[g\Big](\omega)}. \tag{1.183}$$

Proof. The proof of Theorem 1.4.10 can be obtained by replicating the strategy of the proof of Theorem 1.4.9 for Definition 1.4.5. □

Remark 1.4.6. The weighted linear canonical convolution and correlation defined in (1.179) and (1.182) are both dictated by the parameters of the unimodular matrix $M = (A, B; C, D)$. For $M = (0, 1; -1, 0)$, we infer that (1.179) and (1.182) boil down to the classical convolution and correlation in the Fourier domain as defined in (1.75) and (1.88), respectively. On the other hand, for the matrix $M = (\cos\alpha, \sin\alpha; -\sin\alpha, \cos\alpha)$, $\alpha \neq n\pi$, we obtain a new pair of convolution and correlation operations for the fractional Fourier transform.

(C) Modified chirp-free weighted convolution

Here, we intend to present yet another weighted convolution, which enjoys both the commutative and associative properties and also has the advantage that the associated convolution theorem is completely chirp-free.

Definition 1.4.6. Given a unimodular matrix $M = (A, B; C, D)$. Then, for any pair of functions $f, g \in L^2(\mathbb{R})$, the modified chirp-free weighted linear canonical convolution is denoted by $\odot_M$ and is defined as

$$(f \odot_M g)(z) = \int_{\mathbb{R}} f(t) \, g(\sqrt{2}z - t) \, \exp\left\{\frac{iA}{B}\left(\frac{z}{\sqrt{2}} - t\right)^2\right\} dt. \tag{1.184}$$

It is straightforward to verify the commutative and associative properties of the modified chirp-free weighted linear canonical convolution. Below, we shall formulate the convolution theorem corresponding to Definition 1.4.6.

Theorem 1.4.11. *For any $f, g \in L^2(\mathbb{R})$, we have*

$$\mathscr{L}_M\Big[(f \odot_M g)\Big](\omega) = \sqrt{i\pi B}\, \mathscr{L}_M\big[f\big]\left(\frac{\omega}{\sqrt{2}}\right) \mathscr{L}_M\big[g\big]\left(\frac{\omega}{\sqrt{2}}\right). \tag{1.185}$$

Proof. Invoking Definition 1.4.1, we have

$$\mathscr{L}_M\Big[(f \odot_M g)\Big](\omega) = \frac{1}{\sqrt{2\pi i B}} \int_{\mathbb{R}} (f \odot_M g)(z) \exp\left\{\frac{i(Az^2 - 2z\omega + D\omega^2)}{2B}\right\} dz$$

$$= \frac{1}{\sqrt{2\pi i B}} \int_{\mathbb{R}} \left\{\int_{\mathbb{R}} f(t)\, g(\sqrt{2}z - t) \exp\left\{\frac{iA}{B}\left(\frac{z}{\sqrt{2}} - t\right)^2\right\} dt\right\}$$

$$\times \exp\left\{\frac{i(Az^2 - 2z\omega + D\omega^2)}{2B}\right\} dz$$

$$= \frac{1}{\sqrt{2\pi i B}} \int_{\mathbb{R}} \int_{\mathbb{R}} f(t)\, g(\sqrt{2}z - t) \exp\left\{\frac{i(Az^2 - 2z\omega + D\omega^2)}{2B}\right\}$$

$$\times \exp\left\{\frac{iA}{B}\left(\frac{z}{\sqrt{2}} - t\right)^2\right\} dz\, dt. \tag{1.186}$$

The change of variables $\sqrt{2}z = y$ in (1.186) along with the time-shift and frequency modulation property (vii) (Theorem 1.4.4) of the linear canonical transform yields

$$\mathscr{L}_M\Big[(f \odot_M g)\Big](\omega)$$

$$= \frac{1}{2\sqrt{\pi i B}} \int_{\mathbb{R}} \int_{\mathbb{R}} f(t)\, g(y - t) \exp\left\{\frac{iA}{B}\left(\frac{y}{2} - t\right)^2\right\} \exp\left\{\frac{i}{2B}\left(\frac{Ay^2}{2} - \frac{2y\omega}{\sqrt{2}} + D\omega^2\right)\right\} dy\, dt$$

$$= \frac{1}{2\sqrt{\pi i B}} \int_{\mathbb{R}} f(t) \exp\left\{\frac{iAt^2}{B}\right\} \int_{\mathbb{R}} g(y - t) \exp\left\{-\frac{iAyt}{B}\right\} \exp\left\{\frac{iAy^2}{4B}\right\}$$

$$\times \exp\left\{\frac{i}{2B}\left(\frac{Ay^2}{2} - \frac{2y\omega}{\sqrt{2}} + D\omega^2\right)\right\} dy\, dt$$

$$= \frac{1}{\sqrt{2}} \int_{\mathbb{R}} f(t) \exp\left\{\frac{iAt^2}{B}\right\} \exp\left\{\frac{iD\omega^2}{4B}\right\} \left\{\frac{1}{\sqrt{2\pi i B}} \int_{\mathbb{R}} g(y - t) \exp\left\{-\frac{iAyt}{B}\right\}\right.$$

$$\left. \times \exp\left\{\frac{i}{2B}\left(Ay^2 - \frac{2y\omega}{\sqrt{2}} + D\left(\frac{\omega}{\sqrt{2}}\right)^2\right)\right\} dy\right\} dt$$

$$= \sqrt{i\pi B}\, \mathscr{L}_M\big[g\big]\left(\frac{\omega}{\sqrt{2}}\right) \left\{\frac{1}{\sqrt{2\pi i B}} \int_{\mathbb{R}} f(t) \exp\left\{\frac{i}{2B}\left(At^2 - \frac{2t\omega}{\sqrt{2}} + D\left(\frac{\omega}{\sqrt{2}}\right)^2\right)\right\} dt\right\}$$

$$= \sqrt{i\pi B}\, \mathscr{L}_M\big[f\big]\left(\frac{\omega}{\sqrt{2}}\right) \mathscr{L}_M\big[g\big]\left(\frac{\omega}{\sqrt{2}}\right).$$

This completes the proof of Theorem 1.4.11. $\qquad\square$

Remark 1.4.7. The modified chirp-free weighted linear canonical convolution (1.184) reduces to the classical convolution (1.75) upon plugging the unimodular matrix $M = (0, 1; -1, 0)$ in Definition 1.4.6. Nevertheless, for $M = (\cos \alpha, \sin \alpha; -\sin \alpha, \cos \alpha)$, $\alpha \neq n\pi$, the modified chirp-free weighted linear canonical convolution (1.184) yields a new convolution for the fractional Fourier transform.

1.4.3 Applications of linear canonical transform to differential equations

In this subsection, our aim is to study some differential properties of the linear canonical transform defined in (1.164). We initiate the study with the following definition of generalized Schwartz space or generalized Schwartz class:

Definition 1.4.7. The generalized Schwartz class $\mathbb{S}_M(\mathbb{R}) \subseteq L^2(\mathbb{R})$ is defined as

$$\mathbb{S}_M(\mathbb{R}) = \left\{ f(t) \in C^\infty(\mathbb{R}) : \sup_{t \in \mathbb{R}} \left| t^m \mathbb{D}_t^n f(t) \right| < \infty, \; m, n \in \mathbb{N}_0 \right\}, \qquad (1.187)$$

where $\mathbb{N}_0 = \mathbb{N} \cup \{0\}$ and $\mathbb{D}_t^n$ denotes the generalized differential operator given by

$$\mathbb{D}_t^n = \left(\frac{d}{dt} - \frac{iAt}{B} \right)^n. \qquad (1.188)$$

The generalized Schwartz class consists of functions which are infinitely differentiable in the sense of (1.188) and vanish at infinity. The class of functions (1.187) circumscribes both the classical Schwartz class as well as the fractional variant of the Schwartz class. For $M = (0, 1; -1, 0)$, the operator $\mathbb{D}_t^n$ is precisely the ordinary differential operator and the space $\mathbb{S}_M(\mathbb{R})$ boils down to the classical Schwartz space $\mathbb{S}(\mathbb{R})$. On the other hand, for $M = (\cos\alpha, \sin\alpha; -\sin\alpha, \cos\alpha)$, $\alpha \neq k\pi$ with $k \in \mathbb{Z}$, the operator $\mathbb{D}_t^n$ reduces to the fractional differential operator and the corresponding space $\mathbb{S}_M(\mathbb{R})$ is the fractional variant of the Schwartz space denoted by $\mathbb{S}^\alpha(\mathbb{R})$.

Here, it is imperative to mention that for any $f \in \mathbb{S}_M(\mathbb{R})$ the linear canonical transform (1.164) enjoys the following differentiation properties:

$$\mathscr{L}_M \left[t^n f(t) \right](\omega) = \left(D\omega + iB \frac{d}{d\omega} \right)^n \mathscr{L}_M \left[f \right](\omega) \qquad (1.189)$$

and

$$\mathscr{L}_M \left[\frac{d^n}{dt^n} f(t) \right](\omega) = \left(-iC\omega + A \frac{d}{d\omega} \right)^n \mathscr{L}_M \left[f \right](\omega). \qquad (1.190)$$

Example 1.4.1. Consider the Mexican hat function $f(t) = (1 - t^2) e^{-t^2/2}$. Then $f(t)$ can be expressed via the double derivative of the Gaussian function as

$$f(t) = -\frac{d^2}{dt^2} e^{-t^2/2}. \qquad (1.191)$$

In order to compute the linear canonical transform of (1.191), we shall invoke (1.190) and proceed as

$$
\begin{aligned}
\mathscr{L}_M \left[f \right](\omega) &= -\mathscr{L}_M \left[\frac{d^2}{dt^2} \left(e^{-t^2/2} \right) \right](\omega) \\
&= -\left(-iC\omega + A \frac{d}{d\omega} \right)^2 \mathscr{L}_M \left[e^{-t^2/2} \right](\omega) \\
&= -\frac{1}{\sqrt{A + iB}} \left(-iC\omega + A \frac{d}{d\omega} \right)^2 \left[\exp\left\{ \frac{iD\omega^2}{2B} \right\} \exp\left\{ -\frac{i\omega^2}{2B(A + iB)} \right\} \right] \\
&= -\frac{1}{\sqrt{A + iB}} \left(-iC\omega + A \frac{d}{d\omega} \right)^2 \left[\exp\left\{ \left(\frac{iD}{2B} - \frac{i}{2B(A + iB)} \right) \omega^2 \right\} \right].
\end{aligned}
$$

For computational elegance, we set $\left(\dfrac{iD}{2B} - \dfrac{i}{2B(A + iB)} \right) = \gamma$. Then, we observe that

$$
\begin{aligned}
\mathscr{L}_M\big[f\big](\omega) &= -\frac{1}{\sqrt{A + iB}} \left(-iC\omega + A\frac{d}{d\omega} \right)\left(-iC\omega + A\frac{d}{d\omega} \right) e^{\gamma\omega^2} \\
&= \frac{1}{\sqrt{A + iB}} \left[(C\omega)^2 + 4iAC\gamma\omega^2 - (2A\gamma\omega)^2 - 2A^2\gamma \right] e^{\gamma\omega^2}. \qquad (1.192)
\end{aligned}
$$

In the following few propositions, we sequentially examine the properties of the linear canonical transform (1.164) under the influence of the generalized differential operator (1.188).

Proposition 1.4.12. *For any real and unimodular matrix $M = (A, B; C, D)$, the linear canonical kernel $\mathcal{K}_M(t, \omega)$ defined in (1.165) satisfies:*

$$
\mathbb{D}_t^n\big(\mathcal{K}_M(t,\omega)\big) = \left(-\frac{i\omega}{B} \right)^n \mathcal{K}_M(t,\omega), \quad n \in \mathbb{N}_0, \qquad (1.193)
$$

where the generalized differential operator $\mathbb{D}_t^n$ is given by (1.188).

Proof. Differentiating the kernel $\mathcal{K}_M(t,\omega)$ with respect to t yields

$$
\begin{aligned}
\frac{d}{dt}\big(\mathcal{K}_M(t,\omega)\big) &= \frac{1}{\sqrt{2\pi iB}} \frac{d}{dt}\left[\exp\left\{ \frac{i(At^2 - 2t\omega + D\omega^2)}{2B} \right\} \right] \\
&= \frac{1}{\sqrt{2\pi iB}} \exp\left\{ \frac{iD\omega^2}{2B} \right\} \left[\exp\left\{ \frac{i(At^2 - 2t\omega)}{2B} \right\}\left(-\frac{i\omega}{B} \right) \right. \\
&\qquad\qquad\qquad\qquad\qquad \left. + \exp\left\{ \frac{i(At^2 - 2t\omega)}{2B} \right\}\left(\frac{iAt}{B} \right) \right] \\
&= \frac{1}{\sqrt{2\pi iB}} \exp\left\{ \frac{i(At^2 - 2t\omega + D\omega^2)}{2B} \right\}\left(-\frac{i\omega}{B} + \frac{iAt}{B} \right) \\
&= \left(-\frac{i\omega}{B} + \frac{iAt}{B} \right) \mathcal{K}_M(t,\omega),
\end{aligned}
$$

which further implies that

$$
\left(\frac{d}{dt} - \frac{iAt}{B} \right)\mathcal{K}_M(t,\omega) = \left(-\frac{i\omega}{B} \right)\mathcal{K}_M(t,\omega).
$$

By repeatedly executing the above process n-times, we obtain

$$
\left(\frac{d}{dt} - \frac{iAt}{B} \right)^n \mathcal{K}_M(t,\omega) = \left(-\frac{i\omega}{B} \right)^n \mathcal{K}_M(t,\omega).
$$

Equivalently,

$$
\mathbb{D}_t^n\big(\mathcal{K}_M(t,\omega)\big) = \left(-\frac{i\omega}{B} \right)^n \mathcal{K}_M(t,\omega).
$$

This completes the proof of Proposition 1.4.12. $\qquad\qquad\qquad\qquad\qquad\qquad\square$

Proposition 1.4.13. *For any $f \in \mathbb{S}_M(\mathbb{R})$, we have*

$$
\int_{\mathbb{R}} f(t)\, \mathbb{D}_t^n\big(\mathcal{K}_M(t,\omega)\big)\, dt = \int_{\mathbb{R}} \mathcal{K}_M(t,\omega)\, \widetilde{\mathbb{D}}_t^n\big(f(t)\big)\, dt, \quad n \in \mathbb{N}_0, \qquad (1.194)
$$

where

$$
\widetilde{\mathbb{D}}_t = -\left(\frac{d}{dt} + \frac{iAt}{B} \right). \qquad (1.195)
$$

Proof. It suffices to prove the assertion (1.194) for $n = 1$. We note that

$$\int_{\mathbb{R}} f(t) \, \mathbb{D}_t \big(\mathcal{K}_M(t,\omega)\big) \, dt = \int_{\mathbb{R}} f(t) \left\{ \left(\frac{d}{dt} - \frac{iAt}{B}\right) \mathcal{K}_M(t,\omega) \right\} dt$$

$$= \int_{\mathbb{R}} f(t) \left(\frac{d}{dt} \mathcal{K}_M(t,\omega)\right) dt - \int_{\mathbb{R}} \left(\frac{iAt}{B}\right) f(t) \, \mathcal{K}_M(t,\omega) \, dt. \quad (1.196)$$

Integrating the first integral appearing in (1.196) by parts and invoking the condition that f vanishes at infinity, we obtain

$$\int_{\mathbb{R}} f(t) \, \mathbb{D}_t \big(\mathcal{K}_M(t,\omega)\big) \, dt = -\int_{\mathbb{R}} \left(\frac{d}{dt} f(t)\right) \mathcal{K}_M(t,\omega) \, dt - \int_{\mathbb{R}} \left(\frac{iAt}{B}\right) f(t) \, \mathcal{K}_M(t,\omega) \, dt$$

$$= -\int_{\mathbb{R}} \mathcal{K}_M(t,\omega) \left(\frac{d}{dt} + \frac{iAt}{B}\right) f(t) \, dt$$

$$= \int_{\mathbb{R}} \mathcal{K}_M(t,\omega) \, \widetilde{\mathbb{D}}_t \big(f(t)\big) \, dt.$$

Therefore, for any $n \in \mathbb{N}_0$, we can show that

$$\int_{\mathbb{R}} f(t) \, \mathbb{D}_t^n \big(\mathcal{K}_M(t,\omega)\big) \, dt = \int_{\mathbb{R}} \mathcal{K}_M(t,\omega) \, \widetilde{\mathbb{D}}_t^n \big(f(t)\big) \, dt.$$

This completes the proof of Proposition 1.4.13. $\qquad\square$

Proposition 1.4.14. *For any $f \in \mathbb{S}_M(\mathbb{R})$ and $n \in \mathbb{N}_0$, we have*

(i) $\mathscr{L}_M\big[\widetilde{\mathbb{D}}_t^n f(t)\big](\omega) = \left(-\dfrac{i\omega}{B}\right)^n \mathscr{L}_M\big[f\big](\omega),$

(ii) $\mathbb{D}_t^n\Big(\mathscr{L}_M\big[f\big](\omega)\Big) = \mathscr{L}_M\left[\left(-\dfrac{it}{B}\right)^n f(t)\right](\omega).$

Proof. **(i)** Invoking Proposition 1.4.12 abreast of Proposition 1.4.13, we have

$$\mathscr{L}_M\big[\widetilde{\mathbb{D}}_t^n f(t)\big](\omega) = \int_{\mathbb{R}} \big(\widetilde{\mathbb{D}}_t^n f(t)\big) \mathcal{K}_\alpha(t,\omega) \, dt$$

$$= \int_{\mathbb{R}} f(t) \, \mathbb{D}_t^n \big(\mathcal{K}_M(t,\omega)\big) \, dt$$

$$= \int_{\mathbb{R}} \left(-\frac{i\omega}{B}\right)^n f(t) \, \mathcal{K}_M(t,\omega) \, dt$$

$$= \left(-\frac{i\omega}{B}\right)^n \int_{\mathbb{R}} f(t) \, \mathcal{K}_M(t,\omega) \, dt$$

$$= \left(-\frac{i\omega}{B}\right)^n \mathscr{L}_M\big[f\big](\omega).$$

(ii) Since $f \in \mathbb{S}_M(\mathbb{R})$ and the integral defining the linear canonical transform is uniformly convergent for $\omega \in \mathbb{R}$, therefore, we can differentiate within the integral sign, so that

$$\mathbb{D}_\omega^n\Big(\mathscr{L}_M\big[f\big](\omega)\Big) = \int_{\mathbb{R}} f(t) \big(\mathbb{D}_\omega^n \mathcal{K}_M(t,\omega)\big) \, dt$$

$$= \int_{\mathbb{R}} \left(-\frac{it}{B}\right)^n f(t) \, \mathcal{K}_M(t,\omega) \, dt$$

$$= \mathscr{L}_M\left[\left(-\frac{it}{B}\right)^n f(t)\right](\omega).$$

This completes the proof of Proposition 1.4.14.   □

Towards the end of this subsection, our goal is to demonstrate that the linear canonical transform defined in (1.164) can also be invoked in solving certain generalized differential equations. To facilitate the intent, we consider the generalized linear, homogeneous differential equation of order n associated with the linear canonical transform as

$$\left(a_n\widetilde{\mathbb{D}}_t^n + a_{n-1}\widetilde{\mathbb{D}}_t^{n-1} + a_{n-2}\widetilde{\mathbb{D}}_t^{n-2} + \cdots + a_1\widetilde{\mathbb{D}}_t + a_0\right)F(t) = f(t), \tag{1.197}$$

where $\widetilde{\mathbb{D}}_t$ is given by (1.195) and $a_0, a_1, \ldots, a_n$ are constants.

Upon applying the linear canonical transform on both sides of (1.197) and using Proposition 1.4.14, we obtain

$$\left\{a_n\left(-\frac{i\omega}{B}\right)^n + a_{n-1}\left(-\frac{i\omega}{B}\right)^{n-1} + a_{n-2}\left(-\frac{i\omega}{B}\right)^{n-2} + \cdots + a_1\left(-\frac{i\omega}{B}\right) + a_0\right\}\mathscr{L}_M\left[F\right](\omega)$$

$$= \mathscr{L}_M\left[f\right](\omega).$$

Alternatively, we can write the above expression in a more compact form as

$$P_M(\omega)\,\mathscr{L}_M\left[F\right](\omega) = \mathscr{L}_M\left[f\right](\omega),$$

where $P_M(\omega) = \sum_{m=0}^{n} a_m\left(-\frac{i\omega}{B}\right)^m$. Therefore, it follows that

$$\mathscr{L}_M\left[F\right](\omega) = \frac{\mathscr{L}_M\left[f\right](\omega)}{P_M(\omega)}. \tag{1.198}$$

Application of the inverse linear canonical transform (1.169) to (1.198) yields the desired function as

$$F(t) = \mathscr{L}_{M^{-1}}\left(\frac{\mathscr{L}_M\left[f\right](\omega)}{P_M(\omega)}\right)(t). \tag{1.199}$$

Example 1.4.2. Consider the generalized wave equation

$$\frac{\partial^2 F(x,t)}{\partial t^2} = k^2\,\widetilde{\mathbb{D}}_x^2\left(F(x,t)\right), \quad -\infty < x < \infty,\, t > 0 \tag{1.200}$$

with the initial data

$$\left.\begin{array}{c} F(x,0) = f(x) \\[2mm] \dfrac{\partial F(x,0)}{\partial t} = g(x) \end{array}\right\}, \tag{1.201}$$

where k is a constant and $\widetilde{\mathbb{D}}_x$ is given by (1.195).

Application of the linear canonical transform to both sides of the differential equation (1.200) yields

$$\int_{\mathbb{R}} \frac{\partial^2 F(x,t)}{\partial t^2}\,\mathcal{K}_M(x,\omega)\,dx = k^2\int_{\mathbb{R}}\widetilde{\mathbb{D}}_x^2\left(F(x,t)\right)\mathcal{K}_M(x,\omega)\,dx,$$

so that

$$\frac{\partial^2 \mathscr{L}_M[F(x,t)](\omega)}{\partial t^2} = k^2 \int_{\mathbb{R}} F(x,t)\, \mathbb{D}_x^2(\mathcal{K}_M(x,\omega))\, dx = -\left(\frac{k\omega}{B}\right)^2 \mathscr{L}_M[F(x,t)](\omega),$$

where the linear canonical transform of the function $F(x,t)$ is computed with respect to the variable x. Therefore, it follows that

$$\mathscr{L}_M[F(x,t)](\omega) = \mathscr{L}_M[f](\omega)\cos\left(\frac{k\omega t}{B}\right) + \left(\frac{B}{k\,\omega}\right)\mathscr{L}_M[g](\omega)\sin\left(\frac{k\omega t}{B}\right). \tag{1.202}$$

Applying the inverse linear canonical transform (1.169) on both sides of (1.202), we obtain

$$F(x,t) = \mathscr{L}_{M^{-1}}\left(\mathscr{L}_M[f](\omega)\cos\left(\frac{k\omega t}{B}\right)\right)(x)$$

$$+ \mathscr{L}_{M^{-1}}\left(\left(\frac{B}{k\,\omega}\right)\mathscr{L}_M[g](\omega)\sin\left(\frac{k\omega t}{B}\right)\right)(x). \tag{1.203}$$

Setting

$$\left.\begin{array}{l} R(x,t) = \mathscr{L}_{M^{-1}}\left(\mathscr{L}_M[f](\omega)\cos\left(\dfrac{k\omega t}{B}\right)\right)(x) \\[3mm] S(x,t) = \mathscr{L}_{M^{-1}}\left(\left(\dfrac{B}{k\,\omega}\right)\mathscr{L}_M[g](\omega)\sin\left(\dfrac{k\omega t}{B}\right)\right)(x) \end{array}\right\}. \tag{1.204}$$

Then, we need to obtain explicit expressions for the functions $R(x,t)$ and $S(x,t)$ given by (1.204). In order to find a useful estimate for $R(x,t)$, we shall use the formal definitions of linear canonical transform (1.164) and inverse linear canonical transform (1.169) as

$$R(x,t) = \mathscr{L}_{M^{-1}}\left(\mathscr{L}_M[f](\omega)\cos\left(\frac{k\omega t}{B}\right)\right)(x)$$

$$= \frac{1}{\sqrt{2\pi(-i)B}} \int_{\mathbb{R}} \mathscr{L}_M[f](\omega)\cos\left(\frac{k\omega t}{B}\right)\exp\left\{-\frac{i(Ax^2 - 2x\omega + D\omega^2)}{2B}\right\}d\omega$$

$$= \frac{1}{\sqrt{2\pi(-i)B}} \int_{\mathbb{R}}\left\{\frac{1}{\sqrt{2\pi iB}}\int_{\mathbb{R}} f(z)\exp\left\{\frac{i(Az^2 - 2z\omega + D\omega^2)}{2B}\right\}dz\right\}$$

$$\times \cos\left(\frac{k\omega t}{B}\right)\exp\left\{-\frac{i(Ax^2 - 2x\omega + D\omega^2)}{2B}\right\}d\omega$$

$$= \frac{1}{2\pi|B|}\int_{\mathbb{R}}\int_{\mathbb{R}} f(z)\exp\left\{\frac{i(Az^2 - 2z\omega + D\omega^2)}{2B}\right\}\cos\left(\frac{k\omega t}{B}\right)$$

$$\times \exp\left\{-\frac{i(Ax^2 - 2x\omega + D\omega^2)}{2B}\right\}dz\, d\omega$$

$$= \frac{1}{2\pi|B|}\exp\left\{-\frac{iAx^2}{2B}\right\}\int_{\mathbb{R}}\int_{\mathbb{R}} f(z)\cos\left(\frac{k\omega t}{B}\right)\exp\left\{\frac{iAz^2}{2B}\right\}\exp\left\{\frac{2i\omega(x-z)}{2B}\right\}dz\, d\omega$$

$$= \frac{1}{2\pi|B|}\exp\left\{-\frac{iAx^2}{2B}\right\}\int_{\mathbb{R}}\int_{\mathbb{R}} f(z)\cos\left(\frac{k\omega t}{B}\right)\exp\left\{\frac{iAz^2}{2B}\right\}$$

$$\times \exp\left\{\frac{ix\omega}{B}\right\}\exp\left\{-\frac{iz\omega}{B}\right\}dz\, d\omega$$

$$= \frac{1}{2|B|\sqrt{2\pi}}\exp\left\{-\frac{iAx^2}{2B}\right\}\int_{\mathbb{R}}\left[\exp\left\{\frac{i(x+kt)\omega}{B}\right\} + \exp\left\{\frac{i(x-kt)\omega}{B}\right\}\right]$$

$$\times \left\{ \frac{1}{\sqrt{2\pi}} \int_{\mathbb{R}} f(z) \exp\left\{ \frac{iAz^2}{2B} \right\} \exp\left\{ -\frac{iz\omega}{B} \right\} dz \right\} d\omega$$

$$= \frac{1}{2} \exp\left\{ -\frac{iAx^2}{2B} \right\} \frac{1}{\sqrt{2\pi}} \int_{\mathbb{R}} \left[\exp\left\{ \frac{i(x+kt)\omega}{B} \right\} + \exp\left\{ \frac{i(x-kt)\omega}{B} \right\} \right]$$

$$\times \mathscr{F}\left[\exp\left\{ \frac{iAz^2}{2B} \right\} f(z) \right] \left(\frac{\omega}{B} \right) d\left(\frac{\omega}{B} \right)$$

$$= \frac{1}{2} \exp\left\{ -\frac{iAx^2}{2B} \right\} \frac{1}{\sqrt{2\pi}} \int_{\mathbb{R}} \mathscr{F}\left[\exp\left\{ \frac{iAz^2}{2B} \right\} f(z) \right] \left(\frac{\omega}{B} \right) \exp\left\{ \frac{i(x+kt)\omega}{B} \right\} d\left(\frac{\omega}{B} \right)$$

$$+ \frac{1}{2} \exp\left\{ -\frac{iAx^2}{2B} \right\} \frac{1}{\sqrt{2\pi}} \int_{\mathbb{R}} \mathscr{F}\left[\exp\left\{ \frac{iAz^2}{2B} \right\} f(z) \right] \left(\frac{\omega}{B} \right) \exp\left\{ \frac{i(x-kt)\omega}{B} \right\} d\left(\frac{\omega}{B} \right)$$

$$= \frac{1}{2} \exp\left\{ -\frac{iAx^2}{2B} \right\} \left[\exp\left\{ \frac{iA(x+kt)^2}{2B} \right\} f(x+kt) + \exp\left\{ \frac{iA(x-kt)^2}{2B} \right\} f(x-kt) \right].$$

$$(1.205)$$

Similarly, we can obtain

$$S(x,t) = \mathscr{L}_{M^{-1}} \left(\left(\frac{B}{k\,\omega} \right) \mathscr{L}_M \left[g \right](\omega) \sin\left(\frac{k\omega t}{B} \right) \right)(x)$$

$$= \exp\left\{ -\frac{iAx^2}{2B} \right\} \frac{1}{\sqrt{2\pi}} \int_{\mathbb{R}} \left(\frac{B}{k\omega} \right) \exp\left\{ \frac{ix\omega}{B} \right\} \sin\left(\frac{k\omega t}{B} \right)$$

$$\times \mathscr{F}\left[\exp\left\{ \frac{iAz^2}{2B} \right\} g(z) \right] \left(\frac{\omega}{B} \right) d\left(\frac{\omega}{B} \right). \qquad (1.206)$$

Differentiating (1.206) partially on both sides with respect to t, we obtain

$$\frac{\partial S(x,t)}{\partial t} = \exp\left\{ -\frac{iAx^2}{2B} \right\} \frac{1}{\sqrt{2\pi}} \int_{\mathbb{R}} \exp\left\{ \frac{ix\omega}{B} \right\} \cos\left(\frac{k\omega t}{B} \right) \mathscr{F}\left[\exp\left\{ \frac{iAz^2}{2B} \right\} g(z) \right] \left(\frac{\omega}{B} \right) d\left(\frac{\omega}{B} \right)$$

$$= \frac{1}{2} \exp\left\{ -\frac{iAx^2}{2B} \right\} \frac{1}{\sqrt{2\pi}} \int_{\mathbb{R}} \mathscr{F}\left[\exp\left\{ \frac{iAz^2}{2B} \right\} g(z) \right] \left(\frac{\omega}{B} \right) \exp\left\{ \frac{i(x+kt)\omega}{B} \right\} d\left(\frac{\omega}{B} \right)$$

$$+ \frac{1}{2} \exp\left\{ -\frac{iAx^2}{2B} \right\} \frac{1}{\sqrt{2\pi}} \int_{\mathbb{R}} \mathscr{F}\left[\exp\left\{ \frac{iAz^2}{2B} \right\} g(z) \right] \left(\frac{\omega}{B} \right) \exp\left\{ \frac{i(x+kt)\omega}{B} \right\} d\left(\frac{\omega}{B} \right)$$

$$= \frac{1}{2} \exp\left\{ -\frac{iAx^2}{2B} \right\} \left[\exp\left\{ \frac{iA(x+kt)^2}{2B} \right\} g(x+kt) + \exp\left\{ \frac{iA(x-kt)^2}{2B} \right\} g(x-kt) \right].$$

$$(1.207)$$

Upon integrating (1.206) with respect to t, we obtain

$$F_2(x,t) = \frac{1}{2k} \exp\left\{ -\frac{iAx^2}{2B} \right\} \int_{x-kt}^{x+kt} \exp\left\{ \frac{iAy^2}{2B} \right\} g(y)\, dy. \qquad (1.208)$$

Finally, substituting the estimates (1.208) and (1.205) in (1.203), we obtain the solution of the generalized wave equation (1.200) as

$$F(x,t) = \frac{1}{2} \exp\left\{ -\frac{iAx^2}{2B} \right\} \left[\exp\left\{ \frac{iA(x+kt)^2}{2B} \right\} f(x+kt) \right.$$

$$\left. + \exp\left\{ \frac{iA(x-kt)^2}{2B} \right\} f(x-kt) + \frac{1}{k} \int_{x-kt}^{x+kt} \exp\left\{ \frac{iAy^2}{2B} \right\} g(y)\, dy \right]. \qquad (1.209)$$

Example 1.4.3. Consider the generalized heat equation

$$\frac{\partial F(x,t)}{\partial t} = \widetilde{\mathbb{D}}_x^2 \big(F(x,t) \big), \quad -\infty < x < \infty,\ t > 0 \qquad (1.210)$$

with the initial condition $F(x,0) = f(x)$, where $\tilde{\mathbb{D}}_x$ is defined by (1.195).

Applying the linear canonical transform (1.164) on both sides of (1.210) and the invoking Proposition 1.4.14, we obtain

$$\int_{\mathbb{R}} \frac{\partial F(x,t)}{\partial t} \mathcal{K}_M(x,\omega)\,dx = \int_{\mathbb{R}} \tilde{\mathbb{D}}_x^2(F(x,t))\,\mathcal{K}_M(x,\omega)\,dx = -\left(\frac{\omega}{B}\right)^2 \mathscr{L}_M\Big[F(x,t)\Big](\omega),$$

where the linear canonical transform of the function $F(x,t)$ is computed with respect to the variable x. Equivalently, we can write

$$\frac{\partial \mathscr{L}_M\Big[F(x,t)\Big](\omega)}{\partial t} = -\left(\frac{\omega}{B}\right)^2 \mathscr{L}_M\Big[F(x,t)\Big](\omega).$$

Therefore, it follows that

$$\mathscr{L}_M\Big[F(x,t)\Big](\omega) = H(\omega)\exp\left\{-\left(\frac{\omega}{B}\right)^2 t\right\}. \tag{1.211}$$

For $t = 0$, the expression (1.211) yields $\mathscr{L}_M\Big[F(x,0)\Big](\omega) = H(\omega)$. On the other hand, by virtue of the initial condition, it follows that

$$\mathscr{L}_M\Big[F(x,0)\Big](\omega) = \int_{\mathbb{R}} F(x,0)\,\mathcal{K}_M(x,\omega)\,dx = \int_{\mathbb{R}} f(x)\,\mathcal{K}_M(x,\omega)\,dx = \mathscr{L}_M\Big[f\Big](\omega).$$

Hence, we obtain $H(\omega) = \mathscr{L}_M\Big[f\Big](\omega)$, and consequently (1.211) takes the following form:

$$\mathscr{L}_M\Big[F(x,t)\Big](\omega) = \exp\left\{-\left(\frac{\omega}{B}\right)^2 t\right\}\mathscr{L}_M\Big[f\Big](\omega). \tag{1.212}$$

Applying the inverse linear canonical transform defined in (1.169) to (1.212), we obtain

$$F(x,t) = \mathscr{L}_{M^{-1}}\left(\exp\left\{-\left(\frac{\omega}{B}\right)^2 t\right\}\mathscr{L}_M\Big[f\Big](\omega)\right)(x)$$

$$= \frac{1}{\sqrt{2\pi(-i)B}}\int_{\mathbb{R}}\exp\left\{-\left(\frac{\omega}{B}\right)^2 t\right\}\mathscr{L}_M\Big[f\Big](\omega)\exp\left\{-\frac{i(Ax^2 - 2x\omega + D\omega^2)}{2B}\right\}d\omega$$

$$= \frac{1}{2\pi|B|}\int_{\mathbb{R}}\exp\left\{-\left(\frac{\omega}{B}\right)^2 t\right\}\exp\left\{-\frac{i(Ax^2 - 2x\omega + D\omega^2)}{2B}\right\}$$
$$\times\left\{\int_{\mathbb{R}} f(z)\exp\left\{\frac{i(Az^2 - 2z\omega + D\omega^2)}{2B}\right\}dz\right\}d\omega$$

$$= \frac{1}{2\pi|B|}\exp\left\{-\frac{iAx^2}{2B}\right\}\int_{\mathbb{R}}\exp\left\{-\left(\frac{\omega}{B}\right)^2 t\right\}\exp\left\{\frac{i\omega x}{B}\right\}$$
$$\times\left\{\int_{\mathbb{R}} f(z)\exp\left\{-\frac{i\omega z}{B}\right\}\exp\left\{\frac{iAz^2}{2B}\right\}dz\right\}d\omega$$

$$= \frac{1}{\sqrt{2\pi}}\exp\left\{-\frac{iAx^2}{2B}\right\}\int_{\mathbb{R}}\exp\left\{-\left(\frac{\omega}{B}\right)^2 t\right\}\exp\left\{\frac{i\omega x}{B}\right\}$$
$$\times\mathscr{F}\left[\exp\left\{\frac{iAz^2}{2B}\right\}f(z)\right]\left(\frac{\omega}{B}\right)d\left(\frac{\omega}{B}\right)$$

$$= \frac{1}{\sqrt{2\pi}}\exp\left\{-\frac{iAx^2}{2B}\right\}\int_{\mathbb{R}}\exp\left\{\frac{i\omega x}{B}\right\}\left(\frac{1}{\sqrt{2t}}\mathscr{F}\left[\exp\left\{-\frac{z^2}{4t}\right\}\right]\left(\frac{\omega}{B}\right)\right)$$

$$\times \mathscr{F}\left[\exp\left\{\frac{iAz^2}{2B}\right\}f(z)\right]\left(\frac{\omega}{B}\right)d\left(\frac{\omega}{B}\right)$$

$$= \frac{1}{2\pi\sqrt{2t}}\exp\left\{-\frac{iAx^2}{2B}\right\}\int_{\mathbb{R}}\exp\left\{\frac{i\omega x}{B}\right\}$$

$$\times \left(\sqrt{2\pi}\,\mathscr{F}\left[\exp\left\{-\frac{z^2}{4t}\right\}\right]\left(\frac{\omega}{B}\right)\mathscr{F}\left[\exp\left\{\frac{iAz^2}{2B}\right\}f(z)\right]\left(\frac{\omega}{B}\right)\right)d\left(\frac{\omega}{B}\right)$$

$$= \frac{1}{2\pi\sqrt{2t}}\exp\left\{-\frac{iAx^2}{2B}\right\}\int_{\mathbb{R}}\exp\left\{\frac{i\omega x}{B}\right\}$$

$$\times \left(\mathscr{F}\left[\exp\left\{-\frac{z^2}{4t}\right\}*\left(\exp\left\{\frac{iAz^2}{2B}\right\}f(z)\right)\right]\left(\frac{\omega}{B}\right)\right)d\left(\frac{\omega}{B}\right)$$

$$= \frac{1}{\sqrt{4\pi t}}\exp\left\{-\frac{iAx^2}{2B}\right\}\left\{\frac{1}{\sqrt{2\pi}}\int_{\mathbb{R}}\mathscr{F}\left[\exp\left\{-\frac{z^2}{4t}\right\}*\left(\exp\left\{\frac{iAz^2}{2B}\right\}f(z)\right)\right]\left(\frac{\omega}{B}\right)\right.$$

$$\left.\times \exp\left\{\frac{i\omega x}{B}\right\}d\left(\frac{\omega}{B}\right)\right\}$$

$$= \frac{1}{\sqrt{4\pi t}}\exp\left\{-\frac{iAx^2}{2B}\right\}\left[\exp\left\{-\frac{z^2}{4t}\right\}*\left(\exp\left\{\frac{iAz^2}{2B}\right\}f(z)\right)\right]$$

$$= \frac{1}{\sqrt{4\pi t}}\exp\left\{-\frac{iAx^2}{2B}\right\}\int_{\mathbb{R}}f(z)\exp\left\{\frac{iAz^2}{2B}\right\}\exp\left\{-\frac{(y-z)^2}{4t}\right\}dz. \qquad (1.213)$$

The expression (1.213) is the desired solution of the generalized heat equation (1.210).

1.4.4 Uncertainty principle for the linear canonical transform

"Inequality is the cause of all local movements."

-Leonardo da Vinci

It is well-known that the classical Heisenberg's uncertainty principle in the Fourier dominion provides a lower bound for the optimal resolution of a function f and its Fourier transform $\mathscr{F}[f]$; it states that if we try to limit the behavior of the function in its natural domain, we loose control in the Fourier domain. From a signal processing perspective, we can say that the product of the effective widths of a signal in the space and frequency domains has a lower bound and that this bound is reached when the signal is Gaussian. Keeping in view the fact that the linear canonical transform is a parameterized continuum of transforms that embodies several widely used transforms including the Fourier transform, fractional Fourier transform and Fresnel transform, our aim is to derive the Heisenberg-type uncertainty inequality for the linear canonical transform (1.164).

To facilitate the intention, we define the spread of a signal in the space and linear canonical domains as follows:

$$\Delta_{f(t)}^2 = \frac{\int_{\mathbb{R}}|t-t_0|^2|f(t)|^2 dt}{\int_{\mathbb{R}}|f(t)|^2 dt} \quad\text{and}\quad \Delta_{\mathscr{L}_M[f](\omega)}^2 = \frac{\int_{\mathbb{R}}|\omega-\omega_0|^2\left|\mathscr{L}_M[f](\omega)\right|^2 d\omega}{\int_{\mathbb{R}}\left|\mathscr{L}_M[f](\omega)\right|^2 d\omega}, \qquad (1.214)$$

where the first order moments t_0 and ω_0 are respectively, given by

$$t_0 = \frac{\int_{\mathbb{R}} t|f(t)|^2 dt}{\int_{\mathbb{R}} |f(t)|^2 dt} \quad \text{and} \quad \omega_0 = \frac{\int_{\mathbb{R}} \omega \left|\mathscr{L}_M[f](\omega)\right|^2 d\omega}{\int_{\mathbb{R}} \left|\mathscr{L}_M[f](\omega)\right|^2 d\omega}. \tag{1.215}$$

Theorem 1.4.15. *For any $f \in L^2(\mathbb{R})$, the spreads in the spatial and linear canonical domains satisfy the following inequality:*

$$\Delta^2_{f(t)} \cdot \Delta^2_{\mathscr{L}_M[f](\omega)} \geq \frac{|B|^2}{4}, \tag{1.216}$$

with equality if and only if f has the following form:

$$f(t) = K \exp\left\{-\left(\frac{iAt^2}{2B} + a(t - t_0)^2 - i\omega_0 t\right)\right\}, \quad a \in \mathbb{R}^+, K \in \mathbb{C}. \tag{1.217}$$

Proof. Define the function

$$H(\omega) = \frac{1}{\sqrt{2\pi}} \int_{\mathbb{R}} f(t) \exp\left\{\frac{i(At^2 - 2t\omega)}{2B}\right\} dt. \tag{1.218}$$

Then, $H(\omega)$ can also be written as

$$H(\omega) = \sqrt{iB} \exp\left\{-\frac{iD\omega^2}{2B}\right\} \mathscr{L}_M[f](\omega),$$

so that

$$\Delta^2_{\mathscr{L}_M[f](\mathbf{w})} = \frac{\int_{\mathbb{R}} |\omega - \omega_0|^2 |H(\omega)|^2 d\omega}{\int_{\mathbb{R}} |H(\omega)|^2 d\omega} = \Delta^2_{H(\omega)}. \tag{1.219}$$

For $\sigma = \omega/B$, we note that the function $H(\sigma)$ is the Fourier transform of the function $G(t)$:

$$G(t) = f(t) \exp\left\{\frac{iAt^2}{2B}\right\}. \tag{1.220}$$

Using the Heisenberg's uncertainty inequality in the Fourier domain given by (1.108), we obtain

$$\Delta^2_{G(t)} \cdot \Delta^2_{H(\sigma)} \geq \frac{1}{4}. \tag{1.221}$$

Moreover, we observe that

$$\Delta^2_{H(\sigma)} = \frac{\Delta^2_{H(\omega)}}{B^2} = \frac{\Delta^2_{\mathscr{L}_M[f](\omega)}}{|B|^2}. \tag{1.222}$$

In order to compute $\Delta^2_{G(t)}$, we proceed as follows:

$$\Delta^2_{G(t)} = \frac{\int_{\mathbb{R}^n} |t - t_0|^2 \left|f(t) \exp\left\{\frac{iAt^2}{2B}\right\}\right|^2 dt}{\int_{\mathbb{R}} \left|f(t) \exp\left\{\frac{iAt^2}{2B}\right\}\right|^2 dt} = \frac{\int_{\mathbb{R}} |t - t_0|^2 |f(t)|^2 dt}{\int_{\mathbb{R}} |f(t)|^2 dt} = \Delta^2_{f(t)}. \tag{1.223}$$

Finally, using (1.222) and (1.223) in (1.221) yields the following inequality:

$$\Delta^2_{f(t)} \cdot \Delta^2_{\mathscr{L}_M[f](\omega)} \geq \frac{|B|^2}{4}. \tag{1.224}$$

This establishes the Heisenberg's uncertainty inequality for the linear canonical transform (1.164). Note that equality holds in (1.224) provided equality holds in (1.221). Since equality holds in (1.221) for the shifted and modulated Gaussian function: $G(t) = K\,e^{i\omega_0 t}e^{-a(t-t_0)^2}$, $a \in \mathbb{R}^+, K \in \mathbb{C}$, therefore, in view of (1.220), we conclude that equality holds in (1.224) provided the function f has the following form:

$$f(t) = K\exp\left\{-\left(\frac{iAt^2}{2B} + a(t-t_0)^2 - i\omega_0 t\right)\right\}, \quad a \in \mathbb{R}^+, K \in \mathbb{C}. \tag{1.225}$$

This completes the proof of Theorem 1.4.15.   □

To point out the major distinction between the classical uncertainty relation in the Fourier dominion given by (1.108) and the uncertainty inequality (1.216), we recall that the output of any system performing the Fourier transform cannot be narrower than $1/4\Delta^2_{f(t)}$; that is, if f is concentrated about some point in the spatial domain, then f must be spread out in the spectral domain. However, in case of the linear canonical transform, this inverse relation is dictated by the parameter B as the output of any system performing the linear canonical transform is limited by $|B|^2/4\Delta^2_{f(t)}$. This indicates that the spread of the linear canonical transform of a signal f can be made arbitrarily large regardless of the spread of the input signal. Indeed, this is the characteristic feature of linear canonical transform and plays a vital role in its applications, for instance in free-space propagation described by an LCT, the output field can be made arbitrarily wide regardless of the distribution of the input field. For a detailed study on the uncertainty principles pertaining to the linear canonical transform and their applications, we refer the interested reader to [181, 280, 334].

Remark 1.4.8. The uncertainty inequality (1.216) circumscribes both the classical and fractional variants of the Heisenberg's uncertainty principle. For $M = (0,1;-1,0)$, inequality (1.216) is precisely the classical uncertainty inequality (1.108), whereas for $M = (\cos\alpha, \sin\alpha; -\sin\alpha, \cos\alpha)$, $\alpha \neq n\pi$, the inequality (1.216) boils down to the fractional variant of the Hesienberg's uncertainty inequality given by (1.163).

1.5   The Special Affine Fourier Transform

"Not long ago many thought that the mathematical world was created out of analytic functions. It was the Fourier series which disclosed the terra incognita in a second hemisphere."

-Edward V. Vleck

During the culminating years of last century, Abe and Sheridan [1, 2] introduced the notion of special affine Fourier transform (SAFT) in the context of phase-space transforms. The SAFT is an integral transformation associated with a general inhomogeneous lossless linear mapping in phase-space that depends on six parameters independent of the phase-space coordinates. The six-parameters constitute an augmented matrix consisting of a 2×2 unimodular matrix and a real 2×1 augmentation vector. It maps the position x and the wave number k into

$$\begin{pmatrix} x' \\ k' \end{pmatrix} = \begin{pmatrix} A & B \\ C & D \end{pmatrix} \begin{pmatrix} x \\ k \end{pmatrix} + \begin{pmatrix} p \\ q \end{pmatrix}, \tag{1.226}$$

where $AD - BC = 1$. The transformation maps any convex body into another convex body while preserving the area of the body. Such transformations constitute the inhomogeneous special linear group $ISL(2, \mathbb{R})$. In case $p = q = 0$, we obtain the homogeneous special group $SL(2, \mathbb{R})$ corresponding to the linear canonical transformation. Due to the offset by the vector $(p, q)^T$, the SAFT is also referred as the offset linear canonical transform. Owing to the extra degrees of freedom, the special affine Fourier transform can model diverse physical systems, particularly optical systems, at a higher flexibility than the Fourier, fractional Fourier and linear canonical transforms [2, 40, 41, 215, 335].

1.5.1 Definition and basic properties

In this subsection, we present the notion of the special affine Fourier transform and also study some of its fundamental properties.

Definition 1.5.1. For any $f \in L^2(\mathbb{R})$, the special affine Fourier transform with respect to a real, augmented matrix $\Lambda = [M : \lambda]$ is defined by

$$\mathcal{O}_\Lambda\big[f\big](\omega) = \int_{\mathbb{R}} f(t) \, \mathcal{K}_\Lambda(t, \omega) \, dt, \tag{1.227}$$

where the augmented matrix $\Lambda = [M : \lambda]$ is obtained by the offset of a real, 2×2 unimodular matrix $M = \begin{pmatrix} A & B \\ C & D \end{pmatrix}$ with the real 2×1 augmentation vector $\lambda = \begin{pmatrix} p \\ q \end{pmatrix}$ and $\mathcal{K}_\Lambda(t, \omega)$ denotes the "kernel" of the SAFT given by

$$\mathcal{K}_\Lambda(t, \omega) = \begin{cases} \dfrac{1}{\sqrt{2\pi i B}} \exp\left\{ \dfrac{i\left(At^2 + 2t(p - \omega) - 2\omega(Dp - Bq) + D(\omega^2 + p^2)\right)}{2B} \right\}, & B \neq 0 \\[4mm] \sqrt{D} \exp\left\{ \dfrac{iCD(\omega - p)^2}{2} + iq\omega \right\} f\big(D(\omega - p)\big), & B = 0. \end{cases} \tag{1.228}$$

For notational convenience, the augmented matrix $\Lambda = [M : \lambda]$ appearing in Definition 1.5.1 is simply written as $\Lambda = (A, B, C, D : p, q)$. Moreover, Definition 1.5.1 allows us to make the following comments regarding the notion of special affine Fourier transform:

(i) Choosing the augmented matrix as $\Lambda = (0, 1, -1, 0 : p, q)$, the special affine Fourier transform (1.227) yields a new variant of the Fourier transform, namely the offset Fourier transform.

(ii) For $\Lambda = (0, 1, -1, 0 : 0, 0)$, the special affine Fourier transform (1.227), when multiplied by the factor $\sqrt{i}$, boils down to the classical Fourier transform (1.7).

(iii) For $\Lambda = (\cos \alpha, \sin \alpha, -\sin \alpha, \cos \alpha : p, q)$, $\alpha \neq n\pi$, the special affine Fourier transform (1.227), when multiplied by the factor $2i\sqrt{e^{i\alpha}}$, yields a new variant of the fractional Fourier transform, namely the offset fractional Fourier transform.

(iv) For $\Lambda = (\cos \alpha, \sin \alpha, -\sin \alpha, \cos \alpha : 0, 0)$, $\alpha \neq n\pi$, the special affine Fourier transform (1.227), when multiplied by the factor $2i\sqrt{e^{i\alpha}}$, reduces to the ordinary fractional Fourier transform (1.127).

(v) For $\Lambda = (A, B, C, D : 0, 0)$, the special affine Fourier transform (1.227) coincides with the linear canonical transform (1.164).

(vi) Nevertheless, in case the augmented matrix is chosen as $\Lambda = (1, B; 0, 1 : p, q)$, $B \neq 0$, the special affine Fourier transform (1.227) yields the following offset Fresnel transform:

$$\mathcal{O}_\Lambda\big[f\big](\omega) = \frac{1}{\sqrt{2\pi i B}} \int_\mathbb{R} f(t) \, \exp\left\{ \frac{i\big((t - \omega)^2 + 2p(t - \omega) + 2B\omega q + p^2\big)}{2B} \right\} dt. \quad (1.229)$$

(vii) For $\Lambda = (1, B; 0, 1 : 0, 0)$, $B \neq 0$, the special affine Fourier transform (1.227) is precisely the classical Fresnel transform given by (1.166).

(viii) The phase-space transform defined in (1.227) is lossless if and only if the matrix $M = (A, B; C, D)$ is unimodular; that is, $AD - BC = 1$.

(ix) For the case $B = 0$, the special affine Fourier transform (1.227) corresponds to an offset-scaling transformation coupled with offset amplitude and quadratic phase modulation.

(x) For any $f \in L^2(\mathbb{R})$, the special affine Fourier transform (1.227) is expressible as

$$\mathcal{O}_\Lambda\big[f\big](\omega) = \frac{1}{\sqrt{2\pi i B}} \exp\left\{ \frac{i\big(D(\omega^2 + p^2) - 2\omega(Dp - Bq)\big)}{2B} \right\}$$

$$\times \int_\mathbb{R} f(t) \exp\left\{ \frac{iAt^2}{2B} \right\} \exp\left\{ -\frac{i(\omega - p)t}{B} \right\} dt$$

$$= \frac{1}{\sqrt{iB}} \exp\left\{ \frac{i\big(D(\omega^2 + p^2) - 2\omega(Dp - Bq)\big)}{2B} \right\} \mathscr{F}\big[G\big]\left(\frac{\omega - p}{B}\right), \quad (1.230)$$

where

$$G(t) = f(t) \exp\left\{ \frac{iAt^2}{2B} \right\}. \quad (1.231)$$

Therefore, we conclude that the special affine Fourier transform defined in (1.227) performs three actions in unison:

– First, the input is multiplied with a chirp signal; that is, $f(t) \to \exp\left\{ \dfrac{iAt^2}{2B} \right\}$.

– Second, the B^{-1}-scaled and shifted Fourier transform of the chirped signal $G(t)$ is computed.

– Third, the output is multiplied by $\dfrac{1}{\sqrt{iB}} \exp\left\{ \dfrac{i\big(D(\omega^2 + p^2) - 2\omega(Dp - Bq)\big)}{2B} \right\}$.

Hence, the computational load of the special affine Fourier transform is essentially dictated by the classical Fourier transform.

(xi) The special affine Fourier transform (1.227) is in essence, a time-shifted and frequency-modulated version of the linear canonical transform (1.164). Yet, the SAFT enjoys some of its unique properties and has been widely used in quantum physics, optics and signal processing. The SAFT provides a canonical formalism for studying the response of a wide class of physical systems and embodies many integral transforms ranging from classical Fourier transform to the linear canonical transform. Besides, many other optical and signal processing operations can be obtained as particular cases of the SAFT.

(xii) In comparison to the linear canonical transform, the SAFT induces an offset abreast of the twisting operation on the time-frequency plane and the same is lucidly depicted in Figure 1.33.

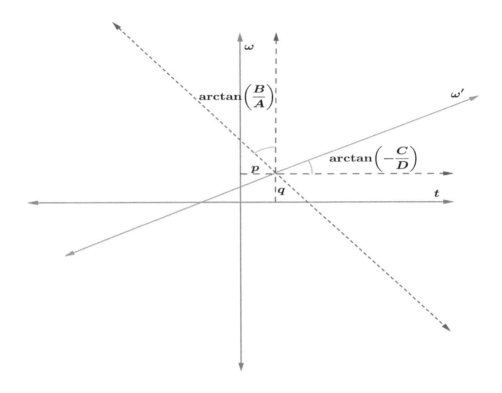

FIGURE 1.33: Offset-twisting operation of SAFT on the time-frequency plane.

In the following theorem, we assemble some elementary properties of the kernel (1.228) associated with the special affine Fourier transform. To facilitate the intent, we denote $\Lambda^{-1} = [M^{-1} : \lambda^{-1}]$, where $M^{-1} = \begin{pmatrix} D & -B \\ -C & A \end{pmatrix}$ and $\lambda^{-1} = \begin{pmatrix} Bq - Dp \\ Cp - Aq \end{pmatrix}$. Moreover, the following notation shall also be frequently used hereon:

$$I_\Lambda = \exp\left\{ \frac{i(CDp^2 + ABq^2 - 2ADpq)}{2} \right\}. \tag{1.232}$$

Theorem 1.5.1. *For $\Lambda^{-1} = [M^{-1} : \lambda^{-1}]$, the SAFT kernel (1.228) satisfies the following properties:*

(i) $\mathcal{K}_{\Lambda^{-1}}(\omega, t) = \overline{I_\Lambda}\,\overline{\mathcal{K}_\Lambda(t, \omega)},$

(ii) $\displaystyle\int_{\mathbb{R}} \mathcal{K}_\Lambda(t, \omega)\,\mathcal{K}_{\Lambda^{-1}}(z, t)\,dt = \overline{I_\Lambda}\,\delta(\omega - z),$

(iii) $\displaystyle\int_{\mathbb{R}} \mathcal{K}_\Lambda(t, \omega)\,\mathcal{K}_{\Lambda^{-1}}(\omega, z)\,d\omega = \overline{I_\Lambda}\,\delta(t - z).$

Proof. For the sake of brevity, we omit the proof. □

Next, we present the Plancherel theorem for the special affine Fourier transform, which foregathers some important characteristic features of the integral operator given by (1.227).

Theorem 1.5.2. (Plancherel Theorem): *For any pair of functions* $f, g \in L^2(\mathbb{R})$, *the following assertions are true:*

(i) *(Inversion Formula): If* $\mathcal{O}_\Lambda[f](\omega)$ *is the special affine Fourier transform corresponding to* $f \in L^2(\mathbb{R})$ *and* $\mathcal{O}_{\Lambda^{-1}}[f](\omega) : L^2(\mathbb{R}) \to L^2(\mathbb{R})$ *denotes the inverse SAFT operator, given by*

$$\mathcal{O}_{\Lambda^{-1}}\left(\mathcal{O}_\Lambda[f](\omega)\right)(t) = I_\Lambda \int_{\mathbb{R}} \mathcal{O}_\Lambda[f](\omega)\, \mathcal{K}_{\Lambda^{-1}}(\omega, t)\, d\omega. \qquad (1.233)$$

Then, we have

$$\mathcal{O}_{\Lambda^{-1}}\left(\mathcal{O}_\Lambda[f](\omega)\right)(t) = f(t). \qquad (1.234)$$

That is, any function $f \in L^2(\mathbb{R})$ *can be reconstructed from the corresponding special affine Fourier transform via the inversion formula (1.233).*

(ii) *(Parseval's Formula): The following orthogonality relation holds:*

$$\left\langle \mathcal{O}_\Lambda[f],\, \mathcal{O}_\Lambda[g] \right\rangle_2 = \left\langle f,\, g \right\rangle_2. \qquad (1.235)$$

That is, whenever the functions f *and* g *are orthogonal, the corresponding special affine Fourier transforms are also orthogonal. In particular, we have*

$$\left\| \mathcal{O}_\Lambda[f] \right\|_2 = \left\| f \right\|_2. \qquad (1.236)$$

The relation (1.236) implies that the total energy of input signal is preserved into the transformed signal.

(iii) *(Isomorphism): The mapping* $f \to \mathcal{O}_\Lambda[f]$ *is a Hilbert space isomorphism from* $L^2(\mathbb{R})$ *onto itself.*

Proof. The proof follows by invoking the fundamental properties of the SAFT kernel given in Theorem 1.5.1. $\qquad\qquad\qquad\qquad\qquad\qquad\qquad\qquad\qquad\qquad\qquad\qquad\qquad\qquad$ □

Below, we investigate upon some of the elementary properties of the special affine Fourier transform (1.227).

Theorem 1.5.3. *Let* $\alpha, \beta \in \mathbb{C}$, $k, \omega_0 \in \mathbb{R}$ *and* $\sigma \in \mathbb{R} \setminus \{0\}$. *Then, for any* $f, g \in L^2(\mathbb{R})$, *the special affine Fourier transform defined in (1.227) satisfies the following properties:*

(i) *Linearity:* $\mathcal{O}_\Lambda[\alpha f + \beta g](\omega) = \alpha\, \mathcal{O}_\Lambda[f](\omega) + \beta\, \mathcal{O}_\Lambda[g](\omega),$

(ii) *Translation:* $\mathcal{O}_\Lambda[f(t-k)](\omega) = \exp\left\{ i\left(Aqk + Ck(\omega - p) - \dfrac{ACk^2}{2} \right) \right\} \mathcal{O}_\Lambda[f](\omega - Ak),$

(iii) *Modulation:* $\mathcal{O}_\Lambda[e^{i\omega_0 t} f(t)](\omega) = \exp\left\{ i\left(D\omega\omega_0 - \omega_0(Dp - Bq) - \dfrac{DB\omega_0^2}{2} \right) \right\}$
$$\times\, \mathcal{O}_\Lambda[f](\omega - B\omega_0),$$

(iv) *Scaling:* $\mathcal{O}_\Lambda\left[f\left(\dfrac{t}{\sigma} \right) \right](\omega) = |\sigma|\, \mathcal{O}_{\Lambda'}[f](\sigma\omega), \quad \Lambda' = \left(A\sigma^2, B, C, D/\sigma^2 : \sigma p, q/\sigma \right),$

(v) *Parity:* $\mathcal{O}_\Lambda\left[f(-t)\right](\omega) = \mathcal{O}_{\Lambda''}\left[f\right](-\omega), \quad \Lambda'' = (A, B, C, D : -p, -q),$

(vi) *Translation and modulation:* $\mathcal{O}_\Lambda\left[e^{i\omega_0 t} f(t-k)\right](\omega) = \mathcal{O}_\Lambda\left[f\right](\omega - B\omega_0 - Ak)$

$$\times \exp\left\{i\left((Ck + D\omega_0)(\omega - p) - BCk\omega_0 + (Ak + B\omega_0)q - \frac{ACk^2}{2} - \frac{DB\omega_0^2}{2}\right)\right\}.$$

Proof. For the sake of brevity, the proof is omitted. □

1.5.2 Special affine convolution and correlation

The convolution and correlation operations pertaining to the special affine Fourier transform are being widely studied mainly due to quest for obtaining practically reliable and efficient convolution structures. In this subsection, we intend to present a pair of convolution and correlation structures associated with the special affine Fourier transform, which are both simple one-dimensional integral expressions and can be easily implemented for sampling and multiplicative filtering in the SAFT domain.

Definition 1.5.2. For any $f, g \in L^2(\mathbb{R})$, the special affine convolution is denoted by $\circledast_\Lambda$ and is defined as

$$(f \circledast_\Lambda g)(z) = \exp\left\{\frac{iDp^2}{2B}\right\} \int_{\mathbb{R}} f(t)\, g(z-t)\, \exp\left\{-\frac{iAt(z-t)}{B}\right\} dt. \qquad (1.237)$$

Next, we look into the elementary properties of the special affine convolution given by the expression (1.237).

Theorem 1.5.4. *For any $f, g, h \in L^2(\mathbb{R})$ and the scalars $k \in \mathbb{R}$, $\sigma \in \mathbb{R} \setminus \{0\}$, the special affine convolution $\circledast_\Lambda$ satisfies the following properties:*

(i) *Commutativity:* $(f \circledast_\Lambda g)(z) = (g \circledast_\Lambda f)(z),$

(ii) *Associativity:* $(f \circledast_\Lambda (g \circledast_\Lambda h))(z) = ((f \circledast_\Lambda g) \circledast_\Lambda h)(z),$

(iii) *Distributivity:* $(f \circledast_\Lambda (g+h))(z) = (f \circledast_\Lambda g)(z) + (f \circledast_\Lambda h)(z),$

(iv) *Translation:* $(f \circledast_\Lambda g)(z-k) = (f(y-k) \circledast_\Lambda G(y))(z), \quad G(y) = e^{iAky/B} g(y),$

(v) *Scaling:* $(f \circledast_\Lambda g)(\sigma z) = |\sigma|\, (f(\sigma y) \circledast_{\Lambda'} g(\sigma y))(z), \quad \Lambda' = (A, B/\sigma^2, \sigma^2 C, D : p, q).$

Proof. The proof can be obtained quite straightforwardly and is, therefore, omitted. □

Following is the convolution theorem associated with the special affine convolution defined in (1.237):

Theorem 1.5.5. *For any $f, g \in L^2(\mathbb{R})$, we have*

$$\mathcal{O}_\Lambda\left[(f \circledast_\Lambda g)\right](\omega) = \sqrt{2\pi i B}\, \exp\left\{-\frac{i(D\omega^2 - 2\omega(Dp - Bq))}{2B}\right\} \mathcal{O}_\Lambda\left[f\right](\omega)\, \mathcal{O}_\Lambda\left[g\right](\omega).$$

$$\qquad (1.238)$$

Proof. In view of Theorem 1.5.3, it follows that for $k, \omega_0 \in \mathbb{R}$, the special affine Fourier transform of the translated and modulated signal $e^{i\omega_0 t} f(t-k)$ is given by

$$\mathcal{O}_\Lambda\left[e^{i\omega_0 t} f(t-k)\right](\omega) = \exp\left\{i\left((Ck + D\omega_0)(\omega - p) - BCk\omega_0 + (Ak + B\omega_0)q - \frac{ACk^2}{2}\right.\right.$$

$$\left.\left. - \frac{DB\omega_0^2}{2}\right)\right\} \mathcal{O}_\Lambda\left[f\right](\omega - B\omega_0 - Ak). \qquad (1.239)$$

Plugging $\omega_0 = -iAk/B$ into the relation (1.239), we obtain

$$\mathcal{O}_\Lambda\left[\exp\left\{-\frac{iAkt}{B}\right\}f(t-k)\right](\omega)$$

$$= \exp\left\{-ik\left(\frac{\omega-p}{B}\right) + iACk^2 - i\left(\frac{ACk^2}{2}\right) - \frac{iAk^2(1+BC)}{2B}\right\}\mathcal{O}_\Lambda\left[f\right](\omega)$$

$$= \exp\left\{-ik\left(\frac{\omega-p}{B}\right) + i\left(\frac{ACk^2}{2}\right) - i\left(\frac{ACk^2}{2}\right) - i\left(\frac{Ak^2}{2B}\right)\right\}\mathcal{O}_\Lambda\left[f\right](\omega)$$

$$= \exp\left\{-ik\left(\frac{\omega-p}{B}\right) - i\left(\frac{Ak^2}{2B}\right)\right\}\mathcal{O}_\Lambda\left[f\right](\omega). \tag{1.240}$$

On the other hand, Definition 1.5.1 implies that

$$\mathcal{O}_\Lambda\left[f\right](\omega)\,\mathcal{O}_\Lambda\left[g\right](\omega)$$

$$= \mathcal{O}_\Lambda\left[f\right](\omega)\exp\left\{\frac{iDp^2}{2B}\right\}\frac{1}{\sqrt{2\pi iB}}\int_\mathbb{R} g(k)$$

$$\times \exp\left\{\frac{i\left(Ak^2 + 2k(p-\omega) - 2\omega(Dp-Bq) + D\omega^2\right)}{2B}\right\}dk$$

$$= \exp\left\{\frac{iDp^2}{2B}\right\}\frac{1}{\sqrt{2\pi iB}}\int_\mathbb{R}\mathcal{O}_\Lambda\left[f\right](\omega)\exp\left\{ik\left(\frac{p-\omega}{B}\right)\right\}g(k)$$

$$\times \exp\left\{\frac{i\left(Ak^2 - 2\omega(Dp-Bq) + D\omega^2\right)}{2B}\right\}dk. \tag{1.241}$$

Using the relation (1.240) in (1.241) yields

$$\mathcal{O}_\Lambda\left[f\right](\omega)\,\mathcal{O}_\Lambda\left[g\right](\omega)$$

$$= \exp\left\{\frac{iDp^2}{2B}\right\}\frac{1}{\sqrt{2\pi iB}}\int_\mathbb{R}\exp\left\{\frac{iAk^2}{2B}\right\}\mathcal{O}_\Lambda\left[\exp\left\{-\frac{iAkt}{B}\right\}f(t-k)\right](\omega)\,g(k)$$

$$\times \exp\left\{\frac{i\left(Ak^2 - 2\omega(Dp-Bq) + D\omega^2\right)}{2B}\right\}dk$$

$$= \exp\left\{\frac{iDp^2}{2B}\right\}\exp\left\{\frac{i\left(D\omega^2 - 2\omega(Dp-Bq)\right)}{2B}\right\}\frac{1}{2\pi iB}\int_\mathbb{R}\int_\mathbb{R} g(k)\,f(t-k)$$

$$\times \exp\left\{-\frac{iAk(t-k)}{B}\right\}\exp\left\{\frac{i\left(At^2 + 2t(p-\omega) - 2\omega(Dp-Bq) + D(\omega^2+p^2)\right)}{2B}\right\}dt\,dk$$

$$= \exp\left\{\frac{i\left(D(\omega^2+p^2) - 2\omega(Dp-Bq)\right)}{2B}\right\}$$

$$\times \frac{1}{\sqrt{2\pi iB}}\int_\mathbb{R}\left\{\int_\mathbb{R} g(k)\,f(t-k)\exp\left\{-\frac{iAk(t-k)}{B}\right\}dk\right\}\mathcal{K}_\Lambda(t,\omega)\,dt$$

$$= \exp\left\{\frac{i\left(D\omega^2 - 2\omega(Dp-Bq)\right)}{2B}\right\}\frac{1}{\sqrt{2\pi iB}}\int_\mathbb{R}\left(g \circledast_\Lambda f\right)(t)\,\mathcal{K}_\Lambda(t,\omega)\,dt. \tag{1.242}$$

Since the special affine convolution operation (1.237) is commutative, therefore, we can rewrite (1.242) as

$$\mathcal{O}_\Lambda\big[f\big](\omega)\,\mathcal{O}_\Lambda\big[g\big](\omega) = \exp\left\{\frac{i\big(D\omega^2 - 2\omega(Dp - Bq)\big)}{2B}\right\} \frac{1}{\sqrt{2\pi iB}} \int_{\mathbb{R}} (f \circledast_\Lambda g)(t)\,\mathcal{K}_\Lambda(t,\omega)\,dt$$

$$= \frac{1}{\sqrt{2\pi iB}} \exp\left\{\frac{i\big(D\omega^2 - 2\omega(Dp - Bq)\big)}{2B}\right\} \mathcal{O}_\Lambda\big[(f \circledast_\Lambda g)\big](\omega). \quad (1.243)$$

Upon reshuffling the factors in (1.243), we get

$$\mathcal{O}_\Lambda\big[(f \circledast_\Lambda g)\big](\omega) = \sqrt{2\pi iB}\,\exp\left\{-\frac{i\big(D\omega^2 - 2\omega(Dp - Bq)\big)}{2B}\right\} \mathcal{O}_\Lambda\big[f\big](\omega)\,\mathcal{O}_\Lambda\big[g\big](\omega).$$

This completes the proof of Theorem 1.5.5. □

Remark 1.5.1. The special affine convolution (1.237) is expressible via the classical convolution (1.75) as

$$(f \circledast_\Lambda g)(z) = \exp\left\{\frac{i(Dp^2 - Az^2)}{2B}\right\} \left(f(t)\exp\left\{\frac{iAt^2}{2B}\right\} * g(t)\exp\left\{\frac{iAt^2}{2B}\right\}\right)(z). \quad (1.244)$$

As a consequence of Theorem 1.5.5, we infer that the special affine convolution of a pair of square integrable functions corresponds to a product of their respective special affine spectra, multiplied by a quadratic phase function. In contrast to this, we shall examine the nature of the convolution (1.237) when the input is a product of two functions.

Theorem 1.5.6. *For any $f, g \in L^2(\mathbb{R})$, we have*

$$\mathcal{O}_\Lambda\big[f(t) \cdot g(t)\big](\omega) = \frac{1}{\sqrt{2\pi|B|}} \exp\left\{\frac{i\big(D\omega^2 - 2\omega(Dp - Bq)\big)}{2B}\right\}$$

$$\times \left(\mathcal{O}_\Lambda\big[g\big](\xi)\exp\left\{-\frac{i\big(D\xi^2 + 2\xi(Bq - Dp)\big)}{2B}\right\} * \mathscr{F}\big[f\big]\left(\frac{\xi}{B}\right)\right)(\omega), \quad (1.245)$$

where $\mathscr{F}$ denotes the classical Fourier transform given by (1.7) and $*$ is the classical convolution operation (1.75).

Proof. Applying Definition 1.5.1 to the product of functions $f(t) \cdot g(t) \in L^2(\mathbb{R})$, we have

$$\mathcal{O}_\Lambda\big[f(t) \cdot g(t)\big](\omega)$$

$$= \int_{\mathbb{R}} f(t)\,g(t)\,\mathcal{K}_\Lambda(t,\omega)\,dt$$

$$= \frac{1}{\sqrt{2\pi iB}} \int_{\mathbb{R}} f(t)\,g(t)\,\exp\left\{\frac{i\big(At^2 + 2t(p - \omega) - 2\omega(Dp - Bq) + D(\omega^2 + p^2)\big)}{2B}\right\}\,dt.$$

$$(1.246)$$

Now, by virtue of the inversion formula (1.233), we have

$$g(t) = I_\Lambda \int_{\mathbb{R}} \mathcal{O}_\Lambda\big[g\big](\xi)\,\mathcal{K}_{\Lambda^{-1}}(\xi,t)\,d\xi$$

$$= \frac{I_\Lambda}{\sqrt{2\pi(-i)B}} \int_{\mathbb{R}} \mathcal{O}_\Lambda\big[g\big](\xi)\exp\left\{-\frac{i\big(D\xi^2 + 2\xi(Bq - Dp - t)\big)}{2B}\right\}$$

$$\times \exp\left\{-\frac{i\big(-2t(A(Bq - Dp) + B(Cp - Aq)) + A(t^2 + (Bq - Dp)^2)\big)}{2B}\right\}d\xi$$

$$= \frac{I_\Lambda}{\sqrt{2\pi(-i)B}} \int_{\mathbb{R}} \mathcal{O}_\Lambda\big[g\big](\xi)\exp\left\{-\frac{i\big(D\xi^2 + 2\xi(Bq - Dp - t)\big)}{2B}\right\}$$

$$\times \exp\left\{-\frac{iA(Bq - Dp)^2}{2B}\right\}\exp\left\{-\frac{i\big(At^2 - 2t(ABq - ADp + BCp - ABq)\big)}{2B}\right\}d\xi$$

$$= \frac{I_\Lambda}{\sqrt{2\pi(-i)B}} \int_{\mathbb{R}} \mathcal{O}_\Lambda\big[g\big](\xi)\exp\left\{-\frac{i\big(D\xi^2 + 2\xi(Bq - Dp - t)\big)}{2B}\right\}$$

$$\times \exp\left\{-\frac{iA(Bq - Dp)^2}{2B}\right\}\exp\left\{-\frac{i\big(At^2 - 2t(-ADp + (AD - 1)p)\big)}{2B}\right\}d\xi$$

$$= \frac{1}{\sqrt{2\pi(-i)B}}\exp\left\{\frac{i\big(CDp^2 + ABq^2 - 2ADpq\big)}{2}\right\}\exp\left\{-\frac{iA(Bq - Dp)^2}{2B}\right\}$$

$$\times \exp\left\{-\frac{i\big(At^2 + 2tp\big)}{2B}\right\}\int_{\mathbb{R}} \mathcal{O}_\Lambda\big[g\big](\xi)\exp\left\{-\frac{i\big(D\xi^2 + 2\xi(Bq - Dp - t)\big)}{2B}\right\}d\xi$$

$$= \frac{1}{\sqrt{2\pi(-i)B}}\exp\left\{-\frac{iDp^2}{2B}\right\}\exp\left\{-\frac{i\big(At^2 + 2tp\big)}{2B}\right\}$$

$$\times \int_{\mathbb{R}} \mathcal{O}_\Lambda\big[g\big](\xi)\exp\left\{-\frac{i\big(D\xi^2 + 2\xi(Bq - Dp - t)\big)}{2B}\right\}d\xi. \tag{1.247}$$

Using (1.247) in (1.246), we get

$$\mathcal{O}_\Lambda\Big[f(t) \cdot g(t)\Big](\omega)$$

$$= \frac{1}{2\pi|B|}\exp\left\{-\frac{iDp^2}{2B}\right\}\int_{\mathbb{R}} f(t)\left\{\exp\left\{-\frac{i\big(At^2 + 2tp\big)}{2B}\right\}\int_{\mathbb{R}} \mathcal{O}_\Lambda\big[g\big](\xi)\right.$$

$$\left.\times \exp\left\{-\frac{i\big(D\xi^2 + 2\xi(Bq - Dp - t)\big)}{2B}\right\}d\xi\right\}$$

$$\times \exp\left\{\frac{i\big(At^2 + 2t(p - \omega) - 2\omega(Dp - Bq) + D(\omega^2 + p^2)\big)}{2B}\right\}dt$$

$$= \frac{1}{2\pi|B|}\exp\left\{-\frac{iDp^2}{2B}\right\}\int_{\mathbb{R}} f(t)\left\{\int_{\mathbb{R}} \mathcal{O}_\Lambda\big[g\big](\xi)\exp\left\{-\frac{i\big(D\xi^2 + 2\xi(Bq - Dp - t)\big)}{2B}\right\}d\xi\right\}$$

$$\times \exp\left\{\frac{i\big(-2t\omega - 2\omega(Dp - Bq) + D(\omega^2 + p^2)\big)}{2B}\right\}dt$$

$$= \frac{1}{2\pi|B|}\exp\left\{\frac{i\big(D\omega^2 - 2\omega(Dp - Bq)\big)}{2B}\right\}\left\{\int_{\mathbb{R}}\left(\int_{\mathbb{R}} f(t)\exp\left\{-\frac{i(\omega - \xi)t}{B}\right\}dt\right)\right.$$

$$\left.\times \mathcal{O}_\Lambda\big[g\big](\xi)\exp\left\{-\frac{i\big(D\xi^2 + 2\xi(Bq - Dp)\big)}{2B}\right\}d\xi\right\}$$

$$= \frac{1}{\sqrt{2\pi|B|}} \exp\left\{ \frac{i(D\omega^2 - 2\omega(Dp - Bq))}{2B} \right\} \int_{\mathbb{R}} \mathcal{O}_\Lambda\big[g\big](\xi) \exp\left\{ -\frac{i(D\xi^2 + 2\xi(Bq - Dp))}{2B} \right\}$$

$$\times \mathscr{F}\big[f\big]\left(\frac{\omega - \xi}{B} \right) d\xi$$

$$= \frac{1}{\sqrt{2\pi|B|}} \exp\left\{ \frac{i(D\omega^2 - 2\omega(Dp - Bq))}{2B} \right\}$$

$$\times \left(\mathcal{O}_\Lambda\big[g\big](\xi) \exp\left\{ -\frac{i(D\xi^2 + 2\xi(Bq - Dp))}{2B} \right\} * \mathscr{F}\big[f\big]\left(\frac{\xi}{B} \right) \right)(\omega).$$

This completes the proof of Theorem 1.5.6. □

Remark 1.5.2. Theorem 1.5.6 asserts that the product of functions in the natural domain leads to a convolution in the spectral domain, as such it is of critical significance for the design and implementation of multiplicative filters beyond the Fourier domain. In the literature, Theorem 1.5.6 is often referred to as product theorem.

Finally, we formulate the notion of special affine correlation and then establish the corresponding correlation theorem.

Definition 1.5.3. For any $f, g \in L^2(\mathbb{R})$, the special affine correlation is denoted by $\otimes_\Lambda$ and is defined as

$$\big(f \otimes_\Lambda g\big)(z) = \exp\left\{ \frac{iDp^2}{2B} \right\} \int_{\mathbb{R}} f(t)\, \overline{g(t - z)} \exp\left\{ -\frac{iAz(z - t)}{B} \right\} dt. \tag{1.248}$$

Remark 1.5.3. The special affine correlation defined in (1.248) can also be expressed in terms of the classical convolution (1.75) as

$$\big(f \otimes_\Lambda g\big)(z) = \exp\left\{ \frac{i(Dp^2 - Az^2)}{2B} \right\} \left(f(t) \exp\left\{ \frac{iAt^2}{2B} \right\} * \overline{g(-t)} \exp\left\{ -\frac{iAt^2}{2B} \right\} \right)(z). \tag{1.249}$$

The correlation theorem corresponding to the special affine correlation given in Definition 1.5.3 is given below:

Theorem 1.5.7. *For any $f, g \in L^2(\mathbb{R})$, we have*

$$\mathcal{O}_\Lambda\big[(f \otimes_\Lambda g)\big](\omega) = \sqrt{2\pi(-i)B} \exp\left\{ \frac{i(D\omega^2 - 2\omega(Dp - Bq))}{2B} \right\} \mathcal{O}_\Lambda\big[f\big](\omega)\, \overline{\mathcal{O}_\Lambda\big[g\big](\omega)}. \tag{1.250}$$

Proof. As a consequence of (1.249), we rewrite the special affine correlation (1.248) as

$$\big(f \otimes_\Lambda g\big)(z)$$
$$= \exp\left\{ \frac{i(Dp^2 - Az^2)}{2B} \right\} \left(f(t) \exp\left\{ \frac{iAt^2}{2B} \right\} * \overline{g(-t)} \exp\left\{ -\frac{iAt^2}{B} \right\} \exp\left\{ \frac{iAt^2}{2B} \right\} \right)(z). \tag{1.251}$$

Then, for the function $h(t) = \overline{g(-t)} \exp\left\{ -\dfrac{iAt^2}{B} \right\}$, the expression (1.251) takes the following form:

$$\big(f \otimes_\Lambda g\big)(z) = \exp\left\{ \frac{i(Dp^2 - Az^2)}{2B} \right\} \left(f(t) \exp\left\{ \frac{iAt^2}{2B} \right\} * h(t) \exp\left\{ \frac{iAt^2}{2B} \right\} \right)(z), \tag{1.252}$$

which has precisely the same form as that of (1.244). Therefore, applying Theorem 1.5.5 to (1.252), we get

$$\mathcal{O}_\Lambda\Big[(f \otimes_\Lambda g)\Big](\omega) = \sqrt{2\pi i B}\,\exp\left\{-\frac{i\big(D\omega^2 - 2\omega(Dp - Bq)\big)}{2B}\right\}\mathcal{O}_\Lambda\Big[f\Big](\omega)\,\mathcal{O}_\Lambda\Big[h\Big](\omega).$$

(1.253)

Now, it remains to compute the special affine Fourier transform corresponding to the function $h(t)$. We proceed as

$$\mathcal{O}_\Lambda\Big[h\Big](\omega)$$

$$= \frac{1}{\sqrt{2\pi i B}}\int_{\mathbb{R}} \overline{g(-t)}\,\exp\left\{-\frac{iAt^2}{B}\right\}\exp\left\{\frac{i\big(At^2 + 2t(p - \omega) - 2\omega(Dp - Bq) + D\omega^2\big)}{2B}\right\}dt$$

$$= \frac{1}{\sqrt{2\pi i B}}\int_{\mathbb{R}} \overline{g(-t)}\,\exp\left\{\frac{i\big(-At^2 + 2t(p - \omega) - 2\omega(Dp - Bq) + D\omega^2\big)}{2B}\right\}dt$$

$$= \frac{1}{\sqrt{2\pi i B}}\int_{\mathbb{R}} \overline{g(y)}\,\exp\left\{\frac{i\big(D\omega^2 - 2\omega(Dp - Bq)\big)}{B}\right\}$$

$$\times \exp\left\{\frac{i\big(-At^2 - 2y(p - \omega) + 2\omega(Dp - Bq) - D\omega^2\big)}{2B}\right\}dy$$

$$= \frac{\sqrt{2\pi(-i)B}}{\sqrt{2\pi i B}}\,\exp\left\{\frac{i\big(D\omega^2 - 2\omega(Dp - Bq)\big)}{B}\right\}\int_{\mathbb{R}} \overline{g(y)\,\mathcal{K}_\Lambda(y,\omega)}\,dy$$

$$= \frac{\sqrt{2\pi(-i)B}}{\sqrt{2\pi i B}}\,\exp\left\{\frac{i\big(D\omega^2 - 2\omega(Dp - Bq)\big)}{B}\right\}\overline{\mathcal{O}_\Lambda\Big[g\Big](\omega)}.$$

(1.254)

Finally, implementing the relation (1.254) in (1.253), we get

$$\mathcal{O}_\Lambda\Big[(f \otimes_\Lambda g)\Big](\omega) = \sqrt{2\pi(-i)B}\,\exp\left\{\frac{i\big(D\omega^2 - 2\omega(Dp - Bq)\big)}{2B}\right\}\mathcal{O}_\Lambda\Big[f\Big](\omega)\,\overline{\mathcal{O}_\Lambda\Big[g\Big](\omega)}.$$

This completes the proof of Theorem 1.5.7.   □

Remark 1.5.4. For $\Lambda = (A, B, C, D : 0, 0)$, both Definition 1.5.2 as well as Definition 1.5.3 yield the linear canonical convolution and correlation operations defined in (1.179) and (1.182), respectively. Moreover, for the augmented matrix $\Lambda = (0, 1, -1, 0 : 0, 0)$, the special affine convolution and correlation operations given by (1.237) and (1.248) reduce to their respective counterparts for the fractional Fourier transform. Nevertheless, for $\Lambda = (0, 1, -1, 0 : 0, 0)$ the convolution and correlation expressions (1.237) and (1.248) boil down to the classical convolution and correlation in the Fourier domain as defined in (1.75) and (1.88), respectively. Apart from the aforementioned cases, many other convolution and correlation operations for the offset variants of the Fourier, fractional Fourier and Fresnel transforms can be obtained from (1.237) and (1.248) by appropriately choosing the augmented matrix $\Lambda = (A, B, C, D : p, q)$.

1.6   The Quadratic-Phase Fourier Transform

"It was, no doubt, partially because of his [Fourier's] very disregard for rigor that he was able to take conceptual steps which were inherently impossible to men of more critical genius."

-Peter G. Dirichlet

Towards the culmination of twentieth century, Saitoh [247, 248, 249] while working on the solution of the heat equation

$$\frac{\partial^2 U(x,t)}{\partial x^2} = \frac{\partial U(x,t)}{\partial t}, \quad x \in \mathbb{R}, \, t \in \mathbb{R}^+ \tag{1.255}$$

with the initial condition $U(x,0) = f(x) \in L^2(\mathbb{R})$ derived a typical result for a novel integral transform arising in the framework of the model (1.255) by using the theory of reproducing kernels. Invoking the classical Fourier transform, it was demonstrated that a solution $U(x,t)$ of (1.255) has the following representation:

$$U_f(x,t) = \frac{1}{\sqrt{4\pi t}} \int_{\mathbb{R}} f(\omega) \, \exp\left\{ -\frac{(x-\omega)^2}{4t} \right\} d\omega. \tag{1.256}$$

Therefore, for any $t > 0$, it was examined that the resulting integral transform $U \to U_f$, $f \in L^2(\mathbb{R})$ can be extended analytically onto $\mathbb{C}$. Inspired by the work of Saitoh, Castro et al. [67, 68] studied certain possibilities for the quadratic Fourier transform by employing a general quadratic function in the exponent of the transform. Keeping in view the contemporary trends of using different chirps in the analysis of finite energy signals, Castro and his colleagues introduced the notion of quadratic-phase Fourier transform which embodies a variety of integral transforms including the Fourier transform, fractional Fourier transform, linear canonical transform and the special affine Fourier transform.

In this section, our goal is to gain deeper insights into the notion of quadratic-phase Fourier transform (QPFT). Primarily, we intent to study all the fundamental properties including Parseval's and inversion formulae, and then formulate some vital results concerning the convolution and correlation in the context of quadratic-phase Fourier transform. The centerpiece of this section is to demonstrate the applications of quadratic-phase Fourier transform to two important aspects of signal processing, viz; sampling and multiplicative filtering.

1.6.1   Definition and basic properties

This subsection aims to provide a stimulus to the notion of quadratic-phase Fourier transform. Below we have the formal definition.

Definition 1.6.1. For a given collection of real parameters $\Omega = (A, B, C, D, E)$, with $B \neq 0$, the quadratic-phase Fourier transform of any $f \in L^2(\mathbb{R})$ is denoted as $\mathcal{Q}_\Omega[f]$ and is defined by

$$\mathcal{Q}_\Omega\big[f\big](\omega) = \frac{1}{\sqrt{2\pi}} \int_{\mathbb{R}} f(t) \, \mathcal{K}_\Omega(t,\omega) \, dt, \tag{1.257}$$

where $\mathcal{K}_\Omega(t,\omega)$ denotes the "kernel" of the quadratic-phase Fourier transform and is given by

$$\mathcal{K}_\Omega(t,\omega) = \exp\left\{ -i\big(At^2 + B\omega t + C\omega^2 + Dt + E\omega\big) \right\}. \tag{1.258}$$

Definition 1.6.1 circumscribes several integral transforms ranging from the classical Fourier to the much recent special affine Fourier transform. Nevertheless, many signal processing operations, such as scaling, shifting, time reversal can also be performed via the transformation (1.257). Here, we make the following comments:

(i) Typically, the constant factors incorporated in the integral operators are considered in view of the final purposes and problems where the operators are used. As is evident from Definition 1.6.1, there is a tangible distinction among these factors in the context of the quadratic-phase Fourier transform and other generalizations of the Fourier transform, such as fractional Fourier transform, linear canonical transform and special affine Fourier transform. In the realm of the quadratic-phase Fourier transform, the factor $1/\sqrt{2\pi}$ is chosen intentionally, since it ensures consequent convenient computations involving the quadratic-phase Fourier integral operator. On the other hand, the constant factors used in the fractional Fourier, linear canonical and special affine Fourier transforms are more convenient in view of the operator theoretic properties of those cases.

(ii) For $A = C = D = E = 0$ and $B = 1$, Definition 1.6.1 boils down to the classical Fourier transform (1.7).

(iii) Taking the parameters as $A = -\cot\theta/2$, $B = \csc\theta$, $C = -\cot\theta/2$ and $D = E = 0$, where $\theta \neq n\pi$, $n \in \mathbb{Z}$. Then, multiplying the integrand in (1.257) with $\sqrt{1 - i\cot\theta}$ yields the θ-angle fractional Fourier transform given by (1.127).

(iv) For $D = E = 0$, we consider the transformations $A \to -A/2B$, $B \to 1/B$, $C \to -C/2B$. Then, after multiplying the integrand with the factor $1/\sqrt{iB}$, the integral transform (1.257) takes the form of the linear canonical transform (1.164).

(v) Shuffling the parameters as $A \to -A/2B$, $B \to 1/B$, $C \to -D/2B$, $D \to -p/B$, $E = (Dp - Bq)/B$ and then multiplying the integrand in (1.257) with $e^{iDp^2/2B}/\sqrt{iB}$, an integral expression similar to the special affine Fourier transform (1.227) is obtained.

In the sequel, we formulate an inversion formula and also obtain an orthogonality relation pertaining to the quadratic-phase Fourier transform (1.257).

Theorem 1.6.1. *If $\mathcal{Q}_\Omega[f](\omega)$ is the quadratic-phase Fourier transform of any $f \in L^2(\mathbb{R})$, then the following inversion formula holds:*

$$f(t) = \mathcal{Q}_\Omega^{-1}\Big(\mathcal{Q}_\Omega\big[f\big](\omega)\Big)(t) := \frac{|B|}{\sqrt{2\pi}} \int_\mathbb{R} \mathcal{Q}_\Omega\big[f\big](\omega)\,\overline{\mathcal{K}_\Omega(t,\omega)}\,d\omega. \qquad (1.259)$$

Proof. Invoking Definition 1.6.1, we have

$$\int_\mathbb{R} \mathcal{Q}_\Omega\big[f\big](\omega)\,\overline{\mathcal{K}_\Omega(t,\omega)}\,d\omega = \int_\mathbb{R} \left\{ \frac{1}{\sqrt{2\pi}} \int_\mathbb{R} f(x)\,\mathcal{K}_\Omega(x,\omega)\,dt \right\} \overline{\mathcal{K}_\Omega(t,\omega)}\,d\omega$$

$$= \frac{1}{\sqrt{2\pi}} \int_\mathbb{R} f(x) \left\{ \int_\mathbb{R} \mathcal{K}_\Omega(x,\omega)\,\overline{\mathcal{K}_\Omega(t,\omega)}\,d\omega \right\} dx$$

$$= \sqrt{2\pi} \int_\mathbb{R} f(x)\,\exp\left\{ i\big(A(t^2 - x^2) + D(t - x)\big) \right\}$$

$$\times \left\{ \frac{1}{2\pi} \int_\mathbb{R} e^{iB\omega(t-x)}\,d\omega \right\} dx$$

$$= \sqrt{2\pi} \int_\mathbb{R} f(x)\,\exp\left\{ i\big(A(t^2 - x^2) + D(t - x)\big) \right\} \left(\frac{\delta(t - x)}{|B|} \right) dx$$

$$= \frac{\sqrt{2\pi}}{|B|}\,f(t).$$

Equivalently,

$$f(t) = \frac{|B|}{\sqrt{2\pi}} \int_{\mathbb{R}} \mathcal{Q}_{\Omega}\big[f\big](\omega)\, \overline{\mathcal{K}_{\Omega}(t,\omega)}\, d\omega.$$

This completes the proof of Theorem 1.6.1. $\square$

Theorem 1.6.2. (Parseval's Formula): *For any pair of functions $f, g \in L^2(\mathbb{R})$, we have*

$$\Big\langle \mathcal{Q}_{\Omega}\big[f\big], \mathcal{Q}_{\Omega}\big[g\big] \Big\rangle_2 = \frac{1}{|B|} \big\langle f, g \big\rangle_2. \tag{1.260}$$

In particular, if $f = g$, then we have

$$\Big\| \mathcal{Q}_{\Omega}\big[f\big] \Big\|_2^2 = \frac{1}{|B|} \big\| f \big\|_2^2. \tag{1.261}$$

Proof. The proof follows directly by taking into account the L.H.S of (1.260) and then proceeding with Definition 1.6.1. $\square$

Remark 1.6.1. For $B = 1$, the quadratic-phase Fourier operator $\mathcal{Q}_{\Omega}$ defines a unitary operator in $L^2(\mathbb{R})$.

In the following theorem, we study some fundamental properties of the quadratic-phase Fourier transform (1.257).

Theorem 1.6.3. *Let $\alpha, \beta \in \mathbb{C}$, $k, \omega_0 \in \mathbb{R}$ and $\lambda \in \mathbb{R} \setminus \{0\}$. Then, for any $f, g \in L^2(\mathbb{R})$ the quadratic-phase Fourier transform defined in (1.257) satisfies the following properties:*

(i) *Linearity:* $\mathcal{Q}_{\Omega}\big[\alpha f + \beta g\big](\omega) = \alpha\, \mathcal{Q}_{\Omega}\big[f\big](\omega) + \beta\, \mathcal{Q}_{\Omega}\big[g\big](\omega)$,

(ii) *Translation:* $\mathcal{Q}_{\Omega}\big[f(t-k)\big](\omega) = \exp\Big\{ i\big((4A^2B^{-2}C - A)k^2 + (4AB^{-1}C - B)\omega k + (2AB^{-1}E - D)k\big)\Big\} \mathcal{Q}_{\Omega}\big[f\big](\omega + 2AB^{-1}k)$,

(iii) *Modulation:* $\mathcal{Q}_{\Omega}\big[e^{i\omega_0 t} f(t)\big](\omega)$
$= \exp\Big\{ i\big(C(B^{-2}\omega_0^2 - 2B^{-1}\omega\omega_0) - EB^{-1}\omega_0\big)\Big\} \mathcal{Q}_{\Omega}\big[f\big](\omega - B^{-1}\omega_0)$,

(iv) *Scaling:* $\mathcal{Q}_{\Omega}\Big[f\Big(\dfrac{t}{\lambda}\Big)\Big](\omega) = |\lambda|\, \mathcal{Q}_{\Omega'}\big[f\big](\omega)$, $\quad \Omega' = (\lambda^2 A, B, \lambda^{-2}C, \lambda D, \lambda^{-1}E)$,

(v) *Parity:* $\mathcal{Q}_{\Omega}\big[f(-t)\big](\omega) = \mathcal{Q}_{\Omega''}\big[f\big](-\omega)$, $\quad \Omega'' = (A, B, C, -D, -E)$,

(vi) *Conjugation:* $\mathcal{Q}_{\Omega}\big[\overline{f}\big](\omega) = \overline{\mathcal{Q}_{-\Omega}\big[f\big](\omega)}$, $\quad -\Omega = (-A, -B, -C, -D, -E)$,

(vii) *Translation and modulation:* $\mathcal{Q}_{\Omega}\big[e^{i\omega_0 t} f(t-k)\big](\omega) = \mathcal{Q}_{\Omega}\big[f\big](\omega - B^{-1}\omega_0 + 2AB^{-1}k)$
$\times \exp\Big\{ i\big((B^{-1}\omega_0 - 2AB^{-1}k)^2 - 2\omega(B^{-1}\omega_0 - 2AB^{-1}k) - EB^{-1}\omega_0 + (2EAB^{-1} - D - B\omega + \omega_0)k - Ak^2\big)\Big\}$.

Proof. To make it brief, we omit the proof. $\square$

1.6.2   Quadratic-phase convolution and correlation

In this subsection, our motive is to introduce the notion of quadratic-phase convolution and correlation and study their fundamental properties. While developing the novel convolution and correlation structures, our primary interest is to obtain the corresponding convolution and correlation theorems which uphold the classical convolution and correlation theorems in the ordinary Fourier domain in the sense that, except for a chirp, the quadratic-phase Fourier transform of the convolution (or correlation) of two functions corresponds to the product of their respective quadratic-phase Fourier transforms.

Definition 1.6.2. For any pair of functions $f, g \in L^2(\mathbb{R})$, the quadratic-phase convolution is denoted by $\circledast_\Omega$ and is defined as

$$\left(f \circledast_\Omega g\right)(z) = \int_{\mathbb{R}} f(t)\, g(z - t)\, e^{-2iAt(t-z)}\, dt. \tag{1.262}$$

Some of the important properties of the quadratic-phase convolution operation given in (1.262) are assembled in the following theorem:

Theorem 1.6.4. *Let $k \in \mathbb{R}$ and $\lambda \in \mathbb{R} \setminus \{0\}$ are scalars. Then for any $f, g, h \in L^2(\mathbb{R})$ the quadratic-phase convolution $\circledast_\Omega$ satisfies the following properties:*

  (i) *Commutativity:* $\left(f \circledast_\Omega g\right)(z) = \left(g \circledast_\Omega f\right)(z),$

  (ii) *Associativity:* $\left(f \circledast_\Omega \left(g \circledast_\Omega h\right)\right)(z) = \left(\left(f \circledast_\Omega g\right) \circledast_\Omega h\right)(z),$

(iii) *Distributivity:* $\left(f \circledast_\Omega (g + h)\right)(z) = \left(f \circledast_\Omega g\right)(z) + \left(f \circledast_\Omega h\right)(z),$

  (iv) *Translation:* $\left(f \circledast_\Omega g\right)(z - k) = e^{-2iAk(z-k)}\left(F_\Omega(\cdot - k) \circledast_\Omega g(\cdot)\right)(z), \quad F_\Omega(t) = e^{2iAkt} f(t),$

  (v) *Scaling:* $\left(f \circledast_\Omega g\right)(\lambda z) = |\lambda|\left(f(\lambda x) \circledast_{\Omega'} g(\lambda x)\right)(z), \quad \Omega' = \left(\lambda^2 A, B, C, D, E\right),$

(vi) *Parity:* $\left(f \circledast_\Omega g\right)(-z) = \left(f(-t) \circledast_\Omega g(-t)\right)(z).$

Proof. For the sake of brevity, we ought to omit the proof.   □

We now establish the convolution theorem associated with the quadratic-phase convolution (1.262).

Theorem 1.6.5. *For any $f, g \in L^2(\mathbb{R})$, we have*

$$\mathcal{Q}_\Omega\left[\left(f \circledast_\Omega g\right)(z)\right](\omega) = \sqrt{2\pi}\, e^{i(C\omega^2 + E\omega)}\, \mathcal{Q}_\Omega\left[f\right](\omega)\, \mathcal{Q}_\Omega\left[g\right](\omega). \tag{1.263}$$

Proof. The proof is accomplished after a direct application of Definition 1.6.1 to (1.262).   □

In analogy to Definition 1.6.2, we now introduce the notion of quadratic-phase correlation operation.

Definition 1.6.3. Given a pair of functions $f, g \in L^2(\mathbb{R})$, the quadratic-phase correlation is denoted by $\star_\Omega$ and is defined as

$$\left(f \star_\Omega g\right)(z) = \int_{\mathbb{R}} f(t)\, \overline{g(t - z)}\, e^{2iAz(z-t)}\, dt. \tag{1.264}$$

Next, our aim is to examine the nature of correlation operation (1.264) in the quadratic-phase Fourier domain. In this direction, we have the following theorem:

Theorem 1.6.6. *For any $f, g \in L^2(\mathbb{R})$, we have*

$$\mathcal{Q}_\Omega\Big[(f \star_\Omega g)(z)\Big](\omega) = \sqrt{2\pi}\, e^{-i(C\omega^2 + E\omega)}\, \mathcal{Q}_\Omega\Big[f\Big](\omega)\, \overline{\mathcal{Q}_\Omega\Big[g\Big](\omega)}. \qquad (1.265)$$

Proof. The assertion (1.265) follows by applying Definition 1.6.1 to Definition 1.6.3, so we omit the formal proof. $\qquad\square$

At the tail end of this subsection, we investigate upon the behaviour of the quadratic-phase Fourier transform (1.257), in case the input is a product of two functions.

Theorem 1.6.7. *The quadratic-phase Fourier transform of the product of two functions $f, g \in L^2(\mathbb{R})$ can be expressed as:*

$$\mathcal{Q}_\Omega\Big[f(t) \cdot g(t)\Big](\omega) = \frac{|B|\, e^{-i(C\omega^2 + E\omega)}}{\sqrt{2\pi}} \left(e^{i(C\eta^2 + E\eta)}\, \mathcal{Q}_\Omega\Big[g\Big](\eta) * \mathscr{F}\Big[f\Big](B\eta) \right)(\omega), \quad (1.266)$$

where $\mathscr{F}$ is the classical Fourier transform defined in (1.7) and $*$ is the corresponding convolution operation given by (1.75).

Proof. As a consequence of Definition 1.6.1 and the associated inversion formula (1.259), it follows that

$$\mathcal{Q}_\Omega\Big[f(t) \cdot g(t)\Big](\omega)$$

$$= \frac{1}{\sqrt{2\pi}} \int_\mathbb{R} f(t)\, g(t)\, \mathcal{K}_\Omega(t, \omega)\, dt$$

$$= \frac{1}{\sqrt{2\pi}} \int_\mathbb{R} f(t)\, g(t)\, \exp\Big\{ -i\big(At^2 + Bt\omega + C\omega^2 + Dt + E\omega\big) \Big\}\, dt$$

$$= \frac{1}{\sqrt{2\pi}} \int_\mathbb{R} f(t) \left\{ \frac{|B|}{\sqrt{2\pi}} \int_\mathbb{R} \mathcal{Q}_\Omega\Big[g\Big](\eta)\, \overline{\mathcal{K}_\Omega(t, \eta)}\, d\eta \right\}$$

$$\times \exp\Big\{ -i\big(At^2 + Bt\omega + C\omega^2 + Dt + E\omega\big) \Big\}\, dt$$

$$= \frac{1}{\sqrt{2\pi}} \int_\mathbb{R} f(t) \left\{ \frac{|B|\, e^{i(At^2 + Dt)}}{\sqrt{2\pi}} \int_\mathbb{R} \mathcal{Q}_\Omega\Big[g\Big](\eta)\, \exp\big\{ i(Bt\eta + C\eta^2 + E\eta) \big\}\, d\eta \right\}$$

$$\times \exp\Big\{ -i\big(At^2 + Bt\omega + C\omega^2 + Dt + E\omega\big) \Big\}\, dt$$

$$= \frac{|B|\, e^{-i(C\omega^2 + E\omega)}}{\sqrt{2\pi}} \int_\mathbb{R} e^{i(C\eta^2 + E\eta)}\, \mathcal{Q}_\Omega\Big[g\Big](\eta) \left\{ \frac{1}{\sqrt{2\pi}} \int_\mathbb{R} f(t)\, e^{-iB(\omega - \eta)t}\, dt \right\} d\eta$$

$$= \frac{|B|\, e^{-i(C\omega^2 + E\omega)}}{\sqrt{2\pi}} \int_\mathbb{R} e^{i(C\eta^2 + E\eta)}\, \mathcal{Q}_\Omega\Big[g\Big](\eta)\, \mathscr{F}\Big[f\Big]\big(B(\omega - \eta)\big)\, d\eta$$

$$= \frac{|B|\, e^{-i(C\omega^2 + E\omega)}}{\sqrt{2\pi}} \left(e^{i(C\eta^2 + E\eta)}\, \mathcal{Q}_\Omega\Big[g\Big](\eta) * \mathscr{F}\Big[f\Big](B\eta) \right)(\omega).$$

This completes the proof of Theorem 1.6.7. $\qquad\square$

Remark 1.6.2. By appropriately choosing the real parameter collection $\Omega = (A, B, C, D, E)$, all the results obtained for the quadratic-phase Fourier transform given in Definition 1.6.1 yield their respective counterparts for all the integral transforms ranging from the classical Fourier transform to the much recent special affine Fourier transform.

1.6.3   Applications of the quadratic-phase Fourier transform

"Almost every problem that you come across is befuddled with all kinds of extraneous data of one sort or another; and if you can bring this problem down into the main issues, you can see more clearly what you are trying to do."

-Claude E. Shannon

From the perspective of applications, the quadratic-phase Fourier transform (1.257) is quite lucrative in the sense that there arise many useful classes of signals, which reveal their characteristics beyond the Fourier domain. Moreover, since the quadratic-phase Fourier transform circumscribes many prominent integral transforms, therefore, all the obtained results can also be carried to the Fourier, fractional Fourier, linear canonical and special affine Fourier transforms. With this prolificacy in hindsight, the goal of this subsection is to demonstrate the applications of the quadratic-phase Fourier transform (1.257) to two important aspects of signal processing, viz; sampling and multiplicative filtering.

(A) Sampling in the quadratic-phase Fourier domain

The sampling theory pertaining to the quadratic-phase Fourier transform is particularly interesting as many chirp-like signals arising in radar and other communication systems are not band-limited in usual Fourier domain but turn to be band-limited in the generalized Fourier domains, in particular in the quadratic-phase Fourier domain. Here, we shall derive the Shannon's sampling theorem pertaining to the quadratic-phase Fourier transform (1.257). To facilitate the narrative, we shall take into consideration the realistic assumption that the parameter collection $\Omega = (A, B, C, D, E)$ in Definition 1.6.1 is chosen such that $B > 0$, which has a tangible influence on simplifying the expressions.

Theorem 1.6.8. *Given a continuous band-limited signal* $f \in L^2(\mathbb{R})$ *with bandwidth U in the quadratic-phase Fourier domain. Then,* f *can be completely reconstructed from the sampled version* $f(nT)$, $n \in \mathbb{Z}$ *via the following reconstruction formula:*

$$f(t) = \exp\left\{i\left(At^2 + Dt\right)\right\} \sum_{n \in \mathbb{Z}} \exp\left\{-i\left(A(nT)^2 + D(nT)\right)\right\} f(nT) \left(\frac{T \sin(B\,U(t - nT))}{\pi(t - nT)}\right).$$

(1.267)

where $T \leq \pi/BU$ *is the sampling period.*

Proof. Let F denotes the uniformly sampled signal obtained from a given continuous signal f; that is,

$$F(t) = f(t) \sum_{n \in \mathbb{Z}} \delta(t - nT) = \sum_{n \in \mathbb{Z}} f(nT)\, \delta(t - nT),$$

(1.268)

where T is the sampling period and $\sum_{n \in \mathbb{Z}} \delta(t - nT)$ is the corresponding uniform impulse train or the Dirac comb. Since the impulse train is integrable and periodic with period T, therefore, application of the Fourier series formula implies that

$$\sum_{n \in \mathbb{Z}} \delta(t - nT) = \frac{1}{T} \sum_{k \in \mathbb{Z}} \exp\left\{\frac{2\pi i k t}{T}\right\}.$$

(1.269)

Next, we shall compute the QPFT of the uniformly sampled signal F given in (1.268). We proceed as

$$\mathcal{Q}_\Omega\big[F\big](\omega) = \frac{1}{\sqrt{2\pi}} \int_{\mathbb{R}} f(t) \sum_{n \in \mathbb{Z}} \delta(t - nT)\, \mathcal{K}_\Omega(t, \omega)\, dt$$

$$= \frac{1}{\sqrt{2\pi}} \sum_{n \in \mathbb{Z}} \int_{\mathbb{R}} f(t) \left\{ \frac{1}{2\pi} \int_{\mathbb{R}} \exp\big\{ i\sigma(t - nT) \big\}\, d\sigma \right\}$$

$$\times \exp\big\{ -i\big(At^2 + Bt\omega + C\omega^2 + Dt + E\omega\big) \big\}\, dt$$

$$= \frac{1}{\sqrt{2\pi}} \sum_{n \in \mathbb{Z}} \int_{\mathbb{R}} f(t) \left\{ \frac{B}{2\pi} \int_{\mathbb{R}} \exp\big\{ iB\lambda(t - nT) \big\}\, d\lambda \right\}$$

$$\times \exp\big\{ -i\big(At^2 + Bt\omega + C\omega^2 + Dt + E\omega\big) \big\}\, dt$$

$$= \frac{B}{2\pi} \sum_{n \in \mathbb{Z}} \int_{\mathbb{R}} e^{-inB\lambda T} \exp\big\{ i\big(C\lambda^2 - 2C\omega\lambda - E\lambda\big) \big\}$$

$$\times \left\{ \frac{1}{\sqrt{2\pi}} \int_{\mathbb{R}} f(t)\, \mathcal{K}_\Omega(t, \omega - \lambda)\, dt \right\} d\lambda$$

$$= \frac{B}{2\pi} \int_{\mathbb{R}} \left(\sum_{n \in \mathbb{Z}} e^{-inB\lambda T} \right) \mathcal{Q}_\Omega\big[f\big](\omega - \lambda) \exp\big\{ i\big(C\lambda^2 - 2C\omega\lambda - E\lambda\big) \big\}\, d\lambda. \qquad (1.270)$$

By using the expansion obtained in (1.269), we can rewrite (1.270) in the following manner:

$$\mathcal{Q}_\Omega\big[F\big](\omega) = \int_{\mathbb{R}} \left(\frac{B}{2\pi} \sum_{n \in \mathbb{Z}} \exp\left\{ -\frac{2\pi in\lambda BT}{2\pi} \right\} \right) \mathcal{Q}_\Omega\big[f\big](\omega - \lambda)$$

$$\times \exp\big\{ i\big(C\lambda^2 - 2C\omega\lambda - E\lambda\big) \big\}\, d\lambda$$

$$= \frac{1}{T} \int_{\mathbb{R}} \sum_{n \in \mathbb{Z}} \delta\left(\lambda - \frac{2\pi n}{BT} \right) \mathcal{Q}_\Omega\big[f\big](\omega - \lambda) \exp\big\{ i\big(C\lambda^2 - 2C\omega\lambda - E\lambda\big) \big\}\, d\lambda$$

$$= \frac{1}{T} \sum_{n \in \mathbb{Z}} \mathcal{Q}_\Omega\big[f\big]\left(\omega - \frac{2\pi n}{BT} \right) \exp\left\{ i\left(C\left(\frac{2\pi n}{BT}\right)^2 - \frac{4\pi n\omega C}{BT} - \frac{2\pi nE}{BT} \right) \right\}. \qquad (1.271)$$

From (1.271), we conclude that $\mathcal{Q}_\Omega\big[F\big](\omega)$ replicates with a period of $2\pi/BT$ along with the phase modulation depending on the harmonic order n. In other words, we can say that the sampling of signal in the spatial domain leads to periodization (phase change) in the quadratic-phase Fourier domain. Furthermore, for $n = 0$, we have $\mathcal{Q}_\Omega\big[F\big](\omega) = \frac{1}{T} \mathcal{Q}_\Omega\big[f\big](\omega)$, which indicates that for $n = 0$, the spectrum $\mathcal{Q}_\Omega\big[F\big](\omega)$ is amplified by a factor $1/T$ without any alteration to the phase. Since f is band-limited in the quadratic-phase Fourier domain, therefore, so is the uniformly sampled signal F and suppose that $[-U, U]$ is the support interval of $\mathcal{Q}_\Omega\big[F\big](\omega)$. Consequently, no overlapping occurs in the spectrum $\mathcal{Q}_\Omega\big[F\big](\omega)$ after sampling, if the alias-free condition holds good; that is, $2\pi/Tb \geq 2U$.

To filter out the replicated spectra from the recovered signal $\mathcal{Q}_\Omega\big[F\big](\omega)$, we consider the low pass filter H whose transfer function is given by

$$\mathcal{Q}_\Omega\big[H\big](\omega) = \begin{cases} T, & |\omega| < \omega_c \\ 0, & |\omega| \geq \omega_c, \end{cases} \qquad (1.272)$$

where ω_c is the cut-off frequency and $\omega_c \in [U, \omega/B - U]$. With Theorem 1.6.5 is hindsight, the befitting candidate for the function on the L.H.S of (1.272) is

$$\mathcal{Q}_\Omega\big[H\big](\omega) = \exp\big\{ i\big(C\omega^2 + E\omega\big) \big\} \mathcal{Q}_\Omega\big[h\big](\omega) = \begin{cases} T, & |\omega| < \omega_c \\ 0, & |\omega| \geq \omega_c, \end{cases} \qquad (1.273)$$

where the function h appearing in (1.273) can be obtained via the inversion formula (1.259) as

$$
\begin{aligned}
h(t) &= \frac{B}{\sqrt{2\pi}} \int_{\mathbb{R}} \mathcal{Q}_\Omega\big[h\big](\omega)\, \overline{\mathcal{K}_\Omega(t,\omega)}\, d\omega \\
&= \frac{B}{\sqrt{2\pi}} \int_{\mathbb{R}} \exp\Big\{ -i\big(C\omega^2 + E\omega\big)\Big\} \mathcal{Q}_\Omega\big[H\big](\omega)\overline{\mathcal{K}_\Omega(t,\omega)}\, d\omega \\
&= \frac{BT}{\sqrt{2\pi}} \exp\Big\{i\big(At^2 + Dt\big)\Big\} \int_{-\omega_c}^{\omega_c} e^{Bt\omega}\, d\omega \\
&= \frac{B}{\sqrt{2\pi}} \exp\Big\{i\big(At^2 + Dt\big)\Big\} \frac{2T\sin(B\omega_c t)}{Bt}.
\end{aligned}
\tag{1.274}
$$

Again invoking Theorem 1.6.5, we get

$$
\begin{aligned}
\mathcal{Q}_\Omega\Big[(F \circledast_\Omega h)(t)\Big](\omega) &= \sqrt{2\pi}\, \exp\Big\{i\big(C\omega^2 + E\omega\big)\Big\} \mathcal{Q}_\Omega\big[F\big](\omega)\, \mathcal{Q}_\Omega\big[h\big](\omega) \\
&= \sqrt{2\pi}\, \mathcal{Q}_\Omega\big[F\big](\omega)\, \mathcal{Q}_\Omega\big[H\big](\omega).
\end{aligned}
\tag{1.275}
$$

Finally, the desired output is obtained by using the inversion formula (1.259) and proceeding as follows:

$$
\begin{aligned}
f(t) &= \frac{B}{\sqrt{2\pi}} \int_{\mathbb{R}} \mathcal{Q}_\Omega\big[F\big](\omega)\, \mathcal{Q}_\Omega\big[H\big](\omega)\, \overline{\mathcal{K}_\Omega(t,\omega)}\, d\omega \\
&= \frac{1}{\sqrt{2\pi}} \left\{ \frac{B}{\sqrt{2\pi}} \int_{\mathbb{R}} \mathcal{Q}_\Omega\Big[(F \circledast_\Omega h)(t)\Big](\omega)\, \overline{\mathcal{K}_\Omega(t,\omega)}\, d\omega \right\} \\
&= \frac{1}{\sqrt{2\pi}} \int_{\mathbb{R}} F(x)\, h(t-x)\, e^{-2iAx(x-t)}\, dx \\
&= \frac{B}{2\pi} \int_{\mathbb{R}} F(x)\, \exp\Big\{i\big(At^2 - Ax^2 + Dt - Dx\big)\Big\} \left(\frac{2T\sin(B\,\omega_c(t-x))}{B(t-x)} \right) dx \\
&= \frac{B}{2\pi} \sum_{n\in\mathbb{Z}} \int_{\mathbb{R}} f(x)\, \delta(x-nT)\, \exp\Big\{i\big(At^2 - Ax^2 + Dt - Dx\big)\Big\} \left(\frac{2T\sin(b\,\omega_c(t-x))}{B(t-x)} \right) dx \\
&= \exp\Big\{i\big(At^2 + Dt\big)\Big\} \sum_{n\in\mathbb{Z}} \exp\Big\{ -i\big(A(nT)^2 + D(nT)\big)\Big\} f(nT) \left(\frac{T\sin(B\omega_c(t-nT))}{\pi(t-nT)} \right).
\end{aligned}
\tag{1.276}
$$

Expression (1.276) serves as the reconstruction formula for the given band-limited signal f. Finally, matching the cut-off frequency ω_c with the bandwidth U of the given signal; that is, $\omega_c = U$, we obtain the desired reconstruction formula:

$$
f(t) = \exp\Big\{i\big(At^2 + Dt\big)\Big\} \sum_{n\in\mathbb{Z}} \exp\Big\{ -i\big(A(nT)^2 + D(nT)\big)\Big\} f(nT) \left(\frac{T\sin(B\,U(t-nT))}{\pi(t-nT)} \right).
$$

This completes the proof of Theorem 1.6.8.   □

Corollary 1.6.9. *Suppose that a continuous signal $f \in L^2(\mathbb{R})$ is band-limited in the quadratic-phase Fourier domain with bandwidth U. Then, f can be critically sampled if the sampling period T satisfies $T = \pi/Ub$ and the reconstruction formula is given by (1.267).*

Remark 1.6.3. By appropriately choosing the collection of real parameters $\Omega = (A, B, C, D, E)$, $B > 0$ in Theorem 1.6.8, we can obtain respective sampling theorems for the Fourier, fractional Fourier, linear canonical and special affine Fourier transforms. Thus, we infer that Theorem 1.6.8 serves as an embodiment of a cluster of sampling theorems.

Example 1.6.1. Consider the analog signal $f(t) = K \operatorname{sinc}(\pi t) \exp\left\{i(\alpha t^2 + \beta t)\right\}$, where $\operatorname{sinc}(t)$ is the unnormalized sinc-function and $\alpha,\ \beta \in \mathbb{R} \setminus \{0\}$, $K \in \mathbb{C} \setminus \{0\}$, which is depicted in Figure 1.34. Firstly, we claim that f is a band-limited in the quadratic-phase Fourier domain. For computational simplicity, we choose $\Omega = (\alpha, B, C, \beta, E)$, $B > 0$, so that

$$
\begin{aligned}
\mathcal{Q}_\Omega\big[f\big](\omega) &= \frac{K}{\sqrt{2\pi}} \exp\left\{i(C\omega^2 + E\omega)\right\} \int_{\mathbb{R}} \frac{\sin \pi t}{\pi t} e^{-iB\omega t}\, dt \\
&= \frac{K}{2\pi} \exp\left\{i(C\omega^2 + E\omega)\right\} \int_{\mathbb{R}} \frac{1}{\sqrt{2\pi}} \left(\frac{2\sin \pi t}{t}\right) e^{-iB\omega t}\, dt \\
&= \frac{K}{2\pi} \exp\left\{i(C\omega^2 + E\omega)\right\} \int_{\mathbb{R}} \mathscr{F}\big(\chi_{[-\pi,\,\pi]}\big)(t)\, e^{-iB\omega t}\, dt \\
&= \frac{K}{\sqrt{2\pi}} \exp\left\{i(C\omega^2 + E\omega)\right\} \left\{ \frac{1}{\sqrt{2\pi}} \int_{\mathbb{R}} \mathscr{F}\big(\chi_{[-\pi,\,\pi]}\big)(t)\, e^{i(-B\omega)t} dt \right\} \\
&= \frac{K}{\sqrt{2\pi}} \exp\left\{i(C\omega^2 + E\omega)\right\} \chi_{\left[-\frac{\pi}{B},\, \frac{\pi}{B}\right]}(\omega), \quad (1.277)
\end{aligned}
$$

where χ is the characteristic function and $\mathscr{F}$ denotes the classical Fourier transform defined in (1.7). From (1.277), we conclude that the quadratic-phase Fourier transform of the given analog signal is supported on the interval $\left[-\frac{\pi}{B}, \frac{\pi}{B}\right]$. Hence, f is indeed a band-limited signal with bandwidth $U = \pi/B$. The quadratic-phase Fourier transform of f is plotted in Figure 1.35. Next, we choose the sampling period T such that $T = \pi/UB$, yielding $T = 1$. Consequently, the sampled version of the given signal is of the following form:

$$
f(nT) = K \operatorname{sinc}(\pi n) \exp\left\{i(\alpha n^2 + \beta n)\right\}, \quad n \in \mathbb{Z}, \quad (1.278)
$$

which is depicted in Figure 1.36. Finally, applying Theorem 1.6.8, we obtain the reconstruction formula for retracting the original signal as

$$
f(t) = K \exp\left\{i(\alpha n^2 + \beta n)\right\} \sum_{n \in \mathbb{Z}} \operatorname{sinc}(\pi n) \operatorname{sinc}(\pi(t - n)). \quad (1.279)
$$

Graphically, the reconstructed signal (1.279) is shown in Figure 1.37.

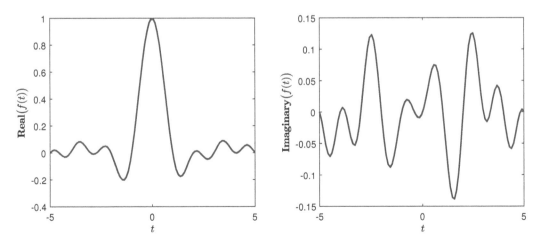

FIGURE 1.34: Real and imaginary parts of the continuous signal $f(t)$ for $\alpha = 0.25$, $\beta = 0.10$ and $K = 1$.

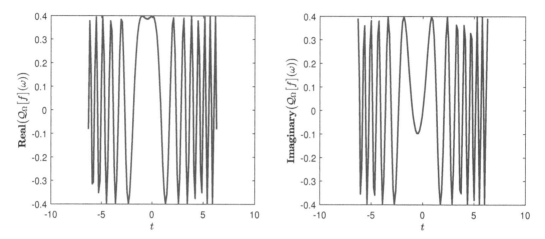

FIGURE 1.35: Real and imaginary parts of the quadratic-phase Fourier transform of f for $\Omega = (0.25, 0.5, 1, 0.10, 1)$ and $K = 1$.

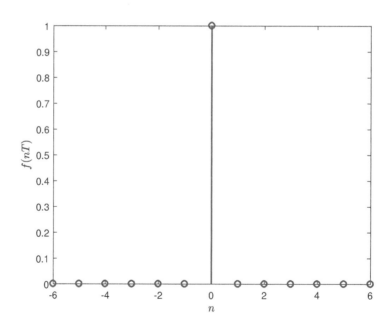

FIGURE 1.36: The sampled signal $f(n)$ for $\alpha = 0.25$, $\beta = 0.10$ and $K = 1$.

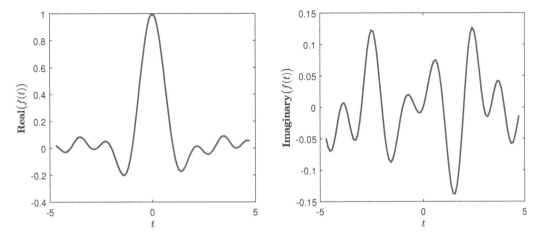

FIGURE 1.37: Real and imaginary parts of the reconstructed signal.

(B) Multiplicative filtering in the quadratic-phase Fourier domain

The notion of filtering is one of the highly acknowledged and widely applied concepts of signal processing. A filter can be regarded as a "black box" that takes the input signal, bites out the unwanted components and then returns the desired output signal. One of the primary concerns during the design of filtering procedures is to minimize the computational complexity while maintaining the efficiency of the filter. Here, we shall demonstrate that the convolution structure (1.262) can be employed to design a computationally efficient multiplicative filter in the quadratic-phase Fourier domain.

The multiplicative filtering procedure in the quadratic-phase Fourier domain is a two-step process: Firstly, the spectrum of the input signal is multiplied by an appropriate filter impulse response and then the invertibility of the quadratic-phase Fourier transform (1.259) is invoked to retract the desired output. If $\mathcal{Q}_\Omega[S_r](\omega)$ denotes the quadratic-phase Fourier transform of the received signal $S_r(t)$ and $\mathcal{Q}_\Omega[H](\omega)$ denotes the frequency response of the filter, then the effect of the multiplicative in Figure 1.38 can be expressed as:

$$S_o(t) = \mathcal{Q}_\Omega^{-1}\Big(\mathcal{Q}_\Omega[S_r](\omega)\,\mathcal{Q}_\Omega[H](\omega)\Big)(t). \tag{1.280}$$

By designing different functions $\mathcal{Q}_\Omega[H](\omega)$ appearing in (1.280), we can achieve different classes of multiplicative filters including band-pass, band-stop, low-pass, high-pass and many others. To illustrate the applicability of the proposed multiplicative filter modelled in Figure 1.38, in filtering out the unwanted signal (noise), we consider a noisy input signal $S_r(t) = f(t) + n(t)$ comprised of the desired signal $f(t)$ and the unwanted signal $n(t)$. We are interested with the quadratic-phase Fourier spectrum, which lives within a region $[\omega_\ell, \omega_h]$ in the quadratic-phase Fourier domain. We take into consideration the reasonable assumptions that the quadratic-phase Fourier transforms of the signals $S_r(t)$, $f(t)$ and $n(t)$ have either no or minimal overlapping. The function $\mathcal{Q}_\Omega[H](\omega)$ can be chosen in accordance to Theorem 1.6.5 as

$$\mathcal{Q}_\Omega[H](\omega) = \sqrt{2\pi}\, e^{i(C\omega^2 + E\omega)}\, \mathcal{Q}_\Omega[h](\omega), \tag{1.281}$$

such that it is constant over the interval $[\omega_\ell, \omega_h]$ and is either zero or of rapid decay outside the interval. Also, the finite impulse response (FIR) for (1.281) can be achieved by any of the standard filtering techniques, such as window design method, frequency sampling method, least MSE (mean square error) method, equiripple method and so on. Then, from Figure 1.38 it is quite evident that the output yields only that part of the spectrum of $f(t)$ which lives over $[\omega_\ell, \omega_h]$. Finally, we can utilize the inversion formula (1.259) to recover the desired signal $f(t)$.

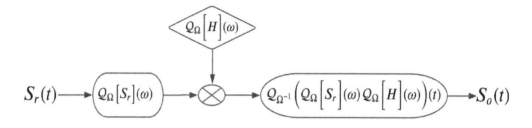

FIGURE 1.38: Multiplicative filter in the quadratic-phase Fourier domain.

Having designed the multiplicative filter in the quadratic-phase Fourier domain, we shall demonstrate that the convolution structure proposed in (1.262) offers an alternative and computationally efficient approach for the design and implementation of the above mentioned filtering procedure. The computational complexity is significantly curtailed in this case as the key components are converted to the Fourier domain and can be candidly handled via the fast Fourier transform (FFT). Note that the quadratic-phase convolution $\circledast_\Omega$ can also be expressed in terms of the classical convolution $*$ given by (1.75) as follows:

$$\left(f \circledast_\Omega h\right)(t) = e^{iAt^2}\left(e^{-iAx^2} f(x) * e^{-iAx^2} h(x)\right)(t). \tag{1.282}$$

A realization of the multiplicative filtering in the quadratic-phase Fourier domain via the convolution form (1.282) is pictorially presented in Figure 1.39. The output of the model is given by

$$S_o(t) = e^{iAt^2}\left(e^{-iAx^2} S_r(x) * \Theta(x)\right)(t). \tag{1.283}$$

In view of the expression (1.282), it is quite evident that the befitting candidate for the convolution function $\Theta(x)$ appearing in (1.283) is $\Theta(x) = e^{-iAx^2} h(x)$. Using the inversion formula (1.259), we can express the function $h(x)$ in terms of the inverse Fourier transform $\mathscr{F}^{-1}$, given by (1.67), as

$$
\begin{aligned}
h(x) &= \frac{|B|}{\sqrt{2\pi}} \int_{\mathbb{R}} \mathcal{Q}_\Omega\big[h\big](\omega)\, \overline{\mathcal{K}_\Omega(x,\omega)}\, d\omega \\
&= \frac{|B|\, e^{i(Ax^2 + Dx)}}{\sqrt{2\pi}} \int_{\mathbb{R}} \mathcal{Q}_\Omega\big[h\big](\omega)\, e^{i(Bx\omega + C\omega^2 + E\omega)}\, d\omega \\
&= \frac{|B|\, e^{i(Ax^2 + Dx)}}{\sqrt{2\pi}} \left\{ \frac{1}{\sqrt{2\pi}} \int_{\mathbb{R}} \mathcal{Q}_\Omega\big[H\big](\omega)\, e^{iBx\omega}\, d\omega \right\} \\
&= \frac{|B|\, e^{i(Ax^2 + Dx)}}{\sqrt{2\pi}}\, \mathscr{F}^{-1}\left(\mathcal{Q}_\Omega\big[H\big](\omega)\right)(Bx). \tag{1.284}
\end{aligned}
$$

Consequently, we obtain

$$\Theta(x) = \frac{|B|\, e^{iDx}}{\sqrt{2\pi}}\, \mathscr{F}^{-1}\left(\mathcal{Q}_\Omega\big[H\big](\omega)\right)(Bx). \tag{1.285}$$

Therefore, the output of the convolution-based multiplicative filter in the quadratic-phase Fourier domain, given in (1.283), takes the following form:

$$S_o(t) = \frac{|B|\, e^{iAt^2}}{\sqrt{2\pi}} \left(e^{-iAx^2} S_r(x) * e^{iDx} \mathscr{F}^{-1}\left(\mathcal{Q}_\Omega\Big[H\Big](\omega)\right)(Bx) \right)(t). \qquad (1.286)$$

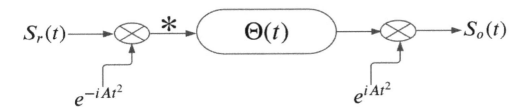

FIGURE 1.39: Multiplicative filter in the quadratic-phase Fourier domain via convolution.

As a consequence of Theorem 1.6.5, we can easily verify that the output of Figure 1.38 is exactly the same as that of the convolution based multiplicative filter. Nevertheless, it is important to notice that the major computational load of (1.286) lies on convolution and computation of the convolution function $\Theta(x)$, which can be efficiently handled via the fast Fourier algorithm. To analyze the computational complexity of the model depicted in Figure 1.39, we shall invoke Theorem 1.2.6 as

$$S_o(t) = e^{iAt^2} \mathscr{F}^{-1}\left(\mathscr{F}\Big[e^{-iAx^2} S_r(x)\Big](\omega)\, \mathcal{Q}_\Omega\Big[H\Big]\left(\frac{\omega + D}{B}\right) \right)(t). \qquad (1.287)$$

From expression (1.287), it is quite evident that the FFT algorithm can be employed to reduce the computational complexity of the convolution-based multiplicative filter in the quadratic-phase Fourier domain. Thus, for a given sample of size N, the computational complexity of this method is $O(N \log_2 N)$.

1.7   The Two-Dimensional Fourier Transform

"The integrals which we have obtained are not only general expressions which satisfy the differential equation, they represent in the most distinct manner the natural effect which is the object of the phenomenon... when this condition is fulfilled, the integral is, properly speaking, the equation of the phenomenon; it expresses clearly the character and progress of it, in the same manner as the finite equation of a line or curved surface makes known all the properties of those forms."

-Joseph Fourier

This section is an exposition on the two-dimensional Fourier transform, which serves as the foundation for the theory of geometrical wavelets to be undertaken in the subsequent chapters of this monograph. To begin with, we shall take into consideration the Hilbert space of square integrable functions on the Euclidean plane, denoted as $L^2(\mathbb{R}^2)$. From a technical perspective, any arbitrary element $f \in L^2(\mathbb{R}^2)$ shall be referred to as a two-dimensional signal or simply signal. The typical example of a two-dimensional signal is an image, because every image can be regarded as a square integrable, complex-valued function defined on the Euclidean plane $\mathbb{R}^2$.

In order to start the formal discourse on the two-dimensional Fourier transform (2D-Fourier transform), we have the following definition:

Definition 1.7.1. For any function $f \in L^2(\mathbb{R}^2)$, the two-dimensional Fourier transform is defined by

$$\mathscr{F}\left[f\right](\mathbf{w}) = \hat{f}(\mathbf{w}) := \frac{1}{2\pi} \int_{\mathbb{R}^2} f(\mathbf{t}) \, e^{-i \mathbf{t}^T \mathbf{w}} \, d\mathbf{t}, \qquad (1.288)$$

where $\mathbf{t} = (t_1, t_2)^T \in \mathbb{R}^2$ and $\mathbf{w} = (\omega_1, \omega_2)^T \in \mathbb{R}^2$.

Remark 1.7.1. The two-dimensional Fourier transform presented in Definition 1.7.1 is also expressible as

$$\mathscr{F}\left[f\right](\mathbf{w}) = \frac{1}{2\pi} \int_{\mathbb{R}^2} f(\mathbf{t}) \, e^{-i \mathbf{t} \cdot \mathbf{w}} \, d\mathbf{t}, \qquad (1.289)$$

where $\mathbf{t} \cdot \mathbf{w}$ corresponds to the usual dot product of $\mathbf{t} = (t_1, t_2) \in \mathbb{R}^2$ and $\mathbf{w} = (\omega_1, \omega_2) \in \mathbb{R}^2$.

In view of Definition 1.7.1, we mark the following points regarding the two-dimensional Fourier transform:

(i) Analogous to usual one-dimensional case, the 2D-Fourier transform (1.288) converts a two-dimensional signal from the spatial domain representation to a frequency domain representation.

(ii) In literature, certain equivalent forms of Definition 1.7.1 can also witnessed. For instance, the 2D-Fourier transform is sometimes expressed as

$$\mathscr{F}\left[f\right](\mathbf{w}) = \int_{\mathbb{R}^2} f(\mathbf{t}) \, e^{-i \mathbf{t}^T \mathbf{w}} \, d\mathbf{t} \qquad (1.290)$$

and the corresponding inversion formula is given by

$$f(\mathbf{t}) = \frac{1}{(2\pi)^2} \int_{\mathbb{R}^2} \mathscr{F}\left[f\right](\mathbf{w}) \, e^{i \mathbf{t}^T \mathbf{w}} \, d\mathbf{w}. \qquad (1.291)$$

However, by doing so, one loses the symmetry of the forward and backward transforms. Yet another elegant approach is to make use of the substitution $\mathbf{w} = 2\pi \mathbf{w}'$, so that the 2D-Fourier transform takes the form

$$\mathscr{F}\left[f\right](\mathbf{w}') = \int_{\mathbb{R}^2} f(\mathbf{t}) \, e^{-2\pi i \mathbf{t}^T \mathbf{w}'} \, d\mathbf{t}, \qquad (1.292)$$

with the inversion formula as

$$f(\mathbf{t}) = \int_{\mathbb{R}^2} \mathscr{F}\left[f\right](\mathbf{w}) \, e^{2\pi i \mathbf{t}^T \mathbf{w}'} \, d\mathbf{w}'. \qquad (1.293)$$

It is pertinent to mention that the definition of two-dimensional Fourier transform given in (1.292) relies on the variable $\mathbf{w}' = (2\pi)^{-1}\mathbf{w}$, which is interpreted as the "frequency," in lieu to the "angular frequency" $\mathbf{w}$ used in (1.288) and (1.290).

(iii) Indeed, the 2D-Fourier transform of any function $f \in L^2(\mathbb{R}^2)$ is defined almost everywhere on $\mathbb{R}^2$ and also belongs to $L^2(\mathbb{R}^2)$.

The Parseval's relation and inversion formula corresponding to Definition 1.7.1 are given by

$$\left\langle \mathscr{F}\big[f\big], \mathscr{F}\big[g\big] \right\rangle_2 = \left\langle f, g \right\rangle_2, \quad \forall\, f, g \in L^2(\mathbb{R}^2) \tag{1.294}$$

and

$$f(\mathbf{t}) = \mathscr{F}^{-1}\left(\mathscr{F}\big[f\big](\mathbf{w})\right)(\mathbf{t}) := \frac{1}{2\pi} \int_{\mathbb{R}^2} \mathscr{F}\big[f\big](\mathbf{w})\, e^{i\mathbf{t}^T \mathbf{w}}\, d\mathbf{w}, \tag{1.295}$$

respectively. Moreover, in analogy to the uncertainty inequality (1.123), the Heisenberg's uncertainty principle associated with the 2D-Fourier transform asserts that for any non-trivial function $f \in L^2(\mathbb{R}^2)$, the following inequality holds:

$$\left\{ \int_{\mathbb{R}^2} |\mathbf{t}|^2 |f(\mathbf{t})|^2 d\mathbf{t} \right\}^{1/2} \left\{ \int_{\mathbb{R}^2} |\mathbf{w}|^2 \left|\mathscr{F}\big[f\big](\mathbf{w})\right|^2 d\mathbf{w} \right\}^{1/2} \geq \big\|f\big\|_2^2. \tag{1.296}$$

Apart from the above described Cartesian coordinate form of the 2D-Fourier transform, the polar variant is of utmost significance. For instance, the notion of curvelets and ripplets to be undertaken in Chapter 3 is heavily reliant on the polar coordinate form of the 2D-Fourier transform. In the remaining part of this section, we shall take a survey of the polar Fourier transform. To facilitate the narrative, we denote the polar coordinates of any $\mathbf{t} \in \mathbb{R}^2$ as (r, θ), where $r \geq 0$ and $\theta \in [0, 2\pi)$.

Definition 1.7.2. For any $f \in L^2(\mathbb{R}^2)$, the polar Fourier transform is defined by

$$\mathscr{F}\big[f(r, \theta)\big](\sigma, \phi) = \frac{1}{2\pi} \int_0^{2\pi} \int_0^\infty f(r, \theta)\, e^{-i\sigma r \cos(\theta - \phi)}\, r\, dr\, d\theta, \tag{1.297}$$

where $r, \sigma \geq 0$ and $\theta, \phi \in [0, 2\pi)$.

The inversion formula corresponding to the above defined polar Fourier transform reads:

$$f(r, \theta) = \frac{1}{2\pi} \int_0^{2\pi} \int_0^\infty \mathscr{F}\big[f(r, \theta)\big](\sigma, \phi)\, e^{i\sigma r \cos(\theta - \phi)}\, \sigma\, d\sigma\, d\phi. \tag{1.298}$$

In order to further explain the notion of polar Fourier transform, we present the following definition:

Definition 1.7.3. A function $f(r, \theta)$ is said to be "separable" in the polar coordinates if it can be expressed as

$$f(r, \theta) = f_r(r)\, f_\theta(\theta), \tag{1.299}$$

where $f_r(r)$ and $f_\theta(\theta)$ denote the radial and angular components, respectively.

A separable function satisfying $f_\theta(\theta) = 1$ is termed as a "circularly" or "radially symmetric" function and the corresponding polar Fourier transform is given by

$$\mathscr{F}\big[f(r, \theta)\big](\sigma, \phi) = \frac{1}{2\pi} \int_0^{2\pi} \int_0^\infty f(r)\, e^{-i\sigma r \cos(\theta - \phi)}\, r\, dr\, d\theta$$

$$= \int_0^\infty r\, f(r) \left\{ \frac{1}{2\pi} \int_0^{2\pi} e^{-i\sigma r \cos(\theta - \phi)}\, d\theta \right\} dr$$

$$= \int_0^\infty r\, f(r)\, J_0(\sigma r)\, dr, \tag{1.300}$$

where the last equality is obtained by using the integral definition of the zeroth-order Bessel function, given by

$$J_0(x) = \frac{1}{2\pi} \int_0^{2\pi} e^{-i\nu \cos x}\, dx, \tag{1.301}$$

which oscillates like a damped cosine. The integral transform given by

$$\mathscr{F}\!\left[f_r\right](\sigma) = \int_0^\infty r\, f(r)\, J_0(\sigma r)\, dr \tag{1.302}$$

is formally called as the "Hankel transform" of order zero or is sometimes also referred as the "Fourier-Bessel transform."

Example 1.7.1. Consider the two-dimensional circularly symmetric function

$$f(r,\theta) = f_r(r) = \mathrm{circ}(r) := \begin{cases} 1, & r < 1 \\ \dfrac{1}{2}, & r = 1 \\ 0, & r > 1. \end{cases} \tag{1.303}$$

In order to compute the polar Fourier transform of the function (1.303), we shall invoke (1.300) and proceed as

$$\mathscr{F}\!\left[f(r,\theta)\right](\sigma,\phi) = \int_0^\infty r\, \mathrm{circ}(r)\, J_0(\sigma r)\, dr = \int_0^1 r\, J_0(\sigma r)\, dr. \tag{1.304}$$

Making use of the substition $\sigma r = r'$, we have

$$\mathscr{F}\!\left[f(r,\theta)\right](\sigma,\phi) = \frac{1}{\sigma^2} \int_0^\sigma J_0(r')\, dr'. \tag{1.305}$$

Also, by virtue of the relationship

$$\int_0^\sigma x\, J_0(x)\, dx = \sigma\, J_1(\sigma), \tag{1.306}$$

we can express the polar Fourier transform of (1.303) as

$$\mathscr{F}\!\left[f(r,\theta)\right](\sigma,\phi) = \frac{J_1(\sigma)}{\sigma}, \tag{1.307}$$

where J_1 is the first-order Bessel function of the first kind, similar to a damped sinusoid. On the other hand, we note that the function $\mathrm{somb}(x) = 2J_1(x)/x$ is the well-known Sombrero (also known as "Bersinc" or "Jinc"). Therefore, from (1.307), we conclude that the Fourier transform of $\mathrm{circ}(r)$ is proportional to a Sombrero function of σ, the radial coordinate in the frequency space. Thus, we can write

$$\mathscr{F}\!\left[\mathrm{circ}(r)\right](\sigma,\phi) = \frac{\mathrm{somb}(\sigma)}{2}. \tag{1.308}$$

The circularly symmetric function (1.303) is depicted in Figure 1.40, whereas the corresponding polar Fourier transform is shown in Figure 1.41.

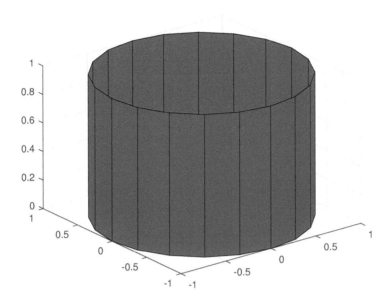

FIGURE 1.40: The 2D-circularly symmetric function (1.303).

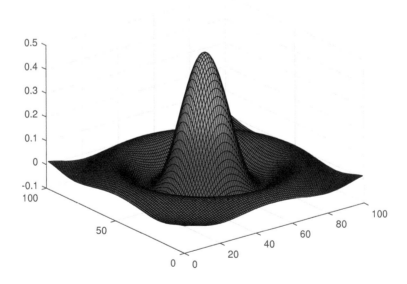

FIGURE 1.41: The polar Fourier transform of 2D-circularly symmetric function.

1.8   Exercises

Exercise 1.8.1. Compute the Fourier transform of the following pulses:

(i) *Pulse centred at origin:*

$$f(t) = \begin{cases} \sin(\omega_0 t), & -\frac{T}{2} \leq t \leq \frac{T}{2}, \quad T > 0 \\ 0, & \text{otherwise}, \end{cases} \tag{1.309}$$

where $\omega_0 = 4n\pi/T$, $n \in \mathbb{Z}$. Also find an expression for the corresponding amplitude spectrum $|\hat{f}(\omega)|$.

(ii) *Pulse centred at T_0:*

$$f(t) = \begin{cases} \sin(\omega_0 t), & -T_0 - \frac{T}{2} \leq t \leq -T_0 + \frac{T}{2}, \quad T > 0 \\ 0, & \text{otherwise}, \end{cases} \tag{1.310}$$

where $\omega_0 = 4n\pi/T$, $n \in \mathbb{Z}$. Choose T_0 such that $|\hat{f}(\omega)|$ is as large as possible for all values of ω. What is the full width at half maximum (FWHM) in this case?

Exercise 1.8.2. Compute the Fourier transform of

$$f(t) = \begin{cases} t e^{-\lambda t}, & t \geq 0, \ \lambda > 0 \\ 0, & \text{otherwise}. \end{cases} \tag{1.311}$$

Is the function (1.311) even, odd or mixed?

Exercise 1.8.3. By virtue of the Plancheral theorem, verify that

$$\int_0^\infty \frac{\sin^2(a\omega)}{\omega^2} \, d\omega = \frac{\pi a}{2}, \quad a > 0. \tag{1.312}$$

Exercise 1.8.4. Find the signals corresponding to the following Fourier transforms:

(i) $\hat{f}(\omega) = \dfrac{1}{7 + i\omega}$,

(ii) $\hat{f}(\omega) = \dfrac{1}{9 + \omega^2}$.

Exercise 1.8.5. Compute the Fourier transform of the impulse train $f(t) = \sum_n \delta(t - nT)$, where T is the period.

Exercise 1.8.6. Determine the Fourier transform of $f(t) = e^{-t/2}g(t)$ in terms of the Fourier transform of $g(t)$. Also, compute the Fourier transform in case g is the unit-step; that is,

$$g(t) = U(t) := \begin{cases} 1, & t \geq 0 \\ 0, & \text{otherwise}. \end{cases} \tag{1.313}$$

Exercise 1.8.7. Compute the Fourier transform corresponding to the function $f(t) = \dfrac{1}{1 + (3t)^2}$.

Exercise 1.8.8. The output $g(t)$ of a linear time invariant (LTI) system is related to the input $f(t)$ by the differential equation

$$\frac{dg(t)}{dt} + 2g(t) = f(t). \tag{1.314}$$

(i) Determine the frequency response $\widehat{H}(\omega) = \dfrac{\hat{g}(\omega)}{\hat{f}(\omega)}$.

(ii) If $f(t) = e^{-t}U(t)$, where $U(t)$ denotes the unit-step, then determine the Fourier transform of the output; that is, $\hat{g}(\omega)$.

(iii) If $\hat{g}(\omega)$ is as obtained in (ii), then find $g(t)$.

Exercise 1.8.9. If $U(t)$ denotes the unit-step, then compute the Fourier transform of $f(t) = e^{-\alpha t}U(t)$, $\alpha > 0$. Use the linearity and scaling properties to derive the Fourier transform of $g(t) = e^{-\alpha|t|} = f(t) + f(-t)$, $\alpha > 0$.

Exercise 1.8.10. If $U(t)$ is the unit-step, then compute the Fourier transform of each of the following signals:

(i) $f(t) = \left(e^{-\alpha t}\cos(\omega_0 t)\right)U(t)$, $\alpha > 0$ and $\omega_0 \in \mathbb{R}$,

(ii) $f(t) = e^{-3|t|}\sin(2t)$,

(iii) $f(t) = \left(\dfrac{\sin(\pi t)}{\pi t}\right)\left(\dfrac{\sin(2\pi t)}{\pi t}\right)$,

(iv) $f(t) = t\,e^{-at^2/2}$, $a > 0$,

(v) $f(t) = e^{-at^2 + 2bt}$, $a, b > 0$,

(vi) $f_n(t) = (-1)^n e^{t^2/2}\dfrac{d^n}{dt^n}\left(e^{-t^2}\right)$,

(vii) $f_n(t) = \dfrac{t^{n-1}}{(n-1)!}e^{-\alpha t}U(t)$,

(viii) $f(t) = \begin{cases} 1 - |t|, & |t| \le 1 \\ 0, & \text{otherwise,} \end{cases}$

(ix) $f(t) = \begin{cases} 1 - t^2, & |t| \le 1 \\ 0, & \text{otherwise.} \end{cases}$

Exercise 1.8.11. Evaluate the convolution expression (1.75) corresponding to a pair of L^2-normalized Gaussian functions given by (1.23) and verify whether the convolution function is L^2-normalized or not. Further, use Theorem 1.2.6 to obtain the Fourier transform of the resultant convolution function.

Exercise 1.8.12. Find the convolution $(f * g)(z)$ corresponding to the following pairs of functions:

(i) $f(t) = e^{at}$, $g(t) = \chi_{[0,\infty)}(t)$, $a \ne 0$,

(ii) $f(t) = \sin(bt)$, $g(t) = e^{-a|t|}$, $a > 0$,

(iii) $f(t) = \chi_{[a,\,b]}(t)$, $g(t) = t^2$, $a, b \in \mathbb{R}$,

(iv) $f(t) = \chi_{[-2,\,2]}(t)$, $g(t) = \chi_{[-2,\,2]}(t)$.

Exercise 1.8.13. Verify the following arguments:

(i) $(f * \delta)(z) = f(z)$,

(ii) $\dfrac{d}{dt}\big((f * g)(z)\big) = \left(\dfrac{df}{dt} * g\right)(z) = \left(f * \dfrac{dg}{dt}\right)(z)$,

(iii) If f and g are both even or both odd, then $(f * g)(z)$ is even,

(iv) If f is even or g is odd, or vice-versa, then $(f * g)(z)$ is odd.

Exercise 1.8.14. Use the Plancheral theorem to solve the below integrals with $a, b > 0$:

(i) $\displaystyle \int_{\mathbb{R}} \frac{dt}{(t^2 + a^2)^2}$,

(ii) $\displaystyle \int_{\mathbb{R}} \frac{e^{-bt^2}}{(t^2 + a^2)}\, dt$,

(iii) $\displaystyle \int_{\mathbb{R}} \frac{\sin(at)}{t(t^2 + b^2)}\, dt$.

Exercise 1.8.15. Compute the Fourier transform of the function

$$f(t) = \begin{cases} t^2 + 4t + 4, & -2 \le t \le -1 \\ 2 - t^2, & -1 \le t \le 1 \\ t^2 - 4t + 4, & 1 \le t \le 2 \\ 0, & \text{otherwise.} \end{cases} \tag{1.315}$$

Exercise 1.8.16. For the function

$$f(t) = \begin{cases} 1, & 0 \le t < 1 \\ 0, & \text{otherwise,} \end{cases} \tag{1.316}$$

show that the self-convolution takes the form

$$(f * f)(z) = \begin{cases} 1 - |z - 1|, & 0 \le z < 2 \\ 0, & \text{otherwise.} \end{cases} \tag{1.317}$$

Exercise 1.8.17. Suppose that the Fourier transform $\hat{f}(\omega)$ corresponding to a signal $f(t)$ vanishes outside the interval $[\omega_1, \omega_2]$. Develop a formula analogous to that of Theorem 1.2.10, where the time-interval of separation is $2\pi/(\omega_1 + \omega_2)$.

Hint: Show that the Fourier transform of $g(t) = \exp\left\{-\dfrac{it(\omega_1 + \omega_2)}{2}\right\} f(t)$ is band-limited to the interval $[-\Omega, \Omega]$ with $\Omega = (\omega_2 - \omega_1)/2$ and then apply Theorem 1.2.10 to the signal $g(t)$.

Exercise 1.8.18. Find the lower bound of the uncertainty product (1.123) for the following functions:

(i) $f(t) = e^{-t^2/2}$,

(ii) $f(t) = e^{-\alpha t^2/2} e^{i\omega_0 t}$, $\alpha > 0$, $\omega_0 \in \mathbb{R}$.

Note that the second term in (ii) is just a complex exponential with frequency ω_0, so it is a rotating phasor, but the first term puts a Gaussian envelope on its amplitude. Thus, it is a kind of localized tone burst with a well defined epoch and frequency.

Exercise 1.8.19. Compute the FrFT, LCT, SAFT and QPFT for the following signals:

(i) *Impulse:* $f(t) = K\,\delta(t)$, $K \in \mathbb{R}$,

(ii) *Harmonic function:* $f(t) = e^{2\pi i t_0 t}$, $t_0 \in \mathbb{R}$,

(iii) *General chirp:* $f(t) = e^{i\pi(at^2 + 2bt)}$, $a, b \in \mathbb{R}$,

(iv) *Hermite Gaussian:* $f_n(t) = \dfrac{2^{1/4}}{\sqrt{2^n n!}} H_n(\sqrt{2\pi}\,t)\, e^{-\pi t^2}$, where $H_n(t)$ is the n-th order Hermite polynomial defined as

$$H_n(t) = (-1)^n e^{t^2/2} \frac{d^n}{dt^n}\left(e^{-t^2/2}\right), \tag{1.318}$$

(v) *General Gaussian:* $f(t) = e^{-\pi(at^2 + 2bt)}$, $a > 0$, $b \in \mathbb{R}$.

Exercise 1.8.20. Compute the two-dimensional Fourier transform of the following functions:

(i) $f(t_1, t_2) = \delta(t_1, t_2) := \delta(t_1)\,\delta(t_2)$,

(ii) $f(t_1, t_2) = \dfrac{\delta(t_1, t_2 - t_0) - \delta(t_1, t_2 + t_0)}{2}$, $t_0 \in \mathbb{R}$,

(iii) $f(r) = \dfrac{1}{2\pi\sigma^2} e^{-r^2/2\sigma^2}$, $\sigma \neq 0$ and $r^2 = t_1^2 + t_2^2$.

2

The Windowed Fourier Transforms

"Motivated by "quantum mechanics," in 1946 the physicist Gabor defined elementary time-frequency atoms as waveforms that have a minimal spread in the time-frequency plane. To measure time-frequency "information" content, he proposed decomposing signals over these elementary atomic waveforms. By showing that such decompositions are closely related to our sensitivity to sounds, and that they exhibit important structures in speech and music recordings, Gabor demonstrated the importance of localized time-frequency signal processing."

-Stéphane G. Mallat

2.1 Introduction

Most of the signals are non-stationary in nature and a complete representation of these signals requires frequency analysis that is local in time, resulting in the time-frequency analysis. The major breakthrough in the context of time-frequency analysis came in the form of the well-known Gabor transform [116], which deals with the decomposition of non-transient signals in terms of time and frequency shifted basis functions, known as "Gabor window functions." With the aid of these window functions, one can analyze the spectral contents of non-transient signals in localized neighbourhoods of time. Owing to the lucid nature and close resemblance with the classical Fourier transform, the Gabor-type transforms have attracted substantial interest during past few decades with numerous applications to various branches of science and engineering, including harmonic analysis, signal and image processing, pseudo-differential operators, sampling theory, quantum mechanics, geophysics, astrophysics and medicine [14, 53, 76, 94, 125, 136, 147, 224, 225, 233].

In case of Fourier transform, the analyzing components are the sinusoidal waves which continue to vibrate throughout the time axis; therefore, it cannot be used for the frequency analysis that is local in time because it requires all previous as well as future information about the signal to evaluate its spectral density at a single frequency component ω. Thus, non-transient signals are not equally well represented under the Fourier transform. To achieve the localization of spectral characteristics of a time varying signal, a window function is introduced into the Fourier transform. A window function $g(t)$ is a function in $L^2(\mathbb{R})$ such that both $g(t)$ and $\hat{g}(\omega)$ have rapid decay, that is, $g(t)$ is well localized in time domain, while $\hat{g}(\omega)$ is well localized in frequency domain. Multiplying a signal $f(t)$ by a window function $g(t)$ before its Fourier transform is computed has the effect of restricting the spectral information of the signal to the domain of influence of the window function. Using the translates of the window function on the time axis to cover the entire time-domain, the signal is analyzed for spectral information in localized neighbourhood's in time. This methodology is formally referred to as "Short-time Fourier transform" (STFT) or "Windowed Fourier transform" or sometimes as "Running window Fourier transform."

DOI: 10.1201/9781003175766-2

Apart from the notion of windowed Fourier transform, many novel time-frequency tools have also appeared over the course of time, which are aimed to harness the advantages of the generalized Fourier transforms, including the fractional Fourier, linear canonical, special affine Fourier and quadratic-phase Fourier transforms [28, 166, 260, 263, 273, 289, 331].

This chapter is exclusively devoted for a detailed analysis of the windowed Fourier transforms. Most importantly, we shall take a survey for analyzing the nature of diverse types of window functions frequently employed in the literature and applications. Besides the fundamental notion of windowed Fourier transform, we shall study the recent ramifications including the windowed fractional Fourier transform, windowed linear canonical transform, windowed special affine Fourier transform and windowed quadratic-phase Fourier transform. Nevertheless, the chapter is concluded with the much recent notion of directional windowed Fourier transform, which opens up a new research prospect in the context of windowed Fourier transforms. For further readings on the subject, we refer to the articles [120, 124, 166, 273, 289, 331] and the references therein.

2.2   The Windowed Fourier Transform

"If we consider a piece of music...and if a note, an *A* for instance, appears once in that piece, Fourier analysis will yield the corresponding frequency with a certain amplitude and a certain phase, without localizing the *A* in time. Clearly the *A* will not be heard at certain instants. Yet the representation is mathematically correct, because the phases of the neighbouring notes conspire to suppress the *A* by interference when it is not heard and to enhance it, again by interference, when it is heard. However, although this conception shows a skilfulness that honours mathematical analysis, one should not hide the fact that it also distorts reality: indeed, when the *A* is not heard, the true reason is that the *A* is not emitted."

-Jean A. Ville

In this section, we shall present a detailed study regarding the windowed Fourier transform and formulate the fundamental properties. The centrepiece of this section is the analysis of different window functions, especially with regard to certain standard parameters such as side-lobe suppression and 3-dB bandwidth of the window functions.

2.2.1   Analysis of window functions

Window functions are used in harmonic analysis to reduce the undesirable effects related to the spectral leakage. They impact on many attributes of a harmonic processor including detestability, resolution, dynamics range and case of implementation. In signal analysis, the use of a particular window function depends upon the requirements in a particular application. As such, it is interesting to analyze the characteristics of window functions including their main-lobes and side-lobes. Any typical window function satisfies the following properties:

(i) It is real, even, non-negative and time-limited.

(ii) The Fourier transform has a main-lobe at the origin and side-lobes on both sides. The side-lobes are decaying with asymptotic attenuation.

As the window functions are taken as even functions, so the corresponding Fourier transforms do not have any imaginary part. Usually in applications, the window functions are non-negative, smooth, bell-shaped curves, however, rectangular pulse, triangular pulse and other functions can also be used as windowing functions. Nevertheless, a more general definition of the window function does not require them to be identically zero outside a specific interval as long as the product of the function multiplied by its argument is square integrable and more specifically, that the function goes sufficiently rapidly towards zero. Here, we shall present a detailed analysis of some prominent window functions and also compare their respective window characteristics. For a lucid comparison of the window qualities, we use logarithmic representations covering equal ranges. That is also the reason why we can't have negative function values. To make sure they do not occur, we shall use the power representation; that is, the squared modulus of the spectrum. For a detailed perspective regarding the notion of windows and their applications, we refer to [57, 117, 121, 161, 168, 224, 233].

Definition 2.2.1. (Window Function): A non-zero function $g \in L^2(\mathbb{R})$ is said to be a "window function" if $t\,g(t) \in L^2(\mathbb{R})$.

The centre (mean) and radius (dispersion about mean) of a window function g are denoted by E_g and Δ_g, respectively, and are defined as

$$E_g = \frac{1}{\|g\|_2^2} \int_{\mathbb{R}} t\,|g(t)|^2 dt \quad \text{and} \quad \Delta_g = \frac{1}{\|g\|_2} \left\{ \int_{\mathbb{R}} (t - E_g)^2 |g(t)|^2 dt \right\}^{1/2}. \tag{2.1}$$

Analogous to (2.1), the centre and radius of $\hat{g}$ are given by

$$E_{\hat{g}} = \frac{1}{\|g\|_2^2} \int_{\mathbb{R}} \omega\,|\hat{g}(\omega)|^2 d\omega \quad \text{and} \quad \Delta_{\hat{g}} = \frac{1}{\|g\|_2} \left\{ \int_{\mathbb{R}} (\omega - E_{\hat{g}})^2 |\hat{g}(\omega)|^2 d\omega \right\}^{1/2}. \tag{2.2}$$

Next, our goal is to exemplify certain window functions, which are of substantial importance in practical aspects of signal processing. Nevertheless, each example shall be followed by a detailed analysis of the window qualities, particularly their respective side-lobe suppression and 3-dB bandwidth.

Example 2.2.1. (The Rectangular Window): The rectangular pulse is one of the simplest window functions in signal processing and is defined by

$$g(t) = \begin{cases} 1, & -\frac{T}{2} \le t \le \frac{T}{2}, \quad T > 0 \\ 0, & \text{otherwise.} \end{cases} \tag{2.3}$$

The centre and radius for the rectangular window function (2.3) are obtained as follows:

$$E_g = \frac{1}{\|g\|_2^2} \int_{\mathbb{R}} t\,|g(t)|^2 dt = \frac{1}{\|g\|_2^2} \int_{-T/2}^{T/2} t\, dt = \frac{1}{T} \left(\frac{t^2}{2} \right)_{-T/2}^{T/2} = \frac{1}{2T} \left(\frac{T^2}{4} - \frac{T^2}{4} \right) = 0$$

and

$$\Delta_g = \frac{1}{\|g\|_2} \left\{ \int_{\mathbb{R}} (t - 0)^2 |g(t)|^2 dt \right\}^{1/2} = \frac{1}{\sqrt{T}} \left\{ \int_{-T/2}^{T/2} t^2\, dt \right\}^{1/2} = \frac{1}{\sqrt{T}} \left(\frac{T^3}{12} \right)^{1/2} = \frac{T}{2\sqrt{3}}.$$

In order to examine the window qualities of the rectangular pulse, we recall that the Fourier transform of (2.3) is obtained in (1.12) and is given by

$$\hat{g}(\omega) = \frac{T}{\sqrt{2\pi}} \left(\frac{\sin\left(\frac{\omega T}{2}\right)}{\left(\frac{\omega T}{2}\right)} \right) = \frac{T}{\sqrt{2\pi}} \operatorname{sinc}\left(\frac{\omega T}{2} \right). \tag{2.4}$$

Consequently, the power representation of the Fourier transform (2.4) is

$$|\hat{g}(\omega)|^2 = \frac{T^2}{2\pi} \left(\frac{\sin\left(\frac{\omega T}{2}\right)}{\left(\frac{\omega T}{2}\right)} \right)^2. \tag{2.5}$$

A disadvantage of the rectangular pulse (2.3) is that it is discontinuous and gives a bad localization in the frequency domain as the spectrum decreases as $1/\omega$. Such limitations are avoided by using more smooth functions. For $T = 1$, the window function $g(t)$ and the power representation $|\hat{g}(\omega)|^2$ (measured in decibels) are plotted in Figure 2.1.

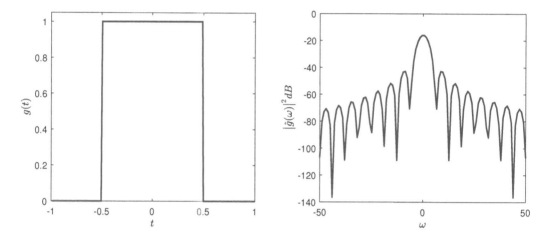

FIGURE 2.1: The rectangular window and its power representation.

The zeros of the function (2.4) or (2.5) are equidistant and are given by $\omega T/2 = \pm n\pi, n \in \mathbb{N}$ or $\omega = \pm 2n\pi/T$. Note that, 0 is not a zero of the function (2.4) as it tends to 1, whenever ω approaches to 0. Next, our motive is to find out the proportion of intensity at the central peak and also get an intuition of how much it gets lost in the sidebands (side-lobes). To achieve this, we need the first zero at $\omega = \pm 2\pi/T$ and compute the following integral:

$$
\begin{aligned}
\int_{-2\pi/T}^{2\pi/T} |\hat{g}(\omega)|^2 \, d\omega &= \int_{-2\pi/T}^{2\pi/T} \frac{T^2}{2\pi} \left(\frac{\sin\left(\frac{\omega T}{2}\right)}{\left(\frac{\omega T}{2}\right)} \right)^2 d\omega \\
&= \int_{-\pi}^{\pi} \left(\frac{T^2}{2\pi} \right) \left(\frac{2}{T} \right) \left(\frac{\sin^2 \eta}{\eta^2} \right) d\eta \\
&= \frac{2T}{\pi} \int_0^{\pi} \frac{\sin^2 \eta}{\eta^2} \, d\eta \\
&= \frac{2T}{\pi} \left\{ \left(-\frac{\sin^2 \eta}{\eta} \right)_0^{\pi} - \int_0^{\pi} \frac{2\sin\eta\cos\eta}{-\eta} \, d\eta \right\} \\
&= \frac{2T}{\pi} \int_0^{\pi} \frac{\sin 2\eta}{\eta} \, d\eta \\
&= \frac{4T}{\pi} \int_0^{\pi} \frac{\sin 2\eta}{2\eta} \, d\eta = \frac{2T}{\pi} \, \mathrm{Si}(2\pi), \tag{2.6}
\end{aligned}
$$

where $\text{Si}(x)$ denotes the integral:

$$\text{Si}(x) = \int_0^x \frac{\sin y}{y} \, dy. \tag{2.7}$$

To compute the total intensity, we shall invoke the Parseval's formula for the Fourier transform (1.70) as

$$\text{Total intensity} = \int_{\mathbb{R}} |\hat{g}(\omega)|^2 d\omega = \int_{\mathbb{R}} |g(t)|^2 dt = \int_{-T/2}^{T/2} dt = T. \tag{2.8}$$

Therefore, the ratio of the intensity at the central peak to the total intensity is given by (2.6) and (2.8) as

$$\frac{\text{Intensity at the central peak}}{\text{Total intensity}} = \frac{2T \, \text{Si}(2\pi)}{\pi} \left(\frac{1}{T}\right) = \frac{2 \, \text{Si}(2\pi)}{\pi} = 0.903. \tag{2.9}$$

From the estimate (2.9), we conclude that approximately 90% of the intensity is in the central peak, whereas the remaining proportion of around 10% is wasted in side-lobes.

Next, we examine the side-lobe suppression and 3-dB bandwidth pertaining to the rectangular window (2.3).

(A) Side-lobe suppression

To determine the height of the first side-lobe, we consider the following differential equation:

$$\frac{d}{d\omega}\left(|\hat{g}(\omega)|\right) = 0 \quad \text{or} \quad \frac{d}{d\omega}\left(\hat{g}(\omega)\right) = 0.$$

That is,

$$\frac{d}{d\omega}\left(\frac{T}{\sqrt{2\pi}} \left(\frac{\sin\left(\frac{\omega T}{2}\right)}{\left(\frac{\omega T}{2}\right)}\right)\right) = 0.$$

As such, we have

$$\frac{T}{\sqrt{2\pi}} \left\{ \frac{\left(\frac{\omega T}{2}\right)\cos\left(\frac{\omega T}{2}\right)\left(\frac{T}{2}\right) - \sin\left(\frac{\omega T}{2}\right)\left(\frac{T}{2}\right)}{\left(\frac{\omega T}{2}\right)^2} \right\} = 0,$$

or equivalently

$$\omega \left(\frac{T}{2}\right)^2 \cos\left(\frac{\omega T}{2}\right) - \left(\frac{T}{2}\right) \sin\left(\frac{\omega T}{2}\right) = 0.$$

The above equation further yields

$$\omega = \left(\frac{2}{T}\right) \tan\left(\frac{\omega T}{2}\right) \quad \text{or} \quad \left(\frac{\omega T}{2}\right) = \tan\left(\frac{\omega T}{2}\right). \tag{2.10}$$

Solving the transcendental equation (2.10) (graphically or using computational software), we get

$$\frac{\omega T}{2} = 0, \quad \frac{\omega T}{2} \approx \pm 10.90412, \ \pm 7.725251, \ \pm 4.49340, \ 14.066193.$$

The smallest possible realistic solution is $\omega T/2 = 4.4934$, which implies that

$$\omega = 2\left(\frac{4.4934}{T}\right) = \frac{8.9868}{T}. \tag{2.11}$$

Using the value of ω given by (2.11) in (2.5) yields

$$\left|\hat{g}\left(\frac{8.9868}{T}\right)\right|^2 = \frac{T^2}{2\pi}\left(\frac{\sin\left(4.4934\right)}{\left(4.4934\right)}\right)^2 = \frac{T^2}{2\pi}\left(-\frac{0.976118}{4.4934}\right)^2 = \frac{T^2}{2\pi}(0.04719). \tag{2.12}$$

Also, for $\omega = 0$, we have

$$\left|\hat{g}\left(0\right)\right|^2 = \frac{T^2}{2\pi}. \tag{2.13}$$

Moreover, upon dividing (2.12) by (2.13), we get

$$\frac{\text{Height of first side-lobe}}{\text{Height of central peak}} = \frac{\left|\hat{g}\left(8.9868/T\right)\right|^2}{\left|\hat{g}\left(0\right)\right|^2} = 0.04719 \tag{2.14}$$

The numeric entity (2.14) gives an intuition of the side-lobe suppression; however, it is customary to express the ratio in decibels (dB). To do so, we have to evaluate ten times the base-10 logarithm of the ratio (2.14) (as per the definition of decibel, a ratio of two measurements of a same type of power quantity can be expressed as a level in decibels by evaluating 10-times the base-10 logarithm). Therefore, the first side-lobe suppression is given by

$$10 \log_{10}\left(0.04719\right) = -13.2 \, \text{dB}. \tag{2.15}$$

(B) 3-dB bandwidth

Since $10 \log_{10}(1/2) = -3.0103 \approx -3$, therefore, the 3-dB bandwidth tells us where the central peak has dropped to half of its height. In order to calculate the 3-dB bandwidth, we simply need to solve the following equation:

$$\frac{\left|\hat{g}(\omega)\right|^2}{\left|\hat{g}(0)\right|^2} = \frac{1}{2}. \tag{2.16}$$

Therefore, we have

$$\frac{T^2}{2\pi}\left(\frac{\sin\left(\frac{\omega T}{2}\right)}{\left(\frac{\omega T}{2}\right)}\right)^2 = \frac{1}{2}\left(\frac{T^2}{2\pi}\right).$$

Or

$$\sin\left(\frac{\omega T}{2}\right) = \frac{1}{\sqrt{2}}\left(\frac{\omega T}{2}\right). \tag{2.17}$$

The solutions of (2.17) are obtained to be

$$\frac{\omega T}{2} \approx -1.39156, \quad \frac{\omega T}{2} = 0, \quad \frac{\omega T}{2} \approx 1.39156.$$

However, the feasible solution is $\omega T/2 = 1.39156$. Thus, the 3-dB bandwidth ω_{3dB} is given by

$$\omega_{3dB} = \frac{2.78312}{T}. \tag{2.18}$$

Therefore, the total width Δ_ω is given by

$$\Delta_\omega = 2\,\omega_{3dB} = \frac{5.56624}{T}. \tag{2.19}$$

This is in fact the slimmest central peak one can get using the Fourier transform. Any other window function will lead to larger 3-dB bandwidths.

Example 2.2.2. (The Triangular Window): The triangular function is a slightly better window function which offers good localization in the frequency domain as the spectrum decreases as $1/\omega^2$. The triangular function is defined by

$$g(t) = \begin{cases} t + \dfrac{T}{2}, & -\dfrac{T}{2} \le t \le 0 \\[2mm] \dfrac{T}{2} - t, & 0 \le t \le \dfrac{T}{2} \\[2mm] 0, & \text{otherwise}, \end{cases} \qquad T > 0. \tag{2.20}$$

Here, we shall compute the Fourier transform of the triangular window by expressing (2.20) as a convolution of the rectangular window functions:

$$g_1(x) = g_2(x) = \begin{cases} 1, & -\dfrac{T}{4} \le x \le \dfrac{T}{4}, \quad T > 0 \\ 0, & \text{otherwise}. \end{cases}$$

Note that

$$\big(g_1 * g_2\big)(t) = \int_{\mathbb{R}} g_1(x)\, g_2(t - x)\, dx,$$

where

$$g_2(t - x) = \begin{cases} 1, & t - \dfrac{T}{4} \le x \le t + \dfrac{T}{4} \\ 0, & \text{otherwise}. \end{cases}$$

Consequently, we have

$$\big(g_1 * g_2\big)(t) = \begin{cases} \displaystyle\int_{-T/4}^{T/4+t} dx, & -\dfrac{T}{2} \le t \le 0 \\[4mm] \displaystyle\int_{t-T/4}^{T/4} dx, & 0 \le t \le \dfrac{T}{2} \end{cases} = \begin{cases} \left(t + \dfrac{T}{4} + \dfrac{T}{4}\right), & -\dfrac{T}{2} \le t \le 0 \\[4mm] \left(\dfrac{T}{4} - t + \dfrac{T}{4}\right), & 0 \le t \le \dfrac{T}{2} \end{cases} = g(t).$$

Thus, the Fourier transform of $g(t)$ can be computed by employing Theorem 1.2.6 together with (2.4) as

$$\hat{g}(\omega) = \sqrt{2\pi}\, \hat{g}_1(\omega)\, \hat{g}_2(\omega)$$

$$= \sqrt{2\pi} \left(\frac{T}{2\sqrt{2\pi}} \frac{\sin\left(\frac{\omega T}{4}\right)}{\left(\frac{\omega T}{4}\right)} \right) \left(\frac{T}{2\sqrt{2\pi}} \frac{\sin\left(\frac{\omega T}{4}\right)}{\left(\frac{\omega T}{4}\right)} \right)$$

$$= \frac{T^2}{4\sqrt{2\pi}} \left(\frac{\sin\left(\frac{\omega T}{4}\right)}{\left(\frac{\omega T}{4}\right)} \right)^2. \tag{2.21}$$

Consequently, the power representation of the Fourier transform (2.21) is given by

$$\left| \hat{g}(\omega) \right|^2 = \frac{1}{2\pi} \left(\frac{T^4}{16} \right) \left(\frac{\sin\left(\frac{\omega T}{4}\right)}{\left(\frac{\omega T}{4}\right)} \right)^4. \tag{2.22}$$

For $T = 1$, the window function (2.20) and the power representation (2.22) are plotted in Figure 2.2.

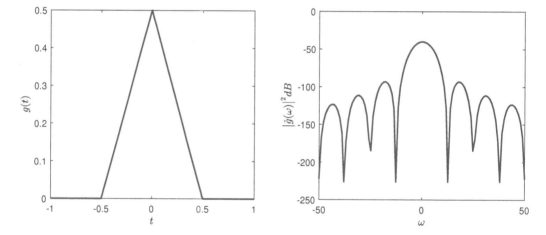

FIGURE 2.2: The triangular window and its power representation.

The zeros of (2.22) are given by $\omega T/4 = \pm n\pi$ or $\omega = \pm 4n\pi/T$, $n \in \mathbb{N}$, which are indeed equidistant, but are twice as much distant as that of the rectangular function (2.3). In order to find the intensity at the central peak, we proceed as follows:

$$\int_{-4\pi/T}^{4\pi/T} \left| \hat{g}(\omega) \right|^2 d\omega = \frac{1}{2\pi} \int_{-4\pi/T}^{4\pi/T} \frac{T^4}{16} \left(\frac{\sin\left(\frac{\omega T}{4}\right)}{\left(\frac{\omega T}{4}\right)} \right)^4 d\omega$$

$$= \frac{4}{2\pi T} \left(\frac{T^4}{16} \right) \int_{-\pi}^{\pi} \left(\frac{\sin\eta}{\eta} \right)^4 d\eta$$

$$= \frac{T^3}{8\pi} (2.08823). \tag{2.23}$$

Also, the total intensity is given by

$$\text{Total intensity} = \int_{\mathbb{R}} \left| \hat{g}(\omega) \right|^2 d\omega$$

$$= \frac{1}{2\pi} \left(\frac{T^4}{16} \right) \int_{\mathbb{R}} \left(\frac{\sin \left(\frac{\omega T}{4} \right)}{\left(\frac{\omega T}{4} \right)} \right)^4 d\omega$$

$$= \frac{4}{2\pi T} \left(\frac{T^4}{16} \right) \int_{\mathbb{R}} \left(\frac{\sin \eta}{\eta} \right)^4 d\eta$$

$$= \frac{T^3}{8\pi} \left(\frac{2\pi}{3} \right) \approx \frac{T^3}{8\pi} (2.0944).$$

Therefore, the ratio of the intensity at the central peak to the total intensity is given by

$$\frac{\text{Intensity at the central peak}}{\text{Total intensity}} = \frac{T^3 (2.0944)}{8\pi} \left(\frac{8\pi}{T^3 (2.0944)} \right) = 0.99705405. \qquad (2.24)$$

From (2.24), we conclude that approximately 99.7% of the intensity is in the central peak and the remaining minute proportion is wasted in the side-lobes.

Next, we analyze the side-lobe suppression and 3-dB bandwidth pertaining to the triangular window.

(A) Side-lobe suppression

To determine the height of the first side-lobe, we consider the following differential equation:

$$\frac{d}{d\omega} \left(|\hat{g}(\omega)| \right) = 0 \quad \text{or} \quad \frac{d}{d\omega} \left(\hat{g}(\omega) \right) = 0.$$

That is;

$$\frac{d}{d\omega} \left(\left(\frac{\sin \left(\frac{\omega T}{4} \right)}{\left(\frac{\omega T}{4} \right)} \right)^2 \right) = 0,$$

so that

$$2 \left(\frac{\sin \left(\frac{\omega T}{4} \right)}{\left(\frac{\omega T}{4} \right)} \right) \left\{ \frac{\left(\frac{T}{4} \right) \left(\frac{\omega T}{4} \right) \cos \left(\frac{\omega T}{4} \right) - \left(\frac{T}{4} \right) \sin \left(\frac{\omega T}{4} \right)}{\left(\frac{\omega T}{4} \right)^2} \right\} = 0$$

or

$$2 \sin \left(\frac{\omega T}{4} \right) \left\{ \left(\frac{T}{4} \right) \left(\frac{\omega T}{4} \right) \cos \left(\frac{\omega T}{4} \right) - \left(\frac{T}{4} \right) \sin \left(\frac{\omega T}{4} \right) \right\} = 0.$$

The above equation further yields

$$2 \left(\frac{T}{4} \right) \left\{ \left(\frac{\omega T}{4} \right) \sin \left(\frac{\omega T}{4} \right) \cos \left(\frac{\omega T}{4} \right) - \sin^2 \left(\frac{\omega T}{4} \right) \right\} = 0.$$

That is,

$$\left(\frac{\omega T}{4} \right) \cos \left(\frac{\omega T}{4} \right) = \sin \left(\frac{\omega T}{4} \right)$$

or

$$\left(\frac{\omega T}{4} \right) = \tan \left(\frac{\omega T}{4} \right). \qquad (2.25)$$

The numerical solutions of the transcendental equation (2.25) are given by

$$\frac{\omega T}{4} = 0, \quad \frac{\omega T}{4} \approx \pm 10.90412, \ \pm 7.725251, \ \pm 4.49340, \ 14.0066193$$

and the smallest possible feasible solution is $\omega T/4 = 4.49340$, which implies that

$$\omega T = 17.9736 \quad \text{or} \quad \omega = \frac{17.9736}{T}. \tag{2.26}$$

Using the numerical estimate given by (2.26) in (2.22), we obtain

$$
\begin{aligned}
\left| \hat{g}\left(\frac{17.9736}{T} \right) \right|^2 &= \frac{1}{2\pi} \left(\frac{T^4}{16} \right) \left(\frac{\sin(4.49340)}{4.49340} \right)^4 \\
&= \frac{1}{2\pi} \left(\frac{T^4}{16} \right) \left(-\frac{0.976118}{4.49340} \right)^4 \\
&= \frac{1}{2\pi} \left(\frac{T^4}{16} \right) (0.04719049)^2 = \frac{1}{2\pi} \left(\frac{T^4}{16} \right) (0.00222694).
\end{aligned}
\tag{2.27}
$$

Also, for $\omega = 0$, we have

$$\left| \hat{g}\left(0 \right) \right|^2 = \frac{1}{2\pi} \left(\frac{T^4}{16} \right). \tag{2.28}$$

Upon dividing (2.27) by (2.28), we obtain

$$\frac{\text{Height of first side-lobe}}{\text{Height of central peak}} = \frac{\left| \hat{g}\left(17.9736/T \right) \right|^2}{\left| \hat{g}\left(0 \right) \right|^2} = 0.00222694. \tag{2.29}$$

Hence, the first side-lobe suppression is given by the following numerical entity:

$$10 \log_{10}(0.00222694) = 10(-2.6523) = -26.523. \tag{2.30}$$

Therefore, we conclude that the height of the first side-lobe is suppressed by -26.523 dB approximately. Here, we emphasize that the first side-lobe suppression pertaining to the triangular window could also have been calculated by using the first side-lobe suppression of the rectangular window function. Since the zeros in case of the triangular window are twice as much distant as that of rectangular window and the first side-lobe suppression of this window is -13.262 dB approximately, hence, the first side-lobe suppression of the triangular window is $2(-13.262) = -26.524$ approximately.

(B) 3-dB bandwidth

Solving the equation $\left| \hat{g}(\omega) \right|^2 = \frac{1}{2} \left| \hat{g}\left(0 \right) \right|^2$ yields

$$\frac{1}{2\pi} \left(\frac{T^4}{16} \right) \left(\frac{\sin\left(\frac{\omega T}{4} \right)}{\left(\frac{\omega T}{4} \right)} \right)^4 = \frac{1}{2} \left(\frac{1}{2\pi} \left(\frac{T^4}{16} \right) \right).$$

That is,

$$\sin\left(\frac{\omega T}{4} \right) = \frac{1}{2^{1/4}} \left(\frac{\omega T}{4} \right). \tag{2.31}$$

The numerical solutions of the transcendental equation (2.31) are given by

$$\frac{\omega T}{4} \approx -1.00191, \quad \frac{\omega T}{4} = 0, \quad \frac{\omega T}{4} \approx 1.00191,$$

with $\omega T/4 = 1.00191$ being the smallest feasible solution. Thus, the 3-dB bandwidth ω_{3dB} is given by

$$\omega_{3dB} = \frac{4(1.00191)}{T} = \frac{4.00764}{T}. \tag{2.32}$$

Therefore, the total width Δ_ω is given by

$$\Delta_\omega = 2\,\omega_{3dB} = \frac{8.015284}{T}. \tag{2.33}$$

Note that the total width of the triangular window function (2.20) in the spectral domain is $\frac{8.01528}{5.56624} = 1.43998103$ times wider than the rectangular window function (2.3).

Example 2.2.3. (The Cosine Window): The triangular window function had a kink when switching on, next kink at the peak and another kink when switching off. Unlike the triangular window, the cosine window avoids the kink at the peak. The cosine window function is defined by

$$g(t) = \begin{cases} \cos\left(\dfrac{\pi t}{T}\right), & -\dfrac{T}{2} \le t \le \dfrac{T}{2}, \quad T > 0 \\ 0, & \text{otherwise.} \end{cases} \tag{2.34}$$

Recall that the Fourier transform corresponding to (2.34) is computed in (1.31), which reads:

$$\hat{g}(\omega) = \frac{T}{\sqrt{2\pi}} \cos\left(\frac{\omega T}{2}\right) \left\{ \frac{1}{(\pi - \omega T)} + \frac{1}{(\pi + \omega T)} \right\}. \tag{2.35}$$

Therefore, we have

$$|\hat{g}(\omega)|^2 = \frac{T^2}{2\pi} \cos^2\left(\frac{\omega T}{2}\right) \left(\frac{2\pi}{\pi^2 - (\omega T)^2}\right)^2. \tag{2.36}$$

For $T = 1$, the cosine window function (2.34) and the power representation of the corresponding Fourier transform given by (2.36) are depicted in Figure 2.3. Furthermore, for $\omega = 0$, we have

$$\hat{g}(0) = \frac{T}{\sqrt{2\pi}} \left(\frac{2\pi}{\pi^2}\right) = \frac{1}{\sqrt{2\pi}} \left(\frac{2T}{\pi}\right).$$

Also,

$$\lim_{\omega T \to \pm\pi} \hat{g}(\omega) = \lim_{\omega T \to \pm\pi} \left\{ \frac{T}{\sqrt{2\pi}} \left(\frac{2\pi \cos\left(\frac{\omega T}{2}\right)}{(\pi - \omega T)(\pi + \omega T)}\right) \right\}$$

$$= \frac{2\pi T}{\sqrt{2\pi}} \lim_{\omega T \to \pm\pi} \left\{ -\frac{\left(\frac{T}{2}\right) \sin\left(\frac{\omega T}{2}\right)}{T(\pi - \omega T) + T(\pi + \omega T)} \right\} = -\frac{T}{2\sqrt{2\pi}}.$$

Thus, we infer that neither $\omega = 0$ nor $\omega = \pm\pi$ is the zero of the function $|\hat{g}(\omega)|$ or $|\hat{g}(\omega)|^2$. In fact, the zeros are given by

$$\frac{\omega T}{2} = \frac{(2n+1)\pi}{2} \quad \text{or} \quad \omega = \frac{(2n+1)\pi}{T}, \quad n \in \mathbb{N},$$

which are spaced at the same distance as in the case of rectangular window. Here, it is worth mentioning that for all practical purposes, the intensity at the central peak is approximately 100%. However, we should have another look at the side-lobes because of the minorities, viz; the chance of detecting additional weak signals. The suppression of the first side-lobe can be calculated by solving

$$\tan\left(\frac{\omega T}{2}\right) = \frac{4\omega T}{\pi^2 - (\omega T)^2},$$

yielding the smallest feasible numerical solution as $\omega \approx 11.87114/T$. This results in a side-lobe suppression of -23 dB, not a lot worse than that of the triangular window. Also, the 3-dB bandwidth is quite better than the triangular window and the total width amounts to

$$\Delta_\omega = \frac{7.47}{T}. \tag{2.37}$$

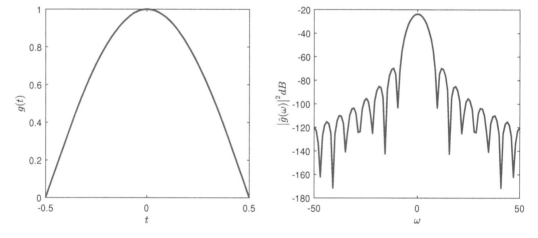

FIGURE 2.3: The cosine window and its power representation.

From the above discussion, we conclude that it is quite lucrative to endorse the quest for more intelligent window functions to meet the diversified needs of signal processing. We continue the analysis of window functions with some of the mostly applied and practically oriented window functions.

Example 2.2.4. (The Hanning Window): Recall that the cosine window function has a kink while switching on (at $-T/2$) and another kink while switching off (at $T/2$). The Austrian meteorologist Julis von Haan $(1839 - 1921)$ introduced a new window function by eliminating the kinks in the cosine window and proposed the $\cos^2$-window which is often referred as the "Hanning window/Haan function" or "Haan smoothing function." The Hanning window function is defined by

$$g(t) = \begin{cases} \dfrac{1}{2}\left(1 + \cos\left(\dfrac{2\pi t}{T}\right)\right) = \cos^2\left(\dfrac{\pi t}{T}\right), & -\dfrac{T}{2} \le t \le \dfrac{T}{2}, \quad T > 0 \\ 0, & \text{otherwise.} \end{cases} \tag{2.38}$$

Recall that the Fourier transform of the Hanning window (2.38) is computed in (1.33) and is given by

$$\hat{g}(\omega) = \frac{T}{4\sqrt{2\pi}} \sin\left(\frac{\omega T}{2}\right) \left\{ \frac{1}{\left(\pi - \frac{\omega T}{2}\right)} - \frac{1}{\left(\pi + \frac{\omega T}{2}\right)} + \frac{2}{\left(\frac{\omega T}{2}\right)} \right\}, \tag{2.39}$$

so that the power representation takes the following form:

$$\left|\hat{g}(\omega)\right|^2 = \frac{T^2}{16(2\pi)} \sin^2\left(\frac{\omega T}{2}\right) \left\{ \frac{1}{\left(\pi - \frac{\omega T}{2}\right)} - \frac{1}{\left(\pi + \frac{\omega T}{2}\right)} + \frac{2}{\left(\frac{\omega T}{2}\right)} \right\}^2. \tag{2.40}$$

For $T = 1$, the window function (2.38) and the power representation (2.40) are plotted in Figure 2.4. Clearly, $\omega = 0$ is not a zero of (2.39) or (2.40) because $\sin\left(\frac{\omega T}{2}\right)/\left(\frac{\omega T}{2}\right) \to 1$ as $\omega \to 0$. Also, $\omega T/2 = \pm\pi$ or $\omega = \pm 2\pi/T$ is not a zero of either (2.39) or (2.40) for the same reasons. In fact, the zeros are given by

$$\frac{\omega T}{2} = \pm n\pi \quad \text{or} \quad \omega = \pm\frac{2n\pi}{T}, \quad n = 2, 3, \ldots.$$

The intensity at the central peak is 100% approximately and the suppression of the first side-lobe is -32 dB. Moreover, the total bandwidth is given by

$$\Delta_\omega = 2\,\omega_{3\,dB} = \frac{9.06}{T}, \tag{2.41}$$

where $\omega_{3\,dB}$ is the usual 3-dB bandwidth. Thus, we observe that the side-lobe suppression for the Hanning window (2.38) is considerably higher than the cosine window (2.34) and the 3-dB bandwidth of the former is also higher than the later. This is actually the effect of increasing the power of the cosine function which in turn results in absorbing more and more zeros near the central peak and thereby corresponds to a gain in the side-lobe suppression. Nevertheless, higher powers of cosine function can also be used to meet the practical needs. For instance, in case of $\cos^3$-window, we have $\Delta_\omega = 10.4/T$, whereas, in case of $\cos^4$-window, the total bandwidth is $\Delta_\omega = 11.66/T$. However, we shall shortly see that there are more intelligent solutions to this problem.

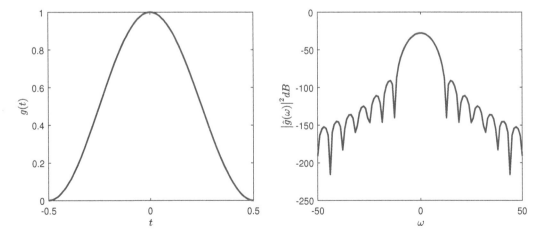

FIGURE 2.4: The Hanning window and its power representation.

Example 2.2.5. (The Hamming Window): Proposed by R.W. Hamming $(1915-1998)$, an American mathematician whose work had many implications for computer engineering and telecommunications, the Hamming window has proved to be a nice alternative to the Hanning window by adjoining some additional parameters to it. For a given real parameter α, the Hamming window function is defined by

$$g(t) = \begin{cases} \alpha + (1-\alpha)\cos^2\left(\dfrac{\pi t}{T}\right), & -\dfrac{T}{2} \le t \le \dfrac{T}{2}, \quad T > 0 \\ 0, & \text{otherwise.} \end{cases} \tag{2.42}$$

In view of (1.35), the Fourier transform corresponding to the Hamming window (2.42) is

$$\hat{g}(\omega) = \frac{T}{4\sqrt{2\pi}}\sin\left(\frac{\omega T}{2}\right)\left\{\frac{1-\alpha}{\left(\pi - \frac{\omega T}{2}\right)} - \frac{1-\alpha}{\left(\pi + \frac{\omega T}{2}\right)} + \frac{2(1+\alpha)}{\left(\frac{\omega T}{2}\right)}\right\}. \tag{2.43}$$

The parameter α involved in the definition of the Hamming window proves to be handy as the same can be used to adjust the side-lobes without affecting on the zeros of (2.43). Practically, $\alpha \approx 0.1$ often proves to be a good choice. Indeed, the zeros are given by

$$\frac{\omega T}{2} = \pm n\pi \quad \text{or} \quad \omega = \pm \frac{2n\pi}{T}, \quad n = 2, 3, \dots.$$

Although, the parameter α has saddled us with the term $\dfrac{T\alpha}{2\sqrt{2\pi}}\left(\dfrac{\sin\left(\frac{\omega T}{2}\right)}{\left(\frac{\omega T}{2}\right)}\right)$, however, it gets added to the side-lobes of the Hamming window. Note that, a squaring of $\hat{g}(\omega)$ is not needed here. On one hand, this will correspond to interference terms of the Fourier transform of the Hamming window, but on the other hand, the same is true for $\hat{g}(\omega)$; here, all we get are positive and negative side-lobes. We note that the absolute values of the side-lobes' heights don't change. For $\alpha = 0.15$ and $T = 1$, the window function (2.42) and $\left|\hat{g}(\omega)\right|^2$ are plotted in Figure 2.5. The first side-lobes are slightly smaller than the second ones. For the optimal parameter $\alpha = 0.08$, the side-lobe suppression is -43 dB and the total bandwidth is $\Delta_\omega = 8.17/T$.

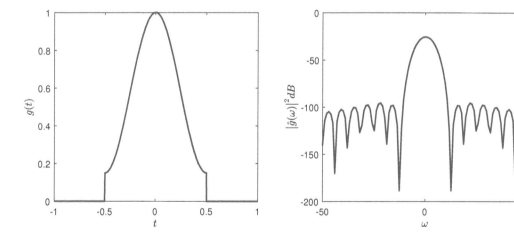

FIGURE 2.5: The Hamming window and its power representation.

Hold the breath, let us look at a literally peculiar window function that creates no side-lobes at all.

Example 2.2.6. (The Triplet Window): For a given parameter $\gamma > 0$, the triplet window function is defined by

$$g(t) = \begin{cases} e^{-\gamma|t|} \cos^2\left(\dfrac{\pi t}{T}\right), & -\dfrac{T}{2} \le t \le \dfrac{T}{2}, \quad T > 0 \\ 0, & \text{otherwise.} \end{cases} \tag{2.44}$$

A lengthy but trivial computation implies that the Fourier transform of the triplet window stands out, as it features oscillating terms (sine and cosine) though there are no more zeros. For $\gamma = 10$ and $T = 1$, the triplet window (2.44) and the corresponding power representation $|\hat{g}(\omega)|^2$ are depicted in Figure 2.6. With regard to the parameter γ, it is imperative to mention that in case γ is chosen to be large enough, then even there won't be any local minima or maxima and the Fourier transform $\hat{g}(\omega)$ decays monotonically. Nevertheless, when the parameter γ is optimized, a total bandwidth of $\Delta_\omega = 9.7/T$ is achieved. Therefore, the idea of re-introducing a spike at $t = 0$ is quite handy, however, there are somewhat better window functions.

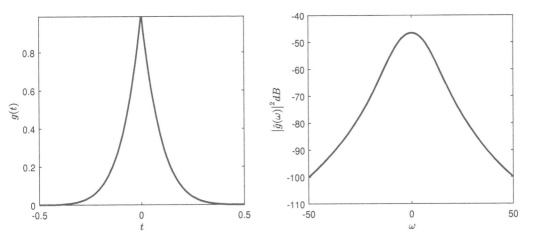

FIGURE 2.6: The triplet window and its power representation.

Example 2.2.7. (The Gaussian Window): The Gaussian window is one of the simplest, yet elegant, window function, which is obtained by truncating the usual Gaussian function as

$$g(t) = \begin{cases} e^{-t^2/2\sigma^2}, & -\dfrac{T}{2} \le t \le \dfrac{T}{2}, \quad T > 0 \\ 0, & \text{otherwise.} \end{cases} \tag{2.45}$$

The Fourier transform corresponding to the Gaussian window (2.45) is be obtained to be

$$\hat{g}(\omega) = \frac{\sigma}{2} \exp\left\{-\frac{\sigma^2\omega^2}{2}\right\} \left\{ \operatorname{erfi}\left(\frac{T}{2\sqrt{2}\,\sigma} - \frac{i\sigma\omega}{\sqrt{2}}\right) + \operatorname{erfi}\left(\frac{T}{2\sqrt{2}\,\sigma} + \frac{i\sigma\omega}{\sqrt{2}}\right) \right\}, \tag{2.46}$$

where $\operatorname{erfi}(\cdot)$ is the complex error function. Note that the error function appears in (2.46) in complex arguments, although together with the conjugate complex argument, $\hat{g}(\omega)$ is real. Choosing $T = 1$ and $\sigma = 1/6$, we plot the Gaussian window (2.45) and the power representation of the Fourier transform $|\hat{g}(\omega)|^2$ in Figure 2.7. A major disadvantage of the

Gaussian window function is that the side-lobes are not decaying monotonically for all values of σ. If σ is chosen sufficiently large, the side-lobes will disappear: The oscillations "creep up" the Gaussian functions' flank shortly before this happens, we get a total width of $\Delta_\omega = 9.06/T$ and a side-lobe suppression of -26 dB.

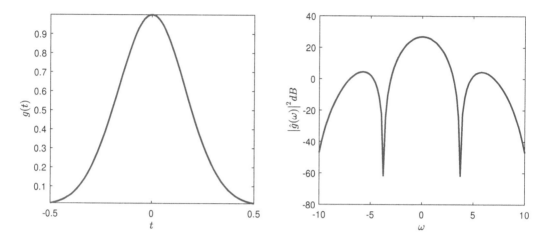

FIGURE 2.7: The Gaussian window and its power representation.

Example 2.2.8. (The Kaiser-Bessel Window): The Kaiser-Bessel window is one of the mostly applied window functions and is defined by

$$g(t) = \begin{cases} \dfrac{I_0\left(\beta\sqrt{1 - \left(\frac{2t}{T}\right)^2}\right)}{I_0(\beta)}, & -\dfrac{T}{2} \le t \le \dfrac{T}{2}, \quad T > 0 \\ 0, & \text{otherwise,} \end{cases} \tag{2.47}$$

where

$$I_0(t) = \sum_{m=0}^{\infty} \frac{1}{(m\,!)^2} \left(\frac{t}{2}\right)^{2m}, \tag{2.48}$$

is the zero-order modified Bessel function of first kind and β is a parameter which may be chosen at will. The Fourier transform of the Kaiser-Bessel window (2.47) is given by

$$\hat{g}(\omega) = \begin{cases} \dfrac{T}{I_0(\beta)\sqrt{2\pi}} \left\{ \dfrac{\sinh\left(\sqrt{\beta^2 - \dfrac{\omega^2 T^2}{4}}\right)}{\sqrt{\beta^2 - \dfrac{\omega^2 T^2}{4}}} \right\}, & \beta > \left|\dfrac{\omega T}{2}\right| \\[3em] \dfrac{T}{I_0(\beta)\sqrt{2\pi}} \left\{ \dfrac{\sin\left(\sqrt{\dfrac{\omega^2 T^2}{4} - \beta^2}\right)}{\sqrt{\dfrac{\omega^2 T^2}{4} - \beta^2}} \right\}, & \beta \le \left|\dfrac{\omega T}{2}\right|. \end{cases} \tag{2.49}$$

An algorithm for the calculation of $I_0(x)$ can be found in [3] as

$$I_0(x) = 1 + (3.5156229)\,t^2 + (3.0899424)\,t^4 + (1.2067492)\,t^6 + (0.2659732)\,t^8 \\ + (0.0360768)\,t^{10} + (0.0045813)\,t^{12} + \varepsilon, \quad |\varepsilon| < 1.6 \times 10^{-7} \Big\}$$

with $t = x/3.75$ for the interval $-3.75 \leq x \leq 3.75$;

$$
\left. \begin{aligned}
x^{1/2} e^{-x} I_0(x) = {} & 0.39894228 + (0.01328592)\,t^{-1} + (0.00225319)\,t^{-2} - (0.00157565)\,t^{-3} \\
& + (0.00916281)\,t^{-4} - (0.02057706)\,t^{-5} + (0.02635537)\,t^{-6} - (0.01647633)\,t^{-7} \\
& + (0.00392377)\,t^{-8} + \varepsilon, \quad |\varepsilon| < 1.9 \times 10^{-7}
\end{aligned} \right\}
$$

with $t = x/3.75$ for the interval $3.75 \leq x < \infty$. Moreover, the zeros are not equidistant as they are located at $(\omega T/2)^2 = (n\pi)^2 + \beta^2$, $n = 1, 2, 3, \ldots$. It is interesting to note that for $\beta = 0$, the Kaiser-Bessel window boils down to the rectangular window function. Practically, the values of β up to $\beta = 9$ are recommended. Choosing $T = 1$ and varying β over a set of values, the Kaiser-Bessel window (2.47) and $|\hat{g}(\omega)|^2$ are plotted in Figures 2.8–2.15. It is pertinent to mention that for $\beta = 9$, the side-lobe suppression is -70 dB whereas the total bandwidth is $\Delta_\omega = 11/T$. Practically, the Kaiser-Bessel window is superior to the Gaussian window function in all aspects.

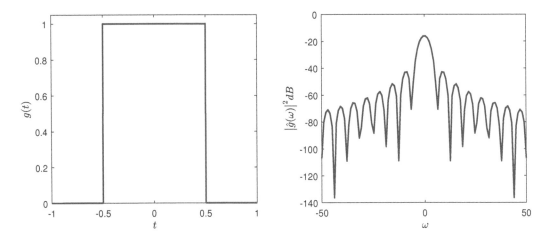

FIGURE 2.8: The Kaiser-Bessel window and its power representation for $\beta = 0$.

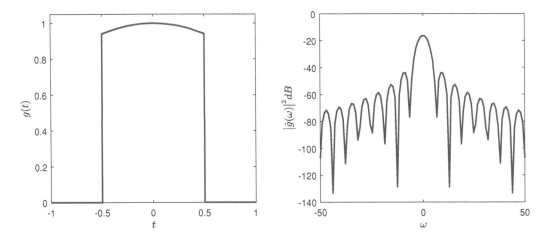

FIGURE 2.9: The Kaiser-Bessel window and its power representation for $\beta = 0.5$.

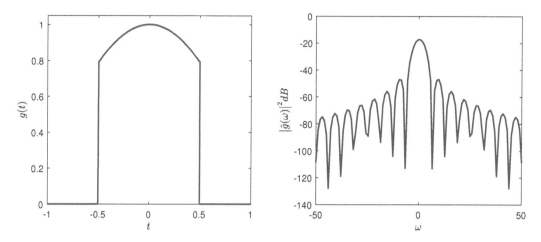

FIGURE 2.10: The Kaiser-Bessel window and its power representation for $\beta = 1$.

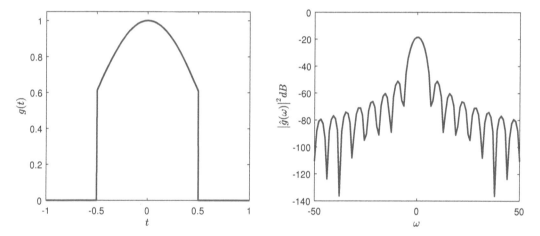

FIGURE 2.11: The Kaiser-Bessel window and its power representation for $\beta = 1.5$.

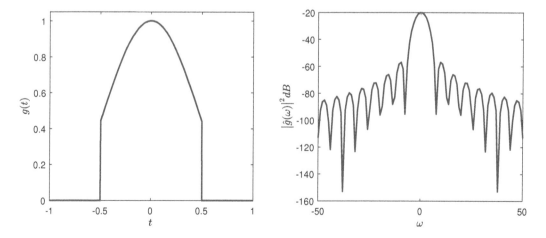

FIGURE 2.12: The Kaiser-Bessel window and its power representation for $\beta = 2$.

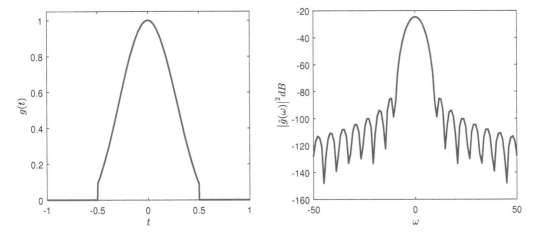

FIGURE 2.13: The Kaiser-Bessel window and its power representation for $\beta = 4$.

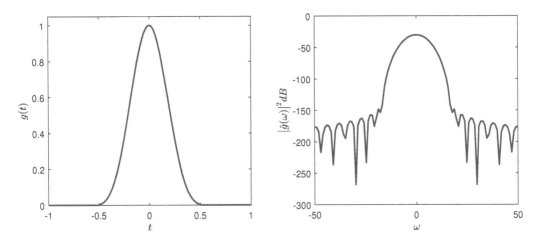

FIGURE 2.14: The Kaiser-Bessel window and its power representation for $\beta = 8$.

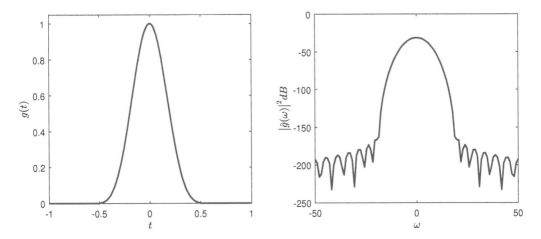

FIGURE 2.15: The Kaiser-Bessel window and its power representation for $\beta = 9$.

2.2.2 Definition and basic properties

In this subsection, we present the formal definition of the windowed Fourier transform (WFT) and study the basic properties thereon.

Definition 2.2.2. Given a window function $g \in L^2(\mathbb{R})$, the windowed Fourier transform of any $f \in L^2(\mathbb{R})$ is denoted by $\mathscr{G}_g[f]$ and is defined as

$$\mathscr{G}_g[f](b,\omega) = \frac{1}{\sqrt{2\pi}} \int_{\mathbb{R}} f(t)\,\overline{g(t-b)}\,e^{-i\omega t}\,dt, \quad b,\omega \in \mathbb{R}. \tag{2.50}$$

In view of Definition 2.2.2, we make the following comments regarding the notion of windowed Fourier transform:

(i) Unlike the usual Fourier transform, the windowed Fourier transform of $f \in L^2(\mathbb{R})$ provides the spectral information of $f(t)$ at a particular instant of time. Besides, the WFT needs to know the signal f only over the interval wherein $g(t-b)$ is non-zero.

(ii) Definition 2.2.2 can also be expressed as

$$\mathscr{G}_g[f](b,\omega) = \frac{1}{\sqrt{2\pi}} \big\langle f, g_{b,\omega} \big\rangle_2 = \frac{1}{\sqrt{2\pi}} \int_{\mathbb{R}} f(t)\,\overline{g_{b,\omega}(t)}\,dt, \tag{2.51}$$

where $g_{b,\omega}(t) = g(t-b)\,e^{i\omega t}$. The members of the family $\{g_{b,\omega}(t);\ b,\omega \in \mathbb{R}\}$ are often referred as "Gabor functions" or "canonical coherent states" in quantum physics.

(iii) The windowed Fourier transform (2.50) is based on the concept of performing Fourier transform over smaller sections of a given signal. Sliding the window function along the spatial domain has the effect of covering up the entire time domain, so that the signal is analyzed for spectral information in localized neighbourhood's in time.

(iv) When the window function g is chosen to be the Gaussian function, the windowed Fourier transform is referred to as the "Gabor transform."

(v) If $\mathcal{T}_b$ and $\mathcal{M}_\omega$ denote the usual translation and modulation operators, then the windowed Fourier transform (2.50) can be alternatively written as

$$\mathscr{G}_g[f](b,\omega) = \frac{1}{\sqrt{2\pi}} \big\langle \mathcal{M}_{-\omega} f, \mathcal{T}_b g \big\rangle_2. \tag{2.52}$$

(vi) The windowed Fourier transform (2.50) can be expressed via the classical convolution operation $*$ given by (1.75) as follows:

$$\begin{aligned}
\mathscr{G}_g[f](b,\omega) &= \frac{1}{\sqrt{2\pi}} \int_{\mathbb{R}} f(t)\,\overline{g(t-b)}\,e^{-i\omega t}\,dt \\
&= \frac{1}{\sqrt{2\pi}} \int_{\mathbb{R}} \mathcal{M}_{-\omega} f(t)\,\overline{\breve{g}(b-t)}\,dt \\
&= \frac{1}{\sqrt{2\pi}} \big((\mathcal{M}_{-\omega} f) * \overline{\breve{g}} \big)(b), \quad \breve{g}(t) = g(-t),
\end{aligned} \tag{2.53}$$

where $\mathcal{M}_\omega$ denotes the usual modulation operator.

In the following proposition, we express the windowed Fourier transform (2.50) via the usual Fourier operator. Subsequently, we demonstrate that indeed the windowed Fourier transform offers a joint resolution of time and frequency components of a signal in the time-frequency plane.

Proposition 2.2.1. *The windowed Fourier transform of any function $f \in L^2(\mathbb{R})$ with respect to a window function $g \in L^2(\mathbb{R})$ satisfies:*

$$\mathscr{G}_g\big[f\big](b,\omega) = e^{-i\omega b}\,\mathscr{F}^{-1}\Big(\hat{f}(\xi)\,\overline{\hat{g}(\omega - \xi)}\Big)(b), \tag{2.54}$$

where $\mathscr{F}^{-1}$ denotes the inverse Fourier transform defined in (1.67).

Proof. In view of (2.51) and the Parseval's formula (1.69) for the Fourier transform, we have

$$\mathscr{G}_g\big[f\big](b,\omega) = \frac{1}{\sqrt{2\pi}}\,\big\langle f,\,g_{b,\omega}\big\rangle_2$$

$$= \frac{1}{\sqrt{2\pi}}\,\big\langle \hat{f},\,\hat{g}_{b,\omega}\big\rangle_2$$

$$= \frac{1}{\sqrt{2\pi}}\int_{\mathbb{R}} \hat{f}(\xi)\,\overline{\hat{g}_{b,\omega}(\xi)}\,d\xi$$

$$= \frac{1}{\sqrt{2\pi}}\int_{\mathbb{R}} \hat{f}(\xi)\left\{\frac{1}{\sqrt{2\pi}}\int_{\mathbb{R}} \overline{g_{b,\omega}(t)\,e^{-i\xi t}}\,dt\right\}d\xi$$

$$= \frac{1}{2\pi}\int_{\mathbb{R}}\int_{\mathbb{R}} \hat{f}(\xi)\,\overline{g(t-b)}\,e^{i(\xi-\omega)t}\,dt\,d\xi$$

$$= \frac{1}{2\pi}\int_{\mathbb{R}} \hat{f}(\xi)\left\{\int_{\mathbb{R}} \overline{g(z)}\,e^{i(\xi-\omega)(b+z)}\,dz\right\}d\xi$$

$$= \frac{1}{2\pi}\int_{\mathbb{R}} \hat{f}(\xi)\,e^{i(\xi-\omega)b}\left\{\int_{\mathbb{R}} \overline{g(z)}\,e^{i(\xi-\omega)z}\,dz\right\}d\xi$$

$$= \frac{e^{-i\omega b}}{\sqrt{2\pi}}\int_{\mathbb{R}} \hat{f}(\xi)\left\{\frac{1}{\sqrt{2\pi}}\int_{\mathbb{R}} \overline{g(z)}\,e^{-i(\omega-\xi)z}\,dz\right\}e^{i\xi b}\,d\xi$$

$$= \frac{e^{-i\omega b}}{\sqrt{2\pi}}\int_{\mathbb{R}} \hat{f}(\xi)\,\overline{\hat{g}(\omega-\xi)}\,e^{i\xi b}\,d\xi$$

$$= e^{-i\omega b}\,\mathscr{F}^{-1}\Big(\hat{f}(\xi)\,\overline{\hat{g}(\omega-\xi)}\Big)(b).$$

This completes the proof of Proposition 2.2.1. $\qquad\square$

The elementary properties of the windowed Fourier transform (2.50) are encapsulated in the following theorem:

Theorem 2.2.2. *If g and h are two window functions and $\alpha,\ \beta \in \mathbb{C}$, $k, \omega_0 \in \mathbb{R}$, $\lambda \in \mathbb{R}\setminus\{0\}$ are scalars, then for any pair of square integrable functions f and s, the windowed Fourier transform (2.50) satisfies the following properties:*

(i) *Linearity:* $\mathscr{G}_g\big[\alpha f + \beta s\big](b,\omega) = \alpha\mathscr{G}_g\big[f\big](a,b) + \beta\mathscr{G}_g\big[s\big](b,\omega),$

(ii) *Anti-linearity:* $\mathscr{G}_{\alpha g+\beta h}\big[f\big](b,\omega) = \overline{\alpha}\,\mathscr{G}_g\big[f\big](b,\omega) + \overline{\beta}\,\mathscr{G}_h\big[f\big](b,\omega),$

(iii) *Time-shift:* $\mathscr{G}_g\big[f(t-k)\big](b,\omega) = e^{-i\omega k}\mathscr{G}_g\big[f\big](b-k,\omega),$

(iv) *Modulation:* $\mathscr{G}_g\big[e^{i\omega_0 t}f(t)\big](b,\omega) = \mathscr{G}_g\big[f\big](b,\omega-\omega_0),$

(v) *Scaling:* $\mathscr{G}_g\big[f(\lambda t)\big](b,\omega) = \dfrac{1}{|\lambda|}\,\mathscr{G}_{\tilde{g}}\big[f\big]\left(\lambda b,\dfrac{\omega}{\lambda}\right),\quad \tilde{g}(t) = g\left(\dfrac{t}{\lambda}\right),$

(vi) *Conjugation:* $\mathscr{G}_g\left[\bar{f}\right](b,\omega) = \overline{\mathscr{G}_{\bar{g}}\left[f\right](b,-\omega)}$,

(vii) *Symmetry:* $\mathscr{G}_g\left[f\right](b,\omega) = e^{-i\omega b}\,\mathscr{G}_{\bar{f}}\left[\bar{g}\right](-b,\omega)$,

(viii) *Translation in window:* $\mathscr{G}_{g(t-k)}\left[f\right](b,\omega) = \mathscr{G}_g\left[f\right](b+k,\omega)$,

(ix) *Scaling in window:* $\mathscr{G}_{g(\lambda t)}\left[f\right](b,\omega) = \dfrac{1}{|\lambda|}\,\mathscr{G}_g\left[\tilde{f}\right]\left(\lambda b, \dfrac{\omega}{\lambda}\right), \quad \tilde{f}(t) = f\left(\dfrac{t}{\lambda}\right).$

Proof. The proof of (i) and (ii) follows directly by using the fact that the inner product $\langle \cdot, \cdot \rangle_2$ is linear in the first component and anti-linear in the second component. Next, we proceed to prove the rest of the properties.

(iii) If the input signal $f \in L^2(\mathbb{R})$ is time-shifted by any $k \in \mathbb{R}$, then we have

$$\mathscr{G}_g\left[f(t-k)\right](b,\omega) = \frac{1}{\sqrt{2\pi}}\int_{\mathbb{R}} f(t-k)\,\overline{g(t-b)}\,e^{-i\omega t}\,dt$$

$$= \frac{1}{\sqrt{2\pi}}\int_{\mathbb{R}} f(z)\,\overline{g(z-(b-k))}\,e^{-i\omega(z+k)}\,dz$$

$$= \frac{e^{-i\omega k}}{\sqrt{2\pi}}\int_{\mathbb{R}} f(z)\,\overline{g(z-(b-k))}\,e^{-i\omega z}\,dz$$

$$= e^{-i\omega k}\,\mathscr{G}_g\left[f\right](b-k,\omega).$$

(iv) After modulating the input signal $f \in L^2(\mathbb{R})$ with $\omega_0 \in \mathbb{R}$, the windowed Fourier transform can be computed as

$$\mathscr{G}_g\left[e^{i\omega_0 t} f(t)\right](b,\omega) = \frac{1}{\sqrt{2\pi}}\int_{\mathbb{R}} e^{i\omega_0 t} f(t)\,\overline{g(t-b)}\,e^{-i\omega t}\,dt$$

$$= \frac{1}{\sqrt{2\pi}}\int_{\mathbb{R}} f(t)\,\overline{g(t-b)}\,e^{-i(\omega-\omega_0)t}\,dt$$

$$= \mathscr{G}_g\left[f\right](b,\omega-\omega_0).$$

(v) For $\lambda \in \mathbb{R} \setminus \{0\}$, we observe that

$$\mathscr{G}_g\left[f(\lambda t)\right](b,\omega) = \frac{1}{\sqrt{2\pi}}\int_{\mathbb{R}} f(\lambda t)\,\overline{g(t-b)}\,e^{-i\omega t}\,dt$$

$$= \frac{1}{\sqrt{2\pi}}\int_{\mathbb{R}} \frac{1}{|\lambda|} f(z)\,\overline{g\left(\frac{z}{\lambda}-b\right)}\,e^{-i\omega(z/\lambda)}\,dz$$

$$= \frac{1}{|\lambda|}\frac{1}{\sqrt{2\pi}}\int_{\mathbb{R}} f(z)\,\overline{g\left(\frac{z-\lambda b}{\lambda}\right)}\,e^{-i(\omega/\lambda)z}\,dz$$

$$= \frac{1}{|\lambda|}\,\mathscr{G}_{\tilde{g}}\left[f\right]\left(\lambda b, \frac{\omega}{\lambda}\right), \quad \tilde{g}(t) = g\left(\frac{t}{\lambda}\right).$$

(vi) Upon complex-conjugating the input, the windowed Fourier transform becomes

$$\mathscr{G}_g\left[\bar{f}\right](b,\omega) = \frac{1}{\sqrt{2\pi}}\int_{\mathbb{R}} \overline{f(t)}\,\overline{g(t-b)}\,e^{-i\omega t}\,dt$$

$$= \overline{\frac{1}{\sqrt{2\pi}}\int_{\mathbb{R}} f(t)\,g(t-b)\,e^{-i(-\omega)t}\,dt}$$

$$= \overline{\mathscr{G}_{\bar{g}}\left[f\right](b,-\omega)}.$$

(vii) Observe that,

$$\mathscr{G}_g\big[f\big](b,\omega) = \frac{1}{\sqrt{2\pi}} \int_{\mathbb{R}} f(t)\,\overline{g(t-b)}\,e^{-i\omega t}\,dt$$

$$= \frac{1}{\sqrt{2\pi}} \int_{\mathbb{R}} f(b+z)\,\overline{g(z)}\,e^{-i\omega(b+z)}\,dz$$

$$= \frac{e^{-i\omega b}}{\sqrt{2\pi}} \int_{\mathbb{R}} f\big(z-(-b)\big)\,\overline{g(z)}\,e^{-i\omega z}\,dz$$

$$= e^{-i\omega b}\,\mathscr{G}_{\bar{f}}\big[\bar{g}\big](-b,\omega).$$

(viii) If the window function $g \in L^2(\mathbb{R})$ is shifted by $k \in \mathbb{R}$, then

$$\mathscr{G}_{g(t-k)}\big[f\big](b,\omega) = \frac{1}{\sqrt{2\pi}} \int_{\mathbb{R}} f(t)\,\overline{g(t-b-k)}\,e^{-i\omega t}\,dt$$

$$= \frac{1}{\sqrt{2\pi}} \int_{\mathbb{R}} f(t)\,\overline{g\big(t-(b+k)\big)}\,e^{-i\omega t}\,dt$$

$$= \mathscr{G}_g\big[f\big](b+k,\omega).$$

(ix) Finally, if the window function $g \in L^2(\mathbb{R})$ is scaled by $\lambda \in \mathbb{R} \setminus \{0\}$, then

$$\mathscr{G}_{g(\lambda t)}\big[f\big](b,\omega) = \frac{1}{\sqrt{2\pi}} \int_{\mathbb{R}} f(t)\,\overline{g\big(\lambda(t-b)\big)}\,e^{-i\omega t}\,dt$$

$$= \frac{1}{\sqrt{2\pi}} \int_{\mathbb{R}} f(t)\,\overline{g(\lambda t - \lambda b)}\,e^{-i\omega t}\,dt$$

$$= \frac{1}{|\lambda|}\frac{1}{\sqrt{2\pi}} \int_{\mathbb{R}} f\left(\frac{z}{\lambda}\right)\,\overline{g(z-\lambda b)}\,e^{-i\omega(z/\lambda)}\,dz$$

$$= \frac{1}{|\lambda|}\,\mathscr{G}_g\big[\tilde{f}\big]\left(\lambda b,\frac{\omega}{\lambda}\right), \quad \tilde{f}(t) = f\left(\frac{t}{\lambda}\right).$$

This completes the proof of Theorem 2.2.2. $\qquad\square$

In the next theorem, we establish an orthogonality relation between two signals and their respective windowed Fourier transforms. Subsequently, we deduce that the windowed Fourier transform (2.50) conserves the energy of the signal from the natural domain to the transformed domain. More precisely, we shall demonstrate that the net concentration of windowed Fourier transform $\mathscr{G}_g\big[f\big](b,\omega)$ in the time-frequency plane is precisely the product of the respective concentrations of the input function $f(t)$ and the window function $g(t)$.

Theorem 2.2.3. *Let $\mathscr{G}_g\big[f\big](b,\omega)$ and $\mathscr{G}_g\big[h\big](b,\omega)$ be the windowed Fourier transforms of the square integrable functions f and h, respectively. Then, we have*

$$\int_{\mathbb{R}}\int_{\mathbb{R}} \mathscr{G}_g\big[f\big](b,\omega)\,\overline{\mathscr{G}_g\big[h\big](b,\omega)}\,db\,d\omega = \big\|g\big\|_2^2 \big\langle f,\,h\big\rangle_2. \qquad (2.55)$$

Proof. Applying Proposition 2.2.1, we have

$$\int_{\mathbb{R}}\int_{\mathbb{R}} \mathscr{G}_g\big[f\big](b,\omega)\,\overline{\mathscr{G}_g\big[h\big](b,\omega)}\,db\,d\omega$$

$$= \int_{\mathbb{R}}\int_{\mathbb{R}} \left\{\frac{e^{-i\omega b}}{\sqrt{2\pi}} \int_{\mathbb{R}} \hat{f}(\xi)\,\overline{\hat{g}(\omega-\xi)}\,e^{i\xi b}\,d\xi\right\} \left\{\frac{e^{i\omega b}}{\sqrt{2\pi}} \int_{\mathbb{R}} \overline{\hat{f}(\eta)}\,\hat{g}(\omega-\eta)\,e^{-i\eta b}\,d\eta\right\} db\,d\omega$$

$$= \frac{1}{2\pi} \int_{\mathbb{R}} \int_{\mathbb{R}} \int_{\mathbb{R}} \int_{\mathbb{R}} \hat{f}(\xi)\, \overline{\hat{f}(\eta)}\, \overline{\hat{g}(\omega - \xi)}\, e^{i\xi b}\, \hat{g}(\omega - \eta)\, e^{-i\eta b}\, db\, d\omega\, d\xi\, d\eta$$

$$= \int_{\mathbb{R}} \int_{\mathbb{R}} \int_{\mathbb{R}} \hat{f}(\xi)\, \overline{\hat{h}(\eta)}\, \overline{\hat{g}(\omega - \xi)}\, \hat{g}(\omega - \eta) \left\{ \frac{1}{2\pi} \int_{\mathbb{R}} e^{i(\xi - \eta)b}\, db \right\} d\omega\, d\xi\, d\eta$$

$$= \int_{\mathbb{R}} \int_{\mathbb{R}} \int_{\mathbb{R}} \hat{f}(\xi)\, \overline{\hat{h}(\eta)}\, \overline{\hat{g}(\omega - \xi)}\, \hat{g}(\omega - \eta)\, \delta(\xi - \eta)\, d\omega\, d\xi\, d\eta$$

$$= \int_{\mathbb{R}} \int_{\mathbb{R}} \hat{f}(\xi)\, \overline{\hat{h}(\xi)}\, \left|\hat{g}(\omega - \xi)\right|^2 d\omega\, d\xi$$

$$= \int_{\mathbb{R}} \hat{f}(\xi)\, \overline{\hat{h}(\xi)} \left\{ \int_{\mathbb{R}} \left|\hat{g}(\omega - \xi)\right|^2 d\omega \right\} d\xi.$$

Invoking translation invariance of the Lebesgue integral, we have

$$\int_{\mathbb{R}} \int_{\mathbb{R}} \mathcal{G}_g\big[f\big](b,\omega)\, \overline{\mathcal{G}_g\big[h\big](b,\omega)}\, db\, d\omega = \left\|g\right\|_2^2 \int_{\mathbb{R}} \hat{f}(\xi)\, \overline{\hat{h}(\xi)}\, d\xi$$

$$= \left\|g\right\|_2^2 \left\langle \hat{f},\, \hat{h} \right\rangle_2$$

$$= \left\|g\right\|_2^2 \left\langle f,\, h \right\rangle_2.$$

This completes the proof of Theorem 2.2.3. □

Corollary 2.2.4. *For $f = h$, the orthogonality relation (2.55) yields the following energy preserving relation:*

$$\int_{\mathbb{R}} \int_{\mathbb{R}} \left| \mathcal{G}_g\big[f\big](b,\omega) \right|^2 db\, d\omega = \left\|f\right\|_2^2 \left\|g\right\|_2^2. \tag{2.56}$$

In particular, for $\left\|g\right\|_2 = 1$, the windowed Fourier transform is an isometry from the space of functions $L^2(\mathbb{R})$ to the space of transformations $L^2(\mathbb{R}^2)$.

The inversion formula pertaining to the windowed Fourier transform (2.50) is obtained in the next theorem. Here, it is imperative to mention that a detailed perspective on different inversion formulae for the windowed Fourier transform can be found in [107].

Theorem 2.2.5. *Let $\mathcal{G}_g\big[f\big](b,\omega)$ be the windowed Fourier transform of the square integrable function f, then we have*

$$f(t) = \frac{1}{\sqrt{2\pi}\, \left\|g\right\|_2^2} \int_{\mathbb{R}} \int_{\mathbb{R}} \mathcal{G}_g\big[f\big](b,\omega)\, g(t - b)\, e^{i\omega t}\, db\, d\omega, \tag{2.57}$$

where equality holds in (2.57) in the almost everywhere sense.

Proof. For the function $h(x) = \delta(x - t)$, $t \in \mathbb{R}$, we have

$$\mathcal{G}_g\big[h\big](b,\omega) = \frac{1}{\sqrt{2\pi}} \int_{\mathbb{R}} \delta(x - t)\, \overline{g(x - b)}\, e^{-i\omega x}\, dx = \frac{1}{\sqrt{2\pi}}\, \overline{g(t - b)}\, e^{-i\omega t}.$$

Consequently, the orthogonality relation (2.55) implies that

$$\int_{\mathbb{R}} \int_{\mathbb{R}} \mathcal{G}_g\big[f\big](b,\omega)\, \overline{\mathcal{G}_g\big[h\big](b,\omega)}\, db\, d\omega = \left\|g\right\|_2^2 \int_{\mathbb{R}} f(x)\, \overline{\delta(x - t)}\, dx = \left\|g\right\|_2^2\, f(t).$$

Equivalently,

$$
\begin{aligned}
f(t) &= \frac{1}{\|g\|_2^2} \int_{\mathbb{R}} \int_{\mathbb{R}} \mathscr{G}_g\big[f\big](b,\omega) \overline{\mathscr{G}_g\big[h\big](b,\omega)} \, db \, d\omega \\
&= \frac{1}{\|g\|_2^2} \int_{\mathbb{R}} \int_{\mathbb{R}} \mathscr{G}_g\big[f\big](b,\omega) \left\{ \frac{1}{\sqrt{2\pi}} g(t-b) \, e^{i\omega t} \right\} db \, d\omega \\
&= \frac{1}{\sqrt{2\pi}\, \|g\|_2^2} \int_{\mathbb{R}} \int_{\mathbb{R}} \mathscr{G}_g\big[f\big](b,\omega) \, g(t-b) \, e^{i\omega t} \, db \, d\omega,
\end{aligned}
$$

where equality holds in the almost everywhere sense. This evidently completes the proof of Theorem 2.2.5. □

Towards the culmination of this subsection, we shall demonstrate that the range of the windowed Fourier transform defined in (2.50) is a reproducing kernel Hilbert space.

Theorem 2.2.6. *Let g be any window function, then the range of the windowed Fourier transform (2.50) is a reproducing kernel Hilbert space in $L^2(\mathbb{R}^2)$ with kernel function:*

$$
\mathscr{K}_g\big(\omega',b';\,\omega,b\big) = \frac{1}{2\pi\|g\|_2^2} \big\langle g_{b',\omega'},\, g_{b,\omega} \big\rangle_2. \tag{2.58}
$$

Proof. Denote $\mathcal{H} = \mathscr{G}_g\big(L^2(\mathbb{R})\big)$. Then, for $F \in \mathcal{H}$, there exists a function $f \in L^2(\mathbb{R})$ such that

$$
F(b,\omega) = \mathscr{G}_g\big[f\big](b,\omega) = \frac{1}{\sqrt{2\pi}} \big\langle f, g_{b,\omega} \big\rangle_2.
$$

Using orthogonality relation (2.55), we obtain

$$
\begin{aligned}
F(b,\omega) &= \frac{1}{\sqrt{2\pi}} \big\langle f, g_{b,\omega} \big\rangle_2 \\
&= \frac{1}{\sqrt{2\pi}\,\|g\|_2^2} \int_{\mathbb{R}} \int_{\mathbb{R}} \mathscr{G}_g\big[f\big](b',\omega') \overline{\mathscr{G}_g\big[g_{b,\omega}\big](b',\omega')} \, db' d\omega' \\
&= \frac{1}{2\pi\|g\|_2^2} \int_{\mathbb{R}} \int_{\mathbb{R}} F(b',\omega') \big\langle g_{b,\omega},\, g_{b',\omega'} \big\rangle_2 \, db' d\omega' \\
&= \frac{1}{2\pi\|g\|_2^2} \int_{\mathbb{R}} \int_{\mathbb{R}} F(b',\omega') \big\langle g_{b',\omega'},\, g_{b,\omega} \big\rangle_2 \, db' d\omega' \\
&= \int_{\mathbb{R}} \int_{\mathbb{R}} F(b',\omega') \, \mathscr{K}_g\big(\omega',b';\,\omega,b\big) \, db' d\omega'.
\end{aligned}
$$

The above argument shows that the range of the windowed Fourier transform is indeed a reproducing kernel Hilbert space in $L^2(\mathbb{R}^2)$ with kernel given by

$$
\mathscr{K}_g\big(\omega',b';\,\omega,b\big) = \frac{1}{2\pi\|g\|_2^2} \big\langle g_{b',\omega'},\, g_{b,\omega} \big\rangle_2.
$$

This completes the proof of Theorem 2.2.6. □

2.2.3   Time-frequency resolution and uncertainty principles

"The indeterminacy relations [uncertainty relations] show first that an accurate knowledge of the parameters, which is needed in classical physics to fix the causal connection, cannot be achieved. A further consequence of the indeterminacy is that also the future behaviour of such an inaccurately known system can be predicted only inaccurately."

-Werner K. Heisenberg

In this subsection, we aim to study the time-frequency resolution and the Heisenberg-type uncertainty principles associated with the windowed Fourier transform (2.50). As a consequence of these uncertainty inequalities, we can compare the localization of the windowed Fourier transform, as a function of either frequency or translation variable, relative to the concentration of either the input function or the Fourier transform of the window function. Towards the culmination, we shall obtain a phase-space uncertainty inequality which elucidates the localization of the windowed Fourier transform in the time-frequency plane.

In order to examine the time-frequency resolution of the windowed Fourier transform, we consider a window function $g \in L^2(\mathbb{R})$ with centre E_g and radius Δ_g. Then, the centre and radius of each time domain analyzing element $g_{b,\omega}(t)$ can be obtained as follows:

$$
\begin{aligned}
E_{g_{b,\omega}} &= \frac{1}{\left\|g_{b,\omega}\right\|_2^2} \int_{\mathbb{R}} t \left|g_{b,\omega}(t)\right|^2 dt \\
&= \frac{1}{\left\|g\right\|_2^2} \int_{\mathbb{R}} t \left|g(t-b)\right|^2 dt \\
&= \frac{1}{\left\|g\right\|_2^2} \int_{\mathbb{R}} (b+z) \left|g(z)\right|^2 dz \\
&= \frac{1}{\left\|g\right\|_2^2} \left\{ b \left\|g\right\|_2^2 + \int_{\mathbb{R}} z \left|g(z)\right|^2 dz \right\} \\
&= b + \int_{\mathbb{R}} z \left|g(z)\right|^2 dz \\
&= b + E_g.
\end{aligned}
$$

Furthermore,

$$
\Delta_{g_{b,\omega}} = \frac{1}{\left\|g\right\|_2} \int_{\mathbb{R}} (t-b-E_g)^2 \left|g(t-b)\right|^2 dt = \frac{1}{\left\|g\right\|_2} \int_{\mathbb{R}} (z-E_g) \left|g(z)\right|^2 dz = \Delta_g.
$$

Consequently, the localization of the windowed Fourier transform (2.50) in the time domain is given by the window

$$
\left[b + E_g - \Delta_g,\, b + E_g + \Delta_g\right], \tag{2.59}
$$

having a width of $2\Delta_g$. By virtue of Proposition 2.2.1, we can recast the windowed Fourier transform in the spectral form as

$$
\mathscr{G}_g\big[f\big](b,\omega) = \frac{e^{-i\omega b}}{\sqrt{2\pi}} \int_{\mathbb{R}} \hat{f}(\xi)\, \overline{\hat{g}(\omega-\xi)}\, e^{i\xi b}\, d\xi.
$$

Therefore, the localization of the windowed Fourier transform (2.50) in the frequency domain is determined by the function $\hat{g}(\omega - \xi)$. Thus, if $E_{\hat{g}}$ and $\Delta_{\hat{g}}$ denote the centre and radius of the window function $\hat{g}$, then $E_{\hat{g}(\omega-\xi)} = \omega + E_{\hat{g}(\xi)}$ and $\Delta_{\hat{g}(\omega-\xi)} = \Delta_{\hat{g}(\xi)}$. Hence, the localization of the windowed Fourier transform in the frequency domain is given by

$$\left[\omega + E_{\hat{g}} - \Delta_{\hat{g}}, \, \omega + E_{\hat{g}} + \Delta_{\hat{g}}\right],\tag{2.60}$$

having a net width of $2\Delta_{\hat{g}}$. Therefore, the joint time-frequency localization of the windowed Fourier transform (2.50) in the time and frequency domains is given by the time-frequency window

$$\left[b + E_g - \Delta_g, \, b + E_g + \Delta_g\right] \times \left[\omega + E_{\hat{g}} - \Delta_{\hat{g}}, \, \omega + E_{\hat{g}} + \Delta_{\hat{g}}\right],\tag{2.61}$$

occupying an area of $4\Delta_g\Delta_{\hat{g}}$. Indeed, (2.61) represent boxes in the time-frequency plane and are formally called as the "Heisenberg's boxes." Owing to the Heisenberg's boxes, the tiling induced by the windowed Fourier transform (2.50) on the time-frequency plane is depicted in Figure 2.16. Moreover, the resolution of the windowed Fourier transform in the time-frequency plane is denoted by R and is defined as

$$R = \frac{1}{\Delta_g\Delta_{\hat{g}}}.\tag{2.62}$$

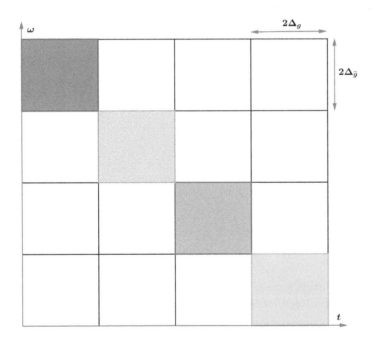

FIGURE 2.16: Time-frequency tiling induced by windowed Fourier transform.

Having presented a detailed perspective of the time-frequency localization characteristics of the windowed Fourier transform, we conclude that a high degree of localization in the time-frequency plane can be obtained by appropriately choosing the window functions with sufficiently narrow spreads in both the time and frequency domains so that the product $\Delta_g\Delta_{\hat{g}}$ is minimized. However, much to the dismay, such a duration-bandwidth product cannot be minimized beyond a certain limit and is dictated by the Heisenberg's uncertainty principle [111, 313].

Next, our aim is to formulate the Heisenberg-type uncertainty principles associated with the windowed Fourier transform defined in (2.50). Recall that, for any $f \in L^2(\mathbb{R})$, the Heisenberg's uncertainty principle in the Fourier domain asserts that

$$\left\{ \int_{\mathbb{R}} t^2 \left| f(t) \right|^2 dt \right\}^{1/2} \left\{ \int_{\mathbb{R}} \xi^2 \left| \hat{f}(\xi) \right|^2 d\xi \right\}^{1/2} \geq \frac{1}{2} \left\| f \right\|_2^2. \tag{2.63}$$

In the following theorem, we obtain a lower bound governing the localization of the windowed Fourier transform $\mathscr{G}_g[f](b, \omega)$, as a function of b, with regard to the localization of the Fourier transform corresponding to the window function g.

Theorem 2.2.7. *If $\mathscr{G}_g[f](b, \omega)$ is the windowed Fourier transform of any $f \in L^2(\mathbb{R})$, then the following uncertainty inequality holds:*

$$\left\{ \int_{\mathbb{R}} \int_{\mathbb{R}} b^2 \left| \mathscr{G}_g[f](b, \omega) \right|^2 db\, d\omega \right\}^{1/2} \left\{ \int_{\mathbb{R}} \xi^2 \left| \hat{g}(\xi) \right|^2 d\xi \right\}^{1/2} \geq \frac{1}{2} \left\| f \right\|_2 \left\| g \right\|_2^2, \tag{2.64}$$

where g is any window function.

Proof. Identifying the windowed Fourier transform $\mathscr{G}_g[f](b, \omega)$ as a function of b and then using (2.63), we obtain

$$\left\{ \int_{\mathbb{R}} b^2 \left| \mathscr{G}_g[f](b, \omega) \right|^2 db \right\}^{1/2} \left\{ \int_{\mathbb{R}} \xi^2 \left| \mathscr{F}\left(\mathscr{G}_g[f](b, \omega) \right)(\xi) \right|^2 d\xi \right\}^{1/2} \geq \frac{1}{2} \left\{ \int_{\mathbb{R}} \left| \mathscr{G}_g[f](b, \omega) \right|^2 db \right\}$$

Integrating the above inequality with respect to the variable ω, followed by an implication of Corollary 2.2.4, we get

$$\int_{\mathbb{R}} \left\{ \int_{\mathbb{R}} b^2 \left| \mathscr{G}_g[f](b, \omega) \right|^2 db \right\}^{1/2} \left\{ \int_{\mathbb{R}} \xi^2 \left| \mathscr{F}\left(\mathscr{G}_g[f](b, \omega) \right)(\xi) \right|^2 d\xi \right\}^{1/2} d\omega$$

$$\geq \frac{1}{2} \int_{\mathbb{R}} \int_{\mathbb{R}} \left| \mathscr{G}_g[f](b, \omega) \right|^2 db\, d\omega$$

$$= \frac{1}{2} \left\| f \right\|_2^2 \left\| g \right\|_2^2.$$

Using the Cauchy-Schwarz inequality, we obtain

$$\frac{1}{2} \left\| f \right\|_2^2 \left\| g \right\|_2^2 \leq \left\{ \int_{\mathbb{R}} \int_{\mathbb{R}} b^2 \left| \mathscr{G}_g[f](b, \omega) \right|^2 db\, d\omega \right\}^{1/2} \left\{ \int_{\mathbb{R}} \int_{\mathbb{R}} \xi^2 \left| \mathscr{F}\left(\mathscr{G}_g[f](b, \omega) \right)(\xi) \right|^2 d\xi\, d\omega \right\}^{1/2}.$$

Employing the convolution form of windowed Fourier transform obtained in (2.53) and then invoking Theorem 1.2.6 together with the application of Theorem 1.2.1, we have

$$\frac{1}{2} \left\| f \right\|_2^2 \left\| g \right\|_2^2 \leq \left\{ \int_{\mathbb{R}} \int_{\mathbb{R}} b^2 \left| \mathscr{G}_g[f](b, \omega) \right|^2 db\, d\omega \right\}^{1/2}$$

$$\times \left\{ \frac{1}{2\pi} \int_{\mathbb{R}} \int_{\mathbb{R}} \xi^2 \left| \mathscr{F}\left[\left((\mathcal{M}_{-\omega} f) * \breve{g} \right)(b) \right](\xi) \right|^2 d\xi\, d\omega \right\}^{1/2}$$

$$= \left\{ \int_{\mathbb{R}} \int_{\mathbb{R}} b^2 \left| \mathscr{G}_g[f](b, \omega) \right|^2 db\, d\omega \right\}^{1/2}$$

$$\times \left\{ \int_{\mathbb{R}} \int_{\mathbb{R}} \xi^2 \left| \mathscr{F}\left[\mathcal{M}_{-\omega} f \right](\xi)\, \mathscr{F}\left[\breve{g} \right](\xi) \right|^2 d\xi\, d\omega \right\}^{1/2}$$

$$= \left\{ \int_{\mathbb{R}} \int_{\mathbb{R}} b^2 \left| \mathscr{G}_g \left[f \right] (b, \omega) \right|^2 db \, d\omega \right\}^{1/2} \left\{ \int_{\mathbb{R}} \xi^2 \left| \hat{g}(\xi) \right|^2 \left(\int_{\mathbb{R}} \left| \hat{f}(\xi + \omega) \right|^2 d\omega \right) d\xi \right\}^{1/2}$$

$$= \left\{ \int_{\mathbb{R}} \int_{\mathbb{R}} b^2 \left| \mathscr{G}_g \left[f \right] (b, \omega) \right|^2 db \, d\omega \right\}^{1/2} \left\{ \left\| \hat{f} \right\|_2^2 \int_{\mathbb{R}} \xi^2 \left| \hat{g}(\xi) \right|^2 d\xi \right\}^{1/2}.$$

Or equivalently,

$$\left\{ \int_{\mathbb{R}} \int_{\mathbb{R}} b^2 \left| \mathscr{G}_g \left[f \right] (b, \omega) \right|^2 db \, d\omega \right\}^{1/2} \left\{ \int_{\mathbb{R}} \xi^2 \left| \hat{g}(\xi) \right|^2 d\xi \right\}^{1/2} \geq \frac{1}{2} \left\| f \right\|_2 \left\| g \right\|_2^2,$$

which is the desired Heisenberg-type uncertainty inequality associated with the windowed Fourier transform. $\square$

Remark 2.2.1. As a consequence of inequality (2.64), we infer that the product of the space localization of the windowed Fourier transform $\mathscr{G}_g \left[f \right] (b, \omega)$ of any signal f and the localization of the Fourier transform corresponding to the window function g is limited by a lower bound $\left\| f \right\|_2 \left\| g \right\|_2^2 / 2$. As such, we note that the uncertainty inequality is dictated by the localization properties of the window function g.

In continuation to the formulation of uncertainty inequalities for the windowed Fourier transform, in the below formulated inequality we compare the localization of the windowed Fourier transform $\mathscr{G}_g \left[f \right] (b, \omega)$, as a function of ω (instead of b), with the localization of the input function f.

Theorem 2.2.8. *If $\mathscr{G}_g \left[f \right] (b, \omega)$ is the windowed Fourier transform of any $f \in L^2(\mathbb{R})$, then the following uncertainty inequality holds:*

$$\left\{ \int_{\mathbb{R}} \int_{\mathbb{R}} \omega^2 \left| \mathscr{G}_g \left[f \right] (b, \omega) \right|^2 db \, d\omega \right\}^{1/2} \left\{ \int_{\mathbb{R}} t^2 |f(t)|^2 dt \right\}^{1/2} \geq \frac{1}{2} \left\| f \right\|_2^2 \left\| g \right\|_2, \qquad (2.65)$$

where g is any window function.

Proof. Using translation invariance of the Lebesgue integral, together with the implication of Definition 2.2.2, we have

$$\left\| g \right\|_2^2 \int_{\mathbb{R}} t^2 \left| f(t) \right|^2 dt = \int_{\mathbb{R}} \int_{\mathbb{R}} t^2 \left| g(t - b) \right|^2 \left| f(t) \right|^2 dt \, db$$

$$= \int_{\mathbb{R}} \int_{\mathbb{R}} t^2 \left| f(t) \, \overline{g(t - b)} \right|^2 dt \, db$$

$$= \int_{\mathbb{R}} \int_{\mathbb{R}} t^2 \left| \mathscr{F}^{-1} \left(\mathscr{G}_g \left[f \right] (b, \omega) \right) (t) \right|^2 dt \, db, \qquad (2.66)$$

where the inverse Fourier transform $\mathscr{F}^{-1}$ in the last equality is computed under the frequency variable ω. As a consequence of the Heisenberg's uncertainty principle in Fourier domain given by (2.63), we can write

$$\left\{ \int_{\mathbb{R}} \omega^2 \left| \mathscr{G}_g \left[f \right] (b, \omega) \right|^2 d\omega \right\}^{1/2} \left\{ \int_{\mathbb{R}} t^2 \left| \mathscr{F}^{-1} \left(\mathscr{G}_g \left[f \right] (b, \omega) \right) (t) \right|^2 dt \right\}^{1/2} \geq \frac{1}{2} \int_{\mathbb{R}} \left| \mathscr{G}_g \left[f \right] (b, \omega) \right|^2 d\omega.$$

Integrating the above inequality on both sides with respect to b and making use of Corollary 2.2.4, we obtain

$$\int_{\mathbb{R}} \left\{ \int_{\mathbb{R}} \omega^2 \left| \mathscr{G}_g \left[f \right] (b, \omega) \right|^2 d\omega \right\}^{1/2} \left\{ \int_{\mathbb{R}} t^2 \left| \mathscr{F}^{-1} \left(\mathscr{G}_g \left[f \right] (b, \omega) \right) (t) \right|^2 dt \right\}^{1/2} db \geq \frac{1}{2} \left\| f \right\|_2^2 \left\| g \right\|_2^2.$$

Applying the Cauchy-Schwarz inequality, the above inequality can be recast as follows:

$$\left\{ \int_{\mathbb{R}} \int_{\mathbb{R}} \omega^2 \left| \mathscr{G}_g \left[f \right] (b,\omega) \right|^2 db\, d\omega \right\}^{1/2} \left\{ \int_{\mathbb{R}} \int_{\mathbb{R}} t^2 \left| \mathscr{F}^{-1} \left(\mathscr{G}_g \left[f \right] (b,\omega) \right)(t) \right|^2 dt\, db \right\}^{1/2} \geq \frac{1}{2} \left\| f \right\|_2^2 \left\| g \right\|_2^2 .$$

Using (2.66) in the above inequality, we obtain

$$\left\{ \int_{\mathbb{R}} \int_{\mathbb{R}} \omega^2 \left| \mathscr{G}_g \left[f \right] (b,\omega) \right|^2 db\, d\omega \right\}^{1/2} \left\{ \left\| g \right\|_2^2 \int_{\mathbb{R}} t^2 \left| f(t) \right|^2 dt \right\}^{1/2} \geq \frac{1}{2} \left\| f \right\|_2^2 \left\| g \right\|_2^2 ,$$

which can be further simplified to

$$\left\{ \int_{\mathbb{R}} \int_{\mathbb{R}} \omega^2 \left| \mathscr{G}_g \left[f \right] (b,\omega) \right|^2 db\, d\omega \right\}^{1/2} \left\{ \int_{\mathbb{R}} t^2 \left| f(t) \right|^2 dt \right\}^{1/2} \geq \frac{1}{2} \left\| f \right\|_2^2 \left\| g \right\|_2 .$$

This completes the proof of Theorem 2.2.8. □

Finally, we have yet another uncertainty inequality, namely the "Phase-space uncertainty principle," associated with the windowed Fourier transform (2.50).

Theorem 2.2.9. *Let g be a window function and* $\mathscr{G}_g [f](b,\omega)$ *be the windowed Fourier transform corresponding to any* $f \in L^2(\mathbb{R})$, *then the following inequality holds:*

$$\left\{ \int_{\mathbb{R}} \int_{\mathbb{R}} \omega^2 \left| \mathscr{G}_g \left[f \right] (b,\omega) \right|^2 db\, d\omega \right\}^{1/2} \left\{ \int_{\mathbb{R}} \int_{\mathbb{R}} b^2 \left| \mathscr{G}_g \left[f \right] (b,\omega) \right|^2 db\, d\omega \right\}^{1/2}$$

$$\times \left\{ \int_{\mathbb{R}} t^2 \left| f(t) \right|^2 dt \right\}^{1/2} \left\{ \int_{\mathbb{R}} \xi^2 \left| \hat{g}(\xi) \right|^2 d\xi \right\}^{1/2} \geq \frac{1}{4} \left\| f \right\|_2^3 \left\| g \right\|_2^3 . \quad (2.67)$$

Proof. The proof follows directly as a consequence Theorem 2.2.7 and Theorem 2.2.8. □

Remark 2.2.2. The phase-space uncertainty principle may be interpreted as follows: The better the localization of the pair $(f, \hat{g})$, the worse is the phase-space localization of the windowed Fourier transform.

2.3   The Windowed Fractional Fourier Transform

"The future cannot be predicted, but futures can be invented. It was man's ability to invent which has made human society what it is."

-Dennis Gabor

Although the fractional Fourier transform has proved to be an elegant alternative to the classical fourier transform by offering representations of signals along a continuum of "fractional" domains making arbitrary angles with the time and frequency domains, however, it is not befitting for an efficient spectral analysis of signals whose fractional frequencies vary with time. This is only due to the fact that the conventional fractional Fourier transform lacks the time localization information. In this section, we shall study a novel ramification of the fractional Fourier transform, coined as the "windowed fractional Fourier transform (WFrFT)," by adequately adjoining a window to an input function prior to the evaluation of the fractional Fourier transform.

2.3.1 Definition and basic properties

In this subsection, we shall be concerned with the formal definition of the windowed fractional Fourier transform and its fundamental properties. Abreast to this, we shall present an illustrative perspective of the time-fractional-frequency resolution.

Definition 2.3.1. Given any window function $g \in L^2(\mathbb{R})$ and $\alpha \neq n\pi$ with $n \in \mathbb{Z}$, the α-angle windowed fractional Fourier transform corresponding to $f \in L^2(\mathbb{R})$ is denoted by $\mathscr{G}_g^\alpha[f]$ and is defined as

$$\mathscr{G}_g^\alpha[f](b,\omega) = \left\langle f, g_{b,\omega}^\alpha \right\rangle_2 = \int_{\mathbb{R}} f(t)\,\overline{g(t-b)}\,\mathcal{K}_\alpha(t,\omega)\,dt, \quad b,\omega \in \mathbb{R}, \tag{2.68}$$

where $\mathcal{K}_\alpha(t,\omega)$ is the usual fractional Fourier kernel given by (1.128) and

$$g_{b,\omega}^\alpha(t) = \sqrt{\frac{1+i\cot\alpha}{2\pi}}\,\exp\left\{ -\frac{i(t^2+\omega^2)\cot\alpha}{2} + i\omega t\,\csc\alpha \right\} g(t-b). \tag{2.69}$$

As a consequence of Definition 2.3.1, we can put forth the following important points concerning the windowed fractional Fourier transform:

(i) Since the fractional Fourier transform leaves the input signal unchanged for $\alpha = 2n\pi$ and induces a reflection for $\alpha = (2n \pm 1)\pi$, where $n \in \mathbb{Z}$, so these two cases are of least importance here and are not dealt with.

(ii) The window function g involved in (2.68) captures the selective portion of the signal f about the point b along the natural domain. Then, the fractional Fourier transform of this portion can be viewed as the instantaneous fractional Fourier domain spectrum of the signal f at b-instant. By sliding the window along the t-axis, we are able to obtain the fractional Fourier domain spectrum at every instant. Thus, the windowed fractional Fourier transform is a two-dimensional representation of a signal which not only allows us to see the fractional Fourier domain frequency contents but allows us to observe how the frequency contents change over time.

(iii) In analogy to the fact that the conventional windowed Fourier transform (2.50) can be written as the joint action of the translation and modulation operators as demonstrated in (2.52), we can revamp the windowed fractional Fourier transform (2.68) as

$$\mathscr{G}_g^\alpha[f](b,\omega)$$
$$= \sqrt{\frac{1-i\cot\alpha}{2\pi}} \int_{\mathbb{R}} f(t)\exp\left\{ \frac{i\omega^2\cot\alpha}{2} - i\omega t\csc\alpha \right\}\exp\left\{ \frac{it^2\cot\alpha}{2} \right\} \overline{g(t-b)}\,dt$$
$$= \sqrt{\frac{1-i\cot\alpha}{2\pi}} \left\langle f, \mathcal{M}_\omega^\alpha \mathcal{T}_b^\alpha g \right\rangle, \tag{2.70}$$

where $\mathcal{T}_b^\alpha$ and $\mathcal{M}_\omega^\alpha$ denote the fractional translation and modulation operators given by

$$\left. \begin{array}{l} \mathcal{T}_b^\alpha g(t) = \exp\left\{ -\frac{it^2\cot\alpha}{2} \right\} g(t-b) \\[4mm] \mathcal{M}_\omega^\alpha g(t) = \exp\left\{ -\frac{i\omega^2\cot\alpha}{2} + i\omega t\csc\alpha \right\} g(t) \end{array} \right\}. \tag{2.71}$$

(iv) In view of the fact that the conventional windowed Fourier transform (2.50) is expressible via the classical convolution operational $*$ as shown in (2.53), we can formulate a couple of new variants of the windowed fractional Fourier transform (2.68) by replacing the classical convolution operation in (2.53) with any of the fractional convolutional convolution operations as defined in (1.141) and (1.146). These convolution-based windowed fractional Fourier transforms are equipped with some of their own fascinating properties, in addition to those imbibed from the fractional Fourier transform.

Next, we have the proposition concerned with the expression of the windowed fractional Fourier transform (2.68) via the spectral representations of the input and window functions. Such an expression plays a key role in understanding the resolving power of windowed fractional Fourier transform abreast of the notion of time-fractional-frequency cells.

Proposition 2.3.1. *Let $\mathscr{G}_g^\alpha[f](b,\omega)$ be the windowed fractional Fourier transform of an arbitrary function $f \in L^2(\mathbb{R})$. Then, we have*

$$\mathscr{G}_g^\alpha[f](b,\omega) = \sqrt{2\pi}\, \mathcal{K}_\alpha(b,\omega) \int_{\mathbb{R}} \mathscr{F}^\alpha[f](\xi)\, \overline{\mathscr{F}[g]\big((\xi-\omega)\csc\alpha\big)}\, \overline{\mathcal{K}_\alpha(b,\xi)}\, d\xi, \qquad (2.72)$$

where $\mathscr{F}$ is the classical Fourier transform defined in (1.7).

Proof. Using Definition 2.3.1 and the Parseval's formula (1.137) for the fractional Fourier transform, we obtain

$$
\begin{aligned}
\mathscr{G}_g^\alpha[f](b,\omega) &= \left\langle f,\, g_{b,\omega}^\alpha \right\rangle_2 \\
&= \left\langle \mathscr{F}^\alpha[f],\, \mathscr{F}^\alpha[g_{b,\omega}^\alpha] \right\rangle_2 \\
&= \int_{\mathbb{R}} \mathscr{F}^\alpha[f](\xi)\, \overline{\mathscr{F}^\alpha[g_{b,\omega}^\alpha](\xi)}\, d\xi. \qquad (2.73)
\end{aligned}
$$

Furthermore, we observe that

$$
\begin{aligned}
\mathscr{F}^\alpha[g_{b,\omega}^\alpha](\xi) &= \sqrt{\frac{1-i\cot\alpha}{2\pi}} \int_{\mathbb{R}} g_{b,\omega}^\alpha(t)\, \exp\left\{\frac{i(t^2+\xi^2)\cot\alpha}{2} - i\xi t\csc\alpha\right\} dt \\
&= \sqrt{\frac{(1-i\cot\alpha)(1+i\cot\alpha)}{4\pi^2}} \int_{\mathbb{R}} g(t-b)\, \exp\left\{-\frac{i(t^2+\omega^2)\cot\alpha}{2} + i\omega t\csc\alpha\right\} \\
&\qquad\qquad\qquad\qquad \times \exp\left\{\frac{i(t^2+\xi^2)\cot\alpha}{2} - i\xi t\csc\alpha\right\} dt \\
&= \sqrt{\frac{(1-i\cot\alpha)(1+i\cot\alpha)}{4\pi^2}} \int_{\mathbb{R}} g(t-b)\, \exp\left\{-\frac{i\omega^2\cot\alpha}{2}\right\} \exp\left\{\frac{i\xi^2\cot\alpha}{2}\right\} \\
&\qquad\qquad\qquad\qquad \times e^{i\omega t\csc\alpha}\, e^{-i\xi t\csc\alpha}\, dt \\
&= \sqrt{\frac{(1-i\cot\alpha)(1+i\cot\alpha)}{4\pi^2}} \int_{\mathbb{R}} g(z)\, \exp\left\{-\frac{i\omega^2\cot\alpha}{2}\right\} \exp\left\{\frac{i\xi^2\cot\alpha}{2}\right\} \\
&\qquad\qquad\qquad\qquad \times e^{i\omega(b+z)\csc\alpha}\, e^{-i\xi(b+z)\csc\alpha}\, dz \\
&= \sqrt{\frac{(1-i\cot\alpha)(1+i\cot\alpha)}{2\pi}} \exp\left\{-\frac{i\omega^2\cot\alpha}{2}\right\} \exp\left\{\frac{i\xi^2\cot\alpha}{2}\right\} e^{i\omega b\csc\alpha}\, e^{-i\xi b\csc\alpha} \\
&\qquad\qquad\qquad\qquad \times \left\{\frac{1}{\sqrt{2\pi}} \int_{\mathbb{R}} g(z)\, e^{-i(\xi-\omega)z\csc\alpha}\, dz\right\} \\
&= \sqrt{\frac{(1-i\cot\alpha)(1+i\cot\alpha)}{2\pi}} \exp\left\{-\frac{i\omega^2\cot\alpha}{2}\right\} \exp\left\{\frac{i\xi^2\cot\alpha}{2}\right\} e^{i\omega b\csc\alpha}\, e^{-i\xi b\csc\alpha} \\
&\qquad\qquad\qquad\qquad \times \mathscr{F}[g]\big((\xi-\omega)\csc\alpha\big). \qquad (2.74)
\end{aligned}
$$

Using the obtained relation (2.74) in (2.73), we get

$$
\mathscr{G}_g^\alpha\Big[f\Big](b,\omega) = \int_{\mathbb{R}} \mathscr{F}^\alpha\Big[f\Big](\xi)\, \overline{\mathscr{F}^\alpha\Big[g_{b,\omega}^\alpha\Big](\xi)}\, d\xi
$$

$$
= \sqrt{\frac{(1 - i\cot\alpha)(1 + i\cot\alpha)}{2\pi}}\, \exp\left\{ \frac{i(\omega^2 + b^2)\cot\alpha}{2} - i\omega b\csc\alpha \right\}
$$

$$
\times \int_{\mathbb{R}} \mathscr{F}^\alpha\Big[f\Big](\xi)\, \overline{\mathscr{F}\Big[g\Big]((\xi - \omega)\csc\alpha)}\, \exp\left\{ -\frac{i(\xi^2 + b^2)\cot\alpha}{2} + i\xi b\csc\alpha \right\} d\xi
$$

$$
= \sqrt{2\pi}\, \mathcal{K}_\alpha(b,\omega) \int_{\mathbb{R}} \mathscr{F}^\alpha\Big[f\Big](\xi)\, \overline{\mathscr{F}\Big[g\Big]((\xi - \omega)\csc\alpha)}\, \overline{\mathcal{K}_\alpha(b,\xi)}\, d\xi.
$$

This completes the proof of Proposition 2.3.1. □

Below, we present a theorem wherein we assemble all fundamental properties of the windowed fractional Fourier transform (2.68).

Theorem 2.3.2. *If g and h are two window functions and c_1, $c_2 \in \mathbb{C}$, $k, \omega_0 \in \mathbb{R}$, $\lambda \in \mathbb{R}\backslash\{0\}$, then for any pair of square integrable functions f and s, the windowed fractional Fourier transform (2.68) satisfies the following properties:*

(i) *Linearity:* $\mathscr{G}_g^\alpha\Big[c_1 f + c_2 s\Big](b,\omega) = c_1 \mathscr{G}_g^\alpha\Big[f\Big](b,\omega) + c_2 \mathscr{G}_g^\alpha\Big[s\Big](b,\omega)$,

(ii) *Anti-linearity:* $\mathscr{G}_{c_1 g + c_2 h}^\alpha\Big[f\Big](b,\omega) = \overline{c_1}\, \mathscr{G}_g^\alpha\Big[f\Big](b,\omega) + \overline{c_2}\, \mathscr{G}_h^\alpha\Big[f\Big](b,\omega)$,

(iii) *Time-shift:* $\mathscr{G}_g^\alpha\Big[f(t-k)\Big](b,\omega) = \exp\left\{ \frac{ik^2\cot\alpha}{2} - i\omega\csc\alpha \right\} \mathscr{G}_g^\alpha\Big[F\Big](b-k,\omega)$, $\quad F(t) = e^{ikt\cot\alpha} f(t)$,

(iv) *Modulation:* $\mathscr{G}_g^\alpha\Big[e^{i\omega_0 t} f(t)\Big](b,\omega) = \exp\left\{ \frac{i(2\omega\omega_0 - \omega_0^2)\cot\alpha}{2} \right\} \mathscr{G}_g^\alpha\Big[f\Big](b,\omega - \omega_0)$,

(v) *Scaling:* $\mathscr{G}_g^\alpha\Big[f(\lambda t)\Big](b,\omega) = \sqrt{\frac{1 - i\cot\alpha}{1 - i\cot\beta}}\, \exp\left\{ \frac{i(\lambda\omega)^2}{2}\left(1 - \frac{\csc^2\alpha}{\lambda^4\csc^2\beta}\right)\cot\beta \right\}$
$\times \mathscr{G}_{G_\lambda}^\alpha\Big[f\Big]\left(\lambda b, \dfrac{\omega\csc\alpha}{\lambda\csc\beta}\right)$, $\quad G_\lambda(t) = \dfrac{1}{|\lambda|}g\left(\dfrac{t}{\lambda}\right)$, $\beta = \arctan(\lambda^2\tan\alpha)$,

(vi) *Symmetry:* $\mathscr{G}_g^\alpha\Big[f\Big](b,\omega) = \exp\left\{ \frac{ib^2\cot\alpha}{2} - i\omega b\csc\alpha \right\} \mathscr{G}_{\tilde{f}}^\alpha\Big[H\Big](-b,\omega)$, $\quad H(t) = e^{-ibt\cot\alpha} g(t)$,

(vii) *Translation in window:* $\mathscr{G}_{g(t-k)}^\alpha\Big[f\Big](b,\omega) = \mathscr{G}_g^\alpha\Big[f\Big](b+k,\omega)$,

(viii) *Scaling in window:* $\mathscr{G}_{g(\lambda t)}^\alpha\Big[f\Big](b,\omega) = \sqrt{\frac{1 - i\cot\alpha}{1 - i\cot\beta}}\, \exp\left\{ \frac{i(\lambda\omega)^2}{2}\left(1 - \frac{\csc^2\alpha}{\lambda^4\csc^2\beta}\right)\cot\beta \right\}$
$\times \mathscr{G}_g^\alpha\Big[F_\lambda\Big]\left(\lambda b, \dfrac{\omega\csc\alpha}{\lambda\csc\beta}\right)$, $\quad F_\lambda(t) = \dfrac{1}{|\lambda|}f\left(\dfrac{t}{\lambda}\right)$, $\beta = \arctan(\lambda^2\tan\alpha)$.

Proof. It is a straightforward computation to verify the assertions made in Theorem 2.3.2, so we omit the proof. □

Below, we formulate an orthogonality relation between a given pair of square integrable functions and the corresponding windowed fractional Fourier transforms.

Theorem 2.3.3. *Let $\mathscr{G}_{g_1}^{\alpha}[f](b,\omega)$ and $\mathscr{G}_{g_2}^{\alpha}[h](b,\omega)$ be the windowed fractional Fourier transforms of any $f, h \in L^2(\mathbb{R})$ with respect to a given pair of window functions $g_1, g_2 \in L^2(\mathbb{R})$, then the following orthogonality relation holds:*

$$\int_{\mathbb{R}} \int_{\mathbb{R}} \mathscr{G}_{g_1}^{\alpha}[f](b,\omega) \overline{\mathscr{G}_{g_2}^{\alpha}[h](b,\omega)} \, db \, d\omega = \left\langle g_2, g_1 \right\rangle_2 \left\langle f, h \right\rangle_2. \tag{2.75}$$

Proof. By virtue of Proposition 2.3.1, we can write

$$\mathscr{G}_{g_1}^{\alpha}[f](b,\omega) = \sqrt{2\pi} \, \mathcal{K}_{\alpha}(b,\omega) \int_{\mathbb{R}} \mathscr{F}^{\alpha}[f](\xi) \, \overline{\mathscr{F}[g_1]\big((\xi - \omega)\csc\alpha\big)} \, \overline{\mathcal{K}_{\alpha}(b,\xi)} \, d\xi$$

and

$$\mathscr{G}_{g_2}^{\alpha}[h](b,\omega) = \sqrt{2\pi} \, \mathcal{K}_{\alpha}(b,\omega) \int_{\mathbb{R}} \mathscr{F}^{\alpha}[h](\eta) \, \overline{\mathscr{F}[g_2]\big((\eta - \omega)\csc\alpha\big)} \, \overline{\mathcal{K}_{\alpha}(b,\eta)} \, d\eta.$$

Therefore, we have

$$\int_{\mathbb{R}} \int_{\mathbb{R}} \mathscr{G}_{g_1}^{\alpha}[f](b,\omega) \overline{\mathscr{G}_{g_2}^{\alpha}[h](b,\omega)} \, db \, d\omega$$

$$= 2\pi \int_{\mathbb{R}} \int_{\mathbb{R}} \left\{ \mathcal{K}_{\alpha}(b,\omega) \int_{\mathbb{R}} \mathscr{F}^{\alpha}[f](\xi) \, \overline{\mathscr{F}[g_1]\big((\xi - \omega)\csc\alpha\big)} \, \overline{\mathcal{K}_{\alpha}(b,\xi)} \, d\xi \right\}$$

$$\times \left\{ \overline{\mathcal{K}_{\alpha}(b,\omega)} \int_{\mathbb{R}} \overline{\mathscr{F}^{\alpha}[h](\eta)} \, \mathscr{F}[g_2]\big((\eta - \omega)\csc\alpha\big) \mathcal{K}_{\alpha}(b,\eta) \, d\eta \right\} db \, d\omega$$

$$= \frac{2\pi}{2\pi|\sin\alpha|} \int_{\mathbb{R}} \int_{\mathbb{R}} \int_{\mathbb{R}} \mathscr{F}^{\alpha}[f](\xi) \, \overline{\mathscr{F}^{\alpha}[h](\eta)} \, \overline{\mathscr{F}[g_1]\big((\xi - \omega)\csc\alpha\big)} \, \mathscr{F}[g_2]\big((\eta - \omega)\csc\alpha\big)$$

$$\times \left\{ \int_{\mathbb{R}} \mathcal{K}_{\alpha}(b,\eta) \, \overline{\mathcal{K}_{\alpha}(b,\xi)} \, db \right\} d\omega \, d\xi \, d\eta$$

$$= \frac{1}{|\sin\alpha|} \int_{\mathbb{R}} \int_{\mathbb{R}} \int_{\mathbb{R}} \mathscr{F}^{\alpha}[f](\xi) \, \overline{\mathscr{F}^{\alpha}[h](\eta)} \, \overline{\mathscr{F}[g_1]\big((\xi - \omega)\csc\alpha\big)} \, \mathscr{F}[g_2]\big((\eta - \omega)\csc\alpha\big)$$

$$\times \, \delta(\eta - \xi) \, d\omega \, d\xi \, d\eta$$

$$= \frac{1}{|\sin\alpha|} \int_{\mathbb{R}} \int_{\mathbb{R}} \mathscr{F}^{\alpha}[f](\xi) \, \overline{\mathscr{F}^{\alpha}[h](\xi)} \, \overline{\mathscr{F}[g_1]\big((\xi - \omega)\csc\alpha\big)} \, \mathscr{F}[g_2]\big((\xi - \omega)\csc\alpha\big) \, d\omega \, d\xi$$

$$= \int_{\mathbb{R}} \mathscr{F}^{\alpha}[f](\xi) \, \overline{\mathscr{F}^{\alpha}[h](\xi)} \left\{ \frac{1}{|\sin\alpha|} \int_{\mathbb{R}} \overline{\mathscr{F}[g_1]\big((\xi - \omega)\csc\alpha\big)} \, \mathscr{F}[g_2]\big((\xi - \omega)\csc\alpha\big) \, d\omega \right\} d\xi$$

$$= \int_{\mathbb{R}} \mathscr{F}^{\alpha}[f](\xi) \, \overline{\mathscr{F}^{\alpha}[h](\xi)} \left\{ \int_{\mathbb{R}} \overline{\mathscr{F}[g_1](\sigma)} \, \mathscr{F}[g_2](\sigma) \, d\sigma \right\} d\omega$$

$$= \left\langle \mathscr{F}[g_2], \mathscr{F}[g_1] \right\rangle_2 \left\langle \mathscr{F}^{\alpha}[f], \mathscr{F}^{\alpha}[h] \right\rangle_2$$

$$= \left\langle g_2, g_1 \right\rangle_2 \left\langle f, h \right\rangle_2.$$

This completes the proof Theorem 2.3.3.   □

Remark 2.3.1. For $f = h$ and $g_1 = g_2 = g$ the orthogonality relation (2.75) yields

$$\int_{\mathbb{R}} \int_{\mathbb{R}} \left| \mathscr{G}_g^{\alpha}[f](b,\omega) \right|^2 db \, d\omega = \left\| f \right\|_2^2 \left\| g \right\|_2^2, \tag{2.76}$$

which gives the net concentration of the windowed fractional Fourier transform in the time-frequency plane. In particular, choosing the window function so that $\|g\|_2 = 1$, we obtain

$$\int_{\mathbb{R}} \int_{\mathbb{R}} \left| \mathscr{G}_g^{\alpha}[f](b,\omega) \right|^2 db \, d\omega = \left\| f \right\|_2^2. \tag{2.77}$$

That is, the total energy of windowed fractional Fourier transform is equal to the total energy of the input signal. In other words, no loss of energy occurs between the natural and transformed domains.

An inversion formula for the windowed fractional Fourier transform (2.68) is given in the following theorem:

Theorem 2.3.4. *Let $\mathscr{G}_{g_1}^{\alpha}[f](b, \omega)$ be the windowed fractional Fourier transform of any $f \in L^2(\mathbb{R})$ with respect to the analyzing window function $g_1 \in L^2(\mathbb{R})$. Then, for any synthesis window function $g_2 \in L^2(\mathbb{R})$, we have*

$$f(t) = \frac{1}{\langle g_2, g_1 \rangle_2} \int_{\mathbb{R}} \int_{\mathbb{R}} \mathscr{G}_{g_1}^{\alpha}[f](b, \omega) \, g_2(t - b) \, \overline{\mathcal{K}_{\alpha}(t, \omega)} \, db \, d\omega, \tag{2.78}$$

with equality holding in the almost everywhere sense.

Proof. Under the given hypothesis and any arbitrary function $h \in L^2(\mathbb{R})$, we can write

$$\langle f, h \rangle_2 = \frac{1}{\langle g_2, g_1 \rangle_2} \int_{\mathbb{R}} \int_{\mathbb{R}} \mathscr{G}_{g_1}^{\alpha}[f](b, \omega) \, \overline{\mathscr{G}_{g_2}^{\alpha}[h](b, \omega)} \, db \, d\omega$$

$$= \frac{1}{\langle g_2, g_1 \rangle_2} \int_{\mathbb{R}} \int_{\mathbb{R}} \mathscr{G}_{g_1}^{\alpha}[f](b, \omega) \left\{ \int_{\mathbb{R}} \overline{h(t)} \, g_{2 b, \omega}^{\alpha}(t) \, dt \right\} db \, d\omega$$

$$= \int_{\mathbb{R}} \left\{ \frac{1}{\langle g_2, g_1 \rangle_2} \int_{\mathbb{R}} \int_{\mathbb{R}} \mathscr{G}_{g_1}^{\alpha}[f](b, \omega) \, g_{2 b, \omega}^{\alpha}(t) \, db \, d\omega \right\} \overline{h(t)} \, dt$$

$$= \int_{\mathbb{R}} \left\{ \frac{1}{\langle g_2, g_1 \rangle_2} \int_{\mathbb{R}} \int_{\mathbb{R}} \mathscr{G}_{g_1}^{\alpha}[f](b, \omega) \, g_2(t - b) \, \overline{\mathcal{K}_{\alpha}(t, \omega)} \, db \, d\omega \right\} \overline{h(t)} \, dt. \tag{2.79}$$

Noting that the function h is arbitrary and using the fundamental property of the inner product, we obtain the desired reconstruction formula as

$$f(t) = \frac{1}{\langle g_2, g_1 \rangle_2} \int_{\mathbb{R}} \int_{\mathbb{R}} \mathscr{G}_{g_1}^{\alpha}[f](b, \omega) \, g_2(t - b) \, \overline{\mathcal{K}_{\alpha}(t, \omega)} \, db \, d\omega, \quad a.e.$$

This completes the proof Theorem 2.3.4.   □

Remark 2.3.2. In case a single window function $g \in L^2(\mathbb{R})$ is applied for both the analysis and synthesis purposes, then the inversion formula (2.78) takes the following form:

$$f(t) = \frac{1}{\|g\|_2^2} \int_{\mathbb{R}} \int_{\mathbb{R}} \mathscr{G}_{g}^{\alpha}[f](b, \omega) \, g(t - b) \, \overline{\mathcal{K}_{\alpha}(t, \omega)} \, db \, d\omega. \tag{2.80}$$

Particularly, if a normalized window is used, then we have

$$f(t) = \int_{\mathbb{R}} \int_{\mathbb{R}} \mathscr{G}_{g}^{\alpha}[f](b, \omega) \, g(t - b) \, \overline{\mathcal{K}_{\alpha}(t, \omega)} \, db \, d\omega. \tag{2.81}$$

However, if the synthesis window g_2 appearing in (2.78) is chosen to be $g_2(t) = \delta(t)$, then we observe that

$$\langle g_2, g_1 \rangle_2 = \int_{\mathbb{R}} \overline{g_1(t)} \, \delta(t - 0) \, dt = \overline{g_1(0)}.$$

Consequently, the inversion formula (2.78) boils down to

$$f(t) = \frac{1}{g_1(0)} \int_{\mathbb{R}} \int_{\mathbb{R}} \mathscr{G}_{g_1}^{\alpha}\big[f\big](b,\omega)\, \delta(t-b)\, \overline{\mathcal{K}_{\alpha}(t,\omega)}\, db\, d\omega$$

$$= \frac{1}{g_1(0)} \int_{\mathbb{R}} \mathscr{G}_{g_1}^{\alpha}\big[f\big](t,\omega)\, \overline{\mathcal{K}_{\alpha}(t,\omega)}\, d\omega, \tag{2.82}$$

provided $g_1(0) \neq 0$, with the equality holding in almost everywhere sense. For the sake of distinction, the inversion formula (2.82) is often referred as the "one-dimensional inversion formula" associated with the windowed fractional Fourier transform.

Towards the end of this subsection, we investigate on the range of the windowed fractional Fourier transform (2.68) and demonstrate that, indeed, it constitutes a reproducing kernel Hilbert space.

Theorem 2.3.5. *A function $f \in L^2(\mathbb{R}^2)$ is the windowed fractional Fourier transform of a certain square integrable function with respect to the window function $g \in L^2(\mathbb{R})$ if and only if it satisfies the following reproduction formula:*

$$f(b',\omega') = \int_{\mathbb{R}} \int_{\mathbb{R}} f(b,\omega)\, \mathscr{K}_g^{\alpha}\big(b,\omega;b',\omega'\big)\, db\, d\omega, \tag{2.83}$$

where

$$\mathscr{K}_g^{\alpha}\big(b,\omega;b',\omega'\big) = \frac{1}{\|g\|_2^2} \Big\langle g(t-b)\, \overline{\mathcal{K}_{\alpha}(t,\omega)},\, g(t-b')\, \overline{\mathcal{K}_{\alpha}(t,\omega')} \Big\rangle_2. \tag{2.84}$$

Proof. Firstly, we suppose that a function $f \in L^2(\mathbb{R}^2)$ is the windowed fractional Fourier transform of $h \in L^2(\mathbb{R})$ with respect to the window function $g \in L^2(\mathbb{R})$; that is, $\mathscr{G}_g^{\alpha}\big[h\big] = f$. Then, we proceed to show that (2.83) holds good. In this direction, we have

$$f(b',\omega') = \mathscr{G}_g^{\alpha}\big[h\big](b',\omega')$$

$$= \int_{\mathbb{R}} h(t)\, \overline{g(t-b')}\, \mathcal{K}_{\alpha}(t,\omega')\, dt$$

$$= \int_{\mathbb{R}} \left\{ \frac{1}{\|g\|_2^2} \int_{\mathbb{R}} \int_{\mathbb{R}} \mathscr{G}_g^{\alpha}\big[h\big](b,\omega)\, g(t-b)\, \overline{\mathcal{K}_{\alpha}(t,\omega)}\, db\, d\omega \right\} \overline{g(t-b')}\, \mathcal{K}_{\alpha}(t,\omega')\, dt$$

$$= \int_{\mathbb{R}} \int_{\mathbb{R}} \mathscr{G}_g^{\alpha}\big[h\big](b,\omega) \left\{ \frac{1}{\|g\|_2^2} \int_{\mathbb{R}} g(t-b)\, \overline{\mathcal{K}_{\alpha}(t,\omega)}\, \overline{g(t-b')}\, \mathcal{K}_{\alpha}(t,\omega')\, dt \right\} db\, d\omega$$

$$= \int_{\mathbb{R}} \int_{\mathbb{R}} \mathscr{G}_g^{\alpha}\big[h\big](b,\omega) \left\{ \frac{1}{\|g\|_2^2} \Big\langle g(t-b)\, \overline{\mathcal{K}_{\alpha}(t,\omega)},\, g(t-b')\, \overline{\mathcal{K}_{\alpha}(t,\omega')} \Big\rangle_2 \right\} db\, d\omega$$

$$= \int_{\mathbb{R}} \int_{\mathbb{R}} f(b,\omega)\, \mathscr{K}_g^{\alpha}\big(b,\omega;b',\omega'\big)\, db\, d\omega. \tag{2.85}$$

On the flip side, we assume that a function $f \in L^2(\mathbb{R}^2)$ satisfies (2.83) and then show that such a function must be the windowed fractional Fourier transform of a certain square integrable function s. We claim that

$$s(t) = \frac{1}{\|g\|_2^2} \int_{\mathbb{R}} \int_{\mathbb{R}} f(b,\omega) \, g(t-b) \, \overline{\mathcal{K}_\alpha(t,\omega)} \, db \, d\omega. \tag{2.86}$$

In order to show that the function $s(t)$ is square integrable, we proceed as

$$\|s\|_2^2 = \int_{\mathbb{R}} s(t) \, \overline{s(t)} \, dt$$

$$= \int_{\mathbb{R}} \left\{ \frac{1}{\|g\|_2^2} \int_{\mathbb{R}} \int_{\mathbb{R}} f(b,\omega) \, g(t-b) \, \overline{\mathcal{K}_\alpha(t,\omega)} \, db \, d\omega \right\}$$

$$\times \left\{ \frac{1}{\|g\|_2^2} \int_{\mathbb{R}} \int_{\mathbb{R}} \overline{f(b',\omega')} \, \overline{g(t-b')} \, \mathcal{K}_\alpha(t,\omega') \, db' \, d\omega' \right\} dt$$

$$= \frac{1}{\|g\|_2^2} \int_{\mathbb{R}} \int_{\mathbb{R}} \int_{\mathbb{R}} \int_{\mathbb{R}} f(b,\omega) \, \overline{f(b',\omega')}$$

$$\times \left\{ \frac{1}{\|g\|_2^2} \int_{\mathbb{R}} g(t-b) \, \overline{\mathcal{K}_\alpha(t,\omega)} \, \overline{g(t-b')} \, \mathcal{K}_\alpha(t,\omega') \, dt \right\} db \, d\omega \, db' \, d\omega'$$

$$= \frac{1}{\|g\|_2^2} \int_{\mathbb{R}} \int_{\mathbb{R}} \int_{\mathbb{R}} \int_{\mathbb{R}} f(b,\omega) \, \overline{f(b',\omega')} \mathcal{H}_g^\alpha(b,\omega;b',\omega') \, db \, d\omega \, db' \, d\omega'$$

$$= \frac{1}{\|g\|_2^2} \int_{\mathbb{R}} \int_{\mathbb{R}} \overline{f(b',\omega')} \left\{ \int_{\mathbb{R}} \int_{\mathbb{R}} f(b,\omega) \, \mathcal{H}_g^\alpha(b,\omega;b',\omega') \, db \, d\omega \right\} db' \, d\omega'$$

$$= \frac{1}{\|g\|_2^2} \int_{\mathbb{R}} \int_{\mathbb{R}} f(b',\omega') \, \overline{f(b',\omega')} \, db' \, d\omega'$$

$$= \frac{1}{\|g\|_2^2} \|f\|_2^2 < \infty. \tag{2.87}$$

As a consequence of (2.87), we infer that the function $s(t)$ defined in (2.86) is indeed square integrable. Finally, we note that the windowed fractional Fourier transform of $s(t)$ is expressible as

$$\mathcal{G}_g^\alpha \big[s\big](b',\omega') = \int_{\mathbb{R}} s(t) \, \overline{g(t-b')} \, \mathcal{K}_\alpha(t,\omega') \, dt$$

$$= \int_{\mathbb{R}} \left\{ \frac{1}{\|g\|_2^2} \int_{\mathbb{R}} \int_{\mathbb{R}} f(b,\omega) \, g(t-b) \, \overline{\mathcal{K}_\alpha(t,\omega)} \, db \, d\omega \right\} \overline{g(t-b')} \, \mathcal{K}_\alpha(t,\omega') \, dt$$

$$= \int_{\mathbb{R}} \int_{\mathbb{R}} f(b,\omega) \left\{ \frac{1}{\|g\|_2^2} \int_{\mathbb{R}} g(t-b) \, \overline{\mathcal{K}_\alpha(t,\omega)} \, \overline{g(t-b')} \, \mathcal{K}_\alpha(t,\omega') \, dt \right\} db \, d\omega$$

$$= \int_{\mathbb{R}} \int_{\mathbb{R}} f(b,\omega) \left\{ \frac{1}{\|g\|_2^2} \Big\langle g(t-b) \, \overline{\mathcal{K}_\alpha(t,\omega)}, \, g(t-b') \, \overline{\mathcal{K}_\alpha(t,\omega')} \Big\rangle_2 \right\} db \, d\omega$$

$$= \int_{\mathbb{R}} \int_{\mathbb{R}} f(b,\omega) \, \mathcal{H}_g^\alpha(b,\omega;b',\omega') \, db \, d\omega$$

$$= f(b',\omega'),$$

where the change in the order of integration is justifiable in view of the well-known Fubini's theorem. This completes the proof of Theorem 2.3.5. $\qquad \square$

Corollary 2.3.6. *The range of the windowed fractional Fourier transform* (2.68) *is a reproducing kernel Hilbert space whose kernel is given by* (2.84).

2.3.2   Time-fractional-frequency resolution

This subsection is entirely devoted to assess the time-fractional-frequency resolution of the windowed fractional Fourier transform (2.68).

Consider a window function $g \in L^2(\mathbb{R})$ with centre E_g and radius Δ_g. In order to obtain the centre and radius of each of the analyzing functions $g_{b,\omega}^\alpha(t)$, we proceed as

$$
\begin{aligned}
E_{g_{b,\omega}^\alpha} &= \frac{1}{\left\|g_{b,\omega}^\alpha\right\|_2^2} \int_{\mathbb{R}} t \left|g_{b,\omega}^\alpha(t)\right|^2 dt \\
&= \frac{1}{\left\|g\right\|_2^2} \int_{\mathbb{R}} t \left|g(t-b)\right|^2 dt \\
&= \frac{1}{\left\|g\right\|_2^2} \int_{\mathbb{R}} (b+z) \left|g(z)\right|^2 dz \\
&= \frac{1}{\left\|g\right\|_2^2} \left\{ b \left\|g\right\|_2^2 + \int_{\mathbb{R}} z \left|g(z)\right|^2 dz \right\} \\
&= b + \int_{\mathbb{R}} z \left|g(z)\right|^2 dz \\
&= b + E_g
\end{aligned}
$$

and

$$
\begin{aligned}
\Delta_{g_{b,\omega}^\alpha} &= \frac{1}{\left\|g\right\|_2} \left\{ \int_{\mathbb{R}} (t-b-E_g)^2 \left|g(t-b)\right|^2 dt \right\}^{1/2} \\
&= \frac{1}{\left\|g\right\|_2} \left\{ \int_{\mathbb{R}} (z-E_g)^2 \left|g(z)\right|^2 dz \right\}^{1/2} \\
&= \Delta_g.
\end{aligned}
$$

Therefore, we infer that the time window of the windowed fractional Fourier transform is given by

$$
\left[b + E_g - \Delta_g, \; b + E_g + \Delta_g \right]. \tag{2.88}
$$

In order to obtain the explicit form of the window in the fractional Fourier domain, we shall use Proposition 2.3.1 and note that the centre of the fractional-frequency domain window $G_\omega^\alpha(\xi) = \hat{g}\big((\xi - \omega)\csc\alpha\big)$ is given by

$$
\begin{aligned}
E_{G_\omega^\alpha(\xi)} &= \frac{1}{\left\|G_\omega^\alpha(\xi)\right\|_2^2} \int_{\mathbb{R}} \xi \left|G_\omega^\alpha(\xi)\right|^2 d\xi \\
&= \frac{|\csc\alpha|}{\left\|\hat{g}\right\|_2^2} \int_{\mathbb{R}} \xi \left|\hat{g}\big((\xi-\omega)\csc\alpha\big)\right|^2 d\xi \\
&= \frac{|\csc\alpha|}{\left\|\hat{g}\right\|_2^2} \int_{\mathbb{R}} \left(\omega + \frac{\eta}{\csc\alpha}\right) \left|\hat{g}(\eta)\right|^2 \frac{d\eta}{|\csc\alpha|}
\end{aligned}
$$

$$= \frac{|\csc\alpha|}{\|\hat{g}\|_2^2} \left\{ \int_{\mathbb{R}} \omega \left|\hat{g}(\eta)\right|^2 \frac{d\eta}{|\csc\alpha|} + \int_{\mathbb{R}} \left(\frac{\eta}{\csc\alpha}\right) \left|\hat{g}(\eta)\right|^2 \frac{d\eta}{|\csc\alpha|} \right\}$$

$$= \omega + \frac{|\csc\alpha|}{\|\hat{g}\|_2^2} \left(\frac{1}{\csc\alpha}\right) \left\{ \int_{\mathbb{R}} \eta \left|\hat{g}(\eta)\right|^2 d\eta \right\} \frac{1}{|\csc\alpha|}$$

$$= \omega + \sin\alpha \, E_{\hat{g}}. \tag{2.89}$$

Also, the radius is given by

$$\Delta_{G_\omega^\alpha(\xi)} = \frac{1}{\|G_\omega^\alpha(\xi)\|_2} \left\{ \int_{\mathbb{R}} (\xi - \omega - \sin\alpha \, E_{\hat{g}})^2 \left|\hat{g}((\xi - \omega)\csc\alpha)\right|^2 d\xi \right\}^{1/2}$$

$$= \frac{|\csc\alpha|^{1/2}}{\|\hat{g}\|_2} \left\{ \sin^2\alpha \int_{\mathbb{R}} ((\xi - \omega)\csc\alpha - E_{\hat{g}})^2 \left|\hat{g}((\xi - \omega)\csc\alpha)\right|^2 d\xi \right\}^{1/2}$$

$$= \frac{|\sin\alpha|}{\|\hat{g}\|_2} \left\{ \int_{\mathbb{R}} ((\xi - \omega)\csc\alpha - E_{\hat{g}})^2 \left|\hat{g}((\xi - \omega)\csc\alpha)\right|^2 d(\xi\csc\alpha) \right\}^{1/2}$$

$$= \frac{|\sin\alpha|}{\|\hat{g}\|_2} \left\{ \int_{\mathbb{R}} (\eta - E_{\hat{g}})^2 \left|\hat{g}(\eta)\right|^2 d\eta \right\}^{1/2}$$

$$= |\sin\alpha| \, \Delta_{\hat{g}}. \tag{2.90}$$

Consequently, the window for localizing the fractional-frequency components is given by

$$\left[\omega + \sin\alpha \, E_{\hat{g}} - |\sin\alpha| \, \Delta_{\hat{g}}, \; \omega + \sin\alpha \, E_{\hat{g}} + |\sin\alpha| \, \Delta_{\hat{g}} \right]. \tag{2.91}$$

Therefore, the joint time and frequency localization of the windowed fractional Fourier transform (2.68) in the time-fractional-frequency plane is given by the window

$$\left[b + E_g - \Delta_g, \; b + E_g + \Delta_g \right] \times \left[\omega + \sin\alpha \, E_{\hat{g}} - |\sin\alpha| \, \Delta_{\hat{g}}, \; \omega + \sin\alpha \, E_{\hat{g}} + |\sin\alpha| \, \Delta_{\hat{g}} \right], \tag{2.92}$$

occupying a net area of $4\Delta_g\Delta_{\hat{g}} |\sin\alpha|$, which is dictated by the fractional parameter α. Keeping in view the intrinsic features of the fractional Fourier transform together with (2.92), we infer that the windowed fractional Fourier transform induces an adjustable, non-rectangular tiling onto the time-frequency plane. More precisely, these adjustable tiles are parallelograms and are called as the "time-fractional-frequency cells." A pictorial illustration of the time-fractional-frequency cells is given in Figure 2.17. Moreover, the time-fractional-frequency resolution of the windowed fractional Fourier transform is denoted by R_α and is defined as the reciprocal value of the duration-bandwidth product; that is,

$$R_\alpha := \frac{1}{\Delta_{g_{b,\omega}^\alpha} \, \Delta_{G_\omega^\alpha(\xi)}} = \frac{1}{\Delta_g \Delta_{\hat{g}} |\sin\alpha|}. \tag{2.93}$$

It is worth noticing that the resolution of the windowed fractional Fourier transform is reliant upon the fractional parameter α and the same can be invoked for resolution adjustments according to the need. In view of the Heisenberg's uncertainty principle, we infer that, for any normalized window function g, the upper bound for the resolution of the windowed fractional Fourier transform (2.68) is $2/|\sin\alpha|$, which is achieved if and only if g is the Gaussian function. It is imperative to mention that, the windowed fractional Fourier transform enjoys higher resolution than the conventional windowed Fourier transform, because

$$R_\alpha = \frac{1}{\Delta_g \Delta_{\hat{g}} |\sin\alpha|} \geq \frac{1}{\Delta_g \Delta_{\hat{g}}} = R, \tag{2.94}$$

where R is the resolution of the usual windowed Fourier transform. Thus, we conclude that the windowed fractional Fourier transform (2.68) is best suited for signals which cannot be efficiently resolved via the conventional windowed Fourier transform, for instance the chirp-like signals. In order to appreciate such nuances, we note that a chirp signal has an oblique support with respect to the time-axis in the usual time-frequency plane. Recall that the conventional windowed Fourier transform induces a rectangular tiling on the time-frequency plane; therefore, the support of any typical chirp signal can be bounded by a rectangular region in the time-frequency plane. Moreover, the coordinates of the centre of the rectangle are the mean-time and mean-frequency, and its two sides are determined by the lengths of the time and frequency localizing windows associated with the windowed Fourier transform. The product of the two sides, which is also a measure of resolution of the windowed Fourier transform in the time-frequency plane, is termed as the time-frequency product. Thus, the support of a chirp signal can be encapsulated in a rectangular region bearing a time-frequency product of $\Delta_g \Delta_{\hat{g}}$. On the other hand, we note that both the mean-time and time-width of windowed fractional Fourier transform are identical with the usual windowed Fourier transform, whereas the mean-frequency and frequency-width are reliant upon the fractional parameter α. By appropriately choosing α, a sense of rotation can be inculcated into the frequency axis of the time-frequency plane, so that the chirp signal has a support parallel to the time-axis, which can be efficiently circumscribed within a parallelogram or a time-fractional-frequency cell occupying an area of $4\Delta_g \Delta_{\hat{g}} |\sin\alpha| \leq 4\Delta_g \Delta_{\hat{g}}$. A pictorial illustration of the aforementioned facts is lucidly presented in Figure 2.18.

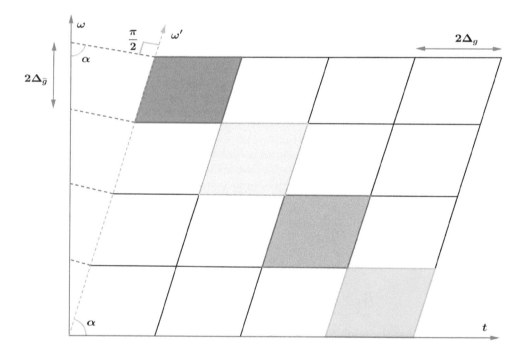

FIGURE 2.17: Time-frequency tiling induced by windowed fractional Fourier transform.

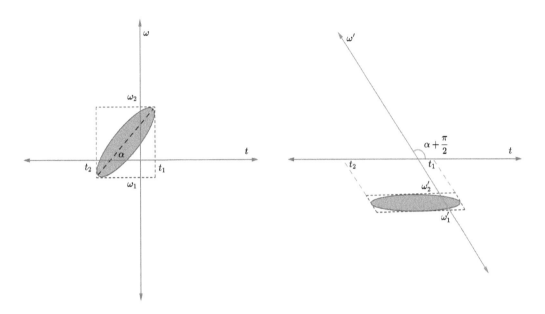

FIGURE 2.18: Working mechanism of windowed fractional Fourier transform for chirps.

2.4 The Windowed Linear Canonical Transform

"In most sciences one generation tears down what another has built, and what one has established, another undoes. In mathematics alone each generation adds a new storey to the old structure."

-Hermann Hankel

In pursuit of localizing the linear canonical spectrum of non-stationary signals, the windowed linear canonical transform has gained a considerable attention and also marked the birth of an all-inclusive time-frequency tool encompassing both the windowed Fourier and windowed fractional Fourier transforms. The windowed linear canonical transform is based on the concept of capturing the linear canonical spectrum of a non-transient signal over restricted temporal sections, confined via a window function. In this section, we shall study the notion of windowed linear canonical transform together with its fundamental properties, including an exclusive discussion on its resolution in the time-frequency plane and the associated uncertainty principles.

2.4.1 Definition and basic properties

This subsection is meant for the formal definition and elementary properties of the windowed linear canonical transform (WLCT). Among the fundamental properties, we formulate an orthogonality relation and inversion formula pertaining to the windowed linear canonical transform. In consonance with the already set notation, we shall write a 2×2 matrix $M = \begin{pmatrix} A & B \\ C & D \end{pmatrix}$ as $M = (A, B; C, D)$.

Definition 2.4.1. For $M = (A, B; C, D) \in SL(2, \mathbb{R})$, the special linear group of degree 2 over $\mathbb{R}$, with $B \neq 0$, the windowed linear canonical transform of any $f \in L^2(\mathbb{R})$ with respect to a window function $g \in L^2(\mathbb{R})$ is denoted by $\mathscr{G}_g^M[f]$ and is defined as

$$\mathscr{G}_g^M[f](b, \omega) = \frac{1}{\sqrt{2\pi i B}} \left\langle f, g_{b,\omega}^M \right\rangle_2 = \int_{\mathbb{R}} f(t) \, \overline{g(t - b)} \, \mathcal{K}_M(t, \omega) \, dt, \quad b, \omega \in \mathbb{R}, \qquad (2.95)$$

where $\mathcal{K}_M(t, \omega)$ is the usual linear canonical kernel defined in (1.165) and $g_{b,\omega}^M(t)$ denote the analyzing functions defined by

$$g_{b,\omega}^M(t) = \exp\left\{ -\frac{i(At^2 - 2t\omega + D\omega^2)}{2B} \right\} g(t - b). \qquad (2.96)$$

In view of Definition 2.4.1, we mark the some important points concerning the notion of windowed linear canonical transform as under:

(i) The quantity $\left| \mathscr{G}_g^M[f](b, \omega) \right|^2$ is termed as the "energy density" of the windowed linear canonical transform. Also, it is important to note that the analyzing functions defined in (2.96) obey the norm equality $\left\| g_{b,\omega}^M \right\|_2^2 = \|g\|_2^2$.

(ii) Infrequently in practice, the pre-factor $1/\sqrt{2\pi i B}$ appearing in (2.95) is replaced with $1/\sqrt{2\pi |B|}$.

(iii) In case the window function $g \in L^2(\mathbb{R})$ is chosen as the Gaussian function, the WLCT (2.95) is referred as the "linear canonical Gabor transform."

(iv) For a fixed b, the windowed linear canonical transform (2.95) can be interpreted as the LCT of the product of a function f with the conjugated and translated window function g; that is,

$$\mathscr{G}_g^M[f](b, \omega) = \mathscr{L}_M\left[f(t) \, \overline{g(t - b)} \right](\omega). \qquad (2.97)$$

Therefore, we infer that the windowed linear canonical transform (2.95) performs the LCT over selective portions of the input signal and can thus be viewed as the instantaneous linear canonical spectrum of the signal f. Upon varying b across the temporal axis, we can ascertain the local information regarding the linear canonical spectrum of the signal.

(v) For the matrix $M = (0, 1; -1, 0)$, the windowed linear canonical transform (2.95) boils down to conventional windowed Fourier transform (2.50), after the former is multiplied with $\sqrt{i}$. In fact, we have the following relationship:

$$\mathscr{G}_g^M[f](b, \omega) = \frac{1}{\sqrt{2\pi i B}} \int_{\mathbb{R}} f(t) \, \overline{g(t - b)} \exp\left\{ \frac{i(At^2 - 2t\omega + D\omega^2)}{2B} \right\} dt$$

$$= \frac{1}{\sqrt{2\pi i B}} \exp\left\{ \frac{iD\omega^2}{2B} \right\} \int_{\mathbb{R}} \exp\left\{ \frac{iAt^2}{2B} \right\} f(t) \, \overline{g(t - b)} \exp\left\{ -\frac{it\omega}{B} \right\} dt$$

$$= \frac{1}{\sqrt{iB}} \exp\left\{ \frac{iD\omega^2}{2B} \right\} \mathscr{G}_g[F]\left(b, \frac{\omega}{B} \right), \qquad (2.98)$$

where $F(t) = \exp\left\{ \dfrac{iAt^2}{2B} \right\} f(t)$. Similarly, an analogous relationship with the windowed fractional Fourier transform (2.68) can be obtained by plugging $M = (\cos\alpha, \sin\alpha; -\sin\alpha, \cos\alpha)$, $\alpha \neq n\pi$.

(vi) For the matrix $M = (1, B; 0, 1)$, we can obtain a novel transform namely, the windowed linear canonical Fresnel transform defined as

$$\mathscr{G}_g^M\big[f\big](b, \omega) = \frac{1}{\sqrt{2\pi i B}} \int_{\mathbb{R}} f(t)\, \overline{g(t - b)} \exp\left\{ \frac{i(t - \omega)^2}{2B} \right\} dt. \tag{2.99}$$

(vii) Since the conventional windowed Fourier transform (2.50) is expressible via the translation and modulation operators as demonstrated in (2.52), therefore, we can recast the windowed linear canonical transform (2.95) as

$$\mathscr{G}_g^M\big[f\big](b, \omega) = \frac{1}{\sqrt{2\pi i B}} \int_{\mathbb{R}} f(t) \exp\left\{ \frac{i(D\omega^2 - 2t\omega)}{2B} \right\} \exp\left\{ \frac{iAt^2}{2B} \right\} \overline{g(t - b)}\, dt$$

$$= \frac{1}{\sqrt{2\pi i B}} \left\langle f,\, \mathcal{M}_\omega^M \mathcal{T}_b^M g \right\rangle, \tag{2.100}$$

where $\mathcal{T}_b^M$ and $\mathcal{M}_\omega^M$ denote the generalized translation and modulation operators given by

$$\left. \begin{aligned} \mathcal{T}_b^M g(t) &= \exp\left\{ -\frac{iAt^2}{2B} \right\} g(t - b) \\ \mathcal{M}_\omega^M g(t) &= \exp\left\{ \frac{i(2t\omega - D\omega^2)}{2B} \right\} g(t) \end{aligned} \right\}. \tag{2.101}$$

(viii) Keeping in view the fact that the conventional windowed Fourier transform (2.50) can be expressed via the classical convolution operation $*$ as shown in (2.53), we can formulate several new variants of the windowed linear canonical transform (2.95) by replacing the classical convolution operation in (2.53) with any of the linear canonical convolutional operations. Such convolution-based windowed linear canonical transforms have the prime advantage of providing the localized information about the linear canonical spectrum at a relatively lower computational cost.

In the following proposition, we obtain a spectral representation of the windowed linear canonical transform defined in (2.95). Such a representation allows us to lucidly examine the localization characteristics of the windowed linear canonical transform.

Proposition 2.4.1. *Let $\mathscr{G}_g^M\big[f\big](b, \omega)$ be the windowed linear canonical transform of any $f \in L^2(\mathbb{R})$, then we have*

$$\mathscr{G}_g^M\big[f\big](b, \omega) = \sqrt{2\pi}\, \mathcal{K}_M(b, \omega) \int_{\mathbb{R}} \mathscr{L}_M\big[f\big](\xi)\, \overline{\mathscr{F}\big[g\big]\left(\frac{\xi - \omega}{B} \right)}\, \overline{\mathcal{K}_M(b, \xi)}\, d\xi, \tag{2.102}$$

where $\mathscr{F}$ is the classical Fourier transform defined in (1.7).

Proof. By virtue of Definition 2.4.1, we have

$$\mathscr{G}_g^M\big[f\big](b, \omega) = \frac{1}{\sqrt{2\pi i B}} \left\langle f,\, g_{b,\omega}^M \right\rangle_2$$

$$= \frac{1}{\sqrt{2\pi i B}} \int_{\mathbb{R}} \mathscr{L}_M\big[f\big](\xi)\, \overline{\mathscr{L}_M\big[g_{b,\omega}^M\big](\xi)}\, d\xi. \tag{2.103}$$

Also, we observe that

$$\mathscr{L}_M\left[g_{b,\omega}^M\right](\xi) = \int_{\mathbb{R}} g_{b,\omega}^M(t)\,\mathcal{K}_M(t,\xi)\,dt$$

$$= \frac{1}{\sqrt{2\pi i B}} \int_{\mathbb{R}} g_{b,\omega}^M(t)\,\exp\left\{\frac{i(At^2 - 2t\xi + D\xi^2)}{2B}\right\}dt$$

$$= \frac{1}{\sqrt{2\pi i B}} \int_{\mathbb{R}} g(t-b)\,\exp\left\{-\frac{i(At^2 - 2t\omega + D\omega^2)}{2B}\right\}\exp\left\{\frac{i(At^2 - 2t\xi + D\xi^2)}{2B}\right\}dt$$

$$= \frac{1}{\sqrt{2\pi i B}} \int_{\mathbb{R}} g(t-b)\,\exp\left\{\frac{i(\omega-\xi)t}{B}\right\}\exp\left\{\frac{iD(\xi^2-\omega^2)}{2B}\right\}dt$$

$$= \frac{1}{\sqrt{2\pi i B}} \int_{\mathbb{R}} g(z)\,\exp\left\{\frac{i(\omega-\xi)(b+z)}{B}\right\}\exp\left\{\frac{iD(\xi^2-\omega^2)}{2B}\right\}dz$$

$$= \exp\left\{\frac{iD(\xi^2-\omega^2)}{2B}\right\}\frac{1}{\sqrt{2\pi i B}} \int_{\mathbb{R}} g(z)\,\exp\left\{\frac{i(\omega-\xi)b}{B}\right\}\exp\left\{\frac{i(\omega-\xi)z}{B}\right\}dz$$

$$= \exp\left\{\frac{iD(\xi^2-\omega^2)+2i(\omega-\xi)b}{2B}\right\}\frac{1}{\sqrt{iB}}\left\{\frac{1}{\sqrt{2\pi}}\int_{\mathbb{R}} g(z)\,\exp\left\{-i\left(\frac{\xi-\omega}{B}\right)\right\}dz\right\}$$

$$= \frac{1}{\sqrt{iB}}\,\exp\left\{\frac{iD(\xi^2-\omega^2)}{2B}\right\}\exp\left\{\frac{2i(\omega-\xi)b}{2B}\right\}\mathscr{F}[g]\left(\frac{\xi-\omega}{B}\right). \tag{2.104}$$

Using (2.104) in (2.103), we obtain

$$\mathscr{G}_g^M\left[f\right](b,\omega) = \frac{1}{\sqrt{2\pi i B}}\,\frac{1}{\sqrt{-iB}}\int_{\mathbb{R}} \mathscr{L}_M\left[f\right](\xi)\,\overline{\mathscr{F}[g]\left(\frac{\xi-\omega}{B}\right)}$$

$$\times \exp\left\{-\frac{iD(\xi^2-\omega^2)}{2B}\right\}\exp\left\{-\frac{2i(\omega-\xi)b}{2B}\right\}d\xi$$

$$= \frac{1}{\sqrt{-iB}}\,\frac{1}{\sqrt{2\pi i B}}\int_{\mathbb{R}} \mathscr{L}_M\left[f\right](\xi)\,\overline{\mathscr{F}[g]\left(\frac{\xi-\omega}{B}\right)}$$

$$\times \exp\left\{-\frac{i(Ab^2 - 2b\xi + D\xi^2)}{2B}\right\}\exp\left\{\frac{i(Ab^2 - 2b\omega + D\omega^2)}{2B}\right\}d\xi$$

$$= \sqrt{2\pi}\,\mathcal{K}_M(b,\omega)\int_{\mathbb{R}} \mathscr{L}_M\left[f\right](\xi)\,\overline{\mathscr{F}[g]\left(\frac{\xi-\omega}{B}\right)}\,\overline{\mathcal{K}_M(b,\xi)}\,d\xi.$$

This completes the proof of Proposition 2.4.1. □

In the following theorem, we assemble some fundamental properties of the windowed linear canonical transform defined in (2.95).

Theorem 2.4.2. *If g and h are two window functions and $\alpha, \beta \in \mathbb{C}$, $k, \omega_0 \in \mathbb{R}$, $\lambda \in \mathbb{R}\setminus\{0\}$, then for any pair of square integrable functions f and s, the windowed linear canonical transform (2.95) satisfies the following properties:*

(i) *Linearity:* $\mathscr{G}_g^M\left[\alpha f + \beta s\right](b,\omega) = \alpha\,\mathscr{G}_g^M\left[f\right](b,\omega) + \beta\,\mathscr{G}_g^M\left[s\right](b,\omega),$

(ii) *Anti-linearity:* $\mathscr{G}_{\alpha g + \beta h}^M\left[f\right](b,\omega) = \overline{\alpha}\,\mathscr{G}_g^M\left[f\right](b,\omega) + \overline{\beta}\,\mathscr{G}_h^M\left[f\right](b,\omega),$

(iii) *Time-shift:* $\mathscr{G}_g^M\left[f(t-k)\right](b,\omega) = \exp\left\{iCk\omega - \frac{iACk^2}{2}\right\}\mathscr{G}_g^M\left[f\right](b-k, \omega - Ak),$

(iv) *Scaling:* $\mathscr{G}_g^M\left[f(\lambda t)\right](b,\omega) = \mathscr{G}_{G_\lambda}^{M'}\left[f\right]\left(\lambda b, \frac{\omega}{\lambda}\right),$ *where $G_\lambda(t) = \frac{1}{|\lambda|}\,g\left(\frac{t}{\lambda}\right)$ and* $M' = (A/\lambda^2, B; C, \lambda^2 D),$

(v) *Modulation:* $\mathscr{G}_g^M\left[e^{i\omega_0 t}f(t)\right](b,\omega) = \exp\left\{iD\omega\omega_0 - \frac{iBD\omega_0^2}{2}\right\}\mathscr{G}_g^M\left[f\right](b, \omega - B\omega_0),$

(vi) *Conjugation:* $\mathscr{G}_g^M\left[\bar{f}\right](b,\omega) = \overline{\mathscr{G}_g^{M^{-1}}\left[f\right](b,\omega)}$, *provided* $A = D$,

(vii) *Symmetry:* $\mathscr{G}_g^M\left[f\right](b,\omega) = \exp\left\{iCb\left(\omega - \dfrac{Ab}{2}\right)\right\}\mathscr{G}_{\bar{f}}^M\left[\bar{g}\right](-b, \omega - Ab)$,

(viii) *Translation in window:* $\mathscr{G}_{g(t-k)}^M\left[f\right](b,\omega) = \mathscr{G}_g^M\left[f\right](b+k, \omega)$,

(ix) *Scaling in window:* $\mathscr{G}_{g(\lambda t)}^M\left[f\right](b,\omega) = \mathscr{G}_g^{M'}\left[F_\lambda\right]\left(\lambda b, \dfrac{\omega}{\lambda}\right)$, *where* $F_\lambda(t) = \dfrac{1}{|\lambda|}f\left(\dfrac{t}{\lambda}\right)$ *and*
$M' = (A/\lambda^2, B; C, \lambda^2 D)$.

Proof. The proof of properties (i) and (ii) follows directly from Definition 2.4.1. Below, we shall prove the remaining properties one by one.

(iii) Upon translating the input function $f \in L^2(\mathbb{R})$ by $k \in \mathbb{R}$, the WLCT is obtained as

$$\mathscr{G}_g^M\left[f(t-k)\right](b,\omega) = \frac{1}{\sqrt{2\pi iB}}\int_{\mathbb{R}} f(t-k)\,\overline{g_{b,\omega}^M(t)}\,dt$$

$$= \frac{1}{\sqrt{2\pi iB}}\int_{\mathbb{R}} f(z)\,\overline{g_{b,\omega}^M(z+k)}\,dz$$

$$= \frac{1}{\sqrt{2\pi iB}}\int_{\mathbb{R}} f(z)\,\overline{g(z+k-b)}$$
$$\times \exp\left\{\frac{i(A(z+k)^2 - 2(z+k)\omega + D\omega^2)}{2B}\right\}dz$$

$$= \frac{1}{\sqrt{2\pi iB}}\int_{\mathbb{R}} f(z)\,\overline{g(z-(b-k))}$$
$$\times \exp\left\{\frac{i(Az^2 + Ak^2 + 2Azk - 2z\omega - 2\omega k + D\omega^2)}{2B}\right\}dz$$

$$= \frac{1}{\sqrt{2\pi iB}}\int_{\mathbb{R}} f(z)\,\overline{g(z-(b-k))}\exp\left\{\frac{iAk^2}{2B}\right\}\exp\left\{-\frac{2ik\omega}{2B}\right\}$$
$$\times \exp\left\{\frac{i(Az^2 - 2z(\omega - Ak) + D(\omega - Ak)^2)}{2B}\right\}$$
$$\times \exp\left\{-\frac{iD((Ak)^2 - 2A\omega k)}{2B}\right\}dz$$

$$= \int_{\mathbb{R}} f(z)\,\overline{g(z-(b-k))}\,\mathcal{K}_M(z, \omega - Ak)$$
$$\times \exp\left\{\frac{i(Ak^2 - 2k\omega - D(Ak)^2 + 2AD\omega k)}{2B}\right\}dz$$

$$= \int_{\mathbb{R}} f(z)\,\overline{g(z-(b-k))}\,\mathcal{K}_M(z, \omega - Ak)$$
$$\times \exp\left\{\frac{i(Ak^2(1 - AD) + 2k\omega(AD - 1))}{2B}\right\}dz$$

$$= \exp\left\{\frac{i(2BCk\omega - ABCk^2)}{2B}\right\}\mathscr{G}_g^M\left[f\right](b-k, \omega - Ak)$$

$$= \exp\left\{iCk\omega - \frac{iACk^2}{2}\right\}\mathscr{G}_g^M\left[f\right](b-k, \omega - Ak),$$

where the last equality is obtained by using the unimodularity $AD - BC = 1$, so that $AD - 1 = BC$.

(iv) In case the input function $f \in L^2(\mathbb{R})$ is scaled by $\lambda \in \mathbb{R} \setminus \{0\}$, then we have

$$\mathscr{G}_g^M\Big[f(\lambda t)\Big](b,\omega) = \frac{1}{\sqrt{2\pi i B}} \int_{\mathbb{R}} f(\lambda t)\, \overline{g_{b,\omega}^M(t)}\, dt$$

$$= \frac{1}{\sqrt{2\pi i B}} \int_{\mathbb{R}} f(z)\, \overline{g_{b,\omega}^M\left(\frac{z}{\lambda}\right)}\, \frac{dz}{|\lambda|}$$

$$= \frac{1}{\sqrt{2\pi i B}} \int_{\mathbb{R}} f(z)\, \overline{g\left(\frac{z}{\lambda} - b\right)} \exp\left\{\frac{i(A(z/\lambda)^2 - 2(z/\lambda)\omega + D\omega^2)}{2B}\right\} \frac{dz}{|\lambda|}$$

$$= \frac{1}{\sqrt{2\pi i B}} \int_{\mathbb{R}} f(z)\, \overline{g\left(\frac{z - \lambda b}{\lambda}\right)}$$

$$\times \exp\left\{\frac{i((A/\lambda^2)z^2 - 2z(\omega/\lambda) + (\lambda^2 D)(\omega/\lambda)^2)}{2B}\right\} \frac{dz}{|\lambda|}$$

$$= \int_{\mathbb{R}} f(z)\, \left\{\frac{1}{|\lambda|}\, \overline{g\left(\frac{z - \lambda b}{\lambda}\right)}\right\} \mathcal{K}_{M'}(z, \omega/\lambda)\, dz$$

$$= \mathscr{G}_{G_\lambda}^{M'}\Big[f\Big]\left(\lambda b, \frac{\omega}{\lambda}\right),$$

where $G_\lambda(t) = \frac{1}{|\lambda|}\, g\left(\frac{t}{\lambda}\right)$ and $M' = (A/\lambda^2, B; C, \lambda^2 D)$. Here, it is interesting to note that the scaling of the input signal causes an inverse scaling of the frequency variable of WLCT, however, such an effect can be neutralized by virtue of the unimodular matrix involved in the windowed linear canonical transform. In fact, we have

$$\mathscr{G}_g^M\Big[f(\lambda t)\Big](b,\omega) = \frac{1}{\sqrt{2\pi i B}} \int_{\mathbb{R}} f(\lambda t)\, \overline{g_{b,\omega}^M(t)}\, dt$$

$$= \frac{1}{\sqrt{2\pi i B}} \int_{\mathbb{R}} f(z)\, \overline{g_{b,\omega}^M\left(\frac{z}{\lambda}\right)}\, \frac{dz}{|\lambda|}$$

$$= \frac{1}{\sqrt{2\pi i B}} \int_{\mathbb{R}} f(z)\, \overline{g\left(\frac{z}{\lambda} - b\right)} \exp\left\{\frac{i(A(z/\lambda)^2 - 2(z/\lambda)\omega + D\omega^2)}{2B}\right\} \frac{dz}{|\lambda|}$$

$$= \frac{1}{\sqrt{2\pi i B}} \int_{\mathbb{R}} f(z)\, \overline{g\left(\frac{z - \lambda b}{\lambda}\right)} \exp\left\{\frac{i(Az^2 - 2z(\lambda\omega) + D(\lambda\omega)^2)}{2\lambda^2 B}\right\} \frac{dz}{|\lambda|}$$

$$= \int_{\mathbb{R}} f(z)\, \overline{g\left(\frac{z - \lambda b}{\lambda}\right)} \mathcal{K}_{M'}(z, \lambda\omega)\, dz$$

$$= \mathscr{G}_{H_\lambda}^{M''}\Big[f\Big](\lambda b, \lambda\omega),$$

where $H_\lambda(t) = g(t/\lambda)$ and $M'' = (A, \lambda^2 B, C/\lambda^2, D)$.

(v) To observe the effect of modulated input on the windowed linear canonical transform, we take $\omega_0 \in \mathbb{R}$ and proceed as

$$\mathscr{G}_g^M\Big[e^{i\omega_0 t} f(t)\Big](b,\omega) = \frac{1}{\sqrt{2\pi i B}} \int_{\mathbb{R}} e^{i\omega_0 t} f(t)\, \overline{g(t - b)} \exp\left\{\frac{i(At^2 - 2t\omega + D\omega^2)}{2B}\right\} dt$$

$$= \frac{1}{\sqrt{2\pi i B}} \int_{\mathbb{R}} f(t)\, \overline{g(t - b)} \exp\left\{\frac{i(At^2 - 2(\omega - B\omega_0)t + D\omega^2)}{2B}\right\} dt$$

$$= \frac{1}{\sqrt{2\pi i B}} \int_{\mathbb{R}} f(t)\, \overline{g(t - b)} \exp\left\{-\frac{i(D(B\omega_0)^2 - 2DB\omega\omega_0)}{2B}\right\}$$

$$\times \exp\left\{\frac{i(At^2 - 2(\omega - B\omega_0)t + D(\omega - B\omega_0)^2)}{2B}\right\} dt$$

$$= \exp\left\{-\frac{iBD(B\omega_0^2 - 2\omega\omega_0)}{2B}\right\} \mathscr{G}_g^M\left[f\right](b, \omega - B\omega_0)$$

$$= \exp\left\{-\frac{iD(B\omega_0^2 - 2\omega\omega_0)}{2}\right\} \mathscr{G}_g^M\left[f\right](b, \omega - B\omega_0)$$

$$= \exp\left\{iD\omega\omega_0 - \frac{iBD\omega_0^2}{2}\right\} \mathscr{G}_g^M\left[f\right](b, \omega - B\omega_0).$$

(vi) In case of the complex conjugated input signal, we have

$$\mathscr{G}_g^M\left[\overline{f}\right](b, \omega) = \int_{\mathbb{R}} \overline{f(t)}\,\overline{g(t-b)}\,\mathcal{K}_M(t, \omega)\,dt$$

$$= \int_{\mathbb{R}} \overline{f(t)\,g(t-b)\,\mathcal{K}_{M^{-1}}(t, \omega)}\,dt$$

$$= \int_{\mathbb{R}} \overline{f(t)\,g(t-b)\,\mathcal{K}_{M^{-1}}(\omega, t)}\,dt$$

$$= \overline{\mathscr{G}_g^{M^{-1}}\left[f\right](b, \omega)},$$

where $\mathcal{K}_{M^{-1}}(t, \omega) = \mathcal{K}_{M^{-1}}(\omega, t)$ follows by making use of the assumption $A = D$.

(vii) To observe the behaviour of WLCT upon switching the input function $f \in L^2(\mathbb{R})$ and the window function $g \in L^2(\mathbb{R})$, we proceed as

$$\mathscr{G}_g^M\left[f\right](b, \omega) = \int_{\mathbb{R}} f(t)\,\overline{g(t-b)}\,\mathcal{K}_M(t, \omega)\,dt$$

$$= \int_{\mathbb{R}} f(z+b)\,\overline{g(z)}\,\mathcal{K}_M(z+b, \omega)\,dz$$

$$= \frac{1}{\sqrt{2\pi iB}} \int_{\mathbb{R}} f(z+b)\,\overline{g(z)}\,\exp\left\{\frac{i(A(z+b)^2 - 2\omega(z+b) + D\omega^2)}{2B}\right\}dz$$

$$= \frac{1}{\sqrt{2\pi iB}} \int_{\mathbb{R}} f(z+b)\,\overline{g(z)}$$

$$\times \exp\left\{\frac{i(Ab^2 + Az^2 + 2Abz - 2\omega z - 2\omega b + D\omega^2)}{2B}\right\}dz$$

$$= \frac{1}{\sqrt{2\pi iB}} \int_{\mathbb{R}} f(z+b)\,\overline{g(z)}\,\exp\left\{\frac{i(Az^2 - 2(\omega - Ab)z + D(\omega - Ab)^2)}{2B}\right\}$$

$$\times \exp\left\{-\frac{i(D(Ab)^2 - 2AD\omega b)}{2B}\right\}\exp\left\{\frac{i(Ab^2 - 2\omega b)}{2B}\right\}dz$$

$$= \int_{\mathbb{R}} f(z+b)\,\overline{g(z)}\,\mathcal{K}_M(z, \omega - Ab)$$

$$\times \exp\left\{\frac{i(Ab^2 - 2\omega b - A^2 Db^2 + 2AD\omega b)}{2B}\right\}dz$$

$$= \exp\left\{\frac{iAb^2(1 - AD)}{2B} + \frac{2i\omega b(AD - 1)}{2B}\right\}$$

$$\times \int_{\mathbb{R}} \overline{g(z)}\,f(z - (-b))\,\mathcal{K}_M(z, \omega - Ab)\,dz$$

$$= \exp\left\{\frac{2iBC\omega b}{2B} - \frac{iABCb^2}{2B}\right\} \mathscr{G}_{\overline{f}}^M\left[\overline{g}\right](-b, \omega - Ab)$$

$$= \exp\left\{iC\omega b - \frac{iACb^2}{2}\right\} \mathscr{G}_{\bar{f}}^M\left[\bar{g}\right](-b, \omega - Ab)$$

$$= \exp\left\{iCb\left(\omega - \frac{Ab}{2}\right)\right\} \mathscr{G}_{\bar{f}}^M\left[\bar{g}\right](-b, \omega - Ab).$$

(viii) In case the window function $g \in L^2(\mathbb{R})$ is shifted along its domain with $k \in \mathbb{R}$, then

$$\mathscr{G}_{g(t-k)}^M\left[f\right](b, \omega) = \int_{\mathbb{R}} f(t)\, \overline{g(t - (b+k))}\, \mathcal{K}_M(t, \omega)\, dt$$

$$= \mathscr{G}_g^M\left[f\right](b + k, \omega).$$

(ix) Upon scaling the window function $g \in L^2(\mathbb{R})$ by $\lambda \in \mathbb{R} \setminus \{0\}$, the WLCT becomes

$$\mathscr{G}_{g(\lambda t)}^M\left[f\right](b, \omega) = \int_{\mathbb{R}} f(t)\, \overline{g(\lambda t - \lambda b))}\, \mathcal{K}_M(t, \omega)\, dt$$

$$= \int_{\mathbb{R}} f\left(\frac{z}{\lambda}\right) \overline{g(z - \lambda b)}\, \mathcal{K}_M(z/\lambda, \omega)\, \frac{dz}{|\lambda|}$$

$$= \frac{1}{\sqrt{2\pi i B}} \int_{\mathbb{R}} f\left(\frac{z}{\lambda}\right) \overline{g(z - \lambda b)} \exp\left\{\frac{i(A(z/\lambda)^2 - 2(z/\lambda)\omega + D\omega^2)}{2B}\right\} \frac{dz}{|\lambda|}$$

$$= \frac{1}{\sqrt{2\pi i B}} \int_{\mathbb{R}} f\left(\frac{z}{\lambda}\right) \overline{g(z - \lambda b)}$$

$$\times \exp\left\{\frac{i((A/\lambda^2)z^2 - 2z(\omega/\lambda) + (\lambda^2 D)(\omega/\lambda)^2)}{2B}\right\} \frac{dz}{|\lambda|}$$

$$= \frac{1}{\sqrt{2\pi i B}} \int_{\mathbb{R}} \left\{\frac{1}{|\lambda|} f\left(\frac{z}{\lambda}\right)\right\} \overline{g(z - \lambda b)}$$

$$\times \exp\left\{\frac{i((A/\lambda^2)z^2 - 2z(\omega/\lambda) + (\lambda^2 D)(\omega/\lambda)^2)}{2B}\right\} dz$$

$$= \int_{\mathbb{R}} F_\lambda(z)\, \overline{g(z - \lambda b)}\, \mathcal{K}_{M'}(z, \omega/\lambda)\, dz$$

$$= \mathscr{G}_g^{M'}\left[F_\lambda\right]\left(\lambda b, \frac{\omega}{\lambda}\right),$$

where $F_\lambda(t) = \dfrac{1}{|\lambda|} f\left(\dfrac{t}{\lambda}\right)$ and $M' = (A/\lambda^2, B; C, \lambda^2 D)$.

This completes the proof of Theorem 2.4.2. $\qquad\qquad\qquad\qquad\qquad\qquad\square$

Next, our aim is to formulate the orthogonality relation and inversion formula concerning the windowed linear canonical transform defined in (2.95). In this direction, we have the following theorem:

Theorem 2.4.3. *Let* $\mathscr{G}_{g_1}^M\left[f\right](b, \omega)$ *and* $\mathscr{G}_{g_2}^M\left[h\right](b, \omega)$ *be the windowed linear canonical transforms of any* $f, h \in L^2(\mathbb{R})$ *with respect to a given pair of window functions* $g_1, g_2 \in L^2(\mathbb{R})$, *then the following orthogonality relation holds:*

$$\int_{\mathbb{R}} \int_{\mathbb{R}} \mathscr{G}_{g_1}^M\left[f\right](b, \omega)\, \overline{\mathscr{G}_{g_2}^M\left[h\right](b, \omega)}\, db\, d\omega = \left\langle g_2, g_1 \right\rangle_2 \left\langle f, h \right\rangle_2. \qquad (2.105)$$

Proof. Making use of Proposition 2.4.1, we observe that

$$\mathscr{G}_{g_1}^M\left[f\right](b, \omega) = \sqrt{2\pi}\, \mathcal{K}_M(b, \omega) \int_{\mathbb{R}} \mathscr{L}_M\left[f\right](\xi)\, \overline{\mathscr{F}\left[g_1\right]\left(\frac{\xi - \omega}{B}\right)}\, \overline{\mathcal{K}_M(b, \xi)}\, d\xi$$

and

$$\mathcal{G}_{g_2}^M \big[h\big](b,\omega) = \sqrt{2\pi}\, \mathcal{K}_M(b,\omega) \int_{\mathbb{R}} \mathcal{L}_M\big[h\big](\eta)\, \overline{\mathscr{F}\big[g_2\big]\left(\frac{\eta-\omega}{B}\right)}\, \overline{\mathcal{K}_M(b,\eta)}\, d\eta.$$

Therefore, we can write

$$\int_{\mathbb{R}}\int_{\mathbb{R}} \mathcal{G}_{g_1}^M\big[f\big](b,\omega)\, \overline{\mathcal{G}_{g_2}^M\big[h\big](b,\omega)}\, db\, d\omega$$

$$= 2\pi \int_{\mathbb{R}}\int_{\mathbb{R}} \left\{ \mathcal{K}_M(b,\omega) \int_{\mathbb{R}} \mathcal{L}_M\big[f\big](\xi)\, \overline{\mathscr{F}\big[g_1\big]\left(\frac{\xi-\omega}{B}\right)}\, \overline{\mathcal{K}_M(b,\xi)}\, d\xi \right\}$$

$$\times \left\{ \overline{\mathcal{K}_M(b,\omega)} \int_{\mathbb{R}} \overline{\mathcal{L}_M\big[h\big](\eta)}\, \mathscr{F}\big[g_2\big]\left(\frac{\eta-\omega}{B}\right)\, \mathcal{K}_M(b,\eta)\, d\eta \right\} db\, d\omega$$

$$= \frac{2\pi}{2\pi|B|} \int_{\mathbb{R}}\int_{\mathbb{R}}\int_{\mathbb{R}} \mathcal{L}_M\big[f\big](\xi)\, \overline{\mathcal{L}_M\big[h\big](\eta)}\, \overline{\mathscr{F}\big[g_1\big]\left(\frac{\xi-\omega}{B}\right)}\, \mathscr{F}\big[g_2\big]\left(\frac{\eta-\omega}{B}\right)$$

$$\times \left\{ \int_{\mathbb{R}} \mathcal{K}_M(b,\eta)\, \overline{\mathcal{K}_M(b,\xi)}\, db \right\} d\omega\, d\xi\, d\eta$$

$$= \frac{1}{|B|} \int_{\mathbb{R}}\int_{\mathbb{R}}\int_{\mathbb{R}} \mathcal{L}_M\big[f\big](\xi)\, \overline{\mathcal{L}_M\big[h\big](\eta)}\, \overline{\mathscr{F}\big[g_1\big]\left(\frac{\xi-\omega}{B}\right)}\, \mathscr{F}\big[g_2\big]\left(\frac{\eta-\omega}{B}\right)\, \delta(\eta-\xi)\, d\omega\, d\xi\, d\eta$$

$$= \frac{1}{|B|} \int_{\mathbb{R}}\int_{\mathbb{R}} \mathcal{L}_M\big[f\big](\xi)\, \overline{\mathcal{L}_M\big[h\big](\xi)}\, \overline{\mathscr{F}\big[g_1\big]\left(\frac{\xi-\omega}{B}\right)}\, \mathscr{F}\big[g_2\big]\left(\frac{\xi-\omega}{B}\right)\, d\omega\, d\xi$$

$$= \int_{\mathbb{R}} \mathcal{L}_M\big[f\big](\xi)\, \overline{\mathcal{L}_M\big[h\big](\xi)}\, \left\{ \frac{1}{|B|} \int_{\mathbb{R}} \overline{\mathscr{F}\big[g_1\big]\left(\frac{\xi-\omega}{B}\right)}\, \mathscr{F}\big[g_2\big]\left(\frac{\xi-\omega}{B}\right)\, d\omega \right\} d\xi$$

$$= \int_{\mathbb{R}} \mathcal{L}_M\big[f\big](\xi)\, \overline{\mathcal{L}_M\big[h\big](\xi)}\, \left\{ \int_{\mathbb{R}} \overline{\mathscr{F}\big[g_1\big](\sigma)}\, \mathscr{F}\big[g_2\big](\sigma)\, d\sigma \right\} d\omega$$

$$= \Big\langle \mathscr{F}\big[g_2\big],\, \mathscr{F}\big[g_1\big]\Big\rangle_2 \Big\langle \mathcal{L}_M\big[f\big],\, \mathcal{L}_M\big[h\big]\Big\rangle_2$$

$$= \Big\langle g_2,\, g_1\Big\rangle_2 \Big\langle f,\, h\Big\rangle_2.$$

This completes the proof Theorem 2.4.3. □

Remark 2.4.1. For $f = h$ and $g_1 = g_2 = g$, the orthogonality relation (2.105) boils down to

$$\int_{\mathbb{R}}\int_{\mathbb{R}} \left| \mathcal{G}_g^M\big[f\big](b,\omega) \right|^2 db\, d\omega = \big\|f\big\|_2^2\, \big\|g\big\|_2^2. \tag{2.106}$$

Since $\left| \mathcal{G}_g^M\big[f\big](b,\omega) \right|^2$ is the energy density of the windowed linear canonical transform, therefore, identity (2.106) implies that the total energy of the transformed signal is essentially distributed between the input signal and the window function. In case, the window function is chosen in a way that $\big\|g\big\|_2 = 1$, we obtain the following energy preserving relation:

$$\int_{\mathbb{R}}\int_{\mathbb{R}} \left| \mathcal{G}_g^M\big[f\big](b,\omega) \right|^2 db\, d\omega = \big\|f\big\|_2^2. \tag{2.107}$$

That is, the total energy of windowed linear canonical transform is exactly equal to the total energy of the input signal.

Following is the inversion formula associated with the windowed linear canonical transform (2.95):

Theorem 2.4.4. *Any function $f \in L^2(\mathbb{R})$ can be reconstructed from the corresponding windowed linear canonical transform $\mathscr{G}_{g_1}^M[f](b, \omega)$ as*

$$f(t) = \frac{1}{\langle g_2, g_1 \rangle_2} \int_{\mathbb{R}} \int_{\mathbb{R}} \mathscr{G}_{g_1}^M[f](b, \omega) \, g_2(t - b) \, \overline{\mathcal{K}_M(t, \omega)} \, db \, d\omega, \qquad (2.108)$$

where $g_2 \in L^2(\mathbb{R})$ is the so called "synthesis window function" and the equality in (2.108) holds in the almost everywhere.

Proof. Using the orthogonality relation (2.105) with respect to any arbitrary function $h \in L^2(\mathbb{R})$, we obtain

$$\cdot \left\langle f, h \right\rangle_2 = \frac{1}{\langle g_2, g_1 \rangle_2} \int_{\mathbb{R}} \int_{\mathbb{R}} \mathscr{G}_{g_1}^M[f](b, \omega) \, \overline{\mathscr{G}_{g_2}^M[h](b, \omega)} \, db \, d\omega$$

$$= \frac{1}{\langle g_2, g_1 \rangle_2} \int_{\mathbb{R}} \int_{\mathbb{R}} \mathscr{G}_{g_1}^M[f](b, \omega) \left\{ \frac{1}{\sqrt{-(2\pi i B)}} \int_{\mathbb{R}} \overline{h(t)} \, g_{2\,b,\omega}^M(t) \, dt \right\} db \, d\omega$$

$$= \frac{1}{\sqrt{-(2\pi i B)}} \int_{\mathbb{R}} \left\{ \frac{1}{\langle g_2, g_1 \rangle_2} \int_{\mathbb{R}} \int_{\mathbb{R}} \mathscr{G}_{g_1}^M[f](b, \omega) \, g_{2\,b,\omega}^M(t) \, db \, d\omega \right\} \overline{h(t)} \, dt$$

$$= \int_{\mathbb{R}} \left\{ \frac{1}{\langle g_2, g_1 \rangle_2} \int_{\mathbb{R}} \int_{\mathbb{R}} \mathscr{G}_{g_1}^M[f](b, \omega) \, g_2(t - b) \, \overline{\mathcal{K}_M(t, \omega)} \, db \, d\omega \right\} \overline{h(t)} \, dt. \qquad (2.109)$$

Since h is arbitrary, therefore from (2.109), it follows that

$$f(t) = \frac{1}{\langle g_2, g_1 \rangle_2} \int_{\mathbb{R}} \int_{\mathbb{R}} \mathscr{G}_{g_1}^M[f](b, \omega) \, g_2(t - b) \, \overline{\mathcal{K}_M(t, \omega)} \, db \, d\omega, \quad a.e.$$

This completes the proof Theorem 2.4.4. $\qquad \qquad \qquad \qquad \qquad \qquad \qquad \qquad \square$

Remark 2.4.2. If a single window function $g \in L^2(\mathbb{R})$ is employed for both the analysis and synthesis purposes; that is, $g_1 = g_2 = g$, then the inversion formula (2.108) can be expressed as

$$f(t) = \frac{1}{\|g\|_2^2} \int_{\mathbb{R}} \int_{\mathbb{R}} \mathscr{G}_g^M[f](b, \omega) \, g(t - b) \, \overline{\mathcal{K}_M(t, \omega)} \, db \, d\omega. \qquad (2.110)$$

Also, if g is a normalized window function, then

$$f(t) = \int_{\mathbb{R}} \int_{\mathbb{R}} \mathscr{G}_g^M[f](b, \omega) \, g(t - b) \, \overline{\mathcal{K}_M(t, \omega)} \, db \, d\omega. \qquad (2.111)$$

Nevertheless, under the special case $g_2(t) = \delta(t)$ and the constraint $g_1(0) \neq 0$, the inversion formula (2.108) boils down to a one-dimensional integral expression

$$f(t) = \frac{1}{g_1(0)} \int_{\mathbb{R}} \mathscr{G}_{g_1}^M[f](t, \omega) \, \overline{\mathcal{K}_M(t, \omega)} \, d\omega, \qquad (2.112)$$

where the equality holds almost everywhere.

As a consequence of the following theorem, we can infer that the range of the windowed linear canonical transform (2.95) constitutes a reproducing kernel Hilbert space.

Theorem 2.4.5. *A function $f \in L^2(\mathbb{R}^2)$ is the windowed linear canonical transform of a certain square integrable function on $\mathbb{R}$ if and only if the following reproduction formula holds:*

$$f(b', \omega') = \int_{\mathbb{R}} \int_{\mathbb{R}} f(b, \omega) \, \mathscr{K}_M(b, \omega; b', \omega') db \, d\omega, \qquad (2.113)$$

where

$$\mathscr{K}_M(b, \omega; b', \omega') = \frac{1}{\|g\|_2^2} \left\langle g(t - b) \, \overline{\mathcal{K}_M(t, \omega)}, \, g(t - b') \, \overline{\mathcal{K}_M(t, \omega')} \right\rangle_2, \qquad (2.114)$$

with $g \in L^2(\mathbb{R})$ being a window function.

Proof. By proceeding in a fashion analogous to the proof of Theorem 2.3.5, we can easily accomplish the proof of Theorem 2.4.5. □

Corollary 2.4.6. *The range of the windowed linear canonical transform (2.95) is a reproducing kernel Hilbert space with the reproducing kernel given by (2.114).*

2.4.2 Resolution of windowed linear canonical transform and uncertainty principle

"An expert is someone who knows some of the worst mistakes that can be made in his subject, and how to avoid them."

-Werner K. Heisenberg

In this subsection, we investigate upon the resolution of the windowed linear canonical transform (2.95) in the time-frequency plane. Besides, we shall also formulate a Heisenberg-type uncertainty inequality associated with the windowed linear canonical transform. Such an inequality sets a lower bound for the localization of windowed linear canonical transform, as a function of frequency variable, in contrast to the localization the input function.

For a given window function $g \in L^2(\mathbb{R})$ with centre E_g and radius Δ_g, the centre and radius of each of the generalized time domain analyzing elements $g_{b,\omega}^M(t)$ are given by

$$
\begin{aligned}
E_{g_{b,\omega}^M} &= \frac{1}{\left\|g_{b,\omega}^M\right\|_2^2} \int_{\mathbb{R}} t \left|g_{b,\omega}^M(t)\right|^2 dt \\
&= \frac{1}{\|g\|_2^2} \int_{\mathbb{R}} t \left|g(t - b)\right|^2 dt \\
&= \frac{1}{\|g\|_2^2} \int_{\mathbb{R}} (b + z) \left|g(z)\right|^2 dz \\
&= b + \int_{\mathbb{R}} z \left|g(z)\right|^2 dz \\
&= b + E_g
\end{aligned}
$$

and

$$\Delta_{g_{b,\omega}^M} = \frac{1}{\|g\|_2} \left\{ \int_{\mathbb{R}} (t - b - E_g)^2 \left|g(t - b)\right|^2 dt \right\}^{1/2} = \frac{1}{\|g\|_2} \left\{ \int_{\mathbb{R}} (z - E_g)^2 \left|g(z)\right|^2 dz \right\}^{1/2} = \Delta_g.$$

Consequently, the localization of the windowed linear canonical transform along the temporal axis is provided by the time-window

$$\left[b + E_g - \Delta_g,\, b + E_g + \Delta_g\right]. \tag{2.115}$$

Next, our goal is to ascertain the localization of linear canonical spectrum via the windowed linear canonical transform. Invoking the spectral representation of the windowed linear canonical transform given in (2.102), we note that the centre and radius of each of the linear canonical domain analyzing elements $G_\omega^M(\xi) = \hat{g}\big(B^{-1}(\xi - \omega)\big)$ are given by

$$\begin{aligned}
E_{G_\omega^M(\xi)} &= \frac{1}{\left\|G_\omega^M(\xi)\right\|_2^2} \int_{\mathbb{R}} \xi \left|G_\omega^M(\xi)\right|^2 d\xi \\
&= \frac{1}{|B|\,\|\hat{g}\|_2^2} \int_{\mathbb{R}} \xi \left|\hat{g}\left(\frac{\xi - \omega}{B}\right)\right|^2 d\xi \\
&= \frac{1}{\|\hat{g}\|_2^2} \int_{\mathbb{R}} (\omega + B\eta) \left|\hat{g}(\eta)\right|^2 d\eta \\
&= \omega + B\,E_{\hat{g}}
\end{aligned} \tag{2.116}$$

and

$$\begin{aligned}
\Delta_{G_\omega^M(\xi)} &= \frac{1}{\left\|G_\omega^M(\xi)\right\|_2} \left\{ \int_{\mathbb{R}} (\xi - \omega - B\,E_{\hat{g}})^2 \left|\hat{g}\left(\frac{\xi - \omega}{B}\right)\right|^2 d\xi \right\}^{1/2} \\
&= \frac{1}{|B|^{1/2}\|\hat{g}\|_2} \left\{ B^2 \int_{\mathbb{R}} (B^{-1}(\xi - \omega) - E_{\hat{g}})^2 \left|\hat{g}\left(\frac{\xi - \omega}{B}\right)\right|^2 d\xi \right\}^{1/2} \\
&= \frac{|B|}{\|\hat{g}\|_2} \left\{ \int_{\mathbb{R}} (B^{-1}(\xi - \omega) - E_{\hat{g}})^2 \left|\hat{g}\left(\frac{\xi - \omega}{B}\right)\right|^2 d(B^{-1}\xi) \right\}^{1/2} \\
&= \frac{|B|}{\|\hat{g}\|_2} \left\{ \int_{\mathbb{R}} (\eta - E_{\hat{g}})^2 \left|\hat{g}(\eta)\right|^2 d\eta \right\}^{1/2} \\
&= |B|\,\Delta_{\hat{g}}.
\end{aligned} \tag{2.117}$$

Therefore, the desired localization of the windowed linear canonical transform in the frequency domain is given by

$$\left[\omega + B\,E_{\hat{g}} - |B|\,\Delta_{\hat{g}},\, \omega + B\,E_{\hat{g}} + |B|\,\Delta_{\hat{g}}\right]. \tag{2.118}$$

Hence, the joint time and frequency localization of the windowed linear canonical transform transform (2.95) in the time-frequency plane is given by the window

$$\left[b + E_g - \Delta_g,\, b + E_g + \Delta_g\right] \times \left[\omega + B\,E_{\hat{g}} - |B|\,\Delta_{\hat{g}},\, \omega + B\,E_{\hat{g}} + |B|\,\Delta_{\hat{g}}\right], \tag{2.119}$$

occupying a net area of $4\Delta_g\Delta_{\hat{g}}\,|B|$, depending upon the matrix parameter B. With the entity (2.119) in hindsight and keeping in view the fact that the linear canonical transform corresponds to a twisting operation, we conclude that the windowed linear canonical transform inculcates a highly flexible tiling onto the time-frequency plane as demonstrated in Figure 2.19. Moreover, the resolution R_M of the windowed linear transform is given by the following inverse relation:

$$R_M := \frac{1}{\Delta_{g_{b,\omega}^M}\,\Delta_{G_\omega^M(\xi)}} = \frac{1}{\Delta_g \Delta_{\hat{g}}\,|B|}, \tag{2.120}$$

which is governed by the arbitrary non-zero real parameter B. Evidently, for any normalized window function g, the classical Heisenberg's uncertainty implies that the upper bound for the resolution of the windowed linear canonical transform (2.95) is $2|B|^{-1}$, which shall be achieved only when the Gaussian window is used. In analogy to the windowed fractional Fourier transform, the windowed linear canonical transform plays a pivotal role in an efficient analysis of chirp-like signals, which are ubiquitous in nature.

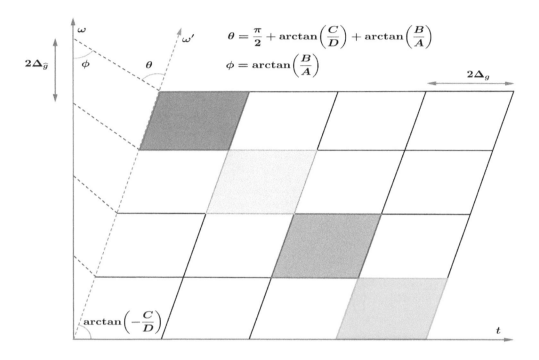

FIGURE 2.19: Time-frequency tiling induced by windowed linear canonical transform.

Next, we recall that for any $f \in L^2(\mathbb{R})$, the Heisenberg's uncertainty principle for the linear canonical transform obtained in Theorem 1.4.15 reads:

$$\left\{ \int_{\mathbb{R}} t^2 |f(t)|^2 dt \right\}^{1/2} \left\{ \int_{\mathbb{R}} \omega^2 \left| \mathscr{L}_M[f](\omega) \right|^2 d\omega \right\}^{1/2} \geq \frac{|B|}{2} \left\{ \int_{\mathbb{R}} |f(t)|^2 dt \right\}. \qquad (2.121)$$

Based upon (2.121), in the following theorem, we shall obtain an uncertainty inequality describing the localization of the windowed linear canonical transform $\mathscr{G}_g^M[f](b,\omega)$, as a function of ω, with respect to the localization of the input function.

Theorem 2.4.7. *For any non-trivial function $f \in L^2(\mathbb{R})$, the windowed linear canonical transform $\mathscr{G}_g^M[f](b,\omega)$ satisfies the following inequality:*

$$\left\{ \int_{\mathbb{R}} \int_{\mathbb{R}} \omega^2 \left| \mathscr{G}_g^M[f](b,\omega) \right|^2 db\, d\omega \right\}^{1/2} \left\{ \int_{\mathbb{R}} t^2 |f(t)|^2 dt \right\}^{1/2} \geq \frac{|B|}{2} \|f\|_2^2 \|g\|_2. \qquad (2.122)$$

Proof. As a consequence of Definition 2.4.1, we can express the windowed linear canonical transform in the following fashion:

$$\mathscr{G}_g^M[f](b,\omega) = \mathscr{L}_M\left[f(t)\,\overline{g(t-b)} \right](\omega), \qquad (2.123)$$

where $\mathscr{L}_M$ denotes the usual linear canonical transform defined in (1.164). Upon taking the windowed linear canonical transform $\mathscr{G}_g^M\left[f\right](b,\omega)$ as a function of ω and applying the inversion formula (1.169) on both sides of (2.123), we get

$$\mathscr{L}_{M^{-1}}\left(\mathscr{G}_g^M\left[f\right](b,\omega)\right)(t) = f(t)\,\overline{g(t-b)}. \tag{2.124}$$

Now, by virtue of the translation invariance of the Lebesgue integral, we obtain

$$\left\|g\right\|_2^2 \int_{\mathbb{R}} t^2 \left|f(t)\right|^2 dt = \int_{\mathbb{R}} t^2 \left|f(t)\right|^2 \left\{\int_{\mathbb{R}} \left|g(t-b)\right|^2 db\right\} dt$$

$$= \int_{\mathbb{R}}\int_{\mathbb{R}} t^2 \left|f(t)\,g(t-b)\right|^2 db\,dt$$

$$= \int_{\mathbb{R}}\int_{\mathbb{R}} t^2 \left|\mathscr{L}_{M^{-1}}\left(\mathscr{G}_g^M\left[f\right](b,\omega)\right)(t)\right|^2 db\,dt. \tag{2.125}$$

Using the Parseval's formula (1.171) for the linear canonical transform, we can rewrite the inequality (2.121) as

$$\left\{\int_{\mathbb{R}} t^2 |f(t)|^2 dt\right\}^{1/2}\left\{\int_{\mathbb{R}}\omega^2\left|\mathscr{L}_M\left[f\right](\omega)\right|^2 d\omega\right\}^{1/2} \geq \frac{|B|}{2}\left\{\int_{\mathbb{R}}\left|\mathscr{L}_M\left[f\right](\omega)\right|^2 d\omega\right\}. \tag{2.126}$$

In view of relation (2.123), we can regard $\mathscr{G}_g^M\left[f\right](b,\omega)$ as the linear canonical transform of the function $f(t)\,\overline{g(t-b)}$, so (2.126) yields the following inequality:

$$\left\{\int_{\mathbb{R}} t^2\left|f(t)\,\overline{g(t-b)}\right|^2 dt\right\}^{1/2}\left\{\int_{\mathbb{R}}\omega^2\left|\mathscr{G}_g^M\left[f\right](b,\omega)\right|^2 d\omega\right\}^{1/2} \geq \frac{|B|}{2}\left\{\int_{\mathbb{R}}\left|\mathscr{G}_g^M\left[f\right](b,\omega)\right|^2 d\omega\right\}. \tag{2.127}$$

Using (2.124), the inequality (2.127) becomes

$$\left\{\int_{\mathbb{R}} t^2\left|\mathscr{L}_{M^{-1}}\left(\mathscr{G}_g^M\left[f\right](b,\omega)\right)(t)\right|^2 dt\right\}^{1/2}\left\{\int_{\mathbb{R}}\omega^2\left|\mathscr{G}_g^M\left[f\right](b,\omega)\right|^2 d\omega\right\}^{1/2}$$
$$\geq \frac{|B|}{2}\left\{\int_{\mathbb{R}}\left|\mathscr{G}_g^M\left[f\right](b,\omega)\right|^2 d\omega\right\}. \tag{2.128}$$

After integrating the above inequality on both sides with respect to b, we apply the Cauchy-Schwarz inequality on the L.H.S to obtain

$$\left\{\int_{\mathbb{R}}\int_{\mathbb{R}} t^2\left|\mathscr{L}_{M^{-1}}\left(\mathscr{G}_g^M\left[f\right](b,\omega)\right)(t)\right|^2 dt\,db\right\}^{1/2}\left\{\int_{\mathbb{R}}\int_{\mathbb{R}}\omega^2\left|\mathscr{G}_g^M\left[f\right](b,\omega)\right|^2 d\omega\,db\right\}^{1/2}$$
$$\geq \frac{|B|}{2}\left\{\int_{\mathbb{R}}\int_{\mathbb{R}}\left|\mathscr{G}_g^M\left[f\right](b,\omega)\right|^2 d\omega\,db\right\}. \tag{2.129}$$

Changing the order of integration in inequality (2.129) yields

$$\left\{\int_{\mathbb{R}}\int_{\mathbb{R}} t^2\left|\mathscr{L}_{M^{-1}}\left(\mathscr{G}_g^M\left[f\right](b,\omega)\right)(t)\right|^2 db\,dt\right\}^{1/2}\left\{\int_{\mathbb{R}}\int_{\mathbb{R}}\omega^2\left|\mathscr{G}_g^M\left[f\right](b,\omega)\right|^2 db\,d\omega\right\}^{1/2}$$
$$\geq \frac{|B|}{2}\left\{\int_{\mathbb{R}}\int_{\mathbb{R}}\left|\mathscr{G}_g^M\left[f\right](b,\omega)\right|^2 db\,d\omega\right\}$$
$$= \frac{|B|}{2}\left\|f\right\|_2^2\left\|g\right\|_2^2. \tag{2.130}$$

Implementing (2.125) in (2.130), we get

$$\left\{ \|g\|_2^2 \int_{\mathbb{R}} t^2 |f(t)|^2 dt \right\}^{1/2} \left\{ \int_{\mathbb{R}} \int_{\mathbb{R}} \omega^2 \left| \mathscr{G}_g^M \left[f \right](b,\omega) \right|^2 db\, d\omega \right\}^{1/2} \geq \frac{|B|}{2} \left\| f \right\|_2^2 \left\| g \right\|_2^2. \quad (2.131)$$

Finally, the inequality (2.131) can be equivalently expressed as

$$\left\{ \int_{\mathbb{R}} \int_{\mathbb{R}} \omega^2 \left| \mathscr{G}_g^M \left[f \right](b,\omega) \right|^2 db\, d\omega \right\}^{1/2} \left\{ \int_{\mathbb{R}} t^2 |f(t)|^2 dt \right\}^{1/2} \geq \frac{|B|}{2} \left\| f \right\|_2^2 \left\| g \right\|_2,$$

which is the desired Heisenberg-type uncertainty inequality for windowed linear canonical transform. This completes the proof of Theorem 2.4.7. □

Remark 2.4.3. The uncertainty inequality (2.122) encapsulates the Heisenberg-type uncertainty principles for both the windowed Fourier and fractional Fourier transforms. For $M = (0, 1; -1, 0)$, inequality (2.122) boils down to the uncertainty inequality (2.65), whereas for $M = (\cos\alpha, \sin\alpha; -\sin\alpha, \cos\alpha)$, $\alpha \neq n\pi$, $n \in \mathbb{Z}$, inequality (2.122) yields the Heisenberg-type uncertainty principle for the windowed fractional Fourier transform defined in (2.68).

2.5 The Windowed Special Affine Fourier Transform

"All the truths of mathematics are linked to each other, and all means of discovering them are equally admissible."

-Adrien-Marie Legendre

The special affine Fourier transform is a significant ramification of the well-known linear canonical transform, which is essentially based on the concept of performing an offset to the linear canonical spectrum of a signal. Here, we shall discuss the notion of windowed special affine Fourier transform (WSAFT), which has the capability of providing local information about the special affine spectrum of a non-transient signal. Primarily, the windowed special affine Fourier transform relies on a typical window function, which restricts the computational domain to the region of influence of the window function. Translating the window function over the temporal axis substantiates that the spectral analysis has been performed for the complete signal. In this section, our goal is to study the fundamental properties of the windowed special affine Fourier transform, including the formulation of the corresponding orthogonality relation and an inversion formula.

Below, we present the formal definition of the windowed special affine Fourier transform. In conformity to the usual notations, we shall denote the augmented matrix obtained by the offset of a real, 2×2 unimodular matrix $M = \begin{pmatrix} A & B \\ C & D \end{pmatrix}$ with the real 2×1 augmentation vector $\lambda = \begin{pmatrix} p \\ q \end{pmatrix}$ by $\Lambda = [M : \lambda]$ and explicitly express it as $\Lambda = (A, B, C, D : p, q)$.

Definition 2.5.1. Given an augmented matrix $\Lambda = (A, B, C, D : p, q)$ with $B \neq 0$, the windowed special affine Fourier transform of any $f \in L^2(\mathbb{R})$ with respect to a window function $g \in L^2(\mathbb{R})$ is denoted by $\mathscr{G}_g^{\Lambda} \left[f \right]$ and is defined as

$$\mathscr{G}_g^{\Lambda} \left[f \right](b,\omega) = \frac{1}{\sqrt{2\pi i B}} \left\langle f, g_{b,\omega}^{\Lambda} \right\rangle_2 = \int_{\mathbb{R}} f(t)\, \overline{g(t-b)}\, \mathcal{K}_{\Lambda}(t,\omega)\, dt, \quad b, \omega \in \mathbb{R}, \quad (2.132)$$

where $\mathcal{K}_\Lambda(t,\omega)$ is the usual kernel as defined in (1.228) and $g^\Lambda_{b,\omega}(t)$ denote the analyzing functions defined by

$$g^\Lambda_{b,\omega}(t) = \exp\left\{-\frac{i\left(At^2 + 2t(p-\omega) - 2\omega(Dp-Bq) + D(\omega^2+p^2)\right)}{2B}\right\} g(t-b). \quad (2.133)$$

Definition 2.5.1 allows us to point out some important observations regarding the notion of windowed special affine Fourier transform:

(i) Indeed, the windowed special affine Fourier transform (2.132) can be regarded as the special affine Fourier transform of the truncated function $f(t)\,\overline{g(t-b)}$; that is,

$$\mathscr{G}^\Lambda_g\left[f\right](b,\omega) = \mathcal{O}_\Lambda\left[f(t)\,\overline{g(t-b)}\right](\omega). \quad (2.134)$$

Moreover, the quantity $\left|\mathscr{G}^\Lambda_g\left[f\right](b,\omega)\right|^2$ represents the energy density of the windowed special affine Fourier transform in the time-frequency plane. Therefore, integrating $\left|\mathscr{G}^\Lambda_g\left[f\right](b,\omega)\right|^2$ with respect to both the time and frequency variables gives the total energy of the windowed special affine Fourier transform.

(ii) Often in practical aspects, somewhat leniency is adopted in dealing with the pre-factor $1/\sqrt{2\pi i B}$ in the sense that it is replaced with $1/\sqrt{2\pi|B|}$.

(iii) Choosing the augmented matrix as $\Lambda = (0,1,-1,0:p,q)$, the windowed special affine Fourier transform (2.132) yields a novel windowed Fourier transform, coined as the "windowed offset Fourier transform."

(iv) Plugging $\Lambda = (0,1,-1,0:0,0)$ into (2.132) and multiplying the integrand with $\sqrt{i}$, Definition 2.5.1 reduces to Definition 2.2.2.

(v) For $\Lambda = (\cos\alpha, \sin\alpha, -\sin\alpha, \cos\alpha : p,q)$, $\alpha \neq n\pi$, the windowed special affine Fourier transform (2.132), when multiplied by the factor $2i\sqrt{e^{i\alpha}}$, yields the offset variant of the windowed fractional Fourier transform.

(vi) For $\Lambda = (\cos\alpha, \sin\alpha, -\sin\alpha, \cos\alpha : p,q)$, $\alpha \neq n\pi$, the windowed special affine Fourier transform (2.132), when multiplied by the factor $2i\sqrt{e^{i\alpha}}$, yields the offset variant of the windowed fractional Fourier transform. On the flip side, choosing $\Lambda = (\cos\alpha, \sin\alpha, -\sin\alpha, \cos\alpha : 0,0)$ and the aforementioned pre-factor in (2.132), Definition 2.5.1 reduces to the windowed fractional Fourier transform (2.68).

(vii) For $\Lambda = (A,B,C,D:0,0)$, the windowed special affine Fourier transform (2.132) precisely boils down to the windowed the linear canonical transform (2.95).

(viii) Choosing $\Lambda = (1,B;0,1:p,q)$, $B \neq 0$, the windowed special affine Fourier transform (2.132) yields the following windowed offset Fresnel transform:

$$\mathscr{G}^\Lambda_g\left[f\right](b,\omega) = \frac{1}{\sqrt{2\pi i B}}\int_\mathbb{R} f(t)\,\overline{g(t-b)}\exp\left\{\frac{i\left((t-\omega)^2 + 2p(t-\omega) + 2B\omega q + p^2\right)}{2B}\right\} dt.$$
$$(2.135)$$

(ix) For $\Lambda = (1,B;0,1:0,0)$, $B \neq 0$, the windowed special affine Fourier transform (2.132) boils down to the windowed linear canonical Fresnel transform defined in (2.99).

(x) Since the conventional windowed Fourier transform (2.50) is expressible via the classical convolution operation $*$ as demonstrated in (2.53), therefore, we can formulate several novel variants of the windowed special affine Fourier transform (2.132) by replacing the classical convolution operation in (2.53) with any of the special affine convolutional operations. Such convolution-based windowed special affine Fourier transforms are distinguished in their own right and have the prime feature of being labour-saving.

Next, we aim to obtain a spectral representation of the windowed special affine Fourier transform defined in (2.132). In this direction, we have the following proposition:

Proposition 2.5.1. *Let $\mathscr{G}_g^\Lambda[f](b,\omega)$ be the windowed special affine Fourier transform of any $f \in L^2(\mathbb{R})$, then we have*

$$\mathscr{G}_g^\Lambda[f](b,\omega) = \sqrt{2\pi}\, \mathcal{K}_\Lambda(b,\omega) \int_{\mathbb{R}} \mathcal{O}_\Lambda[f](\xi)\, \overline{\mathscr{F}[g]\left(\frac{\xi-\omega}{B}\right)}\, \overline{\mathcal{K}_\Lambda(b,\xi)}\, d\xi, \qquad (2.136)$$

where $\mathscr{F}$ is the classical Fourier transform defined in (1.7).

Proof. In view of the orthogonality relation pertaining to the special affine Fourier transform as given in (1.235), Definition 2.5.1 implies that

$$\mathscr{G}_g^\Lambda[f](b,\omega) = \frac{1}{\sqrt{2\pi i B}} \left\langle f, g_{b,\omega}^\Lambda \right\rangle_2$$

$$= \frac{1}{\sqrt{2\pi i B}} \int_{\mathbb{R}} \mathcal{O}_\Lambda[f](\xi)\, \overline{\mathcal{O}_\Lambda[g_{b,\omega}^\Lambda](\xi)}\, d\xi. \qquad (2.137)$$

To facilitate the intent, we proceed as

$$\mathcal{O}_\Lambda[g_{b,\omega}^\Lambda](\omega) = \int_{\mathbb{R}} g_{b,\omega}^\Lambda(t)\, \mathcal{K}_\Lambda(t,\omega)\, dt$$

$$= \frac{1}{\sqrt{2\pi i B}} \int_{\mathbb{R}} g_{b,\omega}^\Lambda(t) \exp\left\{ \frac{i\left(At^2 + 2t(p-\xi) - 2\xi(Dp - Bq) + D(\xi^2 + p^2)\right)}{2B} \right\} dt$$

$$= \frac{1}{\sqrt{2\pi i B}} \int_{\mathbb{R}} g(t-b) \exp\left\{ -\frac{i\left(At^2 + 2t(p-\omega) - 2\omega(Dp - Bq) + D(\omega^2 + p^2)\right)}{2B} \right\}$$

$$\times \exp\left\{ \frac{i\left(At^2 + 2t(p-\xi) - 2\xi(Dp - Bq) + D(\xi^2 + p^2)\right)}{2B} \right\} dt$$

$$= \frac{1}{\sqrt{2\pi i B}} \int_{\mathbb{R}} g(t-b) \exp\left\{ \frac{i\left(2t(\omega-\xi) + 2(\omega-\xi)(Dp - Bq) + D(\xi^2 - \omega^2)\right)}{2B} \right\} dt$$

$$= \frac{1}{\sqrt{2\pi i B}} \int_{\mathbb{R}} g(t-b) \exp\left\{ \frac{i(\omega-\xi)t}{B} \right\} \exp\left\{ \frac{iD(\xi^2 - \omega^2) + 2(\omega-\xi)(Dp - Bq)}{2B} \right\} dt$$

$$= \exp\left\{ \frac{iD(\xi^2 - \omega^2) + 2(\omega-\xi)(Dp - Bq)}{2B} \right\} \frac{1}{\sqrt{2\pi i B}} \int_{\mathbb{R}} g(z) \exp\left\{ \frac{i(\omega-\xi)(b+z)}{B} \right\} dz$$

$$= \exp\left\{ \frac{iD(\xi^2 - \omega^2) + 2(\omega-\xi)(Dp - Bq)}{2B} \right\} \exp\left\{ \frac{2i(\omega-\xi)b}{2B} \right\}$$

$$\times \frac{1}{\sqrt{iB}} \left\{ \frac{1}{\sqrt{2\pi}} \int_{\mathbb{R}} g(z) \exp\left\{ -i\left(\frac{\xi-\omega}{B}\right) \right\} dz \right\}$$

$$= \frac{1}{\sqrt{iB}} \exp\left\{ \frac{iD(\xi^2 - \omega^2) + 2(\omega-\xi)(Dp - Bq)}{2B} \right\} \exp\left\{ \frac{2i(\omega-\xi)b}{2B} \right\} \mathscr{F}[g]\left(\frac{\xi-\omega}{B}\right).$$

$$\qquad (2.138)$$

Using (2.138) in (2.137), we obtain

$$
\mathcal{O}_\Lambda\Big[f\Big](b,\omega) = \frac{1}{\sqrt{2\pi i B}} \frac{1}{\sqrt{-iB}} \int_{\mathbb{R}} \mathcal{O}_\Lambda\Big[f\Big](\xi)\, \overline{\mathscr{F}\Big[g\Big]\left(\frac{\xi-\omega}{B}\right)}
$$
$$
\times \exp\left\{-\frac{iD(\xi^2-\omega^2)+2(\omega-\xi)(Dp-Bq)}{2B}\right\} \exp\left\{-\frac{2i(\omega-\xi)b}{2B}\right\} d\xi
$$

$$
= \frac{1}{\sqrt{-iB}} \frac{1}{\sqrt{2\pi i B}} \int_{\mathbb{R}} \mathcal{O}_\Lambda\Big[f\Big](\xi)\, \overline{\mathscr{F}\Big[g\Big]\left(\frac{\xi-\omega}{B}\right)}
$$
$$
\times \exp\left\{-\frac{i\big(Ab^2+2b(p-\xi)-2\xi(Dp-Bq)+D(\xi^2+p^2)\big)}{2B}\right\}
$$
$$
\times \exp\left\{\frac{i\big(Ab^2+2b(p-\omega)-2\omega(Dp-Bq)+D(\omega^2+p^2)\big)}{2B}\right\} d\xi
$$

$$
= \sqrt{2\pi}\, \mathcal{K}_\Lambda(b,\omega) \int_{\mathbb{R}} \mathcal{O}_\Lambda\Big[f\Big](\xi)\, \overline{\mathscr{F}\Big[g\Big]\left(\frac{\xi-\omega}{B}\right)}\, \overline{\mathcal{K}_\Lambda(b,\xi)}\, d\xi.
$$

This completes the proof of Proposition 2.5.1.   □

With the aim to study the fundamental properties of the windowed special affine Fourier transform (2.132), we have the following theorem:

Theorem 2.5.2. *If g and h are two window functions and $\alpha, \beta \in \mathbb{C}$, $k, \omega_0 \in \mathbb{R}$, $\sigma \in \mathbb{R}\setminus\{0\}$, then for any pair of square integrable functions f and s, the windowed special affine Fourier transform (2.132) satisfies the following properties:*

(i) *Linearity:* $\mathscr{G}_g^\Lambda\Big[\alpha f+\beta s\Big](b,\omega) = \alpha\,\mathscr{G}_g^\Lambda\Big[f\Big](b,\omega) + \beta\,\mathscr{G}_g^\Lambda\Big[s\Big](b,\omega),$

(ii) *Anti-linearity:* $\mathscr{G}_{\alpha g+\beta h}^\Lambda\Big[f\Big](b,\omega) = \overline{\alpha}\,\mathscr{G}_g^\Lambda\Big[f\Big](b,\omega) + \overline{\beta}\,\mathscr{G}_h^\Lambda\Big[f\Big](b,\omega),$

(iii) *Time-shift:* $\mathscr{G}_g^\Lambda\Big[f(t-k)\Big](b,\omega) = \exp\left\{i\left(Aqk+Ck\omega-Ckp-\dfrac{ACk^2}{2}\right)\right\}$
$$
\times \mathscr{G}_g^\Lambda\Big[f\Big](b-k,\omega-Ak),
$$

(iv) *Scaling:* $\mathscr{G}_g^\Lambda\Big[f(\sigma t)\Big](b,\omega) = \mathscr{G}_{G_\sigma}^{\Lambda'}\Big[f\Big]\left(\sigma b,\dfrac{\omega}{\sigma}\right),$ *where* $G_\sigma(t) = \dfrac{1}{|\sigma|}\,g\left(\dfrac{t}{\sigma}\right)$ *and*
$\Lambda' = (A/\sigma^2, B, C, \sigma^2 D : p/\sigma, \sigma q),$

(v) *Modulation:* $\mathscr{G}_g^\Lambda\Big[e^{i\omega_0 t}f(t)\Big](b,\omega) = \exp\left\{i\left(D\omega\omega_0-(Dp-Bq)\omega_0-\dfrac{BD\omega_0^2}{2}\right)\right\}$
$$
\times \mathscr{G}_g^\Lambda\Big[f\Big](b,\omega-B\omega_0),
$$

(vi) *Symmetry:* $\mathscr{G}_g^\Lambda\Big[f\Big](b,\omega) = \exp\left\{i\left(C\omega b-\dfrac{ACb^2}{2}-\dfrac{A(Dp-Bq)b}{2}-\dfrac{bp}{B}\right)\right\}$
$$
\times \mathscr{G}_{\hat{f}}^\Lambda\Big[\bar{g}\Big](-b,\omega-Ab),
$$

(vii) *Translation in window:* $\mathscr{G}_{g(t-k)}^\Lambda\Big[f\Big](b,\omega) = \mathscr{G}_g^\Lambda\Big[f\Big](b+k,\omega),$

(viii) *Scaling in window:* $\mathscr{G}^{\Lambda}_{g(\sigma t)}\left[f\right](b,\omega) = \mathscr{G}^{\Lambda'}_{g}\left[F_{\sigma}\right]\left(\sigma b, \frac{\omega}{\sigma}\right)$, *where* $F_{\sigma}(t) = \frac{1}{|\sigma|} f\left(\frac{t}{\sigma}\right)$ *and*
$\Lambda' = (A/\sigma^2, B, C, \sigma^2 D : p/\sigma, \sigma q)$.

Proof. For the sake of brevity, the proof is omitted. $\qquad\square$

In the next theorem, we establish an orthogonality relation between a given pair of square integrable functions and their respective windowed special affine Fourier transforms.

Theorem 2.5.3. *Let* $\mathscr{G}^{\Lambda}_{g_1}\left[f\right](b,\omega)$ *and* $\mathscr{G}^{\Lambda}_{g_2}\left[h\right](b,\omega)$ *be the windowed special affine transforms of a given pair of functions* $f, h \in L^2(\mathbb{R})$ *with respect to the window functions* $g_1, g_2 \in L^2(\mathbb{R})$, *respectively. Then, the following relation holds:*

$$\int_{\mathbb{R}}\int_{\mathbb{R}} \mathscr{G}^{\Lambda}_{g_1}\left[f\right](b,\omega)\,\overline{\mathscr{G}^{\Lambda}_{g_2}\left[h\right](b,\omega)}\,db\,d\omega = \Big\langle g_2, g_1\Big\rangle_2 \Big\langle f, h\Big\rangle_2. \qquad (2.139)$$

Proof. Firstly, in view of Proposition 2.5.1, we note that the windowed special affine transforms of the given pair of functions $f, h \in L^2(\mathbb{R})$ are expressible as

$$\mathscr{G}^{\Lambda}_{g_1}\left[f\right](b,\omega) = \sqrt{2\pi}\,\mathcal{K}_{\Lambda}(b,\omega) \int_{\mathbb{R}} \mathcal{O}_{\Lambda}\left[f\right](\xi)\,\overline{\mathscr{F}\left[g_1\right]\left(\frac{\xi-\omega}{B}\right)}\,\overline{\mathcal{K}_{\Lambda}(b,\xi)}\,d\xi$$

and

$$\mathscr{G}^{\Lambda}_{g_2}\left[h\right](b,\omega) = \sqrt{2\pi}\,\mathcal{K}_{\Lambda}(b,\omega) \int_{\mathbb{R}} \mathcal{O}_{\Lambda}\left[h\right](\eta)\,\overline{\mathscr{F}\left[g_2\right]\left(\frac{\eta-\omega}{B}\right)}\,\overline{\mathcal{K}_{\Lambda}(b,\eta)}\,d\eta.$$

To obtain the desired orthogonality relation, we shall make use of Theorem 1.5.1 and proceed in the following manner:

$$\int_{\mathbb{R}}\int_{\mathbb{R}} \mathscr{G}^{\Lambda}_{g_1}\left[f\right](b,\omega)\,\overline{\mathscr{G}^{\Lambda}_{g_2}\left[h\right](b,\omega)}\,db\,d\omega$$

$$= 2\pi \int_{\mathbb{R}}\int_{\mathbb{R}} \left\{ \mathcal{K}_{\Lambda}(b,\omega)\int_{\mathbb{R}} \mathcal{O}_{\Lambda}\left[f\right](\xi)\,\overline{\mathscr{F}\left[g_1\right]\left(\frac{\xi-\omega}{B}\right)}\,\overline{\mathcal{K}_{\Lambda}(b,\xi)}\,d\xi \right\}$$

$$\times \left\{ \overline{\mathcal{K}_M(b,\omega)}\int_{\mathbb{R}} \overline{\mathcal{O}_{\Lambda}\left[h\right](\eta)}\,\mathscr{F}\left[g_2\right]\left(\frac{\eta-\omega}{B}\right)\,\mathcal{K}_{\Lambda}(b,\eta)\,d\eta \right\}\,db\,d\omega$$

$$= \frac{2\pi}{2\pi|B|} \int_{\mathbb{R}}\int_{\mathbb{R}}\int_{\mathbb{R}} \mathcal{O}_{\Lambda}\left[f\right](\xi)\,\overline{\mathcal{O}_{\Lambda}\left[h\right](\eta)}\,\overline{\mathscr{F}\left[g_1\right]\left(\frac{\xi-\omega}{B}\right)}\,\mathscr{F}\left[g_2\right]\left(\frac{\eta-\omega}{B}\right)$$

$$\times \left\{ \int_{\mathbb{R}} \mathcal{K}_{\Lambda}(b,\eta)\,\overline{\mathcal{K}_{\Lambda}(b,\xi)}\,db \right\}\,d\omega\,d\xi\,d\eta$$

$$= \frac{2\pi}{2\pi|B|} \int_{\mathbb{R}}\int_{\mathbb{R}}\int_{\mathbb{R}} \mathcal{O}_{\Lambda}\left[f\right](\xi)\,\overline{\mathcal{O}_{\Lambda}\left[h\right](\eta)}\,\overline{\mathscr{F}\left[g_1\right]\left(\frac{\xi-\omega}{B}\right)}\,\mathscr{F}\left[g_2\right]\left(\frac{\eta-\omega}{B}\right)$$

$$\times \left\{ I_{\Lambda}\int_{\mathbb{R}} \mathcal{K}_{\Lambda}(b,\eta)\,\mathcal{K}_{\Lambda^{-1}}(\xi,b)\,db \right\}\,d\omega\,d\xi\,d\eta$$

$$= \frac{1}{|B|} \int_{\mathbb{R}}\int_{\mathbb{R}}\int_{\mathbb{R}} \mathcal{O}_{\Lambda}\left[f\right](\xi)\,\overline{\mathcal{O}_{\Lambda}\left[h\right](\eta)}\,\overline{\mathscr{F}\left[g_1\right]\left(\frac{\xi-\omega}{B}\right)}\,\mathscr{F}\left[g_2\right]\left(\frac{\eta-\omega}{B}\right)\,\delta(\eta-\xi)\,d\omega\,d\xi\,d\eta$$

$$= \frac{1}{|B|} \int_{\mathbb{R}}\int_{\mathbb{R}} \mathcal{O}_{\Lambda}\left[f\right](\xi)\,\overline{\mathcal{O}_{\Lambda}\left[h\right](\xi)}\,\overline{\mathscr{F}\left[g_1\right]\left(\frac{\xi-\omega}{B}\right)}\,\mathscr{F}\left[g_2\right]\left(\frac{\xi-\omega}{B}\right)\,d\omega\,d\xi$$

$$= \int_{\mathbb{R}} \mathcal{O}_{\Lambda}\left[f\right](\xi)\,\overline{\mathcal{O}_{\Lambda}\left[h\right](\xi)} \left\{ \frac{1}{|B|}\int_{\mathbb{R}} \overline{\mathscr{F}\left[g_1\right]\left(\frac{\xi-\omega}{B}\right)}\,\mathscr{F}\left[g_2\right]\left(\frac{\xi-\omega}{B}\right)\,d\omega \right\}\,d\xi$$

$$= \int_{\mathbb{R}} \mathcal{O}_\Lambda\big[f\big](\xi)\, \overline{\mathcal{O}_\Lambda\big[h\big](\xi)} \left\{ \int_{\mathbb{R}} \overline{\mathscr{F}\big[g_1\big](\sigma)}\, \mathscr{F}\big[g_2\big](\sigma)\, d\sigma \right\} d\omega$$

$$= \Big\langle \mathscr{F}\big[g_2\big],\, \mathscr{F}\big[g_1\big] \Big\rangle_2 \Big\langle \mathcal{O}_\Lambda\big[f\big],\, \mathcal{O}_\Lambda\big[h\big] \Big\rangle_2$$

$$= \Big\langle g_2,\, g_1 \Big\rangle_2 \Big\langle f,\, h \Big\rangle_2.$$

This completes the proof Theorem 2.5.3. $\qquad\qquad\qquad\qquad\qquad\qquad\qquad\qquad\square$

Remark 2.5.1. Plugging $f = h$, and $g_1 = g_2 = g$ into the orthogonality relation (2.139) yields the following relation:

$$\int_{\mathbb{R}} \int_{\mathbb{R}} \Big| \mathscr{G}_g^\Lambda\big[f\big](b,\omega) \Big|^2 \, db\, d\omega = \big\| f \big\|_2^2 \, \big\| g \big\|_2^2. \tag{2.140}$$

That is, the total energy of the windowed special affine Fourier transform defined in (2.132) is precisely due the respective energies of the input signal and the window function. In fact, if the window function $g \in L^2(\mathbb{R})$ is normalized; that is, $\|g\|_2 = 1$, then it is evidently of unit energy. In that case, the total energy of the WSAFT exactly matches the input signal, yielding the following energy preserving relation:

$$\int_{\mathbb{R}} \int_{\mathbb{R}} \Big| \mathscr{G}_g^\Lambda\big[f\big](b,\omega) \Big|^2 \, db\, d\omega = \big\| f \big\|_2^2. \tag{2.141}$$

Having obtained the orthogonality relation for the WLCT defined in (2.132), it is imperative to formulate an inversion formula, which allows the retraction of the input signal from the transformed one. In this direction, we have our next theorem.

Theorem 2.5.4. *If $\mathscr{G}_{g_1}^\Lambda\big[f\big](b,\omega)$ is the windowed special affine Fourier transform of any $f \in L^2(\mathbb{R})$ with respect to an analyzing window $g_1 \in L^2(\mathbb{R})$, then for any synthesis window function $g_2 \in L^2(\mathbb{R})$, the following reconstruction formula holds:*

$$f(t) = \frac{1}{\langle g_2, g_1 \rangle_2} \int_{\mathbb{R}} \int_{\mathbb{R}} \mathscr{G}_{g_1}^\Lambda\big[f\big](b,\omega)\, g_2(t-b)\, \overline{\mathcal{K}_\Lambda(t,\omega)}\, db\, d\omega, \tag{2.142}$$

with equality holding in (2.141) in the almost everywhere sense.

Proof. Applying the orthogonality relation (2.139) with respect to any arbitrary function $h \in L^2(\mathbb{R})$ yields

$$\Big\langle f, h \Big\rangle_2 = \frac{1}{\langle g_2, g_1 \rangle_2} \int_{\mathbb{R}} \int_{\mathbb{R}} \mathscr{G}_{g_1}^\Lambda\big[f\big](b,\omega)\, \overline{\mathscr{G}_{g_2}^\Lambda\big[h\big](b,\omega)}\, db\, d\omega$$

$$= \frac{1}{\langle g_2, g_1 \rangle_2} \int_{\mathbb{R}} \int_{\mathbb{R}} \mathscr{G}_{g_1}^\Lambda\big[f\big](b,\omega) \left\{ \frac{1}{\sqrt{-(2\pi i B)}} \int_{\mathbb{R}} \overline{h(t)}\, g_{2b,\omega}^\Lambda(t)\, dt \right\} db\, d\omega$$

$$= \frac{1}{\sqrt{-(2\pi i B)}} \int_{\mathbb{R}} \left\{ \frac{1}{\langle g_2, g_1 \rangle_2} \int_{\mathbb{R}} \int_{\mathbb{R}} \mathscr{G}_{g_1}^\Lambda\big[f\big](b,\omega)\, g_{2b,\omega}^\Lambda(t)\, db\, d\omega \right\} \overline{h(t)}\, dt$$

$$= \int_{\mathbb{R}} \left\{ \frac{1}{\langle g_2, g_1 \rangle_2} \int_{\mathbb{R}} \int_{\mathbb{R}} \mathscr{G}_{g_1}^\Lambda\big[f\big](b,\omega)\, g_2(t-b)\, \overline{\mathcal{K}_\Lambda(t,\omega)}\, db\, d\omega \right\} \overline{h(t)}\, dt$$

$$= \left\langle \frac{1}{\langle g_2, g_1 \rangle_2} \int_{\mathbb{R}} \int_{\mathbb{R}} \mathscr{G}_{g_1}^\Lambda\big[f\big](b,\omega)\, g_2(t-b)\, \overline{\mathcal{K}_\Lambda(t,\omega)}\, db\, d\omega,\, h(t) \right\rangle_2. \tag{2.143}$$

Since h is an arbitrary square integrable function, therefore the inner product relation (2.143) implies that

$$f(t) = \frac{1}{\langle g_2, g_1 \rangle_2} \int_{\mathbb{R}} \int_{\mathbb{R}} \mathscr{G}_{g_1}^{\Lambda}\big[f\big](b, \omega)\, g_2(t - b)\, \overline{\mathcal{K}_{\Lambda}(t, \omega)}\, db\, d\omega, \quad a.e.$$

This completes the proof Theorem 2.5.4. $\square$

Remark 2.5.2. The distinguishing feature of the inversion formula (2.142) is that two different window functions can be brought into play for the analysis and synthesis purposes. However, if a single window function $g \in L^2(\mathbb{R})$ is employed for both the processes; that is, $g_1 = g_2 = g$, then the inversion formula (2.142) becomes

$$f(t) = \frac{1}{\|g\|_2^2} \int_{\mathbb{R}} \int_{\mathbb{R}} \mathscr{G}_g^{\Lambda}\big[f\big](b, \omega)\, g(t - b)\, \overline{\mathcal{K}_{\Lambda}(t, \omega)}\, db\, d\omega. \tag{2.144}$$

Additionally, if the window function g is also normalized, then (2.143) simplifies to the following:

$$f(t) = \int_{\mathbb{R}} \int_{\mathbb{R}} \mathscr{G}_g^{\Lambda}\big[f\big](b, \omega)\, g(t - b)\, \overline{\mathcal{K}_{\Lambda}(t, \omega)}\, db\, d\omega. \tag{2.145}$$

Nevertheless, in case the window functions asserted in Theorem 2.5.4 are chosen in a way that $g_1(0) \neq 0$ and $g_2(t) = \delta(t)$, then the inversion formula (2.142) reduces to a one-dimensional integral expression:

$$f(t) = \frac{1}{g_1(0)} \int_{\mathbb{R}} \mathscr{G}_{g_1}^{\Lambda}\big[f\big](t, \omega)\, \overline{\mathcal{K}_{\Lambda}(t, \omega)}\, d\omega, \tag{2.146}$$

where the equality holds almost everywhere.

In the final theorem of this subsection, we shall present a characterization of the windowed special affine Fourier transform defined in (2.132).

Theorem 2.5.5. *For* $(b, \omega) \in \mathbb{R}^2$, $\mathscr{G}_g^{\Lambda}\big[f\big](b, \omega)$ *is the windowed special affine Fourier transform of a certain signal* $f \in L^2(\mathbb{R})$ *if and only if it satisfies the following reproducing kernel equation:*

$$\mathscr{G}_g^{\Lambda}\big[f\big](b', \omega') = \int_{\mathbb{R}} \int_{\mathbb{R}} \mathscr{G}_g^{\Lambda}\big[f\big](b, \omega)\, \mathscr{K}_g^{\Lambda}(b, \omega; b', \omega')\, db\, d\omega, \quad (b', \omega') \in \mathbb{R}^2, \tag{2.147}$$

where $\mathscr{K}_g^{\Lambda}(b, \omega; b', \omega')$ *denotes the reproducing kernel given by*

$$\mathscr{K}_g^{\Lambda}(b, \omega; b', \omega') = \frac{1}{2\pi|B| \|g\|_2^2} \left\langle g_{b,\omega}^{\Lambda}, g_{b',\omega'}^{\Lambda} \right\rangle_2, \tag{2.148}$$

and $g \in L^2(\mathbb{R})$ *is any arbitrary window function.*

Proof. The proof is a straight away consequence of the inversion formula (2.144) and the fact that

$$\mathscr{G}_g^{\Lambda}\big[f\big](b', \omega') = \frac{1}{\sqrt{2\pi i B}} \int_{\mathbb{R}} f(t)\, \overline{g_{b',\omega'}^{\Lambda}(t)}\, dt$$

$$= \frac{1}{\sqrt{2\pi i B}} \int_{\mathbb{R}} \left\{ \frac{1}{\|g\|_2^2} \int_{\mathbb{R}} \int_{\mathbb{R}} \mathscr{G}_g^{\Lambda}\big[f\big](b, \omega)\, g(t - b)\, \overline{\mathcal{K}_{\Lambda}(t, \omega)}\, db\, d\omega \right\} \overline{g_{b',\omega'}^{\Lambda}(t)}\, dt$$

$$= \frac{1}{2\pi|B|} \int_{\mathbb{R}} \left\{ \frac{1}{\|g\|_2^2} \int_{\mathbb{R}} \int_{\mathbb{R}} \mathscr{G}_g^\Lambda\left[f\right](b,\omega)\, g_{b,\omega}^\Lambda(t)\, db\, d\omega \right\} \overline{g_{b',\omega'}^\Lambda(t)}\, dt$$

$$= \frac{1}{2\pi|B|\,\|g\|_2^2} \int_{\mathbb{R}} \int_{\mathbb{R}} \mathscr{G}_g^\Lambda\left[f\right](b,\omega) \left\{ \int_{\mathbb{R}} g_{b,\omega}^\Lambda(t)\, \overline{g_{b',\omega'}^\Lambda(t)}\, dt \right\} db\, d\omega$$

$$= \frac{1}{2\pi|B|\,\|g\|_2^2} \int_{\mathbb{R}} \int_{\mathbb{R}} \mathscr{G}_g^\Lambda\left[f\right](b,\omega) \left\langle g_{b,\omega}^\Lambda, g_{b',\omega'}^\Lambda \right\rangle_2 \frac{da\, db}{a^2}$$

$$= \int_{\mathbb{R}} \int_{\mathbb{R}} \mathscr{G}_g^\Lambda\left[f\right](b,\omega)\, \mathscr{K}_g^\Lambda(b,\omega;b',\omega')\, db\, d\omega.$$

$$\square$$

Remark 2.5.3. The reproducing kernel (2.148) is a measure of the correlation of the analyzing functions $g_{b,\omega}^\Lambda(t)$ and $g_{b',\omega'}^\Lambda(t)$. As a consequence of (2.147), it follows that the windowed special affine Fourier transform of any $f \in L^2(\mathbb{R})$ about $(b',\omega') \in \mathbb{R}^2$ is always expressible via another $(b,\omega) \in \mathbb{R}^2$ by virtue of the reproducing kernel (2.148). Moreover, the kernel $\mathscr{K}_g^\Lambda(b,\omega;b',\omega')$ is pointwise bounded, because for all $(b,\omega),(b',\omega') \in \mathbb{R}^2$, the implication of Cauchy-Schwarz inequality yields

$$\left| \mathscr{K}_g^\Lambda(b,\omega;b',\omega') \right| = \frac{1}{2\pi|B|\,\|g\|_2^2} \left| \left\langle g_{b,\omega}^\Lambda, g_{b',\omega'}^\Lambda \right\rangle_2 \right|$$

$$\leq \frac{1}{2\pi|B|\,\|g\|_2^2} \left\| g_{b,\omega}^\Lambda \right\|_2 \left\| g_{b',\omega'}^\Lambda \right\|_2$$

$$= \frac{1}{2\pi|B|\,\|g\|_2^2} \|g\|_2^2$$

$$= \frac{1}{2\pi|B|}. \tag{2.149}$$

2.6   The Windowed Quadratic-Phase Fourier Transform

"Before creation, God did just pure mathematics. Then He thought it would be a pleasant change to do some applied."

-John E. Littlewood

This section deals with the notion of windowed quadratic-phase Fourier transform (WQPFT), which is aimed at localizing the spectral components in the quadratic-phase Fourier domain. Such an integral transformation serves as an extreme generalization of the windowed Fourier transforms. Besides an investigation on the fundamental properties, we also establish an orthogonality relation and inversion formula corresponding to WQPFT.

To begin with, we present the formal definition of windowed quadratic-phase Fourier transform followed by the fundamental properties.

Definition 2.6.1. For a given collection of real parameters $\Omega = (A,B,C,D,E)$, with $B \neq 0$, the windowed quadratic-phase Fourier transform of an arbitrary function $f \in L^2(\mathbb{R})$ with respect to a window function $g \in L^2(\mathbb{R})$ is denoted by $\mathscr{G}_g^\Omega\left[f\right]$ and is defined by

$$\mathscr{G}_g^\Omega\left[f\right](b,\omega) = \frac{1}{\sqrt{2\pi}} \left\langle f, g_{b,\omega}^\Omega \right\rangle_2 = \frac{1}{\sqrt{2\pi}} \int_{\mathbb{R}} f(t)\, \overline{g(t-b)}\, \mathcal{K}_\Omega(t,\omega)\, dt, \quad b,\omega \in \mathbb{R}, \tag{2.150}$$

where $\mathcal{K}_\Omega(t,\omega)$ is the usual kernel of the quadratic-phase Fourier transform as given in (1.258), so that the analyzing functions $g^\Omega_{b,\omega}(t)$ are given by

$$g^\Omega_{b,\omega}(t) = g(t-b)\, \exp\left\{ i\left(At^2 + B\omega t + C\omega^2 + Dt + E\omega\right)\right\}. \qquad (2.151)$$

Definition 2.6.1 encompasses several time-frequency tools ranging from the windowed Fourier to the windowed special affine Fourier transforms. Some important special cases are described below:

(i) For $A = C = D = E = 0$ and $B = 1$, Definition 2.6.1 yields the classical windowed Fourier transform (2.50).

(ii) Consider the parameters $A = -\cot\theta/2$, $B = \csc\theta$, $C = -\cot\theta/2$ and $D = E = 0$, where $\theta \neq n\pi$, $n \in \mathbb{Z}$. Then, multiplying the integrand in (2.150) with $\sqrt{1 - i\cot\theta}$ yields the θ-angle windowed fractional Fourier transform as given in (2.68).

(iii) For $D = E = 0$, we consider the transformations $A \to -A/2B$, $B \to 1/B$, $C \to -C/2B$. Then, multiplying the integrand in (2.150) with $1/\sqrt{iB}$, Definition 2.6.1 attains the form of the windowed linear canonical transform (2.95).

(iv) Upon shuffling the parameters as $A \to -A/2B$, $B \to 1/B$, $C \to -D/2B$, $D \to -p/B$, $E = (Dp - Bq)/B$ and then multiplying the integrand in (2.150) with $e^{iDp^2/2B}/\sqrt{iB}$, an integral expression similar to the windowed special affine Fourier transform (2.132) is obtained.

In the following proposition, we shall obtain an alternative representation of the windowed quadratic-phase Fourier transform (2.150), involving the quadratic-phase Fourier transform of the input signal and the classical Fourier transform of the window function.

Proposition 2.6.1. *The windowed quadratic-phase Fourier transform $\mathscr{G}^\Omega_g[f](b,\omega)$ corresponding to any $f \in L^2(\mathbb{R})$ with respect to the window function $g \in L^2(\mathbb{R})$ is expressible as*

$$\mathscr{G}^\Omega_g[f](b,\omega) = \frac{|B|}{\sqrt{2\pi}}\, \overline{\mathcal{K}_\Omega(b,\omega)} \int_{\mathbb{R}} \mathcal{Q}_\Omega[f](\xi)\, \overline{\mathscr{F}[g]\big(B(\omega - \xi)\big)}\, \mathcal{K}_\Omega(b,\xi)\, d\xi, \qquad (2.152)$$

where $\mathcal{Q}_\Omega$ denotes the quadratic-phase Fourier transform (1.257) and $\mathscr{F}$ is the classical Fourier transform defined in (1.7).

Proof. The result is a straightaway consequence of Theorem 1.6.2, so we omit the proof. $\square$

In the following theorem, we assemble some basic properties of the windowed quadratic-phase Fourier transform (2.150).

Theorem 2.6.2. *Let $g, h \in L^2(\mathbb{R})$ be two window functions and $\alpha, \beta \in \mathbb{C}$, $k, \omega_0 \in \mathbb{R}$, $\lambda \in \mathbb{R}\backslash\{0\}$ are scalars. Then, for any pair of functions $f, s \in L^2(\mathbb{R})$, the windowed quadratic-phase Fourier transform (2.150) satisfies the following properties:*

(i) *Linearity:* $\mathscr{G}^\Omega_g[\alpha f + \beta s](b,\omega) = \alpha\, \mathscr{G}^\Omega_g[f](b,\omega) + \beta\, \mathscr{G}^\Omega_g[s](b,\omega),$

(ii) *Anti-linearity:* $\mathscr{G}^\Omega_{\alpha g + \beta h}[f](b,\omega) = \overline{\alpha}\, \mathscr{G}^\Omega_g[f](b,\omega) + \overline{\beta}\, \mathscr{G}^\Omega_h[f](b,\omega),$

(iii) *Time-shift:* $\mathscr{G}^\Omega_g[f(t-k)](b,\omega) = \mathscr{G}^\Omega_g[f](b-k, \omega + 2AB^{-1}k)$
$$\times \exp\left\{ i\big((4A^2B^{-2}C - A)k^2 + (4AB^{-1}C - B)\omega k + (2AB^{-1}E - D)k\big)\right\},$$

(iv) *Scaling:* $\mathscr{G}_g^\Omega\left[f(\lambda t)\right](b,\omega) = \mathscr{G}_{G_\lambda}^{\Omega'}\left[f\right]\left(\lambda b, \frac{\omega}{\lambda}\right)$, *where* $G_\lambda(t) = \frac{1}{|\lambda|}g\left(\frac{t}{\lambda}\right)$ *and*
$\Omega' = (A/\lambda^2, B, \lambda^2 C, D/\lambda, \lambda E)$,

(v) *Modulation:* $\mathscr{G}_g^\Omega\left[e^{i\omega_0 t}f(t)\right](b,\omega) = \exp\left\{i\left(C(B^{-2}\omega_0^2 - 2B^{-1}\omega\omega_0) - EB^{-1}\omega_0\right)\right\}$
$$\times \mathscr{G}_g^\Omega\left[f\right]\left(b, \omega - B^{-1}\omega_0\right),$$

(vi) *Symmetry:* $\mathscr{G}_g^\Omega\left[f\right](b,\omega) = \mathscr{G}_f^\Omega\left[\bar{g}\right]\left(-b, \omega + 2AB^{-1}b\right)$
$$\times \exp\left\{i\left((4A^2B^{-2}C - A)b^2 + (4AB^{-1}C - B)\omega b + (2AB^{-1}E - D)b\right)\right\},$$

(vii) *Translation in window:* $\mathscr{G}_{g(t-k)}^\Omega\left[f\right](b,\omega) = \mathscr{G}_g^\Omega\left[f\right](b+k,\omega)$,

(viii) *Scaling in window:* $\mathscr{G}_{g(\lambda t)}^\Omega\left[f\right](b,\omega) = \mathscr{G}_g^{\Omega'}\left[F_\lambda\right]\left(\lambda b, \frac{\omega}{\lambda}\right)$, *where* $F_\lambda(t) = \frac{1}{|\lambda|}f\left(\frac{t}{\lambda}\right)$ *and*
$\Omega' = (A/\lambda^2, B, \lambda^2 C, D/\lambda, \lambda E)$.

Proof. For the sake of brevity, we omit the proof.   □

In the upcoming theorem, we shall formulate an orthogonality relation associated with the windowed quadratic-phase Fourier transform (2.150). In lieu to the windowed Fourier, fractional Fourier, linear canonical and special affine Fourier transforms, the orthogonality relation pertaining to WQPFT is dictated by the parameter B appearing in Definition 2.6.1.

Theorem 2.6.3. *Let* $\mathscr{G}_{g_1}^\Omega\left[f\right](b,\omega)$ *and* $\mathscr{G}_{g_2}^\Omega\left[h\right](b,\omega)$ *be the windowed quadratic-phase Fourier transforms corresponding to* $f, h \in L^2(\mathbb{R})$ *with respect to the window functions* $g_1, g_2 \in L^2(\mathbb{R})$, *respectively, then we have*

$$\int_\mathbb{R}\int_\mathbb{R} \mathscr{G}_{g_1}^\Omega\left[f\right](b,\omega)\,\overline{\mathscr{G}_{g_2}^\Omega\left[h\right](b,\omega)}\,db\,d\omega = \frac{1}{|B|}\left\langle g_2, g_1\right\rangle_2\left\langle f, h\right\rangle_2. \tag{2.153}$$

Proof. Invoking Proposition 2.6.1, we can express the windowed quadratic-phase Fourier transform of the given pair of functions $f, h \in L^2(\mathbb{R})$ as

$$\mathscr{G}_{g_1}^\Omega\left[f\right](b,\omega) = \frac{|B|}{\sqrt{2\pi}}\,\overline{\mathcal{K}_\Omega(b,\omega)}\int_\mathbb{R}\mathcal{Q}_\Omega\left[f\right](\xi)\,\overline{\mathscr{F}\left[g_1\right]\left(B(\omega-\xi)\right)}\,\mathcal{K}_\Omega(b,\xi)\,d\xi$$

and

$$\mathscr{G}_{g_2}^\Omega\left[h\right](b,\omega) = \frac{|B|}{\sqrt{2\pi}}\,\overline{\mathcal{K}_\Omega(b,\omega)}\int_\mathbb{R}\mathcal{Q}_\Omega\left[h\right](\eta)\,\overline{\mathscr{F}\left[g_2\right]\left(B(\omega-\eta)\right)}\,\mathcal{K}_\Omega(b,\eta)\,d\eta.$$

With Theorem 1.6.2 in hindsight, we can obtain the desired orthogonality relation by proceeding in the following manner:

$$\int_\mathbb{R}\int_\mathbb{R} \mathscr{G}_{g_1}^\Omega\left[f\right](b,\omega)\,\overline{\mathscr{G}_{g_2}^\Omega\left[h\right](b,\omega)}\,db\,d\omega$$

$$= \frac{|B|^2}{2\pi}\int_\mathbb{R}\int_\mathbb{R}\left\{\overline{\mathcal{K}_\Omega(b,\omega)}\int_\mathbb{R}\mathcal{Q}_\Omega\left[f\right](\xi)\,\overline{\mathscr{F}\left[g_1\right]\left(B(\omega-\xi)\right)}\,\mathcal{K}_\Omega(b,\xi)\,d\xi\right\}$$

$$\times\left\{\mathcal{K}_\Omega(b,\omega)\int_\mathbb{R}\overline{\mathcal{Q}_\Omega\left[h\right](\eta)}\,\mathscr{F}\left[g_2\right]\left(B(\omega-\eta)\right)\overline{\mathcal{K}_\Omega(b,\eta)}\,d\eta\right\}db\,d\omega$$

$$= \frac{|B|^2}{2\pi}\int_\mathbb{R}\int_\mathbb{R}\int_\mathbb{R}\mathcal{Q}_\Omega\left[f\right](\xi)\,\overline{\mathcal{Q}_\Omega\left[h\right](\eta)}\,\overline{\mathscr{F}\left[g_1\right]\left(B(\omega-\xi)\right)}\,\mathscr{F}\left[g_2\right]\left(B(\omega-\eta)\right)$$

$$\times \left\{ \int_{\mathbb{R}} \mathcal{K}_\Omega(b,\xi)\, \overline{\mathcal{K}_\Omega(b,\eta)}\, db \right\} d\omega\, d\xi\, d\eta$$

$$= |B| \int_{\mathbb{R}} \int_{\mathbb{R}} \int_{\mathbb{R}} \mathcal{Q}_\Omega\Big[f\Big](\xi)\, \overline{\mathcal{Q}_\Omega\Big[h\Big](\eta)}\, \mathscr{F}\Big[g_1\Big](B(\omega-\xi))\, \overline{\mathscr{F}\Big[g_2\Big](B(\omega-\eta))}$$

$$\times \exp\Big\{ i\big(C(\eta^2-\xi^2)+E(\eta-\xi)\big)\Big\} \left\{ \frac{1}{2\pi} \int_{\mathbb{R}} e^{B(\eta-\xi)b}\, d(Bb) \right\} d\omega\, d\xi\, d\eta$$

$$= |B| \int_{\mathbb{R}} \int_{\mathbb{R}} \int_{\mathbb{R}} \mathcal{Q}_\Omega\Big[f\Big](\xi)\, \overline{\mathcal{Q}_\Omega\Big[h\Big](\eta)}\, \mathscr{F}\Big[g_1\Big](B(\omega-\xi))\, \overline{\mathscr{F}\Big[g_2\Big](B(\omega-\eta))}$$

$$\times \exp\Big\{ i\big(C(\eta^2-\xi^2)+E(\eta-\xi)\big)\Big\}\, \delta(\eta-\xi)\, d\omega\, d\xi\, d\eta$$

$$= |B| \int_{\mathbb{R}} \int_{\mathbb{R}} \mathcal{Q}_\Omega\Big[f\Big](\xi)\, \overline{\mathcal{Q}_\Omega\Big[h\Big](\xi)}\, \mathscr{F}\Big[g_1\Big](B(\omega-\xi))\, \overline{\mathscr{F}\Big[g_2\Big](B(\omega-\xi))}\, d\omega\, d\xi$$

$$= |B| \int_{\mathbb{R}} \mathcal{Q}_\Omega\Big[f\Big](\xi)\, \overline{\mathcal{Q}_\Omega\Big[h\Big](\xi)} \left\{ \int_{\mathbb{R}} \mathscr{F}\Big[g_1\Big](B(\omega-\xi))\, \overline{\mathscr{F}\Big[g_2\Big](B(\omega-\xi))}\, d\omega \right\} d\xi$$

$$= |B| \int_{\mathbb{R}} \mathcal{Q}_\Omega\Big[f\Big](\xi)\, \overline{\mathcal{Q}_\Omega\Big[h\Big](\xi)} \left\{ \frac{1}{|B|} \int_{\mathbb{R}} \mathscr{F}\Big[g_1\Big](\sigma)\, \overline{\mathscr{F}\Big[g_2\Big](\sigma)}\, d\sigma \right\} d\omega$$

$$= \int_{\mathbb{R}} \mathcal{Q}_\Omega\Big[f\Big](\xi)\, \overline{\mathcal{Q}_\Omega\Big[h\Big](\xi)} \left\{ \int_{\mathbb{R}} \mathscr{F}\Big[g_1\Big](\sigma)\, \overline{\mathscr{F}\Big[g_2\Big](\sigma)}\, d\sigma \right\} d\omega$$

$$= \Big\langle \mathscr{F}\Big[g_2\Big],\, \mathscr{F}\Big[g_1\Big] \Big\rangle_2 \Big\langle \mathcal{Q}_\Omega\Big[f\Big],\, \mathcal{Q}_\Omega\Big[h\Big] \Big\rangle_2$$

$$= \frac{1}{|B|} \Big\langle g_2,\, g_1 \Big\rangle_2 \Big\langle f,\, h \Big\rangle_2 .$$

This completes the proof Theorem 2.6.3. $\qquad\square$

Remark 2.6.1. Choosing $f = h$ and $g_1 = g_2 = g$, the orthogonality relation (2.153) comes down to the relation

$$\int_{\mathbb{R}} \int_{\mathbb{R}} \Big| \mathcal{G}_g^\Omega\Big[f\Big](b,\omega) \Big|^2 db\, d\omega = \frac{1}{|B|} \Big\| f \Big\|_2^2 \Big\| g \Big\|_2^2, \tag{2.154}$$

which is evidently somewhat peculiar in the sense that (2.154) is not completely free from the parameters involved in Definition 2.6.1. Therefore, we infer that the energy distribution of the windowed quadratic-phase Fourier transform (2.150) is not equitable between the natural and transformed domains. However, if the window function $g \in L^2(\mathbb{R})$ is chosen in a way that $\|g\|_2^2 = |B|$, then (2.154) precisely yields the energy preserving relation:

$$\int_{\mathbb{R}} \int_{\mathbb{R}} \Big| \mathcal{G}_g^\Omega\Big[f\Big](b,\omega) \Big|^2 db\, d\omega = \Big\| f \Big\|_2^2 . \tag{2.155}$$

Thus, in case of the windowed quadratic-phase Fourier transform, non-normalized window functions can also be used to achieve energy balance between the actual and the transformed domains.

Following is the inversion formula associated with the windowed quadratic-phase Fourier transform defined in (2.150):

Theorem 2.6.4. *For any analysis window $g_1 \in L^2(\mathbb{R})$, let $\mathcal{G}_{g_1}^\Omega\big[f\big](b,\omega)$ be the windowed quadratic-phase Fourier transform corresponding to $f \in L^2(\mathbb{R})$. Then, the signal f can be reconstructed by means of another synthesis window $g_2 \in L^2(\mathbb{R})$ via the following formula:*

$$f(t) = \frac{|B|}{\sqrt{2\pi}\, \langle g_2,\, g_1 \rangle_2} \int_{\mathbb{R}} \int_{\mathbb{R}} \mathcal{G}_{g_1}^\Omega\Big[f\Big](b,\omega)\, g_2(t-b)\, \overline{\mathcal{K}_\Omega(t,\omega)}\, db\, d\omega, \tag{2.156}$$

where equality in (2.156) holds in the almost everywhere sense.

Proof. In view of the orthogonality relation (2.153) and the fundamental properties of the inner product, we can easily accomplish the proof. $\square$

Remark 2.6.2. If a single window function $g \in L^2(\mathbb{R})$ is used for both the analysis and synthesis purposes, then the inversion formula (2.156) boils down to

$$f(t) = \frac{|B|}{\sqrt{2\pi} \, \|g\|_2^2} \int_{\mathbb{R}} \int_{\mathbb{R}} \mathscr{G}_g^{\Omega} \Big[f\Big](b, \omega) \, g(t - b) \, \overline{\mathcal{K}_{\Omega}(t, \omega)} \, db \, d\omega, \qquad (2.157)$$

with equality in the weak sense. Furthermore, in case the window function g satisfies that $\|g\|_2^2 = |B|$, then we have

$$f(t) = \frac{1}{\sqrt{2\pi}} \int_{\mathbb{R}} \int_{\mathbb{R}} \mathscr{G}_g^{\Omega} \Big[f\Big](b, \omega) \, g(t - b) \, \overline{\mathcal{K}_{\Omega}(t, \omega)} \, db \, d\omega. \qquad (2.158)$$

Besides, the above formulated inversion formulae for the windowed quadratic-phase Fourier transform, we can also establish an alternative inversion formula which just involves a one-dimensional integral expression. Note that the windowed quadratic-phase Fourier transform (2.150) can be expressed simply as

$$\mathscr{G}_g^{\Omega} \Big[f\Big](b, \omega) = \mathcal{Q}_{\Omega} \Big[f(t) \, \overline{g(t - b)}\Big](\omega), \qquad (2.159)$$

where $\mathcal{Q}_{\Omega}$ denotes the quadratic-phase Fourier transform as defined in (1.257). Therefore, applying the inversion formula (1.259) pertaining to the quadratic-phase Fourier transform, we get

$$f(t) \, \overline{g(t - b)} = \frac{|B|}{\sqrt{2\pi}} \int_{\mathbb{R}} \mathcal{Q}_{\Omega} \Big[f(t) \, \overline{g(t - b)}\Big](\omega) \, \overline{\mathcal{K}_{\Omega}(t, \omega)} \, d\omega$$

$$= \frac{|B|}{\sqrt{2\pi}} \int_{\mathbb{R}} \mathscr{G}_g^{\Omega} \Big[f\Big](b, \omega) \, \overline{\mathcal{K}_{\Omega}(t, \omega)} \, d\omega. \qquad (2.160)$$

Plugging $b = t$ into (2.160) with the constraint that $g(0) \neq 0$, we obtain the one-dimensional integral expression for the inversion formula of WQPFT as

$$f(t) = \frac{|B|}{\sqrt{2\pi} \, \overline{g(0)}} \int_{\mathbb{R}} \mathscr{G}_g^{\Omega} \Big[f\Big](t, \omega) \, \overline{\mathcal{K}_{\Omega}(t, \omega)} \, d\omega, \qquad (2.161)$$

where the equality holds almost everywhere.

Finally, before we leave this section, our aim is to demonstrate that the range of windowed quadratic-phase Fourier transform is a reproducing kernel Hilbert space, embedded as a subspace in $L^2(\mathbb{R}^2)$. To facilitate the narrative, we shall denote by $\mathcal{H}$ as the range of windowed quadratic-phase Fourier transform defined in (2.150); that is, $\mathcal{H} = \mathscr{G}_g^{\Omega}\big(L^2(\mathbb{R})\big)$.

Theorem 2.6.5. *Any function $F(b, \omega) \in \mathcal{H}$, satisfies the following reproduction formula:*

$$F(b', \omega') = \int_{\mathbb{R}} \int_{\mathbb{R}} F(b, \omega) \, \mathscr{K}_g^{\Omega}(b, \omega; b', \omega') \, db \, d\omega, \quad (b', \omega') \in \mathbb{R}^2, \qquad (2.162)$$

where the reproducing kernel $\mathscr{K}_g^{\Omega}(b, \omega; b', \omega')$ is given by

$$\mathscr{K}_g^{\Omega}(b, \omega; b', \omega') = \frac{|B|}{2\pi \|g\|_2^2} \Big\langle g_{b,\omega}^{\Omega}, \, g_{b',\omega'}^{\Omega} \Big\rangle_2 \qquad (2.163)$$

and is pointwise bounded.

Proof. For any $F \in \mathcal{H}$, we can surely find certain $f \in L^2(\mathbb{R})$, such that $F = \mathscr{G}_g^\Omega[f]$. Therefore, in view of the inversion formula (2.157), it follows that

$$
\begin{aligned}
F(b', \omega') &= \mathscr{G}_g^\Omega\Big[f\Big](b', \omega') \\
&= \frac{1}{\sqrt{2\pi}} \int_\mathbb{R} f(t)\, \overline{g_{b',\omega'}^\Omega(t)}\, dt \\
&= \frac{1}{\sqrt{2\pi}} \int_\mathbb{R} \left\{ \frac{|B|}{\sqrt{2\pi}\left\|g\right\|_2^2} \int_\mathbb{R} \int_\mathbb{R} \mathscr{G}_g^\Omega\Big[f\Big](b,\omega)\, g(t-b)\, \overline{K_\Omega(t,\omega)}\, db\, d\omega \right\} \overline{g_{b',\omega'}^\Omega(t)}\, dt \\
&= \frac{|B|}{2\pi\left\|g\right\|_2^2} \int_\mathbb{R} \left\{ \int_\mathbb{R} \int_\mathbb{R} \mathscr{G}_g^\Omega\Big[f\Big](b,\omega)\, g_{b,\omega}^\Omega(t)\, db\, d\omega \right\} \overline{g_{b',\omega'}^\Omega(t)}\, dt \\
&= \frac{|B|}{2\pi\left\|g\right\|_2^2} \int_\mathbb{R} \int_\mathbb{R} \mathscr{G}_g^\Omega\Big[f\Big](b,\omega) \left\{ \int_\mathbb{R} g_{b,\omega}^\Omega(t)\, \overline{g_{b',\omega'}^\Omega(t)}\, dt \right\} db\, d\omega \\
&= \frac{|B|}{2\pi\left\|g\right\|_2^2} \int_\mathbb{R} \int_\mathbb{R} \mathscr{G}_g^\Omega\Big[f\Big](b,\omega) \left\langle g_{b,\omega}^\Omega, g_{b',\omega'}^\Omega \right\rangle_2 \frac{da\, db}{a^2} \\
&= \int_\mathbb{R} \int_\mathbb{R} F(b,\omega)\, \mathscr{K}_g^\Omega(b,\omega; b', \omega')\, db\, d\omega,
\end{aligned}
$$

where

$$
\mathscr{K}_g^\Omega(b,\omega; b', \omega') = \frac{|B|}{2\pi\left\|g\right\|_2^2} \left\langle g_{b,\omega}^\Omega, g_{b',\omega'}^\Omega \right\rangle_2,
$$

which establishes (2.162). Also, by virtue of the Cauchy-Schwarz inequality, it follows that

$$
\begin{aligned}
\left| \mathscr{K}_g^\Omega(b,\omega; b', \omega') \right| &= \frac{|B|}{2\pi\left\|g\right\|_2^2} \left| \left\langle g_{b,\omega}^\Omega, g_{b',\omega'}^\Omega \right\rangle_2 \right| \\
&\leq \frac{|B|}{2\pi\left\|g\right\|_2^2} \left\| g_{b,\omega}^\Omega \right\|_2 \left\| g_{b',\omega'}^\Omega \right\|_2 \\
&= \frac{|B|}{2\pi},
\end{aligned}
$$

for all $(b,\omega), (b', \omega') \in \mathbb{R}^2$; that is, the kernel $\mathscr{K}_g^\Omega(b,\omega; b', \omega')$ is pointwise bounded. This completes the proof of Theorem 2.6.5. □

Remark 2.6.3. The range of windowed quadratic-phase Fourier transform $\mathcal{H} = \mathscr{G}_g^\Omega(L^2(\mathbb{R}))$ is a reproducing kernel Hilbert space with the reproducing kernel given by (2.163).

2.7 The Directional Windowed Fourier Transform

"[In mathematics] There are two kinds of mistakes. There are fatal mistakes that destroy a theory, but there are also contingent ones, which are useful in testing the stability of a theory."

-Gian-Carlo Rota

The classical windowed Fourier transform is one of the premier signal processing tools, which measures the frequency content of a signal contained in a certain neighbourhood of time by employing a window function to limit the domain of influence of the given non-transient signal. Using the translates of the window function, the signal is analyzed for its entire domain. However, this methodology is not befitting for higher dimensional signals wherein edges, corners and other non-isotropic features are embedded into the signal. To circumvent these limitations, Grafakos and Sensing [124] introduced an elegant ramification of the classical windowed Fourier transform which not only provides the localized frequency content of a signal, but also gives the frequency information along a certain direction or hyperplane (an intrinsic characteristic of the well-known Radon transform). Here, our aim is to present the chronological developments of the directional windowed Fourier transform and also provide an impetus for its extension to generalized Fourier domains. Prior to that, it is imperative to have a bird's-eye view of the notion of multi-dimensional Fourier transform. For any $f \in L^2(\mathbb{R}^n)$, the multi-dimensional Fourier transform is defined as

$$\mathscr{F}\big[f\big](\mathbf{w}) = \frac{1}{(2\pi)^{n/2}} \int_{\mathbb{R}^n} f(\mathbf{t})\, e^{-it\cdot\mathbf{w}}\, d\mathbf{t}, \tag{2.164}$$

where $\mathbf{t}\cdot\mathbf{w}$ denotes the dot product of $\mathbf{t}, \mathbf{w} \in \mathbb{R}^n$. Also, the Parseval's relation and inversion formula corresponding to (2.164) are given by

$$\Big\langle \mathscr{F}\big[f\big], \mathscr{F}\big[g\big] \Big\rangle_2 = \big\langle f, g \big\rangle_2, \quad \forall\, f, g \in L^2(\mathbb{R}^n) \tag{2.165}$$

and

$$f(\mathbf{t}) = \mathscr{F}^{-1}\Big(\mathscr{F}\big[f\big](\mathbf{w}) \Big)(\mathbf{t}) := \frac{1}{(2\pi)^{n/2}} \int_{\mathbb{R}^n} \mathscr{F}\big[f\big](\mathbf{w})\, e^{it\cdot\mathbf{w}}\, d\mathbf{w}, \tag{2.166}$$

respectively.

Typically, for any $f \in L^2(\mathbb{R}^n)$, the multi-dimensional windowed Fourier transform with respect to the window function $g \in L^2(\mathbb{R}^n)$ is defined as the transformation

$$f \to \frac{1}{(2\pi)^{n/2}} \int_{\mathbb{R}^n} f(\mathbf{t})\, \overline{g(\mathbf{t}-\mathbf{b})}\, e^{-i\mathbf{w}\cdot\mathbf{t}}\, d\mathbf{t} = \frac{1}{(2\pi)^{n/2}} \int_{\mathbb{R}^n} f(\mathbf{t})\, \overline{\mathcal{M}_\mathbf{w}\mathcal{T}_\mathbf{b}\, g(\mathbf{t})}\, d\mathbf{t}, \tag{2.167}$$

where "." denotes the usual dot product on $\mathbb{R}^n$ and $\mathcal{T}_\mathbf{b}$, $\mathcal{M}_\mathbf{w}$ denote the multi-dimensional translation and modulation operators acting on $g \in L^2(\mathbb{R}^n)$ as

$$\left. \begin{array}{l} \mathcal{T}_\mathbf{b}\, g(\mathbf{t}) = g(\mathbf{t}-\mathbf{b}), \qquad \mathbf{b} \in \mathbb{R}^n \\ \mathcal{M}_\mathbf{w}\, g(\mathbf{t}) = e^{i\mathbf{w}\cdot\mathbf{t}} g(\mathbf{t}), \quad \mathbf{w} \in \mathbb{R}^n \end{array} \right\}. \tag{2.168}$$

Note that, the joint action of the planer translation and modulation operators as defined in (2.168) is order-dependent in the sense that

$$\mathcal{M}_\mathbf{w}\mathcal{T}_\mathbf{b}\, g(\mathbf{t}) = e^{i\mathbf{w}\cdot\mathbf{t}}\, g(\mathbf{t}-\mathbf{b}) \quad \text{and} \quad \mathcal{T}_\mathbf{b}\mathcal{M}_\mathbf{w}\, g(\mathbf{t}) = e^{i\mathbf{w}\cdot(\mathbf{t}-\mathbf{b})}\, g(\mathbf{t}-\mathbf{b}). \tag{2.169}$$

Consequently, in view of the later ordering, the transformation (2.167) is revamped as follows:

$$f \to \frac{1}{(2\pi)^{n/2}} \int_{\mathbb{R}^n} f(\mathbf{t})\, \overline{g(\mathbf{t}-\mathbf{b})}\, e^{-i\mathbf{w}\cdot(\mathbf{t}-\mathbf{b})}\, d\mathbf{t}, \tag{2.170}$$

which is intimately connected to the usual multi-dimensional windowed Fourier transform, as the transformation (2.167) can be obtained from (2.170) by pulling out the factor $e^{i\mathbf{w}\cdot\mathbf{b}}$ from the integral. With the transformation (2.170) in hindsight, Grafakos and Sensing [124] introduced a directionally sensitive variant of the multi-dimensional windowed Fourier transform. The basic idea was to transform a function f defined on $\mathbb{R}^n$, with respect to an appropriate function g on $\mathbb{R}$, to a function on $\mathbb{S}^{n-1} \times \mathbb{R} \times \mathbb{R}$,

$$(\mathbf{v}, b, \omega) \to \frac{1}{(2\pi)^{n/2}} \int_{\mathbb{R}^n} f(\mathbf{t}) \, \overline{g(\mathbf{v} \cdot \mathbf{t} - b)} \, e^{-i\omega(\mathbf{v}\cdot\mathbf{t}-b)} \, d\mathbf{t}, \qquad (2.171)$$

where $\mathbb{S}^{n-1}$ denotes the unit sphere in $\mathbb{R}^n$. The function given by

$$g_{\mathbf{v},b}(\mathbf{t}) = g(\mathbf{v} \cdot \mathbf{t} - b) \, e^{i\mathbf{u}\cdot(\mathbf{v}\cdot\mathbf{t}-b)} \qquad (2.172)$$

constitutes the analyzing elements and is formally called as the "Gabor ridge function," which behaves like a one-dimensional Gabor function in the direction of $\mathbf{v}$ and is constant in its orthogonal complement. By pulling out $e^{i\omega b}$ from (2.171), Grafakos and Sensing defined a directionally sensitive variant of the multi-dimensional windowed Fourier transform as

$$f \to \frac{1}{(2\pi)^{n/2}} \int_{\mathbb{R}^n} f(\mathbf{t}) \, \overline{g(\mathbf{v} \cdot \mathbf{t} - b)} \, e^{-i\omega(\mathbf{v}\cdot\mathbf{t})} \, d\mathbf{t}. \qquad (2.173)$$

That is, the transformation (2.173) sends a function f defined on $\mathbb{R}^n$ to a function defined on $\mathbb{S}^{n-1} \times \mathbb{R} \times \mathbb{R}$. Interestingly, the directional parameter $\mathbf{v} \in \mathbb{S}^{n-1}$ appears in both the window function and the exponential function defining the modulation operator. This stimulated renewed research interest towards this key problem in the higher dimensional signal processing and prompted the researchers to re-examine the notion of directional windowed Fourier transform.

Notwithstanding the aforementioned approach, Giv [120] introduced another version of the directional windowed Fourier transform by taking into account the fact that the typical multi-dimensional windowed Fourier transform is actually the multi-dimensional Fourier transform of the function $f(\mathbf{t}) \, \mathcal{T}_{\mathbf{b}} \, \overline{g(\mathbf{t})} := f(\mathbf{t}) \, \overline{g(\mathbf{t} - \mathbf{b})}$. This led Giv to re-examine the approach adopted by Grafokos-Sensing [124], following which the directional parameter $\mathbf{v}$ was only inserted into the window function by taking into account a new integral transform which sends a function f defined on $\mathbb{R}^n$ to a function on $\mathbb{S}^{n-1} \times \mathbb{R} \times \mathbb{R}^n$. Formally, we have the following definition of the directional windowed Fourier transform:

Definition 2.7.1. Given a window function $g \in L^2(\mathbb{R})$, the directional windowed Fourier transform of any $f \in L^2(\mathbb{R}^n)$ is denoted by $\mathcal{DG}_g[f]$ and is defined as

$$\mathcal{DG}_g[f](\mathbf{v}, b, \mathbf{w}) = \frac{1}{(2\pi)^{n/2}} \int_{\mathbb{R}^n} f(\mathbf{t}) \, \overline{g(\mathbf{v} \cdot \mathbf{t} - b)} \, e^{-i\mathbf{w}\cdot\mathbf{t}} \, d\mathbf{t}, \quad (\mathbf{v}, b, \mathbf{w}) \in \mathbb{S}^{n-1} \times \mathbb{R} \times \mathbb{R}^n. \qquad (2.174)$$

Below we point out some important features regarding the notion of directional windowed Fourier transform given in Definition 2.7.1.

(i) The transformation $f \to \mathcal{DG}_g[f]$ can be regarded as the multi-dimensional Fourier transform of the truncated and suitably oriented function $f(\mathbf{t}) \, \overline{g(\mathbf{v} \cdot \mathbf{t} - b)}$, where $\mathbf{v} \in \mathbb{S}^{n-1}$ and $b \in \mathbb{R}$. Therefore, (2.174) is legitimately referred to as the "directional windowed Fourier transform."

(ii) The function $g(\mathbf{v}\cdot\mathbf{t}-b)$ appearing in Definition 2.7.1 is called as the "directional window," which behaves like a one-dimensional window in the direction $\mathbf{v}$ and is constant along its orthogonal complement.

(iii) As a function of $\mathbf{w} \in \mathbb{R}^n$, the directional windowed Fourier transform $\mathcal{DG}_g[f](\mathbf{v}, b, \mathbf{w})$ being the multi-dimensional Fourier transform of $f(\mathbf{t})\,\overline{g(\mathbf{v} \cdot \mathbf{t} - b)}$, therefore, if g has a compact support $K \subset \mathbb{R}$, then $\mathcal{DG}_g[f](\mathbf{v}, b, \mathbf{w})$ is concentrated on the set

$$\left\{ \mathbf{t} \in \mathbb{R}^n : \mathbf{v} \cdot \mathbf{t} \in b + K \right\}, \tag{2.175}$$

which is a restricted collection of hyperplanes in $\mathbb{R}^n$ that are perpendicular to $\mathbf{v}$. In particular, if we take $g = \chi_{[0,1]}$ and $n = 2$, then the set (2.175) simply represents the strip $\left\{ \mathbf{t} \in \mathbb{R}^2 : b \le \mathbf{v} \cdot \mathbf{t} \le b + 1 \right\}$, perpendicular to $\mathbf{v} \in \mathbb{S}^1$ and $\mathcal{DG}_g[f](\mathbf{v}, b, \mathbf{w})$ represents the Fourier transform of the restriction of f itself to that strip. Thus, we conclude that the directional windowed Fourier transform (2.174) allows us to take the Fourier transform on sets which are in some sense of "limited width" abreast of a specific direction, which justifies the nomenclature.

(iv) The unit-sphere $\mathbb{S}^{n-1} \subset \mathbb{R}^n$ is equipped with the normalized surface area measure and the integration with respect to this measure is denoted by $d\mathbf{v}$.

(v) The range of the operator $\mathcal{DG}_g$ not only relies upon its domain of definition, but is also heavily dependent upon the properties of the window function g.

In the following theorem, we study some elementary properties of the directional windowed Fourier transform defined in (2.174).

Theorem 2.7.1. *If $g, h \in L^2(\mathbb{R})$ are two one-dimensional window functions and $\alpha, \beta \in \mathbb{C}$, $\mathbf{k}, \mathbf{w}_0 \in \mathbb{R}^n$, $k \in \mathbb{R}$, $\lambda \in \mathbb{R} \setminus \{0\}$, then for any $f, s \in L^2(\mathbb{R}^n)$, the directional windowed Fourier transform (2.174) satisfies the following properties:*

(i) *Linearity:* $\mathcal{DG}_g[\alpha f + \beta s](\mathbf{v}, b, \mathbf{w}) = \alpha\,\mathcal{DG}_g[f](\mathbf{v}, b, \mathbf{w}) + \beta\,\mathcal{DG}_g[s](\mathbf{v}, b, \mathbf{w})$,

(ii) *Anti-linearity:* $\mathcal{DG}_{\alpha g + \beta h}[f](\mathbf{v}, b, \mathbf{w}) = \overline{\alpha}\,\mathcal{DG}_g[f](\mathbf{v}, b, \mathbf{w}) + \overline{\beta}\,\mathcal{DG}_h[f](\mathbf{v}, b, \mathbf{w})$,

(iii) *Time-shift:* $\mathcal{DG}_g[f(\mathbf{t} - \mathbf{k})](\mathbf{v}, b, \mathbf{w}) = e^{-i\mathbf{w} \cdot \mathbf{k}}\,\mathcal{DG}_g[f](\mathbf{v}, b - \mathbf{v} \cdot \mathbf{k}, \mathbf{w})$,

(iv) *Modulation:* $\mathcal{DG}_g[e^{i\mathbf{w}_0 \cdot \mathbf{t}} f(\mathbf{t})](\mathbf{v}, b, \mathbf{w}) = \mathcal{DG}_g[f](\mathbf{v}, b, \mathbf{w} - \mathbf{w}_0)$,

(v) *Scaling:* $\mathcal{DG}_g[f(\lambda \mathbf{t})](\mathbf{v}, b, \mathbf{w}) = \dfrac{1}{|\lambda|^n}\,\mathcal{DG}_{\tilde{g}}[f]\left(\mathbf{v}, \lambda b, \lambda^{-1}\mathbf{w}\right), \quad \tilde{g}(t) = g\left(\dfrac{t}{\lambda}\right)$,

(vi) *Conjugation:* $\mathcal{DG}_g[\overline{f}](\mathbf{v}, b, \mathbf{w}) = \overline{\mathcal{DG}_{\tilde{g}}[f](\mathbf{v}, b, -\mathbf{w})}$,

(vii) *Translation in window:* $\mathcal{DG}_{g(t-k)}[f](\mathbf{v}, b, \mathbf{w}) = \mathcal{DG}_g[f](\mathbf{v}, b + k, \mathbf{w})$,

(viii) *Scaling in window:* $\mathcal{DG}_{g(\lambda t)}[f](\mathbf{v}, b, \mathbf{w}) = \dfrac{1}{|\lambda|^n}\,\mathcal{DG}_g[\tilde{f}](\mathbf{v}, \lambda b, \lambda^{-1}\mathbf{w}), \quad \tilde{f}(\mathbf{t}) = f(\lambda^{-1}\mathbf{t})$.

Proof. The proof can be obtained easily and is therefore omitted. However, it is important to point out that while proving (viii), the scaling factor λ could not be combined with the directional parameter $\mathbf{v}$, because $\mathbf{v} \in \mathbb{S}^{n-1}$, the unit sphere in $\mathbb{R}^n$. $\qquad \square$

Below, we shall formulate an orthogonality relation for the directional windowed Fourier transform defined in (2.174).

Theorem 2.7.2. *Let* $\mathcal{DG}_{g_1}[f](\mathbf{v}, b, \mathbf{w})$ *and* $\mathcal{DG}_{g_2}[h](\mathbf{v}, b, \mathbf{w})$ *be the directional windowed Fourier transforms of any* $f, h \in L^2(\mathbb{R}^n)$ *with respect to the window functions* $g_1, g_2 \in L^2(\mathbb{R})$, *respectively. Then, we have*

$$\int_{\mathbb{S}^{n-1}} \int_{\mathbb{R}} \int_{\mathbb{R}^n} \mathcal{DG}_{g_1}[f](\mathbf{v}, b, \mathbf{w}) \overline{\mathcal{DG}_{g_2}[h](\mathbf{v}, b, \mathbf{w})} \, db \, d\omega = \Big\langle g_2, g_1 \Big\rangle_2 \Big\langle f, h \Big\rangle_2. \qquad (2.176)$$

Proof. Applying Definition 2.7.1 for the given pair of functions $f, h \in L^2(\mathbb{R}^n)$, we obtain

$$\int_{\mathbb{R}^n} \int_{\mathbb{R}} \int_{\mathbb{S}^{n-1}} \mathcal{DG}_{g_1}[f](\mathbf{v}, b, \mathbf{w}) \overline{\mathcal{DG}_{g_2}[h](\mathbf{v}, b, \mathbf{w})} \, d\mathbf{v} \, db \, d\mathbf{w}$$

$$= \int_{\mathbb{R}^n} \int_{\mathbb{R}} \int_{\mathbb{S}^{n-1}} \left\{ \frac{1}{(2\pi)^{n/2}} \int_{\mathbb{R}^n} f(\mathbf{t}) \overline{g_1(\mathbf{v} \cdot \mathbf{t} - b)} \, e^{-i\mathbf{w} \cdot \mathbf{t}} \, d\mathbf{t} \right\}$$
$$\times \left\{ \frac{1}{(2\pi)^{n/2}} \int_{\mathbb{R}^n} \overline{h(\mathbf{z})} \, g_2(\mathbf{v} \cdot \mathbf{z} - b) \, e^{i\mathbf{w} \cdot \mathbf{z}} \, d\mathbf{z} \right\} d\mathbf{v} \, db \, d\mathbf{w}$$

$$= \int_{\mathbb{R}} \int_{\mathbb{S}^{n-1}} \int_{\mathbb{R}^n} \int_{\mathbb{R}^n} f(\mathbf{t}) \overline{g_1(\mathbf{v} \cdot \mathbf{t} - b)} \, \overline{h(\mathbf{z})} \, g_2(\mathbf{v} \cdot \mathbf{z} - b)$$
$$\times \left\{ \frac{1}{(2\pi)^n} \int_{\mathbb{R}^n} e^{i\mathbf{w} \cdot (\mathbf{z} - \mathbf{t})} \, d\mathbf{w} \right\} d\mathbf{t} \, d\mathbf{z} \, d\mathbf{v} \, db$$

$$= \int_{\mathbb{R}} \int_{\mathbb{S}^{n-1}} \int_{\mathbb{R}^n} \int_{\mathbb{R}^n} f(\mathbf{t}) \overline{h(\mathbf{z})} \, \overline{g_1(\mathbf{v} \cdot \mathbf{t} - b)} \, g_2(\mathbf{v} \cdot \mathbf{z} - b) \, \delta(\mathbf{z} - \mathbf{t}) \, d\mathbf{t} \, d\mathbf{z} \, d\mathbf{v} \, db$$

$$= \int_{\mathbb{R}} \int_{\mathbb{S}^{n-1}} \int_{\mathbb{R}^n} f(\mathbf{t}) \overline{h(\mathbf{t})} \, \overline{g_1(\mathbf{v} \cdot \mathbf{t} - b)} \, g_2(\mathbf{v} \cdot \mathbf{t} - b) \, d\mathbf{t} \, d\mathbf{v} \, db$$

$$= \int_{\mathbb{R}^n} f(\mathbf{t}) \overline{h(\mathbf{t})} \left\{ \int_{\mathbb{R}} \int_{\mathbb{S}^{n-1}} \overline{g_1(\mathbf{v} \cdot \mathbf{t} - b)} \, g_2(\mathbf{v} \cdot \mathbf{t} - b) \, d\mathbf{v} \, db \right\} d\mathbf{t}. \qquad (2.177)$$

Now, by virtue of the translation invariance of the Lebesgue integral, it follows that

$$\int_{\mathbb{R}} \overline{g_1(\mathbf{v} \cdot \mathbf{t} - b)} \, g_2(\mathbf{v} \cdot \mathbf{t} - b) \, db = \Big\langle g_2, g_1 \Big\rangle_2.$$

Therefore, noting that the unit-sphere $\mathbb{S}^{n-1} \subset \mathbb{R}^n$ is equipped with the normalized surface area measure, we obtain

$$\int_{\mathbb{S}^{n-1}} \int_{\mathbb{R}} \overline{g_1(\mathbf{v} \cdot \mathbf{t} - b)} \, g_2(\mathbf{v} \cdot \mathbf{t} - b) \, db \, d\mathbf{v} = \Big\langle g_2, g_1 \Big\rangle_2. \qquad (2.178)$$

Hence, after seeking help from the Fubuni's theorem to change the order of integration, implementing (2.178) in (2.177), we get

$$\int_{\mathbb{R}^n} \int_{\mathbb{R}} \int_{\mathbb{S}^{n-1}} \mathcal{DG}_{g_1}[f](\mathbf{v}, b, \mathbf{w}) \overline{\mathcal{DG}_{g_2}[h](\mathbf{v}, b, \mathbf{w})} \, d\mathbf{v} \, db \, d\mathbf{w} = \Big\langle g_2, g_1 \Big\rangle_2 \int_{\mathbb{R}^n} f(\mathbf{t}) \overline{h(\mathbf{t})} \, d\mathbf{t}$$
$$= \Big\langle g_2, g_1 \Big\rangle_2 \Big\langle f, h \Big\rangle_2.$$

This completes the proof Theorem 2.7.2. □

Remark 2.7.1. For $f = h$ and $g_1 = g_2 = g$, the orthogonality relation (2.176) implies that, indeed the energy preserving relation holds for the directional windowed Fourier transform defined in (2.174). Mathematically,

$$\int_{\mathbb{R}^n} \int_{\mathbb{R}} \int_{\mathbb{S}^{n-1}} \Big| \mathcal{DG}_g[f](\mathbf{v}, b, \mathbf{w}) \Big|^2 \, d\mathbf{v} \, db \, d\mathbf{w} = \big\| f \big\|_2^2 \big\| g \big\|_2^2. \qquad (2.179)$$

In case, the window function $g \in L^2(\mathbb{R})$ satisfies $\|g\|_2 = 1$, then (2.179) demonstrates that the operator $\mathcal{DG}_g : L^2(\mathbb{R}^n) \to L^2(\mathbb{S}^{n-1} \times \mathbb{R} \times \mathbb{R}^n)$ is an isometry.

In the final theorem, we shall formulate an inversion formula associated with the directional windowed Fourier transform given in (2.174).

Theorem 2.7.3. *If $\mathcal{D}\mathcal{G}_{g_1}\big[f\big](\mathbf{v}, b, \mathbf{w})$ is the directional windowed Fourier transform of any $f \in L^2(\mathbb{R}^n)$ with respect to an analyzing window $g_1 \in L^2(\mathbb{R})$, then for any synthesis window $g_2 \in L^2(\mathbb{R})$, the following inversion formula holds:*

$$f(\mathbf{t}) = \frac{1}{(2\pi)^{n/2}\langle g_2, g_1\rangle_2} \int_{\mathbb{R}^n} \int_{\mathbb{R}} \int_{\mathbb{S}^{n-1}} \mathcal{D}\mathcal{G}_{g_1}\big[f\big](\mathbf{v}, b, \mathbf{w})\, g_2(\mathbf{v}\cdot\mathbf{t} - b)\, e^{i\mathbf{w}\cdot\mathbf{t}}\, d\mathbf{v}\, db\, d\mathbf{w}, \quad (2.180)$$

where equality in (2.180) holds in the almost everywhere sense.

Proof. Applying the orthogonality relation (2.176) to the pair of functions $f, h \in L^2(\mathbb{R}^n)$, where h is arbitrary, we have

$$\int_{\mathbb{R}^n} f(\mathbf{t})\, \overline{h(\mathbf{t})}\, d\mathbf{t} = \frac{1}{\langle g_2, g_1\rangle_2} \int_{\mathbb{R}^n} \int_{\mathbb{R}} \int_{\mathbb{S}^{n-1}} \mathcal{D}\mathcal{G}_{g_1}\big[f\big](\mathbf{v}, b, \mathbf{w})\, \overline{\mathcal{D}\mathcal{G}_{g_2}\big[h\big](\mathbf{v}, b, \mathbf{w})}\, d\mathbf{v}\, db\, d\mathbf{w}$$

$$= \frac{1}{\langle g_2, g_1\rangle_2} \int_{\mathbb{R}^n} \int_{\mathbb{R}} \int_{\mathbb{S}^{n-1}} \mathcal{D}\mathcal{G}_{g_1}\big[f\big](\mathbf{v}, b, \mathbf{w})$$

$$\times \left\{ \frac{1}{(2\pi)^{n/2}} \int_{\mathbb{R}^n} \overline{h(\mathbf{t})}\, g_2(\mathbf{v}\cdot\mathbf{t} - b)\, e^{i\mathbf{w}\cdot\mathbf{t}}\, d\mathbf{t} \right\} d\mathbf{v}\, db\, d\mathbf{w}$$

$$= \int_{\mathbb{R}^n} \left\{ \frac{1}{(2\pi)^{n/2}\langle g_2, g_1\rangle_2} \int_{\mathbb{R}^n} \int_{\mathbb{R}} \int_{\mathbb{S}^{n-1}} \mathcal{D}\mathcal{G}_{g_1}\big[f\big](\mathbf{v}, b, \mathbf{w})\, g_2(\mathbf{v}\cdot\mathbf{t} - b) \right.$$

$$\left. \times e^{i\mathbf{w}\cdot\mathbf{t}}\, d\mathbf{v}\, db\, d\mathbf{w} \right\} \overline{h(\mathbf{t})}\, d\mathbf{t}. \quad (2.181)$$

Owing to the arbitrariness of the function $h \in L^2(\mathbb{R}^n)$, we obtain the desired inversion formula from (2.181) as

$$f(\mathbf{t}) = \frac{1}{(2\pi)^{n/2}\langle g_2, g_1\rangle_2} \int_{\mathbb{R}^n} \int_{\mathbb{R}} \int_{\mathbb{S}^{n-1}} \mathcal{D}\mathcal{G}_{g_1}\big[f\big](\mathbf{v}, b, \mathbf{w})\, g_2(\mathbf{v}\cdot\mathbf{t} - b)\, e^{i\mathbf{w}\cdot\mathbf{t}}\, d\mathbf{v}\, db\, d\mathbf{w},$$

with equality in the weak sense. This completes the proof Theorem 2.6.4. □

Remark 2.7.2. If the analysis and synthesis processes are carried via a common window function $g \in L^2(\mathbb{R})$, then the inversion formula (2.180) takes the following form:

$$f(\mathbf{t}) = \frac{1}{(2\pi)^{n/2}\|g\|_2^2} \int_{\mathbb{R}^n} \int_{\mathbb{R}} \int_{\mathbb{S}^{n-1}} \mathcal{D}\mathcal{G}_g\big[f\big](\mathbf{v}, b, \mathbf{w})\, g(\mathbf{v}\cdot\mathbf{t} - b)\, e^{i\mathbf{w}\cdot\mathbf{t}}\, d\mathbf{v}\, db\, d\mathbf{w}, \quad (2.182)$$

with equality in the weak sense. Nevertheless, in case a normalized window is chosen; that is, $\|g\|_2 = 1$, then the formula (2.182) boils down to

$$f(\mathbf{t}) = \frac{1}{(2\pi)^{n/2}} \int_{\mathbb{R}^n} \int_{\mathbb{R}} \int_{\mathbb{S}^{n-1}} \mathcal{D}\mathcal{G}_g\big[f\big](\mathbf{v}, b, \mathbf{w})\, g(\mathbf{v}\cdot\mathbf{t} - b)\, e^{i\mathbf{w}\cdot\mathbf{t}}\, d\mathbf{v}\, db\, d\mathbf{w}. \quad (2.183)$$

That is, the multi-dimensional signal $f \in L^2(\mathbb{R}^n)$ can be expressed as a continuous superposition of time-frequency shifts with the directional windowed Fourier transform as the weight function.

Before putting an end to the ongoing discourse, it is worth emphasizing that the inversion formulae obtained for the time-frequency transforms discussed in the present chapter, ranging from the classical windowed Fourier transform to the directional windowed Fourier transform, hold in the weak or almost everywhere sense. However, by following suitable approximation procedures, the corresponding stronger versions of the inversion formulae can also be established. For instance, in the case of classical windowed Fourier transform, this is done in terms of a nested exhausting sequence of subsets of $\mathbb{R} \times \mathbb{R}$. Interestingly, an analogous approach can also be followed for obtaining a stronger version of the inversion formula for the directional windowed Fourier transform [120]. Nevertheless, it is also imperative to mention that the notion of directional windowed Fourier transform paves a way to new research vistas in the context of directional time-frequency analysis. Most importantly, it is of considerable interest to extend this concept to the generalized Fourier domains, including the multi-dimensional analogues of the fractional Fourier, linear canonical and special affine Fourier transforms.

2.8   Exercises

Exercise 2.8.1. Calculate the mean and dispersion of the Triangular window (2.20).

Exercise 2.8.2. Find power representation of the Fourier transform of the truncated Taylor function

$$g(t) = \begin{cases} \dfrac{(1+k)}{2} + \dfrac{(1-k)}{2} \cos\left(\dfrac{2\pi t}{T}\right), & -\dfrac{T}{2} \le t \le \dfrac{T}{2}, \quad T > 0 \\ 0, & \text{otherwise.} \end{cases} \tag{2.184}$$

Determine the side-lobe suppression and 3-dB bandwidth corresponding to (2.184) and show that the range of side-lobe levels varies as a function of $k \in \mathbb{R}$.

Exercise 2.8.3. Find power representation of the Fourier transform of the sum-cosine window

$$g(t) = \begin{cases} (1 - 2a) \cos\left(\dfrac{\pi t}{T}\right) + 2a \cos\left(\dfrac{3\pi t}{T}\right), & -\dfrac{T}{2} \le t \le \dfrac{T}{2}, \quad T > 0 \\ 0, & \text{otherwise,} \end{cases} \tag{2.185}$$

where a is a real constant. Also, determine what value of a is to be chosen to achieve minimum side-lobe level.

Exercise 2.8.4. Show that the sum-cosine window (2.185) can be synthesized from the rectangular pulse (2.3) and the cosine window (2.34).

Exercise 2.8.5. Compute the Fourier transform of the raised-cosine window

$$g(t) = \begin{cases} \dfrac{(1-2a)}{2}\left(1 + \cos\left(\dfrac{2\pi t}{T}\right)\right) + 2a \cos\left(\dfrac{4\pi t}{T}\right), & -\dfrac{T}{2} \le t \le \dfrac{T}{2}, \quad T > 0 \\ 0, & \text{otherwise,} \end{cases} \tag{2.186}$$

and find the optimum value of the parameter $a \in \mathbb{R}$ to achieve minimum side-lobe level. Note that the window function (2.186) is named "raised cosine" as one the functions used is a raised cosine pulse.

Exercise 2.8.6. Discuss the window qualities of the parabolic window

$$g(t) = \begin{cases} 1 - \left(\dfrac{|t|}{2}\right)^2, & -\dfrac{T}{2} \le t \le \dfrac{T}{2}, \quad T > 0 \\ 0, & \text{otherwise.} \end{cases} \tag{2.187}$$

Exercise 2.8.7. Discuss the window qualities of the Papoulis window

$$g(t) = \begin{cases} \dfrac{1}{\pi}\left|\sin\left(\dfrac{2\pi t}{T}\right)\right| + \left(1 - \dfrac{2|t|}{T}\right)\cos\left(\dfrac{2\pi t}{T}\right), & -\dfrac{T}{2} \le t \le \dfrac{T}{2}, \quad T > 0 \\ 0, & \text{otherwise.} \end{cases} \tag{2.188}$$

Exercise 2.8.8. Discuss the window qualities of the Tukey window (Cosine-tapered window)

$$g(t) = \begin{cases} 1, & |t| \le \dfrac{\beta T}{2} \\ 0.5 + 0.5\cos\left(\dfrac{2\pi\left(|t| - \dfrac{\beta T}{2}\right)}{(1-\beta)T}\right), & \dfrac{\beta T}{2} \le |t| \le \dfrac{T}{2} \quad T > 0, \\ 0 & \text{otherwise,} \end{cases} \tag{2.189}$$

where $0 \le \beta < 1$. Also, demonstrate graphically that the window (2.189) evolves from rectangular to Hanning window as the parameter β varies from zero to unity.

Hint: The graphical demonstration can be accomplished via a computational software.

Exercise 2.8.9. Discuss the window qualities of the Parzen window

$$g(t) = \begin{cases} 1 - 6\left(\dfrac{2|t|}{T}\right)^2\left(1 - \dfrac{2|t|}{T}\right), & |t| \le \dfrac{T}{4} \\ 2\left(1 - \dfrac{2|t|}{T}\right)^3, & \dfrac{T}{4} \le |t| \le \dfrac{T}{2} \quad T > 0. \\ 0 & \text{otherwise,} \end{cases} \tag{2.190}$$

Also, compare the efficiency of the Parzen window (2.190) with that of the rectangular window (2.3).

Exercise 2.8.10. Compute the windowed Fourier transform of the function $f(t) = e^{-t^2}$, with respect to the rectangular window (2.3).

Exercise 2.8.11. Compute the windowed Fourier transform of the unit-step

$$U(t) = \begin{cases} 1, & t \ge 0 \\ 0, & \text{otherwise,} \end{cases} \tag{2.191}$$

with respect to the rectangular window (2.3).

Exercise 2.8.12. Compute the windowed Fourier transform of the function $f(t) = \sin(\pi t)$, with respect to the triangular window (2.20).

Exercise 2.8.13. Compute the windowed Fourier transform of the function $f(t) = \cos(\pi t)$, with respect to the cosine window (2.34).

Exercise 2.8.14. Compute the windowed Fourier transform of the function $f(t) = e^{-i\omega_0 t}$, $\omega_0 \in \mathbb{R}$, with respect to the Gaussian window (2.45).

Exercise 2.8.15. Obtain the FrFT, LCT, SAFT and QPFT of the rectangular (2.3), triangular (2.20), Hanning (2.38) and Hamming (2.42) windows. Also, discuss the variations of the respective free parameters.

3

The Wavelet Transforms and Kin

"The wavelets arrive in succession, and each wavelet eventually dies out. The wavelets all have the same basic form and shape, but the strength or impetus of each wavelet is random and uncorrelated with the strength of the other wavelets. Despite the foreordained death of any individual wavelet, the time-series does not die. The reason is that a new wavelet is born each day to take the place of the one that does die on any given day, the time-series is composed of many living wavelets, all of a different age, some young, others old."

-Enders A. Robinson

3.1 Introduction

The concept of "wavelets" can be traced back to the early 1980s as a joint effort of the conglomerate of mathematicians, physicists and engineers. Although the discovery of wavelet transform can be partly attributed to the Russian-American physicist George Zweig (1937− −) who first introduced the cochlear transform while studying the reaction of the ear to sound, however, the French geophysical engineer, Jean Morlet (1931 − 2007) was the first to formally introduce the notion of wavelet transform as a new mathematical tool for seismic signal analysis. The importance of Morlet's work was promptly recognized by the French theoretical physicist Alex Grossmann (1930 − −) who asserted that the Morlet wavelet transform is somewhat similar to the coherent states formalism in quantum mechanics, and developed an exact inversion formula for the wavelet transform. It was as early as 1984 that the joint venture of Morlet and Grossmann led to a rigorous mathematical investigation of the continuous wavelet transforms [126]. Such developments lead to the formulation of wavelet-based expansions for decomposing a function or a signal in a manner analogous to the classical Fourier expansions.

The first major mathematical breakthrough in the theory of wavelets was reported in 1985, when the French pure mathematician Yves Meyer (1939 − −) uncovered a deep connection between the Calderón formula in harmonic analysis and seminal work of Morlet and Grossmann. Building on the knowledge of the Calderón-Zygmund operators and the Littlewood-Paley theory, Meyer laid the foundation of the subject of wavelet analysis. The premiere development in wavelet analysis was reported with the construction of "painless" non-orthogonal wavelet expansion by Daubechies, Grossmann and Meyer in 1986 [90]. In the buffer period of 1985 to 1986, Meyer and Lemarié [209, 210] also proposed the construction of a smooth orthonormal wavelet basis for the Euclidean spaces $\mathbb{R}^n$ which further added flavour to wavelet analysis. Abreast to this, Stéphane Mallat (1962 − −) recognized that some quadratic mirror filters play an important role for the construction of orthogonal wavelet bases generalizing the Haar system. Continuing the work, Meyer and Mallat

DOI: 10.1201/9781003175766-3

realized that the orthogonal wavelet bases could be constructed systematically from a general formalism. Their fruitful collaboration culminated with the remarkable discovery of a new formalism, known as the "multiresolution analysis," for the construction of orthogonal wavelet bases [193, 194]. In 1989, Mallat constructed the wavelet decomposition and reconstruction algorithms by virtue of the novel tool of multiresolution analysis. Mallat's brilliant work served as the pedestal for many subsequent developments, including the construction of orthogonal spline wavelets with exponential decay by Battle and Lemarié, followed by many other developments [195, 196].

The foundations of wavelet analysis laid by Meyer were enough for Ingrid Daubechies $(1954 - -)$ to construct families of compactly supported orthonormal wavelets with some degree of smoothness [84, 85]. Published in 1988, Daubechies' work had a deep influence on wavelet theory as it lucidly explained the connection between the continuous wavelets in $\mathbb{R}^n$ and the discrete wavelets in $\mathbb{Z}^n$ [86, 87]. The idea of discrete wavelets was warmly accepted and was followed with a plethora of research across diverse aspects of digital signal processing. Despite of the fact that the theory of wavelets appeared to be a promising mathematical tool, it was soon realized by researchers that the construction of symmetric, orthogonal and compactly supported wavelets is not a cake walk. To circumvent the difficulties in the construction of such wavelets, in 1992, Cohen [75] initiated the study of bi-orthogonal wavelets. Besides this, Chui and Wang [71, 72, 73] introduced the notion of compactly supported spline wavelets and semi-orthogonal wavelet analysis. On the other hand, Beylkin and his colleagues [38, 39] successfully applied the multiresolution analysis generated by a completely orthogonal scaling function to study a wide variety of integral operators on $\mathbb{L}^2(\mathbb{R})$ by virtue of the matrix resulting from the wavelet basis. Their prolific work further expanded the boundaries of wavelet theory and lead to some remarkable developments in the form of new and efficient algorithms in numerical analysis. As of now, the wavelet-based numerical techniques have gained tremendous attention of researchers working in diverse aspects of science and engineering [175].

The unprecedented progress that the wavelet theory had witnessed over a short span of time was widely acknowledged by the mathematical community and lead the Norwegian Academy of Science and Letters to award the Abel Prize of 2017 to Yves Meyer for his pivotal role in the development of the mathematical theory of wavelets. Nevertheless, another feather was added to the cap of Yves Meyer when the researchers at the Laser Interferometer Gravitational-wave Observatory (LIGO) acknowledged the use of Meyers' work in the detection of gravitational waves. As of now, wavelet theory has gained a respectable status in the scientific, engineering and research communities and the application areas have grown as diverse as signal processing, computer vision, seismology, turbulence, computer graphics, image processing, structures of the galaxies in the Universe, digital communication, pattern recognition, approximation theory, quantum optics, biomedical engineering, sampling theory, matrix theory, operator theory, differential equations, numerical analysis, statistics and multi-scale segmentation of well logs, natural scenes and mammalian visual systems [5, 6, 7, 9, 10, 70, 88, 89, 92, 152, 173, 189, 234, 283, 285, 293]. For further details on wavelets and their applications, we refer to the monographs [13, 45, 47, 56, 71, 87, 94, 95, 114, 123, 145, 148, 154, 160, 216, 226, 231, 302, 303, 304, 316].

In the last couple of decades, the wavelet theory has witnessed many ramifications in pursuit of geometrically efficient waveforms that could perform equally well in dimensions $n > 1$. This led to the development of a new class of waveforms such as steerable wavelets, ridgelets, curvelets, ripplets, shearlets and bendlets. These geometric wavelets are collectively called as "X-lets" and are primarily characterized by their anisotropic, multi-scale nature. The anisotropy of the aforementioned waveforms corresponds to their better directional selectivity and are, hence, also called as "directional wavelets" [20, 103, 171, 301, 308].

In this chapter, our motive is to provide a comprehensive coverage of the theory of wavelet transforms and their kin, including the stockwell transform, two-dimensional (or polar) wavelet transform, ridgelet transform, curvelet transform, ripplet transform, shearlet transform and bendlet transform. The fundamental ideas and results involved in the formulation of the respective integral transforms are discussed in detail, with special attention given to the applications from a signal processing perspective. All the major concepts are illustrated with suitable examples followed by illustrative depictions.

3.2   The Wavelet Transform

"Wavelets are without doubt an exciting and intuitive concept. The concept brings with it a new way of thinking, which is absolutely essential and was entirely missing in previously existing algorithms."

-Yves Meyer

Time-frequency analysis witnessed a giant leap with the birth of the Gabor transform. Although the Gabor transform or short-time Fourier transform (STFT) catered much to the needs of scientific and engineering communities, however, due to the non-dilatable or rigid window functions, they could not perform satisfactorily while analyzing non-transient signals. For instance, if a signal is composed of higher frequencies for shorter duration and lower frequencies for longer duration, then a rigid window function will definitely fail to capture any significant information about the signal. Therefore, in order to find more adequate time-frequency representations, it is desirable to modify the STFT in a fundamentally different way by infusing a sense of adaptability into the time and frequency windows. The only way one can vary the size of the time (or frequency) window for different degrees of localization is by reciprocally varying the size of the frequency (or time) window at the same time, so as to keep the area of the window constant. That is, there must be a trade-off between the windows in the time and frequency domains. In other words, we must have a window function whose width increases in time (reduces in frequency) while resolving the low frequency contents, and decreases in time (increases in frequency) while resolving high frequency contents of a non-transient signal. This is achieved by directly windowing the signal instead of its Fourier transform and its Fourier transform instead of the inverse Fourier transform, and by scaling the window function appropriately to change the window width. Such flexible windows are referred to as "wavelets" and the mathematical tool used for analyzing a given function via wavelets is called as the "wavelet transform."

Nevertheless, one of the promising features of the classical wavelet transform lies in its nice mathematical formulation in the context of group representation theory. The group theoretic approach to the continuous wavelet transform is not only aesthetic, it is also simpler, in that it allows us to understand the underlying deeper mathematical structure in a simple language. Moreover, the group theoretic language also allows a lucid generalization of the continuous wavelet transform to more general manifolds, including the locally compact Abelian groups. In this subsection, we shall present a vigorous treatment to the notion of continuous wavelet transforms, both from analytical as well as group theoretic perspectives.

3.2.1 Wavelet transform: A group theoretic approach

"There is no branch of mathematics, however abstract, which may not some day be applied to phenomena of the real world."

-Nikolai Lobatchevsky

The aim of this subsection is to present an insight into the notion of wavelet transform from a group theoretic perspective. More precisely, we shall formulate the notion of wavelet transforms on locally compact Abelian groups and also discuss some important properties of such wavelet transforms. However, for a detailed perspective on the subject, we refer to [11, 12, 20, 23, 24, 46, 104, 105, 110, 113, 163, 184, 290].

Definition 3.2.1. Given a group G with "$\cdot$" as the binary operation. Then, $(G, \cdot)$ is said to be a topological group if G is a topological space and the mappings $G \times G \ni g \longmapsto g^{-1} \in G$ are continuous, where g^{-1} is the inverse of g.

Remark 3.2.1. The continuity of the map $G \times G \ni (g, h) \to g \cdot h \in G$ means that, for all $g, h \in G$ and any neighbourhood $\mathcal{N}$ of $g \cdot h$, there exists a neighbourhood $\mathcal{N}_1$ of g and a neighbourhood $\mathcal{N}_2$ of h, such that $n_1 \cdot n_2 \in \mathcal{N}$, whenever $n_1 \in \mathcal{N}_1$ and $n_2 \in \mathcal{N}_2$. Thus, the map $G \ni g \to g^{-1} \in G$ is continuous if for all $g \in G$ and any neighbourhood $\mathcal{N}_1$ of g^{-1}, there exists a neighbourhood $\mathcal{N}$ of g, such that $n \in \mathcal{N}$ implies $n^{-1} \in \mathcal{N}_1$.

Before formulating the main results of this subsection, we shall fix some fundamental notations regarding a topological group G. For $H \subseteq G$, we denote

$$H^{-1} = \left\{ h^{-1} : h \in H \right\}. \tag{3.1}$$

For $g_1, g_2 \in G$, the left translate of H by g_1 is denoted by $g_1 H$ and is defined as

$$g_1 H = \left\{ g_1 \cdot h : h \in H \right\}, \tag{3.2}$$

whereas, the right translate of H by g_2 is denoted by $H g_2$ and is given by

$$H g_2 = \left\{ h \cdot g_2 : h \in H \right\}. \tag{3.3}$$

Moreover, if $H_1, H_2 \subseteq G$, then we have

$$H_1 \cdot H_2 = \left\{ h_1 \cdot h_2 : h_1 \in H_1, h_2 \in H_2 \right\}. \tag{3.4}$$

Here, it is imperative to recall some of the separation axioms in the context of modern topology. The separation axioms are denoted by the letter "T" after the German portmanteau *Trennungsaxiom*, which means "separation axiom."

(i) A topological space X is said to be a T_0-space if every pair of distinct points is topologically distinguishable. That is, for any pair of distinct points, at least one of them has a neighbourhood not containing the other.

(ii) A topological space X is said to be a T_1-space if every pair of distinct points is separated in the sense that each lies in a neighbourhood which does not contain the other.

(iii) A Hausdorff space (Separated space) or T_2-space is a topological space in which each pair of distinct points can be separated by open sets in the sense that they have disjoint neighbourhoods. Every Hausdorff space is clearly a T_1-space, and every subspace of Hausdorff space is also a Hausdorff space. Out of the aforementioned separation conditions, the "Hausdorff condition" is the most frequently used and discussed as it implies the uniqueness of limits of sequences, nets and filters. The Hausdorff spaces are named in honour of the Germen mathematician Felix Hausdorff, who is considered to be one of the founders of modern topology.

Theorem 3.2.1. *For any topological group G with topology τ, the following conditions are equivalent:*

(i) G *is a* T_0*−space,*

(ii) G *is a* T_1*−space,*

(iii) G *is a* T_2*−space,*

(iv) $\bigcap_{O \in \tau} O = \{e\}$, *where e is the identity element in G.*

Proof. $(i) \Rightarrow (ii)$. If G is a T_0-space, then we show that G is also T_1-space. For any pair of distinct elements $g_1, g_2 \in G$, there exists an open set O containing either g_1 or g_2, but not both. Without loss of generality, we assume that $g_1 \in O$ and $g_2 \notin O$. Consider the set,

$$V = g_2 \, O^{-1} g_1 = \left\{ g_2 \, n^{-1} g_1 : n \in O \right\}.$$

Then, it is straightforward to observe that V is an open set containing g_2. Note that $g_1 \notin V_1$ and $g_1 = g_2 n^{-1} g_1$, so that $g_2 = n \in O$, which is a clear contradiction. Hence, we conclude that G is indeed a T_1-space.

$(ii) \Rightarrow (iii)$. Assume that G is a T_1-space and $g_1, g_2 \in G$ be any two distinct elements in G. Clearly, $\{g_2^{-1}, g_1\}$ is a subset of G and $H = G - \{g_2^{-1}, g_1\}$ is its complement in G. Observe that H contains the identity element "e" of the group G and for any $h \in H$, we have $h \neq g_2^{-1} g_1$.

Since G is a T_1-space, therefore, there exists a pair of open sets O_1 and O_2, such that $h \in O_1$, $g_2^{-1} g_1 \in O_2$, $h \notin O_2$ and $g_2^{-1} g_1 \notin O_1$. Thus, O_1 is an open neighbourhood of h which is contained in H. Therefore, every point of H is an interior point, hence, H is an open set. Invoking continuity of the multiplication and the fact that $e \cdot e = e$, there exist neighbourhoods U and V of e such that $u \in U$, $v \in V$, which implies $u \cdot v \in H$. Let $W = U \cap U^{-1} \cap V \cap V^{-1}$, then W is evidently an open set containing the identity element "e" and enjoys the properties $W = W^{-1}$ and $W \cdot W^{-1} \subseteq H$. Observe that there is no pair of elements w_1, w_2 in W satisfying $g_1 w_1 = g_2 w_2$, as $g_1 = g_2 w_2 w_1^{-1} \in g_2 H = G - g_1$, which is impossible. Therefore, we conclude that $g_1 W \cap g_2 W = \emptyset$; that is, G is a T_2-space.

$(iii) \Rightarrow (iv)$. Suppose that G is T_2-space. Then, for any $g \in G$ such that $g \neq e$, there exist open sets O_1 and O_2 such that $g \in O_1$, $e \in O_2$ and $O_1 \cap O_2 = \emptyset$. This validates (iv).

$(iv) \Rightarrow (i)$. Finally, suppose that (iv) holds. Let g_1, g_2 be any two distinct elements in G. Then, it is clear that $g_1^{-1} g_2 \neq e$. Thus, there exists an open set O such that $e \in O$ and $g_1^{1} g_2 \notin O$, which implies that $g_1 \in g_1 O$ and $g_2 \notin g_1 O$. This shows that G is a T_0-space, which evidently completes the proof of Theorem 3.2.1. $\qquad \square$

Remark 3.2.2. In case the group $(G, \cdot)$ in Definition 3.2.1 is also a T_2-space, it is formally called as the "Hausdorff group." In what follows, we shall only deal with Hausdorff groups with Theorem 3.2.1 in hindsight.

Definition 3.2.2. A topological group $(G, \cdot)$ is called as a "locally compact group" if the underlying topology is both locally compact and Hausdroff.

Locally compact groups are an important class of topological groups in the sense that many examples of groups that often arise in mathematics are locally compact and such groups have a natural measure called the "Haar measure." This allows one to define integrals of Borel measurable functions on G, so that the fundamental notions of the Fourier and wavelet transforms are developed on the generalized L^p-spaces. In the remaining part of the subsection, we present the concepts of Haar measure, modular function, Haar integral and unitary representations, and then introduce the notion of wavelet transform on locally compact groups.

(A) The Haar measure

Introduced by Alferd Haar in 1933, the Haar measure assigns an invariant volume to subsets of the locally compact groups, consequently defining an integral for functions defined on such groups. The Haar measure is used in many parts of analysis, number theory, group theory, representation theory, statistics, probability theory and ergodic theory.

Given a locally compact group $(G, \cdot)$, the $\sigma-$algebra generated by all open subsets of G is called the "Borel algebra" and any element of the Borel algebra is called a "Borel set." An interesting property of the Borel sets is that both the left and right translates of a subset $H \subseteq G$ by $g \in G$ map Borel sets onto Borel sets. Also, a measure μ on the Borel subsets (Borel measure) of G is called "left translation invariant" if for all Borel subsets $H \subseteq G$, we have $\mu(gH) = \mu(H)$, $g \in G$. Analogously, μ is called as "right translation invariant" if for all Borel subsets $H \subseteq G$, we have $\mu(Hg) = \mu(H)$, $g \in G$.

In order to define the Haar measure on locally compact groups, it is imperative to present the well-known Haar's theorem for locally compact groups. The proof is not necessary for the understanding of subject matter. We are content with the statement of the theorem and the proof is omitted.

Theorem 3.2.2. *Let $(G, \cdot)$ be a locally compact group. There is, upto a positive multiplicative constant, a unique countably additive, non-trivial measure μ on the Borel subsets of G satisfying the following properties:*

(i) *μ is left-translation invariant,*

(ii) *μ is finite on every compact set $K \subseteq G$,*

(iii) *μ is outer regular on all Borel sets $B \subseteq G$; that is,*

$$\mu(B) = \inf \left\{ \mu(U) : B \subseteq U, U \ open \right\},$$

(iv) *μ is inner regular on all sets $U \subseteq G$; that is,*

$$\mu(U) = \inf \left\{ \mu(K) : K \subseteq U, K \ compact \right\}.$$

The measure μ, whose existence is asserted by the Haar's theorem, is formally called as the "left Haar measure" on G. As a consequence of the above cited properties, it can be shown that $\mu(U) > 0$, for any non-empty, open subset $U \subseteq G$. In particular, if G is compact, then $\mu(G)$ is finite and positive, so we can uniquely specify a left Haar measure on G by adding the normalization condition $\mu(G) = 1$.

On the other hand, if the condition (i) in Theorem 3.2.2 is replaced with right translation invariance then the corresponding measure on G is called as the "right Haar measure" and is denoted by v, to mark the distinction. Moreover, if the measure μ is both left and right Haar measure on G, then it is precisely called as the "Haar measure" on G.

It is worth noticing that the left and right Haar measures need not be the same, however, we can obtain an important relationship between the left Haar measure (μ) and the right Haar measure (v) on the locally compact group G. For any Borel set B, consider the inverse element B^{-1} and define $\mu_{-1}(B) = \mu(B^{-1})$. Then, μ_{-1} is indeed a right Haar measure as the right invariance can be shown as

$$\mu_{-1}(Bg) = \mu\left((Bg)^{-1}\right) = \mu\left(g^{-1}B^{-1}\right) = \mu(B^{-1}) = \mu_{-1}(B), \quad \forall g \in G.$$

Using the fact that the right measure is unique, up to a positive multiplicative constant, it follows that there exists a positive constant K such that $\mu(B^{-1}) = \mu_1(B) = Kv(B)$.

(B) Modular function

Let μ be the left Haar measure on a locally compact group G. Then, for all Borel subsets B of G, we define μ_g as the right translate of the left Haar measure μ as

$$\mu_g(B) = \mu(Bg), \quad g \in G. \tag{3.5}$$

Further, it can be easily deduced that μ_g is also a left Haar measure on G. Therefore, by uniqueness, up to a positive constant scaling factor of the Haar measure, there exists a positive number $\Delta(g)$ such that

$$\mu_g = \Delta(g)\,\mu. \tag{3.6}$$

Moreover, if μ' is another left Haar measure on the locally compact group G, then, again by uniqueness, up to a positive constant scaling factor, we can find $K > 0$ such that $\mu' = K\mu$. Hence, by virtue of (3.5) and (3.6), we obtain

$$\mu'_g = K\mu_g = K\Delta(g)\mu = \Delta(g)\mu'. \tag{3.7}$$

From relation (3.7), we infer that the positive function Δ on G is independent of the choice of the left Haar-measure. It is completely determined by the group G and is called as the "modular function" on G.

The modular function $\Delta : G \to \mathbb{R}^+$ is a continuous group homomorphism into the multiplicative group of positive real numbers. A locally compact group G is called "unimodular" if the modular function is identically 1, or equivalently, the left Haar measure is equal to the right Haar measure. As an example, we note that an Abelian, locally compact group G is always unimodular.

(C) The Haar integral

Using the general theory of Lebesgue integration, one can define an integral for all Borel measurable functions f on G. This integral is called as the "Haar integral" and is denoted by $\int_G f\, d\mu$. By virtue of the properties of left Haar measure, for any Haar-integrable function f on G, we have

$$\int_G f(gx)\, d\mu(x) = \int_G f(x)\, d\mu(x), \quad g \in G, \tag{3.8}$$

which is essentially the definition of left-invariance. The analogous result can also be obtained by using the properties of the right Haar measure.

Example 3.2.1. For $a \in \mathbb{R}^+$ and $b \in \mathbb{R}$, consider the set of affine linear transformations of the form $x \to ax + b$, $x \in \mathbb{R}$. Denote,

$$G = \left\{ (a, b) : a \in \mathbb{R}^+, b \in \mathbb{R} \right\} \tag{3.9}$$

and define the binary operation " $\cdot$ " on G as

$$(a_1, b_1) \cdot (a_2, b_2) = (a_1 a_2, a_1 b_2 + b_1). \tag{3.10}$$

We claim that G constitutes a group under the binary operation (3.10). Evidently, $(1, 0)$ is the identity element for G. Also, for any $(a, b) \in G$, we have

$$(a, b) \left(\frac{1}{a}, -\frac{b}{a} \right) = \left(a \cdot \frac{1}{a}, a \left(-\frac{b}{a} \right) + b \right) = (1, 0)$$

and

$$\left(\frac{1}{a}, -\frac{b}{a} \right) (a, b) = \left(\frac{1}{a} \cdot a, \left(\frac{1}{a} \right) b - \frac{b}{a} \right) = (1, 0).$$

That is, for any $(a, b) \in G$, the inverse is given by $(a, b)^{-1} = \left(\frac{1}{a}, -\frac{b}{a} \right)$.

Next, we show that the operation (3.10) is associative. For any (a_1, b_1), (a_2, b_2) and $(a_3, b_3) \in G$, we have

$$\begin{aligned} ((a_1, b_1) \cdot (a_2, b_2)) \cdot (a_3, b_3) &= (a_1 a_2, a_1 b_2 + b_1) \cdot (a_3, b_3) \\ &= ((a_1 a_2) a_3, (a_1 a_2) b_3 + a_1 b_2 + b_1) \\ &= (a_1 a_2 a_3, a_1 a_2 b_3 + a_1 b_2 + b_1) \end{aligned}$$

and

$$\begin{aligned} (a_1, b_1) \cdot ((a_2, b_2) \cdot (a_2, b_3)) &= (a_1, b_1)(a_2 a_3, a_2 b_3 + b_2) \\ &= (a_1(a_2 a_3), a_1(a_2 b_3 + b_2) + b_1) \\ &= (a_1 a_2 a_3, a_1 a_2 b_3 + a_1 b_2 + b_1). \end{aligned}$$

Thus, the binary operation (3.10) is indeed associative. Hence, we conclude that $(G, \cdot)$ is a group and is formally called as the "affine group" on $\mathbb{R}$. In fact, the affine group $(G, \cdot)$ is non-Abelian as the operation " $\cdot$ " is not commutative.

Note that, the affine group $(G, \cdot)$ given by (3.9) can be identified with the right half plane $\mathbb{R}^+ \times \mathbb{R}$, which is locally compact. The left and right Haar measures on $G = \mathbb{R}^+ \times \mathbb{R}$ are given by

$$\mu(B) = \int_B \frac{da\, db}{a^2} \quad \text{and} \quad \nu(B) = \int_B \frac{da\, db}{a}, \tag{3.11}$$

respectively, where B is any Borel subset of $G = \mathbb{R}^+ \times \mathbb{R}$. The assertion (3.11) is justifiable, because if U is any open subset of G, then for fixed $(\alpha, \beta) \in G$, the integration by substitution yields

$$\begin{aligned} \mu((\alpha, \beta) \cdot U) &= \int_{(\alpha, \beta) \cdot U} \frac{da\, db}{a^2} \\ &= \int_U \frac{1}{(\alpha \alpha')^2} \left| (\alpha)(\alpha) - (0)(0) \right| d\alpha' d\beta' \end{aligned}$$

$$= \int_U \frac{d\alpha' d\beta'}{\alpha'^2} = \mu(U).$$

Also, for any fixed $(\alpha', \beta') \in G$, we have

$$\nu(U \cdot (\alpha', \beta')) = \int_{U \cdot (\alpha', \beta')} \frac{da\, db}{a}$$

$$= \int_U \frac{1}{(\alpha\, \alpha')} \left| (\alpha')(1) - (\beta')(0) \right| d\alpha\, d\beta$$

$$= \int_U \frac{d\alpha\, d\beta}{\alpha} = \nu(U).$$

Remark 3.2.3. The affine group $(G, \cdot)$ defined in (3.9) is a Lie group on which the left Haar measure is different from the right Haar measure. Thus, the affine group $(G, \cdot)$ is non-unimodular.

Example 3.2.2. For the general linear group $G = GL(n, \mathbb{R})$, any left Haar measure μ on G is also a right Haar measure and is given by

$$\mu(B) = \int_B \frac{dX}{\left| \det(X) \right|^n}, \tag{3.12}$$

where B is any Borel subset of $G = GL(n, \mathbb{R})$ and dX denotes the Lebesgue measure on $\mathbb{R}^{n \times n}$, identified with set of all $n \times n$ matrices.

(D) Unitary representation

The hindmost goal of this subsection is to gain an intuition regarding the unitary representations of locally compact groups on separable and complex Hilbert spaces. Prior to that, we recall the notions of adjoint and unitary operators on Hilbert spaces.

Definition 3.2.3. If T is a bounded linear operator on a Hilbert space X, then the operator $T^* : X \to X$ defined by

$$\langle Tx, y \rangle = \langle x, T^*y \rangle \tag{3.13}$$

is called as the "adjoint operator" of T.

Definition 3.2.4. A bounded operator T on a Hilbert space X is called a "unitary operator" if $T^*T = TT^* = I$ on X, where I is the identity operator on X.

In the above definition, it is essential that the domain and the range of the operator T be the entire space X. Moreover, it is straightforward to verify that an operator T is unitary if and only if T is invertible and $T^{-1} = T^*$.

Definition 3.2.5. Let G be a locally compact group and X be a separable and complex Hilbert space. If $\mathcal{U}(X)$ denotes the group of all unitary operators on X with respect to the usual composition of mappings, then a homomorphism $\sigma : G \to \mathcal{U}(X)$ is called as a "unitary representation" of G on X.

Here, it is pertinent to mention that the general theory of unitary representations is well-developed in case G is a locally compact group and the representations are strongly continuous; that is, $G \ni g \to \sigma(g)(x) \in X$ is a continuous mapping for all $x \in X$. In that case, the Hilbert space X is called as the "representation space" of $\sigma : G \to \mathcal{U}(X)$ and the dimension of X is called as the "dimension" or "degree" of the map $\sigma : G \to \mathcal{U}(X)$. However, such details go beyond the scope of this subsection and are as such omitted.

Let M be a closed subspace of the separable, complex Hilbert space X. Then, M is said to be invariant with respect to the unitary representation $\sigma : G \to \mathcal{U}(X)$ if $\sigma(g)M \subseteq M$, $g \in G$. It is immediately clear that both $\{0\}$ and X are trivially invariant with respect to any unitary representation. Note that all invariant subspaces are closed by definition. Moreover, a unitary representation $\sigma : G \to \mathcal{U}(X)$ of G on X is said to be irreducible if its only invariant spaces are the trivial subspaces $\{0\}$ and X.

Next, our aim is construct a unitary representation of the affine group $G = \mathbb{R}^+ \times \mathbb{R}$ on the Hilbert space $L^2(\mathbb{R})$. In this direction, we have the following proposition:

Proposition 3.2.3. *For the affine group $G = \mathbb{R}^+ \times \mathbb{R}$ and $(a, b) \in G$, the operator $\sigma(a, b)$ acting on $\psi \in L^2(\mathbb{R})$ as*

$$\sigma(a, b)\,\psi(t) = \frac{1}{\sqrt{a}}\,\psi\left(\frac{t - b}{a}\right) \tag{3.14}$$

is unitary on the Hilbert space $L^2(\mathbb{R})$.

Proof. For (a_1, b_1), $(a_2, b_2) \in G$, we have

$$
\begin{aligned}
\sigma(a_1, b_1)\Big(\sigma(a_2, b_2)\,\psi\Big)(t) &= \frac{1}{\sqrt{a_1}}\,\sigma(a_2, b_2)\left(\psi\left(\frac{t - b_1}{a_1}\right)\right) \\
&= \frac{1}{\sqrt{a_1 a_2}}\,\psi\left(\frac{t - b_1 - a_1 b_2}{a_1 a_2}\right) \\
&= \frac{1}{\sqrt{a_1 a_2}}\,\psi\left(\frac{t - (b_1 + a_1 b_2)}{a_1 a_2}\right) \\
&= \sigma(a_1 a_2, b_1 + a_1 b_2)\,\psi(t) \\
&= \sigma\big((a_1, b_1) \cdot (a_2, b_2)\big)\,\psi(t). \tag{3.15}
\end{aligned}
$$

Equation (3.15) implies that σ is invertible and its inverse is given by

$$\sigma^{-1}(a, b) = \sigma\left(\frac{1}{a}, -\frac{b}{a}\right). \tag{3.16}$$

Moreover, for any pair of functions ψ, ϕ belonging to the Hilbert space $L^2(\mathbb{R})$, we have

$$
\begin{aligned}
\Big\langle \sigma(a, b)\psi,\, \phi \Big\rangle_2 &= \int_{\mathbb{R}} \sigma(a, b)\psi(t)\,\overline{\phi(t)}\, dt \\
&= \frac{1}{\sqrt{a}} \int_{\mathbb{R}} \psi\left(\frac{t - b}{a}\right)\,\overline{\phi(t)}\, dt \\
&= \sqrt{a} \int_{\mathbb{R}} \psi(z)\,\overline{\phi(az + b)}\, dz \\
&= \int_{\mathbb{R}} \psi(z)\,\overline{\sigma^{-1}(a, b)\phi(z)}\, dz \\
&= \Big\langle \psi,\, \sigma^{-1}(a, b)\phi \Big\rangle_2.
\end{aligned}
$$

Therefore, we conclude that σ is indeed a unitary operator and $\sigma\sigma^* = I = \sigma\sigma^{-1}$. $\square$

Proposition 3.2.4. *Let* $\mathcal{U}\big(L^2(\mathbb{R})\big)$ *denotes the group of all unitary operators on* $L^2(\mathbb{R})$ *with respect to the composition of mappings "$\odot$." Then, the map* $\sigma : G \to \mathcal{U}\big(L^2(\mathbb{R})\big)$ *defined in* (3.14) *is a unitary representation of the affine group* $G = \mathbb{R}^+ \times \mathbb{R}$ *on the Hilbert space* $L^2(\mathbb{R})$.

Proof. With Proposition 3.2.3 in hindsight, it remains to show that the map $\sigma : G \to \mathcal{U}\big(L^2(\mathbb{R})\big)$ is a group homomorphism. Let $(a_1, b_1), (a_2, b_2) \in G$. Then, we have

$$
\begin{aligned}
\Big(\sigma(a_1, b_1) \odot \sigma(a_2, b_2)\Big)\psi(t) &= \sigma(a_1, b_1)\Big(\sigma(a_2, b_2)\,\psi\Big)(t) \\
&= \frac{1}{\sqrt{a_1}}\, \sigma(a_2, b_2)\left(\psi\left(\frac{t - b_1}{a_1}\right)\right) \\
&= \frac{1}{\sqrt{a_1 a_2}}\, \psi\left(\frac{t - (b_1 + a_1 b_2)}{a_1 a_2}\right) \\
&= \sigma\big(a_1 a_2, b_1 + a_1 b_2\big)\,\psi(t) \\
&= \sigma\big((a_1, b_1) \cdot (a_2, b_2)\big)\,\psi(t).
\end{aligned}
\tag{3.17}
$$

From relation (3.17), we infer that the map σ is indeed a group homomorphism. Hence, we conclude that $\sigma : G \to \mathcal{U}\big(L^2(\mathbb{R})\big)$ is a unitary representation of the affine group $G = \mathbb{R}^+ \times \mathbb{R}$ on the Hilbert space $L^2(\mathbb{R})$. $\square$

Below, we shall present the notion of square integrable representations on locally compact groups.

Definition 3.2.6. Given a locally compact group G and a Hilbert space X. A unitary representation $\sigma : G \to \mathcal{U}(X)$ is said to be square integrable if there exists a non-zero vector $\psi \in X$ such that

$$
\int_G \big|\langle \psi,\, \sigma(g)\psi \rangle\big|^2 d\mu(g) < \infty.
\tag{3.18}
$$

Any vector $\psi \in X$ satisfying the condition (3.18) is said to be an "admissible vector" for the square integrable representation of G on X.

In the following theorem, we obtain a characterization of admissible vectors for the unitary representation $\sigma : G \to \mathcal{U}(L^2(\mathbb{R}))$ of the affine group G on the Hilbert space of square integrable functions $L^2(\mathbb{R})$.

Theorem 3.2.5. *For any pair of functions* $\psi, \phi \in L^2(\mathbb{R})$, *such that*

$$
C_\psi = 2\pi \int_\mathbb{R} \frac{|\hat{\psi}(\omega)|^2}{|\omega|^2}\, d\omega < \infty,
\tag{3.19}
$$

we have

$$
\int_G \big|\langle \phi,\, \sigma(a, b)\psi \rangle_2\big|^2 d\mu(g) < \infty,
\tag{3.20}
$$

where $\mu(g)$ *denotes the left Haar measure on the affine group* $G = \mathbb{R}^+ \times \mathbb{R}$.

Proof. We shall prove the result for a dense subspace of $L^2(\mathbb{R})$ and then invoke the density argument to extend the result to the whole space. Consider the following dense subspace W of $L^2(\mathbb{R})$:

$$
W = \Big\{ f \in L^2(\mathbb{R}) : \hat{f} \in C_0^\infty(\mathbb{R}) \Big\}.
\tag{3.21}
$$

As a consequence of the Plancherel theorem and the elementary properties of the Fourier transform, we have

$$
\begin{aligned}
\int_G \left| \left\langle \phi, \, \sigma(a,b)\psi \right\rangle_2 \right|^2 d\mu(g) &= \int_{\mathbb{R} \times \mathbb{R}^+} \left| \left\langle \hat{\phi}, \, \sigma(\hat{a,b})\psi \right\rangle_2 \right|^2 \frac{da\,db}{a^2} \\
&= \int_{\mathbb{R} \times \mathbb{R}^+} \left| \int_{\mathbb{R}} \hat{\phi}(\omega)\, \overline{\hat{\psi}(a\omega)}\, e^{i\omega b} \right|^2 \frac{da\,db}{a} \\
&= \int_{\mathbb{R}^+} \left\{ 2\pi \int_{\mathbb{R}} \left| \mathscr{F}^{-1}\left(\hat{\phi}(\omega)\, \overline{\hat{\psi}(a\omega)} \right)(b) \right|^2 db \right\} \frac{da}{a} \\
&= 2\pi \int_{\mathbb{R} \times \mathbb{R}^+} \left| \hat{\phi}(\omega) \right|^2 \left| \hat{\psi}(a\omega) \right|^2 \frac{da\,d\omega}{a} \\
&= \int_{\mathbb{R}} \left| \hat{\phi}(\omega) \right|^2 \left\{ 2\pi \int_{\mathbb{R}^+} \left| \hat{\psi}(a\omega) \right|^2 \frac{da}{a} \right\} d\omega.
\end{aligned}
\tag{3.22}
$$

In order to simplify the second integral on the R.H.S of (3.22), we substitute $a\omega = \eta$, so that

$$
\int_{\mathbb{R}^+} \left| \hat{\psi}(a\omega) \right|^2 \frac{da}{a} = \int_{\mathbb{R}} \frac{\left| \hat{\psi}(\eta) \right|^2}{|\eta|}\, d\eta.
$$

Thus, we have

$$
\int_G \left| \left\langle \phi, \, \sigma(a,b)\psi \right\rangle_2 \right|^2 d\mu(g) = \int_{\mathbb{R}} \left| \hat{\phi}(\omega) \right|^2 \left\{ 2\pi \int_{\mathbb{R}} \frac{\left| \hat{\psi}(\eta) \right|^2}{|\eta|}\, d\eta \right\} d\omega = C_\psi \left\| \phi \right\|_2^2,
\tag{3.23}
$$

for all $\phi \in W$. Since W is a dense subspace of $L^2(\mathbb{R})$, therefore, there exists a sequence of functions $\{\phi_k\}_{k=1}^{\infty}$ in W such that $\phi_k \to \phi$ in $L^2(\mathbb{R})$ as $k \to \infty$. Therefore, by virtue of (3.23), we conclude that $\left\{ \left\langle \phi_k, \, \sigma(\cdot,\cdot)\psi \right\rangle_2 \right\}_{k=1}^{\infty}$ is a Cauchy sequence in $L^2(G)$. Hence, there exists a function $h \in L^2(G)$ such that

$$
\left\langle \phi_k, \, \sigma(\cdot,\cdot)\psi \right\rangle_2 \to h \quad \text{as } k \to \infty.
\tag{3.24}
$$

Thus, we can find a subsequence of $\left\{ \left\langle \phi_k, \sigma(\cdot,\cdot)\psi \right\rangle_2 \right\}_{k=1}^{\infty}$ which converges to h almost everywhere on G. Ignoring the threat of causing typographical confusion, we shall again denote such a sequence by $\left\{ \left\langle \phi_k, \sigma(\cdot,\cdot)\psi \right\rangle_2 \right\}_{k=1}^{\infty}$. Using the fact that $\phi_k \to \phi$, we infer that $\left\langle \phi_k, \, \sigma(a,b)\psi \right\rangle_2 \to \left\langle \phi, \, \sigma(a,b)\psi \right\rangle_2$, $\forall\, (a,b) \in G$ as $k \to \infty$. Hence, we arrive at

$$
\int_G \left| \left\langle \phi_k, \sigma(a,b)\psi \right\rangle_2 \right|^2 d\mu(g) \to \int_G \left| \left\langle \phi, \sigma(a,b)\psi \right\rangle_2 \right|^2 d\mu(g).
\tag{3.25}
$$

On the other hand, by virtue of the Plancheral theorem and the fact that $\phi_k \to \phi$ in $L^2(\mathbb{R})$ as $k \to \infty$, we have

$$
\int_G \left| \left\langle \phi_k, \sigma(a,b)\psi \right\rangle_2 \right|^2 d\mu(g) \to \int_{\mathbb{R}} \left| \hat{\phi}(\omega) \right|^2 \left\{ 2\pi \int_{\mathbb{R}} \frac{\left| \hat{\psi}(\eta) \right|^2}{|\eta|}\, d\eta \right\} d\omega = C_\psi \left\| \phi \right\|_2^2.
\tag{3.26}
$$

By combining (3.25) and (3.26), we obtain the desired result. $\square$

Corollary 3.2.6. *The unitary representation $\sigma : G \to \mathcal{U}(L^2(\mathbb{R}))$ of the affine group G on the Hilbert space $L^2(\mathbb{R})$ defined in (3.14) is square integrable.*

As a consequence of Theorem 3.2.5, we note that the set of all those functions in $L^2(\mathbb{R})$ which satisfy the condition (3.19) is particularly interesting in the sense that the representation $\sigma : G \to \mathcal{U}(L^2(\mathbb{R}))$ defined in (3.14) is square integrable. In the physical literature, such functions are formally called as "coherent states" of the representation $\sigma : G \to \mathcal{U}(L^2(\mathbb{R}))$ or are simply referred as "wavelets." In this direction, we have the following definition of a wavelet:

Definition 3.2.7. A function $\psi \in L^2(\mathbb{R})$ is said to be an "admissible wavelet" if the following condition holds:

$$C_\psi = 2\pi \int_{\mathbb{R}} \frac{|\hat{\psi}(\omega)|^2}{|\omega|} \, d\omega < \infty. \tag{3.27}$$

Condition (3.27) is called as the "admissibility condition."

Finally, we are in a position to introduce the notion of continuous wavelet transform on the space of square integrable functions $L^2(\mathbb{R})$.

Definition 3.2.8. Let $\psi \in L^2(\mathbb{R})$ be an admissible wavelet for the square integrable representation $\sigma : G \to \mathcal{U}(L^2(\mathbb{R}))$ of the affine group G on the Hilbert space $L^2(\mathbb{R})$ defined in (3.14). Then, for any $f \in L^2(\mathbb{R})$, the continuous wavelet transform is defined by

$$\mathscr{W}_\psi\big[f\big](a,b) = \Big\langle f, \sigma(a,b)\,\psi \Big\rangle_2 = \frac{1}{\sqrt{a}} \int_{\mathbb{R}} f(t)\, \overline{\psi\left(\frac{t-b}{a}\right)} \, dt, \tag{3.28}$$

where $(a,b) \in G = \mathbb{R}^+ \times \mathbb{R}$.

Remark 3.2.4. Definition 3.2.8 implies that the continuous wavelet transform is indeed a square integrable representation of the affine group $G = \mathbb{R}^+ \times \mathbb{R}$ on the Hilbert space of square integrable functions $L^2(\mathbb{R})$.

3.2.2 Wavelet transform: An analytical perspective

"Today the boundaries between mathematics and signal and image processing have faded, and mathematics has benefited from the rediscovery of wavelets by experts from other disciplines. The detour through signal and image processing was the most direct path leading from Haar basis to Daubechies's wavelets."

-Yves Meyer

Having formulated the notion of wavelet transform in the abstract setting, we shall proceed with the analytical perspective of the wavelet transform. From the above discourse, it is quite evident that a good starting point of the analytical theory of wavelets is the notion of an admissible wavelet as given in Definition 3.2.7.

The admissibility condition (3.27) guarantees the existence of an inversion formula for the continuous wavelet transform. From this condition, it follows that $\hat{\psi}(\omega) \to 0$ as $\omega \to 0$. Indeed, if $\hat{\psi}(\omega)$ is continuous, then $\hat{\psi}(0) = 0$; that is, $\int_{\mathbb{R}} \psi(t)\, dt = 0$, which implies that ψ must be an oscillatory function with zero mean. Thus, a wavelet ψ is by necessity an oscillating function, real or complex-valued and in fact, should have good time localization properties so that it looks like a "small wave." That is why ψ is named as a " wavelet."

In addition to the admissibility condition, there are other properties that may be useful in particular applications. For instance, restrictions on the support of ψ and of $\hat{\psi}$ or ψ may be required to have a certain number of vanishing moments that represent the regularity of the wavelet functions and ability of the wavelet transform to capture localized information. A wavelet $\psi(t)$ has n-vanishing moments if the following condition is satisfied:

$$m_k = \int_{\mathbb{R}} t^k \psi(t)\, dt = 0, \quad k = 0, 1, \ldots, n. \tag{3.29}$$

Or equivalently,

$$\left[\frac{d^k \hat{\psi}(\omega)}{d\omega^k} \right]_{\omega=0} = 0, \quad \text{for} \quad k = 0, 1, \ldots, n. \tag{3.30}$$

The number of vanishing moments is directly related to the regularity of any wavelet. Thus, a more regular wavelet has a greater number of vanishing moments. Another desirable property of wavelets is the so-called "localization property" that helps to capture the localized effects of a signal in both the time and frequency domains. Localization and regularity (vanishing moments) are inversely related to each other. Hence, wavelets with a large number of vanishing moments result in more flatness when the frequency ω is small.

Example 3.2.3. (The Haar Wavelet): The Haar wavelet is the first known wavelet proposed by the Hungarian mathematician Alfréd Haar $(1885 - 1933)$ in 1910 [135]. The Haar wavelet, being an odd rectangular pulse pair, is the simplest and oldest orthonormal wavelet with compact support. It is defined by

$$\psi(t) = \begin{cases} 1, & 0 \le t < \dfrac{1}{2} \\ -1, & \dfrac{1}{2} \le t < 1 \\ 0, & \text{otherwise.} \end{cases} \tag{3.31}$$

For the Haar wavelet (3.31), it can be easily verified that

$$\int_{\mathbb{R}} \psi(t)\, dt = 0 \quad \text{and} \quad \int_{\mathbb{R}} \left| \psi(t) \right|^2 dt = 1.$$

It is pertinent to mention that the Haar wavelet (3.31) is very well localized in the time domain, but it is not continuous. Moreover, the Fourier transform of the Haar wavelet is given by

$$\hat{\psi}(\omega) = \frac{i}{\sqrt{2\pi}} \exp\left\{ -\frac{i\omega}{2} \right\} \frac{\sin^2\left(\dfrac{\omega}{4}\right)}{\left(\dfrac{\omega}{4}\right)}. \tag{3.32}$$

By virtue of (3.32), it can be shown that indeed the function (3.31) satisfies the admissibility condition (3.27) as

$$\int_{\mathbb{R}} \frac{\left| \hat{\psi}(\omega) \right|^2}{|\omega|}\, d\omega = \frac{16}{2\pi} \int_{\mathbb{R}} |\omega|^{-3} \left| \sin\left(\frac{\omega}{4}\right) \right|^4 d\omega < \infty. \tag{3.33}$$

Both $\psi(t)$ as well as the real and imaginary parts of $\hat{\psi}(\omega)$ are plotted in Figure 3.1. These figures indicate that the Haar wavelet has good time localization but poor frequency localization. The function $\left| \hat{\psi}(\omega) \right|$ is even, attains its maximum at the frequency $\omega_0 \sim 4.662$, and decays slowly at the rate of ω^{-1} as $\omega \to \infty$, which means that it does not have compact support in the frequency domain. Indeed, the discontinuity of ψ causes a slow decay of $\hat{\psi}$ as $\omega \to \infty$. The discontinuous nature of the Haar wavelet poses severe limitations in many applications. However, the Haar wavelet is one of the most fundamental examples that elegantly illustrates major features of the general wavelet theory.

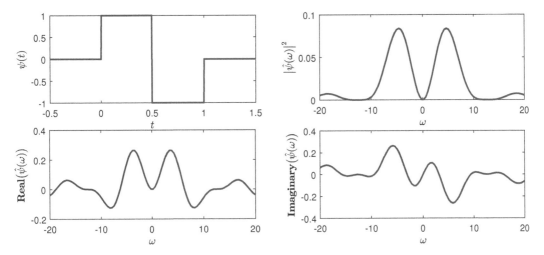

FIGURE 3.1: The Haar wavelet and its Fourier transform.

In the following theorem, we demonstrate that the convolution operation (1.75) offers a lucid method for constructing a new wavelet by appropriately convolving an existing wavelet with a bounded integrable function.

Theorem 3.2.7. *For any wavelet ψ and a bounded integrable function ϕ, the convolution function $(\phi * \psi)$ is a wavelet.*

Proof. Using Definition 1.2.2, we have

$$
\int_{\mathbb{R}} \left| (\phi * \psi)(t) \right|^2 dt = \int_{\mathbb{R}} \left| \int_{\mathbb{R}} \phi(x)\, \psi(t-x)\, dx \right|^2 dt
$$

$$
\leq \int_{\mathbb{R}} \left(\int_{\mathbb{R}} |\phi(x)| |\psi(t-x)|\, dx \right)^2 dt
$$

$$
= \int_{\mathbb{R}} \left(\int_{\mathbb{R}} |\phi(x)|^{1/2} |\psi(t-x)|\, |\phi(x)|^{1/2}\, dx \right)^2 dt
$$

$$
\leq \int_{\mathbb{R}} \left\{ \left(\int_{\mathbb{R}} |\phi(x)|\, |\psi(t-x)|^2 dx \right) \left(\int_{\mathbb{R}} |\phi(x)|\, dx \right) \right\} dt
$$

$$
= \left\| \phi \right\|_1 \int_{\mathbb{R}} \int_{\mathbb{R}} |\phi(x)|\, |\psi(t-x)|^2 dt\, dx
$$

$$
= \left\| \phi \right\|_1^2 \left\| \psi \right\|_2^2 < \infty.
$$

Thus, we infer that $(\phi * \psi) \in L^2(\mathbb{R})$. Moreover, as a consequence of Theorem 1.2.6, we have

$$
\int_{\mathbb{R}} \frac{\left| \mathscr{F}\left[(\phi * \psi) \right](\omega) \right|^2}{|\omega|}\, d\omega = 2\pi \int_{\mathbb{R}} \frac{\left| \hat{\phi}(\omega)\, \hat{\psi}(\omega) \right|^2}{|\omega|}\, d\omega
$$

$$
= 2\pi \int_{\mathbb{R}} \frac{\left| \hat{\psi}(\omega) \right|^2}{|\omega|} \left| \hat{\phi}(\omega) \right|^2 d\omega
$$

$$
\leq \sup \left(\left| \hat{\phi}(\omega) \right|^2 \right) \left\{ 2\pi \int_{\mathbb{R}} \frac{\left| \hat{\psi}(\omega) \right|^2}{|\omega|}\, d\omega \right\} < \infty.
$$

Hence, we conclude that the convolution function $(\phi * \psi)$ is a wavelet. $\qquad \square$

Example 3.2.4. Convoluting the Haar wavelet (3.31) with the function ϕ defined as

$$\phi(t) = \begin{cases} 0, & t < 0 \\ 1, & 0 \le t \le 1 \\ 0, & t \ge 1 \end{cases} \tag{3.34}$$

yields the wavelet shown in Figure 3.2.

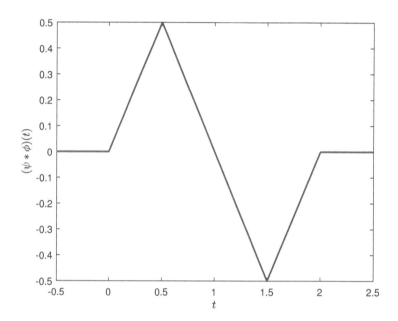

FIGURE 3.2: The wavelet obtained via convolution.

Example 3.2.5. (The Morlet Wavelet). The Morlet wavelet (or Gabor wavelet) is comprised of a complex exponential (carrier) and a Gaussian window (envelope). For a given real parameter ω_0, the Morlet wavelet is defined by

$$\psi(t) = \exp\left(i\omega_0 t - \frac{t^2}{2}\right). \tag{3.35}$$

By virtue of the standard Gaussian integral

$$\int_{\mathbb{R}} e^{-at^2 + bt}\, dt = \sqrt{\frac{\pi}{a}}\, e^{b^2/4a},$$

the Fourier transform corresponding to (3.35) can be easily computed to be

$$\hat{\psi}(\omega) = \exp\left\{-\frac{(\omega - \omega_0)^2}{2}\right\}. \tag{3.36}$$

The parameter ω_0 appearing in the Morlet wavelet allows a trade-off between spatial and frequency resolutions. This wavelet is closely related to both the human audio and visual perceptions and is of particular significance in medical imaging and music. For instance, in magnetic resonance spectroscopy imaging, the Morlet wavelet offers an intuitive bridge between the temporal and frequency information which can efficiently elucidate the complex head trauma spectra. Nevertheless, in the context of music, the Morlet wavelet is quite apt for capturing short bursts of repeating and alternating music notes with a clear start and end time for each note. For $\omega_0 = 7$, the real and imaginary parts of the wavelet (3.35) are plotted in Figure 3.3, whereas the Fourier transform (3.36) is depicted in Figure 3.4.

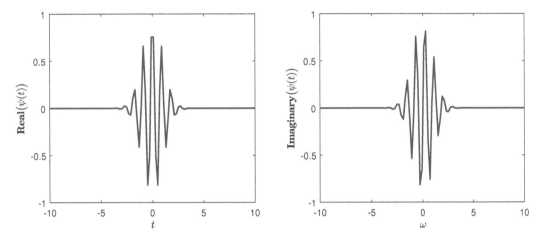

FIGURE 3.3: Real and imaginary parts of the Morlet wavelet.

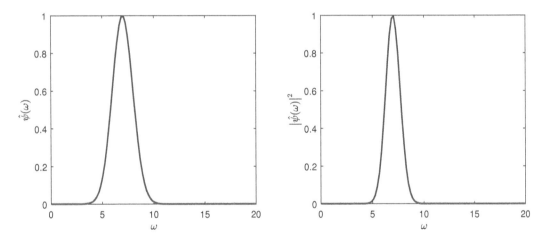

FIGURE 3.4: The Fourier transform of the Morlet wavelet.

"I found my wavelets by trial and error; there was no underlying concept."

-Yves Meyer

Besides the aforementioned classes of wavelets, we present a couple of other examples on the Shannon and Meyer wavelets. These wavelets have the intrinsic feature of being constructed via the celebrated tool of multiresolution analysis (MRA), however, we shall omit the construction formalism as the same is out of the scope of this monograph. For a detailed perspective on the subject, we refer the interested reader to the monographs [87, 94, 95, 145, 193].

Example 3.2.6. (The Shannon Wavelet). The Shannon wavelet is named after the American mathematician Claude E. Shannon (1916 − 2001), who is also known as the "father of information theory." The name is derived from the fact that the recipes for the construction of this wavelet are the sinc-functions, which constitute essence of the celebrated "Shannon's sampling theorem." Mathematically, the Shannon wavelet (or sinc wavelet) is defined as

$$\psi(t) = \left\{ \frac{\sin\left(\dfrac{\pi t}{2}\right)}{\left(\dfrac{\pi t}{2}\right)} \right\} \cos\left(\frac{3\pi t}{2}\right). \tag{3.37}$$

The Shannon wavelet belongs to the class of infinitely differentiable functions (C^∞) and is compactly supported in the frequency domain with the corresponding Fourier transform as

$$\hat{\psi}(\omega) = \begin{cases} \dfrac{1}{\sqrt{2\pi}}, & \pi < |\omega| < 2\pi \\ 0, & \text{otherwise.} \end{cases} \tag{3.38}$$

However, the wavelet (3.37) is not compactly supported in the spatial domain as it decreases slowly at infinity, due to the discontinuity of the Fourier transform (3.38). Practically, the Shannon wavelet (3.37) is applied for evaluation of the radar cross section of the conducting and resistive surfaces. It can also quite handy in situations demanding the analysis of functions ranging in multi-frequency bands. For a pictorial illustration, the Shannon wavelet (3.37) and the corresponding Fourier transform (3.38) are shown in Figure 3.5.

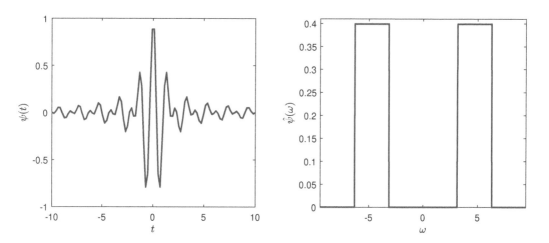

FIGURE 3.5: The Shannon wavelet and its Fourier transform.

Example 3.2.7. (The Meyer Wavelet). Proposed by the French mathematician Yves Meyer (1939 − −) in 1986, the Meyer wavelet is one of the marvellous wavelets available in the literature. The Meyer wavelet is a continuous, orthogonal wavelet and is defined in the frequency domain as

$$\hat{\psi}(\omega) = \begin{cases} \dfrac{1}{\sqrt{2\pi}}\, e^{i\omega/2} \sin\left(\dfrac{\pi}{2}\, \nu\left(\dfrac{3|\omega|}{2\pi} - 1\right)\right), & \dfrac{2\pi}{3} \le |\omega| \le \dfrac{4\pi}{3} \\ \dfrac{1}{\sqrt{2\pi}}\, e^{i\omega/2} \cos\left(\dfrac{\pi}{2}\, \nu\left(\dfrac{3|\omega|}{4\pi} - 1\right)\right), & \dfrac{4\pi}{3} \le |\omega| \le \dfrac{8\pi}{3} \\ 0, & \text{otherwise,} \end{cases} \tag{3.39}$$

where ν is a smooth function (C^k or C^∞) satisfying

$$\nu(t) = \begin{cases} 0, & t \leq 0 \\ 1, & t \geq 1, \end{cases} \tag{3.40}$$

with the additional property that $\nu(t) + \nu(1 - t) = 1$. It is imperative to mention that the regularity of $\hat{\psi}$ is same as that of the constituent function ν. A typical choice for the function ν is

$$\nu(t) = \begin{cases} 0, & t \leq 0 \\ t, & 0 < t < 1 \\ 1, & t \geq 1. \end{cases} \tag{3.41}$$

However, there are much better options for choosing the function ν with enhanced regularity, for instance one may also choose $\nu(t) = t^4(35 - 84t + 70t^2 - 20t^3)$, $0 \leq t \leq 1$. In the spatial domain, the Meyer wavelet $\psi(t)$ is not compactly supported and the asymptotic decay is fast when the degree of smoothness k is large. As a result, the Meyer wavelets are generally implemented in the Fourier domain. However, unlike the Shannon wavelet, the Fourier transform of the Meyer wavelet is smooth, which provides a comparatively much faster asymptotic decay in time. As of now, the Meyer wavelet has been successfully applied in diverse aspects of signal processing, such as in adaptive filters, fractal random fields and multi-fault classification. The graphical demonstration of the Meyer wavelet $\psi(t)$ and the amplitude spectrum $|\hat{\psi}(\omega)|$ is given in Figure 3.6.

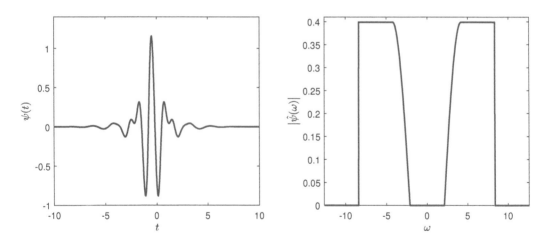

FIGURE 3.6: The Meyer wavelet $\psi(t)$ and $|\hat{\psi}(\omega)|$.

"A generating function is a clothesline on which we hang up a sequence of numbers for display."

-Herbert S. Wilf

To follow-up the subsequent developments, for $a \in \mathbb{R}^+$ and $b \in \mathbb{R}$, we define the translation operator $\mathcal{T}_b$ and the dilation operator $\mathcal{D}_a$ acting on a function $\psi \in L^2(\mathbb{R})$ as

$$\mathcal{T}_b\psi(t) = \psi(t - b) \quad \text{and} \quad \mathcal{D}_a\psi(t) = \frac{1}{\sqrt{a}}\psi\left(\frac{t}{a}\right).$$

Based on the idea that the unitary operator $\sigma(a,b)$ as defined in (3.14) can be viewed as the joint action of the above defined translation and dilation operators on $\psi \in L^2(\mathbb{R})$, we define the following family of functions:

$$\mathscr{F}_\psi(a,b) := \left\{ \psi_{a,b}(t) = \mathcal{D}_a \mathcal{T}_b \psi(t) = \frac{1}{\sqrt{a}} \psi\left(\frac{t-b}{a}\right) ; \ a \in \mathbb{R}^+ \text{ and } b \in \mathbb{R} \right\}. \tag{3.42}$$

Note that $\psi \in L^2(\mathbb{R})$ suffices to conclude that $\psi_{a,b}(t) \in L^2(\mathbb{R})$. In fact, we have the norm equality

$$\left\|\psi_{a,b}\right\|_2^2 = \frac{1}{a} \int_{\mathbb{R}} \left| \psi\left(\frac{t-b}{a}\right)\right|^2 dt = \int_{\mathbb{R}} |\psi(x)|^2 dx = \left\|\psi\right\|_2^2. \tag{3.43}$$

Moreover, the Fourier transform of $\psi_{a,b}(t)$ is given by

$$\begin{aligned}
\hat{\psi}_{a,b}(\omega) &= \frac{1}{\sqrt{2\pi}} \int_{\mathbb{R}} \frac{1}{\sqrt{a}} \psi\left(\frac{t-b}{a}\right) e^{-i\omega t}\, dt \\
&= \sqrt{a}\, e^{-ib\omega} \left\{ \frac{1}{\sqrt{2\pi}} \int_{\mathbb{R}} \psi(z)\, e^{-i(a\omega)z}\, dz \right\} \\
&= \sqrt{a}\, \hat{\psi}(a\omega)\, e^{-ib\omega}.
\end{aligned} \tag{3.44}$$

The generating function $\psi(t)$ appearing in (3.42) and satisfying the admissibility condition (3.27) is called as the "mother wavelet," whereas the obtained functions $\psi_{a,b}(t)$ are called as "daughter wavelets" or simply as "wavelets." Moreover, the parameter a is the "scaling parameter," which measures the degree of compression or scale, and b is the "translation parameter," which determines the time location of the wavelet. Clearly, wavelets $\psi_{a,b}(t)$ generated by the mother wavelet ψ are somewhat similar to the Gabor functions $g_{b,\omega}(t) = g(t-b)\, e^{i\omega t}$ which can be considered as musical notes that oscillate at the frequency ω inside the envelope described by $|g(t-b)|$ as a function of t. For $0 < a < 1$, the wavelet $\psi_{a,b}(t)$ given by (3.42) is the compressed version (smaller support in time-domain) of the mother wavelet $\psi(t)$ and corresponds mainly to higher frequencies. On the flip side, for higher values of a the daughter wavelets are stretched versions of the mother wavelet and correspond to lower frequencies. Thus, wavelets have time-widths adapted to their frequencies. It may be noted that the resolution of wavelets at different scales varies in the time and frequency domains as governed by the Heisenberg uncertainty principle.

Example 3.2.8. (The Mexican Hat Wavelet). The Mexican hat wavelet is defined by the second derivative of a Gaussian function as

$$\psi(t) = \left(1 - t^2\right)e^{-t^2/2} = -\frac{d^2}{dt^2}\left(e^{-t^2/2}\right) = \psi_{1,0}(t). \tag{3.45}$$

The Fourier transform corresponding to (3.45) is given by

$$\hat{\psi}(\omega) = \hat{\psi}_{1,0}(\omega) = \sqrt{2\pi}\, \omega^2 e^{-\omega^2/2}. \tag{3.46}$$

The function (3.45) clearly satisfies the admissibility condition and has two vanishing moments. The Mexican hat wavelet is endowed with excellent localization in time and frequency domains. Unlike the Haar wavelet, the Mexican hat wavelet is a C^∞-function. A typical pair of daughter wavelets obtained from the mother wavelet (3.45) are given by $\psi_{\frac{3}{2},-2}(t)$ and $\psi_{\frac{1}{4},\sqrt{2}}(t)$. A graphical demonstration of the Mexican hat wavelet (3.45) and the corresponding Fourier transform (3.46) is given in Figure 3.7. Moreover, the daughter wavelets $\psi_{\frac{3}{2},-2}(t)$

and $\psi_{\frac{1}{4},\sqrt{2}}(t)$ are depicted in Figure 3.8. Here, it is pertinent to mention that other classes of wavelets with more vanishing moments are obtained by taking into account the higher derivatives of the Gaussian function:

$$\psi^{(m)}(t) = \left(\frac{1}{i}\frac{d}{dt}\right)^m e^{-t^2/2}, \qquad \widehat{\psi^{(m)}}(\omega) = \sqrt{2\pi}\,\omega^m e^{-\omega^2/2}.$$

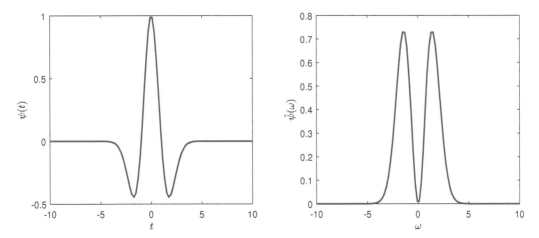

FIGURE 3.7: The Mexican hat wavelet and its Fourier transform.

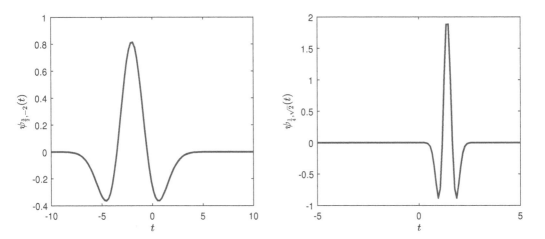

FIGURE 3.8: Two daughter wavelets of the Mexican hat wavelet.

From the above discussion, it is clear that the notion of wavelet transform can be liberated from group theoretic conceptions by taking into account the family of functions $\mathscr{F}_\psi(a,b)$ defined in (3.42). Formally, we have the following definition of wavelet transform:

Definition 3.2.9. Given an admissible wavelet $\psi \in L^2(\mathbb{R})$, the wavelet transform of any $f \in L^2(\mathbb{R})$ is defined as

$$\mathscr{W}_\psi\big[f\big](a,b) = \int_{\mathbb{R}} f(t)\,\overline{\psi_{a,b}(t)}\,dt, \quad a \in \mathbb{R}^+,\ b \in \mathbb{R}, \tag{3.47}$$

where $\psi_{a,b}(t)$ is given by (3.42).

Definition 3.2.9 allows us to make the following comments:

(i) The kernel $\psi_{a,b}(t)$ in (3.47) plays the analogous role of the kernel $e^{-i\omega t}$ in the Fourier transform. However, unlike the Fourier transform, the kernel of the wavelet transform is a two-parameter family of functions determining the scale and location of the wavelet.

(ii) Both the wavelet transform as well as the Fourier transform are linear transforms. However, in contrast to the Fourier transform, the wavelet transform is not a single transform, but a group of transforms determined by the nature of the underlying wavelet.

(iii) For a fixed scale a, the wavelet transform $\mathscr{W}_\psi[f](a,b)$, as a function of b, represents the detailed information contained in the signal $f(t)$ at scale a. In case negative scales are used, the dilation factor $a^{-1/2}$ appearing in (3.42) is to be replaced with $|a|^{-1/2}$.

(iv) If the mother wavelet $\psi(t)$ possesses n-vanishing moments, then expanding the function $f(t)$ in (3.47) via Taylor's series at $t = b$, we obtain

$$\mathscr{W}_\psi\big[f\big](a,b) = \bigg[f(b) \int_{\mathbb{R}} \overline{\psi_{a,0}(t-b)}\, dt + f^{(1)}(b) \int_{\mathbb{R}} (t-b)\, \overline{\psi_{a,0}(t-b)}\, dt +$$

$$\frac{f^{(2)}(b)}{2!} \int_{\mathbb{R}} (t-b)^2\, \overline{\psi_{a,0}(t-b)}\, dt + \cdots + \frac{f^{(n)}(b)}{n!} \int_{\mathbb{R}} (t-b)^n\, \overline{\psi_{a,0}(t-b)}\, dt + \cdots \bigg].$$

$$(3.48)$$

By moment condition (3.29), it follows that the first n-terms of (3.48) vanish, and hence, they do not contribute to $\mathscr{W}_\psi[f](a,b)$.

Proposition 3.2.8. *If $\mathscr{W}_\psi[f](a,b)$ is the continuous wavelet transform of any $f \in L^2(\mathbb{R})$, then we have*

$$\mathscr{W}_\psi\big[f\big](a,b) = \sqrt{a} \int_{\mathbb{R}} \hat{f}(\omega)\, \overline{\hat{\psi}(a\omega)}\, e^{ib\omega}\, d\omega. \qquad (3.49)$$

Proof. Invoking the Parseval's formula (1.69) together with (3.44), we get

$$\mathscr{W}_\psi\big[f\big](a,b) = \int_{\mathbb{R}} f(t)\, \overline{\psi_{a,b}(t)}\, dt$$

$$= \int_{\mathbb{R}} \hat{f}(\omega)\, \overline{\hat{\psi}_{a,b}(t)}\, d\omega$$

$$= \sqrt{a} \int_{\mathbb{R}} \hat{f}(\omega)\, \overline{\hat{\psi}(a\omega)}\, e^{ib\omega}\, d\omega.$$

This completes the proof of Proposition 3.2.8.   □

Remark 3.2.5. By considering the continuous wavelet transform $\mathscr{W}_\psi[f](a,b)$ as a function of b, we can compute the corresponding Fourier transform by virtue of (3.49) as

$$\mathscr{F}\left(\mathscr{W}_\psi\big[f\big](a,b)\right)(\omega) = \sqrt{2\pi a}\, \hat{f}(\omega)\, \overline{\hat{\psi}(a\omega)}. \qquad (3.50)$$

Next, our aim is to study the resolution of the wavelet transform (3.47) in the time-frequency plane. From the Definition 3.2.9 and Proposition 3.2.8, it is quite evident that the localizing behaviour of the wavelet transform is completely determined by the analyzing functions or wavelets $\psi_{a,b}(t)$. Thus, if the wavelets are supported in time domain or frequency domain, then the wavelet transform $\mathscr{W}_\psi[f](a,b)$ is accordingly supported in the respective domains. For a clear mathematical description of the preceding arguments, let E_ψ and $E_{\hat\psi}$ denote the centre (mean) of $\psi(t)$ and $\hat\psi(\omega)$, respectively. Then, by definition, we have

$$E_\psi = \frac{1}{\|\psi\|_2^2} \int_{\mathbb{R}} t \, |\psi(t)|^2 dt \quad \text{and} \quad E_{\hat\psi} = \frac{1}{\|\hat\psi\|_2^2} \int_{\mathbb{R}} \omega \, |\hat\psi(\omega)|^2 d\omega. \quad (3.51)$$

Moreover, the dispersion about mean (radius) of $\psi(t)$ and $\hat\psi(\omega)$ is given by

$$\Delta_\psi = \left\{ \frac{1}{\|\psi\|_2^2} \int_{\mathbb{R}} (t - E_\psi) |\psi(t)|^2 dt \right\}^{1/2} \quad \text{and} \quad \Delta_{\hat\psi} = \left\{ \frac{1}{\|\hat\psi\|_2^2} \int_{\mathbb{R}} (\omega - E_{\hat\psi}) |\hat\psi(\omega)|^2 d\omega \right\}^{1/2}. \quad (3.52)$$

Consequently, the centre and radius of the wavelet $\psi_{a,b}(t)$ are given by

$$\begin{aligned}
E_{\psi_{a,b}} &= \frac{1}{\|\psi_{a,b}\|_2^2} \int_{\mathbb{R}} t \, |\psi_{a,b}(t)|^2 dt \\
&= \frac{1}{\|\psi\|_2^2} \int_{\mathbb{R}} t \, |\psi_{a,b}(t)|^2 dt \\
&= a E_\psi + b
\end{aligned} \quad (3.53)$$

and

$$\Delta_{\psi_{a,b}} = \left\{ \frac{1}{\|\psi_{a,b}\|_2^2} \int_{\mathbb{R}} (t - a E_\psi - b) |\psi_{a,b}(t)|^2 dt \right\}^{1/2} = a \Delta_\psi. \quad (3.54)$$

From the expressions (3.53) and (3.54), we conclude that the wavelet transform $\mathscr{W}_\psi[f](a,b)$ gives the localized information of the signal $f(t)$ in the time window

$$\left[a E_\psi + b - a \Delta_\psi, \; a E_\psi + b + a \Delta_\psi \right] \quad (3.55)$$

having width $2a\Delta_\psi$, which is directly proportional to the scale a. Furthermore, by virtue of (3.49), we infer that the wavelet transform $\mathscr{W}_\psi[f](a,b)$ also gives the localized information of $\hat{f}(\omega)$ in the frequency window

$$\left[\frac{E_{\hat\psi}}{a} - \frac{\Delta_{\hat\psi}}{a}, \; \frac{E_{\hat\psi}}{a} + \frac{\Delta_{\hat\psi}}{a} \right] \quad (3.56)$$

having width $2a^{-1}\Delta_{\hat\psi}$, which is inversely proportional to the scale a. Since the inverse scale is interpreted as the frequency, therefore, we infer that for lower frequencies, the time window (3.55) expands, whereas the frequency window (3.56) shrinks, thereby providing a sharper frequency resolution. On the flip side, for higher frequencies, the time window (3.55) shrinks, whereas the frequency window (3.56) expands, thereby providing a sharper time resolution. In view of the above discourse, we conclude that the continuous wavelet transform $\mathscr{W}_\psi[f](a,b)$ defined in (3.47) offers the joint time-frequency information of a signal within the window

$$\left[a E_\psi + b - a \Delta_\psi, \; a E_\psi + b + a \Delta_\psi \right] \times \left[\frac{E_{\hat\psi}}{a} - \frac{\Delta_{\hat\psi}}{a}, \; \frac{E_{\hat\psi}}{a} + \frac{\Delta_{\hat\psi}}{a} \right]. \quad (3.57)$$

having a net area of $4\Delta_\psi\Delta_{\hat\psi}$, which is independent of the scale a and is governed by the Heisenberg's uncertainty principle $\Delta_\psi\Delta_{\hat\psi} \geq 1/2$. For different values of the translation and scaling parameters, the windows given by (3.57) describe rectangular regions in the time-frequency plane, formally called as the "time-frequency tiles." The characteristic feature of these time-frequency tiles is that they enjoy variable geometry in the sense that the tiles become increasingly wider for capturing lower frequencies and tend to be increasingly narrower for capturing the higher frequencies. For a neat elucidation, the effective tiling onto the time-frequency plane due to the wavelet transform (3.47) is demonstrated in Figure 3.9.

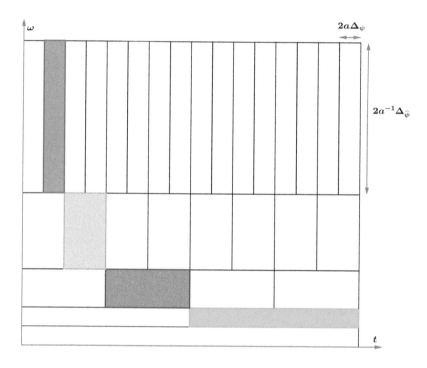

FIGURE 3.9: Time-frequency tiling induced by the wavelet transform.

In the following theorem, we assemble some fundamental properties of the wavelet transform defined in (3.47).

Theorem 3.2.9. *Let $\alpha, \beta \in \mathbb{C}$, $k \in \mathbb{R}$, $c > 0$ are scalars and ψ, ϕ be any two admissible wavelets, then for $f, g \in L^2(\mathbb{R})$, we have*

(i) *Linearity:* $\mathscr{W}_\psi\big[\alpha f + \beta g\big](a, b) = \alpha\,\mathscr{W}_\psi\big[f\big](a, b) + \beta\,\mathscr{W}_\psi\big[g\big](a, b),$

(ii) *Translation:* $\mathscr{W}_\psi\big[T_k f\big](a, b) = \mathscr{W}_\psi\big[f\big](a, b - k),$

(iii) *Dilation:* $\mathscr{W}_\psi\big[D_c f\big](a, b) = \mathscr{W}_\psi\big[f\big]\left(\dfrac{a}{c}, \dfrac{b}{c}\right),$

(iv) *Symmetry:* $\mathscr{W}_\psi\big[f\big](a, b) = \overline{\mathscr{W}_f\big[\psi\big]}\left(\dfrac{1}{a}, -\dfrac{b}{a}\right),$

(v) *Parity:* $\mathscr{W}_{P\psi}\big[Pf\big](a, b) = \mathscr{W}_\psi\big[f\big](a, -b),$ *where* $P\big(f(t)\big) = f(-t),$

(vi) *Anti-linearity:* $\mathscr{W}_{\alpha\psi+\beta\phi}\big[f\big](a,b)(a,b) = \overline{\alpha}\,\mathscr{W}_{\psi}\big[f\big](a,b) + \overline{\beta}\,\mathscr{W}_{\phi}\big[f\big](a,b),$

(vii) *Translation in wavelet:* $\mathscr{W}_{T_k\psi}\big[f\big](a,b) = \mathscr{W}_{\psi}\big[f\big](a,b+ak),$

(viii) *Dilation in wavelet:* $\mathscr{W}_{D_c\psi}\big[f\big](a,b) = \mathscr{W}_{\psi}\big[f\big](ac,b).$

Proof. The proof of the theorem is straightforward and is as such omitted. $\qquad\square$

The wavelet transform of a one-dimensional signal is a two-dimensional time-scale representation. In the following theorem, we shall demonstrate that the continuous wavelet representations of a pair of signals admit an orthogonality relation with the signals. That is, if the signals f and g are orthogonal in $L^2(\mathbb{R})$, then the corresponding wavelet transforms $\mathscr{W}_{\psi}[f](a,b)$ and $\mathscr{W}_{\phi}[g](a,b)$ are also orthogonal in $L^2(\mathbb{R}^+ \times \mathbb{R})$.

Theorem 3.2.10. *Let $\mathscr{W}_{\psi}[f](a,b)$ and $\mathscr{W}_{\phi}[g](a,b)$ be the wavelet transforms of a pair of functions f and g with respect to the wavelets ψ and ϕ, respectively. Then, the following orthogonality relation holds:*

$$\int_{\mathbb{R}}\int_{\mathbb{R}^+} \mathscr{W}_{\psi}\big[f\big](a,b)\,\overline{\mathscr{W}_{\phi}\big[g\big](a,b)}\,\frac{da\,db}{a^2} = C_{\psi,\phi}\big\langle f,\,g\big\rangle_2, \tag{3.58}$$

where $C_{\psi,\phi}$ is given by the cross admissibility condition:

$$C_{\psi,\phi} = 2\pi \int_{\mathbb{R}} \frac{\overline{\hat{\psi}(\omega)}\,\hat{\phi}(\omega)}{|\omega|}\,d\omega < \infty. \tag{3.59}$$

Proof. Using Proposition 3.2.8, we can express the wavelet transforms of f and g as follows:

$$\mathscr{W}_{\psi}\big[f\big](a,b) = \sqrt{a} \int_{\mathbb{R}} \hat{f}(\omega)\,\overline{\hat{\psi}(a\omega)}\,e^{ib\omega}\,d\omega \tag{3.60}$$

and

$$\mathscr{W}_{\phi}\big[g\big](a,b) = \sqrt{a} \int_{\mathbb{R}} \hat{g}(\eta)\,\overline{\hat{\phi}(a\eta)}\,e^{ib\eta}\,d\eta. \tag{3.61}$$

Thus, as a consequence of (3.60) and (3.61) together with a justifiable change in the order of integration, we have

$$\int_{\mathbb{R}}\int_{\mathbb{R}^+} \mathscr{W}_{\psi}\big[f\big](a,b)\,\overline{\mathscr{W}_{\psi}\big[g\big](a,b)}\,\frac{da\,db}{a^2}$$

$$= \int_{\mathbb{R}}\int_{\mathbb{R}^+} \left\{\sqrt{a}\int_{\mathbb{R}} \hat{f}(\omega)\,\overline{\hat{\psi}(a\omega)}\,e^{ib\omega}\,d\omega\right\}\left\{\sqrt{a}\int_{\mathbb{R}} \overline{\hat{g}(\eta)}\,\hat{\phi}(a\eta)\,e^{-ib\eta}\,d\eta\right\}\frac{da\,db}{a^2}$$

$$= 2\pi \int_{\mathbb{R}}\int_{\mathbb{R}}\int_{\mathbb{R}^+} \hat{f}(\omega)\,\overline{\hat{g}(\eta)}\,\overline{\hat{\psi}(a\omega)}\,\hat{\phi}(a\eta)\left\{\frac{1}{2\pi}\int_{\mathbb{R}} e^{ib(\omega-\eta)}\,db\right\}\frac{da\,d\omega\,d\eta}{a}$$

$$= 2\pi \int_{\mathbb{R}}\int_{\mathbb{R}}\int_{\mathbb{R}^+} \hat{f}(\omega)\,\overline{\hat{g}(\eta)}\,\overline{\hat{\psi}(a\omega)}\,\hat{\phi}(a\eta)\,\delta(\omega-\eta)\frac{da\,d\omega\,d\eta}{a}$$

$$= 2\pi \int_{\mathbb{R}}\int_{\mathbb{R}^+} \hat{f}(\omega)\,\overline{\hat{g}(\omega)}\,\overline{\hat{\psi}(a\omega)}\,\hat{\phi}(a\omega)\frac{da\,d\omega}{a}$$

$$= \int_{\mathbb{R}} \hat{f}(\omega)\,\overline{\hat{g}(\omega)}\left\{2\pi \int_{\mathbb{R}^+} \overline{\hat{\psi}(a\omega)}\,\hat{\phi}(a\omega)\frac{da}{a}\right\}d\omega. \tag{3.62}$$

Simplifying the second integral on the R.H.S of (3.62) by substituting $a\omega = \xi$, we get

$$2\pi \int_{\mathbb{R}^+} \overline{\hat{\psi}(a\omega)}\, \hat{\phi}(a\omega) \frac{da}{a} = 2\pi \int_{\mathbb{R}^+} \frac{\overline{\hat{\psi}(\xi)}\, \hat{\phi}(\xi)}{|\xi|}\, d\xi = C_{\psi,\phi}. \tag{3.63}$$

Using (3.63) in (3.62), we obtain the desired result as

$$\int_{\mathbb{R}} \int_{\mathbb{R}^+} \mathscr{W}_\psi\big[f\big](a,b)\, \overline{\mathscr{W}_\psi\big[g\big](a,b)}\, \frac{da\, db}{a^2} = C_{\psi,\phi} \big\langle f,\, g \big\rangle_2.$$

$\square$

Remark 3.2.6. Choosing $\psi = \phi$ in Theorem 3.2.10, we obtain the following orthogonality relation:

$$\int_{\mathbb{R}} \int_{\mathbb{R}^+} \mathscr{W}_\psi\big[f\big](a,b)\, \overline{\mathscr{W}_\psi\big[g\big](a,b)}\, \frac{da\, db}{a^2} = C_\psi \big\langle f,\, g \big\rangle_2, \tag{3.64}$$

where C_ψ is given by the usual admissibility condition (3.27). Moreover, choosing $f = g$ in (3.64) yields the energy preserving relation:

$$\int_{\mathbb{R}} \int_{\mathbb{R}^+} \big|\mathscr{W}_\psi\big[f\big](a,b)\big|^2\, \frac{da\, db}{a^2} = C_\psi \big\|f\big\|_2^2, \tag{3.65}$$

which indicates that for $C_\psi = 1$ the continuous wavelet transform defined in (3.47) is an isometry from the space of signals $L^2(\mathbb{R})$ to the space of of transforms $L^2(\mathbb{R}^+ \times \mathbb{R})$.

The prime significance of the orthogonality relation (3.58) lies in the fact that it allows a simple reconstruction of a signal from the corresponding wavelet transform by virtue of the "resolution of identity" formula.

Theorem 3.2.11. *Let ψ and ϕ be any pair of wavelets, then any function $f \in L^2(\mathbb{R})$ can be reconstructed from the corresponding wavelet transform $\mathscr{W}_\psi[f](a,b)$ via the "resolution of identity" formula:*

$$f(t) = \frac{1}{C_{\psi,\phi}} \int_{\mathbb{R}} \int_{\mathbb{R}^+} \mathscr{W}_\psi\big[f\big](a,b)\, \phi_{a,b}(t)\, \frac{da\, db}{a^2}, \qquad a.e., \tag{3.66}$$

where $C_{\psi,\phi} \neq 0$ is given by (3.59).

Proof. For any arbitrary function $g \in L^2(\mathbb{R})$, the orthogonality relation (3.58) yields

$$C_{\psi,\phi} \big\langle f,\, g \big\rangle_2 = \int_{\mathbb{R}} \int_{\mathbb{R}^+} \mathscr{W}_\psi\big[f\big](a,b)\, \overline{\mathscr{W}_\phi\big[g\big](a,b)}\, \frac{da\, db}{a^2}$$

$$= \int_{\mathbb{R}} \int_{\mathbb{R}^+} \mathscr{W}_\psi\big[f\big](a,b) \left\{ \int_{\mathbb{R}} \overline{g(t)}\, \phi_{a,b}(t)\, dt \right\} \frac{da\, db}{a^2}$$

$$= \int_{\mathbb{R}} \left\{ \int_{\mathbb{R}} \int_{\mathbb{R}^+} \mathscr{W}_\psi\big[f\big](a,b)\, \phi_{a,b}(t)\, \frac{da\, db}{a^2} \right\} \overline{g(t)}\, dt$$

$$= \left\langle \int_{\mathbb{R}} \int_{\mathbb{R}^+} \mathscr{W}_\psi\big[f\big](a,b)\, \phi_{a,b}(t)\, \frac{da\, db}{a^2},\, g(t) \right\rangle_2. \tag{3.67}$$

Since the function g is chosen arbitrarily, therefore, we obtain the following "resolution of identity" formula:

$$f(t) = \frac{1}{C_{\psi,\phi}} \int_{\mathbb{R}} \int_{\mathbb{R}^+} \mathscr{W}_\psi\big[f\big](a,b)\, \phi_{a,b}(t)\, \frac{da\, db}{a^2},$$

with convergence of the integral "in the weak sense." This completes the proof of Theorem 3.2.11. $\square$

Remark 3.2.7. Choosing $\psi = \phi$ in Theorem 3.2.11, we obtain the following "resolution of identity" formula:

$$f(t) = \frac{1}{C_\psi} \int_{\mathbb{R}} \int_{\mathbb{R}^+} \mathscr{W}_\psi\big[f\big](a,b)\, \psi_{a,b}(t)\, \frac{da\,db}{a^2}, \qquad a.e., \tag{3.68}$$

where $C_\psi \neq 0$ is given by the usual admissibility condition (3.27).

Remark 3.2.8. As a consequence of Theorem 3.2.11, we observe that although the wavelet transform is computed with respect to the mother wavelet ψ, it can be reconstructed using the wavelets of any other mother wavelet ϕ, provided $C_{\psi,\phi} \neq 0$. For the sake of distinction, the wavelet ψ is called as "analyzing wavelet," whereas the wavelet ϕ is called as "reconstruction wavelet." Moreover, up to a multiplicative constant, the reconstruction formula is exactly the same as (3.68) in which the same wavelet ψ is used both for analyzing and reconstructing purposes. This indicates, that in some sense, analysis and reconstruction are independent of the choice of the mother wavelet.

In the following theorem, our motive is to show that the convergence of the inversion formula (3.66) also holds in a slightly stronger sense.

Theorem 3.2.12. *Let ψ and ϕ be any pair of analyzing and reconstruction wavelets, then for any $f \in L^2(\mathbb{R})$, we have*

$$\lim_{A_1 \to 0; A_2, B \to \infty} \left\| f(t) - C_{\psi,\phi}^{-1} \int_{|b| \le B} \int_{A_1 \le a \le A_2} \mathscr{W}_\psi\big[f\big](a,b)\, \phi_{a,b}(t)\, \frac{da\,db}{a^2} \right\|_2 = 0, \tag{3.69}$$

where $C_{\psi,\phi} \neq 0$ is given by (3.59).

Proof. For any arbitrary function $g \in L^2(\mathbb{R})$, the orthogonality relation (3.58) yields

$$\left| \left\langle \int_{|b| \le B} \int_{A_1 \le a \le A_2} \mathscr{W}_\psi\big[f\big](a,b)\, \phi_{a,b}(t)\, \frac{da\,db}{a^2}, g(t) \right\rangle_2 \right|$$

$$= \left| \int_{\mathbb{R}} \left\{ \int_{|b| \le B} \int_{A_1 \le a \le A_2} \mathscr{W}_\psi\big[f\big](a,b)\, \phi_{a,b}(t)\, \frac{da\,db}{a^2} \right\} \overline{g(t)}\, dt \right|$$

$$= \left| \int_{|b| \le B} \int_{A_1 \le a \le A_2} \mathscr{W}_\psi\big[f\big](a,b)\, \big\langle \phi_{a,b}(t), g(t) \big\rangle_2\, \frac{da\,db}{a^2} \right|$$

$$\le \int_{|b| \le B} \int_{A_1 \le a \le A_2} \big\| f \big\|_2\, \big\| \psi_{a,b} \big\|_2\, \big\| \phi_{a,b} \big\|_2\, \big\| g \big\|_2\, \frac{da\,db}{a^2}$$

$$\le \int_{|b| \le B} \int_{A_1 \le a \le A_2} \big\| f \big\|_2\, \big\| \psi \big\|_2\, \big\| \phi \big\|_2\, \big\| g \big\|_2\, \frac{da\,db}{a^2}$$

$$= 2B \left(\frac{1}{A_1} - \frac{1}{A_2} \right) \big\| f \big\|_2\, \big\| \psi \big\|_2\, \big\| \phi \big\|_2\, \big\| g \big\|_2 < \infty. \tag{3.70}$$

Therefore, by virtue of Riesz' lemma, we conclude that the integral in the first component of the inner product on the L.H.S of (3.70) is well defined. Thus, we have

$$\left\| f(t) - C_{\psi,\phi}^{-1} \int_{|b| \le B} \int_{A_1 \le a \le A_2} \mathscr{W}_\psi\big[f\big](a,b)\, \phi_{a,b}(t)\, \frac{da\,db}{a^2} \right\|_2$$

$$= \sup_{\|g\|_2 = 1} \left| \left\langle f(t) - C_{\psi,\phi}^{-1} \int_{|b| \le B} \int_{A_1 \le a \le A_2} \mathscr{W}_\psi\big[f\big](a,b)\, \phi_{a,b}(t)\, \frac{da\,db}{a^2}, g(t) \right\rangle \right|$$

$$\leq \sup_{\|g\|_2=1} \left| C_{\psi,\phi}^{-1} \int_{|b|\geq B} \int_{A_2 \leq a \leq A_1} \mathscr{W}_\psi \big[f\big](a,b) \, \overline{\mathscr{W}_\phi \big[g\big](a,b)} \, \frac{da\,db}{a^2} \right|$$

$$\leq \sup_{\|g\|_2=1} \left\{ C_{\psi,\phi}^{-1} \int_{|b|\geq B} \int_{A_2 \leq a \leq A_1} \left| \mathscr{W}_\psi \big[f\big](a,b) \right| \frac{da\,db}{a^2} \right\}^{1/2} \left\{ C_{\psi,\phi}^{-1} \int_{\mathbb{R}} \int_{\mathbb{R}^+} \left| \mathscr{W}_\psi \big[g\big](a,b) \right|^2 \frac{da\,db}{a^2} \right\}^{1/2}.$$

$$(3.71)$$

As a consequence of Theorem 3.2.10, the expression in the second pair of brackets in (3.71) is equal to $\|g\|_2^2 = 1$ and the expression in the first pair of brackets in (3.71) converges to zero as $A_1 \to 0$ and $A_2, B \to \infty$, because the infinite integral converges. Therefore, we have

$$\lim_{A_1 \to 0; A_2, B \to \infty} \left\| f(t) - C_{\psi,\phi}^{-1} \int_{|b|\leq B} \int_{A_1 \leq a \leq A_2} \mathscr{W}_\psi \big[f\big](a,b) \, \phi_{a,b}(t) \, \frac{da\,db}{a^2} \right\|_2 = 0,$$

which shows that any $f \in L^2(\mathbb{R})$ can be arbitrarily well approximated by a superposition of wavelets. This completes the proof of Theorem 3.2.12. $\qquad\square$

Remark 3.2.9. For a detailed study on other variations of the inversion formula (3.66) associated with the wavelet transform, we refer the interested reader to the monograph [87].

Next, we shall present a characterization of range of the wavelet transform defined in (3.47). Subsequently, we shall exhibit the reproducing kernel Hilbert space underlying the continuous wavelet transform.

Theorem 3.2.13. *A function* $f \in L^2(\mathbb{R}^+ \times \mathbb{R})$ *is the wavelet transform of a certain square integrable function on* $\mathbb{R}$ *if and only if it satisfies the following reproduction formula:*

$$f(c,d) = \frac{1}{C_\psi} \int_{\mathbb{R}} \int_{\mathbb{R}^+} f(a,b) \left\langle \psi_{a,b}, \psi_{c,d} \right\rangle_2 \frac{da\,db}{a^2}, \tag{3.72}$$

where $C_\psi \neq 0$ *is given by* (3.27).

Proof. Consider a function f in the range of the continuous wavelet transform; that is, $f \in \mathscr{W}_\psi(L^2(\mathbb{R}))$. Then, there exists a function $h \in L^2(\mathbb{R})$, such that $\mathscr{W}_\psi[h] = f$. Thus, we have

$$f(c,d) = \mathscr{W}_\psi \big[h\big](c,d)$$

$$= \int_{\mathbb{R}} h(t) \, \overline{\psi_{c,d}(t)} \, dt$$

$$= \int_{\mathbb{R}} \left\{ \frac{1}{C_\psi} \int_{\mathbb{R}} \int_{\mathbb{R}^+} \mathscr{W}_\psi \big[h\big](a,b) \, \psi_{a,b}(t) \, \frac{da\,db}{a^2} \right\} \overline{\psi_{c,d}(t)} \, dt$$

$$= \frac{1}{C_\psi} \int_{\mathbb{R}} \int_{\mathbb{R}^+} \mathscr{W}_\psi \big[h\big](a,b) \left\{ \int_{\mathbb{R}} \psi_{a,b}(t) \, \overline{\psi_{c,d}(t)} \, dt \right\} \frac{da\,db}{a^2}$$

$$= \frac{1}{C_\psi} \int_{\mathbb{R}} \int_{\mathbb{R}^+} \mathscr{W}_\psi \big[h\big](a,b) \left\langle \psi_{a,b}, \psi_{c,d} \right\rangle_2 \frac{da\,db}{a^2}$$

$$= \frac{1}{C_\psi} \int_{\mathbb{R}} \int_{\mathbb{R}^+} f(a,b) \left\langle \psi_{a,b}, \psi_{c,d} \right\rangle_2 \frac{da\,db}{a^2}. \tag{3.73}$$

Conversely, suppose that a function $f \in L^2(\mathbb{R}^+ \times \mathbb{R})$ satisfies (3.72). Then, we have to show that $f \in \mathscr{W}_\psi(L^2(\mathbb{R}))$; that is, there must exist a function $g \in L^2(\mathbb{R})$, such that $\mathscr{W}_\psi[g] = f$. We claim that

$$g(t) = \frac{1}{C_\psi} \int_{\mathbb{R}} \int_{\mathbb{R}^+} f(a,b) \, \psi_{a,b}(t) \, \frac{da\,db}{a^2}. \tag{3.74}$$

To show that g is indeed square integrable, we proceed as

$$\|g\|_2^2 = \int_{\mathbb{R}} g(t)\, \overline{g(t)}\, dt$$

$$= \int_{\mathbb{R}} \left\{ \frac{1}{C_\psi} \int_{\mathbb{R}} \int_{\mathbb{R}^+} f(a,b)\, \psi_{a,b}(t)\, \frac{da\, db}{a^2} \right\} \left\{ \frac{1}{C_\psi} \int_{\mathbb{R}} \int_{\mathbb{R}^+} \overline{f(a',b')}\, \overline{\psi_{a',b'}(t)}\, \frac{da'\, db'}{a'^2} \right\} dt$$

$$= \frac{1}{C_\psi} \int_{\mathbb{R}} \int_{\mathbb{R}^+} \left\{ \frac{1}{C_\psi} \int_{\mathbb{R}} \int_{\mathbb{R}^+} f(a,b) \left\langle \psi_{a,b},\, \psi_{a',b'} \right\rangle_2 \frac{da\, db}{a^2} \right\} \overline{f(a',b')}\, \frac{da'\, db'}{a'^2}$$

$$= \frac{1}{C_\psi} \int_{\mathbb{R}} \int_{\mathbb{R}^+} f(a',b')\, \overline{f(a',b')}\, \frac{da'\, db'}{a'^2}$$

$$= \frac{1}{C_\psi} \|f\|_2^2 < \infty. \tag{3.75}$$

Finally, upon making a change in the order of integration, it follows that the function $f \in L^2(\mathbb{R}^+ \times \mathbb{R})$ belongs to the range of the wavelet transform as

$$\mathscr{W}_\psi\big[g\big](c,d) = \int_{\mathbb{R}} g(t)\, \overline{\psi_{c,d}(t)}\, dt$$

$$= \int_{\mathbb{R}} \left\{ \frac{1}{C_\psi} \int_{\mathbb{R}} \int_{\mathbb{R}^+} f(a,b)\, \psi_{a,b}(t)\, \frac{da\, db}{a^2} \right\} \overline{\psi_{c,d}(t)}\, dt$$

$$= \frac{1}{C_\psi} \int_{\mathbb{R}} \int_{\mathbb{R}^+} f(a,b) \left\langle \psi_{a,b},\, \psi_{c,d} \right\rangle_2 \frac{da\, db}{a^2}$$

$$= f(c,d).$$

This completes the proof of Theorem 3.2.13. □

The image of the wavelet transform, that is $\mathscr{W}_\psi(L^2(\mathbb{R}))$, constitutes a closed subspace of the Hilbert space $L^2(\mathbb{R}^+ \times \mathbb{R};\, da\, db/a^2)$. In fact, the argument in Theorem 3.2.13 indicates that the range of the wavelet transform is a reproducing kernel Hilbert space. Formally, we have the following corollary:

Corollary 3.2.14. *The continuous wavelet transform defined in* (3.47) *is a reproducing kernel Hilbert space embedded as a subspace in* $L^2(\mathbb{R}^2)$ *with kernel given by*

$$K_\psi(a,b;c,d) = C_\psi^{-1} \left\langle \psi_{a,b},\, \psi_{c,d} \right\rangle_2. \tag{3.76}$$

3.3 The Stockwell Transform

"Wavelets aren't well adapted to music and speech. Yet standard windowed Fourier analysis is incompatible with orthogonality and, because of its fixed window size, lacks the flexibility of wavelets."

-Yves Meyer

Time-frequency analysis is a pivotal tool in signal analysis which is concerned with how the frequency of a function (or signal) behaves in time, and it has evolved into a widely recognized applied discipline of signal processing. The inception of the wavelet transform completely revolutionized the field of time-frequency analysis as wavelets capture the local details of non-transient signals at a wide range of scales. This progressive resolution captures more information than the short-time Fourier transform (STFT), however, unlike the STFT the detail that is measured is not directly analogs to the frequency and phase of the Fourier transform. In this direction, R.G. Stockwell [281] introduced the notion of Stockwell transform as a bridge between the STFT and the wavelet transform. By adopting the progressive resolution of wavelets (using wider windows for lower frequencies and narrower windows for higher frequencies), it is able to resolve a wider range of frequencies and changes in frequencies than the ordinary STFT. By using a Fourier-like basis and maintaining a phase of zero about the time $t = 0$ as in the STFT, Fourier based analysis could be performed locally. Such unique features of the Stockwell transform make it a highly valuable tool in signal processing and is one of the hottest research areas of the contemporary era. The major distinction between the Stockwell and wavelet transforms can be ascertained from the fact that, technically, the wavelet transform is a time-scale transform which is inept for providing appropriate phase information, whereas, the Stockwell transform is a time-frequency transform and offers the absolutely referenced phase information of a signal. The loss of phase information in the wavelet transform is due to the reason that as the wavelet is shifted, the phase of the wavelet changes with respect to the origin. On the other hand, the Stockwell transform is unique in that it provides frequency-dependent resolution while maintaining a direct relationship with the Fourier spectrum. These advantages of the Stockwell transform are due to the fact that the modulating sinusoids are fixed with respect to the time axis, whereas the localizing scalable window dilates and translates. A detailed comparison between the Stockwell and wavelet transforms can be found in [118, 281].

3.3.1 Definition and basic properties

In this subsection, our motive is to study the Stockwell transform abreast of its fundamental properties. To facilitate this, we shall introduce three elementary operators on the space of square integrable functions $L^2(\mathbb{R})$. For $a \in \mathbb{R} \setminus \{0\}$ and $b \in \mathbb{R}$, let $\mathscr{D}_a, \mathcal{M}_a$ and $\mathcal{T}_b$ denote the dilation, modulation and translation operators acting on $\psi \in L^2(\mathbb{R})$ as

$$(\mathscr{D}_a \psi)(t) = |a|\, \psi(at), \quad (\mathcal{M}_a \psi)(t) = e^{iat} \psi(t) \quad \text{and} \quad (\mathcal{T}_b \psi)(t) = \psi(t - b), \tag{3.77}$$

respectively. We define a family of analyzing functions by employing the above defined elementary operators on a window function $\psi \in L^2(\mathbb{R})$ as

$$\mathscr{F}_\psi(a, b) := \left\{ \psi_{a,b}(t) = \mathcal{M}_a \mathcal{T}_b \mathscr{D}_a \psi(t) = |a|\, e^{iat}\, \psi\big(a(t - b)\big);\ a \in \mathbb{R} \setminus \{0\},\ b \in \mathbb{R} \right\}. \tag{3.78}$$

Having constructed a new family of analyzing functions defined in (3.78), we shall present the formal definition of the Stockwell transform.

Definition 3.3.1. For a given window function $\psi \in L^2(\mathbb{R})$, the Stockwell transform of any function $f \in L^2(\mathbb{R})$ with respect to ψ is denoted by $\mathcal{S}_\psi\big[f\big]$ and is defined as

$$\mathcal{S}_\psi\big[f\big](a, b) = \frac{1}{\sqrt{2\pi}} \Big\langle f,\, \mathcal{M}_a \mathcal{T}_b \mathscr{D}_a \psi \Big\rangle_2 = \frac{1}{\sqrt{2\pi}} \int_\mathbb{R} f(t)\, \overline{\psi_{a,b}(t)}\, dt, \tag{3.79}$$

where $a \in \mathbb{R} \setminus \{0\}$, $b \in \mathbb{R}$ and $\psi_{a,b}(t)$ is given by (3.78).

Remark 3.3.1. An easy computation shows that the continuous Stockwell transform defined in (3.79) can be expressed in terms of the classical convolution operation $*$ as

$$\mathcal{S}_\psi\big[f\big](a, b) = \frac{1}{\sqrt{2\pi}} \left((\mathcal{M}_{-a} f) * \big(\mathscr{D}_a \overline{\check{\psi}}\big) \right)(b), \qquad \check{\psi}(t) = \psi(-t). \tag{3.80}$$

A notable feature in the Stockwell transform (3.79) is the normalizing factor in the dilation operator, which is $|a|$ instead of $|a|^{-1/2}$ as is the case with the wavelet transform. This distinction is the mathematical underpinning of the notion of "absolutely referenced phase information" which mainly distinguishes the continuous Stockwell transform from the wavelet transform. Apart from these major distinctions, the Stockwell transform shares a nice bond with the wavelet transform via the following expression:

$$
\begin{aligned}
\mathcal{S}_\psi\big[f\big](a,b) &= \frac{|a|}{\sqrt{2\pi}} \int_{\mathbb{R}} f(t)\, \overline{\psi\big(a(t-b)\big)}\, e^{-iat}\, dt \\
&= \frac{|a|}{\sqrt{2\pi}} \int_{\mathbb{R}} f(t)\, \overline{\psi\big(a(t-b)\big)}\, e^{-ia(t-b)}\, e^{-iab}\, dt \\
&= \frac{|a|^{1/2}}{\sqrt{2\pi}} \frac{e^{-iab}}{|a|^{-1/2}} \int_{\mathbb{R}} f(t)\, \overline{\psi\big(a(t-b)\big)\, e^{ia(t-b)}}\, dt \\
&= \frac{|a|^{1/2}}{\sqrt{2\pi}}\, e^{-iab} \mathcal{W}_\phi\big[f\big] \left(\frac{1}{a}, b\right),
\end{aligned}
\tag{3.81}
$$

where $\phi(t) = e^{it}\psi(t)$.

The importance of the expression (3.81) lies in the phase correction term e^{-iab} which appears due to the phase function e^{-iat} occurring in Definition 3.3.1. For this reason, the Stockwell transform is also referred as the "phase-corrected wavelet transform." It is pertinent to mention that the continuous Stockwell transform is a reminiscent of the well-known Morlet wavelet transform. That is, if $\phi(t)$ denotes the Morlet wavelet defined by

$$
\phi(t) = e^{-t^2/2} e^{it}.
\tag{3.82}
$$

Then,

$$
\begin{aligned}
\mathcal{W}_\phi\big[f\big] \left(\frac{1}{a}, b\right) &= |a|^{1/2} \int_{\mathbb{R}} f(t)\, \overline{\phi\big(a(t-b)\big)}\, dt \\
&= |a|^{1/2} \int_{\mathbb{R}} f(t)\, e^{-ia(t-b)}\, e^{-a(t-b)^2/2}\, dt,
\end{aligned}
\tag{3.83}
$$

which further yields

$$
\begin{aligned}
\mathcal{W}_\phi\big[f\big] \left(\frac{1}{a}, b\right) &= |a|^{1/2} e^{iab} \int_{\mathbb{R}} f(t)\, e^{-a(t-b)^2/2}\, e^{-iat}\, dt \\
&= |a|^{-1/2} \sqrt{2\pi}\, e^{iab} \mathcal{S}_\psi\big[f\big](a,b),
\end{aligned}
\tag{3.84}
$$

where $\psi(t) = e^{-t^2/2}$. Thus, for the Morlet wavelet $\phi(t)$, we have

$$
\mathcal{S}_\psi\big[f\big](a,b) = \frac{|a|^{1/2}}{\sqrt{2\pi}\, e^{iab}} \mathcal{W}_\phi\big[f\big] \left(\frac{1}{a}, b\right),
\tag{3.85}
$$

where $\phi(t) = e^{it}\psi(t)$. A major distinction between the Morlet wavelet transform given by (3.83) and the Stockwell transform is that the phase function e^{-iat}, appearing inside the integral defining the Stockwell transform, picks out the frequency to be localized but is not translated along the time axis as is the case with the Morlet wavelet transform (3.83). This distinction is more evident in real life applications where the signal f and the window ψ are both real-valued functions. Therefore, the information about the phase $\arg\big(\mathcal{S}_\psi\big[f\big](a,b)\big)$ of the Stockwell transform $\mathcal{S}_\psi\big[f\big](a,b)$ at time b and frequency a comes from the term e^{-iat}

at time $b = 0$. On the other hand, in case of the Morlet wavelet transform, the phase information is obtained by referencing the windowed signal $f(t) e^{-a(t-b)^2/2}$ with respect to $e^{-ia(t-b)}$. This distinction is put into words as "the Stockwell transform gives the absolutely referenced phase information." Such distinguishing features of the Stockwell transform are fruitful in many practical application ranging over diverse aspects of science and engineering, particularly geophysics, astrophysics, quantum mechanics and medical imaging [33, 47, 118, 281, 282, 311].

In the following theorem, we shall present some fundamental properties of the continuous Stockwell transform defined in (3.79).

Theorem 3.3.1. *Let ψ, $\phi \in L^2(\mathbb{R})$ be a given pair of window functions and α, $\beta \in \mathbb{C}$, $k \in \mathbb{R}$, $c \in \mathbb{R} \setminus \{0\}$ are scalars. For any $f \in L^2(\mathbb{R})$, the continuous Stockwell transform defined in (3.79) satisfies the following properties:*

(i) *Linearity:* $\mathcal{S}_\psi\big[\alpha f + \beta g\big](a, b) = \alpha \, \mathcal{S}_\psi\big[f\big](a, b) + \beta \, \mathcal{S}_\psi\big[g\big](a, b),$

(ii) *Anti-linearity:* $\mathcal{S}_{\alpha\psi+\beta\phi}\big[f\big](a, b) = \overline{\alpha} \, \mathcal{S}_\psi\big[f\big](a, b) + \overline{\beta} \, \mathcal{S}_\phi\big[f\big](a, b),$

(iii) *Translation:* $\mathcal{S}_\psi\big[(\mathcal{T}_k f)\big](a, b) = e^{-iak} \mathcal{S}_\psi\big[f\big](a, b - k),$

(iv) *Dilation:* $\mathcal{S}_\psi\big[(\mathcal{D}_c f)\big](a, b) = |c| \, \mathcal{S}_\psi\big[f\big]\left(\dfrac{a}{c}, bc\right),$

(v) *Parity:* $\mathcal{S}_\psi\big[(Pf)\big](a, b) = \mathcal{S}_\psi\big[f\big](-a, -b),$ *where* $P\big(f(t)\big) = f(-t),$

(vi) *Conjugation:* $\mathcal{S}_\psi\big[\overline{f}\big](a, b) = \overline{\mathcal{S}_{\check{\psi}}\big[f\big](-a, b)}, \; \check{\psi}(t) = \psi(-t).$

Proof. For the sake of brevity, we omit the proof of Theorem 3.3.1.   □

Below, we shall examine the effect of the continuous Stockwell transform defined in (3.79) on a convoluted input.

Proposition 3.3.2. *For any pair of functions f, $g \in L^2(\mathbb{R})$, we have*

$$\mathcal{S}_\psi\big[(f * g)\big](a, b) = \int_{\mathbb{R}} f(x) \, \mathcal{S}_\psi\big[g\big](a, b - x) \, e^{-iax} \, dx. \tag{3.86}$$

Proof. Using Definition 3.3.1, followed by a change in the order of integration, we have

$$\mathcal{S}_\psi\big[(f * g)\big](a, b) = \frac{1}{\sqrt{2\pi}} \int_{\mathbb{R}} (f * g)(t) \, \overline{\psi_{a,b}(t)} \, dt$$

$$= \frac{|a|}{\sqrt{2\pi}} \int_{\mathbb{R}} \left\{ \int_{\mathbb{R}} f(x) \, g(t - x) \, dx \right\} \overline{\psi\big(a(t - b)\big)} \, e^{-iat} \, dt$$

$$= \frac{|a|}{\sqrt{2\pi}} \int_{\mathbb{R}} \int_{\mathbb{R}} f(x) \, g(t - x) \, \overline{\psi\big(a(t - b)\big)} \, e^{-iat} \, dx \, dt$$

$$= \frac{|a|}{\sqrt{2\pi}} \int_{\mathbb{R}} f(x) \left\{ \int_{\mathbb{R}} g(t - x) \, \overline{\psi\big(a(t - b)\big)} \, e^{-iat} \, dt \right\} dx$$

$$= \frac{|a|}{\sqrt{2\pi}} \int_{\mathbb{R}} f(x) \left\{ \int_{\mathbb{R}} g(z) \, \overline{\psi\big(a(z + x - b)\big)} \, e^{-ia(z+x)} \, dz \right\} dx$$

$$= \int_{\mathbb{R}} f(x) \left\{ \frac{|a|}{\sqrt{2\pi}} \int_{\mathbb{R}} g(z) \, \overline{\psi\big(a(z - (b - x))\big)} \, e^{-iaz} \, dz \right\} e^{-iax} \, dx$$

$$= \int_{\mathbb{R}} f(x) \, \mathcal{S}_\psi\big[g\big](a, b - x) \, e^{-iax} \, dx.$$

This completes the proof of Proposition 3.3.2.   □

If the Stockwell transform (3.79) is indeed a representation of the local spectrum, one would expect a simple operation of averaging the local spectra over time to give the Fourier spectrum. This pivotal property of the Stockwell transform is justified in the following proposition:

Proposition 3.3.3. *Let $\psi \in L^2(\mathbb{R})$ be a window function satisfying*

$$\int_{\mathbb{R}} \psi(t)\, dt = 1. \tag{3.87}$$

Then, for any $f \in L^2(\mathbb{R})$, we have

$$\int_{\mathbb{R}} S_\psi\big[f\big](a,b)\, db = \hat{f}(a) \tag{3.88}$$

Proof. Invoking Definition 3.3.1, we have

$$
\begin{aligned}
\int_{\mathbb{R}} S_\psi\big[f\big](a,b)\, db &= \int_{\mathbb{R}} \left\{ \frac{|a|}{\sqrt{2\pi}} \int_{\mathbb{R}} f(t)\, \overline{\psi\big(a(t-b)\big)}\, e^{-iat}\, dt \right\} db \\
&= \frac{|a|}{\sqrt{2\pi}} \int_{\mathbb{R}} f(t) \left\{ \int_{\mathbb{R}} \overline{\psi\big(a(t-b)\big)}\, db \right\} e^{-iat}\, dt \\
&= \frac{|a|}{\sqrt{2\pi}} \int_{\mathbb{R}} f(t) \left\{ \int_{\mathbb{R}} \overline{\psi(z)}\, \frac{dz}{|a|} \right\} e^{-iat}\, dt \\
&= \frac{1}{\sqrt{2\pi}} \int_{\mathbb{R}} f(t) \left\{ \int_{\mathbb{R}} \overline{\psi(z)}\, dz \right\} e^{-iat}\, dt \\
&= \frac{1}{\sqrt{2\pi}} \int_{\mathbb{R}} f(t)\, e^{-iat}\, dt \\
&= \hat{f}(a).
\end{aligned}
$$

This completes the proof of Proposition 3.3.3. $\qquad\square$

Theorem 3.3.4. *For any $f \in L^2(\mathbb{R})$, the continuous Stockwell transform $S_\psi\big[f\big](a,b)$ defined in (3.79) can be expressed as*

$$S_\psi\big[f\big](a,b) = e^{-iab}\, \mathscr{F}^{-1}\left[\hat{f}(\omega)\, \overline{\hat{\psi}\left(\frac{\omega-a}{a} \right)} \right](b). \tag{3.89}$$

Proof. By virtue of Parseval's formula (1.69), for any $f \in L^2(\mathbb{R})$, we have

$$
\begin{aligned}
S_\psi\big[f\big](a,b) &= \frac{1}{\sqrt{2\pi}} \big\langle f,\, \psi_{a,b} \big\rangle_2 \\
&= \frac{1}{\sqrt{2\pi}} \big\langle \hat{f},\, \hat{\psi}_{a,b} \big\rangle_2 \\
&= \frac{1}{\sqrt{2\pi}} \int_{\mathbb{R}} \hat{f}(\omega)\, \overline{\hat{\psi}_{a,b}(\omega)}\, d\omega. \tag{3.90}
\end{aligned}
$$

Note that,

$$
\begin{aligned}
\hat{\psi}_{a,b}(\omega) &= \frac{1}{\sqrt{2\pi}} \int_{\mathbb{R}} \psi_{a,b}(t)\, e^{-i\omega t}\, dt \\
&= \frac{|a|}{\sqrt{2\pi}} \int_{\mathbb{R}} \psi\big(a(t-b)\big)\, e^{iat}\, e^{-i\omega t}\, dt \\
&= e^{i(a-\omega)b}\, \hat{\psi}\left(\frac{\omega}{a} - 1 \right). \tag{3.91}
\end{aligned}
$$

Plugging (3.91) into (3.90), we obtain

$$
\begin{aligned}
\mathcal{S}_\psi\big[f\big](a,b) &= \frac{1}{\sqrt{2\pi}}\int_{\mathbb{R}} \hat{f}(\omega)\, e^{-i(a-\omega)b}\, \overline{\hat{\psi}\left(\frac{\omega}{a}-1\right)}\, d\omega \\
&= \frac{e^{-iab}}{\sqrt{2\pi}}\int_{\mathbb{R}} \hat{f}(\omega)\, \overline{\hat{\psi}\left(\frac{\omega}{a}-1\right)}\, e^{i\omega b}\, d\omega \\
&= e^{-iab}\,\mathscr{F}^{-1}\left[\hat{f}(\omega)\, \overline{\hat{\psi}\left(\frac{\omega-a}{a}\right)}\right](b).
\end{aligned}
$$

This completes the proof of Theorem 3.3.4.   □

Remark 3.3.2. As a consequence of (3.80) together with an implication of Theorem 1.2.6, we can compute the Fourier transform of $\mathcal{S}_\psi\big[f\big](a,b)$, when regarded as a function of b, as

$$
\mathscr{F}\left(\mathcal{S}_\psi\big[f\big](a,b)\right)(\omega) = \hat{f}(\omega+a)\, \overline{\hat{\psi}\left(\frac{\omega}{a}\right)}. \tag{3.92}
$$

Lemma 3.3.5. *For any window function* $\psi \in L^2(\mathbb{R})$ *and* $a \in \mathbb{R}\setminus\{0\}$*, we have*

$$
\int_{\mathbb{R}} \left|\hat{\psi}\left(\frac{\omega-a}{a}\right)\right|^2 \frac{da}{|a|} = \int_{\mathbb{R}} \frac{\left|\hat{\psi}(\eta-1)\right|^2}{|\eta|}\, d\eta. \tag{3.93}
$$

Proof. Observe that,

$$
\int_{\mathbb{R}} \left|\hat{\psi}\left(\frac{\omega-a}{a}\right)\right|^2 \frac{da}{|a|} = \int_{\mathbb{R}} \left|\hat{\psi}\left(\frac{\omega}{a}-1\right)\right|^2 \frac{da}{|a|}.
$$

Making use of the substitution $\omega/a = 1/\eta$, so that $a = \eta\omega$, we have

$$
\int_{\mathbb{R}} \left|\hat{\psi}\left(\frac{\omega-a}{a}\right)\right|^2 \frac{da}{|a|} = \int_{\mathbb{R}} \frac{\left|\hat{\psi}(\eta-1)\right|^2}{|\eta|}\, d\eta,
$$

which establishes the desired result.   □

Theorem 3.3.6. *If* $\mathcal{S}_\psi\big[f\big](a,b)$ *and* $\mathcal{S}_\psi\big[g\big](a,b)$ *are the continuous Stockwell transforms corresponding to the functions* f *and* g *belonging to* $L^2(\mathbb{R})$*, then*

$$
\int_{\mathbb{R}}\int_{\mathbb{R}} \mathcal{S}_\psi\big[f\big](a,b)\, \overline{\mathcal{S}_\psi\big[g\big](a,b)}\, \frac{da\,db}{|a|} = C_\psi\big\langle f,\, g\big\rangle_2, \tag{3.94}
$$

where the window function ψ *is such that*

$$
C_\psi = \int_{\mathbb{R}} \frac{\left|\hat{\psi}(\omega-1)\right|^2}{|\omega|}\, d\omega < \infty. \tag{3.95}
$$

Proof. As a consequence of Theorem 3.3.4, the continuous Stockwell transform of the given pair of functions f and g can be expressed as

$$
\mathcal{S}_\psi\big[f\big](a,b) = \frac{e^{-iab}}{\sqrt{2\pi}}\int_{\mathbb{R}} \hat{f}(\omega)\, \overline{\hat{\psi}\left(\frac{\omega-a}{a}\right)}\, e^{i\omega b}\, d\omega \tag{3.96}
$$

and

$$
\mathcal{S}_\psi\big[g\big](a,b) = \frac{e^{-iab}}{\sqrt{2\pi}}\int_{\mathbb{R}} \hat{g}(\xi)\, \overline{\hat{\psi}\left(\frac{\xi-a}{a}\right)}\, e^{i\xi b}\, d\xi. \tag{3.97}
$$

Therefore, we have

$$
\int_{\mathbb{R}} \int_{\mathbb{R}} \mathcal{S}_{\psi}\big[f\big](a,b)\, \overline{\mathcal{S}_{\psi}\big[g\big](a,b)}\, \frac{da\,db}{|a|}
$$

$$
= \int_{\mathbb{R}} \int_{\mathbb{R}} \left\{ \frac{e^{-iab}}{\sqrt{2\pi}} \int_{\mathbb{R}} \hat{f}(\omega)\, \hat{\psi}\left(\frac{\omega - a}{a}\right) e^{i\omega b}\, d\omega \right\} \left\{ \frac{e^{iab}}{\sqrt{2\pi}} \int_{\mathbb{R}} \overline{\hat{g}(\xi)}\, \overline{\hat{\psi}\left(\frac{\xi - a}{a}\right)}\, e^{-i\xi b}\, d\xi \right\} \frac{da\,db}{|a|}
$$

$$
= \int_{\mathbb{R}} \int_{\mathbb{R}} \int_{\mathbb{R}} \hat{f}(\omega)\, \overline{\hat{g}(\xi)}\, \hat{\psi}\left(\frac{\omega - a}{a}\right) \overline{\hat{\psi}\left(\frac{\xi - a}{a}\right)} \left\{ \frac{1}{2\pi} \int_{\mathbb{R}} e^{i(\omega - \xi)b}\, db \right\} \frac{da\,d\omega\,d\xi}{|a|}
$$

$$
= \int_{\mathbb{R}} \int_{\mathbb{R}} \int_{\mathbb{R}} \hat{f}(\omega)\, \overline{\hat{g}(\xi)}\, \hat{\psi}\left(\frac{\omega - a}{a}\right) \overline{\hat{\psi}\left(\frac{\xi - a}{a}\right)}\, \delta(\omega - \xi)\, \frac{da\,d\omega\,d\xi}{|a|}
$$

$$
= \int_{\mathbb{R}} \int_{\mathbb{R}} \int_{\mathbb{R}} \hat{f}(\omega)\, \overline{\hat{g}(\xi)}\, \hat{\psi}\left(\frac{\omega - a}{a}\right) \overline{\hat{\psi}\left(\frac{\xi - a}{a}\right)}\, \delta(\omega - \xi)\, da\,d\omega\,d\xi
$$

$$
= \int_{\mathbb{R}} \int_{\mathbb{R}} \hat{f}(\omega)\, \overline{\hat{g}(\omega)} \left| \hat{\psi}\left(\frac{\omega - a}{a}\right) \right|^2 \frac{da\,d\omega}{|a|}
$$

$$
= \int_{\mathbb{R}} \hat{f}(\omega)\, \overline{\hat{g}(\omega)} \left\{ \int_{\mathbb{R}} \left| \hat{\psi}\left(\frac{\omega - a}{a}\right) \right|^2 \frac{da}{|a|} \right\} d\omega. \tag{3.98}
$$

Invoking Lemma 3.3.5, we can rewrite (3.98) in the following form:

$$
\int_{\mathbb{R}} \int_{\mathbb{R}} \mathcal{S}_{\psi}\big[f\big](a,b)\, \overline{\mathcal{S}_{\psi}\big[g\big](a,b)}\, da\,db = \int_{\mathbb{R}} \hat{f}(\omega)\, \overline{\hat{g}(\omega)} \left\{ \int_{\mathbb{R}} \frac{|\hat{\psi}(\eta - 1)|^2}{|\eta|}\, d\eta \right\} d\omega
$$

$$
= C_{\psi} \int_{\mathbb{R}} \hat{f}(\omega)\, \overline{\hat{g}(\omega)}\, d\omega
$$

$$
= C_{\psi} \left\langle \hat{f},\, \hat{g} \right\rangle_2
$$

$$
= C_{\psi} \left\langle f,\, g \right\rangle_2.
$$

This completes the proof of Theorem 3.3.6. □

Remark 3.3.3. For $f = g$, Theorem 3.3.6 yields the following energy preserving relation for the continuous Stockwell transform defined in (3.79):

$$
\int_{\mathbb{R}} \int_{\mathbb{R}} \left| \mathcal{S}_{\psi}\big[f\big](a,b) \right|^2 \frac{da\,db}{|a|} = C_{\psi} \big\| f \big\|_2^2. \tag{3.99}
$$

The expression (3.99) shows that, except for the factor C_{ψ}, the continuous Stockwell transform is an isometry from $L^2(\mathbb{R})$ to $L^2(\mathbb{R}^2)$.

Next, our aim is to formulate a reconstruction formula associated with the continuous Stockwell transform (3.79).

Theorem 3.3.7. *Any $f \in L^2(\mathbb{R})$ can be reconstructed from the corresponding continuous Stockwell transform via the formula:*

$$
f(t) = \frac{1}{\sqrt{2\pi} C_{\psi}} \int_{\mathbb{R}} \int_{\mathbb{R}} \mathcal{S}_{\psi}\big[f\big](a,b)\, \psi_{a,b}(t)\, \frac{da\,db}{|a|}, \quad a.e., \tag{3.100}
$$

where $C_{\psi} \neq 0$ is given by (3.95).

Proof. Choose the function $g(t) = \delta(t - x)$, $x \in \mathbb{R}$. Then, we have

$$\mathcal{S}_\psi\big[g\big](a,b) = \frac{1}{\sqrt{2\pi}} \int_{\mathbb{R}} \delta(t - x) \, \overline{\psi_{a,b}(x)} \, dx = \frac{1}{\sqrt{2\pi}} \, \overline{\psi_{a,b}(t)}. \tag{3.101}$$

Therefore, by virtue of the orthogonality relation (3.94), we can write

$$\frac{1}{\sqrt{2\pi}} \int_{\mathbb{R}} \int_{\mathbb{R}} \mathcal{S}_\psi\big[f\big](a,b) \, \psi_{a,b}(t) \, \frac{da\,db}{|a|} = C_\psi \int_{\mathbb{R}} f(x) \, \delta(t - x) \, dx = C_\psi \, f(t), \tag{3.102}$$

so that the inversion formula takes the following form:

$$f(t) = \frac{1}{\sqrt{2\pi} C_\psi} \int_{\mathbb{R}} \int_{\mathbb{R}} \mathcal{S}_\psi\big[f\big](a,b) \, \psi_{a,b}(t) \, \frac{da\,db}{|a|}, \quad a.e.$$

This completes the proof of Theorem 3.3.7.   □

Towards the end of this section, we shall present a characterization of range of the continuous Stockwell transform defined in (3.79). As a consequence of such a characterization, we could easily infer that the range of the Stockwell transform constitutes a reproducing kernel Hilbert space embedded as a subspace in $L^2(\mathbb{R})$. Subsequently, we intend to exhibit the reproducing kernel Hilbert space underlying the Stockwell transform.

Theorem 3.3.8. *A function $f \in L^2(\mathbb{R}^2)$ is the Stockwell transform of a certain square integrable function on $\mathbb{R}$ if and only if it satisfies the following reproduction formula:*

$$f(c,d) = \frac{1}{2\pi \, C_\psi} \int_{\mathbb{R}} \int_{\mathbb{R}} f(a,b) \, \mathcal{K}_\psi\big(a,b : c,d\big) \, \frac{da\,db}{|a|}, \tag{3.103}$$

where $C_\psi \neq 0$ is defined in (3.95) and the reproducing kernel is given by

$$\mathcal{K}_\psi\big(a,b : c,d\big) = \big\langle \psi_{a,b}, \, \psi_{c,d} \big\rangle_2. \tag{3.104}$$

Proof. With the proof of Theorem 3.2.13 in hindsight, we can easily formulate the proof of Theorem 3.3.8.   □

As a consequence of Theorem 3.3.8, we can conclude that the image of the Stockwell transform constitutes a reproducing kernel Hilbert space. In fact, the image $\mathcal{S}_\psi\big(L^2(\mathbb{R})\big)$ constitutes only a closed subspace, not all of $L^2\big(\mathbb{R}^2; da\,db/2\pi|a|C_\psi\big)$. Formally, we have the following corollary:

Corollary 3.3.9. *The Stockwell transform defined in (3.79) is a reproducing kernel Hilbert space embedded as a subspace in $L^2\big(\mathbb{R}^2; da\,db/2\pi|a|C_\psi\big)$ with kernel given by*

$$\mathcal{K}_\psi\big(a,b;c,d\big) = \big\langle \psi_{a,b}, \, \psi_{c,d} \big\rangle_2. \tag{3.105}$$

3.4   The Two-Dimensional Wavelet Transform

"For most one-dimensional signals, speed isn't essential; today you can compute very fast with specialized processors. But this becomes completely false when you talk about pictures in two dimensions, not to mention higher dimensions; then there are things that you just can't do without fast algorithms."

-Alex Grossmann

In Section 3.2, we thoroughly discussed the continuous wavelet transform in the Hilbert space $L^2(\mathbb{R})$ and demonstrated that, in essence, the whole wavelet theory in the simpler one-dimensional case is firmly grounded in the group theory. Indeed, the wavelet transform and all its properties may be entirely derived from an appropriate representation of the affine group $G = \mathbb{R}^+ \times \mathbb{R}$. Moreover, the square integrability of the representation ensures the validity of the derivation, in particular the possibility of inverting the wavelet transform. Nevertheless, it was quite conceivable that the one-dimensional wavelet transform projects a signal onto the wavelet functions $\psi_{a,b}(t)$ obtained by suitable translation and dilation of the mother wavelet $\psi(t)$ by b and a, respectively. That is, the one-dimensional continuous wavelet transform is completely determined by the elementary operations of translations and dilations of the line. In analogy to this, the development of two-dimensional wavelet theory can be carried out via two approaches: The group theoretic approach and the analytical approach. The group theoretic approach relies on the most natural extension of the classical affine group, known as the "polar affine group." In lieu to this, the analytical approach is characterized by the elementary operations of translations, dilations and rotation of the Euclidean plane $\mathbb{R}^2$.

This section deals with a systematic study of the two-dimensional wavelet transform both from the group theoretic and analytical perspectives. Abreast of studying all fundamental properties of the two-dimensional wavelet transform, the notions of isotropic, directional and multi-directional wavelets are also discussed in great detail. Finally, some important representations of the two-dimensional wavelet transform and their respective partial energy densities are also studied.

3.4.1 Two-dimensional wavelet transform: A group theoretic approach

"Mathematicians have various ways of judging the merits of new theorems and constructions. One very important criterion is esthetic-some developments just feel right, fitting, and beautiful. Just as in other venues where beauty or esthetics are discussed, taste plays an important role in this, but I think I am not alone in being especially excited when apparently different fields suddenly meet in a new concept, a new understanding. It is often of the sparks of such encounters that our esthetic enjoyment of mathematics is born. Another important criterion for according merit to some particular piece of mathematics is the extent to which it can be useful in applications; this is the criterion almost exclusively used by non mathematicians."

-Ingrid Daubechies

This subsection is completely devoted to formulate the notion of two-dimensional wavelet transform via a unitary and square integrable representation of the underlying polar affine group.

(A) The polar affine group

Consider the upper half space in $\mathbb{R}^3$ given by $\mathbb{U} = \{(a,\mathbf{b}) : a > 0, \mathbf{b} = (b_1, b_2) \in \mathbb{R}^2\}$ and define the set $\mathbb{PU} = \mathbb{U} \times [0, 2\pi)$; that is,

$$\mathbb{PU} = \Big\{(a,\mathbf{b},\theta) : a > 0, \mathbf{b} = (b_1, b_2) \in \mathbb{R}^2 \text{ and } \theta \in [0, 2\pi)\Big\}, \tag{3.106}$$

equipped with the binary operation " $\cdot$ " defined as

$$(a,\mathbf{b},\theta) \cdot (a',\mathbf{b}',\theta') = (aa', \mathbf{b} + aR_\theta\mathbf{b}', \theta + \theta'), \tag{3.107}$$

where R_θ denotes the planer rotation operator acting on any $\mathbf{t} = (t_1, t_2)^T \in \mathbb{R}^2$ as follows:

$$R_\theta \mathbf{t} = \begin{pmatrix} \cos\theta & -\sin\theta \\ \sin\theta & \cos\theta \end{pmatrix} \begin{pmatrix} t_1 \\ t_2 \end{pmatrix}.$$

Then, it can be easily verified that $(1, \mathbf{0}, 0)$ is the identity element for the binary operation (3.107). Also, for any $(a, \mathbf{b}, \theta) \in \mathbb{PU}$, we have

$$(a, \mathbf{b}, \theta) \cdot \left(\frac{1}{a}, -\frac{1}{a} R_{-\theta} \mathbf{b}, -\theta \right) = \left(a \left(\frac{1}{a} \right), \mathbf{b} + a R_\theta \left(-\frac{1}{a} R_{-\theta} \mathbf{b} \right), \theta - \theta \right) = (1, \mathbf{0}, 0)$$

and

$$\left(\frac{1}{a}, -\frac{1}{a} R_{-\theta} \mathbf{b}, -\theta \right) \cdot (a, \mathbf{b}, \theta) = \left(\left(\frac{1}{a} \right) a, \left(-\frac{1}{a} R_{-\theta} \mathbf{b} \right) + \left(\frac{1}{a} \right) R_{-\theta} \mathbf{b}, \theta - \theta \right) = (1, \mathbf{0}, 0),$$

which shows that the binary operation (3.107) is indeed invertible. Moreover, for any trio of elements $(a, \mathbf{b}, \theta)$, $(a', \mathbf{b}', \theta')$ and $(a'', \mathbf{b}'', \theta'')$ belonging to $\mathbb{PU}$, we have

$$\begin{aligned}
(a, \mathbf{b}, \theta) \cdot \big((a', \mathbf{b}', \theta') \cdot (a'', \mathbf{b}'', \theta'') \big) &= (a, \mathbf{b}, \theta) \cdot \big(a'a'', \mathbf{b}' + a' R_{\theta'} \mathbf{b}'', \theta' + \theta'' \big) \\
&= \big(aa'a'', \mathbf{b} + a R_\theta \mathbf{b}' + aa' R_{\theta+\theta'} \mathbf{b}'', \theta + \theta' + \theta'' \big) \\
&= \big((aa')a'', (\mathbf{b} + a R_\theta \mathbf{b}') + aa' R_{\theta+\theta'} \mathbf{b}'', (\theta + \theta') + \theta'' \big) \\
&= (aa', \mathbf{b} + a R_\theta \mathbf{b}', \theta + \theta') \cdot (a'', aa' R_{\theta+\theta'} \mathbf{b}'', \theta'') \\
&= \big((a, \mathbf{b}, \theta) \cdot ((a', \mathbf{b}', \theta')) \big) \cdot (a'', \mathbf{b}'', \theta'').
\end{aligned}$$

Hence, we conclude that $(\mathbb{PU}, \cdot)$ is a group, known as the "polar affine group." It can be easily observed that the polar affine group is non-Abelian under the binary operation defined in (3.107).

In the following proposition, we shall obtain the left and right Haar measures on the polar affine group $(\mathbb{PU}, \cdot)$ defined in (3.106).

Proposition 3.4.1. *The left and right Haar measures on $(\mathbb{PU}, \cdot)$ are, respectively, given by*

$$d\mu = \frac{da\, d\mathbf{b}\, d\theta}{a^3} \qquad and \qquad d\nu = \frac{da\, d\mathbf{b}\, d\theta}{a}. \tag{3.108}$$

Proof. Let f be any integrable function on $(\mathbb{PU}, \cdot)$ with respect to $d\mu$. Then, for $(a', \mathbf{b}', \theta') \in \mathbb{PU}$, we have

$$\int_{\mathbb{PU}} f\big((a', \mathbf{b}', \theta') \cdot (a, \mathbf{b}, \theta) \big) d\mu = \int_{\mathbb{R}^+} \int_{\mathbb{R}^2} \int_0^{2\pi} f\big(a'a, \mathbf{b}' + a' R_{\theta'} \mathbf{b}, \theta' + \theta \big) \frac{da\, d\mathbf{b}\, d\theta}{a^3}.$$

Making use of the substitutions $a'a = p$, $\mathbf{b}' + a' R_{\theta'} \mathbf{b} = \mathbf{q}$ and $\theta' + \theta = \phi$, we obtain

$$\int_{\mathbb{PU}} f\big((a', \mathbf{b}', \theta') \cdot (a, \mathbf{b}, \theta) \big) d\mu = \int_{\mathbb{R}^+} \int_{\mathbb{R}^2} \int_0^{2\pi} f(p, \mathbf{q}, \phi) \frac{dp\, d\mathbf{q}\, d\phi}{p^3}.$$

Therefore, we conclude that $d\mu$ is the left Haar measure on the polar affine group $(\mathbb{PU}, \cdot)$.

Proceeding in a manner analogous to above, it can be demonstrated that $d\nu$ is indeed the right Haar measure on the polar affine group $(\mathbb{PU}, \cdot)$. $\square$

Remark 3.4.1. The polar affine group $(\mathbb{PU}, \cdot)$ is a Lie group on which the left Haar measure is different from the right Haar measure. Thus, $(\mathbb{PU}, \cdot)$ is a non-unimodular group.

(B) Unitary representation of the polar affine group

Here, our main aim is to construct a unitary representation of the polar affine group $(\mathbb{PU}, \cdot)$ defined in (3.106) on the Hilbert space $L^2(\mathbb{R}^2)$.

Proposition 3.4.2. *Let* $(\mathbb{PU}, \cdot)$ *denotes the polar affine group defined in (3.106). Then, for* $(a, \mathbf{b}, \theta) \in \mathbb{PU}$, *the operator* $\sigma(a, \mathbf{b}, \theta)$ *defined by*

$$\sigma(a, \mathbf{b}, \theta)\, \psi(\mathbf{t}) = \frac{1}{a}\, \psi\left(\frac{R_{-\theta}(\mathbf{t} - \mathbf{b})}{a}\right), \tag{3.109}$$

where $\psi \in L^2(\mathbb{R}^2)$, *is unitary on the Hilbert space* $L^2(\mathbb{R}^2)$.

Proof. Let $(a, \mathbf{b}, \theta), (a', \mathbf{b}', \theta') \in \mathbb{PU}$. Then, we have

$$\sigma(a, \mathbf{b}, \theta)\Big(\sigma(a', \mathbf{b}', \theta')\,\psi\Big)(\mathbf{t}) = \frac{1}{a}\,\sigma(a', \mathbf{b}', \theta')\,\psi\left(\frac{R_{-\theta}(\mathbf{t} - \mathbf{b})}{a}\right)$$

$$= \frac{1}{aa'}\,\psi\left(\frac{R_{-\theta'} R_{-\theta}(\mathbf{t} - \mathbf{b}) - aR_{-\theta'}\mathbf{b}'}{aa'}\right)$$

$$= \frac{1}{aa'}\,\psi\left(\frac{R_{-(\theta+\theta')}\big(\mathbf{t} - (\mathbf{b} + aR_\theta \mathbf{b}')\big)}{aa'}\right)$$

$$= \sigma(aa',\, \mathbf{b} + aR_\theta \mathbf{b}',\, \theta + \theta')\,\psi(\mathbf{t})$$

$$= \sigma\big((a, \mathbf{b}, \theta) \cdot (a', \mathbf{b}', \theta')\big)\,\psi(\mathbf{t}). \tag{3.110}$$

From (3.110), it is quite evident that the operator σ is invertible and its inverse is given by

$$\sigma^{-1}(a, \mathbf{b}, \theta) = \sigma\left(\frac{1}{a},\, -\frac{1}{a}R_{-\theta}\mathbf{b},\, -\theta\right). \tag{3.111}$$

Moreover, for any pair of functions ψ, ϕ belonging to the Hilbert space $L^2(\mathbb{R}^2)$, we have

$$\Big\langle \sigma(a, \mathbf{b}, \theta)\psi,\, \phi \Big\rangle_2 = \int_{\mathbb{R}^2} \sigma(a, \mathbf{b}, \theta)\psi(\mathbf{t})\, \overline{\phi(\mathbf{t})}\, d\mathbf{t}$$

$$= \frac{1}{a}\int_{\mathbb{R}^2} \psi\left(\frac{R_{-\theta}(\mathbf{t} - \mathbf{b})}{a}\right) \overline{\phi(\mathbf{t})}\, d\mathbf{t}$$

$$= a\int_{\mathbb{R}^2} \psi(\mathbf{z})\, \overline{\phi\left(aR_\theta \mathbf{z} + \mathbf{b}\right)}\, d\mathbf{z}$$

$$= \int_{\mathbb{R}^2} \psi(\mathbf{z})\, \overline{\sigma^{-1}(a, \mathbf{b}, \theta)\phi(\mathbf{z})}\, d\mathbf{z}$$

$$= \Big\langle \psi,\, \sigma^{-1}(a, \mathbf{b}, \theta)\phi \Big\rangle_2.$$

Hence, we conclude that σ is indeed a unitary operator and $\sigma\sigma^* = I = \sigma\sigma^{-1}$. $\qquad\square$

Proposition 3.4.3. *Let* $\mathcal{U}(L^2(\mathbb{R}^2))$ *denotes the group of all unitary operators on* $L^2(\mathbb{R}^2)$ *with respect to the composition of mappings "* $\odot$ *." Then, the map* $\sigma : \mathbb{PU} \to \mathcal{U}(L^2(\mathbb{R}^2))$ *defined in (3.109) is a unitary representation of the polar affine group* $(\mathbb{PU}, \cdot)$, *defined in (3.109), on the Hilbert space* $L^2(\mathbb{R}^2)$.

Proof. In order to accomplish the proof, it remains to show that the map $\sigma : \mathbb{PU} \to \mathcal{U}(L^2(\mathbb{R}^2))$ given by (3.109) is a group homomorphism. For any pair of elements $(a, \mathbf{b}, \theta)$ and $(a', \mathbf{b}', \theta')$ belonging to the group $(\mathbb{PU}, \cdot)$, we have

$$\Big(\sigma(a, \mathbf{b}, \theta) \odot \sigma(a', \mathbf{b}', \theta')\Big)\psi(\mathbf{t}) = \sigma(a, \mathbf{b}, \theta)\Big(\sigma(a', \mathbf{b}', \theta')\,\psi\Big)(\mathbf{t})$$

$$= \frac{1}{aa'}\,\psi\left(\frac{R_{-(\theta+\theta')}\big(\mathbf{t} - (\mathbf{b} + aR_\theta \mathbf{b}')\big)}{aa'}\right)$$

$$= \sigma(aa',\, \mathbf{b} + aR_\theta \mathbf{b}',\, \theta + \theta')\,\psi(\mathbf{t})$$

$$= \sigma\big((a, \mathbf{b}, \theta) \cdot (a', \mathbf{b}', \theta')\big)\,\psi(\mathbf{t}). \tag{3.112}$$

From the expression (3.112), it is clear that the map $\sigma : \mathbb{PU} \to \mathcal{U}(L^2(\mathbb{R}^2))$ preserves the group operations and is thus a group homomorphism. Hence, we conclude that $\sigma : \mathbb{PU} \to \mathcal{U}(L^2(\mathbb{R}^2))$ is a unitary representation of the polar affine group $(\mathbb{PU}, \cdot)$ on the Hilbert space of square integrable functions $L^2(\mathbb{R}^2)$. $\qquad\square$

In the following proposition, we aim to demonstrate that the rotation parameter $\theta \in [0, 2\pi)$ involved in (3.109) plays a significant role in concluding that the representation $\sigma : \mathbb{PU} \to \mathcal{U}(L^2(\mathbb{R}^2))$ is also irreducible.

Proposition 3.4.4. *The representation $\sigma : \mathbb{PU} \to \mathcal{U}(L^2(\mathbb{R}^2))$ defined in (3.109) is irreducible.*

Proof. Let M be a non-zero closed subspace of $L^2(\mathbb{R}^2)$ which is invariant with respect to the representation $\sigma : \mathbb{PU} \to \mathcal{U}(L^2(\mathbb{R}^2))$ defined in (3.109). For any non-zero element $g \in M$, we have

$$\Big\{\sigma(a, \mathbf{b}, \theta)\,g : \ (a, \mathbf{b}, \theta) \in \mathbb{PU}\Big\} \subseteq M. \tag{3.113}$$

Let $f \in L^2(\mathbb{R}^2)$ be such that $f \in M^\perp$, the orthogonal complement of M. Then, for all $(a, \mathbf{b}, \theta) \in \mathbb{PU}$, we have

$$\int_{\mathbb{R}^2} f(\mathbf{t})\,\overline{\sigma(a, \mathbf{b}, \theta)\,g(\mathbf{t})}\,d\mathbf{t} = \Big\langle f,\, \sigma(a, \mathbf{b}, \theta)\Big\rangle_2 = 0. \tag{3.114}$$

By virtue of the Parseval's formula (1.294) for the two-dimensional Fourier transform, we can express (3.114) as

$$\Big\langle \hat{f},\, \sigma(a, \hat{\mathbf{b}}, \theta)\,g(\omega)\Big\rangle_2 = 0,$$

which further yields

$$a \int_{\mathbb{R}^2} e^{i\mathbf{b}\cdot\mathbf{w}}\,\hat{f}(\mathbf{w})\,\overline{\hat{g}(aR_{-\theta}\mathbf{w})}\,d\mathbf{w} = 0. \tag{3.115}$$

From the expression (3.115), it follows that $\hat{f}(\mathbf{w})\,\overline{\hat{g}(aR_{-\theta}\mathbf{w})} = 0$, for almost all $\mathbf{w} \in \mathbb{R}^2$. Note that the joint action of rotations and dilation on $\mathbb{R}^2$ is transitive. Therefore, if the support of $\hat{g}$ is a patch (for instance, a disc or an ellipse) in the $\mathbf{w}$-plane, then the support of $\hat{g}(aR_{-\theta}\mathbf{w})$ will cover the entire plane when a and θ vary over their respective ranges. Thus, we must have $\hat{f}(\mathbf{w}) = 0$, almost everywhere; that is, $f = 0$. Hence, we conclude that $M = L^2(\mathbb{R}^2)$. That is, the only invariant subspaces of $L^2(\mathbb{R}^2)$ under the representation $\sigma : \mathbb{PU} \to \mathcal{U}(L^2(\mathbb{R}^2))$ are either the full space or the null space. In other words, the representation $\sigma : \mathbb{PU} \to \mathcal{U}(L^2(\mathbb{R}^2))$ defined in (3.109) is irreducible. $\qquad\square$

Finally, our motive is to show that the representation (3.109) is square integrable in view of Definition 3.2.6. As such, we shall obtain a characterization of admissible vectors for the unitary, irreducible representation $\sigma : \mathbb{PU} \to \mathcal{U}(L^2(\mathbb{R}^2))$ of the polar affine group on the Hilbert space $L^2(\mathbb{R}^2)$.

Theorem 3.4.5. *For any pair of functions $\psi, \phi \in L^2(\mathbb{R}^2)$ such that*

$$C_\psi = (2\pi)^2 \int_{\mathbb{R}} \frac{|\hat{\psi}(\mathbf{w})|^2}{|\mathbf{w}|^2} \, d\omega < \infty, \tag{3.116}$$

we have

$$\int_{\mathbb{PU}} \left| \left\langle \phi, \, \sigma(a,\mathbf{b},\theta)\psi \right\rangle_2 \right|^2 d\mu < \infty, \tag{3.117}$$

where $d\mu$ denotes the left Haar measure on the polar affine affine group.

Proof. Consider the following dense subspace of $L^2(\mathbb{R}^2)$:

$$W = \left\{ f \in L^2(\mathbb{R}^2) : \hat{f} \in C_0^\infty(\mathbb{R}^2) \right\}. \tag{3.118}$$

In order to accomplish the proof, it suffices to prove the result for the subspace W and then use the density argument to extend the result to the full space. Applying the Parseval's relation (1.294) for the two-dimensional Fourier transform, we can write

$$
\begin{aligned}
\int_{\mathbb{PU}} \left| \left\langle \phi, \sigma(a,\mathbf{b},\theta)\psi \right\rangle_2 \right|^2 d\mu &= \int_0^{2\pi} \int_{\mathbb{R}^2} \int_{\mathbb{R}^+} \left| \left\langle \hat{\phi}, \sigma(a,\hat{\mathbf{b}},\theta)\psi \right\rangle_2 \right|^2 \frac{da \, d\mathbf{b} \, d\theta}{a^3} \\
&= \int_0^{2\pi} \int_{\mathbb{R}^2} \int_{\mathbb{R}^+} \left| \int_{\mathbb{R}^2} \hat{\phi}(\mathbf{w}) \overline{\hat{\psi}(aR_{-\theta}\mathbf{w})} \, e^{i\mathbf{w}\cdot\mathbf{b}} d\mathbf{w} \right|^2 \frac{da \, d\mathbf{b} \, d\theta}{a} \\
&= \int_0^{2\pi} \int_{\mathbb{R}^+} \left\{ (2\pi)^2 \int_{\mathbb{R}^2} \left| \mathscr{F}^{-1}\left(\hat{\phi}(\mathbf{w}) \overline{\hat{\psi}(aR_{-\theta}\mathbf{w})} \right)(\mathbf{b}) \right|^2 d\mathbf{b} \right\} \frac{da \, d\theta}{a} \\
&= (2\pi)^2 \int_0^{2\pi} \int_{\mathbb{R}^+} \int_{\mathbb{R}^2} \left| \hat{\phi}(\mathbf{w}) \right|^2 \left| \hat{\psi}(aR_{-\theta}\mathbf{w}) \right|^2 \frac{d\mathbf{w} \, da \, d\theta}{a} \\
&= \int_{\mathbb{R}^2} \left| \hat{\phi}(\mathbf{w}) \right|^2 \left\{ (2\pi)^2 \int_0^{2\pi} \int_{\mathbb{R}^+} \left| \hat{\psi}(aR_{-\theta}\mathbf{w}) \right|^2 \frac{da \, d\theta}{a} \right\} d\mathbf{w}.
\end{aligned}
\tag{3.119}
$$

Making use of the substitution $aR_{-\theta}\mathbf{w} = \eta$ into the inner integral on the R.H.S of the expression (3.119), we get

$$\int_0^{2\pi} \int_{\mathbb{R}^+} \left| \hat{\psi}(aR_{-\theta}\mathbf{w}) \right|^2 \frac{da \, d\theta}{a^2} = \int_{\mathbb{R}^2} \frac{|\hat{\psi}(\eta)|^2}{|\eta|^2} \, d\eta,$$

so that

$$\int_{\mathbb{PU}} \left| \left\langle \phi, \sigma(a,\mathbf{b},\theta)\psi \right\rangle_2 \right|^2 d\mu = \int_{\mathbb{R}^2} \left| \hat{\phi}(\mathbf{w}) \right|^2 \left\{ (2\pi)^2 \int_{\mathbb{R}^2} \frac{|\hat{\psi}(\eta)|^2}{|\eta|^2} \, d\eta \right\} d\mathbf{w} = C_\psi \left\| \phi \right\|_2^2, \tag{3.120}$$

for all $\phi \in W$. Since W is a dense subspace of $L^2(\mathbb{R}^2)$, therefore, there exists a sequence of functions $\{\phi_k\}_{k=1}^\infty$ in W such that $\phi_k \to \phi$ in $L^2(\mathbb{R}^2)$ as $k \to \infty$. Thus, as a consequence of

(3.120), we infer that $\left\{\left\langle \phi_k, \sigma(\cdot, \cdot, \cdot)\psi\right\rangle_2\right\}_{k=1}^{\infty}$ is a Cauchy sequence in $L^2(\mathbb{PU})$. Hence, there exists a function $h \in L^2(\mathbb{PU})$ such that

$$\left\langle \phi_k, \sigma(\cdot, \cdot, \cdot)\psi\right\rangle_2 \to h \quad \text{as } k \to \infty. \tag{3.121}$$

Thus, we can find a subsequence of the aforementioned Cauchy sequence, again denoted by $\left\{\left\langle \phi_k, \sigma(\cdot, \cdot, \cdot)\psi\right\rangle_2\right\}_{k=1}^{\infty}$, which converges to h almost everywhere on $\mathbb{PU}$. Using the fact that $\phi_k \to \phi$, we infer that $\left\langle \phi_k, \sigma(a, \mathbf{b}, \theta)\psi\right\rangle_2 \to \left\langle \phi, \sigma(a, \mathbf{b}, \theta)\psi\right\rangle_2$, $\forall\, (a, \mathbf{b}, \theta) \in \mathbb{PU}$ as $k \to \infty$. Consequently, we obtain

$$\int_{\mathbb{PU}} \left|\left\langle \phi_k, \sigma(a, \mathbf{b}, \theta)\psi\right\rangle_2\right|^2 d\mu \to \int_{\mathbb{PU}} \left|\left\langle \phi, \sigma(a, \mathbf{b}, \theta)\psi\right\rangle_2\right|^2 d\mu. \tag{3.122}$$

On the other hand, a joint implication of the Parseval's formula (1.294) and the fact that $\phi_k \to \phi$ in $L^2(\mathbb{R}^2)$ as $k \to \infty$ yields

$$\int_{\mathbb{PU}} \left|\left\langle \phi_k, \sigma(a, \mathbf{b}, \theta)\psi\right\rangle_2\right|^2 d\mu \to \int_{\mathbb{R}^2} \left|\hat{\phi}(\mathbf{w})\right|^2 \left\{(2\pi)^2 \int_{\mathbb{R}^2} \frac{|\hat{\psi}(\eta)|^2}{|\eta|^2}\, d\eta\right\} d\mathbf{w} = C_\psi \left\|\phi\right\|_2^2. \tag{3.123}$$

Then, the desired result follows by combining (3.122) and (3.123). $\square$

Corollary 3.4.6. *The unitary, irreducible representation* $\sigma : \mathbb{PU} \to \mathcal{U}\big(L^2(\mathbb{R}^2)\big)$ *of the polar affine group* $\big(\mathbb{PU}, \cdot\big)$ *on the Hilbert space* $L^2(\mathbb{R}^2)$ *defined in (3.109) is square integrable.*

Analogous to the classical one-dimensional case, Theorem 3.4.5 demonstrates that the set of all those functions in $L^2(\mathbb{R}^2)$ satisfying the condition (3.116) is particularly interesting in the sense that the unitary and irreducible representation $\sigma : \mathbb{PU} \to \mathcal{U}\big(L^2(\mathbb{R}^2)\big)$ defined in (3.109) is square integrable. Such functions are formally called as "coherent states" of the representation $\sigma : \mathbb{PU} \to \mathcal{U}\big(L^2(\mathbb{R}^2)\big)$ of the polar affine group $\big(\mathbb{PU}, \cdot\big)$ on the Hilbert space $L^2(\mathbb{R}^2)$. More precisely, such functions are also called as the "two-dimensional wavelets" or "polar wavelets." In this direction, we have the following definition of a two-dimensional or polar wavelet:

Definition 3.4.1. A function $\psi \in L^2(\mathbb{R}^2)$ is said to be an "admissible two-dimensional wavelet" (or "polar wavelet") if the following condition holds:

$$C_\psi = (2\pi)^2 \int_{\mathbb{R}^2} \frac{|\hat{\psi}(\mathbf{w})|^2}{|\mathbf{w}|^2}\, d\mathbf{w} < \infty. \tag{3.124}$$

Condition (3.124) is called as the "admissibility condition."

Based on Definition 3.4.1, we introduce the notion of two-dimensional wavelet transform, which may also be referred as the "polar wavelet transform."

Definition 3.4.2. Let $\psi \in L^2(\mathbb{R}^2)$ be an admissible two-dimensional wavelet (or polar wavelet) for the irreducible, square integrable representation $\sigma : \mathbb{PU} \to \mathcal{U}\big(L^2(\mathbb{R}^2)\big)$ as defined in (3.109). For any $f \in L^2(\mathbb{R}^2)$, the two-dimensional wavelet transform (or polar wavelet transform) is defined as

$$\mathscr{W}_\psi\big[f\big](a, \mathbf{b}, \theta) = \left\langle f, \sigma(a, \mathbf{b}, \theta)\,\psi\right\rangle_2 = \frac{1}{a}\int_{\mathbb{R}^2} f(\mathbf{t})\, \overline{\psi\left(\frac{R_{-\theta}(\mathbf{t} - \mathbf{b})}{a}\right)}\, d\mathbf{t}, \tag{3.125}$$

where $(a, \mathbf{b}, \theta) \in \mathbb{PU}$.

Remark 3.4.2. Definition 3.4.2 implies that the continuous wavelet transform is indeed an irreducible and square integrable representation of the polar affine group $\big(\mathbb{PU}, \cdot\big)$ on the Hilbert space $L^2(\mathbb{R}^2)$.

3.4.2 Two-dimensional wavelet transform: An analytical perspective

"One cannot expect any serious understanding of what wavelet analysis means without a deep knowledge of the corresponding operator theory."

-Yves Meyer

In this subsection, we shall rigorously study the two-dimensional wavelet transform (or polar wavelet transform) from an analytical perspective. Recall that the classical one-dimensional wavelet transform projects a signal onto the daughter wavelets obtained by applying all the elementary operations of a line, namely translation and dilation, to the mother wavelet. In an analogy to this, the two-dimensional continuous wavelet transform can be obtained by taking into consideration all the three elementary operations of the plane, viz; translation, dilation and rotation.

As in the one-dimensional setting, the analytical approach of the two-dimensional wavelet transform relies on the admissibility condition (3.124). Here, it is worth emphasizing that if ψ is regular enough, then the admissibility condition (3.124) implies that

$$\hat{\psi}(\mathbf{0}) = \mathbf{0} \iff \int_{\mathbb{R}^2} \psi(\mathbf{t}) \, d\mathbf{t} = \mathbf{0}. \tag{3.126}$$

Intuitively, as in the one-dimensional case, the expression (3.126) means that a wavelet must be an oscillating function with zero mean. Strictly speaking, the condition (3.126) is only necessary, but in fact it is almost sufficient as a slightly stronger condition than that of usual integrability is to be imposed on the function ψ (for more details see [87]).

Below, we shall define a set of three elementary operators on the Euclidean plane, which shall serve as the pedestal for studying the two-dimensional wavelet transform from an analytical perspective.

(i) *Rigid translation:* For $\mathbf{b} \in \mathbb{R}^2$, the translation operator acting on $\psi \in L^2(\mathbb{R}^2)$ is denoted by $\mathcal{T}_{\mathbf{b}}$ and is defined as

$$(\mathcal{T}_{\mathbf{b}}\psi)(\mathbf{t}) = \psi(\mathbf{t} - \mathbf{b}). \tag{3.127}$$

(ii) *Planer dilation:* For $a > 0$, the dilation operator acting on $\psi \in L^2(\mathbb{R}^2)$ is denoted by $\mathcal{D}_a$ and is defined as

$$(\mathcal{D}_a\psi)(\mathbf{t}) = \frac{1}{a} \, \psi\left(\frac{\mathbf{t}}{a}\right). \tag{3.128}$$

(iii) *Planer rotation:* For $\theta \in [0, 2\pi)$, the rotation operator acting on $\psi \in L^2(\mathbb{R}^2)$ is denoted by R_θ and is defined as

$$(R_\theta\psi)(\mathbf{t}) = \psi\left(R_{-\theta}\mathbf{t}\right), \tag{3.129}$$

where R_θ represents the 2×2 rotation matrix effecting a planer rotation of θ-radians in anti-clockwise direction and is given by

$$R_\theta = \begin{pmatrix} \cos\theta & -\sin\theta \\ \sin\theta & \cos\theta \end{pmatrix}.$$

The above defined translation, dilation and rotation operators obey the following commutation rules:

$$\left. \begin{array}{l} T_{\mathbf{b}} D_a = D_a T_{a^{-1}\mathbf{b}} \\ T_{\mathbf{b}} R_\theta = R_\theta T_{R_{-\theta}\mathbf{b}} \\ R_\theta D_a = D_a R_\theta \end{array} \right\} . \tag{3.130}$$

Moreover, we note that the planer dilation and rotation operators given by (3.128) and (3.129), respectively, can be combined to perform a joint action via a new operator $S(a, \theta)$ defined as

$$S(a, \theta)\psi(\mathbf{t}) = a^{-1} \psi \left(S^{-1}(a, \theta)\mathbf{t} \right), \tag{3.131}$$

where

$$S(a, \theta) = \begin{pmatrix} a\cos\theta & -a\sin\theta \\ a\sin\theta & a\cos\theta \end{pmatrix}. \tag{3.132}$$

By virtue of (3.132), the joint influence of the translation, dilation and rotation operators on the function $\psi \in L^2(\mathbb{R}^2)$ can be expressed in the following compact form:

$$R_\theta D_a T_{\mathbf{b}} \, \psi(\mathbf{t}) = S(a, \theta) \, \psi(\mathbf{t} - \mathbf{b}). \tag{3.133}$$

Having presented a healthy overview of the elementary geometric operations on the Euclidean plane $\mathbb{R}^2$, we are now in a position to introduce the notion of two-dimensional or polar wavelet transform. Note that the operator σ defined in (3.109) can be viewed as the joint action of the translation, dilation and rotation operators defined in (3.127)–(3.129). Therefore, for any $\psi \in L^2(\mathbb{R}^2)$ satisfying the admissibility condition (3.124), we define a family of two-dimensional daughter wavelets as

$$\mathfrak{F}_\psi(a, \mathbf{b}, \theta) := \left\{ \psi_{a,\mathbf{b},\theta}(\mathbf{t}) = R_\theta D_a T_{\mathbf{b}} \, \psi(\mathbf{t}); \ a \in \mathbb{R}^+, \ \mathbf{b} \in \mathbb{R}^2 \text{ and } \theta \in [0, 2\pi) \right\}. \tag{3.134}$$

The generating function ψ appearing in (3.134) is called as the "two-dimensional" or "polar" mother wavelet. Moreover, we observe that

$$\left\| \psi_{a,\mathbf{b},\theta}(\mathbf{t}) \right\|_2^2 = \int_{\mathbb{R}^2} \left| R_\theta D_a T_{\mathbf{b}} \, \psi(\mathbf{t}) \right|^2 d\mathbf{t} = \int_{\mathbb{R}^2} \left| \psi(\mathbf{x}) \right|^2 d\mathbf{x} = \left\| \psi \right\|_2^2. \tag{3.135}$$

Also, the two-dimensional Fourier transform corresponding to $\psi_{a,\mathbf{b},\theta}(\mathbf{t})$ is given by

$$\hat{\psi}_{a,\mathbf{b},\theta}(\mathbf{w}) = a \, e^{-i\mathbf{b}\cdot\mathbf{w}} \hat{\psi} \left(a \, R_{-\theta} \, \mathbf{w} \right). \tag{3.136}$$

Next, we show that the linear span of the four-parameter family of functions given by (3.134) constitutes a dense subspace of $L^2(\mathbb{R}^2)$.

Proposition 3.4.7. *The linear span of the family $\mathfrak{F}_\psi(a, \mathbf{b}, \theta)$ defined in (3.134) is a dense subspace of $L^2(\mathbb{R}^2)$.*

Proof. Clearly $\mathfrak{F}_\psi(a, \mathbf{b}, \theta) \subseteq L^2(\mathbb{R}^2)$. In order to accomplish the proof, it suffices to show that the only vector in $L^2(\mathbb{R}^2)$ which is orthogonal to every member of $\mathfrak{F}_\psi(a, \mathbf{b}, \theta)$ is the zero vector. Let $f \in L^2(\mathbb{R}^2)$ be such that

$$\left\langle f, \psi_{a,\mathbf{b},\theta} \right\rangle_2 = 0. \tag{3.137}$$

As a consequence of the Parseval's formula (1.294) for the two-dimensional Fourier transform together with an implication of (3.136), we get

$$a \int_{\mathbb{R}} \hat{f}(\mathbf{w}) \, \overline{\hat{\psi}\,(a\,R_{-\theta}\,\mathbf{w})} \, e^{i\mathbf{b}\cdot\mathbf{w}} \, d\mathbf{w} = 0. \tag{3.138}$$

By virtue of the identity (3.138), we infer that $\hat{f}(\mathbf{w}) \, \overline{\hat{\psi}\,(a\,R_{-\theta}\,\mathbf{w})} = 0$, almost everywhere, for all $a \in \mathbb{R}^+$, $\mathbf{b} \in \mathbb{R}^2$ and $\theta \in [0, 2\pi)$. Since the joint action of planer rotations and dilations is transitive, therefore, varying the parameters a and θ over their respective ranges, the support of $\hat{\psi}\,(a\,R_{-\theta}\,\mathbf{w})$ will cover the entire plane. Therefore, we must have $\hat{f}(\mathbf{w}) = 0$ almost everywhere. Hence, we conclude that $f = 0$, which evidently completes the proof. $\qquad\square$

Remark 3.4.3. As a consequence of the Proposition 3.4.7, we note that any square integrable function on $\mathbb{R}^2$ is uniquely determined by its projections on the elements of the family $\mathfrak{F}_\psi(a, \mathbf{b}, \theta)$ defined in (3.134). This justifies the following definition of the two-dimensional or polar wavelet transform:

Definition 3.4.3. For any $f \in L^2(\mathbb{R}^2)$, the two-dimensional wavelet transform with respect to the two-dimensional mother wavelet $\psi \in L^2(\mathbb{R}^2)$ is defined by

$$\mathscr{W}_\psi\big[f\big](a, \mathbf{b}, \theta) = \big\langle f, \psi_{a,\mathbf{b},\theta} \big\rangle_2 = \frac{1}{a} \int_{\mathbb{R}^2} f(\mathbf{t}) \, \overline{\psi\left(\frac{R_{-\theta}(\mathbf{t} - \mathbf{b})}{a}\right)} \, d\mathbf{t}, \tag{3.139}$$

where $\psi_{a,\mathbf{b},\theta}(\mathbf{t})$ is given by (3.134).

Here, it is imperative to make the following important points regarding the above definition of the two-dimensional wavelet transform:

(i) The factor a^{-1} appearing outside the integral in Definition 3.4.3 is only meant to uphold the mathematical elegance in the sense that the corresponding dilation operator $\mathcal{D}_a$ is unitary. Often in practice, a factor a^{-2} is used instead of the factor a^{-1}, so as to enhance the high-frequency part of the signal, which makes the singularities more conspicuous.

(ii) Unlike the one-dimensional continuous wavelet transform (3.47), the distinguishing feature of Definition 3.4.3 is that we can take into account the negative dilations $a < 0$ by including a rotation of $\theta = \pi$ and applying the usual positive dilation $a > 0$.

Proposition 3.4.8. *If $\mathscr{W}_\psi\big[f\big](a, \mathbf{b}, \theta)$ is the two-dimensional wavelet transform of any $f \in L^2(\mathbb{R}^2)$, then we have*

$$\mathscr{F}\left(\mathscr{W}_\psi\big[f\big](a, \mathbf{b}, \theta)\right)\mathbf{w} = 2\pi a \, \hat{f}(\mathbf{w}) \, \overline{\hat{\psi}\,(a\,R_{-\theta}\,\mathbf{w})}, \tag{3.140}$$

where the Fourier transform on the L.H.S of (3.140) is computed with respect to $\mathbf{b}$.

Proof. Invoking the Parseval's formula (1.294) for the two-dimensional Fourier transform together with (3.136), we get

$$\begin{aligned}
\mathscr{W}_\psi\big[f\big](a, \mathbf{b}, \theta) &= \big\langle f, \psi_{a,\mathbf{b},\theta} \big\rangle_2 \\
&= \big\langle \hat{f}, \hat{\psi}_{a,\mathbf{b},\theta} \big\rangle_2 \\
&= a \int_{\mathbb{R}^2} \hat{f}(\mathbf{w}) \, \overline{\hat{\psi}\,(a\,R_{-\theta}\,\mathbf{w})} \, e^{i\mathbf{b}\cdot\mathbf{w}} \, d\mathbf{w} \\
&= 2\pi a \, \mathscr{F}^{-1}\left(\hat{f}(\mathbf{w}) \, \overline{\hat{\psi}\,(a\,R_{-\theta}\,\mathbf{w})}\right)\mathbf{b}. \tag{3.141}
\end{aligned}$$

Identifying $\mathscr{W}_\psi[f](a, \mathbf{b}, \theta)$ as a function of $\mathbf{b}$ and applying the two-dimensional Fourier transform on both sides of (3.141), we get

$$\mathscr{F}\left(\mathscr{W}_\psi[f](a, \mathbf{b}, \theta)\right)\mathbf{w} = 2\pi a\,\hat{f}(\mathbf{w})\,\overline{\hat{\psi}(a\,R_{-\theta}\,\mathbf{w})}.$$

This completes the proof of Proposition 3.4.8. □

Next, we present some elementary properties of the two-dimensional wavelet transform defined in (3.139). We note that the two-dimensional wavelet transform enjoys invariance under all the three elementary operations of rigid translations, planer dilations and planer rotations. These invariance properties play a vital role in the applications of the two-dimensional wavelet transform, including fractional analysis, inflation properties and so on.

Theorem 3.4.9. *Let $\alpha, \beta \in \mathbb{C}$, $\mathbf{k} \in \mathbb{R}^2$, $c > 0$ and $\theta' \in [0, 2\pi)$. Then, for a pair of two-dimensional wavelets ψ and ϕ, and any $f, g \in L^2(\mathbb{R}^2)$, the wavelet transform defined in (3.139) satisfies the following properties:*

(i) *Linearity:* $\mathscr{W}_\psi\big[\alpha f + \beta g\big](a, \mathbf{b}, \theta) = \alpha\,\mathscr{W}_\psi\big[f\big](a, \mathbf{b}, \theta) + \beta\,\mathscr{W}_\psi\big[g\big](a, \mathbf{b}, \theta),$

(ii) *Translation:* $\mathscr{W}_\psi\big[T_\mathbf{k}f\big](a, \mathbf{b}, \theta) = \mathscr{W}_\psi\big[f\big](a, \mathbf{b} - \mathbf{k}, \theta),$

(iii) *Dilation:* $\mathscr{W}_\psi\big[D_c f\big](a, \mathbf{b}, \theta) = \dfrac{1}{c}\,\mathscr{W}_\psi\big[f\big]\left(\dfrac{a}{c}, c^{-1}\mathbf{b}, \theta\right),$

(iv) *Symmetry:* $\mathscr{W}_\psi\big[f\big](a, \mathbf{b}, \theta) = \overline{\mathscr{W}_f\big[\psi\big]\left(\dfrac{1}{a}, -a^{-1}R_{-\theta}\mathbf{b}, -\theta\right)},$

(v) *Parity:* $\mathscr{W}_\psi\big[Pf\big](a, \mathbf{b}, \theta) = -\mathscr{W}_\psi\big[f\big](-a, -\mathbf{b}, \theta),$ *where* $Pf(\mathbf{t}) = f(-\mathbf{t}),$

(vi) *Rotation:* $\mathscr{W}_\psi\big[R_{\theta'}f\big](a, \mathbf{b}, \theta) = \mathscr{W}_\psi\big[f\big](a, R_{-\theta'}\mathbf{b}, \theta - \theta'),$

(vii) *Anti-linearity:* $\mathscr{W}_{\alpha\psi+\beta\phi}\big[f\big](a, \mathbf{b}, \theta) = \overline{\alpha}\,\mathscr{W}_\psi\big[f\big](a, \mathbf{b}, \theta) + \overline{\beta}\,\mathscr{W}_\phi\big[f\big](a, \mathbf{b}, \theta),$

(viii) *Translation in wavelet:* $\mathscr{W}_{T_\mathbf{k}\psi}\big[f\big](a, \mathbf{b}, \theta) = \mathscr{W}_\psi\big[f\big](a, \mathbf{b} + aR_\theta\mathbf{k}, \theta),$

(ix) *Dilation in wavelet:* $\mathscr{W}_{D_c\psi}\big[f\big](a, \mathbf{b}, \theta) = \mathscr{W}_\psi\big[f\big](ac, \mathbf{b}, \theta),$

(x) *Rotation in wavelet:* $\mathscr{W}_{R_{\theta'}\psi}\big[f\big](a, \mathbf{b}, \theta) = \mathscr{W}_\psi\big[f\big](a, \mathbf{b}, \theta + \theta').$

Proof. The proof of Theorem 3.4.9 is left as an exercise to the reader. □

The wavelet transform defined in (3.139) is a four-dimensional representation determining the scale, position and orientation of a given two-dimensional signal. The following theorem exhibits an orthogonality relation between a given pair of signals and their respective two-dimensional wavelet transforms. More precisely, we shall demonstrate that the orthogonality of signals in the space $L^2(\mathbb{R}^2)$ implies and is implied by the orthogonality of their respective wavelet transforms in the space $L^2\big(\mathbb{R}^+ \times \mathbb{R}^2 \times [0, 2\pi)\big)$.

Theorem 3.4.10. *Let* $\mathscr{W}_\psi[f](a, \mathbf{b}, \theta)$ *and* $\mathscr{W}_\phi[g](a, \mathbf{b}, \theta)$ *be the two-dimensional wavelet transforms of a given pair of functions* f *and* g *with respect to the two-dimensional wavelets* ψ *and* ϕ, *respectively. Then, the following orthogonality relation holds:*

$$\int_0^{2\pi} \int_{\mathbb{R}^2} \int_{\mathbb{R}^+} \mathscr{W}_\psi[f](a, \mathbf{b}, \theta) \overline{\mathscr{W}_\phi[g](a, \mathbf{b}, \theta)} \frac{da\,d\mathbf{b}\,d\theta}{a^3} = C_{\psi,\phi} \left\langle f, g \right\rangle_2, \quad (3.142)$$

where $C_{\psi,\phi}$ *is given by the cross admissibility condition:*

$$C_{\psi,\phi} = (2\pi)^2 \int_{\mathbb{R}^2} \frac{\overline{\hat{\psi}(\mathbf{w})}\,\hat{\phi}(\mathbf{w})}{|\mathbf{w}|^2}\,d\mathbf{w} < \infty. \quad (3.143)$$

Proof. Using Proposition 3.4.8, we can express the two-dimensional wavelet transforms of f and g as follows:

$$\mathscr{W}_\psi[f](a, \mathbf{b}, \theta) = a \int_{\mathbb{R}^2} \hat{f}(\mathbf{w}) \overline{\hat{\psi}(a\,R_{-\theta}\,\mathbf{w})}\, e^{i\mathbf{b}\cdot\mathbf{w}}\,d\mathbf{w} \quad (3.144)$$

and

$$\mathscr{W}_\phi[g](a, \mathbf{b}, \theta) = a \int_{\mathbb{R}^2} \hat{g}(\mathbf{w}') \overline{\hat{\phi}(a\,R_{-\theta}\,\mathbf{w}')}\, e^{i\mathbf{b}\cdot\mathbf{w}'}\,d\mathbf{w}'. \quad (3.145)$$

Invoking a change in the order of integration, we can express the L.H.S of (3.142) as

$$\int_0^{2\pi} \int_{\mathbb{R}^2} \int_{\mathbb{R}^+} \mathscr{W}_\psi[f](a, \mathbf{b}, \theta) \overline{\mathscr{W}_\phi[g](a, \mathbf{b}, \theta)} \frac{da\,d\mathbf{b}\,d\theta}{a^3}$$

$$= \int_0^{2\pi} \int_{\mathbb{R}^2} \int_{\mathbb{R}^+} \left\{ a \int_{\mathbb{R}^2} \hat{f}(\mathbf{w}) \overline{\hat{\psi}(a\,R_{-\theta}\,\mathbf{w})}\, e^{i\mathbf{b}\cdot\mathbf{w}}\,d\mathbf{w} \right\}$$

$$\times \left\{ a \int_{\mathbb{R}^2} \overline{\hat{g}(\mathbf{w}')}\, \hat{\phi}(a\,R_{-\theta}\,\mathbf{w}')\, e^{-i\mathbf{b}\cdot\mathbf{w}'}\,d\mathbf{w}' \right\} \frac{da\,d\mathbf{b}\,d\theta}{a^3}$$

$$= \int_{\mathbb{R}^2} \int_{\mathbb{R}^2} \int_{\mathbb{R}^+} \int_0^{2\pi} \hat{f}(\mathbf{w}) \overline{\hat{g}(\mathbf{w}')}\, \hat{\phi}(a\,R_{-\theta}\,\mathbf{w}') \overline{\hat{\psi}(a\,R_{-\theta}\,\mathbf{w})} \left\{ \int_{\mathbb{R}^2} e^{-i\mathbf{b}\cdot(\mathbf{w}-\mathbf{w}')}d\mathbf{b} \right\} \frac{d\theta\,da\,d\mathbf{w}\,d\mathbf{w}'}{a}$$

$$= (2\pi)^2 \int_{\mathbb{R}^2} \int_{\mathbb{R}^2} \int_{\mathbb{R}^+} \int_0^{2\pi} \hat{f}(\mathbf{w}) \overline{\hat{g}(\mathbf{w}')}\, \hat{\phi}(a\,R_{-\theta}\,\mathbf{w}') \overline{\hat{\psi}(a\,R_{-\theta}\,\mathbf{w})}$$

$$\times \left\{ \frac{1}{(2\pi)^2} \int_{\mathbb{R}^2} e^{-i\mathbf{b}\cdot(\mathbf{w}-\mathbf{w}')}d\mathbf{b} \right\} \frac{d\theta\,da\,d\mathbf{w}\,d\mathbf{w}'}{a}$$

$$= (2\pi)^2 \int_{\mathbb{R}^2} \int_{\mathbb{R}^2} \int_{\mathbb{R}^+} \int_0^{2\pi} \hat{f}(\mathbf{w}) \overline{\hat{g}(\mathbf{w}')}\, \hat{\phi}(a\,R_{-\theta}\,\mathbf{w}') \overline{\hat{\psi}(a\,R_{-\theta}\,\mathbf{w})}\, \delta(\mathbf{w}-\mathbf{w}') \frac{d\theta\,da\,d\mathbf{w}\,d\mathbf{w}'}{a}$$

$$= (2\pi)^2 \int_{\mathbb{R}^2} \int_{\mathbb{R}^+} \int_0^{2\pi} \hat{f}(\mathbf{w}) \overline{\hat{g}(\mathbf{w})}\, \hat{\phi}(a\,R_{-\theta}\,\mathbf{w}) \overline{\hat{\psi}(a\,R_{-\theta}\,\mathbf{w})} \frac{d\theta\,da\,d\mathbf{w}}{a}$$

$$= \int_{\mathbb{R}^2} \hat{f}(\mathbf{w}) \overline{\hat{g}(\mathbf{w})} \left\{ (2\pi)^2 \int_{\mathbb{R}^+} \int_0^{2\pi} \overline{\hat{\psi}(a\,R_{-\theta}\,\mathbf{w})}\, \hat{\phi}(a\,R_{-\theta}\,\mathbf{w}) \frac{d\theta\,da}{a} \right\} d\mathbf{w}. \quad (3.146)$$

Using the polar coordinates $\mathbf{w} = (r, \theta')$, where $r = |\mathbf{w}|$ and $0 \le \theta' < 2\pi$, the integral within the braces on the R.H.S of (3.146) can be simplified as

$$(2\pi)^2 \int_{\mathbb{R}^+} \int_0^{2\pi} \overline{\hat{\psi}(a\,R_{-\theta}\,\mathbf{w})}\, \hat{\phi}(a\,R_{-\theta}\,\mathbf{w}) \frac{d\theta\,da}{a}$$

$$= (2\pi)^2 \int_{\mathbb{R}^+} \int_0^{2\pi} \overline{\hat{\psi}(ar, \theta' - \theta)}\, \hat{\phi}(ar, \theta' - \theta) \frac{d\theta\,da}{a}$$

$$= (2\pi)^2 \int_{\mathbb{R}^+} \int_0^{2\pi} \overline{\hat{\psi}(r', \theta'')}\, \hat{\phi}(r', \theta'')\, \frac{d\theta'\, dr'}{r'}$$

$$= (2\pi)^2 \int_{\mathbb{R}^+} \int_0^{2\pi} \frac{\overline{\hat{\psi}(\mathbf{w}'')}\, \hat{\phi}(\mathbf{w}'')}{|\mathbf{w}''|^2}\, d\mathbf{w}''$$

$$= C_{\psi,\phi}. \tag{3.147}$$

Implementation of (3.147) in (3.146) yields the desired orthogonality relation as

$$\int_0^{2\pi} \int_{\mathbb{R}^2} \int_{\mathbb{R}^+} \mathscr{W}_\psi\big[f\big](a,\mathbf{b},\theta)\, \overline{\mathscr{W}_\phi\big[g\big](a,\mathbf{b},\theta)}\, \frac{da\, d\mathbf{b}\, d\theta}{a^3} = C_{\psi,\phi} \left\langle f,\, g \right\rangle_2.$$

$\square$

Remark 3.4.4. Choosing $\psi = \phi$ in Theorem 3.4.10, we obtain the orthogonality relation:

$$\int_0^{2\pi} \int_{\mathbb{R}^2} \int_{\mathbb{R}^+} \mathscr{W}_\psi\big[f\big](a,\mathbf{b},\theta)\, \overline{\mathscr{W}_\psi\big[g\big](a,\mathbf{b},\theta)}\, \frac{da\, d\mathbf{b}\, d\theta}{a^3} = C_\psi \left\langle f,\, g \right\rangle_2, \tag{3.148}$$

where C_ψ is given by the usual admissibility condition (3.124). Moreover, choosing $f = g$ in (3.148) yields the energy preserving relation:

$$\int_0^{2\pi} \int_{\mathbb{R}^2} \int_{\mathbb{R}^+} \left|\mathscr{W}_\psi\big[f\big](a,\mathbf{b},\theta)\right|^2 \frac{da\, d\mathbf{b}\, d\theta}{a^3} = C_\psi \left\|f\right\|_2^2, \tag{3.149}$$

which indicates that for $C_\psi = 1$ the two-dimensional wavelet transform defined in (3.139) turns to be an isometry from the space of signals $L^2(\mathbb{R}^2)$ to the space of transformations $L^2\big(\mathbb{R}^+ \times \mathbb{R}^2 \times [0, 2\pi)\big)$.

As a consequence of Theorem 3.4.10, we obtain a "resolution of identity" formula, which allows the reconstruction of input signal from the corresponding two-dimensional wavelet transform. In this direction, we have the following theorem:

Theorem 3.4.11. *Let* ψ *and* ϕ *be a given pair of two-dimensional wavelets, then any function* $f \in L^2(\mathbb{R}^2)$ *can be reconstructed from the corresponding wavelet transform* $\mathscr{W}_\psi\big[f\big](a,\mathbf{b},\theta)$ *via the "resolution of identity" formula:*

$$f(\mathbf{t}) = \frac{1}{C_{\psi,\phi}} \int_0^{2\pi} \int_{\mathbb{R}^2} \int_{\mathbb{R}^+} \mathscr{W}_\psi\big[f\big](a,\mathbf{b},\theta)\, \phi_{a,\mathbf{b},\theta}(\mathbf{t})\, \frac{da\, d\mathbf{b}\, d\theta}{a^3}, \qquad a.e., \tag{3.150}$$

where $C_{\psi,\phi} \neq 0$ *is given by* (3.143).

Proof. For an arbitrary function $g \in L^2(\mathbb{R}^2)$, the orthogonality relation (3.142) yields

$$\left\langle f,\, g \right\rangle_2 = \frac{1}{C_{\psi,\phi}} \int_0^{2\pi} \int_{\mathbb{R}^2} \int_{\mathbb{R}^+} \mathscr{W}_\psi\big[f\big](a,\mathbf{b},\theta)\, \overline{\mathscr{W}_\psi\big[g\big](a,\mathbf{b},\theta)}\, \frac{da\, d\mathbf{b}\, d\theta}{a^3}$$

$$= \frac{1}{C_{\psi,\phi}} \int_0^{2\pi} \int_{\mathbb{R}^2} \int_{\mathbb{R}^+} \mathscr{W}_\psi\big[f\big](a,\mathbf{b},\theta) \left\{ \int_{\mathbb{R}^2} \overline{g(\mathbf{t})}\, \phi_{a,\mathbf{b},\theta}(\mathbf{t})\, d\mathbf{t} \right\} \frac{da\, d\mathbf{b}\, d\theta}{a^3}$$

$$= \int_{\mathbb{R}^2} \left\{ \frac{1}{C_{\psi,\phi}} \int_0^{2\pi} \int_{\mathbb{R}^2} \int_{\mathbb{R}^+} \mathscr{W}_\psi\big[f\big](a,\mathbf{b},\theta)\, \phi_{a,\mathbf{b},\theta}(\mathbf{t})\, \frac{da\, d\mathbf{b}\, d\theta}{a^3} \right\} \overline{g(\mathbf{t})}\, d\mathbf{t}$$

$$= \left\langle \frac{1}{C_{\psi,\phi}} \int_0^{2\pi} \int_{\mathbb{R}^2} \int_{\mathbb{R}^+} \mathscr{W}_\psi\big[f\big](a,\mathbf{b},\theta)\, \phi_{a,\mathbf{b},\theta}(\mathbf{t})\, \frac{da\, d\mathbf{b}\, d\theta}{a^3},\, g(\mathbf{t}) \right\rangle_2. \tag{3.151}$$

Using the fundamental properties of the inner product in (3.151), we obtain the desired reconstruction formula as

$$f(\mathbf{t}) = \frac{1}{C_{\psi,\phi}} \int_0^{2\pi} \int_{\mathbb{R}^2} \int_{\mathbb{R}^+} \mathscr{W}_\psi\big[f\big](a, \mathbf{b}, \theta)\, \phi_{a,\mathbf{b},\theta}(\mathbf{t})\, \frac{da\, d\mathbf{b}\, d\theta}{a^3},$$

with convergence of the integral "in the weak sense." This completes the proof of Theorem 3.4.11. $\qquad\square$

Remark 3.4.5. Choosing $\psi = \phi$ in Theorem 3.4.11, we obtain the following "resolution of identity" formula:

$$f(\mathbf{t}) = \frac{1}{C_\psi} \int_0^{2\pi} \int_{\mathbb{R}^2} \int_{\mathbb{R}^+} \mathscr{W}_\psi\big[f\big](a, \mathbf{b}, \theta)\, \psi_{a,\mathbf{b},\theta}(\mathbf{t})\, \frac{da\, d\mathbf{b}\, d\theta}{a^3}, \quad a.e., \tag{3.152}$$

where $C_\psi \neq 0$ is given by the usual admissibility condition (3.124). Moreover, in case ψ is rotation invariant, then the reconstruction formula (3.152) further simplifies to

$$f(\mathbf{t}) = \frac{2\pi}{C_\psi} \int_{\mathbb{R}^2} \int_{\mathbb{R}^+} \mathscr{W}_\psi\big[f\big](a, \mathbf{b})\, \psi_{a,\mathbf{b}}(\mathbf{t})\, \frac{da\, d\mathbf{b}}{a^3}, \quad a.e. \tag{3.153}$$

Remark 3.4.6. In analogy to the classical one-dimensional wavelet transform, Theorem 3.4.11 asserts that a distinct pair of two-dimensional wavelets can be employed for the analyzing and reconstruction processes. For instance, in case the reconstruction wavelet ϕ is chosen to be the Dirac delta function, then the reconstruction formula (3.150) becomes

$$f(\mathbf{t}) = \frac{1}{C_{\psi,\delta}} \int_0^{2\pi} \int_{\mathbb{R}^2} \int_{\mathbb{R}^+} \mathscr{W}_\psi\big[f\big](a, \mathbf{b}, \theta)\, \delta\left(\frac{R_{-\theta}(\mathbf{t} - \mathbf{b})}{a}\right) \frac{da\, d\mathbf{b}\, d\theta}{a^4}$$

$$= \frac{1}{C_{\psi,\delta}} \int_0^{2\pi} \int_{\mathbb{R}^+} \mathscr{W}_\psi\big[f\big](a, \mathbf{t}, \theta)\, \frac{da\, d\theta}{a^2}, \quad a.e., \tag{3.154}$$

where $C_{\psi,\delta} \neq 0$ is given

$$C_{\psi,\delta} = (2\pi)^2 \int_{\mathbb{R}^2} \frac{\overline{\hat{\psi}(\mathbf{w})}}{|\mathbf{w}|^2}\, d\mathbf{w} < \infty. \tag{3.155}$$

Again, if the analyzing wavelet ψ is rotation invariant, then the reconstruction formula (3.153) can be expressed in an extremely compact form as

$$f(\mathbf{t}) = \frac{2\pi}{C_{\psi,\delta}} \int_{\mathbb{R}^+} \mathscr{W}_\psi\big[f\big](a, \mathbf{t})\, \frac{da}{a^2}, \quad a.e. \tag{3.156}$$

In the following theorem, we obtain a characterization of the range of two-dimensional wavelet transform, which demonstrates the existence of the reproducing kernel for the two-dimensional wavelet transform defined in (3.139).

Theorem 3.4.12. *A function $f \in L^2\big(\mathbb{R}^+ \times \mathbb{R}^2 \times [0, 2\pi)\big)$ is the two-dimensional wavelet transform of a certain square integrable function on $\mathbb{R}^2$ if and only if it satisfies the following reproduction formula:*

$$f(a', \mathbf{b}', \theta') = \frac{1}{C_\psi} \int_0^{2\pi} \int_{\mathbb{R}^2} \int_{\mathbb{R}^+} f(a, \mathbf{b}, \theta) \big\langle \psi_{a,\mathbf{b},\theta},\, \psi_{a',\mathbf{b}',\theta'} \big\rangle_2 \frac{da\, d\mathbf{b}\, d\theta}{a^3}, \tag{3.157}$$

where $C_\psi \neq 0$ is given by the usual admissibility condition (3.124).

Proof. Suppose that $f \in \mathscr{W}_\psi\big(L^2(\mathbb{R}^2)\big)$; that is, there exists a function $h \in L^2(\mathbb{R}^2)$, such that $\mathscr{W}_\psi[h] = f$. Then, we have

$$
\begin{aligned}
f(a', \mathbf{b}', \theta') &= \mathscr{W}_\psi\big[h\big](a', \mathbf{b}', \theta') \\[6pt]
&= \int_{\mathbb{R}^2} h(\mathbf{t})\, \overline{\psi_{a', \mathbf{b}', \theta'}}\, d\mathbf{t} \\[6pt]
&= \int_{\mathbb{R}^2} \left\{ \frac{1}{C_\psi} \int_0^{2\pi} \int_{\mathbb{R}^2} \int_{\mathbb{R}^+} \mathscr{W}_\psi\big[h\big](a, \mathbf{b}, \theta)\, \psi_{a, \mathbf{b}, \theta}(\mathbf{t})\, \frac{da\, d\mathbf{b}\, d\theta}{a^3} \right\} \overline{\psi_{a', \mathbf{b}', \theta'}}\, d\mathbf{t} \\[6pt]
&= \frac{1}{C_\psi} \int_0^{2\pi} \int_{\mathbb{R}^2} \int_{\mathbb{R}^+} \mathscr{W}_\psi\big[h\big](a, \mathbf{b}, \theta) \left\{ \int_{\mathbb{R}^2} \psi_{a, \mathbf{b}, \theta}(\mathbf{t})\, \overline{\psi_{a', \mathbf{b}', \theta'}(\mathbf{t})}\, d\mathbf{t} \right\} \frac{da\, d\mathbf{b}\, d\theta}{a^3} \\[6pt]
&= \frac{1}{C_\psi} \int_0^{2\pi} \int_{\mathbb{R}^2} \int_{\mathbb{R}^+} f(a, \mathbf{b}, \theta) \big\langle \psi_{a, \mathbf{b}, \theta},\ \psi_{a', \mathbf{b}', \theta'} \big\rangle_2 \frac{da\, d\mathbf{b}\, d\theta}{a^3}, \qquad (3.158)
\end{aligned}
$$

which establishes the necessary implication. In order to prove the reverse implication, we suppose that a function $f \in L^2\big(\mathbb{R}^+ \times \mathbb{R}^2 \times [0, 2\pi)\big)$ satisfies the reproduction formula (3.157). Then, we must show the existence of a function $g \in L^2(\mathbb{R}^2)$, such that $\mathscr{W}_\psi[g] = f$. We claim that,

$$
g(\mathbf{t}) = \frac{1}{C_\psi} \int_0^{2\pi} \int_{\mathbb{R}^2} \int_{\mathbb{R}^+} f(a, \mathbf{b}, \theta)\, \psi_{a, \mathbf{b}, \theta}(\mathbf{t})\, \frac{da\, d\mathbf{b}\, d\theta}{a^3}. \qquad (3.159)
$$

Indeed, the function g is square integrable as

$$
\begin{aligned}
\big\| g \big\|_2^2 &= \int_{\mathbb{R}^2} g(\mathbf{t}) \overline{g(\mathbf{t})}\, d\mathbf{t} \\[6pt]
&= \int_{\mathbb{R}^2} \left\{ \frac{1}{C_\psi} \int_0^{2\pi} \int_{\mathbb{R}^2} \int_{\mathbb{R}^+} f(a, \mathbf{b}, \theta)\, \psi_{a, \mathbf{b}, \theta}(\mathbf{t})\, \frac{da\, d\mathbf{b}\, d\theta}{a^3} \right\} \\[6pt]
&\qquad\qquad \times \left\{ \frac{1}{C_\psi} \int_0^{2\pi} \int_{\mathbb{R}^2} \int_{\mathbb{R}^+} \overline{f(a', \mathbf{b}', \theta')}\, \overline{\psi_{a', \mathbf{b}', \theta'}(\mathbf{t})}\, \frac{da'\, d\mathbf{b}'\, d\theta'}{a'^3} \right\} d\mathbf{t} \\[6pt]
&= \frac{1}{C_\psi} \int_0^{2\pi} \int_{\mathbb{R}^2} \int_{\mathbb{R}^+} \left\{ \frac{1}{C_\psi} \int_0^{2\pi} \int_{\mathbb{R}^2} \int_{\mathbb{R}^+} f(a, \mathbf{b}, \theta) \big\langle \psi_{a, \mathbf{b}, \theta},\ \psi_{a', \mathbf{b}', \theta'} \big\rangle_2 \frac{da\, d\mathbf{b}\, d\theta}{a^3} \right\} \\[6pt]
&\qquad\qquad\qquad\qquad\qquad\qquad\qquad\qquad\qquad \times \overline{f(a', \mathbf{b}', \theta')}\, \frac{da'\, d\mathbf{b}'\, d\theta'}{a'^3} \\[6pt]
&= \frac{1}{C_\psi} \int_0^{2\pi} \int_{\mathbb{R}^2} \int_{\mathbb{R}^+} f(a', \mathbf{b}', \theta')\, \overline{f(a', \mathbf{b}', \theta')}\, \frac{da'\, d\mathbf{b}'\, d\theta'}{a'^3} \\[6pt]
&= \frac{1}{C_\psi} \big\| f \big\|_2^2 < \infty. \qquad\qquad\qquad\qquad\qquad\qquad\qquad\qquad\qquad\qquad (3.160)
\end{aligned}
$$

Finally, altering the order of integration, we get

$$
\begin{aligned}
\mathscr{W}_\psi\big[g\big](a', \mathbf{b}', \theta') &= \int_{\mathbb{R}^2} g(\mathbf{t}) \overline{\psi_{a', \mathbf{b}', \theta'}(\mathbf{t})}\, d\mathbf{t} \\[6pt]
&= \int_{\mathbb{R}^2} \left\{ \frac{1}{C_\psi} \int_0^{2\pi} \int_{\mathbb{R}^2} \int_{\mathbb{R}^+} f(a, \mathbf{b}, \theta)\, \psi_{a, \mathbf{b}, \theta}(\mathbf{t})\, \frac{da\, d\mathbf{b}\, d\theta}{a^3} \right\} \overline{\psi_{a', \mathbf{b}', \theta'}(\mathbf{t})}\, d\mathbf{t} \\[6pt]
&= \frac{1}{C_\psi} \int_0^{2\pi} \int_{\mathbb{R}^2} \int_{\mathbb{R}^+} f(a, \mathbf{b}, \theta) \big\langle \psi_{a, \mathbf{b}, \theta},\ \psi_{a', \mathbf{b}', \theta'} \big\rangle_2 \frac{da\, d\mathbf{b}\, d\theta}{a^3} \\[6pt]
&= f(a', \mathbf{b}', \theta').
\end{aligned}
$$

This completes the proof of Theorem 3.4.12. $\square$

Corollary 3.4.13. *The projection from* $L^2\left(\mathbb{R}^+ \times \mathbb{R}^2 \times [0, 2\pi)\right)$ *onto the range of the two-dimensional wavelet transform* $\mathscr{W}_\psi\left(L^2(\mathbb{R}^2)\right)$ *is an integral operator whose reproducing kernel* $K_\psi\left(a, \mathbf{b}, \theta; a', \mathbf{b}', \theta'\right)$ *is given by*

$$K_\psi\left(a, \mathbf{b}, \theta; a', \mathbf{b}', \theta'\right) = C_\psi^{-1}\left\langle \psi_{a,\mathbf{b},\theta}, \, \psi_{a',\mathbf{b}',\theta'} \right\rangle_2. \tag{3.161}$$

Corollary 3.4.14. *The family of wavelets given by* (3.134) *generates a resolution of identity in the sense that if* I *denotes the identity operator on* $L^2(\mathbb{R}^2)$, *then*

$$C_\psi^{-1} \int_0^{2\pi} \int_{\mathbb{R}^2} \int_{\mathbb{R}^+} \left\langle \psi_{a,\mathbf{b},\theta}, \, \psi_{a,\mathbf{b},\theta} \right\rangle_2 \frac{da \, d\mathbf{b} \, d\theta}{a^3} = I. \tag{3.162}$$

Remark 3.4.7. From (3.161), it is evident that the reproducing kernel $K_\psi\left(a, \mathbf{b}, \theta; a', \mathbf{b}', \theta'\right)$ is simply obtained by computing the wavelet transform of the wavelet itself; that is, the autocorrelation function of the wavelet. Therefore, the reproducing kernel leads to the notion of "correlation length," which determines the region of influence of a given wavelet in the parameter space determined by $a \in \mathbb{R}^+, \mathbf{b} \in \mathbb{R}^2$ and $\theta \in [0, 2\pi)$. As such, it plays an important role in the "wavelet calibration," which deals with the determination of the capabilities of a given wavelet and evaluating the performances of the two-dimensional or polar wavelet transform. For a detailed information on wavelet calibration, we refer the interested reader to the monograph [20].

3.4.3 Different classes of two-dimensional wavelets

"You play with the width of the wavelet in order to catch the rhythm of the signal."

-Yves Meyer

After a sound mathematical foundation of the two-dimensional wavelet transform, our next motive is to gain deeper insights into some of the characteristic features of two-dimensional wavelets. As it is well-known that choosing an appropriate wavelet is crucial in the applications of wavelet transforms to diverse aspects of science and engineering. For instance, in image processing one is often interested in using the wavelets which are better at detecting the oriented features of an image; that is, regions where the amplitude is regular along one direction and has a sharp variation along the perpendicular direction. Therefore, while analyzing the signals one is inclined to choose a wavelet which best suits the purpose. In what follows, we shall study three classes of two-dimensional wavelets, viz; isotropic, directional and multi-directional wavelets.

(A) Isotropic wavelets

The isotropic wavelets are suitable for performing the point-wise analysis; that is, when the directional features are either not present or are not relevant in the signal. Being rotationally invariant, the angular parameter θ has no influence on the wavelet. The most familiar example of the isotropic wavelets is the 2D-Mexican Hat wavelet, which is defined as the Laplacian of a Gaussian function

$$\psi(\mathbf{t}) = \left(2 - |\mathbf{t}|^2\right) e^{-\frac{1}{2}|\mathbf{t}|^2} = -\Delta \, e^{-\frac{1}{2}|\mathbf{t}|^2}. \tag{3.163}$$

For this reason, the Mexican hat wavelet (3.163) is often referred as the "LOG wavelet" (Laplacian-of-Gaussian). Moreover, the Fourier transform of (3.163) is given by

$$\hat{\psi}(\mathbf{w}) = |\mathbf{w}|^2 \, e^{-\frac{1}{2}|\mathbf{w}|^2}. \tag{3.164}$$

The Mexican hat wavelet (3.163) is real, rotation invariant with vanishing moments of order up to 1. The 2D-Mexican Hat wavelet (3.163) and the corresponding Fourier transform are depicted in Figure 3.10.

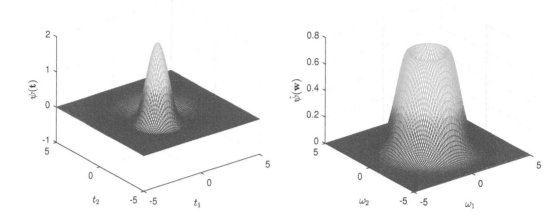

FIGURE 3.10: The 2D-Mexican hat wavelet and its Fourier transform.

As the Mexican wavelet is obtained by applying the first order Laplacian to a Gaussian, thus one may also use the higher order Laplacians to construct other wavelets as

$$\psi^{(n)}(\mathbf{t}) = (-\Delta)^n \, e^{-\frac{1}{2}|\mathbf{t}|^2}. \tag{3.165}$$

For increasing values of n, the wavelets (3.165) enjoy more and more vanishing moments and are, therefore, sensitive to increasingly sharper details. In certain practical applications, an additional parameter σ pertaining to the width of the Gaussian function is also incorporated into (3.163) as

$$\psi(\mathbf{t}) = -\Delta \, e^{-\frac{\sigma^2}{2}|\mathbf{t}|^2}. \tag{3.166}$$

The Fourier transform corresponding to (3.166) reads:

$$\hat{\psi}(\mathbf{w}) = \frac{|\mathbf{w}|^2}{\sigma^2} \, e^{-\frac{1}{2\sigma^2}|\mathbf{w}|^2}. \tag{3.167}$$

Although the parameter σ is redundant as the Gaussian can be dilated to an arbitrary width, however, it is handy for fixing the central frequency explicitly.

Besides the Mexican hat wavelet, many other isotropic wavelets have been constructed to meet the needs of signal analysts. For instance, the "Pet hat" wavelet is defined in the frequency space as follows:

$$\hat{\psi}(\mathbf{w}) = \begin{cases} \cos^2\left(\dfrac{\pi}{2}\log_2\dfrac{|\mathbf{w}|}{2\pi}\right), & \pi < |\mathbf{w}| < 4\pi \\ 0, & \text{otherwise.} \end{cases} \tag{3.168}$$

Compared to the Mexican hat wavelet, the Pet hat wavelet has a better resolving power in scale and is as such befitting for analyzing astrophysical images according to their characteristic scale. The Pet hat wavelet (3.168) is depicted in Figure 3.11.

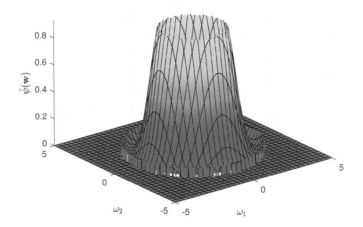

FIGURE 3.11: The Pet hat wavelet (3.168).

Yet another isotropic wavelet, which is also continuous is the "Halo wavelet" defined in the frequency domain as

$$\hat{\psi}(\mathbf{w}) = c\, e^{-(|\mathbf{w}|^2 - |\mathbf{w}_0|^2)}, \tag{3.169}$$

for some real parameter c. The Halo wavelet is real and selects the angular region $|\mathbf{w}| \cong |\mathbf{w}_0|$. For $c = 1$ and $c = 5$, the Halo wavelet (3.169) is shown in Figure 3.12.

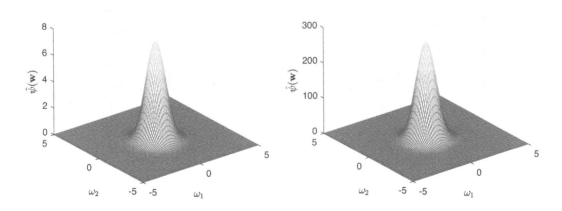

FIGURE 3.12: The Halo wavelet (3.171).

Finally, we shall present another important class of isotropic wavelets, formally called as the "difference wavelets." The difference wavelets are obtained by taking the difference of two positive functions in the following fashion:

$$\psi(\mathbf{t}) = \frac{1}{a_1}\big(\phi_{a_1, \mathbf{b}_1, \theta_1}(\mathbf{t})\big) - \frac{1}{a_2}\big(\phi_{a_2, \mathbf{b}_2, \theta_2}(\mathbf{t})\big), \tag{3.170}$$

where ϕ is a wavelet. In pursuit of the construction of isotropic wavelets via the procedure described in (3.170), the only possibility is to take the difference between a single isotropic function ϕ and a contracted version of the latter; that is, the particular case where only scale factors differ in (3.170). Indeed, if ϕ is smooth, non-negative, integrable and square integrable function whose all moments are of order 1, vanishing at the origin, the function $\psi(\mathbf{t})$ given by the relations

$$\psi(\mathbf{t}) = \alpha^{-2}\,\phi\big(\alpha^{-1}\mathbf{t}\big) - \phi(\mathbf{t}) \quad \text{and} \quad \hat{\psi}(\mathbf{w}) = \hat{\phi}(\alpha\mathbf{w}) - \hat{\phi}(\mathbf{w}), \qquad (3.171)$$

where $0 < \alpha < 1$, is easily seen to be a wavelet satisfying the condition (3.126). As ϕ is typically a smoothing function, the wavelet $\psi(\mathbf{t})$ is often called as "DOS wavelet" (Difference-of-Smoothings). In practice, one of the commonly used difference wavelet is the DOG wavelet (Difference-of-Gaussians) which is obtained by taking $\phi(\mathbf{t})$ to be an isotropic Gaussian:

$$\psi(\mathbf{t}) = \frac{1}{2\alpha^2}\,e^{-\frac{1}{2\alpha^2}|\mathbf{t}|^2} - e^{-\frac{1}{2}|\mathbf{t}|^2}, \quad 0 < \alpha < 1. \qquad (3.172)$$

The DOG wavelet (3.172) serves as a nice alternative to the Mexican hat wavelet and for $\alpha = 1.6$ their shapes are almost similar. For $\alpha = 0.2$ and $\alpha = 0.5$, the DOG wavelet given by (3.172) is depicted in Figure 3.13.

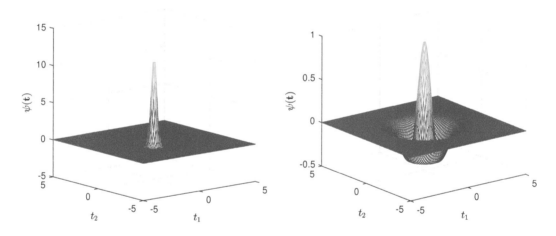

FIGURE 3.13: The DOG wavelet (3.172).

Before concluding the discourse on isotropic wavelets, it is pertinent to mention that the difference wavelets are of substantial importance for denoising applications, unsharp masking, automatic target recognition and so on [20].

(B) Directional wavelets

One of the key problems in multi-dimensional signal analysis is the detection of oriented features, such as segments, edges, vector field and so on. Thus, the isotropic wavelets are inept in such situations and one needs a directionally sensitive wavelet. One of the simplest approaches to instil directional sensitivity into a wavelet is to modify an isotropic wavelet, for instance the Maxican hat wavelet, by simply stretching it. Mathematically, the stretching is achieved by replacing $\mathbf{t}$ with $A\mathbf{t}$ in (3.163), where $A = \mathrm{diag}\big(\epsilon^{-1/2}, 1\big)$ and $\epsilon \geq 1$ is the anisotropy parameter. However, such wavelets are of little use in practice as they perform poorly and are shown to lack a needful directional sensitivity no matter how large the anisotropy parameter ϵ is chosen. Thus, we conclude that stretching an isotropic wavelet is not a feasible option to incorporate directional selectivity to a wavelet. The next step in the queue is to make use of the directional derivative. As a precursor, we note that the first derivative of the Gaussian function; that is,

$$\psi^{(1)}(\mathbf{t}) = \frac{d}{d\mathbf{t}} e^{-|\mathbf{t}|^2/\sigma^2} \tag{3.173}$$

is feasible for detecting the edges oriented in one direction, and it suffices to rotate it to get an edge detector that is sensitive to an arbitrary direction. In addition, many different values of the parameter σ can also be chosen which amounts to varying the scale.

Although the directional derivative wavelets are reasonably capable of performing directional filtering, however, they are by far not sufficient because their angular selectivity is rather poor. As such, there is a dire requirement of other directional wavelets, so it is imperative to formally characterize the notion of directional wavelets. In this regard, we have the following definition:

Definition 3.4.4. A wavelet $\psi(\mathbf{t})$ is said to be "directional" if the effective support of the corresponding Fourier transform $\hat{\psi}(\mathbf{w})$ is contained in a convex cone with apex at the origin. Moreover, if the support of $\hat{\psi}(\mathbf{w})$ is contained in a finite union of such disjoint cones, then $\psi(\mathbf{t})$ is called as "multi-directional."

Since Definition 3.4.4 may sound counter-intuitive, so it requires a word of justification. As a consequence Proposition 3.4.8, it is quite evident that the two-dimensional wavelet acts as a multiplicative filter in the Fourier domain. Suppose that a given signal $f(\mathbf{t}) = f(t_1, t_2)$ is highly oriented, for instance, a long segment along t_1-axis. Then, clearly the corresponding Fourier transform $\hat{f}(\mathbf{w}) = \hat{f}(\omega_1, \omega_2)$ will be a long segment along the ω_2-axis. For an efficient detection of such a signal, it is desirable to have a wavelet $\psi(\mathbf{t})$ which is supported in a narrow cone in the $\mathbf{w}$-space. Then, the wavelet transform is negligible unless $\hat{\psi}(\mathbf{w})$ is essentially aligned onto $\hat{f}(\mathbf{w})$, because the directional selectivity demands the restriction of the support of $\hat{\psi}(\mathbf{w})$ not $\psi(\mathbf{t})$. The corresponding standard practice in signal processing is to design an adequate filter in the frequency domain, for instance, high pass, low pass, band pass and so on. Nevertheless, in many applications, such as magnetic resonance imaging, the data is acquired in $\mathbf{w}$-space and the image space is obtained after a Fourier transform: here again directional filtering takes place in $\mathbf{w}$-space.

According to Definition 3.4.4, the non-isotropic Mexican hat wavelet is not directional, since the corresponding Fourier transform is supported at the origin irrespective of the anisotropic parameter ϵ. On the flip side, the Morlet wavelet could serve as a nice prototype of a directional wavelet. Mathematically, the two-dimensional Morlet wavelet is defined as

$$\psi_M(\mathbf{t}) = \exp\left(i\mathbf{w}_0 \cdot \mathbf{t}\right) \exp\left(-\frac{|A\mathbf{t}|^2}{2}\right) - \exp\left(-\frac{|A^{-1}\mathbf{w}_0|^2}{2}\right) \exp\left(-\frac{|A\mathbf{t}|^2}{2}\right). \tag{3.174}$$

Moreover, the Fourier transform corresponding to the two-dimensional Morlet wavelet (3.174) is given by

$$\hat{\psi}_M(\mathbf{w}) = \sqrt{\epsilon} \left\{ \exp\left(-\frac{|A^{-1}(\mathbf{w} - \mathbf{w}_0)|^2}{2}\right) - \exp\left(-\frac{|A^{-1}\mathbf{w}_0|^2}{2}\right) \exp\left(-\frac{|A^{-1}\mathbf{w}|^2}{2}\right) \right\}. \tag{3.175}$$

The parameter $\mathbf{w}_0$ involved in (3.174) and (3.175) is called as the "wave-vector" and $A = \text{diag}(\epsilon^{-1/2}, 1), \epsilon \geq 1$ is the 2×2 "anisotropy matrix." For $\epsilon = 2$ and $\mathbf{w}_0 = (1, 1)$, the real and imaginary parts of the two-dimensional Morlet wavelet (3.174) are shown in Figure 3.14, whereas the corresponding Fourier transform (3.175) is depicted in Figure 3.15.

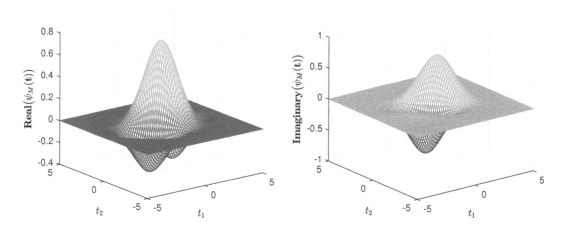

FIGURE 3.14: Real and imaginary parts of the two-dimensional Morlet wavelet.

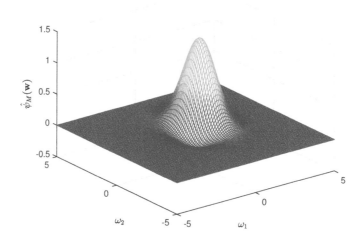

FIGURE 3.15: The Fourier transform of the two-dimensional Morlet wavelet.

Note that the subtracted terms on the R.H.S of (3.174) and (3.175) are actually the correction terms, which enforce the admissibility condition $\hat{\psi}_M(\mathbf{0}) = 0$. However, since the correction terms are numerically negligible for $|\mathbf{w}_0| \geq 5.6$, one may drop these terms altogether, whenever feasible. In that case, taking $\epsilon = 1$ yields the Gabor function, given by

$$\psi_G(\mathbf{t}) = \exp\left(i\mathbf{w}_0 \cdot \mathbf{t}\right) \exp\left(-\frac{|\mathbf{t}|^2}{2}\right). \tag{3.176}$$

Owing to its computational simplicity, the Gabor function (3.176) has become increasingly popular in the image processing literature. One of the distinguishing features of the Gabor function is that it is befitting for the modelling of human vision. This is because a large fraction of cells in the primary visual cortex of primates (including humans) have a receptive field that resembles a Gabor function. For $\mathbf{w}_0 = (6, 0)$, the real and imaginary parts of the Gabor function (3.176) are depicted in Figure 3.16.

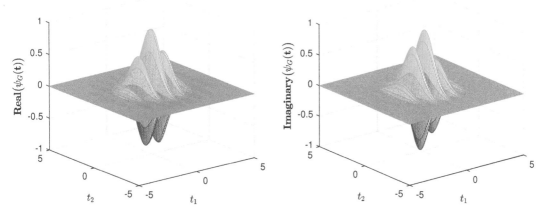

FIGURE 3.16: Real and imaginary parts of the Gabor function.

It is interesting to note that the Gabor function (3.176) has the qualitative behaviour expected from a wavelet as it is well localized in both the natural domain as well as in the frequency domain. However, strictly speaking, the Gabor function is not admissible. On the other hand, the function (3.174) is always admissible, but for small values of $|\mathbf{w}_0|$, it is not an efficient wavelet as the corresponding Fourier transform essentially consists of two disjoint pieces. Thus, we conclude that the Morlet wavelet (3.176) is a reliable wavelet for the case when $|\mathbf{w}_0|$ is sufficiently large. For $\epsilon = 2$ and $\mathbf{w}_0 = (0, 6)$, the real and imaginary parts of the truncated Morlet wavelet obtained by pulling out the correction term in (3.174) are eloquently shown in Figure 3.17, whereas the corresponding Fourier transform is displayed in Figure 3.18.

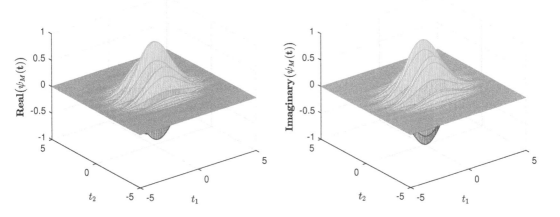

FIGURE 3.17: Real and imaginary parts of the truncated Morlet wavelet.

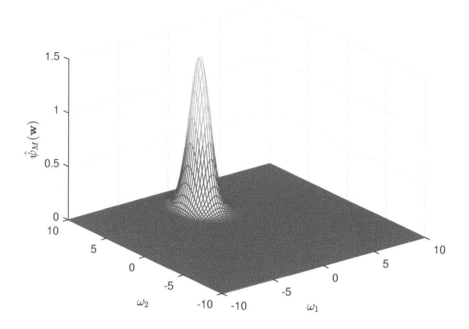

FIGURE 3.18: The Fourier transform of the truncated Morlet wavelet.

The Morlet wavelet is complex-valued and the modulus of the truncated wavelet obtained from (3.174) (ignoring the correction term) is a Gaussian elongated in the t_1-direction if $\epsilon > 1$ and its phase is constant along the direction orthogonal to the wave vector $\mathbf{w}_0$ and linear in $\mathbf{t}$, $\mathrm{mod}(2\pi/|\mathbf{w}_0|)$, along the direction of $\mathbf{w}_0$. Therefore, plotting the phase of the truncated wavelet as a function of $\mathbf{t}$, we obtain a succession of straight lines perpendicular to the wave vector $\mathbf{w}_0$, whose intensity varies periodically and linearly from 0 to 2π. In contrast to the usual one-dimensional case, the main feature of the Morlet wavelet lies in its inherent directional selectivity, entirely contained in its phase. As such, the Morlet wavelet is a promising tool for the detection of oriented features in higher dimensional signals. By virtue of the fact that the wavelet transform is essentially the convolution of the input signal with the dilated wavelet, it can be observed that the Morlet wavelet smooths the signal in all directions but detects the sharp transitions in the direction orthogonal to the wave vector $\mathbf{w}_0$.

Moreover, the effective support of $\hat{\psi}_M(\mathbf{w}) = \hat{\psi}_M(\omega_1, \omega_2)$ is an ellipse centred about $\mathbf{w}_0$ and elongated along the ω_2 direction, thus contained in a convex cone. Since the ratio of the axes of the ellipse is $\epsilon^{1/2}$, we infer that the cone tends to become narrower with increasing ϵ. Clearly, the Morlet wavelet $\psi_M(\mathbf{t}) = \psi_M(t_1, t_2)$ is befitting to detect singularities in the t_1-direction and its angular selectivity increases with $|\mathbf{w}_0|$ and with the anisotropic parameter ϵ. The optimal selectivity is achieved by choosing the wave vector $\mathbf{w}_0$ parallel to the major axis of the ellipse in the frequency plane; that is, $\mathbf{w}_0 = (0, 1)$. In that case, the truncated Morlet wavelet $\psi_M(\mathbf{t})$ becomes

$$\psi_G(\mathbf{t}) = \exp\left(it_2\right) \exp\left(-\frac{1}{2}\left(\frac{t_1^2}{\epsilon} + t_2^2\right)\right), \quad \mathbf{t} = (t_1, t_2). \tag{3.177}$$

For $\epsilon = 5$, the real and imaginary parts of the wavelet (3.177) are depicted in Figure 3.19.

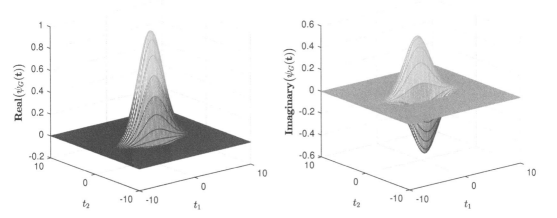

FIGURE 3.19: Real and imaginary parts of the truncated Morlet wavelet (3.177).

Nevertheless, to cater the needs of signal analysts, many variants of the basic wavelets are often designed. For instance, it is well known that the Mexican hat wavelet is good for detecting discontinuities, such as edges in images, but it lacks the sense of directionality. On the other hand, the Morlet wavelet performs well in directional selectivity but mostly selective in frequency. As such the novel wavelets may be defined by combing the respective merits of these wavelets. One of the prominent wavelets obtained from such a construction is the Gabor (or modulated) Mexican hat wavelet defined as

$$\psi_{GM}(\mathbf{t}) = \psi_{GM}(t_1, t_2)$$

$$= -\left(\epsilon \frac{\partial^2}{\partial t_1^2} + \frac{\partial^2}{\partial t_2^2} \right) \left[\exp\left(ik_0 t_2 \right) \exp\left\{ -\frac{1}{2} \left(\frac{t_1^2}{\epsilon} + t_2^2 \right) \right\} \right]$$

$$= \left(2 - \frac{t_1^2}{\epsilon} - \left(t_2 - ik_0 \right)^2 \right) \exp\left(ik_0 t_2 \right) \exp\left\{ -\frac{1}{2} \left(\frac{t_1^2}{\epsilon} + t_2^2 \right) \right\}, \quad \epsilon \geq 1, \quad (3.178)$$

where k_0 is given by choosing the wave vector $\mathbf{w}_0 = (0, k_0)$ in the truncated Morlet wavelet. It is worthwhile to note that no correction term is needed here to enforce the admissibility as the function $\psi_{GM}(\mathbf{t})$ is admissible in its own right. Moreover, the Fourier transform corresponding to (3.178) is given by

$$\hat{\psi}_{GM}(\mathbf{w}) = \hat{\psi}_{GM}(\omega_1, \omega_2)$$

$$= \sqrt{\epsilon} \left(\epsilon \omega_1^2 + \omega_2^2 \right) \exp\left\{ -\frac{1}{2} \left(\epsilon \omega_1^2 + \left(\omega_2 - k_0 \right)^2 \right) \right\}. \quad (3.179)$$

Choosing $\epsilon = 10$ and $k_0 = 1$, the real and imaginary parts of the modulated Mexican hat wavelet (3.178) are shown in Figure 3.20. Besides, the Fourier transform (3.179) corresponding to the modulated Mexican hat wavelet (3.178) is depicted in Figure 3.21. It is worth mentioning that the wavelet (3.178) does not require any correction term, because it is admissible in its own right. From the signal processing perspective, the distinguishing characteristics of the modulated Mexican hat wavelet (3.178) are that, it is efficient for detecting edges even in the presence of heavy noise and is also known for its applications in character recognition. For a detailed information regarding the notion of directional wavelets together

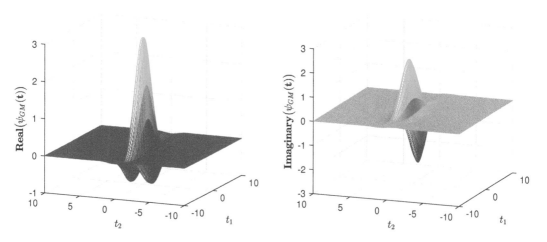

FIGURE 3.20: Real and imaginary parts of the modulated Mexican hat wavelet.

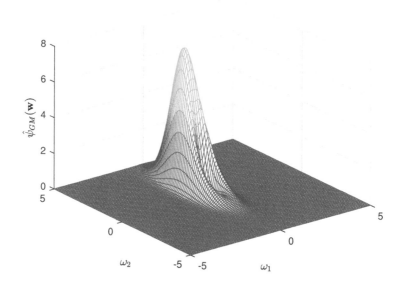

FIGURE 3.21: The Fourier transform of the modulated Mexican hat wavelet.

with their applications to different aspects of signal processing, the interested reader is referred to [19, 20, 22].

(C) Multi-directional wavelets

A multi-directional wavelet with n-fold symmetry is constructed by appropriately superimposing n-rotated copies of a directional wavelet $\psi(\mathbf{t})$. Mathematically,

$$\psi_n(\mathbf{t}) = \frac{1}{n}\sum_{\ell=0}^{n-1}\psi\left(R_{-\theta_\ell}\mathbf{t}\right), \quad \theta_\ell = \frac{2\pi\ell}{n}, \quad \text{where} \quad \ell =, 1, 2, \ldots, n-1. \tag{3.180}$$

As an example, choosing $\psi(\mathbf{t})$ as the Gabor (or truncated Morlet) wavelet and $n = 4$, we obtain the following real wavelet with 4-fold symmetry:

$$\psi_{4M}(\mathbf{t}) = \psi_{4M}(t_1, t_2) = \frac{(\cos k_0 t_1 + \cos k_0 t_2)}{2}\exp\left\{-\frac{(t_1^2 + t_2^2)}{2}\right\}. \tag{3.181}$$

For $k_0 = 1$, the wavelet (3.181) is shown in Figure 3.22. The wavelet (3.181) filters out all features which are not primarily horizontal or vertical. In analogy to this, one can obtain wavelets with symmetry 6 or 10, which may find applications in biological problems or the analysis of quasi-crystals, respectively. In general, the multi-directional wavelets are of substantial importance in pattern recognition.

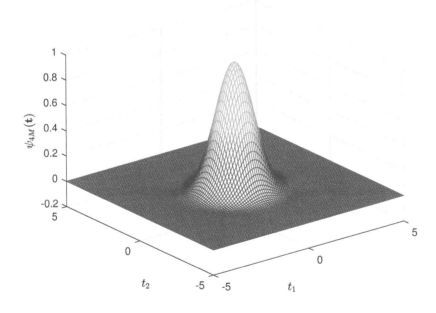

FIGURE 3.22: The multi-directional wavelet $\psi_{4M}(\mathbf{t})$.

In addition to the directional wavelets, many other directional representation systems have been introduced in the literature to capture the geometric features embedded in higher dimensional signals. Among a series of directional representation systems introduced in the open literature, the notable developments include ridgelets, curvelets, ripplets, shearlets and bendlets. Such geometrical representation systems give birth to new integral transforms, which shall be studied in the remaining portion of the ongoing Chapter 3.

3.4.4 Representation classes and partial energy densities of two-dimensional wavelet transform

"Mathematics is the most exact science, and its conclusions are capable of absolute proof. But this is so only because mathematics does not attempt to draw absolute conclusions. All mathematical truths are relative, conditional."

-Charles P. Steinmetz

In this subsection, our aim is to focus on different representations pertaining to the two-dimensional wavelet transform (3.139). Besides, it is also of considerable interest to look into the partial energy densities for certain representations of the two-dimensional wavelet transform.

(A) Representation classes

As a consequence of Definition 3.4.3, we observe that the two-dimensional or polar wavelet transform $\mathscr{W}_\psi[f](a, \mathbf{b}, \theta)$ can be visualized as four-parameter representation of a signal $f \in L^2(\mathbb{R}^2)$. As such, a simultaneous visualization of the two-dimensional wavelet transform in all the four variables is hardly possible. Therefore, in order to obtain a sound visualization of the two-dimensional wavelet transform some variables are either fixed or eliminated, so that one is restricted to a particular section of the parameter space determined by $a \in \mathbb{R}^+, \mathbf{b} \in \mathbb{R}^2$ and $\theta \in [0, 2\pi)$. This gives rise to several representations associated with the two-dimensional wavelet transform.

(i) *The position representation:* In case the scale and angular variables a and θ are fixed in (3.139); that is, $\mathscr{W}_\psi[f](a, \mathbf{b}, \theta)$ is considered as a function of the translation variable $\mathbf{b}$, the two-dimensional wavelet transform yields a position representation system $\mathscr{W}_\psi[f](\cdot, \mathbf{b}, \cdot)$ which amounts to taking a set of snap-shots one for each (a, θ). The position representation system is the standard representation which is widely used in image processing: Detection of position, shape and contours of objects, pattern recognition, image filtering and so on.

(ii) *The scale-angle representation:* When the translation variable $\mathbf{b}$ is fixed; that is, $\mathscr{W}_\psi[f](a, \mathbf{b}, \theta)$ is considered as a function of scale a and angle θ, the two-dimensional wavelet transform boils down to a scale-angle representation. The scale angle representation $\mathscr{W}_\psi[f](a, \cdot, \theta)$ serves as a keyhole about $\mathbf{b}$ from which all the scales and directions are observed once. The scale-angle representation is particularly fruitful in case the underlying wavelet is directionally selective. Such representations are important in fractional analysis, wherein the scaling-behaviour and angular selection play a key role.

The aforementioned representations of the two-dimensional wavelet transform are complementary in the sense that a complete information of $\mathscr{W}_\psi[f](a, \mathbf{b}, \theta)$ in all the four variables can be obtained by a joint understanding of both these representations. Besides, another class of representations is obtained by fixing any two members of the four parameters $(a, |\mathbf{b}|, \phi, \theta)$, where $\mathbf{b} = (|\mathbf{b}|, \phi)$ are the polar coordinates of $\mathbf{b}$ in which $|\mathbf{b}|$ and ϕ are interpreted as range and aspect/perception angle, respectively.

(iii) *The scale-perception-angle representation:* By fixing the range $|\mathbf{b}|$ and the angle θ, the two-dimensional wavelet transform defined in (3.139) yields a representation at all scales a and all perception angles ϕ. Such a representation is formally known as the "scale-perception-angle representation."

(iv) *The range-anisotropy-angle representation:* Such a representation is obtained by fixing the scale a and the perception angle ϕ, thus provides an analysis at all ranges $|\mathbf{b}|$ and all anisotropy angles θ.

(v) *The scale-range representation:* In case the perception angle ϕ and the anisotropy angle θ are fixed, then the two-dimensional wavelet transform defined in (3.139) provides an analysis at all scales a and all ranges $|\mathbf{b}|$. Such a representation is known as "scale-range representation."

(vi) *The perception-angle-anisotropy representation:* This representation is obtained by fixing the scale a and the range $|\mathbf{b}|$, thus providing an analysis of all perception angles along all directions. It is worth emphasizing that this representation is different from the aforementioned representations mainly due to the reason that the restricted parameter space is compact as it is a torus.

Yet another class of representations of the two-dimensional wavelet transform defined in (3.139) is obtained by fixing one variable and letting the other three to vary over their respective ranges. From the applications perspective, the following pair of representations are important:

(vii) *The position-scale representation:* In case the underlying wavelet is directionally insensitive, then the two-dimensional wavelet transform $\mathscr{W}_\psi[f](a, \mathbf{b}, \theta)$ is essentially a function of a and $\mathbf{b}$ alone. Such a representation is known as "position-scale representation." These representations are useful for the detection of coherent structures which survive through a whole range of scales. The position-scale representation systems are particularly useful in astrophysics, turbulence, fluid dynamics and so on.

(viii) *The position-anisotropy representation:* Such a representation of the two-dimensional wavelet transform (3.139) is obtained by fixing the scale a and viewing $\mathscr{W}_\psi[f](a, \mathbf{b}, \theta)$ as a function of translation variable $\mathbf{b}$ and anisotropy angle θ. If the anisotropy angle θ is plotted on the vertical axis of a three-dimensional graph, this directional representation system means that the plane $\theta = \theta_0$ selects all features that live in the corresponding line of sight. Similarly, an angular sector of opening $\Delta\theta$ is represented by a horizontal slice with thickness $\Delta\theta$. Such representation systems may offer distinct advantages over the conventional ones.

Apart from the above illustrated process of visualizing the two-dimensional wavelet transform by fixing some elements of the parameter space determined by $a \in \mathbb{R}^+, \mathbf{b} \in \mathbb{R}^2$ and $\theta \in [0, 2\pi)$, there is an alternative approach to do so by integrating out some variables using the part of the natural measure $da\, d\mathbf{b}\, d\theta/a^3$. This approach leads to the notion of partial energy densities as the integral of $\left|\mathscr{W}_\psi[f](a, \mathbf{b}, \theta)\right|^2$ over the entire parameter space under the natural measure $da\, d\mathbf{b}\, d\theta/a^3$ is interpreted as the total energy of the signal. This approach is fruitful in situations wherein the variables to be eliminated (typically scale a) take only discrete values. As the previously mentioned approach of fixing some variables make some sense only if the variables in question take arbitrary values in a continuous range. For instance, if the input signal has significant features ranging over the discrete set of scales $\{a_j : j \in \mathbb{Z}\}$, then the corresponding properties can be visualized only if we choose one of the scale values a_j. Otherwise, nothing will be seen, and the transform is useless. A typical example of such a situation is the problem of dilation symmetry in patterns. In such situations, clearly one should not fix the scale variable, but integrate over all the scale part da/a^3 of the natural measure $da\, db\, d\theta/a^3$. The same process can be repeated for the other variables or any combination of variables under consideration.

(B) Partial energy densities

Here, we examine the partial energy densities associated with certain representations of the two-dimensional wavelet transform defined in (3.139).

(i) *Position energy density:* The position energy density is obtained by integrating $\left|\mathscr{W}_\psi[f](a, \mathbf{b}, \theta)\right|^2$ over all scales and orientations and is given by

$$\int_0^{2\pi} \int_{\mathbb{R}^+} \left|\mathscr{W}_\psi[f](a, \mathbf{b}, \theta)\right|^2 \frac{da\, d\theta}{a^3}. \tag{3.182}$$

The position energy density (3.182) is mostly used for the detection and recognition of objects in forward-looking infra-red radar (FLIR) imagery.

(ii) *Scale-angle energy density:* The scale-angle energy density is obtained by integrating $\left|\mathscr{W}_\psi[f](a, \mathbf{b}, \theta)\right|^2$ over all the positions $\mathbf{b}$ and is given by

$$\int_{\mathbb{R}^2} \left|\mathscr{W}_\psi[f](a, \mathbf{b}, \theta)\right|^2 d\mathbf{b}. \tag{3.183}$$

The energy density (3.183) is also referred as "scale-angle measure" or the "scale-angle spectrum" of the signal f and gives the distribution of energy at different scales and directions. The scale-angle energy density is used for distinguishing a class of objects according to their size and orientation or more precisely in target classification in FLIR imagery.

Another variant of the scale-angle energy density is the "relative scale-angle spectrum" which is obtained by normalizing (3.183) over the range of θ; that is,

$$\frac{\int_{\mathbb{R}^2} \left|\mathscr{W}_\psi[f](a, \mathbf{b}, \theta)\right|^2 d\mathbf{b}}{\int_0^{2\pi} \left(\int_{\mathbb{R}^2} \left|\mathscr{W}_\psi[f](a, \mathbf{b}, \theta)\right|^2 d\mathbf{b}\right) d\theta}. \tag{3.184}$$

The relative scale-angle spectrum gives the relative distribution of the signal energy along different directions at a particular scale with respect to the total energy at that scale.

Besides the partial energy densities described above, one can define other partial energy densities corresponding to other choices of the variables which are irrelevant to a particular situation. However, the above defined energy densities are the most important ones and are often used in practice.

3.5   The Ridgelet Transform

"Wavelet transform has provided not only a wealth of new mathematical results, but also a common language and rallying call for researchers in a remarkably wide variety of fields: mathematicians working in harmonic analysis because of the special properties of wavelet bases; mathematical physicists because of the implications for time-frequency or phase-space analysis and relationships to concepts of renormalisation; digital signal processors because of connections with multirate filtering, quadrature mirror filters, and subband coding; image processors because of applications in pyramidal image representation and compression; researchers in computer vision who have used scale-space for some time; researchers in stochastic processes interested in self-similar processes, noise, and fractals; speech processors interested in efficient representation, event extraction and mimicking the human auditory system. And the list goes on."

-Ingrid Daubechies

In the pursuit to develop more efficient signal processing tools, the theory of wavelet transforms has witnessed a giant leap with the inception of novel, modified versions of the traditional two-dimensional wavelet transform. Some of the prominent ramifications include the ridgelet transform, curvelet transform, ripplet transform, shearlet transform, bendlet transform and so on [32, 59, 62, 78, 98, 103, 171, 176, 183]. In the remaining part of the ongoing chapter, we shall exclusively explore these novel geometric wavelet transforms in chronological order. Here, we initiate our endeavour by studying the ridgelet transform. In order to develop a sound mathematical theory of the ridgelet transform, it is imperative to take a survey of the well-known Radon transform.

3.5.1 Radon transform

"We live in an age in which mathematics plays a more and more important role, to the extent that it is hard to think of an aspect of human life to which it either has not provided, or does not have the potential to provide, crucial insights. Mathematics is the language in which quantitative models of the world around us are described. As subjects become more understood, they become more mathematical. A good example is medicine, where the Radon transform is what makes X-ray tomography work, where statistics form the basis of evaluating the success or failure of treatments, and where mathematical models of organs such as the heart, of tumor growth, and of nerve impulses are of key importance."

-John M. Ball

The Radon transform dates back to 1917 with the seminal work of Johann Radon $(1887 - 1956)$ to construct a bivariate function from its integrals over all straight lines in the plane [143]. Furthermore, Radon also generalized his work by reconstructing a bivariate function from its integrals over smooth curves. The Radon transform has found numerous applications across diverse aspects of science and engineering including medical imaging, astronomy, crystallography, electron microscopy, geophysics, material science and optics [32, 91, 93]. It is pertinent to mention that the Radon transform has also been heavily employed in computer assisted tomography, which deals with the problem of finding the internal structure of an object by observations or projections. Mathematically, the Radon transform of a bivariate function $f(\mathbf{x})$ is defined as

$$\Re\big[f\big](t,\theta) = \int_{\mathbb{R}^2} f(\mathbf{x})\,\delta(\Lambda_\theta \cdot \mathbf{x} - t)\,d\mathbf{x}, \qquad (3.185)$$

where $\mathbf{x} = (x_1, x_2) \in \mathbb{R}^2$, $t \in \mathbb{R}$ and $\theta \in [0, 2\pi)$. Also, $\delta(\cdot)$ denotes the well-known Dirac function and $\Lambda_\theta = (\cos\theta, \sin\theta)$ is the unit vector in the direction θ. In literature, the integral transform (3.185) is also referred as the "X-ray transform." The name derives from the fact that in medical imaging, the integral along a line represents a measurement of the intensity of the X-ray beam at the detector after passing through the object being radio-graphed.

In the following theorem, we assemble some fundamental properties of the Radon transform (3.185).

Theorem 3.5.1. *Let $\alpha, \beta \in \mathbb{C}$, $\mathbf{k} \in \mathbb{R}^2$, $c > 0$ and $\theta' \in [0, 2\pi)$. Then, for any pair of functions $f, g \in L^2(\mathbb{R}^2)$, the Radon transform (3.185) satisfies the following properties:*

(i) *Linearity:* $\mathfrak{R}\big[\alpha f + \beta g\big](t, \theta) = \alpha \, \mathfrak{R}\big[f\big](t, \theta) + \beta \, \mathfrak{R}\big[g\big](t, \theta),$

(ii) *Translation:* $\mathfrak{R}\big[\mathcal{T}_{\mathbf{k}} f\big](t, \theta) = \mathfrak{R}\big[f\big](t - \Lambda_\theta \cdot \mathbf{k}, \theta),$

(iii) *Planer dilation:* $\mathfrak{R}\big[\mathcal{D}_c f\big](t, \theta) = \dfrac{1}{c} \mathfrak{R}\big[f\big]\left(\dfrac{t}{c}, \theta\right),$

(iv) *Parity:* $\mathfrak{R}\big[Pf\big](t, \theta) = -\mathfrak{R}\big[f\big](t, \theta),$ *where* $Pf(\mathbf{x}) = f(-\mathbf{x}),$

(v) *Rotation:* $\mathfrak{R}\big[R_{\theta'} f\big](t, \theta) = \mathfrak{R}\big[f\big](t, \theta - \theta').$

Proof. **(i)** By virtue of (3.185), we observe that

$$\mathfrak{R}\big[\alpha f + \beta g\big](t, \theta) = \int_{\mathbb{R}^2} \big(\alpha \, f(\mathbf{x}) + \beta \, g(\mathbf{x})\big) \delta(\Lambda_\theta \cdot \mathbf{x} - t) \, d\mathbf{x}$$

$$= \alpha \int_{\mathbb{R}^2} f(\mathbf{x}) \, \delta(\Lambda_\theta \cdot \mathbf{x} - t) \, d\mathbf{x} + \beta \int_{\mathbb{R}^2} g(\mathbf{x}) \, \delta(\Lambda_\theta \cdot \mathbf{x} - t) \, d\mathbf{x}$$

$$= \alpha \, \mathfrak{R}\big[f\big](t, \theta) + \beta \, \mathfrak{R}\big[g\big](t, \theta).$$

(ii) For $\mathbf{k} \in \mathbb{R}^2$, we have

$$\mathfrak{R}\big[\mathcal{T}_{\mathbf{k}} f\big](t, \theta) = \int_{\mathbb{R}^2} f(\mathbf{x} - \mathbf{k}) \, \delta(\Lambda_\theta \cdot \mathbf{x} - t) \, d\mathbf{x}$$

$$= \int_{\mathbb{R}^2} f(\mathbf{z}) \, \delta\big(\Lambda_\theta \cdot (\mathbf{z} + \mathbf{k}) - t\big) \, d\mathbf{z}$$

$$= \int_{\mathbb{R}^2} f(\mathbf{z}) \, \delta\big(\Lambda_\theta \cdot \mathbf{z} - (t - \Lambda_\theta \cdot \mathbf{k})\big) \, d\mathbf{z}$$

$$= \mathfrak{R}\big[f\big](t - \Lambda_\theta \cdot \mathbf{k}, \theta).$$

(iii) For $c > 0$, applying the dilation operator $\mathcal{D}_c$ to the input signal, we have

$$\mathfrak{R}\big[\mathcal{D}_c f\big](t, \theta) = \frac{1}{c} \int_{\mathbb{R}^2} f\left(\frac{\mathbf{x}}{c}\right) \delta(\Lambda_\theta \cdot \mathbf{x} - t) \, d\mathbf{x}$$

$$= \int_{\mathbb{R}^2} f(\mathbf{y}) \, \delta(c \, \Lambda_\theta \cdot \mathbf{y} - t) \, d\mathbf{y}$$

$$= \int_{\mathbb{R}^2} f(\mathbf{y}) \, \delta\left(c \left(\Lambda_\theta \cdot \mathbf{y} - \frac{t}{c}\right)\right) \, d\mathbf{y}$$

$$= \frac{1}{c} \int_{\mathbb{R}^2} f(\mathbf{y}) \, \delta\left(\Lambda_\theta \cdot \mathbf{y} - \frac{t}{c}\right) \, d\mathbf{y}$$

$$= \frac{1}{c} \mathfrak{R}\big[f\big]\left(\frac{t}{c}, \theta\right).$$

(iv) To observe the behaviour of the Radon transform under the reflection of the input signal, we proceed as

$$\Re\big[Pf\big](t,\theta) = \int_{\mathbb{R}^2} f(-\mathbf{x})\,\delta(\Lambda_\theta \cdot \mathbf{x} - t)\,d\mathbf{x}$$

$$= \int_{\mathbb{R}^2} f(\mathbf{y})\,\delta\big(\Lambda_\theta \cdot (-\mathbf{y}) - t\big)\,d\mathbf{y}$$

$$= \int_{\mathbb{R}^2} f(\mathbf{y})\,\delta\big(-(\Lambda_\theta \cdot \mathbf{y} + t)\big)\,d\mathbf{y}$$

$$= -\int_{\mathbb{R}^2} f(\mathbf{y})\,\delta(\Lambda_\theta \cdot \mathbf{y} + t)\,d\mathbf{y}$$

$$= -\Re\big[f\big](t,\theta).$$

(v) For $\theta' \in [0, 2\pi)$, we have

$$\Re\big[R_{\theta'} f\big](t,\theta) = \int_{\mathbb{R}^2} f\big(R_{-\theta'}\mathbf{x}\big)\,\delta(\Lambda_\theta \cdot \mathbf{x} - t)\,d\mathbf{x}$$

$$= \int_{\mathbb{R}^2} f(\mathbf{y})\,\delta\big(\Lambda_\theta \cdot (R_{\theta'}\mathbf{y}) - t\big)\,d\mathbf{y}$$

$$= \int_{\mathbb{R}^2} f(\mathbf{y})\,\delta(\Lambda_{\theta-\theta'} \cdot \mathbf{y} - t)\,d\mathbf{y}$$

$$= \Re\big[f\big](t,\theta - \theta').$$

This completes the proof of Theorem 3.5.1. □

The Radon transform (3.185) maps a bivariate function f to another unique function $\Re[f](t,\theta)$, whose domain is the set of all lines in $\mathbb{R}^2$. From a practical point of view, the major problem is to reconstruct the function f from the transformed function $\Re[f](t,\theta)$. This is called as the "reconstruction (inverse) problem" and has a definite solution based on the general mathematical theory. However, we shall demonstrate that the two-dimensional wavelet transform (3.139) can also be invoked to obtain a reconstruction formula for the Radon transform defined in (3.185). To illustrate this fact, consider the analyzing wavelet

$$\psi(\mathbf{t}) = \delta(\mathbf{u} \cdot \mathbf{t}), \tag{3.186}$$

which is a Dirac distribution along the line $\mathbf{u} \cdot \mathbf{t}$, that is perpendicular to the unit vector $\mathbf{u}$. The wavelet (3.186) is very singular, but if we choose an appropriate synthesis wavelet $\phi(\mathbf{t})$, then the cross admissibility condition:

$$C_{\psi,\phi} = (2\pi)^2 \int_{\mathbb{R}^2} \frac{\overline{\hat{\psi}(\mathbf{w})}\,\hat{\phi}(\mathbf{w})}{|\mathbf{w}|^2} < \infty, \tag{3.187}$$

allows the use of such entities. Upon applying the usual translation operator $\mathcal{T}_\mathbf{b}$, $\mathbf{b} \in \mathbb{R}^2$, dilation operator $\mathcal{D}_a$, $a \in \mathbb{R}^+$ and the clockwise rotation operator $R_{-\theta}$, $\theta \in [0, 2\pi)$ on the mother wavelet $\psi(\mathbf{t})$ given by (3.186), we obtain a family of daughter wavelets $\psi_{a,\mathbf{b},\theta}(\mathbf{t})$ as

$$\psi_{a,\mathbf{b},\theta}(\mathbf{t}) = R_{-\theta}\mathcal{D}_a\mathcal{T}_\mathbf{b}\,\psi(\mathbf{t})$$

$$= R_{-\theta}\mathcal{D}_a\mathcal{T}_\mathbf{b}\big(\delta(\mathbf{u} \cdot \mathbf{t})\big)$$

$$= R_{-\theta}\mathcal{D}_a\big(\delta(\mathbf{u} \cdot (\mathbf{t} - \mathbf{b}))\big)$$

$$= \frac{1}{a} \delta \left(\frac{R_\theta \mathbf{u} \cdot (\mathbf{t} - \mathbf{b})}{a} \right)$$

$$= a \, \delta \left(R_\theta \mathbf{u} \cdot \mathbf{t} - R_\theta \mathbf{u} \cdot \mathbf{b} \right). \tag{3.188}$$

The reason that we have employed the clockwise rotation operator $R_{-\theta}$ in (3.188) instead of the usual anti-clockwise rotation operator R_θ is to infuse the angular compatibity between the two-dimensional wavelet transform and the Radon transform. Therefore, by virtue of (3.188), we can express the two-dimensional wavelet transform as

$$\mathscr{W}_\psi \big[f \big] (a, \mathbf{b}, \theta) = \int_{\mathbb{R}^2} f(\mathbf{t}) \, \overline{\psi_{a,\mathbf{b},\theta}(\mathbf{t})} \, d\mathbf{t}$$

$$= a \int_{\mathbb{R}^2} f(\mathbf{t}) \, \delta \left(R_\theta \mathbf{u} \cdot \mathbf{t} - R_\theta \mathbf{u} \cdot \mathbf{b} \right) d\mathbf{t}$$

$$= a \, \mathfrak{R} \big[f \big] (R_\theta \mathbf{u} \cdot \mathbf{b}, \theta), \tag{3.189}$$

which is the desired relationship between the two-dimensional wavelet transform given by (3.139) and the Radon transform. Exploiting the inversion formula (3.150) for the two-dimensional wavelet transform, we can obtain a reconstruction formula for the Radon transform (3.185) as follows:

$$f(\mathbf{t}) = \frac{1}{C_{\psi,\phi}} \int_0^{2\pi} \int_{\mathbb{R}^2} \int_{\mathbb{R}^+} \mathfrak{R} \big[f \big] (R_\theta \mathbf{u} \cdot \mathbf{b}, \theta) \, \phi_{a,\mathbf{b},\theta}(\mathbf{t}) \, \frac{da \, d\mathbf{b} \, d\theta}{a^2}, \tag{3.190}$$

where the convergence holds almost everywhere. Thus, by choosing an appropriate synthesis wavelet ϕ we can revert the original function f from the corresponding Radon transform. Nevertheless, Holschreider [150] demonstrated that the inversion formula (3.190) holds strongly in $L^2(\mathbb{R}^2)$ provided the synthesis wavelet ϕ satisfies the following condition:

$$\int_{\mathbb{R}} \frac{\hat{\phi}(\xi \mathbf{u})}{\xi^2} \, d\xi < \infty. \tag{3.191}$$

3.5.2   Ridgelet transform: Definition and basic properties

The ridgelet transform deals with an efficient capturing/resolving of the line singularities in two-dimensional signals. A ridgelet behaves like a one-dimensional wavelet in a specific direction represented by the unit vector Λ_θ of orientation θ and remains constant along the orthogonal direction. A ridgelet can be translated along its oscillating direction and can also be scaled to obtain a family of ridgelets. Mathematically, let $\psi \in L^2(\mathbb{R})$ be a classical one-dimensional wavelet, then for $a \in \mathbb{R}^+$, $b \in \mathbb{R}$ and $\theta \in [0, 2\pi)$, the family of ridgelets is denoted by $\mathfrak{F}_\psi(a, b, \theta)$ and is is defined as

$$\mathfrak{F}_\psi(a, b, \theta) := \left\{ \psi_{a,b,\theta}(\mathbf{t}) = a^{-1/2} \, \psi \left(\frac{\Lambda_\theta \cdot \mathbf{t} - b}{a} \right); \ a \in \mathbb{R}^+, b \in \mathbb{R}, \theta \in [0, 2\pi) \right\}, \tag{3.192}$$

where Λ_θ denotes the unit-vector of orientation θ. From (3.192), it is quite evident that ridgelets can be roughly visualized as two-dimensional wavelets with the major distinction of being constant along a preferred direction and are thus never admissible with regard to the usual definition of admissibility. However, we have a slightly modified definition of the notion of an admissible ridgelet as given below:

Definition 3.5.1. Any function $\psi \in L^2(\mathbb{R})$ satisfying the condition:

$$C_\psi = (2\pi)^2 \int_{\mathbb{R}} \frac{\left| \hat{\psi}(\omega) \right|^2}{|\omega|^2} \, d\omega < \infty, \tag{3.193}$$

is said to be an "admissible ridgelet."

With the notion of ridgelets at hand, particularly admissible ridgelets, we are ready to introduce the formal definition of the ridgelet transform.

Definition 3.5.2. For any $f \in L^2(\mathbb{R}^2)$, the ridgelet transform is denoted by $R_\psi[f]$ and is defined as

$$R_\psi[f](a,b,\theta) = \left\langle f, \psi_{a,b,\theta} \right\rangle_2 = \int_{\mathbb{R}^2} f(\mathbf{t}) \, \overline{\psi_{a,b,\theta}(\mathbf{t})} \, d\mathbf{t}, \qquad (3.194)$$

where $\psi_{a,b,\theta}(\mathbf{t})$ is given by (3.192).

Remark 3.5.1. Unlike the usual two-dimensional wavelet transform, the ridgelet transform defined in (3.194) is determined by a total of three parameters, viz; $a \in \mathbb{R}^+$, $b \in \mathbb{R}$ and $\theta \in [0, 2\pi)$. This is justified by the fact that the ridgelet behaves like a one-dimensional wavelet in a specific direction and is constant along the orthogonal direction.

One of the distinguishing features of the ridgelet transform (3.194) is that it intertwines the well-known Radon and wavelet transforms. Next, our objective is to demonstrate the coupling between the Radon and ridgelet transforms. Prior to that, we study some fundamental properties of the ridgelet transform.

Theorem 3.5.2. *Let* $\alpha, \beta \in \mathbb{C}$, $\mathbf{k} \in \mathbb{R}^2$, $k \in \mathbb{R}$, $c > 0$ *and* $\theta' \in [0, 2\pi)$. *Then, for any pair of functions* $f, g \in L^2(\mathbb{R}^2)$, *the ridgelet transform* (3.194) *satisfies the following properties:*

(i) *Linearity:* $R_\psi[\alpha f + \beta g](a,b,\theta) = \alpha R_\psi[f](a,b,\theta) + \beta R_\psi[g](a,b,\theta)$,

(ii) *Translation:* $R_\psi[T_{\mathbf{k}} f](a,b,\theta) = R_\psi[f](a, b - \Lambda_\theta \cdot \mathbf{k}, \theta)$,

(iii) *Planer dilation:* $R_\psi[D_c f](a,b,\theta) = \dfrac{1}{\sqrt{c}} R_\psi[f]\left(\dfrac{a}{c}, \dfrac{b}{c}, \theta\right)$,

(iv) *Parity:* $R_{P\psi}[Pf](a,b,\theta) = R_\psi[f](a, -b, \theta)$, *where* $Pf(\mathbf{t}) = f(-\mathbf{t})$,

(v) *Rotation:* $R_\psi[R_{\theta'} f](a,b,\theta) = R_\psi[f](a, b, \theta - \theta')$,

(vi) *Anti-linearity:* $R_{\alpha\psi + \beta\phi}[f](a,b,\theta) = \overline{\alpha} R_\psi[f](a,b,\theta) + \overline{\beta} R_\phi[f](a,b,\theta)$,

(vii) *Translation in ridgelet:* $R_{T_k \psi}[f](a,b,\theta) = R_\psi[f](a, b + ak, \theta)$,

(viii) *Dilation in ridgelet:* $R_{D_c \psi}[f](a,b,\theta) = R_\psi[f](ac, b, \theta)$.

Proof. (i) Given any pair of functions $f, g \in L^2(\mathbb{R}^2)$, we observe that

$$R_\psi[\alpha f + \beta g](a,b,\theta) = \left\langle \alpha f + \beta g, \psi_{a,b,\theta} \right\rangle_2$$
$$= \alpha \left\langle f, \psi_{a,b,\theta} \right\rangle_2 + \beta \left\langle g, \psi_{a,b,\theta} \right\rangle_2$$
$$= \alpha R_\psi[f](a,b,\theta) + \beta R_\psi[g](a,b,\theta).$$

(ii) For $\mathbf{k} \in \mathbb{R}^2$, we have

$$R_\psi\left[\mathcal{T}_\mathbf{k} f\right](a,b,\theta) = \frac{1}{\sqrt{a}} \int_{\mathbb{R}^2} f(\mathbf{t} - \mathbf{k})\, \overline{\psi\left(\frac{\Lambda_\theta \cdot \mathbf{t} - b}{a}\right)}\, dt$$

$$= \frac{1}{\sqrt{a}} \int_{\mathbb{R}^2} f(\mathbf{z})\, \overline{\psi\left(\frac{\Lambda_\theta \cdot (\mathbf{z} + \mathbf{k}) - b}{a}\right)}\, dt$$

$$= \frac{1}{\sqrt{a}} \int_{\mathbb{R}^2} f(\mathbf{z})\, \overline{\psi\left(\frac{\Lambda_\theta \cdot \mathbf{z} - (b - \Lambda_\theta \cdot \mathbf{k})}{a}\right)}\, dt$$

$$= R_\psi\left[f\right](a, b - \Lambda_\theta \cdot \mathbf{k}, \theta).$$

(iii) To observe the behaviour of ridgelet transform under the planer-dilated input, we have

$$R_\psi\left[\mathcal{D}_c f\right](a,b,\theta) = \frac{1}{c\sqrt{a}} \int_{\mathbb{R}^2} f\left(\frac{\mathbf{t}}{c}\right) \overline{\psi\left(\frac{\Lambda_\theta \cdot \mathbf{t} - b}{a}\right)}\, dt$$

$$= \frac{1}{\sqrt{a}} \int_{\mathbb{R}^2} f(\mathbf{y})\, \overline{\psi\left(\frac{c\,\Lambda_\theta \cdot \mathbf{y} - b}{a}\right)}\, dy$$

$$= \frac{1}{\sqrt{(a/c)c}} \int_{\mathbb{R}^2} f(\mathbf{y})\, \overline{\psi\left(\frac{\Lambda_\theta \cdot \mathbf{y} - b/c}{a/c}\right)}\, dy$$

$$= \frac{1}{\sqrt{c}}\, R_\psi\left[f\right]\left(\frac{a}{c}, \frac{b}{c}, \theta\right).$$

(iv) We observe that,

$$R_{P\psi}\left[Pf\right](a,b,\theta) = \frac{1}{\sqrt{a}} \int_{\mathbb{R}^2} f(-\mathbf{t})\, \overline{\psi\left(-\frac{\Lambda_\theta \cdot \mathbf{t} - b}{a}\right)}\, dt$$

$$= \frac{1}{\sqrt{a}} \int_{\mathbb{R}^2} f(\mathbf{y})\, \overline{\psi\left(-\frac{\Lambda_\theta \cdot (-\mathbf{y}) - b}{a}\right)}\, dy$$

$$= \frac{1}{\sqrt{a}} \int_{\mathbb{R}^2} f(\mathbf{y})\, \overline{\psi\left(\frac{\Lambda_\theta \cdot \mathbf{y} + b}{a}\right)}\, dy$$

$$= R_\psi\left[f\right](a, -b, \theta).$$

(v) For $\theta' \in [0, 2\pi)$, we have

$$R_\psi\left[R_{\theta'} f\right](a,b,\theta) = \frac{1}{\sqrt{a}} \int_{\mathbb{R}^2} f(R_{-\theta'}\mathbf{t})\, \overline{\psi\left(\frac{\Lambda_\theta \cdot \mathbf{t} - b}{a}\right)}\, dt$$

$$= \frac{1}{\sqrt{a}} \int_{\mathbb{R}^2} f(\mathbf{y})\, \overline{\psi\left(\frac{\Lambda_\theta \cdot (R_{\theta'}\mathbf{y}) - b}{a}\right)}\, dy$$

$$= \frac{1}{\sqrt{a}} \int_{\mathbb{R}^2} f(\mathbf{y})\, \overline{\psi\left(\frac{\Lambda_{\theta-\theta'} \cdot \mathbf{y} - b}{a}\right)}\, dy$$

$$= R_\psi\left[f\right](a, b, \theta - \theta').$$

(vi) Since the inner product is anti-linear in the second component, therefore, for any $f \in L^2(\mathbb{R}^2)$ and a given pair of wavelets $\psi, \phi \in L^2(\mathbb{R})$, we have

$$R_{\alpha\psi+\beta\phi}\left[f\right](a,b,\theta) = \Big\langle f,\, \alpha\, \psi_{a,b,\theta} + \beta\, \phi_{a,b,\theta} \Big\rangle_2$$

$$= \overline{\alpha} \left\langle f, \psi_{a,b,\theta} \right\rangle_2 + \overline{\beta} \left\langle f, \phi_{a,b,\theta} \right\rangle_2$$

$$= \overline{\alpha} \, R_\psi \big[f\big](a,b,\theta) + \overline{\beta} \, R_\phi \big[f\big](a,b,\theta).$$

(vii) For $k \in \mathbb{R}$, we note that

$$R_{\mathcal{T}_k \psi} \big[f\big](a,b,\theta) = \frac{1}{\sqrt{a}} \int_{\mathbb{R}^2} f(\mathbf{t}) \, \overline{\psi\left(\frac{\Lambda_\theta \cdot \mathbf{t} - b}{a} - k\right)} \, d\mathbf{t}$$

$$= \frac{1}{\sqrt{a}} \int_{\mathbb{R}^2} f(\mathbf{t}) \, \overline{\psi\left(\frac{\Lambda_\theta \cdot \mathbf{t} - (b + ak)}{a}\right)} \, d\mathbf{t}$$

$$= R_\psi \big[f\big](a, b + ak, \theta).$$

(viii) For a given wavelet ψ, taking $\mathcal{D}_c \psi$, $c > 0$ as the generating function yields:

$$R_{\mathcal{D}_c \psi} \big[f\big](a,b,\theta) = \frac{1}{\sqrt{ac}} \int_{\mathbb{R}^2} f(\mathbf{t}) \, \overline{\psi\left(\frac{\Lambda_\theta \cdot \mathbf{t} - b}{ac}\right)} \, d\mathbf{t}$$

$$= R_\psi \big[f\big](ac, b, \theta).$$

This completes the proof of Theorem 3.5.2. $\qquad\qquad\square$

As the next endeavour, we intend to show that the wavelet transform bridges the gap between the Radon and ridgelet transforms. More precisely, we shall uncover the fundamental fact that the ridgelet transform (3.194) can be expressed as the wavelet transform of the Radon transform (3.185), when considered as a function of $t \in \mathbb{R}$. We note that,

$$\mathscr{W}_\psi \Big[\mathfrak{R}\big[f\big](t,\theta)\Big](a,b) = \frac{1}{\sqrt{a}} \int_{\mathbb{R}} \mathfrak{R}\big[f\big](t,\theta) \, \overline{\psi\left(\frac{t - b}{a}\right)} \, dt$$

$$= \frac{1}{\sqrt{a}} \int_{\mathbb{R}} \left\{ \int_{\mathbb{R}^2} f(\mathbf{x}) \, \delta(\Lambda_\theta \cdot \mathbf{x} - t) \, d\mathbf{x} \right\} \overline{\psi\left(\frac{t - b}{a}\right)} \, dt$$

$$= \frac{1}{\sqrt{a}} \int_{\mathbb{R}} \int_{\mathbb{R}^2} f(\mathbf{x}) \, \delta(\Lambda_\theta \cdot \mathbf{x} - t) \, \overline{\psi\left(\frac{t - b}{a}\right)} \, d\mathbf{x} \, dt$$

$$= \frac{1}{\sqrt{a}} \int_{\mathbb{R}^2} f(\mathbf{x}) \, \overline{\psi\left(\frac{\Lambda_\theta \cdot \mathbf{x} - b}{a}\right)} \, d\mathbf{x}$$

$$= R_\psi \big[f\big](a,b,\theta), \tag{3.195}$$

which is the desired relationship shared by the wavelet, Radon and ridgelet transforms. It is interesting to note that in case the given bivariate function has singularity along a single line, then the Radon coefficients at this angle will be constant. As such, because of the vanishing moments of the wavelet ψ, the continuous ridgelet coefficients $R_\psi \big[f\big](a,b,\theta)$ will also vanish for this direction.

In the following proposition, we shall obtain an important relationship between the Fourier transform and the ridgelet transform.

Proposition 3.5.3. *For any $f \in L^2(\mathbb{R}^2)$, the ridgelet transform defined in (3.194) can be expressed via the Fourier transform as follows:*

$$R_\psi \big[f\big](a,b,\theta) = \sqrt{2\pi a} \int_{\mathbb{R}} \hat{f}(\omega \Lambda_\theta) \, \overline{\hat{\psi}(a\omega)} \, e^{ib\omega} \, d\omega. \tag{3.196}$$

Proof. Invoking the fundamental relationship between the ridgelet and Radon transforms obtained in (3.195), we have

$$R_\psi\big[f\big](a,b,\theta) = \mathscr{W}_\psi\big[\mathfrak{R}\big[f\big](t,\theta)\big](a,b)$$

$$= \frac{1}{\sqrt{2\pi}} \int_\mathbb{R} \sqrt{a}\, \mathscr{F}\big[\mathfrak{R}\big[f\big](t,\theta)\big](\omega)\, \overline{\hat{\psi}(a\omega)}\, e^{ib\omega}\, d\omega. \qquad (3.197)$$

We observe that

$$\mathscr{F}\Big(\mathfrak{R}\big[f\big](t,\theta)\Big)(\omega) = \frac{1}{\sqrt{2\pi}} \int_\mathbb{R} \mathfrak{R}\big[f\big](t,\theta)\, e^{-i\omega t}\, dt$$

$$= \frac{1}{\sqrt{2\pi}} \int_\mathbb{R} \left\{ \int_{\mathbb{R}^2} f(\mathbf{x})\, \delta(\Lambda_\theta \cdot \mathbf{x} - t)\, d\mathbf{x} \right\} e^{-i\omega t}\, dt$$

$$= \frac{1}{\sqrt{2\pi}} \int_{\mathbb{R}^2} f(\mathbf{x}) \left\{ \int_\mathbb{R} \delta(\Lambda_\theta \cdot \mathbf{x} - t)\, e^{-i\omega t}\, dt \right\} d\mathbf{x}. \qquad (3.198)$$

Making use of the substitution $\Lambda_\theta \cdot \mathbf{x} - t = z$ into the second integral on the R.H.S of equation (3.198), we obtain

$$\mathscr{F}\Big(\mathfrak{R}\big[f\big](t,\theta)\Big)(\omega) = \frac{1}{\sqrt{2\pi}} \int_{\mathbb{R}^2} f(\mathbf{x}) \left\{ \int_\mathbb{R} \delta(z)\, e^{-i\omega(\Lambda_\theta \cdot \mathbf{x} - z)}\, dz \right\} d\mathbf{x}$$

$$= \frac{1}{\sqrt{2\pi}} \int_{\mathbb{R}^2} f(\mathbf{x}) \left\{ \int_\mathbb{R} \delta(z)\, e^{i\omega z}\, dz \right\} e^{-i\omega \Lambda_\theta \cdot \mathbf{x}}\, d\mathbf{x}$$

$$= \int_{\mathbb{R}^2} f(\mathbf{x})\, e^{-i\omega \Lambda_\theta \cdot \mathbf{x}}\, d\mathbf{x}$$

$$= 2\pi\, \hat{f}(\omega\Lambda_\theta). \qquad (3.199)$$

Substituting (3.199) in (3.197), we obtain

$$R_\psi\big[f\big](a,b,\theta) = \sqrt{2\pi a} \int_\mathbb{R} \hat{f}(\omega\Lambda_\theta)\, \overline{\hat{\psi}(a\omega)}\, e^{ib\omega}\, d\omega.$$

This completes the proof of Proposition 3.5.3   □

Next, our motive is to demonstrate that the ridgelet transform (3.194) is an isometric map from $L^2(\mathbb{R}^2)$ to $L^2(\mathbb{R}^+ \times \mathbb{R} \times [0, 2\pi))$.

Theorem 3.5.4. *For any pair of functions $f, g \in L^2(\mathbb{R}^2)$, we have*

$$\int_0^{2\pi} \int_\mathbb{R} \int_{\mathbb{R}^+} R_\psi\big[f\big](a,b,\theta)\, \overline{R_\phi\big[g\big](a,b,\theta)}\, \frac{da\, db\, d\theta}{a^3} = C_{\psi,\phi} \big\langle f, g \big\rangle_2, \qquad (3.200)$$

where $C_{\psi,\phi}$ is given by the cross admissibility condition:

$$C_{\psi,\phi} = (2\pi)^2 \int_\mathbb{R} \frac{\overline{\hat{\psi}(\omega)}\, \hat{\phi}(\omega)}{|\omega|^2}\, d\omega < \infty. \qquad (3.201)$$

Proof. For any pair of functions $f, g \in L^2(\mathbb{R}^2)$, the implication of Proposition 3.5.3 yields

$$\int_0^{2\pi} \int_\mathbb{R} \int_{\mathbb{R}^+} R_\psi\big[f\big](a,b,\theta)\, \overline{R_\phi\big[g\big](a,b,\theta)}\, \frac{da\, db\, d\theta}{a^3}$$

$$= \int_0^{2\pi} \int_\mathbb{R} \int_{\mathbb{R}^+} \left\{ \sqrt{2\pi a} \int_\mathbb{R} \hat{f}(\omega\Lambda_\theta)\, \overline{\hat{\psi}(a\omega)}\, e^{ib\omega}\, d\omega \right\}$$

$$\times \left\{ \sqrt{2\pi a} \int_{\mathbb{R}} \overline{\hat{g}(\eta \Lambda_\theta)} \, \hat{\phi}(a\eta) \, e^{-ib\eta} \, d\eta \right\} \frac{da \, db \, d\theta}{a^3}$$

$$= (2\pi) \int_0^{2\pi} \int_{\mathbb{R}^+} \int_{\mathbb{R}} \int_{\mathbb{R}} \hat{f}(\omega \Lambda_\theta) \, \overline{\hat{g}(\eta \Lambda_\theta)} \, \overline{\hat{\psi}(a\omega)} \, \hat{\phi}(a\eta) \left\{ \int_{\mathbb{R}} e^{ib(\omega - \eta)} \, db \right\} \frac{d\omega \, d\eta \, da \, d\theta}{a^2}$$

$$= (2\pi)^2 \int_0^{2\pi} \int_{\mathbb{R}^+} \int_{\mathbb{R}} \int_{\mathbb{R}} \hat{f}(\omega \Lambda_\theta) \, \overline{\hat{g}(\eta \Lambda_\theta)} \, \overline{\hat{\psi}(a\omega)} \, \hat{\phi}(a\eta) \, \delta(\omega - \eta) \frac{d\omega \, d\eta \, da \, d\theta}{a^2}$$

$$= (2\pi)^2 \int_0^{2\pi} \int_{\mathbb{R}^+} \int_{\mathbb{R}} \hat{f}(\omega \Lambda_\theta) \, \overline{\hat{g}(\omega \Lambda_\theta)} \, \overline{\hat{\psi}(a\omega)} \, \hat{\phi}(a\omega) \frac{d\omega \, da \, d\theta}{a^2}$$

$$= (2\pi)^2 \int_0^{2\pi} \int_{\mathbb{R}} \hat{f}(\omega \Lambda_\theta) \, \overline{\hat{g}(\omega \Lambda_\theta)} \left\{ \int_{\mathbb{R}^+} \overline{\hat{\psi}(a\omega)} \, \hat{\phi}(a\omega) \frac{da}{a^2} \right\} d\omega \, d\theta$$

$$= \int_0^{2\pi} \int_{\mathbb{R}} \hat{f}(\omega \Lambda_\theta) \, \overline{\hat{g}(\omega \Lambda_\theta)} \left\{ (2\pi)^2 \int_{\mathbb{R}} \frac{\overline{\hat{\psi}(\xi)} \, \hat{\phi}(\xi)}{|\xi|^2} \, d\xi \right\} |\omega| \, d\omega \, d\theta$$

$$= C_{\psi,\phi} \int_0^{2\pi} \int_{\mathbb{R}} \hat{f}(\omega \Lambda_\theta) \, \overline{\hat{g}(\omega \Lambda_\theta)} |\omega| \, d\omega \, d\theta$$

$$= C_{\psi,\phi} \left\langle \hat{f}, \hat{g} \right\rangle_2$$

$$= C_{\psi,\phi} \left\langle f, g \right\rangle_2.$$

This completes the proof of Theorem 3.5.4. □

Remark 3.5.2. For $f = g$ and $\psi = \phi$, Theorem 3.5.4 yields

$$\int_0^{2\pi} \int_{\mathbb{R}} \int_{\mathbb{R}^+} \left| R_\psi[f](a, b, \theta) \right|^2 \frac{da \, db \, d\theta}{a^3} = C_\psi \left\| f \right\|_2^2, \qquad (3.202)$$

where C_ψ is given by (3.193). If we choose the wavelet function ψ such that $C_\psi = 1$, then the energy preserving relation (3.202) implies that the ridgelet transform (3.194) is an isometry from $L^2(\mathbb{R}^2)$ to $L^2(\mathbb{R}^+ \times \mathbb{R} \times [0, 2\pi))$.

Next, our motive is to develop an inversion formula for the ridgelet transform defined in (3.194). In this direction, we have the next theorem.

Theorem 3.5.5. *Any function $f \in L^2(\mathbb{R}^2)$ can be reverted from the corresponding ridgelet transform $R_\psi[f](a, b, \theta)$ as follows:*

$$f(\mathbf{t}) = \frac{1}{C_{\psi,\phi}} \int_0^{2\pi} \int_{\mathbb{R}} \int_{\mathbb{R}^+} R_\psi[f](a, b, \theta) \, \phi_{a,b,\theta}(\mathbf{t}) \frac{da \, db \, d\theta}{a^3}, \quad a.e., \qquad (3.203)$$

where $C_{\psi,\phi} \neq 0$ is given by the cross admissibility condition (3.201).

Proof. Choosing $g(\mathbf{x}) = \delta(\mathbf{x} - \mathbf{t})$, $\mathbf{t} \in \mathbb{R}^2$, the corresponding ridgelet transform is given by

$$R_\phi[g](a, b, \theta) = \frac{1}{\sqrt{a}} \int_{\mathbb{R}^2} \delta(\mathbf{x} - \mathbf{t}) \, \overline{\phi\left(\frac{\Lambda_\theta \cdot \mathbf{x} - b}{a} \right)} \, d\mathbf{x} = \frac{1}{\sqrt{a}} \, \overline{\phi\left(\frac{\Lambda_\theta \cdot \mathbf{t} - b}{a} \right)}. \qquad (3.204)$$

Moreover, we observe that

$$\left\langle f, g \right\rangle_2 = \int_{\mathbb{R}^2} f(\mathbf{x}) \, \delta(\mathbf{x} - \mathbf{t}) \, d\mathbf{t} = f(\mathbf{t}). \qquad (3.205)$$

Using (3.204) and (3.205) in the orthogonality relation (3.200), we get

$$f(\mathbf{t}) = \frac{1}{C_{\psi,\phi}} \int_0^{2\pi} \int_{\mathbb{R}} \int_{\mathbb{R}^+} \frac{1}{\sqrt{a}} R_\psi \big[f\big](a,b,\theta)\, \phi\left(\frac{\Lambda_\theta \cdot \mathbf{t} - b}{a}\right) \frac{da\, db\, d\theta}{a^3}$$

$$= \frac{1}{C_{\psi,\phi}} \int_0^{2\pi} \int_{\mathbb{R}} \int_{\mathbb{R}^+} R_\psi \big[f\big](a,b,\theta)\, \phi_{a,b,\theta}(\mathbf{t}) \frac{da\, db\, d\theta}{a^3}, \quad a.e.,$$

which is the desired reconstruction formula and completes the proof of Theorem 3.5.5. $\square$

Before wrapping up the ongoing discourse on the ridgelet transform, we formulate a Heisenberg-type uncertainty inequality which governs the simultaneous localization of the Fourier and ridgelet transforms of a square integrable function.

Theorem 3.5.6. *If $R_\psi\big[f\big](a,b,\theta)$ is the ridgelet transform of any non-trivial function $f \in L^2(\mathbb{R}^2)$, then the following uncertainty inequality holds:*

$$\left\{ \int_0^{2\pi} \int_{\mathbb{R}} \int_{\mathbb{R}^+} b^2 \left| R_\psi\big[f\big](a,b,\theta) \right|^2 \frac{da\, db\, d\theta}{a^3} \right\}^{1/2} \left\{ \int_{\mathbb{R}^2} |\mathbf{w}|^2 |\hat{f}(\mathbf{w})|^2 d\mathbf{w} \right\}^{1/2} \geq \frac{\sqrt{C_\psi}}{2} \|f\|_2^2. \tag{3.206}$$

Proof. For any $g \in L^2(\mathbb{R})$, the Heisenberg's uncertainty inequality for the one-dimensional Fourier transform reads:

$$\left\{ \int_{\mathbb{R}} t^2 |g(t)|^2 dt \right\}^{1/2} \left\{ \int_{\mathbb{R}} \omega^2 \left| \mathscr{F}\big[g\big](\omega) \right|^2 d\omega \right\}^{1/2} \geq \frac{1}{2} \left\{ \int_{\mathbb{R}} |g(t)|^2 dt \right\}. \tag{3.207}$$

Identifying $R_\psi\big[f\big](a,b,\theta)$ as a function of b and then invoking inequality (3.207), we obtain

$$\left\{ \int_{\mathbb{R}} b^2 \left| R_\psi\big[f\big](a,b,\theta) \right|^2 db \right\}^{1/2} \left\{ \int_{\mathbb{R}} \omega^2 \left| \mathscr{F}\left(R_\psi\big[f\big](a,b,\theta) \right)(\omega) \right|^2 d\omega \right\}^{1/2}$$

$$\geq \frac{1}{2} \left\{ \int_{\mathbb{R}} \left| R_\psi\big[f\big](a,b,\theta) \right|^2 db \right\}. \tag{3.208}$$

Integrating (3.208) with respect to the measure $da\, d\theta/a^3$ yields

$$\int_0^{2\pi} \int_{\mathbb{R}^+} \left\{ \int_{\mathbb{R}} b^2 \left| R_\psi\big[f\big](a,b,\theta) \right|^2 db \right\}^{1/2} \left\{ \int_{\mathbb{R}} \omega^2 \left| \mathscr{F}\left(R_\psi\big[f\big](a,b,\theta) \right)(\omega) \right|^2 d\omega \right\}^{1/2} \frac{da\, d\theta}{a^3}$$

$$\geq \frac{1}{2} \left\{ \int_0^{2\pi} \int_{\mathbb{R}^+} \int_{\mathbb{R}} \left| R_\psi\big[f\big](a,b,\theta) \right|^2 \frac{db\, da\, d\theta}{a^3} \right\}. \tag{3.209}$$

Employing the Cauchy-Schwarz inequality followed by a change in the order of integration, the inequality (3.209) can be expressed as

$$\left\{ \int_0^{2\pi} \int_{\mathbb{R}} \int_{\mathbb{R}^+} b^2 \left| R_\psi\big[f\big](a,b,\theta) \right|^2 \frac{da\, db\, d\theta}{a^3} \right\}^{1/2}$$

$$\times \left\{ \int_0^{2\pi} \int_{\mathbb{R}} \int_{\mathbb{R}^+} \omega^2 \left| \mathscr{F}\left(R_\psi\big[f\big](a,b,\theta) \right)(\omega) \right|^2 \frac{da\, d\omega\, d\theta}{a^3} \right\}^{1/2} \geq \frac{C_\psi}{2} \|f\|_2^2. \tag{3.210}$$

As a consequence of (3.196), we can express the ridgelet transform $R_\psi\big[f\big](a,b,\theta)$ via the classical one-dimensional Fourier transform as

$$R_\psi\big[f\big](a,b,\theta) = \sqrt{2\pi a} \int_{\mathbb{R}} \hat{f}(\omega\Lambda_\theta)\, \overline{\hat{\psi}(a\omega)}\, e^{ib\omega}\, d\omega$$

$$= 2\pi \sqrt{a}\, \mathscr{F}^{-1}\left(\hat{f}(\omega\Lambda_\theta)\, \overline{\hat{\psi}(a\omega)}\right)(b),$$

so that

$$\mathscr{F}\left(R_\psi\Big[f\Big](a,b,\theta)\right)(\omega) = 2\pi\sqrt{a}\, \hat{f}(\omega\Lambda_\theta)\, \overline{\hat{\psi}(a\omega)}. \tag{3.211}$$

Using (3.211) in (3.210), we get

$$\left\{\int_0^{2\pi}\int_{\mathbb{R}}\int_{\mathbb{R}+} b^2\left|R_\psi\Big[f\Big](a,b,\theta)\right|^2 \frac{da\,db\,d\theta}{a^3}\right\}^{1/2}$$

$$\times \left\{(2\pi)^2\int_0^{2\pi}\int_{\mathbb{R}}\int_{\mathbb{R}+}\omega^2\left|\hat{f}(\omega\Lambda_\theta)\right|^2\left|\hat{\psi}(a\omega)\right|^2\frac{da\,d\omega\,d\theta}{a^2}\right\}^{1/2} \geq \frac{C_\psi}{2}\left\|f\right\|_2^2. \tag{3.212}$$

Upon simplifying the second expression on the L.H.S of (3.212), we obtain

$$\int_0^{2\pi}\int_{\mathbb{R}}\int_{\mathbb{R}+}\omega^2\left|\hat{f}(\omega\Lambda_\theta)\right|^2\left|\hat{\psi}(a\omega)\right|^2\frac{da\,d\omega\,d\theta}{a^3}$$

$$= \int_0^{2\pi}\int_{\mathbb{R}}\omega^2\left|\hat{f}(\omega\Lambda_\theta)\right|^2\left\{(2\pi)^2\int_{\mathbb{R}+}\left|\hat{\psi}(a\omega)\right|^2\frac{da}{a^2}\right\}d\omega\,d\theta$$

$$= C_\psi\int_{\mathbb{R}^2}|\mathbf{w}|^2|\hat{f}(\mathbf{w})|^2 d\mathbf{w}, \tag{3.213}$$

where the last integral is obtained by making use of the substitution $\mathbf{w} = (\omega_1,\omega_2) = (\omega\cos\theta,\ \omega\sin\theta)$ and noting that the Jacobian of the transformation is equal to ω. Finally, implementing (3.213) in (3.212), we obtain the following inequality:

$$\left\{\int_0^{2\pi}\int_{\mathbb{R}}\int_{\mathbb{R}+} b^2\left|R_\psi\Big[f\Big](a,b,\theta)\right|^2\frac{da\,db\,d\theta}{a^3}\right\}^{1/2}\left\{\int_{\mathbb{R}^2}|\mathbf{w}|^2|\hat{f}(\mathbf{w})|^2 d\mathbf{w}\right\}^{1/2} \geq \frac{\sqrt{C_\psi}}{2}\left\|f\right\|_2^2,$$

which is the desired Heisenberg-type inequality. This completes the proof of Theorem 3.5.6. □

3.6 The Curvelet and Ripplet Transforms

"The research worker, in his efforts to express the fundamental laws of Nature in mathematical form, should strive mainly for mathematical beauty. He should take simplicity into consideration in a subordinate way to beauty. ... It often happens that the requirements of simplicity and beauty are the same, but where they clash the latter must take precedence."

-Paul A. Dirac

In this section, our goal is to present a heuristic discussion on the fundamental notions of the curvelet and ripplet transforms, which constitute a premier class of integral transforms in the context of multi-scale analysis of multi-dimensional signals. To facilitate a smooth transition from curvelet to ripplet transforms, we subdivide this section into two subsections.

3.6.1   Curvelet transform

Although the wavelet transforms have provided the bottom line for representing multi-dimensional signals in a simple and insightful way, however, these representations are not optimal. An illustration to the argument can be provided by taking into account the fact that, a smooth curve is in essence a one-dimensional object; therefore, it is a terrible waste of time or bits to represent it via analyzers designed for genuine two-dimensional objects. A nice alternative to this approach is provided by ridgelets, which are good at resolving the line singularities. Unlike the usual wavelets, the ridgelet functions employ an anisotropic refinement mechanism. For instance, if a signal contains a linear singularity, then the orientation of the ridgelet can be tuned to that of the line and can be arbitrarily refined in the orthogonal direction to obtain a nice approximation of the singularity. However, ridgelets alone are not sufficient to deal efficiently with curved singularities, but intuitively a correct localization of ridgelet elements might be handy. In this direction, Candès and Donoho [62, 63] introduced the notion of curvelets as a novel multi-scale system particularly well suited for curved singularities. The intrinsic multi-scale and anisotropic nature of curvelet waveforms leads to optimally sparse representations of objects which display curve-punctuated smoothness; that is, smoothness except for discontinuity along a general curve with bounded curvature. Another remarkable property of curvelets is that they elegantly model the geometry of wave propagation; curvelets may be viewed as coherent waveforms with enough frequency localization to behave like waves but at the same time, with sufficient spatial localization to behave like particles [64]. For more about curvelets and their applications, we refer to the monographs [60, 65, 103, 191, 199, 277, 286].

In this subsection, we develop the mathematical foundations of the theory of curvelets and then study the fundamental properties of the curvelet transform. It is worth emphasizing that unlike the classical two-dimensional wavelets, curvelets are obtained by applying parabolic dilations to a given waveform. For $a > 0$, the parabolic dilations of any bivariate function $f(t_1, t_2)$ are given by $f(\sqrt{a}t_1, at_2)$, so-called because they leave invariant the parabola $t_2 = t_1^2$. Thus, in case of parabolic dilations, the dilation is always twice as powerful in one fixed direction as in the orthogonal direction.

Consider the frequency plane $\mathbb{R}^2$ and let (r, ϕ), $r \geq 0$, $\phi \in [0, 2\pi)$ denote the polar coordinates of an arbitrary point $\mathbf{w} = (\omega_1, \omega_2) \in \mathbb{R}^2$. We choose a pair of window functions $W : (0, \infty) \to (0, \infty)$, called as "radial window," and $V : (-\infty, \infty) \to (0, \infty)$, called as "angular window," satisfying the following admissibility conditions:

$$\int_0^\infty |W(r)|^2 \frac{dr}{r} = 1, \qquad \operatorname{supp}(W) \subseteq \left(\frac{1}{2}, 2\right) \tag{3.214}$$

and

$$(2\pi)^2 \int_{-1}^1 |V(\phi)|^2 d\phi = 1, \qquad \operatorname{supp}(V) \subseteq [-1, 1]. \tag{3.215}$$

The window functions satisfying (3.214) and (3.215) are used to construct a family of complex-valued waveforms adopted to scale $a > 0$, location $\mathbf{b} \in \mathbb{R}^2$ and orientation $\theta \in [0, 2\pi)$ (or $(-\pi, \pi)$ according to convenience). For a fixed scale $a \in (0, a_0)$, where $a_0 < \pi^2$, the basic curvelet $\Psi_a : \mathbb{R}^2 \to \mathbb{C}$ is defined via the polar Fourier transform as

$$\mathscr{F}\big[\Psi_a\big](r, \phi) = a^{3/4} W(ar) V\left(\frac{\phi}{\sqrt{a}}\right), \tag{3.216}$$

where $\mathscr{F}$ denotes the two-dimensional Fourier transform:

$$\mathscr{F}\big[f\big](\mathbf{w}) = \frac{1}{2\pi} \int_{\mathbb{R}^2} f(\mathbf{t}) \, e^{-i\mathbf{t}^T \mathbf{w}} \, d\mathbf{t}, \tag{3.217}$$

which is expressible via the polar coordinates as

$$\mathscr{F}\big[f\big](r,\phi) = \frac{1}{2\pi} \int_0^{2\pi} \int_0^{\infty} f(\rho,\eta) \, \exp\big\{ -i\rho r \cos(\eta - \phi) \big\} \, \rho \, d\rho \, d\eta. \tag{3.218}$$

Consequently, the family of analyzing waveforms $\Psi_{a,\mathbf{b},\theta}(\mathbf{t})$, called "curvelets," is generated by translation and rotation of the basic element $\Psi_a(\mathbf{t})$; that is,

$$\Psi_{a,\mathbf{b},\theta}(\mathbf{t}) = \Psi_a \left(R_\theta(\mathbf{t} - \mathbf{b}) \right), \quad \mathbf{t} \in \mathbb{R}^2, \tag{3.219}$$

where

$$R_\theta = \begin{pmatrix} \cos\theta & -\sin\theta \\ \sin\theta & \cos\theta \end{pmatrix}$$

denotes the usual 2×2 rotation matrix effecting the planar rotation by θ radians. From (3.216), we note that the support of the basic element Ψ_a in the frequency domain is a polar wedge governed by the respective supports of the radial and angular windows. The scaling in the radial and angular windows is parabolic in nature with ϕ being the "thin" variable. The coarsest scale a_0 is fixed once for all and must obey $a_0 < \pi^2$. Thus, we infer that the curvelet waveforms (3.219) are essentially the result of directional parabolic dilations and they become increasingly needle like at fine scales, obeying $width \approx length^2$. For a pictorial illustration of the aforementioned facts, the wedge-tiling inculcated by the curvelet waveforms onto the frequency plane is depicted in Figure 3.23.

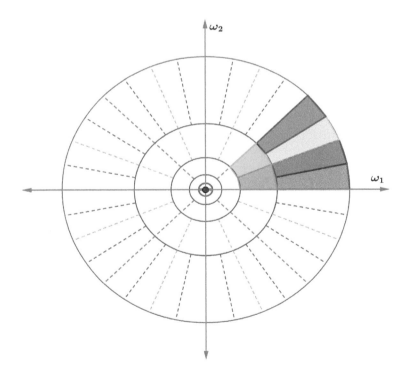

FIGURE 3.23: Wedge-tiling on the frequency plane due to curvelet waveforms.

After formulating the notion of curvelet waveforms, we have the formal definition of the curvelet transform.

Definition 3.6.1. For any bivariate function $f \in L^2(\mathbb{R}^2)$, the curvelet transform is denoted by $\Gamma_\Psi[f]$ and is defined as

$$\Gamma_\Psi[f](a, \mathbf{b}, \theta) = \int_{\mathbb{R}^2} f(\mathbf{t}) \, \overline{\Psi_{a,\mathbf{b},\theta}(\mathbf{t})} \, d\mathbf{t}, \qquad (3.220)$$

where $a \in (0, a_0)$ with $a_0 < \pi^2$, $\mathbf{b} \in \mathbb{R}^2$, $\theta \in [0, 2\pi)$ and $\Psi_{a,\mathbf{b},\theta}(\mathbf{t})$ is given by (3.219).

Proposition 3.6.1. *For any $f \in L^2(\mathbb{R}^2)$, we have*

$$\mathscr{F}\left(\Gamma_\Psi[f](a, \mathbf{b}, \theta)\right)(\mathbf{w}) = 2\pi \, \mathscr{F}[f](\mathbf{w}) \, \mathscr{F}[\Psi_a](R_\theta \mathbf{w}) \,. \qquad (3.221)$$

Proof. Using the definition of Fourier transform in Cartesian coordinates (3.217), we have

$$\mathscr{F}\left[\Psi_{a,\mathbf{b},\theta}\right](\mathbf{w}) = \frac{1}{2\pi} \int_{\mathbb{R}^2} \Psi_{a,\mathbf{b},\theta}(\mathbf{t}) \, e^{-i \mathbf{t}^T \mathbf{w}} \, d\mathbf{t}$$

$$= \frac{1}{2\pi} \int_{\mathbb{R}^2} \Psi_a\left(R_\theta(\mathbf{t} - \mathbf{b})\right) e^{-i \mathbf{t}^T \mathbf{w}} \, d\mathbf{t}$$

$$= \frac{1}{2\pi} \int_{\mathbb{R}^2} \Psi_a(\mathbf{z}) \, e^{-i(\mathbf{b} + R_{-\theta}\mathbf{z})^T \mathbf{w}} \, d\mathbf{z}$$

$$= \frac{e^{-i\mathbf{b}^T \mathbf{w}}}{2\pi} \int_{\mathbb{R}^2} \Psi_a(\mathbf{z}) \, e^{-i \mathbf{z}^T (R_\theta \mathbf{w})} \, d\mathbf{z}$$

$$= e^{-i\mathbf{b}^T \mathbf{w}} \mathscr{F}[\Psi_a](R_\theta \mathbf{w}) \,. \qquad (3.222)$$

In order to translate the above expression into polar coordinates, let (σ, μ), (ρ, η) and (r, ϕ) be the polar coordinates of the variables $\mathbf{b}$, $\mathbf{t}$ and $\mathbf{w}$, respectively. Then, we have

$$\mathscr{F}\left[\Psi_{a,\mathbf{b},\theta}\right](r, \phi) = e^{-ir\sigma \cos(\mu-\phi)} \mathscr{F}[\Psi_a](r, \phi - \theta)$$

$$= e^{-ir\sigma \cos(\mu-\phi)} a^{3/4} W(ar) \, V\left(\frac{\phi - \theta}{\sqrt{a}}\right). \qquad (3.223)$$

Invoking the well-known Parseval's formula in polar coordinates, we obtain

$$\Gamma_\Psi[f](a, \mathbf{b}, \theta) = \left\langle f, \, \Psi_{a,\mathbf{b},\theta} \right\rangle_2$$

$$= a^{3/4} \int_0^{2\pi} \int_0^\infty e^{ir\sigma \cos(\mu-\phi)} \mathscr{F}[f](r, \phi) \, W(ar) \, V\left(\frac{\phi - \theta}{\sqrt{a}}\right) r \, dr \, d\phi. \qquad (3.224)$$

Reverting the expression (3.224) into Cartesian coordinates yields the following relation:

$$\Gamma_\Psi[f](a, \mathbf{b}, \theta) = \int_{\mathbb{R}^2} \mathscr{F}[f](\mathbf{w}) \overline{\mathscr{F}[\Psi_a](R_\theta \mathbf{w})} \, e^{i\mathbf{b}^T \mathbf{w}} \, d\mathbf{w}$$

$$= 2\pi \, \mathscr{F}^{-1}\left(\mathscr{F}[f](\mathbf{w}) \overline{\mathscr{F}[\Psi_a](R_\theta \mathbf{w})}\right)(\mathbf{b}). \qquad (3.225)$$

Applying the Fourier transform on both sides of (3.225), we obtain the desired result as

$$\mathscr{F}\left(\Gamma_\Psi[f](a, \mathbf{b}, \theta)\right)(\mathbf{w}) = 2\pi \, \mathscr{F}[f](\mathbf{w}) \, \mathscr{F}[\Psi_a](R_\theta \mathbf{w}) \,.$$

This completes the proof of Proposition 3.6.1.   □

Next, we shall analyze the support and oscillatory behaviour of the curvelet waveforms. From (3.222), we observe that supp $\left(\mathscr{F}\left[\Psi_{a,b,\theta}\right](\mathbf{w})\right) = \text{supp}\left(\mathscr{F}\left[\Psi_a\right](R_\theta \mathbf{w})\right)$. That is, the support of curvelet waveforms $\Psi_{a,b,\theta}(\mathbf{t})$ in the Fourier domain is completely independent of the translation parameter $\mathbf{b}$ and for any $\theta \in [0, 2\pi)$, the support matches with the support of $\mathscr{F}\left[\Psi_a\right](\mathbf{w})$ with an effective planar rotation of θ radians in clockwise direction. Moreover, we note that the curvelet functions are complex-valued, in general; however, one can construct the real counterparts by replacing the function $W(r)$ appearing in (3.216) with $W(r) + W(-r)$. In that case $\left(\mathscr{F}\left[\Psi_{a,b,\theta}\right](\mathbf{w})\right)$ is supported on two polar wedges being symmetric with respect origin.

In the time domain, the effective support of the curvelet waveforms is governed by the uncertainty inequality in the Fourier domain. Since the curvelet functions $\Psi_{a,b,\theta}(\mathbf{t})$ have compact support in the frequency domain, therefore, the well-known Heisenberg's uncertainty principle implies that they cannot enjoy compact support in the time-domain. Also, for large $|\mathbf{t}|$, the decay of the functions $\Psi_{a,b,\theta}(\mathbf{t})$ depends upon the smoothness of $\mathscr{F}\left[\Psi_{a,b,\theta}\right](\mathbf{w})$; the smoother the $\mathscr{F}\left[\Psi_{a,b,\theta}\right](\mathbf{w})$, the faster the decay. Moreover, by definition $\mathscr{F}\left[\Psi_a\right](\mathbf{w})$ is supported away from the vertical axis $\omega_1 = 0$, but near the horizontal axis $\omega_2 = 0$. Hence, for smaller values of $a < a_0$, the basic waveform $\Psi_a(\mathbf{t})$ is less oscillatory in t_2 direction (with frequency about $\sqrt{a}$) and very oscillatory in t_1 direction (containing frequencies of about $1/2a$). The essential support of the amplitude spectrum of Ψ_a is a rectangle given by $[-\pi/2a, \pi/2a] \times [-\pi/\sqrt{a}, \pi/\sqrt{a}]$ and its decay outside this rectangle depends up on the smoothness of the corresponding Fourier transform. Consequently, the support of the curvelet waveforms $\Psi_{a,b,\theta}(\mathbf{t})$ is the aforementioned rectangle rotated by an angle of θ radians and translated by $R_\theta \mathbf{b}$.

Below, we shall present the formal reconstruction formula associated with the curvelet transform. We note that the said reconstruction formula is valid for high frequency signals. The analogue for low-frequency signals will be dealt afterwards. To facilitate the narrative, we have the following definition of two-dimensional convolution operation:

Definition 3.6.2. Given any two functions $f, g \in L^2(\mathbb{R}^2)$, the two-dimensional convolution operation is denoted by $\circledast$ and is defined as

$$\left(f \circledast g\right)(\mathbf{z}) = \int_{\mathbb{R}^2} f(\mathbf{t})\, g(\mathbf{z} - \mathbf{t})\, d\mathbf{t}. \tag{3.226}$$

In view of (3.217), the convolution theorem associated with the two-dimensional convolution defined in (3.226) reads:

$$\mathscr{F}\left[(f \circledast g)\right](\mathbf{w}) = 2\pi\, \mathscr{F}\left[f\right](\mathbf{w})\, \mathscr{F}\left[g\right](\mathbf{w}). \tag{3.227}$$

Theorem 3.6.2. *For any $f \in L^2(\mathbb{R}^2)$ satisfying $\mathscr{F}\left[f\right](\mathbf{w}) = 0$, $\forall\, |\mathbf{w}| < 2/a_0$, $a_0 < \pi^2$, the reconstruction formula for the curvelet transform $\Gamma_\Psi\left[f\right](a, \mathbf{b}, \theta)$ defined in (3.220) is given by*

$$f(\mathbf{t}) = \int_0^{2\pi} \int_{\mathbb{R}^2} \int_0^{a_0} \Gamma_\Psi\left[f\right](a, \mathbf{b}, \theta)\, \Psi_{a,b,\theta}(\mathbf{t})\, \frac{da\, d\mathbf{b}\, d\theta}{a^3}, \tag{3.228}$$

where the radial and angular windows W and V satisfy their respective admissibility conditions (3.214) and (3.215).

Proof. We note that the curvelet transform $\Gamma_\Psi\left[f\right](a, \mathbf{b}, \theta)$ defined in (3.220) can be expressed via the convolution (3.226) as

$$\Gamma_\Psi\left[f\right](a, \mathbf{b}, \theta) = \int_{\mathbb{R}^2} f(\mathbf{t})\, \overline{\Psi_a\left(R_\theta(\mathbf{t} - \mathbf{b})\right)}\, d\mathbf{t}$$

$$= \int_{\mathbb{R}^2} f(\mathbf{t}) \, \overline{\tilde{\Psi}_{a,0,\theta}(-(\mathbf{b}-\mathbf{t}))} \, d\mathbf{t}$$

$$= (f \circledast \tilde{\Psi}_{a,0,\theta})(\mathbf{b}), \qquad \tilde{\Psi}(\mathbf{t}) = \overline{\Psi(-\mathbf{t})}. \tag{3.229}$$

Next, we define a function

$$F_{a,\theta}(\mathbf{t}) = \int_{\mathbb{R}^2} \Gamma_{\Psi}\Big[f\Big](a,\mathbf{b},\theta) \, \Psi_{a,\mathbf{b},\theta}(\mathbf{t}) \, d\mathbf{b}. \tag{3.230}$$

Invoking (3.226), we can express (3.230) as follows:

$$F_{a,\theta}(\mathbf{t}) = \Big((f \circledast \tilde{\Psi}_{a,0,\theta})(\mathbf{b}) \circledast \Psi_{a,0,\theta}(\mathbf{b})\Big)(\mathbf{t}). \tag{3.231}$$

Applying the convolution theorem (3.227), we can compute the Fourier transform of the function $F_{a,\theta}(\mathbf{t})$ as

$$\mathscr{F}\Big[F_{a,\theta}\Big](\mathbf{w}) = 2\pi \, \mathscr{F}\Big[(f \circledast \tilde{\Psi}_{a,0,\theta})\Big](\mathbf{w}) \, \mathscr{F}\Big[\Psi_{a,0,\theta}\Big](\mathbf{w})$$

$$= (2\pi)^2 \mathscr{F}\Big[f\Big](\mathbf{w}) \left|\mathscr{F}\Big[\Psi_{a,0,\theta}\Big](\mathbf{w})\right|^2. \tag{3.232}$$

Consequently, we have

$$\int_0^{a_0} \int_0^{2\pi} \mathscr{F}\Big[F_{a,\theta}\Big](\mathbf{w}) \, \frac{d\theta \, da}{a^3} = (2\pi)^2 \mathscr{F}\Big[f\Big](\mathbf{w}) \int_0^{a_0} \int_0^{2\pi} \left|\mathscr{F}\Big[\Psi_{a,0,\theta}\Big](\mathbf{w})\right|^2 \frac{d\theta \, da}{a^3}. \tag{3.233}$$

We now evaluate the integral on the R.H.S of (3.233). To do so, we shall use the polar coordinates of $\mathbf{w}$ and invoke the admissibility conditions (3.214) and (3.215). For $r \geq 2/a_0$, $a_0 < \pi^2$, we have

$$(2\pi)^2 \int_0^{a_0} \int_0^{2\pi} \left|\mathscr{F}\Big[\Psi_{a,0,\theta}\Big](r,\phi)\right|^2 \frac{d\theta \, da}{a^3} = (2\pi)^2 \int_0^{a_0} \int_0^{2\pi} \left|W(ar)\right|^2 \left|V\left(\frac{\phi-\theta}{\sqrt{a}}\right)\right|^2 \frac{d\theta \, da}{a^{3/2}}$$

$$= \int_0^{a_0} \left|W(ar)\right|^2 \left\{(2\pi)^2 \int_0^{2\pi} \left|V\left(\frac{\phi-\theta}{\sqrt{a}}\right)\right|^2 d\theta\right\} \frac{da}{a^{3/2}}$$

$$= \int_0^{a_0} \left|W(ar)\right|^2 \frac{da}{a}$$

$$= \int_0^{a_0 r} \left|W(r')\right|^2 \frac{dr'}{r'} = 1. \tag{3.234}$$

Implementing (3.234) in (3.233), we obtain

$$\mathscr{F}\Big[f\Big](\mathbf{w}) = \int_0^{a_0} \int_0^{2\pi} \mathscr{F}\Big[F_{a,\theta}\Big](\mathbf{w}) \, \frac{d\theta \, da}{a^3}. \tag{3.235}$$

That is,

$$f(\mathbf{t}) = \int_0^{2\pi} \int_{\mathbb{R}^2} \int_0^{a_0} \Gamma_{\Psi}\Big[f\Big](a,\mathbf{b},\theta) \, \Psi_{a,\mathbf{b},\theta}(\mathbf{t}) \, \frac{da \, d\mathbf{b} \, d\theta}{a^3}.$$

This completes the proof of Theorem 3.6.2.   □

Theorem 3.6.3. *For any* $f \in L^2(\mathbb{R}^2)$ *satisfying* $\mathscr{F}\big[f\big](\mathbf{w}) = 0$, $\forall \, |\mathbf{w}| < 2/a_0$, $a_0 < \pi^2$, *we have*

$$\left\|\Gamma_{\Psi}\Big[f\Big](a,\mathbf{b},\theta)\right\|_2^2 = \left\|f\right\|_2^2. \tag{3.236}$$

That is, the total energy of the given signal is preserved from the natural domain $L^2(\mathbb{R}^2)$ *to transformed domain* $L^2\big((0, a_0) \times \mathbb{R}^2 \times [0, 2\pi)\big)$, *where* $a_0 < \pi^2$.

Proof. In view of the Fourier domain representation (3.227) of the convolution (3.226), we obtain

$$
\begin{aligned}
\left\| \Gamma_\Psi\big[f\big](a,\mathbf{b},\theta) \right\|_2^2 &= \int_0^{2\pi} \int_{\mathbb{R}^2} \int_0^{a_0} \left| \Gamma_\Psi\big[f\big](a,\mathbf{b},\theta) \right|^2 \frac{da\,d\mathbf{b}\,d\theta}{a^3} \\
&= \int_0^{2\pi} \int_{\mathbb{R}^2} \int_0^{a_0} \left| (f \circledast \tilde{\Psi}_{a,0,\theta})(\mathbf{b}) \right|^2 \frac{da\,d\mathbf{b}\,d\theta}{a^3} \\
&= \int_0^{2\pi} \int_{\mathbb{R}^2} \int_0^{a_0} \left| \mathscr{F}\left[(f \circledast \tilde{\Psi}_{a,0,\theta}) \right](\mathbf{w}) \right|^2 \frac{da\,d\mathbf{w}\,d\theta}{a^3} \\
&= (2\pi)^2 \int_0^{2\pi} \int_{\mathbb{R}^2} \int_0^{a_0} \left| \mathscr{F}\big[f\big](\mathbf{w}) \right|^2 \left| \mathscr{F}\big[\Psi_{a,0,\theta}\big](\mathbf{w}) \right|^2 \frac{da\,d\mathbf{w}\,d\theta}{a^3} \\
&= \int_{\mathbb{R}^2} \left| \mathscr{F}\big[f\big](\mathbf{w}) \right|^2 \left\{ (2\pi)^2 \int_0^{a_0} \int_0^{2\pi} \left| \mathscr{F}\big[\Psi_{a,0,\theta}\big](\mathbf{w}) \right|^2 \frac{d\theta\,da}{a^3} \right\} d\mathbf{w} \\
&= \int_{\mathbb{R}^2} \left| \mathscr{F}\big[f\big](\mathbf{w}) \right|^2 d\mathbf{w} \\
&= \left\| f \right\|_2^2,
\end{aligned}
$$

which evidently completes the proof of Theorem 3.6.3. □

Remark 3.6.1. From the norm equality (3.236), we infer that the curvelet transform defined in (3.220) is an isometry from the space of those signals in $L^2(\mathbb{R}^2)$ satisfying $\mathscr{F}\big[f\big](\mathbf{w}) = 0, \forall\, |\mathbf{w}| < 2/a_0,\ a_0 < \pi^2$ to the space of transforms $L^2((0,a_0) \times \mathbb{R}^2 \times [0,2\pi))$.

Remark 3.6.2. The reference measure involved in both Theorem 3.6.2 as well as Theorem 3.6.3 can be thought of as follows:

$$
\frac{da\,d\mathbf{b}\,d\theta}{a^3} = \left(\frac{da}{a} \right) \left(\frac{d\mathbf{b}}{a^{3/2}} \right) \left(\frac{d\theta}{a^{1/2}} \right),
$$

suggesting that the range of $\mathbf{b}$ can be viewed as divided into unit cells of side a by $\sqrt{a}$, so that the area is $a^{3/2}$, the range of θ is naturally viewed as divided into unit cells (intervals) of side $\sqrt{a}$ and the range of $\log(a)$ has unit-cells of side 1.

In the next instance, we note that the reconstruction formula (3.228) is concerned for those signals $f \in L^2(\mathbb{R}^2)$ satisfying $\mathscr{F}\big[f\big](\mathbf{w}) = 0, \forall\, |\mathbf{w}| < 2/a_0,\ a_0 < \pi^2$. In order to have a complete reconstruction formula, we need to take care of the other frequency components as well. To facilitate the narrative, we consider an arbitrary square integrable function f on $\mathbb{R}^2$ and define

$$
\begin{aligned}
(T_1 f)(\mathbf{t}) &= \int_0^{2\pi} \int_{\mathbb{R}^2} \int_0^{a_0} \Gamma_\Psi\big[f\big](a,\mathbf{b},\theta)\, \Psi_{a,\mathbf{b},\theta}(\mathbf{t}) \frac{da\,d\mathbf{b}\,d\theta}{a^3} \\
&= \int_0^{2\pi} \int_0^{a_0} \left((f \circledast \tilde{\Psi}_{a,0,\theta})(\mathbf{b}) \circledast \Psi_{a,0,\theta}(\mathbf{b}) \right)(\mathbf{t}) \frac{da\,d\theta}{a^3} \quad (3.237)
\end{aligned}
$$

and

$$
(T_0 f)(\mathbf{t}) = f(\mathbf{t}) - (T_1 f)(\mathbf{t}). \quad (3.238)
$$

Here, we note that

$$
\mathscr{F}\big[(T_1 f)\big](\mathbf{w}) = (2\pi)^2 \int_0^{2\pi} \int_0^{a_0} \mathscr{F}\big[f\big](\mathbf{w}) \left| \mathscr{F}\big[\Psi_{a,0,\theta}\big](\mathbf{w}) \right|^2 \frac{da\,d\theta}{a^3}
$$

$$= \mathscr{F}\left[f\right](\mathbf{w}) \int_0^{a_0|\mathbf{w}|} \left|W(a)\right|^2 \frac{da}{a}$$

$$= (2\pi)^2 \, \mathscr{F}\left[f\right](\mathbf{w}) \left(\mathscr{F}\left[\Omega\right](\mathbf{w})\right)^2, \tag{3.239}$$

where $\left(\mathscr{F}\left[\Omega\right](\mathbf{w})\right)^2 = \dfrac{1}{(2\pi)^2} \displaystyle\int_0^{a_0|\mathbf{w}|} \left|W(a)\right|^2 \dfrac{da}{a}$.

Furthermore, using the additivity of the Fourier transform, we can express the Fourier transform of (3.238) as

$$\mathscr{F}\left[(T_0 f)\right](\mathbf{w}) = \mathscr{F}\left[f\right](\mathbf{w}) - \mathscr{F}\left[(T_1 f)\right](\mathbf{w})$$

$$= \mathscr{F}\left[f\right](\mathbf{w}) \left(1 - (2\pi)^2 \left(\mathscr{F}\left[\Omega\right](\mathbf{w})\right)^2\right)$$

$$= (2\pi)^2 \mathscr{F}\left[f\right](\mathbf{w}) \left(\frac{1}{(2\pi)^2} - \left(\mathscr{F}\left[\Omega\right](\mathbf{w})\right)^2\right)$$

$$= (2\pi)^2 \mathscr{F}\left[f\right](\mathbf{w}) \left(\mathscr{F}\left[\Phi\right](\mathbf{w})\right)^2, \tag{3.240}$$

where $\left(\mathscr{F}\left[\Phi\right](\mathbf{w})\right)^2 = \dfrac{1}{(2\pi)^2} - \left(\mathscr{F}\left[\Omega\right](\mathbf{w})\right)^2$.

By virtue of the assertion made in (3.227), we conclude from (3.237) and (3.238) that

$$(T_0 f)(\mathbf{t}) = \left(f \circledast \Phi \circledast \Phi\right)(\mathbf{t}) \quad \text{and} \quad (T_1 f)(\mathbf{t}) = \left(f \circledast \Omega \circledast \Omega\right)(\mathbf{t}). \tag{3.241}$$

Moreover, we note that

$$(2\pi)^2 \left[\left(\mathscr{F}\left[\Omega\right](\mathbf{w})\right)^2 + \left(\mathscr{F}\left[\Phi\right](\mathbf{w})\right)^2\right] = 1. \tag{3.242}$$

Also,

$$\mathscr{F}\left[\Phi\right](\mathbf{w}) = 0 \quad |\mathbf{w}| > 2/a_0, \quad \text{and} \quad \mathscr{F}\left[\Phi\right](\mathbf{w}) = 1/2\pi, \quad |\mathbf{w}| < 1/2a_0. \tag{3.243}$$

Finally, we define the father wavelet $\Phi_{\mathbf{b}}(\mathbf{t}) = \Phi(\mathbf{t} - \mathbf{b})$, so that

$$(T_0 f)(\mathbf{t}) = \int_{\mathbb{R}^2} \left\langle f, \, \Phi_{\mathbf{b}} \right\rangle_2 \Phi_{\mathbf{b}}(\mathbf{t}) \, d\mathbf{b}. \tag{3.244}$$

Consequently, (3.244) implies that

$$f(\mathbf{t}) = \int_{\mathbb{R}^2} \left\langle f, \, \Phi_{\mathbf{b}} \right\rangle_2 \Phi_{\mathbf{b}}(\mathbf{t}) \, d\mathbf{b} + \int_0^{2\pi} \int_{\mathbb{R}^2} \int_0^{a_0} \Gamma_\Psi\left[f\right](a, \mathbf{b}, \theta) \, \Psi_{a,\mathbf{b},\theta}(\mathbf{t}) \, \frac{da \, d\mathbf{b} \, d\theta}{a^3}.$$

Hence, we conclude that the complete reconstruction formula for the curvelet transform (3.220) is composed of both curvelet waveforms and isotropic father wavelets. The above discussion can be summarized into the following theorem:

Theorem 3.6.4. *For any $f \in L^2(\mathbb{R}^2)$, the reproducing formula for the curvelet transform $\Gamma_\Psi\left[f\right](a, \mathbf{b}, \theta)$ defined in (3.220) is given by*

$$f(\mathbf{t}) = \int_{\mathbb{R}^2} \left\langle f, \, \Phi_{\mathbf{b}} \right\rangle_2 \Phi_{\mathbf{b}}(\mathbf{t}) \, d\mathbf{b} + \int_0^{2\pi} \int_{\mathbb{R}^2} \int_0^{a_0} \Gamma_\Psi\left[f\right](a, \mathbf{b}, \theta) \, \Psi_{a,\mathbf{b},\theta}(\mathbf{t}) \, \frac{da \, d\mathbf{b} \, d\theta}{a^3}, \tag{3.245}$$

where the radial and angular windows W and V satisfy their respective admissibility conditions (3.214) and (3.215).

In continuation to the quest for obtaining novel uncertainty inequalities, here we formulate a Heisenberg-type uncertainty inequality for the curvelet transform (3.220). Such an inequality allows us to compare the localization of the Fourier transform of a function f with the corresponding curvelet transform $\Gamma_\Psi[f](a, \mathbf{b}, \theta)$, regarded as a function of the planer translation variable $\mathbf{b}$.

Theorem 3.6.5. *If $\Gamma_\Psi[f](a, \mathbf{b}, \theta)$ is the curvelet transform of any non-trivial function $f \in L^2(\mathbb{R}^2)$ and satisfying $\mathscr{F}[f](\mathbf{w}) = 0, \forall |\mathbf{w}| < 2/a_0, \ a_0 < \pi^2$, then the following uncertainty inequality holds:*

$$\left\{ \int_0^{2\pi} \int_{\mathbb{R}^2} \int_0^{a_0} |\mathbf{b}|^2 \big|\Gamma_\Psi[f](a, \mathbf{b}, \theta)\big|^2 \frac{da\, d\mathbf{b}\, d\theta}{a^3} \right\}^{1/2} \left\{ \int_{\mathbb{R}^2} |\mathbf{w}|^2 \big|\mathscr{F}[f](\mathbf{w})\big|^2 d\mathbf{w} \right\}^{1/2} \geq \big\|f\big\|_2^2.$$
(3.246)

Proof. For the given square integrable function f on $\mathbb{R}^2$, we shall recast the Heisenberg's uncertainty inequality (1.296) associated with the two-dimensional Fourier transform as:

$$\left\{ \int_{\mathbb{R}^2} |\mathbf{t}|^2 |f(\mathbf{t})|^2 d\mathbf{t} \right\}^{1/2} \left\{ \int_{\mathbb{R}^2} |\mathbf{w}|^2 \big|\mathscr{F}[f](\mathbf{w})\big|^2 d\mathbf{w} \right\}^{1/2} \geq \left\{ \int_{\mathbb{R}^2} |f(\mathbf{t})|^2 d\mathbf{t} \right\}.$$
(3.247)

Identifying $\Gamma_\Psi[f](a, \mathbf{b}, \theta)$ as a function of the translation variable $\mathbf{b}$ and invoking the inequality (3.247), so that

$$\left\{ \int_{\mathbb{R}^2} |\mathbf{b}|^2 \big|\Gamma_\Psi[f](a, \mathbf{b}, \theta)\big|^2 d\mathbf{b} \right\}^{1/2} \left\{ \int_{\mathbb{R}^2} |\mathbf{w}|^2 \big|\mathscr{F}\big(\Gamma_\Psi[f](a, \mathbf{b}, \theta)\big)(\mathbf{w})\big|^2 d\mathbf{w} \right\}^{1/2}$$
$$\geq \left\{ \int_{\mathbb{R}^2} \big|\Gamma_\Psi[f](a, \mathbf{b}, \theta)\big|^2 d\mathbf{b} \right\}.$$
(3.248)

Integrating (3.248) with respect to the measure $da\, d\theta/a^3$, we obtain

$$\int_0^{2\pi} \int_0^{a_0} \left\{ \int_{\mathbb{R}^2} |\mathbf{b}|^2 \big|\Gamma_\Psi[f](a, \mathbf{b}, \theta)\big|^2 d\mathbf{b} \right\}^{1/2}$$
$$\times \left\{ \int_{\mathbb{R}^2} |\mathbf{w}|^2 \big|\mathscr{F}\big(\Gamma_\Psi[f](a, \mathbf{b}, \theta)\big)(\mathbf{w})\big|^2 d\mathbf{w} \right\}^{1/2} \frac{da\, d\theta}{a^3}$$
$$\geq \left\{ \int_0^{2\pi} \int_0^{a_0} \int_{\mathbb{R}^2} \big|\Gamma_\Psi[f](a, \mathbf{b}, \theta)\big|^2 \frac{d\mathbf{b}\, da\, d\theta}{a^3} \right\}.$$
(3.249)

As a consequence of the Cauchy-Schwarz inequality and the norm equality (3.236), the above inequality can be expressed as

$$\left\{ \int_0^{2\pi} \int_{\mathbb{R}^2} \int_0^{a_0} |\mathbf{b}|^2 \big|\Gamma_\Psi[f](a, \mathbf{b}, \theta)\big|^2 \frac{da\, d\mathbf{b}\, d\theta}{a^3} \right\}^{1/2}$$
$$\times \left\{ \int_0^{2\pi} \int_{\mathbb{R}^2} \int_0^{a_0} |\mathbf{w}|^2 \big|\mathscr{F}\big(\Gamma_\Psi[f](a, \mathbf{b}, \theta)\big)(\mathbf{w})\big|^2 \frac{da\, d\mathbf{w}\, d\theta}{a^3} \right\}^{1/2} \geq \big\|f\big\|_2^2.$$
(3.250)

Invoking Proposition 3.6.1, then employing the polar coordinates (r, ϕ) for $\mathbf{w} \in \mathbb{R}^2$, together with the implication of (3.234), we observe that

$$\int_0^{2\pi} \int_{\mathbb{R}^2} \int_0^{a_0} |\mathbf{w}|^2 \left| \mathscr{F}\left(\Gamma_\Psi\left[f\right](a, \mathbf{b}, \theta)\right)(\mathbf{w}) \right|^2 \frac{da\, d\mathbf{w}\, d\theta}{a^3}$$

$$= (2\pi)^2 \int_0^{2\pi} \int_{\mathbb{R}^2} \int_0^{a_0} |\mathbf{w}|^2 \left| \mathscr{F}\left[f\right](\mathbf{w}) \right|^2 \left| \mathscr{F}\left[\Psi_a\right](R_\theta \mathbf{w}) \right|^2 \frac{da\, d\mathbf{w}\, d\theta}{a^3}$$

$$= (2\pi)^2 \int_{\mathbb{R}^2} |\mathbf{w}|^2 \left| \mathscr{F}\left[f\right](\mathbf{w}) \right|^2 \left\{ \int_0^{2\pi} \int_0^{a_0} \left| \mathscr{F}\left[\Psi_{a,0,\theta}\right](\mathbf{w}) \right|^2 \frac{da\, d\theta}{a^3} \right\} d\mathbf{w}$$

$$= \int_0^{2\pi} \int_0^\infty r^2 \left| \mathscr{F}\left[f\right](r, \phi) \right|^2 \left\{ (2\pi)^2 \int_0^{2\pi} \int_0^{a_0} \left| \mathscr{F}\left[\Psi_{a,0,\theta}\right](r, \phi) \right|^2 \frac{da\, d\theta}{a^3} \right\} r\, dr\, d\phi$$

$$= \int_0^{2\pi} \int_0^\infty r^2 \left| \mathscr{F}\left[f\right](r, \phi) \right|^2 r\, dr\, d\phi$$

$$= \int_{\mathbb{R}^2} |\mathbf{w}|^2 \left| \mathscr{F}\left[f\right](\mathbf{w}) \right|^2 d\mathbf{w}. \tag{3.251}$$

Plugging (3.251) in (3.250), we obtain the Heisenberg-type uncertainty inequality for the curvelet transform (3.220) as

$$\left\{ \int_0^{2\pi} \int_{\mathbb{R}^2} \int_0^{a_0} |\mathbf{b}|^2 \left| \Gamma_\Psi\left[f\right](a, \mathbf{b}, \theta) \right|^2 \frac{da\, d\mathbf{b}\, d\theta}{a^3} \right\}^{1/2} \left\{ \int_{\mathbb{R}^2} |\mathbf{w}|^2 \left| \mathscr{F}\left[f\right](\mathbf{w}) \right|^2 d\mathbf{w} \right\}^{1/2} \geq \left\| f \right\|_2^2.$$

This completes the proof of Theorem 3.6.5.   □

Before concluding the discourse on the curvelet transform, it is of critical significance to demonstrate the construction of curvelets via an example. Note that, the construction of basic curvelet waveforms is governed by the radial and angular window functions W and V satisfying the conditions (3.214) and (3.215). Therefore, our core objective is to illustrate the construction of the admissible window functions W and V. In this direction, we have the following illustrative example:

Example 3.6.1. Consider the following window functions:

$$W(r) = \begin{cases} \cos\left[\frac{\pi}{2}\left(\nu(5 - 6r)\right)\right], & 2/3 \leq r \leq 5/6 \\ 1, & 5/6 \leq r \leq 4/3 \\ \cos\left[\frac{\pi}{2}\left(\nu(3r - 4)\right)\right], & 4/3 \leq r \leq 5/3 \\ 0, & \text{otherwise,} \end{cases} \tag{3.252}$$

and

$$V(\phi) = \begin{cases} \dfrac{1}{2\pi} \cos\left[\frac{\pi}{2}\left(\nu(-3\phi - 1)\right)\right], & -2/3 \leq \phi \leq -1/3 \\ \dfrac{1}{2\pi}, & -1/3 \leq \phi \leq 1/3 \\ \dfrac{1}{2\pi} \cos\left[\frac{\pi}{2}\left(\nu(3\phi - 1)\right)\right], & 1/3 \leq \phi \leq 2/3 \\ 0, & \text{otherwise,} \end{cases} \tag{3.253}$$

where ν is a smooth function, such that

$$\nu(y) = \begin{cases} 0, & y \leq 0 \\ 1, & y \geq 1, \end{cases} \quad \text{and} \quad \nu(y) + \nu(1 - y) = 1, \quad y \in \mathbb{R}. \tag{3.254}$$

Certain choices of the function ν include $\nu(y) = y$ over $[0, 1]$ or even smoother polynomials like $\nu(y) = 3y^2 - 2y^3$ and $\nu(y) = y^5 - 5y^4 + 5y^3$. We note that the smoothness of the window functions W and V is governed by the function ν. The smoother the function ν is, the smoother are the window functions and consequently faster is the decay of curvelets. A typical example of a sufficiently smooth function is given below:

$$\nu(y) = \begin{cases} 0, & y \leq 0 \\ \dfrac{\alpha(y-1)}{\alpha(y-1) + \alpha(y)}, & 0 < y < 1 \\ 1, & y \geq 1, \end{cases} \quad \text{where} \quad \alpha(y) = \exp\left\{ -\dfrac{1}{(1+y)^2} - \dfrac{1}{(1-y)^2} \right\}.$$

$$(3.255)$$

We claim that the window functions (3.252) and (3.253) satisfy the following conditions:

$$\sum_{j=-\infty}^{\infty} |W(2^j r)|^2 = 1 \quad \text{and} \quad \sum_{\ell=-\infty}^{\infty} |V(\phi - \ell)|^2 = 1. \tag{3.256}$$

Since $\text{supp}(V) \subseteq [-2/3, \, 2/3]$, therefore, for any fixed $\phi \in \mathbb{R}$, the sum $\sum_{\ell=-\infty}^{\infty} |V(\phi - \ell)|^2$ contains only two non-vanishing terms, and for $\phi \in [1/3, \, 2/3]$, we have

$$(2\pi)^2 \sum_{\ell=-\infty}^{\infty} |V(\phi - \ell)|^2 = (2\pi)^2 \Big(|V(\phi)|^2 + |V(\phi - 1)|^2 \Big)$$

$$= \cos^2\left[\frac{\pi}{2}(\nu(3\phi - 1))\right] + \cos^2\left[\frac{\pi}{2}(\nu(-3\phi + 2))\right]$$

$$= \cos^2\left[\frac{\pi}{2}(\nu(x))\right] + \cos^2\left[\frac{\pi}{2}(\nu(1 - x))\right]$$

$$= \cos^2\left[\frac{\pi}{2}(\nu(x))\right] + \cos^2\left[\frac{\pi}{2}(1 - \nu(x))\right]$$

$$= \cos^2\left[\frac{\pi}{2}(\nu(x))\right] + \sin^2\left[\frac{\pi}{2}(\nu(x))\right]$$

$$= 1. \tag{3.257}$$

Again, since, $\text{supp}(W) \subset [1/2, \, 2]$, it follows that $\text{supp}(W(2^j r)) \subseteq [2^{-j-1}, \, 2^{1-j}]$, where $j \in \mathbb{Z}$. Consequently, the sum $\sum_{j \in \mathbb{Z}} |W(2^j r)|^2$ has only two non-vanishing terms corresponding to $r \in [1/2, \, 1]$, viz; $|W(r)|^2$ and $|W(2r)|^2$. Thus, for $r \in [1/2, \, 1]$, we have

$$\sum_{j=-\infty}^{\infty} |W(2^j r)|^2 = \Big(|W(r)|^2 + |W(2r)|^2 \Big)$$

$$= \begin{cases} 1, & 1/2 \leq r \leq 2/3 \\ \cos^2\left[\frac{\pi}{2}(\nu(6r - 4))\right] + \cos^2\left[\frac{\pi}{2}(\nu(5 - 6r))\right], & 2/3 \leq r \leq 5/6 \\ 1, & 5/6 \leq r \leq 1 \\ 0, & \text{otherwise.} \end{cases}$$

$$(3.258)$$

Moreover, we observe that

$$\cos^2\left[\frac{\pi}{2}(\nu(6r - 4))\right] + \cos^2\left[\frac{\pi}{2}(\nu(5 - 6r))\right] = \cos^2\left[\frac{\pi}{2}(\nu(z))\right] + \cos^2\left[\frac{\pi}{2}(\nu(1 - z))\right]$$

$$= \cos^2\left[\frac{\pi}{2}(\nu(z))\right] + \cos^2\left[\frac{\pi}{2}(1 - \nu(z))\right]$$

$$= \cos^2\left[\frac{\pi}{2}(\nu(z))\right] + \sin^2\left[\frac{\pi}{2}(\nu(z))\right]$$

$$= 1. \tag{3.259}$$

Using the identity (3.259) in (3.258), we obtain

$$\sum_{j=-\infty}^{\infty} \left| W\left(2^j r\right) \right|^2 = 1. \tag{3.260}$$

Finally, if we choose $\ln 2\, W'(r) = W(r)$, then we shall demonstrate that the window functions W' and V satisfy the admissibility conditions (3.214) and (3.215). We proceed as

$$1 = (2\pi)^2 \sum_{\ell=-\infty}^{\infty} \left| V(\phi - \ell) \right|^2 = (2\pi)^2 \int_0^1 \sum_{\ell=-\infty}^{\infty} \left| V(\phi - \ell) \right|^2 d\phi = (2\pi)^2 \int_{-\infty}^{\infty} \left| V(\phi) \right|^2 d\phi. \tag{3.261}$$

Also, for $r \in (0, \infty)$, we take $r = 2^x$, $x \in (-\infty, \infty)$. Then, we observe that

$$
\begin{aligned}
1 &= \sum_{j=-\infty}^{\infty} \left| W\left(2^j r\right) \right|^2 = B^2 \sum_{j=-\infty}^{\infty} \left| W\left(2^{j+x}\right) \right|^2 = B^2 \int_0^1 \sum_{j=-\infty}^{\infty} \left| W\left(2^{j+x}\right) \right|^2 dx \\
&= \sum_{j=-\infty}^{\infty} \frac{1}{\ln 2} \int_{2^j}^{2^{j+1}} \left| W(y) \right|^2 \frac{dy}{y} \\
&= \sum_{j=-\infty}^{\infty} \int_{2^j}^{2^{j+1}} \left| W'(y) \right|^2 \frac{dy}{y} \\
&= \int_0^{\infty} \left| W'(y) \right|^2 \frac{dy}{y}.
\end{aligned} \tag{3.262}
$$

From the above discourse, it follows that the window functions $W'(r)$ and $V(\phi)$ can be effectively employed for the construction of curvelet waveforms.

3.6.2 Ripplet transform

"Mathematical analysis is as extensive as nature itself; it defines all perceptible relations, measures times, spaces, forces, temperatures; this difficult science is formed slowly, but it preserves every principle which it has once acquired; it grows and strengthens itself incessantly in the midst of the many variations and errors of the human mind."

-Joseph Fourier

The directional representation system of curvelets has gained considerable attention with the versatile capabilities to deal with certain interesting phenomena occurring along curved edges in higher dimensional signals. Unlike the classical wavelets, curvelets obey a parabolic scaling law and can achieve higher anisotropic directionality for resolving two-dimensional singularities occurring along smooth curves. However, the parabolic scaling of the curvelet waveforms is not optimal for resolving the critical geometrical features in higher dimensional signals. The ripplet transform is a refinement of the ordinary curvelet transform relying upon a modified form of the curvelet waveforms with non-parabolic scaling law for resolving two-dimensional singularities. These modified waveforms are known as "ripplets" and are obtained by adding a couple of additional parameter to the classical curvelet waveforms for controlling the effective support and degree of the waveforms [318].

For $r \geq 0$ and $\phi \in [0, 2\pi)$, the basic ripplet waveforms are defined via the polar Fourier transform as

$$\mathscr{F}\left[\Upsilon_{a,\mathbf{0},0}\right](r, \phi) = \frac{a^{(1+d)/2d}}{\sqrt{c}} W(ar) V\left(\frac{a^{1/d}\phi}{ca}\right), \quad c > 0, \; d > 1, \tag{3.263}$$

where $a \in (0, a_0)$, $a_0 < \pi^2$ and the window functions W and V are called as "radial" and "angular" windows, satisfying the admissibility conditions (3.214) and (3.215), respectively. Moreover, the parameters c and d appearing in (3.263) are used to control the effective support and degree of the ripplet waveforms. The introduction of support c and degree d provides an enhanced anisotropy capability for representing singularities along arbitrarily shaped curves. Consequently, the family of complex-valued ripplet waveforms adopted to scale $a > 0$, location $\mathbf{b} \in \mathbb{R}^2$ and orientation $\theta \in [0, 2\pi)$ (or $(-\pi, \pi)$ according to convenience) are defined by

$$\Upsilon_{a,\mathbf{b},\theta}(\mathbf{t}) = \Upsilon_{a,\mathbf{0},0}\left(R_\theta(\mathbf{t} - \mathbf{b})\right), \quad \mathbf{t} \in \mathbb{R}^2, \tag{3.264}$$

where R_θ denotes the usual rotation matrix effecting the planar rotation by θ radians in anti-clockwise direction. In the spatial domain, the functions $\Upsilon_{a,\mathbf{b},\theta}(\mathbf{t})$ have ripple-like shapes, so they are named as "ripplets." The effective support of the ripplets is determined by the major axis, referred as "effective length" pointing in the direction of ripplet and the minor axis, referred as "effective width" which is orthogonal to the major axis. Moreover, the effective region satisfies $width \approx c \times (length)^d$, which plays a crucial role in optimizing the resolving power of ripplets for capturing singularities along arbitrary curves. It is worth noticing that for $c = 1$ and $d = 2$, the ripplet waveforms (3.264) reduce to the usual curvelet waveforms (3.219). As of now, the ripplet waveforms have proved to be of substantial importance for diverse aspects of image processing, for instance image compression, medical diagnosis and feature extraction [159, 318, 319].

Based on the family of ripplet waveforms (3.264), we have the formal definition of the ripplet transform.

Definition 3.6.3. For any function $f \in L^2(\mathbb{R}^2)$, the ripplet transform is denoted by $\Re_\Upsilon\left[f\right]$ and is defined as

$$\Re_\Upsilon\left[f\right](a, \mathbf{b}, \theta) = \int_{\mathbb{R}^2} f(\mathbf{t}) \, \overline{\Upsilon_{a,\mathbf{b},\theta}(\mathbf{t})} \, d\mathbf{t}, \tag{3.265}$$

where $a \in (0, a_0)$ with $a_0 < \pi^2$, $\mathbf{b} \in \mathbb{R}^2$, $\theta \in [0, 2\pi)$ and $\Upsilon_{a,\mathbf{b},\theta}(\mathbf{t})$ is given by (3.264).

Here it is imperative to point out that the reconstruction formula associated with the ripplet transform defined in (3.265) can be obtained in a manner analogous to Theorem 3.6.4. Thus, we infer that the complete reconstruction formula for the ripplet transform is also composed of both ripplet waveforms and isotropic father wavelets. More precisely, we have the following theorem:

Theorem 3.6.6. *If $\Re_\Upsilon\left[f\right](a, \mathbf{b}, \theta)$ is the ripplet transform of any $f \in L^2(\mathbb{R}^2)$, then there exists a band-limited, purely radial function $\Phi_\mathbf{b} \in L^2(\mathbb{R}^2)$ with $\Phi_\mathbf{b}(\mathbf{t}) = \Phi(\mathbf{t} - \mathbf{b})$, such that*

$$f(\mathbf{t}) = \int_{\mathbb{R}^2} \left\langle f, \, \Phi_\mathbf{b} \right\rangle_2 \Phi_\mathbf{b}(\mathbf{t}) \, d\mathbf{b} + \int_0^{2\pi} \int_{\mathbb{R}^2} \int_0^{a_0} \Re_\Upsilon\left[f\right](a, \mathbf{b}, \theta) \, \Upsilon_{a,\mathbf{b},\theta}(\mathbf{t}) \, \frac{da \, d\mathbf{b} \, d\theta}{a^3}, \tag{3.266}$$

where the radial and angular windows W and V satisfy their respective admissibility conditions (3.214) and (3.215).

3.7   The Shearlet Transforms

"Beauty is the first test: there is no permanent place in the world for ugly mathematics."

-Godfrey H. Hardy

Shearlets are the outcome of a series of multiscale methods such as wavelets, ridgelets, curvelets, ripplets and many others introduced during the last few decades with the aim to achieve optimally sparse approximations for higher dimensional signals by employing the basis elements with much higher directional sensitivity and various shapes [171]. Unlike the classical wavelets, shearlets are non-isotropic in nature, they offer optimally sparse representations, they allow compactly supported analyzing elements, they are associated with fast decomposition algorithms and they provide a unified treatment of continuum and digital data. However, similar to the wavelets, they are an affine-like system of well localized waveforms at various scales, locations and orientations; that is, they are generated by dilating and translating a single generating function, where the dilation matrix is the product of a parabolic scaling matrix and a shear matrix and hence, they are a specific type of composite dilation wavelets. The importance of shearlet transforms has been widely acknowledged and since their inception, they have emerged as one of the most effective frameworks for representing multidimensional data ranging over the areas of signal and image processing, remote sensing, data compression, and several others, where the detection of directional structures of the analyzed signals play a role [17, 32, 78, 81, 82, 101, 128, 129, 130, 131, 158, 170, 171, 172, 299].

This section is exclusively meant for a detailed investigation of the theory of shearlet transforms. Primarily, we shall study the notion of shearlet transform from a group theoretic perspective and then proceed with the conventional analytical approach. Subsequently, we study the most important ramification of the continuous shearlet transform, namely, the cone-adapted shearlet transform. Following the convention, we shall rely on the definition of the two-dimensional Fourier transform as given in (1.292).

3.7.1   Shearlet transform: A group theoretic approach

"Mathematics is the most beautiful and most powerful creation of the human spirit."

-Stefan Banach

In this subsection, we begin the formal study of the shearlet transform from a group theoretic perspective. More precisely, we shall demonstrate that the shearlet transform can be obtained via a unitary and square integrable representation of a special group that we refer to as the "shearlet group."

(A) The shearlet group

Consider the set $\mathbb{S} = \mathbb{R}^+ \times \mathbb{R} \times \mathbb{R}^2$, which can be explicitly expressed as

$$\mathbb{S} = \left\{ (a, s, \mathbf{t}) \middle| \ a \in \mathbb{R}^+, s \in \mathbb{R}, \mathbf{t} \in \mathbb{R}^2 \right\}. \tag{3.267}$$

Suppose that the set $\mathbb{S}$ is equipped with the binary operation "$\cdot$" defined as

$$\left(a, s, \mathbf{t}\right) \cdot \left(a', s', \mathbf{t}'\right) = \left(aa', s + s'\sqrt{a}, \mathbf{t} + S_s A_a \mathbf{t}'\right), \tag{3.268}$$

where the matrices A_a and S_s are given by

$$A_a = \begin{pmatrix} a & 0 \\ 0 & \sqrt{a} \end{pmatrix} \quad \text{and} \quad S_s = \begin{pmatrix} 1 & s \\ 0 & 1 \end{pmatrix}. \tag{3.269}$$

Our aim is to show that the set $\mathbb{S}$ constitutes a group under the binary operation given by (3.268). Clearly, $(1, 0, 0)$ is the identity element for the set $\mathbb{S}$. Also, for any $(a, s, \mathbf{t}) \in \mathbb{S}$, we observe that

$$(a, s, \mathbf{t}) \cdot \left(\frac{1}{a}, \frac{-s}{\sqrt{a}}, -A_a^{-1} S_s^{-1} \mathbf{t} \right) = \left(a \left(\frac{1}{a} \right), s + \left(\frac{-s}{\sqrt{a}} \right) \sqrt{a}, \mathbf{t} + S_s A_a \left(-A_a^{-1} S_s^{-1} \mathbf{t} \right) \right)$$
$$= (1, 0, \mathbf{0}).$$

Moreover, we have

$$S_{-s/\sqrt{a}} A_{1/a} = \begin{pmatrix} 1 & -s/\sqrt{a} \\ 0 & 1 \end{pmatrix} \begin{pmatrix} 1/a & 0 \\ 0 & 1/\sqrt{a} \end{pmatrix}$$
$$= \begin{pmatrix} 1/a & -s/a \\ 0 & 1/\sqrt{a} \end{pmatrix}$$
$$= \begin{pmatrix} 1/a & 0 \\ 0 & 1/\sqrt{a} \end{pmatrix} \begin{pmatrix} 1 & -s \\ 0 & 1 \end{pmatrix}$$
$$= A_a^{-1} S_s^{-1},$$

so that

$$\left(\frac{1}{a}, \frac{-s}{\sqrt{a}}, -A_a^{-1} S_s^{-1} \mathbf{t} \right) \cdot (a, s, \mathbf{t})$$
$$= \left(\left(\frac{1}{a} \right) a, \left(\frac{-s}{\sqrt{a}} \right) + s \left(\frac{1}{\sqrt{a}} \right), (-A_a^{-1} S_s^{-1} \mathbf{t}) + S_{-s/\sqrt{a}} A_{1/a} \mathbf{t} \right)$$
$$= (1, 0, \mathbf{0}).$$

Hence, we conclude that any arbitrary element $(a, s, \mathbf{t}) \in \mathbb{S}$ is invertible and the inverse is given by $(1/a, -s/\sqrt{a}, -A_a^{-1} S_s^{-1} \mathbf{t})$. In order to show that the binary operation (3.268) is associative, we choose any trio $(a, s, \mathbf{t}), (a', s', \mathbf{t}'), (a'', s'', \mathbf{t}'') \in \mathbb{S}$ and proceed as follows:

$$((a, s, \mathbf{t}) \cdot (a', s', \mathbf{t}')) \cdot (a'', s'', \mathbf{t}'')$$
$$= (aa', s + s'\sqrt{a}, \mathbf{t} + S_s A_a \mathbf{t}') \cdot (a'', s'', \mathbf{t}'')$$
$$= \left((aa')a'', (s + s'\sqrt{a}) + s''\sqrt{aa'}, (\mathbf{t} + S_s A_a \mathbf{t}') + S_{(s+s'\sqrt{a})} A_{aa'} \mathbf{t}'' \right). \tag{3.270}$$

We observe that

$$s + s'\sqrt{a} + s''\sqrt{aa'} = s + \sqrt{a}s' + (s''\sqrt{a'})\sqrt{a}$$
$$= s + \sqrt{a}s' + \sqrt{a}(s''\sqrt{a'})$$
$$= s + \sqrt{a}(s' + s''\sqrt{a'}) \tag{3.271}$$

and

$$S_{s+s'\sqrt{a}} A_{aa'} = \begin{pmatrix} 1 & s + s'\sqrt{a} \\ 0 & 1 \end{pmatrix} \begin{pmatrix} aa' & 0 \\ 0 & \sqrt{aa'} \end{pmatrix}$$
$$= \begin{pmatrix} aa' & \sqrt{aa'}(s + s'\sqrt{a}) \\ 0 & \sqrt{aa'} \end{pmatrix}$$

$$= \begin{pmatrix} a & s\sqrt{a} \\ 0 & \sqrt{a} \end{pmatrix} \begin{pmatrix} a' & s'\sqrt{a'} \\ 0 & \sqrt{a'} \end{pmatrix}$$

$$= \begin{pmatrix} 1 & s \\ 0 & 1 \end{pmatrix} \begin{pmatrix} a & 0 \\ 0 & \sqrt{a} \end{pmatrix} \begin{pmatrix} 1 & s' \\ 0 & 1 \end{pmatrix} \begin{pmatrix} a' & 0 \\ 0 & \sqrt{a'} \end{pmatrix}$$

$$= S_s A_a S_{s'} A_{a'}. \tag{3.272}$$

Using (3.271) and (3.272) in (3.270), we obtain

$$\big((a, s, \mathbf{t}) \cdot (a', s', \mathbf{t}')\big) \cdot (a'', s'', \mathbf{t}'')$$

$$= \Big(a(a'a''), s + \sqrt{a}(s' + s''\sqrt{a'}, \mathbf{t} + (S_s A_a \mathbf{t}') + (S_s A_a)(S_{s'} A_{'a})\mathbf{t}''\Big)$$

$$= \Big(a(a'a''), s + \sqrt{a}(s' + s''\sqrt{a'}, \mathbf{t} + S_s A_a\big(\mathbf{t}' + (S_{s'} A_{a'})\mathbf{t}''\big)\Big)$$

$$= (a, s, \mathbf{t}) \cdot \big((a', s', \mathbf{t}') \cdot (a'', s'', \mathbf{t}'')\big).$$

Hence, we conclude that $(\mathbb{S}, \cdot)$ is a group and is formally called as the "shearlet group." Moreover, it can be easily verified that the shearlet group is non-Abelian under the binary operation (3.268).

Next, we shall demonstrate that the shearlet group $(\mathbb{S}, \cdot)$ is isomorphic to the locally compact group $G \times \mathbb{R}^2$, where $G = \{M_{sa} = S_s A_a : s \in \mathbb{R}, a \in \mathbb{R}^+\}$, with multiplication defined by $(M, \mathbf{t}) \cdot (M', \mathbf{t}') = (MM', \mathbf{t} + M\mathbf{t}')$. Define a map $\mu : \mathbb{S} \to G \times \mathbb{R}^2$ as

$$\mu(a, s, \mathbf{t}) = \big(S_s A_a, \mathbf{t}\big) = \big(M_{sa}, \mathbf{t}\big). \tag{3.273}$$

Then, by virtue of relation (3.272), we have

$$\mu\big((a, s, \mathbf{t}) \cdot (a', s', \mathbf{t}')\big) = \mu\big(aa', s + s'\sqrt{a}, \mathbf{t} + S_s A_a \mathbf{t}'\big)$$

$$= \big(S_{s+s'\sqrt{a}} A_{aa'}, \mathbf{t} + S_s A_a \mathbf{t}'\big)$$

$$= (S_s A_a S_{s'} A_{a'}, \mathbf{t} + S_s A_a \mathbf{t}')$$

$$= \mu(a, s, \mathbf{t}) \cdot \mu(a', s', \mathbf{t}'). \tag{3.274}$$

Thus, μ is a homomorphism from $\mathbb{S}$ to $G \times \mathbb{R}^2$. Also, for any pair of distinct elements $(a, s, \mathbf{t}), (a', s', \mathbf{t}') \in \mathbb{S}$, we have

$$\mu\big(a, s, \mathbf{t}\big) = \big(S_s A_a, \mathbf{t}\big) = \big(M_{sa}, \mathbf{t}\big) \quad \text{and} \quad \mu\big(a', s', \mathbf{t}'\big) = \big(S_{s'} A_{a'}, \mathbf{t}'\big) = \big(M_{s'a'}, \mathbf{t}\big);$$

that is, $\mu\big(a, s, \mathbf{t}\big) \neq \mu\big(a', s', \mathbf{t}'\big)$. Therefore, we infer that the map $\mu : \mathbb{S} \to G \times \mathbb{R}^2$ is one-to-one. Also, for any $(M_{sa}, \mathbf{t}) \in G \times \mathbb{R}^2$, there exists $(a, s, \mathbf{t}) \in \mathbb{S}$ such that $\mu(a, s, \mathbf{t}) = (S_s A_a, \mathbf{t}) = (M_{sa}, \mathbf{t})$; that is, $\mu : \mathbb{S} \to G \times \mathbb{R}^2$ is an onto map. Hence, we conclude that $\mu : \mathbb{S} \to G \times \mathbb{R}^2$ is an isomorphic map and $\mathbb{S} \approx G \times \mathbb{R}^2$.

Remark 3.7.1. The shearlet group $(\mathbb{S}, \cdot)$ is a subgroup of the full group of motions $GL_2(\mathbb{R}) \times \mathbb{R}^2$ under the binary operation $(M, \mathbf{t}) \cdot (M', \mathbf{t}') = (MM', \mathbf{t} + M\mathbf{t}')$.

In the following proposition, we obtain the left and right Haar measures on the shearlet group $(\mathbb{S}, \cdot)$ defined in (3.267).

Proposition 3.7.1. *The left and right Haar measures on the shearlet group $(\mathbb{S}, \cdot)$ are, respectively, given by*

$$d\mu = \frac{da\, ds\, d\mathbf{t}}{a^3} \qquad and \qquad d\nu = \frac{da\, ds\, d\mathbf{t}}{a}. \tag{3.275}$$

Proof. For any integrable function f defined on the shearlet group $(\mathbb{S}, \cdot)$, we have

$$\int_{\mathbb{S}} f\left((a', s', \mathbf{t}') \cdot (a, s, \mathbf{t})\right) d\mu = \int_{\mathbb{R}^+} \int_{\mathbb{R}} \int_{\mathbb{R}^2} f\left(a'a, s' + s\sqrt{a'}, \mathbf{t}' + S_{s'} A_{a'} \mathbf{t}\right) \frac{da\, ds\, d\mathbf{t}}{a^3}. \quad (3.276)$$

Making use of the substitutions $a'a = \tilde{a}$, $s' + s\sqrt{a'} = \tilde{s}$ and $\mathbf{t}' + S_{s'} A_{a'} \mathbf{t} = \tilde{\mathbf{t}}$ yields

$$\int_{\mathbb{R}^+} \int_{\mathbb{R}} \int_{\mathbb{R}^2} f\left(a'a, s' + s\sqrt{a'}, \mathbf{t}' + S_{s'} A_{a'} \mathbf{t}\right) \frac{da\, ds\, d\mathbf{t}}{a^3} = \int_{\mathbb{R}^+} \int_{\mathbb{R}} \int_{\mathbb{R}^2} f\left(\tilde{a}, \tilde{s}, \tilde{\mathbf{t}}\right) \frac{d\tilde{a}\, d\tilde{s}\, d\tilde{\mathbf{t}}}{\tilde{a}^3}. \quad (3.277)$$

From relations (3.276) and (3.277), we conclude that $d\mu = da\, ds\, d\mathbf{t}/a^3$ is the left Haar measure on the shearlet group $(\mathbb{S}, \cdot)$. Also, using the right operation on the shearlet group $(\mathbb{S}, \cdot)$, we have

$$\int_{\mathbb{S}} f\left((a, s, \mathbf{t}) \cdot (a', s', \mathbf{t}')\right) d\nu = \int_{\mathbb{R}^+} \int_{\mathbb{R}} \int_{\mathbb{R}^2} f\left(aa', s + s'\sqrt{a}, \mathbf{t} + S_s A_a \mathbf{t}'\right) \frac{da\, ds\, d\mathbf{t}}{a}. \quad (3.278)$$

Putting $aa' = \tilde{a}$, $s + s'\sqrt{a} = \tilde{s}$ and $\mathbf{t} + S_s A_a \mathbf{t}' = \tilde{\mathbf{t}}$ into the integrand on the R.H.S of (3.277), we obtain

$$\int_{\mathbb{R}^+} \int_{\mathbb{R}} \int_{\mathbb{R}^2} f\left(aa', s + s'\sqrt{a}, \mathbf{t} + S_s A_a \mathbf{t}'\right) \frac{da\, ds\, d\mathbf{t}}{a} = \int_{\mathbb{R}^+} \int_{\mathbb{R}} \int_{\mathbb{R}^2} f\left(\tilde{a}, \tilde{s}, \tilde{\mathbf{t}}\right) \frac{d\tilde{a}\, d\tilde{s}\, d\tilde{\mathbf{t}}}{\tilde{a}}. \quad (3.279)$$

Therefore, from (3.278) and (3.279) we infer that $d\nu = da\, ds\, d\mathbf{t}/a$ is the right Haar-measure on the shearlet group $(\mathbb{S}, \cdot)$ defined in (3.267). $\qquad \square$

Remark 3.7.2. The shearlet group $(\mathbb{S}, \cdot)$ is non-unimodular as the left Haar measure is different from the right Haar measure.

(B) Unitary representation of the shearlet group

Here, our primary goal is to construct a unitary and square integrable representation of the shearlet group $(\mathbb{S}, \cdot)$ on the Hilbert space $L^2(\mathbb{R}^2)$. To begin with, we have the following proposition:

Proposition 3.7.2. *Let $\psi \in L^2(\mathbb{R}^2)$ and $(a, s, \mathbf{t}) \in (\mathbb{S}, \cdot)$, then the operator $\sigma : \mathbb{S} \to L^2(\mathbb{R}^2)$ defined by*

$$\sigma(a, s, \mathbf{t})\, \psi(\mathbf{x}) = a^{-3/4} \psi\left(A_a^{-1} S_s^{-1}(\mathbf{x} - \mathbf{t})\right) \quad (3.280)$$

is unitary on the Hilbert space $L^2(\mathbb{R}^2)$.

Proof. Let $(a, s, \mathbf{t}), (a', s', \mathbf{t}') \in \mathbb{S}$. Then, we have

$$\sigma(a, s, \mathbf{t})\left(\sigma(a', s', \mathbf{t}')\, \psi\right)(\mathbf{x}) = a^{-3/4}\, \sigma(a', s', \mathbf{t}')\, \psi\left(A_a^{-1} S_s^{-1}(\mathbf{x} - \mathbf{t})\right)$$
$$= (aa')^{-3/4} \psi\left(A_{a'}^{-1} S_{s'}^{-1}\left(A_a^{-1} S_s^{-1}(\mathbf{x} - \mathbf{t}) - \mathbf{t}'\right)\right). \quad (3.281)$$

Observe that,

$$A_{a'}^{-1} S_{s'}^{-1} A_a^{-1} S_s^{-1} = \begin{pmatrix} 1/a' & 0 \\ 0 & 1/\sqrt{a'} \end{pmatrix} \begin{pmatrix} 1 & -s' \\ 0 & 1 \end{pmatrix} \begin{pmatrix} 1 & -s \\ 0 & 1 \end{pmatrix}$$
$$= \begin{pmatrix} 1/a' & -s'/a' \\ 0 & 1/\sqrt{a'} \end{pmatrix} \begin{pmatrix} 1/a & -s/a \\ 0 & 1/\sqrt{a} \end{pmatrix}$$

$$= \begin{pmatrix} \dfrac{1}{aa'} & \dfrac{-s}{aa'} - \dfrac{s'}{a'\sqrt{a}} \\ 0 & \dfrac{1}{\sqrt{aa'}} \end{pmatrix}$$

$$= A_{aa'}^{-1} S_{s+s'\sqrt{a}}^{-1}. \tag{3.282}$$

Using (3.282) in (3.281), we obtain

$$\sigma(a, s, \mathbf{t}) \Big(\sigma(a', s', \mathbf{t}') \, \psi \Big)(\mathbf{x}) = (aa')^{-3/4} \, \psi \big(A_{aa'}^{-1} S_{s+s'\sqrt{a}}^{-1} (\mathbf{x} - (\mathbf{t} + S_s A_a \mathbf{t}'))\big)$$

$$= \sigma \big((a, s, \mathbf{t}) \cdot (a', s', \mathbf{t}') \big) \psi(\mathbf{x}). \tag{3.283}$$

From expression (3.283), it is quite evident that the operator σ defined in (3.280) is invertible and the inverse is given by

$$\sigma^{-1}(a, s, \mathbf{t}) = \sigma \left(\frac{1}{a}, -\frac{s}{\sqrt{a}}, -A_a^{-1} S_s^{-1} \mathbf{t} \right). \tag{3.284}$$

To complete the proof, it remains to show that the operator (3.284) is the adjoint operator corresponding to (3.280). Thus, for any pair of functions ψ, ϕ belonging to the Hilbert space $L^2(\mathbb{R}^2)$, we have

$$\Big\langle \sigma(a, s, \mathbf{t}) \psi, \, \phi \Big\rangle_2 = \int_{\mathbb{R}^2} \sigma(a, s, \mathbf{t}) \psi(\mathbf{x}) \, \overline{\phi(\mathbf{x})} \, d\mathbf{x}$$

$$= a^{-3/4} \int_{\mathbb{R}^2} \psi \big(A_a^{-1} S_s^{-1} (\mathbf{x} - \mathbf{t}) \big) \, \overline{\phi(\mathbf{x})} \, d\mathbf{x}$$

$$= a^{3/2} \int_{\mathbb{R}^2} \psi(\mathbf{z}) \, \overline{\phi \left(S_s A_a \mathbf{z} + \mathbf{t} \right)} \, d\mathbf{z}$$

$$= \int_{\mathbb{R}^2} \psi(\mathbf{z}) \, \overline{\sigma^{-1}(a, \mathbf{b}, \theta) \phi(\mathbf{z})} \, d\mathbf{z}$$

$$= \Big\langle \psi, \, \sigma^{-1}(a, s, \mathbf{t}) \phi \Big\rangle_2.$$

Hence, we conclude that $\sigma : \mathbb{S} \to L^2(\mathbb{R}^2)$ is indeed a unitary operator. $\square$

Proposition 3.7.3. *Let* $\mathcal{U}\big(L^2(\mathbb{R}^2)\big)$ *denotes the group of all unitary operators on* $L^2(\mathbb{R}^2)$ *with respect to the composition of mappings "* $\odot$ *." Then, the map* $\sigma : \mathbb{S} \to \mathcal{U}\big(L^2(\mathbb{R}^2)\big)$ *defined in (3.280) is a unitary representation of the shearlet group* $(\mathbb{S}, \cdot)$ *on the Hilbert space* $L^2(\mathbb{R}^2)$.

Proof. To accomplish the proof, we have to show that the map $\sigma : \mathbb{S} \to L^2(\mathbb{R}^2)$ given by (3.280) is a group homomorphism. For any pair of elements $(a, s, \mathbf{t})$ and $(a', s', \mathbf{t}')$ belonging to the shearlet group $(\mathbb{S}, \cdot)$, we have

$$\Big(\sigma(a, s, \mathbf{t}) \odot \sigma(a', s', \mathbf{t}') \Big) \psi(\mathbf{x}) = \sigma(a, s, \mathbf{t}) \Big(\sigma(a', s', \mathbf{t}') \, \psi \Big)(\mathbf{t})$$

$$= (aa')^{-3/4} \psi \big(A_{a'}^{-1} S_{s'}^{-1} \big(A_a^{-1} S_s^{-1} (\mathbf{x} - \mathbf{t}) - \mathbf{t}' \big) \big)$$

$$= \sigma(aa', s + s'\sqrt{a}, \mathbf{t} + S_s A_a \mathbf{t}') \, \psi(\mathbf{x})$$

$$= \sigma \big((a, s, \mathbf{t}) \cdot (a', s', \mathbf{t}') \big) \, \psi(\mathbf{x}). \tag{3.285}$$

From the expression (3.285), it is clear that the map $\sigma : \mathbb{S} \to \mathcal{U}\big(L^2(\mathbb{R}^2)\big)$ preserves the group operations and is thus a group homomorphism. Hence, we conclude that $\sigma : \mathbb{S} \to \mathcal{U}\big(L^2(\mathbb{R}^2)\big)$ is a unitary representation of the shearlet group $(\mathbb{S}, \cdot)$ on the Hilbert space $L^2(\mathbb{R}^2)$. $\square$

Theorem 3.7.4. *For any pair of functions* $\psi, \phi \in L^2(\mathbb{R}^2)$, *we have*

$$\int_{\mathbb{S}} \left| \left\langle \phi, \sigma(a, s, \mathbf{t}) \psi \right\rangle_2 \right|^2 d\mu = \int_{\mathbb{R}} \int_0^\infty \left| \hat{f}(\omega_1, \omega_2) \right|^2 \left\{ \int_0^\infty \int_{\mathbb{R}} \frac{|\hat{\psi}(\xi_1, \xi_2)|^2}{\xi_1^2} d\xi_2 \, d\xi_1 \right\} d\omega_1 d\omega_2$$

$$+ \int_{\mathbb{R}} \int_{-\infty}^0 \left| \hat{f}(\omega_1, \omega_2) \right|^2 \left\{ \int_{-\infty}^0 \int_{\mathbb{R}} \frac{|\hat{\psi}(\xi_1, \xi_2)|^2}{\xi_1^2} d\xi_2 \, d\xi_1 \right\} d\omega_1 d\omega_2.$$

$$(3.286)$$

Proof. In the first instance, we note that the integrand on the L.H.S of (3.286) can be expressed via the classical two-dimensional convolution operation $\circledast$, given in (3.226), as

$$\left\langle \phi, \sigma(a, s, \mathbf{t}) \psi \right\rangle_2 = \int_{\mathbb{R}^2} \phi(\mathbf{x}) \, \overline{\sigma(a, s, \mathbf{t}) \psi(\mathbf{x})} \, d\mathbf{x}$$

$$= a^{-3/4} \int_{\mathbb{R}^2} \phi(\mathbf{x}) \, \overline{\psi\left(A_a^{-1} S_s^{-1}(\mathbf{x} - \mathbf{t})\right)} \, d\mathbf{x}$$

$$= a^{-3/4} \int_{\mathbb{R}^2} \phi(\mathbf{x}) \, \overline{\psi\left(A_a^{-1} S_s^{-1}(\mathbf{t} - \mathbf{x})\right)} \, d\mathbf{x}$$

$$= \int_{\mathbb{R}^2} \phi(\mathbf{x}) \, \sigma(a, s, \mathbf{0}) \overline{\check{\psi}}(\mathbf{t} - \mathbf{x}) \, d\mathbf{x},$$

$$= \left(\phi \circledast \sigma(a, s, \mathbf{0}) \overline{\check{\psi}} \right)(\mathbf{t}), \quad \check{\psi}(\mathbf{x}) = \psi(-\mathbf{x}). \qquad (3.287)$$

In view of the definition of two-dimensional Fourier transform given by (1.292), we can express (3.287) as follows:

$$\int_{\mathbb{S}} \left| \left\langle \phi, \sigma(a, s, \mathbf{t}) \psi \right\rangle_2 \right|^2 d\mu = \int_{\mathbb{R}^2} \int_{\mathbb{R}} \int_{\mathbb{R}^+} \left| \left(\phi \circledast \sigma(a, s, \mathbf{0}) \overline{\check{\psi}} \right)(\mathbf{t}) \right|^2 \frac{da \, ds \, d\mathbf{t}}{a^3}$$

$$= \int_{\mathbb{R}^2} \int_{\mathbb{R}} \int_{\mathbb{R}^+} \left| \mathscr{F}\left(\phi \circledast \sigma(a, s, \mathbf{0}) \overline{\check{\psi}} \right)(\mathbf{w}) \right|^2 \frac{da \, ds \, d\mathbf{w}}{a^3}$$

$$= \int_{\mathbb{R}^2} \int_{\mathbb{R}} \int_{\mathbb{R}^+} \left| \mathscr{F}[\phi](\mathbf{w}) \, \mathscr{F}\left[\sigma(a, s, \mathbf{0}) \overline{\check{\psi}} \right](\mathbf{w}) \right|^2 \frac{da \, ds \, d\mathbf{w}}{a^3}. \quad (3.288)$$

In order to simplify the R.H.S of (3.288), we proceed as

$$\mathscr{F}\left[\sigma(a, s, \mathbf{0}) \overline{\check{\psi}} \right](\mathbf{w}) = \int_{\mathbb{R}^2} \sigma(a, s, \mathbf{0}) \overline{\check{\psi}}(\mathbf{x}) \, e^{-2\pi i \mathbf{x}^T \mathbf{w}} \, d\mathbf{x}$$

$$= a^{-3/4} \int_{\mathbb{R}^2} \overline{\psi\left(- A_a^{-1} S_s^{-1} \mathbf{x} \right)} \, e^{-2\pi i \mathbf{x}^T \cdot \mathbf{w}} \, d\mathbf{x}$$

$$= a^{3/4} \int_{\mathbb{R}^2} \overline{\psi(\mathbf{z})} \, e^{-2\pi i (-S_s A_a \mathbf{z})^T \mathbf{w}} \, d\mathbf{z}$$

$$= a^{3/4} \int_{\mathbb{R}^2} \overline{\psi(\mathbf{z})} \, e^{-2\pi i \mathbf{z}^T (A_a^T S_s^T \mathbf{w})} \, d\mathbf{z}$$

$$= a^{3/4} \overline{\mathscr{F}[\psi]\left(A_a^T S_s^T \mathbf{w} \right)}. \qquad (3.289)$$

Implementing (3.289) in (3.288), we obtain

$$\int_{\mathbb{S}} \left| \left\langle \phi, \sigma(a, s, \mathbf{t}) \psi \right\rangle_2 \right|^2 d\mu = \int_{\mathbb{R}^2} \int_{\mathbb{R}} \int_{\mathbb{R}^+} \left| \mathscr{F}[\phi](\mathbf{w}) \right|^2 \left| \mathscr{F}[\psi]\left(A_a^T S_s^T \mathbf{w} \right) \right|^2 \frac{da \, ds \, d\mathbf{w}}{a^{3/2}}. \quad (3.290)$$

Now, we observe that

$$A_a^T S_s^T \omega = \begin{pmatrix} a & 0 \\ 0 & \sqrt{a} \end{pmatrix} \begin{pmatrix} 1 & 0 \\ s & 1 \end{pmatrix} \begin{pmatrix} \omega_1 \\ \omega_2 \end{pmatrix}$$

$$= \begin{pmatrix} a & 0 \\ s\sqrt{a} & \sqrt{a} \end{pmatrix} \begin{pmatrix} \omega_1 \\ \omega_2 \end{pmatrix}$$

$$= \left(a\,\omega_1,\, s\sqrt{a}\,\omega_1 + \sqrt{a}\,\omega_2 \right). \tag{3.291}$$

Using (3.291) in (3.290), we get

$$\int_{\mathbb{S}} \left| \left\langle \phi, \sigma(a,s,\mathbf{t})\psi \right\rangle_2 \right|^2 d\mu = \int_{\mathbb{R}^+} \int_{\mathbb{R}} \int_{\mathbb{R}} \int_{\mathbb{R}} \left| \hat{f}(\omega_1,\omega_2) \right|^2 \left| \hat{\psi}\left(a\,\omega_1,\, \sqrt{a}(s\omega_1 + \omega_2) \right) \right|^2 \frac{ds\,d\omega_1\,d\omega_2\,da}{a^{3/2}}. \tag{3.292}$$

Making appropriate substitutions in the R.H.S of (3.292) yields the desired result as follows:

$$\int_{\mathbb{S}} \left| \left\langle \phi, \sigma(a,s,\mathbf{t})\psi \right\rangle_2 \right|^2 d\mu = \int_{\mathbb{R}} \int_0^\infty \int_0^\infty \int_{\mathbb{R}} \left| \hat{f}(\omega_1,\omega_2) \right|^2 \left| \hat{\psi}(a\,\omega_1, \xi_2) \right|^2 \frac{d\xi_2\,da\,d\omega_1 d\omega_2}{a^2 \omega_1}$$

$$- \int_{\mathbb{R}} \int_{-\infty}^0 \int_0^\infty \int_{\mathbb{R}} \left| \hat{f}(\omega_1,\omega_2) \right|^2 \left| \hat{\psi}(a\,\omega_1, \xi_2) \right|^2 \frac{d\xi_2\,da\,d\omega_1 d\omega_2}{a^2 \omega_1}$$

$$= \int_{\mathbb{R}} \int_0^\infty \left| \hat{f}(\omega_1,\omega_2) \right|^2 \left\{ \int_0^\infty \int_{\mathbb{R}} \frac{\left| \hat{\psi}(\xi_1,\xi_2) \right|^2}{\xi_1^2} d\xi_2\,d\xi_1 \right\} d\omega_1 d\omega_2$$

$$+ \int_{\mathbb{R}} \int_{-\infty}^0 \left| \hat{f}(\omega_1,\omega_2) \right|^2 \left\{ \int_{-\infty}^0 \int_{\mathbb{R}} \frac{\left| \hat{\psi}(\xi_1,\xi_2) \right|^2}{\xi_1^2} d\xi_2\,d\xi_1 \right\} d\omega_1 d\omega_2.$$

This completes the proof of Theorem 3.7.4.   □

Below, we present an important corollary regarding the square integrability of the unitary representation $\sigma : \mathbb{S} \to \mathcal{U}\left(L^2(\mathbb{R}^2)\right)$ defined in (3.280).

Corollary 3.7.5. *For any pair of functions $\psi, \phi \in L^2(\mathbb{R}^2)$ such that*

$$\int_{\mathbb{R}} \int_{\mathbb{R}} \frac{\left| \hat{\psi}(\omega_1,\omega_2) \right|^2}{\omega_1^2} d\omega_1\,d\omega_2 < \infty, \tag{3.293}$$

we have

$$\int_{\mathbb{S}} \left| \left\langle \phi, \sigma(a,s,\mathbf{t})\psi \right\rangle_2 \right|^2 d\mu < \infty, \tag{3.294}$$

where $d\mu$ denotes the left Haar measure on the shearlet group $(\mathbb{S}, \cdot)$ defined in (3.267). That is, the unitary representation $\sigma : \mathbb{S} \to \mathcal{U}\left(L^2(\mathbb{R}^2)\right)$ of the shearlet group $(\mathbb{S}, \cdot)$ on the Hilbert space $L^2(\mathbb{R}^2)$ is square integrable.

By virtue of Corollary 3.7.5, we infer that the set of all those functions in $L^2(\mathbb{R}^2)$ satisfying the condition (3.293) is particularly interesting in the sense that the unitary representation $\sigma : \mathbb{S} \to \mathcal{U}\left(L^2(\mathbb{R}^2)\right)$ turns to be square integrable. In this direction, we have the following definition of an admissible shearlet:

Definition 3.7.1. A function $\psi \in L^2(\mathbb{R}^2)$ satisfying the condition:

$$\int_{\mathbb{R}} \int_{\mathbb{R}} \frac{\left| \hat{\psi}(\omega_1,\omega_2) \right|^2}{\omega_1^2} d\omega_1\,d\omega_2 < \infty, \tag{3.295}$$

is called an "admissible shearlet." Condition (3.295) is called as the "admissibility condition."

The admissibility condition (3.295) plays a central role in the development of the theory of shearlet transforms as it usually yields a resolution of identity which in turn allows the reconstruction of the input signal from the transformed one. Based on Definition 3.7.1, we shall introduce the notion of abstract shearlet transform.

Definition 3.7.2. Let $\psi \in L^2(\mathbb{R}^2)$ be an admissible shearlet for the square integrable representation $\sigma : \mathbb{S} \to \mathcal{U}(L^2(\mathbb{R}^2))$ of the shearlet group $(\mathbb{S}, \cdot)$ on the Hilbert space $L^2(\mathbb{R}^2)$ defined in (3.280). Then, for any $f \in L^2(\mathbb{R}^2)$, the abstract shearlet transform is defined as

$$\mathcal{S}_\psi\big[f\big](a, s, \mathbf{t}) = \Big\langle f,\, \sigma(a, s, \mathbf{t})\, \psi \Big\rangle_2 = a^{-3/4} \int_{\mathbb{R}^2} f(\mathbf{x})\, \overline{\psi\big(A_a^{-1} S_s^{-1}(\mathbf{x} - \mathbf{t})\big)}\, d\mathbf{x}, \qquad (3.296)$$

where $(a, s, \mathbf{t}) \in \mathbb{S}$.

Remark 3.7.3. Definition 3.7.2 implies that, indeed, the shearlet transform is a unitary, square integrable representation of the shearlet group $(\mathbb{S}, \cdot)$ on the Hilbert space $L^2(\mathbb{R}^2)$.

Proposition 3.7.6. *Given an admissible shearlet* $\psi \in L^2(\mathbb{R}^2)$, *denote*

$$C_\psi^+ = \int_{\mathbb{R}} \int_0^\infty \frac{\big|\hat{\psi}(\omega_1, \omega_2)\big|^2}{\omega_1^2}\, d\omega_1\, d\omega_2 \quad and \quad C_\psi^- = \int_{\mathbb{R}} \int_{-\infty}^0 \frac{\big|\hat{\psi}(\omega_1, \omega_2)\big|^2}{\omega_1^2}\, d\omega_1\, d\omega_2. \qquad (3.297)$$

If $C_\psi^+ = C_\psi^- = C_\psi$, *then the shearlet transform defined in* (3.296) *is a* C_ψ-*multiple of an isometry.*

Proof. Using Theorem 3.7.4, the result follows immediately. □

3.7.2 Shearlet transform: An analytical perspective

"It is always the case, with mathematics, that a little direct experience of thinking over things on your own can provide a much deeper understanding than merely reading about them."

-Roger Penrose

In this subsection, our aim is liberate the concept of shearlet transform from the abstract notions and heuristically examine the theory purely from an analytical perspective.

Despite the spectacular success of classical wavelets in the analysis of univariate data, they tend to lose their efficacy when dealing with multivariate data. In fact, the wavelet representations are optimal for approximating data with pointwise singularities only and cannot handle equally well distributed singularities, such as singularities along curves. Such limitations of the traditional multiscale systems have stimulated a flurry of activity among researchers and lead to the development of several directional representation systems including two-dimensional wavelets, curvelets, ripplets and so on. One of the severe drawbacks of the aforementioned directional representation systems is that they are reliant upon rotations, which do not preserve the digital lattice and thereby prevent a direct transition from the continuum to the digital setting.

Shearlets emerged as part of an extensive research activity developed over the last decade with the aim to create a new generation of analysis and processing tools for massive and higher dimensional data, which could go beyond the limitations of traditional Fourier and wavelet systems. This approach was derived within a larger class of affine-like systems-the so called composite wavelets-as a truly multivariate extension of the wavelet framework.

One of the distinctive features of shearlets is the use of shearing to control directional selectivity, in contrast to rotation used in the construction of two-dimensional wavelets, curvelets, ripplets and so on. This is a fundamentally different concept, since it allows shearlet systems to be derived from a single or finite set of generators, and it also ensures a unified treatment of the continuum and digital world due to the fact that the shear matrix preserves the integer lattice [171].

The construction of shearlet waveforms requires a combination of an appropriate scaling operator to generate elements at different scales, an orthogonal operator to change their orientations, and a translation operator to displace these elements over the plane. For $a \in \mathbb{R}^+$, we choose the parabolic scaling matrix:

$$A_a = \begin{pmatrix} a & 0 \\ 0 & \sqrt{a} \end{pmatrix}. \tag{3.298}$$

Based on the parabolic scaling matrix (3.298), let $\mathcal{D}_{A_a}$ denotes the dilation operator acting on $\psi \in L^2(\mathbb{R}^2)$ as

$$\big(\mathcal{D}_{A_a}\psi\big)(\mathbf{x}) = |\det A_a|^{-1/2}\psi(A_a^{-1}\mathbf{x}). \tag{3.299}$$

The use of parabolic scaling is quite justifiable from the fact that such a scaling law is best adapted to obtain optimally sparse approximations of two-dimensional signals endowed with anisotropic features, abreast of upholding the mathematical elegance.

Next, we require an orthogonal transformation to change the orientations of the waveforms. It is pertinent to mention that the most obvious choice of using a rotation operator is eluded due to the reason that rotations destroy the structure of the integer lattice $\mathbb{Z}^2$ whenever the rotation angle is different from $0, \pm\pi/2, \pm\pi, \pm 3\pi/2$. This issue becomes a serious problem for the transition from the continuum to the digital setting. As an alternative orthogonal transformation, we choose the shearing operator $\mathscr{S}_{S_s}$ based on the shearing matrix:

$$S_s = \begin{pmatrix} 1 & s \\ 0 & 1 \end{pmatrix} \tag{3.300}$$

and acting on $\psi \in L^2(\mathbb{R}^2)$ as

$$\big(\mathscr{S}_{S_s}\psi\big)(\mathbf{x}) = \psi(S_s^{-1}\mathbf{x}). \tag{3.301}$$

The shearing matrix (3.300) parameterizes the orientations using the variable s associated with the slopes rather than the angles, and has the advantage of leaving the integer lattice invariant, provided s is an integer.

Finally, in order to displace the waveforms over the Euclidean plane $\mathbb{R}^2$, we choose the usual planer translation operator $\mathcal{T}_{\mathbf{t}}$, $\mathbf{t} \in \mathbb{R}^2$ acting on the function $\psi \in L^2(\mathbb{R}^2)$ in the following fashion:

$$\big(\mathcal{T}_{\mathbf{t}}\psi\big)(\mathbf{x}) = \psi(\mathbf{x} - \mathbf{t}). \tag{3.302}$$

In pursuit of constructing a new family of analysis and processing tools for multi-variate data, it is of crucial importance to choose an appropriate generating function $\psi \in L^2(\mathbb{R}^2)$. With the notion of abstract shearlet transform in hindsight, it is imperative to choose the generating function in such a way that it satisfies the admissibility condition (3.295). Also, note that the operator σ defined in (3.280) can be viewed as the joint action of the above defined dilation, shearing and translation operators. Therefore, for any $\psi \in L^2(\mathbb{R}^2)$ satisfying the admissibility condition (3.295), we define a family of functions as

$$\mathfrak{F}_\psi(a, s, \mathbf{t}) := \left\{ \psi_{a,s,\mathbf{t}}(\mathbf{x}) = \mathcal{T}_\mathbf{t} \mathcal{D}_{A_a} \mathscr{S}_{S_s} \psi(\mathbf{x}); \; a \in \mathbb{R}^+, s \in \mathbb{R} \text{ and } \mathbf{t} \in \mathbb{R}^2 \right\}. \qquad (3.303)$$

The generating function ψ in (3.303) is called as the "shearlet." Moreover, it is worth noticing that the analyzing elements of the family (3.303) satisfy the following norm equality:

$$\left\| \psi_{a,s,\mathbf{t}}(\mathbf{x}) \right\|_2^2 = \int_{\mathbb{R}^2} \left| \mathcal{T}_\mathbf{t} \mathcal{D}_{A_a} \mathscr{S}_{S_s} \psi(\mathbf{x}) \right|^2 d\mathbf{x} = \int_{\mathbb{R}^2} \left| \psi(\mathbf{x}) \right|^2 d\mathbf{x} = \left\| \psi \right\|_2^2. \qquad (3.304)$$

Based on the family of analyzing shearlets (3.303), we have the following definition of shearlet transform:

Definition 3.7.3. The shearlet transform of any $f \in L^2(\mathbb{R}^2)$ with respect to the admissible shearlet $\psi \in L^2(\mathbb{R}^2)$ is defined by

$$\mathcal{S}_\psi \big[f \big](a, s, \mathbf{t}) = \big\langle f, \psi_{a,s,\mathbf{t}} \big\rangle_2 = |\det A_a|^{-1/2} \int_{\mathbb{R}^2} f(\mathbf{x}) \overline{\psi\big(A_a^{-1} S_s^{-1}(\mathbf{x} - \mathbf{t})\big)} \, d\mathbf{x}, \qquad (3.305)$$

where $\psi_{a,s,\mathbf{t}}(\mathbf{x})$ is given by (3.303).

Below, we present some important points regarding the above definition of the continuous shearlet transform.

(i) Unlike the two-dimensional wavelet transform, the dilation operator used in Definition 3.7.3 is based on the parabolic scaling law; that is, the dilation is always doubly effective in one fixed direction as in the orthogonal direction.

(ii) The shearing matrix involved in Definition 3.7.3 parameterizes the orientations using the variable s associated with the slopes in lieu of the angles used in the two-dimensional wavelet transform.

Remark 3.7.4. The shearlet transform of any $f \in L^2(\mathbb{R}^2)$ with respect to the shearlet $\psi \in L^2(\mathbb{R}^2)$ can be expressed via the two-dimensional convolution operation (3.226) as

$$\mathcal{S}_\psi \big[f \big](a, s, \mathbf{t}) = \int_{\mathbb{R}^2} f(\mathbf{x}) \, \overline{\psi_{a,s,\mathbf{t}}(\mathbf{x})} \, d\mathbf{x}$$

$$= \int_{\mathbb{R}^2} f(\mathbf{x}) \, \overline{\breve{\psi}_{a,s,\mathbf{0}}(\mathbf{t} - \mathbf{x})} \, d\mathbf{x}$$

$$= \big(f \circledast \overline{\breve{\psi}}_{a,s,\mathbf{0}} \big)(\mathbf{t}), \quad \breve{\psi}(\mathbf{x}) = \psi(-\mathbf{x}). \qquad (3.306)$$

Having presented the formal definition of the continuous shearlet transform as given by (3.305), we shall demonstrate the construction of admissible shearlets in the following illustrative example:

Example 3.7.1. The process of construction of admissible shearlets is somewhat straightforward. Essentially, any function $\psi \in L^2(\mathbb{R}^2)$ such that the corresponding Fourier transform $\hat{\psi}$ is compactly supported away from origin is an admissible shearlet. One of the important examples of such functions is the well-known classical shearlet, defined by

$$\hat{\psi}(\mathbf{w}) = \hat{\psi}(\omega_1, \omega_2) = \hat{\psi}_1(\omega_1) \, \hat{\psi}_2 \left(\frac{\omega_2}{\omega_1} \right), \qquad (3.307)$$

where $\psi_1 \in L^2(\mathbb{R})$ is such that $\hat{\psi}_1 \in C^\infty(\mathbb{R})$, the class of smooth functions, with supp $(\hat{\psi}_1) \subseteq [-1/2, -1/16] \cup [1/16, 1/2]$ and satisfies the discrete wavelet Claderón condition:

$$\sum_{j\in\mathbb{Z}} \left| \hat{\psi}_1\left(2^{-j}\omega_1\right) \right|^2 = 1 \quad \text{a.e.} \quad \omega_1 \in \mathbb{R}. \tag{3.308}$$

Moreover, $\psi_2 \in L^2(\mathbb{R})$ is chosen to be bump function in the sense that

$$\sum_{k=-1}^{1} \left| \hat{\psi}_2(\omega_1 + k) \right|^2 = 1 \quad \text{a.e.} \quad \omega_1 \in [-1, 1], \tag{3.309}$$

with $\hat{\psi}_2 \in C^\infty(\mathbb{R})$ and supp $(\hat{\psi}_2) \subseteq [-1, 1]$. Therefore, the classical shearlet ψ defined via (3.307) is wavelet-like along one axis and bump-like along the other axis.

There are several choices for choosing the functions ψ_1 and ψ_2 satisfying the conditions (3.308) and (3.309). Below, we shall construct such a pair of functions. Consider a real valued function $\alpha : \mathbb{R} \to \mathbb{R}$ defined as

$$\alpha(x_1) = \begin{cases} 0, & x_1 < 0 \\ 35x_1^4 - 84x_1^5 + 70x_1^6 - 20x_1^7 & 0 \le x_1 \le 1 \\ 1, & x_1 > 1. \end{cases} \tag{3.310}$$

Note that the function α is symmetric about $(1/2, 1/2)$ and is the same function used in the construction of the well-known Meyer wavelet. Define another function $\beta : \mathbb{R} \to \mathbb{R}$ as

$$\beta(x_2) = \begin{cases} \sin\left(\frac{\pi}{2}\alpha(|x_2| - 1)\right), & 1 \le |x_2| \le 2 \\ \cos\left(\frac{\pi}{2}\alpha\left(\frac{1}{2}|x_2| - 1\right)\right), & 2 \le |x_2| \le 4 \\ 0, & \text{otherwise.} \end{cases} \tag{3.311}$$

The function β is positive, real-valued symmetric with compact support on the interval $[-4, -1] \cup [1, 4]$. For a pictorial illustration, the functions α and β are shown in Figure 3.24.

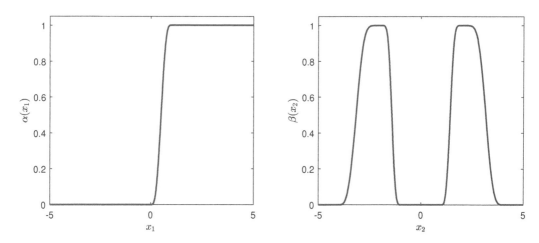

FIGURE 3.24: The shearlet recipes $\alpha(x_1)$ and $\beta(x_2)$.

Using the functions α and β, we construct the function ψ_1 via the corresponding Fourier transform as

$$\hat{\psi}_1(\omega_1) = \sqrt{\beta^2(\omega_1) + \beta^2(2\omega_1)}. \tag{3.312}$$

Then, we observe that the function $\hat{\psi}_1$ has compact support over $[-4, -1/2]\cup[1/2, 4]$. Next, we define another function ψ_2 via the corresponding Fourier transform as

$$\hat{\psi}_2(\omega_2) = \begin{cases} \sqrt{\alpha(1+\omega_2)}, & \omega_2 \leq 0 \\ \sqrt{\alpha(1-\omega_2)}, & \omega_2 > 0. \end{cases} \tag{3.313}$$

Consequently, the classical shearlet is obtained by substituting (3.312) and (3.313) in (3.307). The functions (3.312) and (3.313) together with the resultant shearlet in the frequency domain are depicted in Figures 3.25–3.27. Moreover, for an intuitive understanding of the effect of shearlets on the frequency plane, the corresponding tiling on the frequency plane in depicted Figure 3.28.

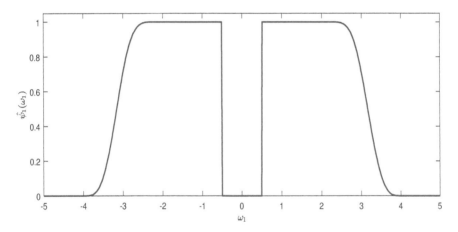

FIGURE 3.25: The function $\hat{\psi}_1(\omega_1)$.

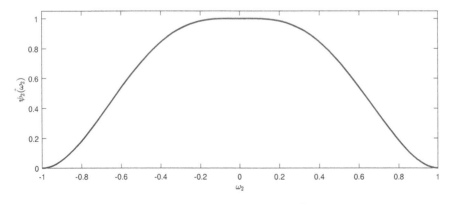

FIGURE 3.26: The function $\hat{\psi}_2(\omega_2)$.

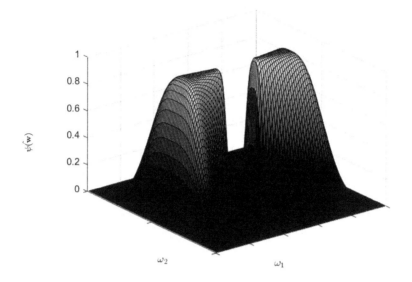

FIGURE 3.27: The classical shearlet (3.307).

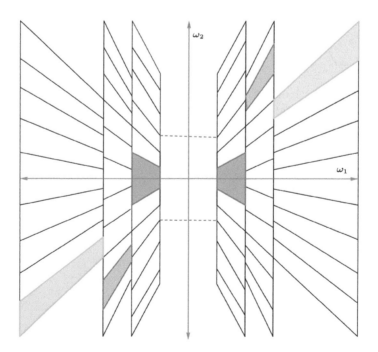

FIGURE 3.28: Tiling on the frequency plane due to shearlets.

In the following proposition, we express the shearlet transform (3.305) via the two-dimensional Fourier transform defined in (1.292).

Proposition 3.7.7. *Let* $\psi \in L^2(\mathbb{R}^2)$ *be an admissible shearlet. Then, for any* $f \in L^2(\mathbb{R}^2)$, *the continuous shearlet transform* $\mathcal{S}_\psi[f](a,s,\mathbf{t})$ *defined in (3.305) can be expressed as:*

$$\mathscr{F}\Big(\mathcal{S}_\psi[f](a,s,\mathbf{t})\Big)(\mathbf{w}) = a^{3/4}\,\mathscr{F}[f](\mathbf{w})\,\overline{\mathscr{F}[\psi](A_a^T S_s^T \mathbf{w})}, \qquad (3.314)$$

where the Fourier transform on L.H.S of (3.314) is computed with respect to the translation variable $\mathbf{t}$.

Proof. Using the convolution expression of the continuous shearlet transform obtained in (3.306) and then applying the two-dimensional Fourier transform given by (1.292), we obtain

$$\mathscr{F}\Big(\mathcal{S}_\psi[f](a,s,\mathbf{t})\Big)(\mathbf{w}) = \mathscr{F}\Big[(f \circledast \check{\bar{\psi}}_{a,s,0})(\mathbf{t})\Big](\mathbf{w})$$
$$= \mathscr{F}[f](\mathbf{w})\,\mathscr{F}\Big[\check{\bar{\psi}}_{a,s,0}\Big](\mathbf{w}). \qquad (3.315)$$

Observe that,

$$\mathscr{F}\Big[\check{\bar{\psi}}_{a,s,0}\Big](\mathbf{w}) = \int_{\mathbb{R}^2} \check{\bar{\psi}}_{a,s,0}(\mathbf{x})\, e^{-2\pi i \mathbf{x}^T \mathbf{w}}\, d\mathbf{x}$$
$$= a^{-3/4} \int_{\mathbb{R}^2} \overline{\psi\big(-A_a^{-1} S_s^{-1} \mathbf{x}\big)}\, e^{-2\pi i \mathbf{x}^T \mathbf{w}}\, d\mathbf{x}$$
$$= a^{3/4} \int_{\mathbb{R}^2} \overline{\psi(\mathbf{z})}\, e^{-2\pi i (-S_s A_a \mathbf{z})^T \mathbf{w}}\, d\mathbf{z}$$
$$= a^{3/4} \int_{\mathbb{R}^2} \overline{\psi(\mathbf{z})}\, e^{-2\pi i \mathbf{z}^T (A_a^T S_s^T \mathbf{w})}\, d\mathbf{z}$$
$$= a^{3/4}\,\overline{\mathscr{F}[\psi](A_a^T S_s^T \mathbf{w})}. \qquad (3.316)$$

Plugging (3.316) in (3.315), we obtain the desired expression as

$$\mathscr{F}\Big(\mathcal{S}_\psi[f](a,s,\mathbf{t})\Big)(\mathbf{w}) = a^{3/4}\,\mathscr{F}[f](\mathbf{w})\,\overline{\mathscr{F}[\psi](A_a^T S_s^T \mathbf{w})}.$$

This completes the proof of Proposition 3.7.7. $\qquad\qquad\Box$

In the following theorem, we present some fundamental properties of the shearlet transform (3.305).

Theorem 3.7.8. *If* ψ *and* ϕ *are any two shearlets, then for the scalars* $\alpha, \beta \in \mathbb{C},\ \mathbf{k} \in \mathbb{R}^2,\ \lambda \in \mathbb{R}^+$ *and any pair of functions* $f, g \in L^2(\mathbb{R}^2)$, *the shearlet transform (3.305) satisfies the following properties:*

(i) *Linearity:* $\mathcal{S}_\psi[\alpha f + \beta g](a,s,\mathbf{t}) = \alpha\,\mathcal{S}_\psi[f](a,s,\mathbf{t}) + \beta\,\mathcal{S}_\psi[g](a,s,\mathbf{t}),$

(ii) *Anti-linearity:* $\mathcal{S}_{\alpha\psi+\beta\phi}[f](a,s,\mathbf{t}) = \overline{\alpha}\,\mathcal{S}_\psi[f](a,s,\mathbf{t}) + \overline{\beta}\,\mathcal{S}_\phi[f](a,s,\mathbf{t}),$

(iii) *Translation:* $\mathcal{S}_\psi[\mathcal{T}_\mathbf{k} f](a,s,\mathbf{t}) = \mathcal{S}_\psi[f](a,s,\mathbf{t}-\mathbf{k}),$

(iv) *Translation in shearlet:* $\mathcal{S}_{\mathcal{T}_\mathbf{k}\psi}[f](a,s,\mathbf{t}) = \mathcal{S}_\psi[f](a,s,\mathbf{t}+S_s A_a \mathbf{k}),$

(v) *Scaling:* $\mathcal{S}_\psi\big[f(\lambda\mathbf{x})\big](a,s,\mathbf{t}) = \dfrac{1}{\lambda}\,\mathcal{S}_{\check{\psi}}\big[f\big](a,s,\lambda\mathbf{t})$, *where* $\check{\psi}(\mathbf{x}) = \psi\left(\lambda^{-1}\mathbf{x}\right)$,

(vi) *Reflection:* $\mathcal{S}_\psi\big[f(-\mathbf{x})\big](a,s,\mathbf{t}) = \mathcal{S}_{\check{\psi}}\big[f\big](a,s,-\mathbf{t})$, *where* $\check{\psi}(\mathbf{x}) = \psi(-\mathbf{x})$.

Proof. The proofs of (i) and (ii) are straightforward and follow by virtue of the elementary properties of the inner product. Moreover, the assertion (vi) can also be verified easily. We proceed to prove the rest of the properties of shearlet transform.

(iii) For any $\mathbf{k} \in \mathbb{R}^2$, we observe that

$$\mathcal{S}_\psi\big[\mathcal{T}_\mathbf{k}f\big](a,s,\mathbf{t}) = \int_{\mathbb{R}^2} f(\mathbf{x}-\mathbf{k})\,\overline{\psi_{a,s,\mathbf{t}}(\mathbf{x})}\,d\mathbf{x}$$

$$= |\det A_a|^{-1/2}\int_{\mathbb{R}^2} f(\mathbf{x}-\mathbf{k})\,\overline{\psi\big(A_a^{-1}S_s^{-1}(\mathbf{x}-\mathbf{t})\big)}\,d\mathbf{x}$$

$$= |\det A_a|^{-1/2}\int_{\mathbb{R}^2} f(\mathbf{z})\,\overline{\psi\big(A_a^{-1}S_s^{-1}(\mathbf{z}-(\mathbf{t}-\mathbf{k}))\big)}\,d\mathbf{z}$$

$$= \mathcal{S}_\psi\big[f\big](a,s,\mathbf{t}-\mathbf{k}).$$

(iv) In case the shearlet $\psi \in L^2(\mathbb{R}^2)$ is shifted by any $\mathbf{k} \in \mathbb{R}^2$, then we have

$$\mathcal{S}_{\mathcal{T}_\mathbf{k}\psi}\big[f\big](a,s,\mathbf{t}) = \int_{\mathbb{R}^2} f(\mathbf{x})\,\overline{(\mathcal{T}_\mathbf{k}\psi)_{a,s,\mathbf{t}}(\mathbf{x})}\,d\mathbf{x}$$

$$= |\det A_a|^{-1/2}\int_{\mathbb{R}^2} f(\mathbf{x})\,\overline{(\mathcal{T}_\mathbf{k}\psi)\big(A_a^{-1}S_s^{-1}(\mathbf{x}-\mathbf{t})\big)}\,d\mathbf{x}$$

$$= |\det A_a|^{-1/2}\int_{\mathbb{R}^2} f(\mathbf{x})\,\overline{\psi\big(A_a^{-1}S_s^{-1}(\mathbf{x}-\mathbf{t})-\mathbf{k}\big)}\,d\mathbf{x}$$

$$= |\det A_a|^{-1/2}\int_{\mathbb{R}^2} f(\mathbf{x})\,\overline{\psi\big(A_a^{-1}S_s^{-1}(\mathbf{x}-\mathbf{t}-S_sA_a\mathbf{k})\big)}\,d\mathbf{x}$$

$$= |\det A_a|^{-1/2}\int_{\mathbb{R}^2} f(\mathbf{x})\,\overline{\psi\big(A_a^{-1}S_s^{-1}(\mathbf{x}-(\mathbf{t}+S_sA_a\mathbf{k}))\big)}\,d\mathbf{x}$$

$$= \mathcal{S}_\psi\big[f\big](a,s,\mathbf{t}+S_sA_a\mathbf{k}).$$

(v) For any $\lambda \in \mathbb{R}^+$, we have

$$\mathcal{S}_\psi\big[f(\lambda\mathbf{x})\big](a,s,\mathbf{t}) = \int_{\mathbb{R}^2} f(\lambda\mathbf{x})\,\overline{\psi_{a,s,\mathbf{t}}(\mathbf{x})}\,d\mathbf{x}$$

$$= |\det A_a|^{-1/2}\int_{\mathbb{R}^2} f(\lambda\mathbf{x})\,\overline{\psi\big(A_a^{-1}S_s^{-1}(\mathbf{x}-\mathbf{t})\big)}\,d\mathbf{x}$$

$$= |\det A_a|^{-1/2}\int_{\mathbb{R}^2} f(\mathbf{z})\,\overline{\check{\psi}\big(A_a^{-1}S_s^{-1}(\mathbf{z}-\lambda\mathbf{t})\big)}\,\frac{d\mathbf{z}}{\lambda}$$

$$= \frac{1}{\lambda}\,\mathcal{S}_{\check{\psi}}\big[f\big](a,s,\lambda\mathbf{t}), \quad \text{where } \check{\psi}(\mathbf{x}) = \psi\left(\lambda^{-1}\mathbf{x}\right).$$

This completes the proof of Theorem 3.7.8. □

In order to study some other important properties of the shearlet transform (3.305), we have the following main definition:

Definition 3.7.4. A pair of shearlets (ψ, ϕ) is said to be an "admissible pair" if the following condition holds:

$$\int_{\mathbb{R}} \int_{\mathbb{R}} \frac{\overline{\hat{\psi}(\omega_1, \omega_2)}\, \hat{\phi}(\omega_1, \omega_2)}{\omega_1^2}\, d\omega_1\, d\omega_2 < \infty. \tag{3.317}$$

Condition (3.317) is called as the "cross admissibility condition."

Remark 3.7.5. For $\psi = \phi$, the cross admissibility condition (3.317) boils down to the usual admissibility condition (3.295).

Next, our aim is to obtain an orthogonality relation between a given pair of signals and the corresponding shearlet transforms as defined in (3.305). More precisely, we shall demonstrate that a pair of signals is orthogonal in the space $L^2(\mathbb{R}^2)$ if and only if their corresponding shearlet transforms are orthogonal in the transformed space $L^2(\mathbb{R}^+ \times \mathbb{R} \times \mathbb{R}^2)$. To facilitate the intent, we firstly set the following notation with regard to an admissible pair (ψ, ϕ) given by (3.317):

$$\left. \begin{aligned} C_{\psi,\phi}^+ &= \int_{\mathbb{R}} \int_0^\infty \frac{\overline{\hat{\psi}(\omega_1, \omega_2)}\, \hat{\phi}(\omega_1, \omega_2)}{\omega_1^2}\, d\omega_1\, d\omega_2 \\ C_{\psi,\phi}^- &= \int_{\mathbb{R}} \int_{-\infty}^0 \frac{\overline{\hat{\psi}(\omega_1, \omega_2)}\, \hat{\phi}(\omega_1, \omega_2)}{\omega_1^2}\, d\omega_1\, d\omega_2 \end{aligned} \right\}. \tag{3.318}$$

In case $C_{\psi,\phi}^+ = C_{\psi,\phi}^-$, the common entity shall be denoted by $C_{\psi,\phi}$; that is, we shall follow the notation $C_{\psi,\phi} = C_{\psi,\phi}^+ = C_{\psi,\phi}^-$.

Theorem 3.7.9. *Let $\mathcal{S}_\psi[f](a, s, \mathbf{t})$ and $\mathcal{S}_\phi[g](a, s, \mathbf{t})$ be the shearlet transforms of a given pair of functions f and g with respect to the admissible pair (ψ, ϕ), respectively. Then, the following orthogonality relation holds:*

$$\int_{\mathbb{R}^2} \int_{\mathbb{R}} \int_{\mathbb{R}^+} \mathcal{S}_\psi[f](a, s, \mathbf{t})\, \overline{\mathcal{S}_\phi[g](a, s, \mathbf{t})}\, \frac{da\, ds\, d\mathbf{t}}{a^3} = C_{\psi,\phi}\big\langle f, g \big\rangle_2, \tag{3.319}$$

provided $C_{\psi,\phi}^+ = C_{\psi,\phi}^- = C_{\psi,\phi}$.

Proof. By virtue of the expression (3.314), we can express the shearlet transforms of the given pair of functions f and g with respect to the shearlet ψ and ϕ as follows:

$$\mathcal{S}_\psi[f](a, s, \mathbf{t}) = a^{3/4} \int_{\mathbb{R}^2} \hat{f}(\mathbf{w})\, \overline{\hat{\psi}(A_a^T S_s^T \mathbf{w})}\, e^{2\pi i \mathbf{t}^T \mathbf{w}}\, d\mathbf{w} \tag{3.320}$$

and

$$\mathcal{S}_\psi[g](a, s, \mathbf{t}) = a^{3/4} \int_{\mathbb{R}^2} \hat{g}(\mathbf{w}')\, \overline{\hat{\phi}(A_a^T S_s^T \mathbf{w}')}\, e^{2\pi i \mathbf{t}^T \mathbf{w}'}\, d\mathbf{w}'. \tag{3.321}$$

As a consequence of (3.320) and (3.321), we have

$$\int_{\mathbb{R}^2} \int_{\mathbb{R}} \int_{\mathbb{R}^+} \mathcal{S}_\psi[f](a, s, \mathbf{t})\, \overline{\mathcal{S}_\phi[g](a, s, \mathbf{t})}\, \frac{da\, ds\, d\mathbf{t}}{a^3}$$

$$= \int_{\mathbb{R}^2} \int_{\mathbb{R}} \int_{\mathbb{R}^+} \left\{ a^{3/4} \int_{\mathbb{R}^2} \hat{f}(\mathbf{w})\, \overline{\hat{\psi}(A_a^T S_s^T \mathbf{w})}\, e^{2\pi i \mathbf{t}^T \mathbf{w}}\, d\mathbf{w} \right\}$$

$$\times \left\{ a^{3/4} \int_{\mathbb{R}^2} \overline{\hat{g}(\mathbf{w}')}\, \hat{\phi}(A_a^T S_s^T \mathbf{w}')\, e^{-2\pi i \mathbf{t}^T \mathbf{w}'}\, d\mathbf{w}' \right\} \frac{da\, ds\, d\mathbf{t}}{a^3}$$

$$= \int_{\mathbb{R}^2} \int_{\mathbb{R}^2} \int_{\mathbb{R}} \int_{\mathbb{R}^+} \hat{f}(\mathbf{w}) \, \overline{\hat{g}(\mathbf{w}')} \, \hat{\phi}\left(A_a^T S_s^T \mathbf{w}'\right) \, \overline{\hat{\psi}\left(A_a^T S_s^T \mathbf{w}\right)} \left\{ \int_{\mathbb{R}^2} e^{-2\pi i \mathbf{t} \cdot (\mathbf{w} - \mathbf{w}')} d\mathbf{t} \right\} \frac{da \, ds \, d\mathbf{w} \, d\mathbf{w}'}{a^{3/2}}$$

$$= \int_{\mathbb{R}^2} \int_{\mathbb{R}^2} \int_{\mathbb{R}} \int_{\mathbb{R}^+} \hat{f}(\mathbf{w}) \, \overline{\hat{g}(\mathbf{w}')} \, \hat{\phi}\left(A_a^T S_s^T \mathbf{w}'\right) \, \overline{\hat{\psi}\left(A_a^T S_s^T \mathbf{w}\right)} \, \delta(\mathbf{w} - \mathbf{w}') \frac{da \, ds \, d\mathbf{w} \, d\mathbf{w}'}{a^{3/2}}$$

$$= \int_{\mathbb{R}^2} \int_{\mathbb{R}} \int_{\mathbb{R}^+} \hat{f}(\mathbf{w}) \, \overline{\hat{g}(\mathbf{w})} \, \hat{\phi}\left(A_a^T S_s^T \mathbf{w}\right) \, \overline{\hat{\psi}\left(A_a^T S_s^T \mathbf{w}\right)} \frac{da \, ds \, d\mathbf{w}}{a^{3/2}}. \tag{3.322}$$

For $\mathbf{w} = (\omega_1, \omega_2)^T$, we can simplify the R.H.S of (3.322) by making appropriate substitutions in the following manner:

$$\int_{\mathbb{R}^+} \int_{\mathbb{R}^2} \int_{\mathbb{R}} \hat{f}(\mathbf{w}) \, \overline{\hat{g}(\mathbf{w})} \, \hat{\phi}\left(A_a^T S_s^T \mathbf{w}\right) \, \overline{\hat{\psi}\left(A_a^T S_s^T \mathbf{w}\right)} \frac{da \, ds \, d\mathbf{w}}{a^{3/2}}$$

$$= \int_{\mathbb{R}^+} \int_{\mathbb{R}} \int_{\mathbb{R}} \int_{\mathbb{R}} \hat{f}(\omega_1, \omega_2) \overline{\hat{g}(\omega_1, \omega_2)} \, \hat{\phi}\left(a\omega_1, s\sqrt{a}\omega_1 + \sqrt{a}\omega_2\right)$$

$$\times \overline{\hat{\psi}\left(a\omega_1, s\sqrt{a}\omega_1 + \sqrt{a}\omega_2\right)} \frac{ds \, d\omega_1 \, d\omega_2 \, da}{a^{3/2}}$$

$$= \int_{\mathbb{R}} \int_0^\infty \int_0^\infty \int_{\mathbb{R}} \hat{f}(\omega_1, \omega_2) \overline{\hat{g}(\omega_1, \omega_2)} \, \hat{\phi}(a\omega_1, \xi_2) \overline{\hat{\psi}(a\omega_1, \xi_2)} \frac{d\xi_2 \, da \, d\omega_1 \, d\omega_2}{a^2 \omega_1}$$

$$- \int_{\mathbb{R}} \int_{-\infty}^0 \int_0^\infty \int_{\mathbb{R}} \hat{f}(\omega_1, \omega_2) \overline{\hat{g}(\omega_1, \omega_2)} \, \hat{\phi}(a\omega_1, \xi_2) \overline{\hat{\psi}(a\omega_1, \xi_2)} \frac{d\xi_2 \, da \, d\omega_1 \, d\omega_2}{a^2 \omega_1}$$

$$= \int_{\mathbb{R}} \int_0^\infty \hat{f}(\omega_1, \omega_2) \overline{\hat{g}(\omega_1, \omega_2)} \left\{ \int_0^\infty \int_{\mathbb{R}} \frac{\hat{\phi}(\xi_1, \xi_2) \overline{\hat{\psi}(\xi_1, \xi_2)}}{\xi_1^2} d\xi_2 \, d\xi_1 \right\} d\omega_1 \, d\omega_2$$

$$+ \int_{\mathbb{R}} \int_{-\infty}^0 \hat{f}(\omega_1, \omega_2) \overline{\hat{g}(\omega_1, \omega_2)} \left\{ \int_{-\infty}^0 \int_{\mathbb{R}} \frac{\hat{\phi}(\xi_1, \xi_2) \overline{\hat{\psi}(\xi_1, \xi_2)}}{\xi_1^2} d\xi_2 \, d\xi_1 \right\} d\omega_1 \, d\omega_2$$

$$= C_{\psi,\phi}^+ \int_{\mathbb{R}} \int_0^\infty \hat{f}(\omega_1, \omega_2) \overline{\hat{g}(\omega_1, \omega_2)} \, d\omega_1 \, d\omega_2 + C_{\psi,\phi}^- \int_{\mathbb{R}} \int_{-\infty}^0 \hat{f}(\omega_1, \omega_2) \overline{\hat{g}(\omega_1, \omega_2)} \, d\omega_1 \, d\omega_2. \tag{3.323}$$

Furthermore, using the hypothesis that $C_{\psi,\phi}^+ = C_{\psi,\phi}^- = C_{\psi,\phi}$ and invoking the Parseval's formula for two-dimensional Fourier transform, we can simplify (3.322) as

$$\int_{\mathbb{R}^+} \int_{\mathbb{R}^2} \int_{\mathbb{R}} \hat{f}(\mathbf{w}) \, \overline{\hat{g}(\mathbf{w})} \, \hat{\phi}\left(A_a^T S_s^T \mathbf{w}\right) \, \overline{\hat{\psi}\left(A_a^T S_s^T \mathbf{w}\right)} \frac{da \, ds \, d\mathbf{w}}{a^{3/2}}$$

$$= C_{\psi,\phi} \left\{ \int_{\mathbb{R}} \int_0^\infty \hat{f}(\omega_1, \omega_2) \overline{\hat{g}(\omega_1, \omega_2)} \, d\omega_1 \, d\omega_2 + \int_{\mathbb{R}} \int_{-\infty}^0 \hat{f}(\omega_1, \omega_2) \overline{\hat{g}(\omega_1, \omega_2)} \, d\omega_1 \, d\omega_2 \right\}$$

$$= C_{\psi,\phi} \left\{ \int_{\mathbb{R}} \int_{\mathbb{R}} \hat{f}(\omega_1, \omega_2) \overline{\hat{g}(\omega_1, \omega_2)} \, d\omega_1 \, d\omega_2 \right\}$$

$$= C_{\psi,\phi} \int_{\mathbb{R}^2} \hat{f}(\mathbf{w}) \overline{\hat{g}(\mathbf{w})} \, d\mathbf{w}$$

$$= C_{\psi,\phi} \left\langle \hat{f}, \hat{g} \right\rangle_2$$

$$= C_{\psi,\phi} \left\langle f, g \right\rangle_2. \tag{3.324}$$

Finally, implementing (3.324) in (3.322), we obtain the desired result as

$$\int_{\mathbb{R}^2} \int_{\mathbb{R}} \int_{\mathbb{R}^+} \mathcal{S}_\psi\big[f\big](a, s, \mathbf{t}) \, \overline{\mathcal{S}_\phi\big[g\big](a, s, \mathbf{t})} \frac{da \, ds \, d\mathbf{t}}{a^3} = C_{\psi,\phi} \left\langle f, g \right\rangle_2.$$

This completes the proof of Theorem 3.7.9. $\square$

Remark 3.7.6. For $\psi = \phi$, Theorem 3.7.9 yields the below version of the orthogonality relation:

$$\int_{\mathbb{R}^2} \int_{\mathbb{R}} \int_{\mathbb{R}^+} \mathcal{S}_\psi \big[f\big](a, s, \mathbf{t}) \, \overline{\mathcal{S}_\phi \big[g\big](a, s, \mathbf{t})} \, \frac{da \, ds \, d\mathbf{t}}{a^3} = C_\psi \big\langle f, g \big\rangle_2, \qquad (3.325)$$

where $C_\psi = C_\psi^+ = C_\psi^-$ and the quantities C_ψ^+, C_ψ^- are given by (3.297). Moreover, for $f = g$ in (3.325), we obtain the energy preserving relation associated with the shearlet transform as

$$\int_{\mathbb{R}^2} \int_{\mathbb{R}} \int_{\mathbb{R}^+} \Big| \mathcal{S}_\psi \big[f\big](a, s, \mathbf{t}) \Big|^2 \frac{da \, ds \, d\mathbf{t}}{a^3} = C_\psi \big\| f \big\|_2^2. \qquad (3.326)$$

Thus, choosing the shearlet ψ in such a way that $C_\psi = 1$, the shearlet transform (3.305) is an isometry from the space of signals $L^2(\mathbb{R}^2)$ to the space of of transforms $L^2(\mathbb{R}^+ \times \mathbb{R} \times \mathbb{R}^2)$.

Corollary 3.7.10. *The family of shearlets given by (3.303) generates a resolution of identity in the sense that if I denotes the identity operator on $L^2(\mathbb{R}^2)$, then*

$$C_\psi^{-1} \int_{\mathbb{R}^2} \int_{\mathbb{R}} \int_{\mathbb{R}^+} \big\langle \psi_{a,s,\mathbf{t}}, \, \psi_{a,s,\mathbf{t}} \big\rangle_2 \, \frac{da \, ds \, d\mathbf{t}}{a^3} = I. \qquad (3.327)$$

We now present a reconstruction formula associated with the shearlet transform defined in (3.305).

Theorem 3.7.11. *Let (ψ, ϕ) be an admissible pair of shearlets with $C_{\psi,\phi} = 1$ and $\{g_n\}_{n=1}^\infty$ be an approximation identity of even, square integrable functions on $\mathbb{R}^2$. Then, for any $f \in L^2(\mathbb{R}^2)$ such that*

$$f_n(\mathbf{x}) = \int_{\mathbb{R}^2} \int_{\mathbb{R}} \int_{\mathbb{R}^+} \mathcal{S}_\psi \big[f\big](a, s, \mathbf{t}) \, (g_n \circledast \phi_{a,s,\mathbf{t}})(\mathbf{x}) \, \frac{da \, ds \, d\mathbf{t}}{a^3}, \qquad (3.328)$$

we have $\lim\limits_{n \to \infty} \big\| f - f_n \big\|_2 = 0$.

Proof. Since each g_n is an even function and for $C_{\psi,\phi} = 1$, the continuous shearlet transform $\mathcal{S}_\psi \big[f\big](a, s, \mathbf{t})$ is an isometry, therefore, by using the definition of two-dimensional convolution operation $\circledast$ given in (3.226), it follows that

$$\begin{aligned}
\big(f \circledast g_n\big)(\mathbf{x}) &= \int_{\mathbb{R}^2} f(\mathbf{z}) \, g_n(\mathbf{x} - \mathbf{z}) \, d\mathbf{z} \\
&= \Big\langle f, \, \overline{T_\mathbf{x} g_n} \Big\rangle_2 \\
&= \Big\langle \mathcal{S}_\psi \big[f\big](a, s, \mathbf{t}), \, \overline{\mathcal{S}_\phi \big[T_\mathbf{x} g_n\big](a, s, \mathbf{t})} \Big\rangle_2 \\
&= \int_{\mathbb{R}^2} \int_{\mathbb{R}} \int_{\mathbb{R}^+} \mathcal{S}_\psi \big[f\big](a, s, \mathbf{t}) \, \overline{\Big\langle g_n(\cdot - \mathbf{x}), \, \phi_{a,s,\mathbf{t}}(\cdot) \Big\rangle_2} \, \frac{da \, ds \, d\mathbf{t}}{a^3} \\
&= \int_{\mathbb{R}^2} \int_{\mathbb{R}} \int_{\mathbb{R}^+} \mathcal{S}_\psi \big[f\big](a, s, \mathbf{t}) \, (g_n \circledast \phi_{a,s,\mathbf{t}})(\mathbf{x}) \, \frac{da \, ds \, d\mathbf{t}}{a^3}. \qquad (3.329)
\end{aligned}$$

But, $\{g_n\}_{n=1}^\infty$ being an approximation identity, the expression (3.329) demonstrates that $\lim\limits_{n \to \infty} \big\| f - f_n \big\|_2 = 0$. This completes the proof of Theorem 3.7.11. $\qquad \square$

Remark 3.7.7. Theorem 3.7.11 clearly implies that a distinct pair of shearlets ψ and ϕ satisfying $C_{\psi,\phi} = 1$ can be used for the analysis and synthesis processes, respectively. Moreover, choosing $\psi = \phi$ in (3.328), we obtain the following "resolution of identity" formula:

$$f(\mathbf{x}) = \int_{\mathbb{R}^2} \int_{\mathbb{R}} \int_{\mathbb{R}^+} \mathcal{S}_\psi \big[f\big](a, s, \mathbf{t}) \, (g_n \circledast \psi_{a,s,\mathbf{t}})(\mathbf{x}) \, \frac{da \, ds \, d\mathbf{t}}{a^3}. \qquad (3.330)$$

In the following theorem, we obtain a characterization of range of the continuous shearlet transform, which demonstrates the existence of the reproducing kernel for the continuous shearlet transform defined in (3.305).

Theorem 3.7.12. *A function $f \in L^2(\mathbb{R}^+ \times \mathbb{R} \times \mathbb{R}^2)$ is the continuous shearlet transform of a certain square integrable function on $\mathbb{R}^2$ with respect to the shearlet $\psi \in L^2(\mathbb{R}^2)$ satisfying $C_\psi = 1$ if and only if it satisfies the following reproduction formula:*

$$f(a', s', \mathbf{t}') = \int_{\mathbb{R}^2} \int_{\mathbb{R}} \int_{\mathbb{R}^+} f(a, s, \mathbf{t}) \left\langle (g_n \circledast \psi_{a,s,\mathbf{t}}), \psi_{a',s',\mathbf{t}'} \right\rangle_2 \frac{da\, ds\, d\mathbf{t}}{a^3}, \tag{3.331}$$

where $\{g_n\}_{n=1}^{\infty}$ is an approximation identity of even, square integrable functions on $\mathbb{R}^2$ and $C_\psi = C_\psi^+ = C_\psi^-$, with the quantities C_ψ^+, C_ψ^- being given by (3.297).

Proof. Suppose that the function $f \in L^2(\mathbb{R}^+ \times \mathbb{R} \times \mathbb{R}^2)$ belongs to the range of the shearlet transform. Then, there must exist some $h \in L^2(\mathbb{R}^2)$, such that $\mathcal{S}_\psi[h] = f$. Thus, we have

$$
\begin{aligned}
f(a', s', \mathbf{t}') &= \mathcal{S}_\psi\big[h\big](a', s', \mathbf{t}') \\
&= \int_{\mathbb{R}^2} h(\mathbf{x}) \, \overline{\psi_{a',s',\mathbf{t}'}(\mathbf{x})} \, d\mathbf{x} \\
&= \int_{\mathbb{R}^2} \left\{ \int_{\mathbb{R}^2} \int_{\mathbb{R}} \int_{\mathbb{R}^+} \mathcal{S}_\psi\big[h\big](a, s, \mathbf{t}) \, (g_n \circledast \psi_{a,s,\mathbf{t}})(\mathbf{x}) \, \frac{da\, ds\, d\mathbf{t}}{a^3} \right\} \overline{\psi_{a',s',\mathbf{t}'}(\mathbf{x})} \, d\mathbf{x} \\
&= C_\psi^{-1} \int_{\mathbb{R}^2} \int_{\mathbb{R}} \int_{\mathbb{R}^+} \mathcal{S}_\psi\big[h\big](a, s, \mathbf{t}) \left\{ \int_{\mathbb{R}^2} (g_n \circledast \psi_{a,s,\mathbf{t}})(\mathbf{x}) \, \overline{\psi_{a',s',\mathbf{t}'}(\mathbf{x})} \, d\mathbf{x} \right\} \frac{da\, ds\, d\mathbf{t}}{a^3} \\
&= C_\psi^{-1} \int_{\mathbb{R}^2} \int_{\mathbb{R}} \int_{\mathbb{R}^+} f(a, s, \mathbf{t}) \left\langle (g_n \circledast \psi_{a,s,\mathbf{t}}), \psi_{a',s',\mathbf{t}'} \right\rangle_2 \frac{da\, ds\, d\mathbf{t}}{a^3}. \tag{3.332}
\end{aligned}
$$

Conversely, suppose that a function $f \in L^2(\mathbb{R}^+ \times \mathbb{R} \times \mathbb{R}^2)$ satisfies the reproduction formula (3.331). Then, we have to show that there exists a function $g \in L^2(\mathbb{R}^2)$, such that $\mathcal{S}_\psi[g] = f$. The obvious candidate for such a function is

$$g(\mathbf{x}) = \int_{\mathbb{R}^2} \int_{\mathbb{R}} \int_{\mathbb{R}^+} f(a, s, \mathbf{t}) \, (g_n \circledast \psi_{a,s,\mathbf{t}})(\mathbf{x}) \, \frac{da\, ds\, d\mathbf{t}}{a^3}, \tag{3.333}$$

which is indeed square integrable on $\mathbb{R}^2$. Finally, we note that

$$
\begin{aligned}
\mathcal{S}_\psi\big[g\big](a', s', \mathbf{t}') &= \int_{\mathbb{R}^2} g(\mathbf{x}) \, \overline{\psi_{a',s',\mathbf{t}'}(\mathbf{x})} \, d\mathbf{x} \\
&= \int_{\mathbb{R}^2} \left\{ \int_{\mathbb{R}^2} \int_{\mathbb{R}} \int_{\mathbb{R}^+} f(a, s, \mathbf{t}) \, (g_n \circledast \psi_{a,s,\mathbf{t}})(\mathbf{x}) \, \frac{da\, ds\, d\mathbf{t}}{a^3} \right\} \overline{\psi_{a',s',\mathbf{t}'}(\mathbf{x})} \, d\mathbf{x} \\
&= \int_{\mathbb{R}^2} \int_{\mathbb{R}} \int_{\mathbb{R}^+} f(a, s, \mathbf{t}) \left\langle (g_n \circledast \psi_{a,s,\mathbf{t}}), \psi_{a',s',\mathbf{t}'} \right\rangle_2 \frac{da\, ds\, d\mathbf{t}}{a^3} \\
&= f(a', s', \mathbf{t}').
\end{aligned}
$$

This completes the proof of Theorem 3.7.12. $\square$

Corollary 3.7.13. *The projection from $L^2(\mathbb{R}^+ \times \mathbb{R} \times \mathbb{R}^2)$ onto the range of the continuous shearlet transform defined in (3.305) is an integral operator whose reproducing kernel $K_\psi(a, s, \mathbf{t}; a', s', \mathbf{t}')$ is given by*

$$K_\psi(a, s, \mathbf{t}; a', s', \mathbf{t}') = C_\psi^{-1} \left\langle (g_n \circledast \psi_{a,s,\mathbf{t}}), \psi_{a',s',\mathbf{t}'} \right\rangle_2, \tag{3.334}$$

where $\{g_n\}_{n=1}^{\infty}$ is an approximation identity of even, square integrable functions on $\mathbb{R}^2$ and $C_\psi = C_\psi^+ = C_\psi^-$, with the quantities C_ψ^+, C_ψ^- being given by (3.297).

3.7.3 Visualization of the shearlet coefficients

"In every mathematical investigation, the question will arise whether we can apply our mathematical results to the real world."

-Vladimir I. Arnold

The shearlet transform is unique in emphasizing different attributes of signals and is advantageous over the classical integral transforms as it faithfully provides information about the directionality within an image. In this subsection, we examine the directional properties of the continuous shearlet transform defined in (3.305). More precisely, our aim is to visualize the behaviour of the shearlet coefficients while dealing with a specific class of problems in texture analysis, including the edge detection and tunable detection of orientations in images.

Consider an input signal $f \in L^2(\mathbb{R}^2)$, then the shearlet coefficients $\langle f, \psi_{a,s,\mathbf{t}} \rangle_2$ can be regarded as a *pattern matching task*, wherein the pattern ψ is scaled by $a \in \mathbb{R}^+$, sheared by $s \in \mathbb{R}$ and translated by $\mathbf{t} \in \mathbb{R}^2$. In case the signal f is similar to $\psi_{a,s,\mathbf{t}}$, when being restricted to the support of $\psi_{a,s,\mathbf{t}}$, the absolute value of the inner product $\langle f, \psi_{a,s,\mathbf{t}} \rangle_2$ is expected to be large. In order to visualize this behaviour, we firstly select specific values for a and s with regard to a given signal f and then display the quantity $|\langle f, \psi_{a,s,\mathbf{t}} \rangle_2|$ as a function of $\mathbf{t} \in \mathbb{R}^2$; that is, as an image, denoted by $I(\psi, a, s)$. With the aforementioned details in hindsight, we infer that the image intensity should be large for those values of $\mathbf{t} \in \mathbb{R}^2$, where the signal f is similar to $\psi_{a,s,\mathbf{t}}$, when the former is restricted to the support of $\psi_{a,s,\mathbf{t}}$. In this sense, the intensity peaks of the image $I(\psi, a, s)$ should indicate, where scale a and shear s are present in the original signal $f \in L^2(\mathbb{R}^2)$.

Here, it is imperative to mention that the admissibility condition (3.295) facilitates the invertibility of the transformation $f \to \langle f, \psi_{a,s,\mathbf{t}} \rangle_2$. For applications like those discussed above, this condition is not *a priori* necessary. Therefore, we shall analyze the image $I(\psi, a, s)$ for simple functions ψ which do not satisfy the admissibility condition (3.295), but allow for an easy interpretation of the function $I(\psi, a, s)$. In this direction, we choose

$$\psi_1(x_1, x_2) = \psi_H(x_1)\,\chi_{(-1/2,\,1/2)}(x_2), \tag{3.335}$$

where

$$\psi_H(x_1) = \begin{cases} 1, & -\dfrac{1}{2} \le x_1 < 0 \\ -1 & 0 \le x_1 \le \dfrac{1}{2} \\ 0, & \text{otherwise} \end{cases} \tag{3.336}$$

is the shifted version of the one-dimensional Haar-wavelet and $\chi_{(-1/2,\,1/2)}(x_2)$ denotes the characteristic function of the interval $(-1/2, 1/2)$. Further, we choose another function as the second radial derivative of a Gaussian function; that is,

$$\psi_2(x_1, x_2) = e^{-4r^2}\left(32r^2 - 4\right), \quad r^2 = x_1^2 + x_2^2. \tag{3.337}$$

Then, we observe that both the functions given by (3.335) and (3.337) have zero mean and are thus capable of resolving the edges and boundaries occurring due to the spatial occlusion between different objects in the signal $f \in L^2(\mathbb{R}^2)$. Obviously ψ_1 will respond on vertical edges, whereas ψ_2 is isotropic.

For gaining an intuition of the above cited procedure, we choose two types of input signals: A square and a circle as displayed in Figure 3.29. In order to investigate, how $I(\psi, a, s)$ responds to shear, we compute and display $I(\psi, a, s)$ for sheared versions of the input signal f; that is, for $f_{a', s', \mathbf{t}'}(\mathbf{x})$, $\mathbf{x} \in \mathbb{R}^2$ with $a' = 1$ and $\mathbf{t}' = \mathbf{0}$. The reason for taking the scale to be 1 lies in the fact that we want to focus only on the shear. For notational convenience, we denote the image simply as $f_{s'}(\mathbf{x})$. In all the plots, the gray-level coding is scaled; that is, the smallest value is displayed black and the largest value is displayed white. In Figure 3.30, the original square is sheared with $s' = 1$ and $I(\psi_1, 1, 0)$ is displayed; that is, ψ_1 is not sheared as $s = 0$. From Figure 3.31, we observe the expected directional sensitivity of ψ_1 (horizontal edges are not detected) and, moreover, we also note that the shearlet coefficients are not well localized. On the other hand, in Figure 3.31, we choose $s = 1$; that is, ψ_1 is sheared by the same factor as the original image. In this case, the shearlet coefficients are obviously very well localized, thus indicating that the shearlet transform is indeed able to detect shear. Nevertheless, Figures 3.32–3.33 essentially illustrate the same phenomenon for the circle instead of the square. Since the function ψ_2 is isotropic, therefore all edges are detected, independent from their orientation.

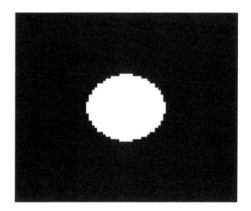

FIGURE 3.29: Input signals: A square and a circle.

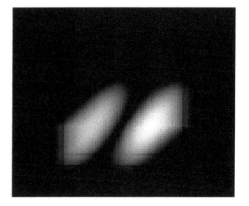

FIGURE 3.30: Sheared square f_1 and $I(\psi_1, 1, 0)$ ("wrong" shear factor).

FIGURE 3.31: Sheared square f_1 and $I(\psi_1, 1, 1)$ ("matching" shear factor).

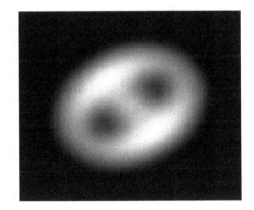

FIGURE 3.32: Sheared circle $f_{0.5}$ and $I(\psi_2, 1, 0)$ ("wrong" shear factor).

FIGURE 3.33: Sheared circle $f_{0.5}$ and $I(\psi_2, 1, 0.5)$ ("matching" shear factor).

3.7.4 Cone-adapted shearlet transform

"To those who do not know mathematics it is difficult to get across a real feeling as to the beauty, the deepest beauty, of nature... If you want to learn about nature, to appreciate nature, it is necessary to understand the language that she speaks in."

-Richard P. Feynman

The construction of shearlets discussed in the preceding subsections is distinguishable in the sense that it is deeply interlinked with an elegant group structure, which plays a significant role in diverse mathematical developments pertaining to the continuous shearlet transform. However, a major issue with this construction is that the associated shearlet systems are directionally biased: The more concentrated a function is along an axis, the more of its information is perceptible in the shearlet coefficients $\mathcal{S}_\psi[f](a, s, \mathbf{t})$ as $s \to \infty$. This problem is specifically evident in case the Fourier transform of the analyzing function is concentrated along one axis of the domain, as the function can only be detected in the shearlet domain for $s \to \infty$. Therefore, the directionally biased behaviour of the shearlet transform can pose serious limitations in certain applications. One possible solution to address the issue is to split the frequency plane $\mathbb{R}^2 = \{\mathbf{w} = (\omega_1, \omega_2)^T\}$ into four cones C_1, C_2, C_3, C_4 and cut out the low frequency region by a square centred around the origin in the frequency plane, denoted by C_0. Such a partition of the frequency plane is depicted in Figure 3.34.

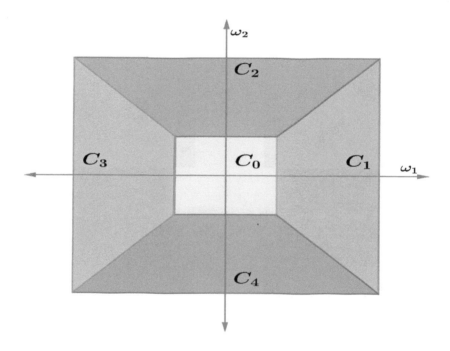

FIGURE 3.34: Conic-partition of the frequency plane.

Note that, within each cone, the shearing variable s varies only over a finite range, thus producing elements whose orientations are distributed more uniformly. Also, the low-frequency box C_0 cut out from the frequency plane, as shown in Figure 3.34, is explicitly given by

$$C_0 = \left\{ \mathbf{w} = (\omega_1, \omega_2)^T : |\omega_1| \le 1, |\omega_2| \le 1 \right\}. \tag{3.338}$$

Based on the aforementioned partitioning of the frequency plane, a variant of the usual shearlet systems is defined by taking into account a triplet of square integrable functions $\phi, \psi, \tilde{\psi}$ and constructing the systems

$$
\left.
\begin{aligned}
\mathcal{F}_1(\phi) &:= \left\{ \phi_{\mathbf{t}'}(\mathbf{x}) = \phi(\mathbf{x} - \mathbf{t}'); \ \mathbf{t}' \in \mathbb{R}^2 \right\} \\
\mathcal{F}_2(\psi) &:= \left\{ \psi_{a,s,\mathbf{t}}(\mathbf{x}) = a^{-3/4} \psi\left(A_a^{-1} S_s^{-1}(\mathbf{x} - \mathbf{t})\right); \ a \in (0,1], \ |s| \leq 1 + \sqrt{a}, \ \mathbf{t} \in \mathbb{R}^2 \right\} \\
\mathcal{F}_3(\tilde{\psi}) &:= \left\{ \tilde{\psi}_{\tilde{a},\tilde{s},\tilde{\mathbf{t}}}(\mathbf{x}) = \tilde{a}^{-3/4} \psi\left(\tilde{A}_{\tilde{a}^{-1}} S_{\tilde{s}}^{-1}(\mathbf{x} - \tilde{\mathbf{t}})\right); \ \tilde{a} \in (0,1], \ |\tilde{s}| \leq 1 + \sqrt{\tilde{a}}, \ \tilde{\mathbf{t}} \in \mathbb{R}^2 \right. \\
&\qquad\qquad\qquad\qquad\qquad\qquad\qquad \text{and } \tilde{A}_{\tilde{a}} = \operatorname{diag}\left(\sqrt{\tilde{a}}, \tilde{a}\right)\Big\}
\end{aligned}
\right\}.
$$

$$(3.339)$$

Consequently, the cone-adapted shearlet system is defined by

$$
\mathcal{F}_{\mathrm{cone}}(\phi, \psi, \tilde{\psi}) = \mathcal{F}_1(\phi) \cup \mathcal{F}_2(\psi) \cup \mathcal{F}_3(\tilde{\psi}). \tag{3.340}
$$

The function ϕ is often referred as the "scaling function" of the cone-adapted shearlet system and is chosen to have compact support in the Fourier domain about the origin, which in turn ensures that the associated system $\mathcal{F}_1(\phi)$ is suited for the low-frequency region $C_0 = \left\{ \mathbf{w} = (\omega_1, \omega_2)^T : |\omega_1|, |\omega_2| \leq 1 \right\}$. Moreover, by choosing ψ to be classical shearlet, the associated shearlet system $\mathcal{F}_2(\psi)$ is suited for the horizontal cones $C_1 \cup C_3 = \left\{ \mathbf{w} = (\omega_1, \omega_2)^T : |\omega_2/\omega_1| \leq 1, |\omega_1| > 1 \right\}$. Also, $\tilde{\psi}$ can be chosen likewise with the roles of ω_1 and ω_2 reversed; that is, $\tilde{\psi}(\omega_1, \omega_2) = \psi(\omega_2, \omega_1)$ so that the corresponding shearlet system $\mathcal{F}_3(\tilde{\psi})$ is befitting for the vertical cones $C_2 \cup C_4 = \left\{ \mathbf{w} = (\omega_1, \omega_2)^T : |\omega_2/\omega_1| > 1, |\omega_1| > 1 \right\}$.

Having presented an illustrative construction of the cone-adapted shearlet systems, we are in a position to formally introduce the notion of the cone-adapted shearlet transform. Prior to that, we set the following notation:

$$
\mathbb{S}_{\mathrm{cone}} = \left\{ (a, s, \mathbf{t}) : a \in (0,1], \ |s| \leq 1 + \sqrt{a}, \ \mathbf{t} \in \mathbb{R}^2 \right\}. \tag{3.341}
$$

Definition 3.7.5. For any $f \in L^2(\mathbb{R}^2)$, the cone-adapted shearlet transform with respect to $\phi, \psi, \tilde{\psi} \in L^2(\mathbb{R}^2)$ is defined as

$$
\mathcal{S}_{\phi,\psi,\tilde{\psi}}\big[f\big]\left(\mathbf{t}', (a, s, \mathbf{t}), (\tilde{a}, \tilde{s}, \tilde{\mathbf{t}})\right) = \left(\langle f, \phi_{\mathbf{t}'}\rangle_2, \langle f, \psi_{a,s,\mathbf{t}}\rangle_2, \langle f, \tilde{\psi}_{\tilde{a},\tilde{s},\tilde{\mathbf{t}}}\rangle_2\right), \tag{3.342}
$$

where $\left(\mathbf{t}', (a, s, \mathbf{t}), (\tilde{a}, \tilde{s}, \tilde{\mathbf{t}})\right) \in \mathbb{R}^2 \times \mathbb{S}_{\mathrm{cone}} \times \mathbb{S}_{\mathrm{cone}}$ and the shearlet systems $\phi_{\mathbf{t}'}, \psi_{a,s,\mathbf{t}}, \tilde{\psi}_{\tilde{a},\tilde{s},\tilde{\mathbf{t}}}$ are given by (3.339).

With the formal definition of the cone-adapted shearlet transform at hand, our motive is to demonstrate the construction of cone-adapted shearlets by taking into consideration the notion of classical shearlets described in the previous subsection. Let $\chi_{C_i}, i = 1, 2, 3, 4$ denotes the characteristic function over the i-th cone, that is

$$
\chi_{C_i}(\mathbf{w}) = \begin{cases} 1, & \mathbf{w} \in C_i \\ 0, & \text{otherwise,} \end{cases} \qquad i = 1, 2, 3, 4. \tag{3.343}
$$

Then, a shearlet is assigned to a particular cone by multiplying it with an appropriate characteristic function. For instance, if $\psi, \tilde{\psi}$ are two classical shearlets assigned to the horizontal and vertical cones $C_1 \cup C_3$ and $C_2 \cup C_4$, respectively. Subsequently, we can express the complete shearlet system as

$$\left.\begin{aligned}
\hat{\psi}(\omega_1, \omega_2) &= \hat{\psi}_1(\omega_1)\,\hat{\psi}_2\left(\frac{\omega_2}{\omega_1}\right)\chi_{C_1 \cup C_3} \\
\hat{\tilde{\psi}}(\omega_1, \omega_2) &= \hat{\tilde{\psi}}_1(\omega_1)\,\hat{\tilde{\psi}}_2\left(\frac{\omega_2}{\omega_1}\right)\chi_{C_2 \cup C_4}
\end{aligned}\right\}, \tag{3.344}$$

where $\psi_1, \tilde{\psi}_1, \psi_2, \tilde{\psi}_2$ are chosen in accordance to the definition of the classical shearlet given in previous subsection. Since the horizontal and vertical cones can be treated similarly by interchanging the roles of ω_1 and ω_2, therefore, the shearlets (3.344) can be simply chosen as

$$\left.\begin{aligned}
\hat{\psi}(\omega_1, \omega_2) &= \hat{\psi}_1(\omega_1)\,\hat{\psi}_2\left(\frac{\omega_2}{\omega_1}\right)\chi_{C_1 \cup C_3} \\
\hat{\tilde{\psi}}(\omega_2, \omega_1) &= \hat{\psi}(\omega_2, \omega_1)\,\chi_{C_2 \cup C_4} \\
&= \hat{\psi}_1(\omega_2)\,\hat{\psi}_2\left(\frac{\omega_1}{\omega_2}\right)\chi_{C_2 \cup C_4}
\end{aligned}\right\}. \tag{3.345}$$

The shearlets given by (3.345) are the desired "cone-adapted shearlets." The tiling on the frequency plane induced by the cone-adapted shearlets is shown in Figure 3.35.

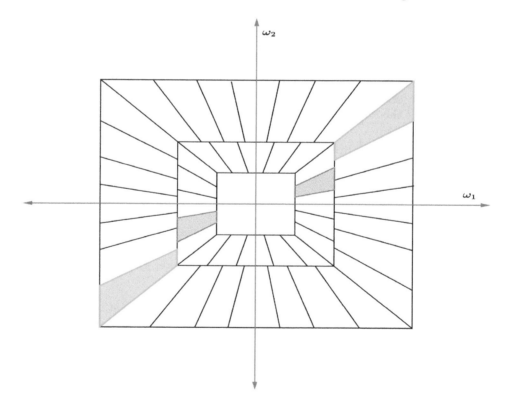

FIGURE 3.35: Tiling on the frequency plane due to cone-adapted shearlets.

Finally, we shall conclude this subsection with a fundamental result demonstrating that a square integrable function f is continuously reproduced by using isotropic window functions at coarse scales, and two sets of continuous shearlet systems at fine scales: One set corresponding to the horizontal cones and another set corresponding to the vertical cones. The prime advantage of this construction over the classical case, where the shear variable "s" attains any real value, is that $|s| \leq 2$. Such properties are of critical significance in the characterization of singularities arising in higher dimensional signals [170].

Theorem 3.7.14. *Let $\phi \in L^2(\mathbb{R}^2)$ be the scaling function and $\psi, \tilde{\psi} \in L^2(\mathbb{R}^2)$ be two classical shearlets, such that*

$$\left|\hat{\phi}(\mathbf{w})\right| + \chi_{C_1 \cup C_3}(\mathbf{w}) \left(\int_0^1 \left|\hat{\psi}(a\omega_1)\right|^2 \frac{da}{a}\right) + \chi_{C_2 \cup C_4}(\mathbf{w}) \left(\int_0^1 \left|\hat{\psi}(\tilde{a}\omega_2)\right|^2 \frac{d\tilde{a}}{\tilde{a}}\right) = 1, \quad (3.346)$$

almost everywhere $\mathbf{w} = (\omega_1, \omega_2)^T \in \mathbb{R}^2$. Then, for any $f \in L^2(\mathbb{R}^2)$, we have

$$\int_{\mathbb{R}^2} \left|\left\langle f, \phi_{\mathbf{t}'}\right\rangle_2\right|^2 d\mathbf{t}' + \int_0^1 \int_{|s|\leq 2} \int_{\mathbb{R}^2} \left|\left\langle \mathscr{F}^{-1}\left(\hat{f}(\mathbf{w})\chi_{C_1 \cup C_3}(\mathbf{w})\right), \psi_{a,s,\mathbf{t}}\right\rangle_2\right|^2 \frac{d\mathbf{t}\, ds\, da}{a^3}$$

$$+ \int_0^1 \int_{|\tilde{s}|\leq 2} \int_{\mathbb{R}^2} \left|\left\langle \mathscr{F}^{-1}\left(\hat{f}(\mathbf{w})\chi_{C_2 \cup C_4}(\mathbf{w})\right), \tilde{\psi}_{\tilde{a},\tilde{s},\tilde{\mathbf{t}}}\right\rangle_2\right|^2 \frac{d\tilde{\mathbf{t}}\, d\tilde{s}\, d\tilde{a}}{\tilde{a}^3} = \left\|f\right\|_2^2. \quad (3.347)$$

Proof. For the scaling function $\phi \in L^2(\mathbb{R}^2)$, we have

$$\int_{\mathbb{R}^2} \left|\left\langle f, \phi_{\mathbf{t}'}\right\rangle_2\right|^2 d\mathbf{t}' = \int_{\mathbb{R}^2} \left|\left\langle \hat{f}, \hat{\phi}_{\mathbf{t}'}\right\rangle_2\right|^2 d\mathbf{t}'$$

$$= \int_{\mathbb{R}^2} \left|\int_{\mathbb{R}^2} \hat{f}(\mathbf{w})\, \overline{\hat{\phi}_{\mathbf{t}'}(\mathbf{w})}\, d\mathbf{w}\right|^2 d\mathbf{t}'$$

$$= \int_{\mathbb{R}^2} \left|\int_{\mathbb{R}^2} \hat{f}(\mathbf{w})\, \overline{\hat{\phi}(\mathbf{w})}\, e^{2\pi i \mathbf{t}'^T \mathbf{w}}\, d\mathbf{w}\right|^2 d\mathbf{t}'$$

$$= \int_{\mathbb{R}^2} \left|\mathscr{F}^{-1}\left(\hat{f}(\mathbf{w})\, \overline{\hat{\phi}(\mathbf{w})}\right)(\mathbf{t}')\right|^2 d\mathbf{t}'$$

$$= \left\|\mathscr{F}^{-1}\left(\hat{f}(\mathbf{w})\, \overline{\hat{\phi}(\mathbf{w})}\right)(\mathbf{t}')\right\|_2^2$$

$$= \left\|\hat{f}(\mathbf{w})\, \overline{\hat{\phi}(\mathbf{w})}\right\|_2^2$$

$$= \int_{\mathbb{R}^2} \left|\hat{f}(\mathbf{w})\right|^2 \left|\hat{\phi}(\mathbf{w})\right|^2 d\mathbf{w}. \quad (3.348)$$

Moreover, we observe that

$$\int_{\mathbb{S}_{\text{cone}}} \left|\left\langle \mathscr{F}^{-1}\left(\hat{f}(\mathbf{w})\chi_{C_1 \cup C_3}(\mathbf{w})\right), \psi_{a,s,\mathbf{t}}\right\rangle_2\right|^2 \frac{d\mathbf{t}\, ds\, da}{a^3}$$

$$= \int_0^1 \int_{|s|\leq 2} \int_{\mathbb{R}^2} \left|\int_{\mathbb{R}^2} \hat{f}(\mathbf{w})\chi_{C_1 \cup C_3}(\mathbf{w})\, \overline{\hat{\psi}_{a,s,\mathbf{t}}(\mathbf{w})}\, d\mathbf{w}\right|^2 \frac{d\mathbf{t}\, ds\, da}{a^3}$$

$$= \int_0^1 \int_{|s|\leq 2} \int_{\mathbb{R}^2} \left|\int_{\mathbb{R}^2} \hat{f}(\mathbf{w})\chi_{C_1 \cup C_3}(\mathbf{w})\, a^{3/2}\, \overline{\hat{\psi}\left(A_a^T S_s^T \mathbf{w}\right)}\, e^{2\pi i \mathbf{t}^T \mathbf{w}}\, d\mathbf{w}\right|^2 \frac{d\mathbf{t}\, ds\, da}{a^3}$$

$$= \int_0^1 \int_{|s|\leq 2} \int_{\mathbb{R}^2} \left|\mathscr{F}^{-1}\left(\hat{f}(\mathbf{w})\chi_{C_1 \cup C_3}(\mathbf{w})\, \overline{\hat{\psi}\left(A_a^T S_s^T \mathbf{w}\right)}\right)(\mathbf{t})\right|^2 \frac{d\mathbf{t}\, ds\, da}{a^{3/2}}$$

$$= \int_0^1 \int_{|s| \le 2} \left\| \mathscr{F}^{-1} \left(\hat{f}(\mathbf{w}) \, \chi_{C_1 \cup C_3}(\mathbf{w}) \, \overline{\hat{\psi} \left(A_a^T S_s^T \mathbf{w} \right)} \right) (\mathbf{t}) \right\|^2 \frac{ds \, da}{a^{3/2}}$$

$$= \int_0^1 \int_{|s| \le 2} \left\| \hat{f}(\mathbf{w}) \, \chi_{C_1 \cup C_3}(\mathbf{w}) \, \overline{\hat{\psi} \left(A_a^T S_s^T \mathbf{w} \right)} \right\|^2 \frac{ds \, da}{a^{3/2}}$$

$$= \int_0^1 \int_{|s| \le 2} \int_{\mathbb{R}^2} \left| \hat{f}(\mathbf{w}) \right|^2 \chi_{C_1 \cup C_3}(\mathbf{w}) \left| \hat{\psi} \left(A_a^T S_s^T \mathbf{w} \right) \right|^2 \frac{d\mathbf{w} \, ds \, da}{a^{3/2}}$$

$$= \int_{\mathbb{R}^2} \left| \hat{f}(\mathbf{w}) \right|^2 \chi_{C_1 \cup C_3}(\mathbf{w}) \left\{ \int_0^1 \int_{|s| \le 2} \left| \hat{\psi} \left(A_a^T S_s^T \mathbf{w} \right) \right|^2 \frac{ds \, da}{a^{3/2}} \right\} d\mathbf{w}. \qquad (3.349)$$

Now, for $\mathbf{w} = (\omega_1, \omega_2)^T \in C_1 \cup C_3$, we have

$$\int_0^1 \int_{|s| \le 2} \left| \hat{\psi} \left(A_a^T S_s^T \mathbf{w} \right) \right|^2 \frac{ds \, da}{a^{3/2}} = \int_0^1 \int_{-2}^2 \left| \hat{\psi}(a\omega_1) \right|^2 \left| \hat{\psi}_2 \left(a^{-1/2} \left(\frac{\omega_2}{\omega_1} - s \right) \right) \right|^2 \frac{ds \, da}{a^{3/2}}$$

$$= \int_0^1 \left| \hat{\psi}(a\omega_1) \right|^2 \left\{ \int_{\frac{1}{\sqrt{a}} \left(\frac{\omega_2}{\omega_1} - 2 \right)}^{\frac{1}{\sqrt{a}} \left(\frac{\omega_2}{\omega_1} + 2 \right)} \left| \hat{\psi}_2(s) \right|^2 ds \right\} \frac{da}{a^{3/2}}$$

$$= \int_0^1 \left| \hat{\psi}(a\omega_1) \right|^2 \frac{da}{a}. \qquad (3.350)$$

Implementing (3.350) in (3.349), we obtain

$$\int_{\mathbb{S}_{cone}} \left| \left\langle \mathscr{F}^{-1} \left(\hat{f}(\mathbf{w}) \, \chi_{C_1 \cup C_3}(\mathbf{w}) \right), \psi_{a,s,\mathbf{t}} \right\rangle_2 \right|^2 \frac{d\mathbf{t} \, ds \, da}{a^3}$$

$$= \int_{\mathbb{R}^2} \left| \hat{f}(\mathbf{w}) \right|^2 \chi_{C_1 \cup C_3}(\mathbf{w}) \left\{ \int_0^1 \left| \hat{\psi}(a\omega_1) \right|^2 \frac{da}{a} \right\} d\mathbf{w}. \qquad (3.351)$$

By proceeding in a similar manner for the function $\tilde{\psi} \in L^2(\mathbb{R}^2)$ and noting that $\tilde{\psi}(\omega_1, \omega_2) = \psi(\omega_2, \omega_1)$, we obtain

$$\int_0^1 \int_{|\tilde{s}| \le 2} \int_{\mathbb{R}^2} \left| \left\langle \mathscr{F}^{-1} \left(\hat{f}(\mathbf{w}) \, \chi_{C_2 \cup C_4}(\mathbf{w}) \right), \tilde{\psi}_{\tilde{a},\tilde{s},\tilde{\mathbf{t}}} \right\rangle_2 \right|^2 \frac{d\tilde{\mathbf{t}} \, d\tilde{s} \, d\tilde{a}}{\tilde{a}^3}$$

$$= \int_{\mathbb{R}^2} \left| \hat{f}(\mathbf{w}) \right|^2 \chi_{C_2 \cup C_4}(\mathbf{w}) \left\{ \int_0^1 \left| \hat{\psi}(\tilde{a}\omega_2) \right|^2 \frac{d\tilde{a}}{\tilde{a}} \right\} d\mathbf{w}. \qquad (3.352)$$

Upon adding (3.348), (3.351) and (3.352), we get the desired result as

$$\int_{\mathbb{R}^2} \left| \left\langle f, \phi_{\mathbf{t}'} \right\rangle_2 \right|^2 d\mathbf{t}' + \int_0^1 \int_{|s| \le 2} \int_{\mathbb{R}^2} \left| \left\langle \mathscr{F}^{-1} \left(\hat{f}(\mathbf{w}) \, \chi_{C_1 \cup C_3}(\mathbf{w}) \right), \psi_{a,s,\mathbf{t}} \right\rangle_2 \right|^2 \frac{d\mathbf{t} \, ds \, da}{a^3}$$

$$+ \int_0^1 \int_{|\tilde{s}| \le 2} \int_{\mathbb{R}^2} \left| \left\langle \mathscr{F}^{-1} \left(\hat{f}(\mathbf{w}) \, \chi_{C_2 \cup C_4}(\mathbf{w}) \right), \tilde{\psi}_{\tilde{a},\tilde{s},\tilde{\mathbf{t}}} \right\rangle_2 \right|^2 \frac{d\tilde{\mathbf{t}} \, d\tilde{s} \, d\tilde{a}}{\tilde{a}^3}$$

$$= \int_{\mathbb{R}^2} \left| \hat{f}(\mathbf{w}) \right| \left| \hat{\phi}(\mathbf{w}) \right| d\mathbf{w} + \int_{\mathbb{R}^2} \left| \hat{f}(\mathbf{w}) \right|^2 \chi_{C_1 \cup C_3}(\mathbf{w}) \left\{ \int_0^1 \left| \hat{\psi}(a\omega_1) \right|^2 \frac{da}{a} \right\} d\mathbf{w}$$

$$+ \int_{\mathbb{R}^2} \left| \hat{f}(\mathbf{w}) \right|^2 \chi_{C_2 \cup C_4}(\mathbf{w}) \left\{ \int_0^1 \left| \hat{\psi}(\tilde{a}\omega_2) \right|^2 \frac{d\tilde{a}}{\tilde{a}} \right\} d\mathbf{w}$$

$$= \int_{\mathbb{R}^2} \left| \hat{f}(\mathbf{w}) \right|^2 \left\{ \left| \hat{\phi}(\mathbf{w}) \right| + \chi_{C_1 \cup C_3}(\mathbf{w}) \left(\int_0^1 \left| \hat{\psi}(a\omega_1) \right|^2 \frac{da}{a} \right) \right.$$

$$\left. + \chi_{C_2 \cup C_4}(\mathbf{w}) \left(\int_0^1 \left| \hat{\psi}(\tilde{a}\omega_2) \right|^2 \frac{d\tilde{a}}{\tilde{a}} \right) \right\} d\mathbf{w}$$

$$= \int_{\mathbb{R}^2} \left| \hat{f}(\mathbf{w}) \right|^2 d\mathbf{w}$$

$$= \left\| \hat{f} \right\|_2^2$$

$$= \left\| f \right\|_2^2.$$

This completes the proof of Theorem 3.7.14. □

3.8 The Bendlet Transform

"How can it be that mathematics, being after all a product of human thought independent of experience, is so admirably adapted to the objects of reality? Is human reason, then, without experience, merely by taking thought, able to fathom the properties of real things."

-Albert Einstein

The analysis of multi-dimensional data is one of the pivotal as well contemporary areas of research in science and engineering. One of the prime interests in multi-dimensional signal analysis is to efficiently capture the edges and corners in signals appearing due to the spatial occlusion between different objects. For example, in medical imaging curves separate bones and different kinds of soft tissue; and in geoscience the boundaries between different geological strata correspond to discontinuities in the measured data. Therefore, the key problem in multi-dimensional signal analysis is to extract and characterize the relevant and directional information regarding the occurrence of curves and boundaries in signals. In this context, many directional representation systems have been developed and successfully employed from time to time including the previously discussed polar wavelets, ridgelets, curvelets, ripplets and shearlets. Yet another directional representation system has been recently introduced by Christian Lessig and his colleagues coined as "bendlets" [176]; a shearlet like system that is based on anisotropic scaling, translation, shearing and bending of a compactly supported generator. Despite having much similarities with the well-known shearlets, one of the major distinguishing features of the bendlets is that they allow a precise characterization of location, orientation and curvature of discontinuities in higher dimensional signals. In this section, we investigate on the mathematical theory of bendlet transforms and study the fundamental properties, including orthogonality relation, inversion formula and obtain a characterization of the range of bendlet transform. Besides, we also invoke the bendlet transform for the integral representations of the solutions of Laplace and wave equations.

3.8.1 Definition and basic properties

In this subsection, we focus on some mathematical aspects of the theory of bendlets; a novel directional representation system that allows for the precise classification of both local orientation as well as curvature in higher dimensional signals. As an extension of shearlets, the bendlets are constructed by enhancing the adaptivity of shearlets by incorporating bending as another degree of freedom for the analyzing elements. This enables that the elements can align with the local bending of the boundary curves, which is the key for the classification of curvature. Since shearing has linear and bending has quadratic dependence on spatial coordinates, bendlets can be considered as a second-order shearlet system. Indeed, they are a special case of shearlet systems of arbitrary orders. Unlike, the shearlet systems, the bendlets cannot be obtained by the action of a group on the generating function, the reason being that the generalized shearing operation used in case of shearlet systems of arbitrary orders yields a composition that does not satisfy associativity.

For $a \in \mathbb{R}^+$ and $\alpha \in [0,1]$, let $\mathcal{D}_{A_{a,\alpha}}$ denote the scaling operator acting on a square integrable function $\psi \in L^2(\mathbb{R}^2)$ as

$$\left(\mathcal{D}_{A_{a,\alpha}} \psi \right)(\mathbf{x}) = |\det A_{a,\alpha}|^{-1/2} \, \psi \left(A_{a,\alpha}^{-1} \mathbf{x} \right), \quad \text{where } \mathbf{x} = (x_1, x_2)^T \in \mathbb{R}^2, \; A_{a,\alpha} = \begin{pmatrix} a & 0 \\ 0 & a^\alpha \end{pmatrix}.$$

The parameter α appearing in the scaling matrix $A_{a,\alpha}$ determines the scaling anisotropy. For example, $\alpha = 1$ corresponds to isotropic scaling, $\alpha = 1/2$ leads to parabolic scaling whereas $\alpha = 0$ determines the pure directional scaling. Also, for $\ell \in \mathbb{N}$ and $\mathbf{r} = (r_1, r_2, \ldots, r_\ell) \in \mathbb{R}^\ell$, let $\mathcal{S}_{S_{r,\ell}}$ denotes the generalized shearing operator acting on $\psi \in L^2(\mathbb{R}^2)$ as

$$\left(\mathcal{S}_{S_{r,\ell}} \psi \right)(\mathbf{x}) = \psi \left(S_{-r,\ell} \, \mathbf{x} \right), \quad \text{where } S_{-r,\ell} \, \mathbf{x} = \begin{pmatrix} 1 & \sum_{m=1}^{\ell} r_m x_2^{m-1} \\ 0 & 1 \end{pmatrix} \begin{pmatrix} x_1 \\ x_2 \end{pmatrix}.$$

For $\ell = 1$, the operator $\mathcal{S}_{S_{r,\ell}}$ reduces to the ordinary shearing operator, whereas $\ell = 2$ yields an operator implementing shearing and bending. Moreover, for $\mathbf{t} = (t_1, t_2)^T \in \mathbb{R}^2$, let $\mathcal{T}_{\mathbf{t}}$ denotes the translation operator acting on $\psi \in L^2(\mathbb{R}^2)$ as $(\mathcal{T}_{\mathbf{t}} \psi)(\mathbf{x}) = \psi(\mathbf{x} - \mathbf{t})$. Then, the combined action of the above defined scaling, generalized shearing and translation operators; that is, $\mathcal{D}_{A_{a,\alpha}}$, $\mathcal{S}_{S_{r,\ell}}$ and $\mathcal{T}_{\mathbf{t}}$, on a given generating function $\psi \in L^2(\mathbb{R}^2)$ yields a new family of analyzing functions $\mathscr{B}_\psi^{\ell,\alpha}$ called the "ℓ-th order α-shearlet systems." Mathematically, we write it as

$$\mathscr{B}_\psi^{\ell,\alpha} := \left\{ \psi_{a,\mathbf{r},\mathbf{t}}(\mathbf{x}) = \left(\mathcal{D}_{A_{a,\alpha}} \mathcal{S}_{S_{r,\ell}} \mathcal{T}_{\mathbf{t}} \psi \right)(\mathbf{x}) = |\det A_{a,\alpha}|^{-1/2} \, \psi \left(A_{a,\alpha}^{-1} S_{-r,\ell} \, (\mathbf{x} - \mathbf{t}) \right) \right\}. \tag{3.353}$$

Having constructed a family of ℓ-th order α-shearlets, the associated ℓ-th order α-shearlet transform of any $f \in L^2(\mathbb{R}^2)$ is defined as

$$\mathbb{S}_\psi^{(\ell,\alpha)} \big[f \big](a, \mathbf{r}, \mathbf{t}) = \big\langle f, \psi_{a,\mathbf{r},\mathbf{t}} \big\rangle_2 = \int_{\mathbb{R}^2} f(\mathbf{x}) \, \overline{\psi_{a,\mathbf{r},\mathbf{t}}}(\mathbf{x}) \, d\mathbf{x}. \tag{3.354}$$

It is worth noticing that for $\ell = 1$, the ℓ-th order α-shearlet transform (3.354) boils down to the classical shearlet transform. Moreover, for $\ell \geq 2$, the parameter $\mathbf{r}$ appearing in (3.354) describes a joint action of the shearing and bending which in turn corresponds to sparse representations of curved discontinuities. By increasing the order ℓ, the associated transform provides additional information about the input signals and results in finer characterization of curved discontinuities. For $\ell = 2$, the associated second-order transform is called as the "bendlet transform." The formal definition of the bendlet transform is given below:

Definition 3.8.1. For any function $f \in L^2(\mathbb{R}^2)$, the bendlet transform with respect to the analyzing function $\psi \in L^2(\mathbb{R}^2)$ is denoted by $\mathbb{B}_\psi^{(\alpha)}[f]$ and is defined as

$$\mathbb{B}_\psi^{(\alpha)}[f](a, s, b, \mathbf{t}) = \left\langle f, \psi_{a,s,b,\mathbf{t}} \right\rangle_2 = \int_{\mathbb{R}^2} f(\mathbf{x}) \, \overline{\psi_{a,s,b,\mathbf{t}}}(\mathbf{x}) \, d\mathbf{x}, \qquad (3.355)$$

where $\psi_{a,s,b,\mathbf{t}}(\mathbf{x})$ is given by (3.353) with $\mathbf{r} = (s, b)$, $\ell = 2$, so that s takes the role of shearing and b corresponds to bending of the generating function $\psi \in L^2(\mathbb{R}^2)$.

Remark 3.8.1. The bendlet transform (3.355) can be expressed via the two-dimensional convolution operation $\circledast$ given in (3.226) as

$$\mathbb{B}_\psi^{(\alpha)}[f](a, s, b, \mathbf{t}) = \left(f \circledast \check{\overline{\psi}}_{a,s,b,0}\right)(\mathbf{t}), \quad \check{\psi}(\mathbf{x}) = \psi(-\mathbf{x}). \qquad (3.356)$$

In the following theorem, we study some elementary properties of the bendlet transform defined in (3.355).

Theorem 3.8.1. *For the scalars $\beta, \gamma \in \mathbb{C}$, $\mathbf{k} \in \mathbb{R}^2$, $\lambda \in \mathbb{R}^+$ and any pair of functions $f, g \in L^2(\mathbb{R}^2)$, the bendlet transform (3.355) satisfies the following properties:*

(i) *Linearity:* $\mathbb{B}_\psi^{(\alpha)}[\beta f + \gamma g](a, s, b, \mathbf{t}) = \beta \, \mathbb{B}_\psi^{(\alpha)}[f](a, s, b, \mathbf{t}) + \gamma \, \mathbb{B}_\psi^{(\alpha)}[g](a, s, b, \mathbf{t}),$

(ii) *Anti-linearity:* $\mathbb{B}_{\beta\psi + \gamma\phi}^{(\alpha)}[f](a, s, b, \mathbf{t}) = \bar{\beta} \, \mathbb{B}_\psi^{(\alpha)}[f](a, s, b, \mathbf{t}) + \bar{\gamma} \, \mathbb{B}_\phi^{(\alpha)}[f](a, s, b, \mathbf{t}),$

(iii) *Translation:* $\mathbb{B}_\psi^{(\alpha)}[\mathcal{T}_\mathbf{k} f](a, s, b, \mathbf{t}) = \mathbb{B}_\psi^{(\alpha)}[f](a, s, b, \mathbf{t} - \mathbf{k}),$

(iv) *Translation in bendlet:* $\mathbb{B}_{\mathcal{T}_\mathbf{k}\psi}[f](a, s, b, \mathbf{t}) = \mathbb{B}_\psi^{(\alpha)}[f](a, s, b, \mathbf{t} + S_{(s,b)} A_{a,\alpha}\mathbf{k}),$

(v) *Scaling:* $\mathbb{B}_\psi^{(\alpha)}[f(\lambda\mathbf{x})](a, s, b, \mathbf{t}) = \dfrac{1}{\lambda} \, \mathbb{B}_{\tilde{\psi}}^{(\alpha)}[f](a, s, b, \lambda\mathbf{t}), \; \tilde{\psi}(\mathbf{x}) = \psi\left(\lambda^{-1}\mathbf{x}\right),$

(vi) *Reflection:* $\mathbb{B}_\psi^{(\alpha)}[f(-\mathbf{x})](a, s, b, \mathbf{t}) = -\mathbb{B}_{\tilde{\psi}}^{(\alpha)}[f](a, s, b, -\mathbf{t}).$

Proof. The proofs of (i) and (ii) are straightforward and follow directly by using the fact that the inner product is linear in first component and anti-linear in second component.

(iii) For any $\mathbf{k} \in \mathbb{R}^2$, we observe that

$$\begin{aligned}
\mathbb{B}_\psi^{(\alpha)}[\mathcal{T}_\mathbf{k} f](a, s, b, \mathbf{t}) &= \int_{\mathbb{R}^2} f(\mathbf{x} - \mathbf{k}) \, \overline{\psi_{a,s,b,\mathbf{t}}}(\mathbf{x}) \, d\mathbf{x} \\
&= |\det A_{a,\alpha}|^{-1/2} \int_{\mathbb{R}^2} f(\mathbf{x} - \mathbf{k}) \, \overline{\psi\left(A_{a,\alpha}^{-1} S_{-(s,b)} \, (\mathbf{x} - \mathbf{t})\right)} \, d\mathbf{x} \\
&= |\det A_{a,\alpha}|^{-1/2} \int_{\mathbb{R}^2} f(\mathbf{z}) \, \overline{\psi\left(A_{a,\alpha}^{-1} S_{-(s,b)} \, (\mathbf{z} - (\mathbf{t} - \mathbf{k}))\right)} \, d\mathbf{z} \\
&= \mathbb{B}_\psi^{(\alpha)}[f](a, s, b, \mathbf{t} - \mathbf{k}).
\end{aligned}$$

(iv) Upon translating the generating function $\psi \in L^2(\mathbb{R}^2)$ by any $\mathbf{k} \in \mathbb{R}^2$, we have

$$\mathbb{B}_{\mathcal{T}_{\mathbf{k}}\psi}\Big[f\Big](a,s,b,\mathbf{t}) = \int_{\mathbb{R}^2} f(\mathbf{x})\,\overline{(\mathcal{T}_{\mathbf{k}}\psi)_{a,s,b,\mathbf{t}}(\mathbf{x})}\,d\mathbf{x}$$

$$= |\det A_{a,\alpha}|^{-1/2} \int_{\mathbb{R}^2} f(\mathbf{x})\,\overline{(\mathcal{T}_{\mathbf{k}}\psi)\left(A_{a,\alpha}^{-1}S_{-(s,b)}\,(\mathbf{x}-\mathbf{t})\right)}\,d\mathbf{x}$$

$$= |\det A_{a,\alpha}|^{-1/2} \int_{\mathbb{R}^2} f(\mathbf{x})\,\overline{\psi\left(A_{a,\alpha}^{-1}S_{-(s,b)}\,(\mathbf{x}-\mathbf{t})-\mathbf{k}\right)}\,d\mathbf{x}$$

$$= |\det A_{a,\alpha}|^{-1/2} \int_{\mathbb{R}^2} f(\mathbf{x})\,\overline{\psi\left(A_{a,\alpha}^{-1}S_{-(s,b)}\,(\mathbf{x}-\mathbf{t}-S_{(s,b)}A_{a,\alpha}\mathbf{k})\right)}\,d\mathbf{x}$$

$$= |\det A_{a,\alpha}|^{-1/2} \int_{\mathbb{R}^2} f(\mathbf{x})\,\overline{\psi\left(A_{a,\alpha}^{-1}S_{-(s,b)}\,(\mathbf{x}-(\mathbf{t}+S_{(s,b)}A_{a,\alpha}\mathbf{k}))\right)}\,d\mathbf{x}$$

$$= \mathbb{B}_{\psi}^{(\alpha)}\Big[f\Big](a,s,b,\mathbf{t}+S_{(s,b)}A_{a,\alpha}\mathbf{k}).$$

(**v**) For any $\lambda \in \mathbb{R}^{+}$, we have

$$\mathbb{B}_{\psi}^{(\alpha)}\Big[f(\lambda\mathbf{x})\Big](a,s,b,\mathbf{t}) = \int_{\mathbb{R}^2} f(\lambda\mathbf{x})\,\overline{\psi_{a,s,b,\mathbf{t}}(\mathbf{x})}\,d\mathbf{x}$$

$$= |\det A_{a,\alpha}|^{-1/2} \int_{\mathbb{R}^2} f(\lambda\mathbf{x})\,\overline{\psi\left(A_{a,\alpha}^{-1}S_{-(s,b)}\,(\mathbf{x}-\mathbf{t})\right)}\,d\mathbf{x}$$

$$= |\det A_{a,\alpha}|^{-1/2} \int_{\mathbb{R}^2} f(\mathbf{z})\,\overline{\tilde{\psi}\left(A_{a,\alpha}^{-1}S_{-(s,b)}\,(\mathbf{z}-\lambda\mathbf{t})\right)}\,\frac{d\mathbf{z}}{\lambda}$$

$$= \frac{1}{\lambda}\,\mathbb{B}_{\tilde{\psi}}^{(\alpha)}\Big[f\Big](a,s,b,\lambda\mathbf{t}), \quad \text{where}\ \ \tilde{\psi}(\mathbf{x}) = \psi\left(\lambda^{-1}\mathbf{x}\right).$$

(**vi**) To observe the behaviour of the bendlet transform under the reflection of the input signal, we proceed as

$$\mathbb{B}_{\psi}^{(\alpha)}\Big[f(-\mathbf{x})\Big](a,s,b,\mathbf{t}) = \int_{\mathbb{R}^2} f(-\mathbf{x})\,\overline{\psi_{a,s,b,\mathbf{t}}(\mathbf{x})}\,d\mathbf{x}$$

$$= |\det A_{a,\alpha}|^{-1/2} \int_{\mathbb{R}^2} f(-\mathbf{x})\,\overline{\psi\left(A_{a,\alpha}^{-1}S_{-(s,b)}\,(\mathbf{x}-\mathbf{t})\right)}\,d\mathbf{x}$$

$$= -|\det A_{a,\alpha}|^{-1/2} \int_{\mathbb{R}^2} f(\mathbf{z})\,\overline{\psi\left(A_{a,\alpha}^{-1}S_{-(s,b)}\,(-\mathbf{z}-\mathbf{t})\right)}\,d\mathbf{z}$$

$$= -|\det A_{a,\alpha}|^{-1/2} \int_{\mathbb{R}^2} f(\mathbf{z})\,\overline{\psi\left(-A_{a,\alpha}^{-1}S_{-(s,b)}\,(\mathbf{z}+\mathbf{t})\right)}\,d\mathbf{z}$$

$$= -|\det A_{a,\alpha}|^{-1/2} \int_{\mathbb{R}^2} f(\mathbf{z})\,\overline{\check{\psi}\left(A_{a,\alpha}^{-1}S_{-(s,b)}\,(\mathbf{z}+\mathbf{t})\right)}\,d\mathbf{z}$$

$$= -\mathbb{B}_{\check{\psi}}^{(\alpha)}\Big[f\Big](a,s,b,-\mathbf{t}), \quad \text{where}\ \ \check{\psi}(\mathbf{x}) = \psi(-\mathbf{x}).$$

This completes the proof of Theorem 3.8.1.   □

Next, we intend to compute the two-dimensional Fourier transform of the bendlet transform $\mathbb{B}_{\psi}^{(\alpha)}\big[f\big](a,s,b,\mathbf{t})$, when considered as a function of the translation variable $\mathbf{t}$. Such a relationship shall be invoked in the formulation of orthogonality relation for the bendlet transform defined in (3.355). In this direction, we have the following lemma:

Lemma 3.8.2. *Let* $\mathbb{B}_\psi^{(\alpha)}[f](a, s, b, \mathbf{t})$ *be the bendlet transform of the function* $f \in L^2(\mathbb{R}^2)$. *Then, the following formula holds:*

$$\mathscr{F}\left(\mathbb{B}_\psi^{(\alpha)}[f](a, s, b, \mathbf{t})\right)(\mathbf{w}) = a^{\frac{(1+\alpha)}{2}} \, \mathscr{F}[f](\mathbf{w}) \, \overline{\mathscr{F}[\psi]\left(A_{a,\alpha}^T S_{(s,b)}^T \mathbf{w}\right)}, \qquad (3.357)$$

where $\mathscr{F}$ *denotes the two-dimensional Fourier transform defined in* (1.292).

Proof. For the generating function $\psi \in L^2(\mathbb{R}^2)$, we have

$$\begin{aligned}
\mathscr{F}\left[\psi_{a,s,b,\mathbf{t}}(\mathbf{x})\right](\mathbf{w}) &= \int_{\mathbb{R}^2} \psi_{a,s,b,\mathbf{t}}(\mathbf{x}) \, e^{-2\pi i \mathbf{x}^T \mathbf{w}} d\mathbf{x} \\
&= |\det A_{a,\alpha}|^{-1/2} \int_{\mathbb{R}^2} \psi\left(A_{a,\alpha}^{-1} S_{-(s,b)} (\mathbf{x} - \mathbf{t})\right) e^{-2\pi i \mathbf{x}^T \mathbf{w}} d\mathbf{x} \\
&= a^{\frac{(1+\alpha)}{2}} \int_{\mathbb{R}^2} \psi(\mathbf{z}) \, e^{-2\pi i \left(\mathbf{t} + S_{(s,b)} A_{a,\alpha} \mathbf{z}\right)^T \mathbf{w}} d\mathbf{z} \\
&= a^{\frac{(1+\alpha)}{2}} e^{-2\pi i \mathbf{t}^T \mathbf{w}} \int_{\mathbb{R}^2} \psi(\mathbf{z}) \, e^{-2\pi i \mathbf{z}^T A_{a,\alpha}^T S_{(s,b)}^T \mathbf{w}} d\mathbf{z} \\
&= a^{\frac{(1+\alpha)}{2}} e^{-2\pi i \mathbf{t}^T \mathbf{w}} \mathscr{F}[\psi]\left(A_{a,\alpha}^T S_{(s,b)}^T \mathbf{w}\right). \qquad (3.358)
\end{aligned}$$

Then, by virtue of the Parseval's formula for the two-dimensional Fourier transform together with the implication of (3.358), we have

$$\begin{aligned}
\mathbb{B}_\psi^{(\alpha)}[f](a, s, b, \mathbf{t}) &= \left\langle f, \psi_{a,s,b,\mathbf{t}} \right\rangle_2 \\
&= \left\langle \mathscr{F}[f], \mathscr{F}[\psi_{a,s,b,\mathbf{t}}] \right\rangle_2 \\
&= \int_{\mathbb{R}^2} \mathscr{F}[f](\mathbf{w}) \, \overline{\mathscr{F}[\psi_{a,s,b,\mathbf{t}}](\mathbf{w})} \, d\mathbf{w} \\
&= \int_{\mathbb{R}^2} \mathscr{F}[f](\mathbf{w}) \, a^{\frac{(1+\alpha)}{2}} e^{2\pi i \mathbf{t}^T \mathbf{w}} \, \overline{\mathscr{F}[\psi]\left(A_{a,\alpha}^T S_{(s,b)}^T \mathbf{w}\right)} \, d\mathbf{w} \\
&= a^{\frac{(1+\alpha)}{2}} \, \mathscr{F}^{-1}\left(\mathscr{F}[f](\mathbf{w}) \, \overline{\mathscr{F}[\psi]\left(A_{a,\alpha}^T S_{(s,b)}^T \mathbf{w}\right)}\right)(\mathbf{t}). \qquad (3.359)
\end{aligned}$$

Taking the Fourier transform on both sides of (3.359) yields the desired result as

$$\mathscr{F}\left(\mathbb{B}_\psi^{(\alpha)}[f](a, s, b, \mathbf{t})\right)(\mathbf{w}) = a^{\frac{(1+\alpha)}{2}} \, \mathscr{F}[f](\mathbf{w}) \, \overline{\mathscr{F}[\psi]\left(A_{a,\alpha}^T S_{(s,b)}^T \mathbf{w}\right)}.$$

This completes the proof of Lemma 3.8.2. $\qquad\qquad\qquad\qquad\qquad\qquad\qquad$ □

In the next instance, we prove the Rayleigh's theorem for the bendlet transform. As a consequence of the Rayleigh's theorem, we shall demonstrate that the orthogonality of bendlet transforms of two signals is both implied by and implies the orthogonality of the input signals. Moreover, such a theorem also yields a resolution of identity for the bendlet transform defined in (3.355).

Theorem 3.8.3. *If* $\mathbb{B}_\psi^{(\alpha)}[f](a, s, b, \mathbf{t})$ *and* $\mathbb{B}_\psi^{(\alpha)}[g](a, s, b, \mathbf{t})$ *are the bendlet transforms of a given pair of signals* $f, g \in L^2(\mathbb{R}^2)$ *with respect to the generating function* $\psi \in L^2(\mathbb{R}^2)$, *then the following orthogonality relation holds:*

$$\int_{\mathbb{R}^2 \times \mathbb{R} \times \mathbb{R} \times \mathbb{R}^+} \mathbb{B}_\psi^{(\alpha)}[f](a, s, b, \mathbf{t}) \, \overline{\mathbb{B}_\psi^{(\alpha)}[g](a, s, b, \mathbf{t})} \, \frac{da \, ds \, db \, d\mathbf{t}}{a^{2(1+\alpha)}} = C_\psi \left\langle f, g \right\rangle_2, \qquad (3.360)$$

provided

$$C_\psi := \int_{\mathbb{R}\times\mathbb{R}\times\mathbb{R}^+} \left| \hat{\psi}\left(A_{a,\alpha}^T S_{(s,b)}^T \mathbf{w}\right) \right|^2 \frac{da\,ds\,db}{a^{(1+\alpha)}} < \infty, \quad a.e. \quad \mathbf{w} \in \mathbb{R}^2. \tag{3.361}$$

Proof. Invoking Lemma 3.8.2, the bendlet transforms of the given pair of functions $f, g \in L^2(\mathbb{R}^2)$ can be expressed as

$$\mathbb{B}_\psi^{(\alpha)}\big[f\big](a,s,b,\mathbf{t}) = a^{\frac{(1+\alpha)}{2}} \int_{\mathbb{R}^2} \hat{f}(\mathbf{w})\, \overline{\hat{\psi}\left(A_{a,\alpha}^T S_{(s,b)}^T \mathbf{w}\right)}\, e^{2\pi i \mathbf{t}^T \mathbf{w}}\, d\mathbf{w} \tag{3.362}$$

and

$$\mathbb{B}_\psi^{(\alpha)}\big[g\big](a,s,b,\mathbf{t}) = a^{\frac{(1+\alpha)}{2}} \int_{\mathbb{R}^2} \hat{g}(\mathbf{w}')\, \overline{\hat{\psi}\left(A_{a,\alpha}^T S_{(s,b)}^T \mathbf{w}'\right)}\, e^{2\pi i \mathbf{t}^T \mathbf{w}'}\, d\mathbf{w}'. \tag{3.363}$$

Thus, as a consequence of (3.362) and (3.363), we can write

$$\int_{\mathbb{R}^2\times\mathbb{R}\times\mathbb{R}\times\mathbb{R}^+} \mathbb{B}_\psi^{(\alpha)}\big[f\big](a,s,b,\mathbf{t})\, \overline{\mathbb{B}_\psi^{(\alpha)}\big[g\big](a,s,b,\mathbf{t})}\, \frac{da\,ds\,db\,d\mathbf{t}}{a^{2(1+\alpha)}}$$

$$= \int_{\mathbb{R}^2\times\mathbb{R}\times\mathbb{R}\times\mathbb{R}^+} \left\{ a^{\frac{(1+\alpha)}{2}} \int_{\mathbb{R}^2} \hat{f}(\mathbf{w})\, \overline{\hat{\psi}\left(A_{a,\alpha}^T S_{(s,b)}^T \mathbf{w}\right)}\, e^{2\pi i \mathbf{t}^T \mathbf{w}}\, d\mathbf{w} \right\}$$

$$\times \left\{ a^{\frac{(1+\alpha)}{2}} \int_{\mathbb{R}^2} \overline{\hat{g}(\mathbf{w}')}\, \hat{\psi}\left(A_{a,\alpha}^T S_{(s,b)}^T \mathbf{w}'\right)\, e^{-2\pi i \mathbf{t}^T \mathbf{w}'}\, d\mathbf{w}' \right\} \frac{da\,ds\,db\,d\mathbf{t}}{a^{2(1+\alpha)}}$$

$$= \int_{\mathbb{R}^2\times\mathbb{R}\times\mathbb{R}\times\mathbb{R}^+} a^{(1+\alpha)} \left\{ \int_{\mathbb{R}^2} \hat{f}(\mathbf{w})\, \overline{\hat{\psi}\left(A_{a,\alpha}^T S_{(s,b)}^T \mathbf{w}\right)}\, e^{2\pi i \mathbf{t}^T \mathbf{w}}\, d\mathbf{w} \right\}$$

$$\times \left\{ \int_{\mathbb{R}^2} \overline{\hat{g}(\mathbf{w}')}\, \hat{\psi}\left(A_{a,\alpha}^T S_{(s,b)}^T \mathbf{w}'\right)\, e^{-2\pi i \mathbf{t}^T \mathbf{w}'}\, d\mathbf{w}' \right\} \frac{da\,ds\,db\,d\mathbf{t}}{a^{2(1+\alpha)}}$$

$$= \int_{\mathbb{R}\times\mathbb{R}\times\mathbb{R}^+} \left\{ \int_{\mathbb{R}^2} \int_{\mathbb{R}^2} \hat{f}(\mathbf{w})\, \overline{\hat{g}(\mathbf{w}')}\, \hat{\psi}\left(A_{a,\alpha}^T S_{(s,b)}^T \mathbf{w}'\right)\, \overline{\hat{\psi}\left(A_{a,\alpha}^T S_{(s,b)}^T \mathbf{w}\right)}\, d\mathbf{w}\, d\mathbf{w}' \right\}$$

$$\times \left\{ \int_{\mathbb{R}^2} e^{-2\pi i \mathbf{t}^T (\mathbf{w}-\mathbf{w}')}\, d\mathbf{t} \right\} \frac{da\,ds\,db}{a^{(1+\alpha)}}$$

$$= \int_{\mathbb{R}\times\mathbb{R}\times\mathbb{R}^+} \left\{ \int_{\mathbb{R}^2} \int_{\mathbb{R}^2} \hat{f}(\mathbf{w})\, \overline{\hat{g}(\mathbf{w}')}\, \hat{\psi}\left(A_{a,\alpha}^T S_{(s,b)}^T \mathbf{w}'\right)\, \overline{\hat{\psi}\left(A_{a,\alpha}^T S_{(s,b)}^T \mathbf{w}\right)}\, d\mathbf{w}\, d\mathbf{w}' \right\}$$

$$\times \delta\left(\mathbf{w} - \mathbf{w}'\right) \frac{da\,ds\,db}{a^{(1+\alpha)}}$$

$$= \int_{\mathbb{R}^2} \hat{f}(\mathbf{w})\, \overline{\hat{g}(\mathbf{w})} \left\{ \int_{\mathbb{R}\times\mathbb{R}\times\mathbb{R}^+} \left| \hat{\psi}\left(A_{a,\alpha}^T S_{(s,b)}^T \mathbf{w}\right) \right|^2 \frac{da\,ds\,db}{a^{(1+\alpha)}} \right\} d\mathbf{w}$$

$$= C_\psi \big\langle \hat{f}, \hat{g} \big\rangle_2$$

$$= C_\psi \big\langle f, g \big\rangle_2.$$

This completes the proof of Theorem 3.8.3. $\square$

Remark 3.8.2. The condition (3.361) is known as the "admissibility condition" pertaining to the generating function $\psi \in L^2(\mathbb{R}^2)$.

Remark 3.8.3. For $f = g$, the orthogonality relation (3.360) yields

$$\int_{\mathbb{R}^2\times\mathbb{R}\times\mathbb{R}\times\mathbb{R}^+} \left| \mathbb{B}_\psi^{(\alpha)}\big[f\big](a,s,b,\mathbf{t}) \right|^2 \frac{da\,ds\,db\,d\mathbf{t}}{a^{2(1+\alpha)}} = C_\psi \big\| f \big\|_2^2. \tag{3.364}$$

Therefore, choosing the generating function $\psi \in L^2(\mathbb{R}^2)$ in such a way that $C_\psi = 1$, the expression (3.364) implies that

$$\left\| \mathbb{B}_\psi^{(\alpha)} \big[f \big] (a, s, b, \mathbf{t}) \right\|_2^2 = \| f \|_2^2. \tag{3.365}$$

That is, the bendlet transform given by (3.355) is an isometry from the space $L^2(\mathbb{R}^2)$ to the transformed space $L^2\big(\mathbb{R}^+ \times \mathbb{R} \times \mathbb{R} \times \mathbb{R}^2\big)$.

In the forthcoming theorem, we formulate an inversion formula for the bendlet transform defined in (3.355).

Theorem 3.8.4. *Any function $f \in L^2(\mathbb{R}^2)$ can be reconstructed from the corresponding bendlet transform $\mathbb{B}_\psi^{(\alpha)} \big[f \big] (a, s, b, \mathbf{t})$ via the reconstruction formula:*

$$f(\mathbf{x}) = \frac{1}{C_\psi} \int_{\mathbb{R}^2 \times \mathbb{R} \times \mathbb{R} \times \mathbb{R}^+} \mathbb{B}_\psi^{(\alpha)} \big[f \big] (a, s, b, \mathbf{t}) \, \psi_{a,s,b,\mathbf{t}}(\mathbf{x}) \, \frac{da \, ds \, db \, d\mathbf{t}}{a^{2(1+\alpha)}}, \quad a.e., \tag{3.366}$$

where C_ψ is given by (3.361).

Proof. For any arbitrary square integrable function $g \in L^2(\mathbb{R}^2)$, the orthogonality relation (3.360) yields

$$\begin{aligned}
\big\langle f, g \big\rangle_2 &= \frac{1}{C_\psi} \Big\langle \mathbb{B}_\psi^{(\alpha)} \big[f \big] (a, s, b, \mathbf{t}), \, \mathbb{B}_\psi^{(\alpha)} \big[f \big] (a, s, b, \mathbf{t}) \Big\rangle_2 \\
&= \frac{1}{C_\psi} \int_{\mathbb{R}^2 \times \mathbb{R} \times \mathbb{R} \times \mathbb{R}^+} \mathbb{B}_\psi^{(\alpha)} \big[f \big] (a, s, b, \mathbf{t}) \left\{ \int_{\mathbb{R}^2} \overline{g(\mathbf{x})} \, \psi_{a,s,b,\mathbf{t}}(\mathbf{x}) \, d\mathbf{x} \right\} \frac{da \, ds \, db \, d\mathbf{t}}{a^{2(1+\alpha)}} \\
&= \frac{1}{C_\psi} \int_{\mathbb{R}^2} \left\{ \int_{\mathbb{R}^2 \times \mathbb{R} \times \mathbb{R} \times \mathbb{R}^+} \mathbb{B}_\psi^{(\alpha)} \big[f \big] (a, s, b, \mathbf{t}) \, \psi_{a,s,b,\mathbf{t}}(\mathbf{x}) \, \frac{da \, ds \, db \, d\mathbf{t}}{a^{2(1+\alpha)}} \right\} \overline{g(\mathbf{x})} \, d\mathbf{x} \\
&= \frac{1}{C_\psi} \left\langle \int_{\mathbb{R}^2 \times \mathbb{R} \times \mathbb{R} \times \mathbb{R}^+} \mathbb{B}_\psi^{(\alpha)} \big[f \big] (a, s, b, \mathbf{t}) \, \psi_{a,s,b,\mathbf{t}}(\mathbf{x}) \, \frac{da \, ds \, db \, d\mathbf{t}}{a^{2(1+\alpha)}}, \, g(\mathbf{x}) \right\rangle_2. \tag{3.367}
\end{aligned}$$

Since g is arbitrary, therefore it follows that

$$f(\mathbf{x}) = \frac{1}{C_\psi} \int_{\mathbb{R}^2 \times \mathbb{R} \times \mathbb{R} \times \mathbb{R}^+} \mathbb{B}_\psi^{(\alpha)} \big[f \big] (a, s, b, \mathbf{t}) \, \psi_{a,s,b,\mathbf{t}}(\mathbf{x}) \, \frac{da \, ds \, db \, d\mathbf{t}}{a^{2(1+\alpha)}} \quad a.e.,$$

which is the desired reconstruction formula. $\qquad \square$

In the next theorem, we intend to study the pointwise convergence of the reconstruction formula (3.366). We note that the pointwise convergence is guaranteed under the assumptions that $f \in L^1 \cap L^2(\mathbb{R}^2)$ and the generating function ψ satisfies the admissibility condition (3.361). To facilitate the intent, we have the following proposition:

Proposition 3.8.5. *Suppose that the generating function $\psi \in L^2(\mathbb{R}^2)$ satisfies the admissibility condition (3.361). For any $f \in L^2(\mathbb{R}^2)$ and $A_2 > A_1 > 0$, $B > 0$, $C > 0$, define*

$$f_{A_1, A_2 : B : C}(\mathbf{x}) = \frac{1}{C_\psi} \int_{A_1}^{A_2} \int_{-B}^{B} \int_{-C}^{C} \int_{\mathbb{R}^2} \mathbb{B}_\psi^{(\alpha)} \big[f \big] (a, s, b, \mathbf{t}) \, \psi_{a,s,b,\mathbf{t}}(\mathbf{x}) \, \frac{d\mathbf{t} \, db \, ds \, da}{a^{2(1+\alpha)}}. \tag{3.368}$$

Then, $f_{A_1, A_2 : B : C}$ is uniformly continuous on $\mathbb{R}^2$ and the corresponding Fourier transform is given by

$$\widehat{f}_{A_1, A_2 : B : C}(\mathbf{w}) = \frac{1}{C_\psi} \hat{f}(\mathbf{w}) \left\{ \int_{A_1}^{A_2} \int_{-B}^{B} \int_{-C}^{C} \left| \hat{\psi}\big(A_{a,\alpha}^T S_{(s,b)}^T \mathbf{w}\big) \right|^2 \frac{db \, ds \, da}{a^{(1+\alpha)}} \right\}. \tag{3.369}$$

Proof. Applying the well-known Cauchy-Schwarz inequality, we can write

$$\int_{A_1}^{A_2} \int_{-B}^{B} \int_{-C}^{C} \int_{\mathbb{R}^2} \left| \mathbb{B}_{\psi}^{(\alpha)}[f](a,s,b,\mathbf{t}) \, \psi_{a,s,b,\mathbf{t}}(\mathbf{x}) \right| \frac{d\mathbf{t} \, db \, ds \, da}{a^{2(1+\alpha)}}$$

$$\leq \int_{A_1}^{A_2} \int_{-B}^{B} \int_{-C}^{C} \left\{ \int_{\mathbb{R}^2} \left| \mathbb{B}_{\psi}^{(\alpha)}[f](a,s,b,\mathbf{t}) \right|^2 d\mathbf{t} \right\}^{1/2} \left\{ \int_{\mathbb{R}^2} \left| \psi_{a,s,b,\mathbf{t}}(\mathbf{x}) \right|^2 d\mathbf{t} \right\}^{1/2} \frac{db \, ds \, da}{a^{2(1+\alpha)}}$$

$$\leq \int_{A_1}^{A_2} \int_{-B}^{B} \int_{-C}^{C} \left\{ \int_{\mathbb{R}^2} \left| \mathbb{B}_{\psi}^{(\alpha)}[f](a,s,b,\mathbf{t}) \right|^2 d\mathbf{t} \right\}^{1/2} \left\| \psi \right\|_2 \frac{db \, ds \, da}{a^{2(1+\alpha)}}$$

$$= \int_{A_1}^{A_2} \int_{-B}^{B} \int_{-C}^{C} \left\{ \int_{\mathbb{R}^2} \left| a^{\frac{(1+\alpha)}{2}} \hat{f}(\mathbf{w}) \overline{\hat{\psi}(A_{a,\alpha}^T S_{(s,b)}^T \mathbf{w})} \right|^2 d\mathbf{w} \right\}^{1/2} \left\| \psi \right\|_2 \frac{db \, ds \, da}{a^{2(1+\alpha)}}$$

$$\leq \left\{ \int_{A_1}^{A_2} \int_{-B}^{B} \int_{-C}^{C} \int_{\mathbb{R}^2} a^{(1+\alpha)} \left| \hat{f}(\mathbf{w}) \right|^2 \left| \hat{\psi}(A_{a,\alpha}^T S_{(s,b)}^T \mathbf{w}) \right|^2 \frac{d\mathbf{w} \, db \, ds \, da}{a^{2(1+\alpha)}} \right\}^{1/2}$$

$$\times \left\{ \int_{A_1}^{A_2} \int_{-B}^{B} \int_{-C}^{C} \left\| \psi \right\|_2^2 \frac{db \, ds \, da}{a^{2(1+\alpha)}} \right\}^{1/2}$$

$$= \left\| \psi \right\|_2 \left\{ \frac{4BC \left(A_1^{-(1+2\alpha)} - A_2^{-(1+2\alpha)} \right)}{1+2\alpha} \right\}$$

$$\times \left\{ \int_{\mathbb{R}^2} \left| \hat{f}(\mathbf{w}) \right|^2 \left(\int_{A_1}^{A_2} \int_{-B}^{B} \int_{-C}^{C} \left| \hat{\psi}(A_{a,\alpha}^T S_{(s,b)}^T \mathbf{w}) \right|^2 \frac{db \, ds \, da}{a^{(1+\alpha)}} \right) d\mathbf{w} \right\}^{1/2}$$

$$\leq \sqrt{C_\psi} \left\| f \right\|_2 \left\| \psi \right\|_2 \left\{ \frac{4BC \left(A_1^{-(1+2\alpha)} - A_2^{-(1+2\alpha)} \right)}{1+2\alpha} \right\} < \infty. \tag{3.370}$$

By virtue of inequality (3.370), we conclude that the function $f_{A_1,A_2:B:C}$ is well defined on $\mathbb{R}^2$. In order to exhibit the uniform continuity of the function $f_{A_1,A_2:B:C}$, we choose any $\mathbf{x}, \mathbf{x}' \in \mathbb{R}^2$ and proceed as follows:

$$\left| f_{A_1,A_2:B:C}(\mathbf{x}) - f_{A_1,A_2:B:C}(\mathbf{x}') \right|$$

$$= \left| \frac{1}{C_\psi} \int_{A_1}^{A_2} \int_{-B}^{B} \int_{-C}^{C} \int_{\mathbb{R}^2} \mathbb{B}_{\psi}^{(\alpha)}[f](a,s,b,\mathbf{t}) \left(\psi_{a,s,b,\mathbf{t}}(\mathbf{x}) - \psi_{a,s,b,\mathbf{t}}(\mathbf{x}') \right) \frac{d\mathbf{t} \, db \, ds \, da}{a^{2(1+\alpha)}} \right|$$

$$\leq \frac{1}{C_\psi} \left\{ \int_{A_1}^{A_2} \int_{-B}^{B} \int_{-C}^{C} \int_{\mathbb{R}^2} a^{(1+\alpha)} \left| \hat{f}(\mathbf{w}) \right|^2 \left| \hat{\psi}(A_{a,\alpha}^T S_{(s,b)}^T \mathbf{w}) \right|^2 \frac{d\mathbf{w} \, db \, ds \, da}{a^{2(1+\alpha)}} \right\}^{1/2}$$

$$\times \left\{ \int_{A_1}^{A_2} \int_{-B}^{B} \int_{-C}^{C} \int_{\mathbb{R}^2} \left| \psi_{a,s,b,\mathbf{t}}(\mathbf{x}) - \psi_{a,s,b,\mathbf{t}}(\mathbf{x}') \right|^2 \frac{d\mathbf{t} \, db \, ds \, da}{a^{2(1+\alpha)}} \right\}^{1/2}$$

$$\leq \frac{1}{\sqrt{C_\psi}} \left\| f \right\|_2 \left\{ \int_{A_1}^{A_2} \int_{-B}^{B} \int_{-C}^{C} \int_{\mathbb{R}^2} a^{-(1+\alpha)} \left| \psi \left(A_{a,\alpha}^{-1} S_{-\mathbf{r},\ell} \, \mathbf{x} - A_{a,\alpha}^{-1} S_{-\mathbf{r},\ell} \, \mathbf{t} \right) \right. \right.$$

$$\left. \left. - \psi \left(A_{a,\alpha}^{-1} S_{-\mathbf{r},\ell} \, \mathbf{x}' - A_{a,\alpha}^{-1} S_{-\mathbf{r},\ell} \, \mathbf{t} \right) \right|^2 \frac{d\mathbf{t} \, db \, ds \, da}{a^{2(1+\alpha)}} \right\}^{1/2}$$

$$= \frac{\left\| f \right\|_2}{\sqrt{C_\psi}} \left\{ \int_{A_1}^{A_2} \int_{-B}^{B} \int_{-C}^{C} \left\| \psi \left(A_{a,\alpha}^{-1} S_{-\mathbf{r},\ell} \, \mathbf{x} - \cdot \right) - \psi \left(A_{a,\alpha}^{-1} S_{-\mathbf{r},\ell} \, \mathbf{x}' - \cdot \right) \right\|_2^2 \frac{db \, ds \, da}{a^{2(1+\alpha)}} \right\}^{1/2}.$$

$$\tag{3.371}$$

From inequality (3.371), it is quite evident that $|f_{A_1,A_2:B:C}(\mathbf{x}) - f_{A_1,A_2:B}(\mathbf{x}')| \to 0$, whenever $|\mathbf{x} - \mathbf{x}'| \to 0$. Hence, we conclude that the function $f_{A_1,A_2:B:C}(\mathbf{x})$ defined in (3.368) is uniformly continuous on $\mathbb{R}^2$. Finally, for any arbitrary function $g \in L^2(\mathbb{R}^2)$, we have

$$
\left\langle f_{A_1,A_2:B:C},\, g \right\rangle_2
$$

$$
= \int_{\mathbb{R}^2} \left\{ \frac{1}{C_\psi} \int_{A_1}^{A_2} \int_{-B}^{B} \int_{-C}^{C} \int_{\mathbb{R}^2} \mathbb{B}_\psi^{(\alpha)}\big[f\big](a,s,b,\mathbf{t})\, \psi_{a,s,b,\mathbf{t}}(\mathbf{x})\, \frac{d\mathbf{t}\, db\, ds\, da}{a^{2(1+\alpha)}} \right\} \overline{g(\mathbf{x})}\, d\mathbf{x}
$$

$$
= \frac{1}{C_\psi} \int_{A_1}^{A_2} \int_{-B}^{B} \int_{-C}^{C} \int_{\mathbb{R}^2} \mathbb{B}_\psi^{(\alpha)}\big[f\big](a,s,b,\mathbf{t}) \left\{ \int_{\mathbb{R}^2} \psi_{a,s,b,\mathbf{t}}(\mathbf{x})\, \overline{g(\mathbf{x})}\, d\mathbf{x} \right\} \frac{d\mathbf{t}\, db\, ds\, da}{a^{2(1+\alpha)}}
$$

$$
= \frac{1}{C_\psi} \int_{A_1}^{A_2} \int_{-B}^{B} \int_{-C}^{C} \int_{\mathbb{R}^2} \mathbb{B}_\psi^{(\alpha)}\big[f\big](a,s,b,\mathbf{t})\, \overline{\mathbb{B}_\psi^{(\alpha)}\big[g\big](a,s,b,\mathbf{t})}\, \frac{d\mathbf{t}\, db\, ds\, da}{a^{2(1+\alpha)}}
$$

$$
= \frac{1}{C_\psi} \int_{A_1}^{A_2} \int_{-B}^{B} \int_{-C}^{C} \int_{\mathbb{R}^2} \left\{ a^{\frac{(1+\alpha)}{2}} \int_{\mathbb{R}^2} \hat{f}(\mathbf{w})\, \overline{\hat{\psi}\big(A_{a,\alpha}^T S_{(s,b)}^T \mathbf{w}\big)}\, e^{2\pi i \mathbf{t}^T \mathbf{w}}\, d\mathbf{w} \right\}
$$
$$
\times \left\{ a^{\frac{(1+\alpha)}{2}} \int_{\mathbb{R}^2} \overline{\hat{g}(\mathbf{w}')}\, \hat{\psi}\big(A_{a,\alpha}^T S_{(s,b)}^T \mathbf{w}'\big)\, e^{-2\pi i \mathbf{t}^T \mathbf{w}'}\, d\mathbf{w}' \right\} \frac{d\mathbf{t}\, db\, ds\, da}{a^{2(1+\alpha)}}
$$

$$
= \frac{1}{C_\psi} \int_{A_1}^{A_2} \int_{-B}^{B} \int_{-C}^{C} \int_{\mathbb{R}^2} \int_{\mathbb{R}^2} \hat{f}(\mathbf{w})\, \overline{\hat{g}(\mathbf{w}')}\, \overline{\hat{\psi}\big(A_{a,\alpha}^T S_{(s,b)}^T \mathbf{w}\big)}\, \hat{\psi}\big(A_{a,\alpha}^T S_{(s,b)}^T \mathbf{w}'\big)
$$
$$
\times \left\{ \int_{\mathbb{R}^2} e^{2\pi i \mathbf{t}^T (\mathbf{w}-\mathbf{w}')}\, d\mathbf{t} \right\} \frac{d\mathbf{w}\, d\mathbf{w}'\, db\, ds\, da}{a^{(1+\alpha)}}
$$

$$
= \frac{1}{C_\psi} \int_{A_1}^{A_2} \int_{-B}^{B} \int_{-C}^{C} \int_{\mathbb{R}^2} \int_{\mathbb{R}^2} \hat{f}(\mathbf{w})\, \overline{\hat{g}(\mathbf{w}')}\, \overline{\hat{\psi}\big(A_{a,\alpha}^T S_{(s,b)}^T \mathbf{w}\big)}\, \hat{\psi}\big(A_{a,\alpha}^T S_{(s,b)}^T \mathbf{w}'\big)
$$
$$
\times\, \delta(\mathbf{w}-\mathbf{w}')\, \frac{d\mathbf{w}\, d\mathbf{w}'\, db\, ds\, da}{a^{(1+\alpha)}}
$$

$$
= \frac{1}{C_\psi} \int_{A_1}^{A_2} \int_{-B}^{B} \int_{-C}^{C} \int_{\mathbb{R}^2} \hat{f}(\mathbf{w})\, \overline{\hat{g}(\mathbf{w})}\, \left| \hat{\psi}\big(A_{a,\alpha}^T S_{(s,b)}^T \mathbf{w}\big) \right|^2 \frac{d\mathbf{w}\, db\, ds\, da}{a^{(1+\alpha)}}
$$

$$
= \frac{1}{C_\psi} \int_{\mathbb{R}^2} \hat{f}(\mathbf{w}) \left\{ \int_{A_1}^{A_2} \int_{-B}^{B} \int_{-C}^{C} \left| \hat{\psi}\big(A_{a,\alpha}^T S_{(s,b)}^T \mathbf{w}\big) \right|^2 \frac{db\, ds\, da}{a^{(1+\alpha)}} \right\} \overline{\hat{g}(\mathbf{w})}\, d\mathbf{w}. \qquad (3.372)
$$

Invoking the Parseval's formula on L.H.S of (3.372) and then using the fundamental properties of the inner product, we deduce that

$$
\widehat{f}_{A_1,A_2:B:C}(\mathbf{w}) = \frac{1}{C_\psi}\, \hat{f}(\mathbf{w}) \left\{ \int_{A_1}^{A_2} \int_{-B}^{B} \int_{-C}^{C} \left| \hat{\psi}\big(A_{a,\alpha}^T S_{(s,b)}^T \mathbf{w}\big) \right|^2 \frac{db\, ds\, da}{a^{(1+\alpha)}} \right\}.
$$

This completes the proof of Proposition 3.8.5. $\qquad\qquad\square$

Theorem 3.8.6. *Suppose that the generating function $\psi \in L^2(\mathbb{R}^2)$ satisfies the admissibility condition (3.361). Then, for any $f \in L^1 \cap L^2(\mathbb{R}^2)$, we have*

$$
\lim_{\substack{A_1 \to 0,\, A_2 \to \infty \\ B,\, C \to \infty}} \big\| f - f_{A_1,A_2:B:C} \big\|_\infty = 0 \quad \text{and} \quad \lim_{\substack{A_1 \to 0,\, A_2 \to \infty \\ B,\, C \to \infty}} \big\| f - f_{A_1,A_2:B:C} \big\|_2 = 0. \qquad (3.373)
$$

Proof. Invoking Proposition 3.8.5 together with the implication of Parseval's formula for the Fourier transform, we have

$$\left\|f - f_{A_1, A_2:B:C}\right\|_\infty$$

$$\leq \left\|f - f_{A_1, A_2:B:C}\right\|_1$$

$$= \left\|\hat{f} - \hat{f}_{A_1, A_2:B:C}\right\|_1$$

$$= \left\|\hat{f}(\mathbf{w}) - \frac{1}{C_\psi}\, \hat{f}(\mathbf{w}) \left\{\int_{A_1}^{A_2} \int_{-B}^{B} \int_{-C}^{C} \left|\hat{\psi}\left(A_{a,\alpha}^T S_{(s,b)}^T \mathbf{w}\right)\right|^2 \frac{db\,ds\,da}{a^{(1+\alpha)}}\right\}\right\|_1$$

$$= \left\|\hat{f}(\mathbf{w}) \left(1 - \frac{1}{C_\psi}\int_{A_1}^{A_2} \int_{-B}^{B} \int_{-C}^{C} \left|\hat{\psi}\left(A_{a,\alpha}^T S_{(s,b)}^T \mathbf{w}\right)\right|^2 \frac{db\,ds\,da}{a^{(1+\alpha)}}\right)\right\|_1$$

$$= \int_{\mathbb{R}^2} |\hat{f}(\mathbf{w})| \left|1 - \frac{1}{C_\psi}\int_{A_1}^{A_2} \int_{-B}^{B} \int_{-C}^{C} \left|\hat{\psi}\left(A_{a,\alpha}^T S_{(s,b)}^T \mathbf{w}\right)\right|^2 \frac{db\,ds\,da}{a^{(1+\alpha)}}\right| d\mathbf{w}. \qquad (3.374)$$

Letting $A_1 \to 0$, $A_2 \to \infty$, $B, C \to \infty$ and noting that $\psi \in L^2(\mathbb{R}^2)$ satisfies the admissibility condition (3.361), we obtain

$$\lim_{\substack{A_1 \to 0,\, A_2 \to \infty \\ B, C \to \infty}} \left|1 - \frac{1}{C_\psi}\int_{A_1}^{A_2} \int_{-B}^{B} \int_{-C}^{C} \left|\hat{\psi}\left(A_{a,\alpha}^T S_{(s,b)}^T \mathbf{w}\right)\right|^2 \frac{db\,ds\,da}{a^{(1+\alpha)}}\right| = 0. \qquad (3.375)$$

Finally, using the dominated convergence theorem in (3.374), we conclude that

$$\lim_{\substack{A_1 \to 0,\, A_2 \to \infty \\ B, C \to \infty}} \left\|f - f_{A_1, A_2:B:C}\right\|_\infty = 0.$$

Proceeding in a manner analogous to the above case, we can show that

$$\lim_{\substack{A_1 \to 0,\, A_2 \to \infty \\ B, C \to \infty}} \left\|f - f_{A_1, A_2:B:C}\right\|_2 = 0.$$

This completes the proof of Theorem 3.8.6. $\square$

Towards the end of this subsection, we present a theorem providing a characterization of range of the bendlet transform (3.355).

Theorem 3.8.7. *Any function* $f \in L^2\left(\mathbb{R}^+ \times \mathbb{R} \times \mathbb{R} \times \mathbb{R}^2\right)$ *is the bendlet transform of a certain signal with respect to the generating function* $\psi \in L^2(\mathbb{R}^2)$ *satisfying (3.361) if and only if the following reproduction property holds:*

$$f(a', s', b', \mathbf{t}') = \frac{1}{C_\psi} \int_{\mathbb{R}^2 \times \mathbb{R} \times \mathbb{R} \times \mathbb{R}^+} f(a, s, b, \mathbf{t}) \left\langle \psi_{a,s,b,\mathbf{t}}, \, \psi_{a',s',b',\mathbf{t}'} \right\rangle_2 \frac{da\,ds\,db\,d\mathbf{t}}{a^{2(1+\alpha)}}. \qquad (3.376)$$

Proof. Firstly, suppose that a function $f \in L^2\left(\mathbb{R}^+ \times \mathbb{R} \times \mathbb{R} \times \mathbb{R}^2\right)$ is the bendlet transform of the function $g \in L^2(\mathbb{R}^2)$ with respect to the admissible generator $\psi \in L^2(\mathbb{R}^2)$. It remains to show that (3.376) holds good. We proceed as

$$f(a', s', b', \mathbf{t}')$$

$$= \mathbb{B}_\psi^{(\alpha)}\left[g\right](a', s', b', \mathbf{t}')$$

$$= \int_{\mathbb{R}^2} g(\mathbf{x})\, \overline{\psi_{a',s',b',\mathbf{t}'}(\mathbf{x})}\, d\mathbf{x}$$

$$= \int_{\mathbb{R}^2} \left\{ \frac{1}{C_\psi} \int_{\mathbb{R}^2 \times \mathbb{R} \times \mathbb{R} \times \mathbb{R}^+} \mathbb{B}_\psi^{(\alpha)} \big[g\big] (a, s, b, \mathbf{t}) \, \psi_{a,s,b,\mathbf{t}}(\mathbf{x}) \, \frac{da \, ds \, db \, d\mathbf{t}}{a^{2(1+\alpha)}} \right\} \overline{\psi_{a',s',b',\mathbf{t}'}}(\mathbf{x}) \, d\mathbf{x}$$

$$= \frac{1}{C_\psi} \int_{\mathbb{R}^2 \times \mathbb{R} \times \mathbb{R} \times \mathbb{R}^+} \mathbb{B}_\psi^{(\alpha)} \big[g\big] (a, s, b, \mathbf{t}) \left\{ \int_{\mathbb{R}^2} \psi_{a,s,b,\mathbf{t}}(\mathbf{x}) \, \overline{\psi_{a',s',b',\mathbf{t}'}}(\mathbf{x}) \, d\mathbf{x} \right\} \frac{da \, ds \, db \, d\mathbf{t}}{a^{2(1+\alpha)}}$$

$$= \frac{1}{C_\psi} \int_{\mathbb{R}^2 \times \mathbb{R} \times \mathbb{R} \times \mathbb{R}^+} f(a, s, b, \mathbf{t}) \left\langle \psi_{a,s,b,\mathbf{t}}, \, \psi_{a',s',b',\mathbf{t}'} \right\rangle_2 \frac{da \, ds \, db \, d\mathbf{t}}{a^{2(1+\alpha)}}. \tag{3.377}$$

Conversely, suppose that a function $f \in L^2\left(\mathbb{R}^+ \times \mathbb{R} \times \mathbb{R} \times \mathbb{R}^2\right)$ satisfies (3.376). Then, we have to show that there exists a function $g \in L^2(\mathbb{R}^2)$ such that $\mathbb{B}_\psi^{(\alpha)}[g] = f$. We claim that

$$g(\mathbf{x}) = \frac{1}{C_\psi} \int_{\mathbb{R}^2 \times \mathbb{R} \times \mathbb{R} \times \mathbb{R}^+} f(a, s, b, \mathbf{t}) \, \psi_{a,s,b,\mathbf{t}}(\mathbf{x}) \, \frac{da \, ds \, db \, d\mathbf{t}}{a^{2(1+\alpha)}} \tag{3.378}$$

is an apt candidate for the aforesaid function. A straightforward computation demonstrates that

$$\big\|g\big\|_2 = C_\psi^{-1} \big\|f\big\|_2 < \infty.$$

That is, the function g defined in (3.378) is square integrable. Finally, we observe that

$$\mathbb{B}_\psi^{(\alpha)} \big[g\big] (a', s', b', \mathbf{t}')$$

$$= \int_{\mathbb{R}^2} g(\mathbf{x}) \, \overline{\psi_{a',s',b',\mathbf{t}'}}(\mathbf{x}) \, d\mathbf{x}$$

$$= \int_{\mathbb{R}^2} \left\{ \frac{1}{C_\psi} \int_{\mathbb{R}^2 \times \mathbb{R} \times \mathbb{R} \times \mathbb{R}^+} f(a, s, b, \mathbf{t}) \, \psi_{a,s,b,\mathbf{t}}(\mathbf{x}) \, \frac{da \, ds \, db \, d\mathbf{t}}{a^{2(1+\alpha)}} \right\} \overline{\psi_{a',s',b',\mathbf{t}'}}(\mathbf{x}) \, d\mathbf{x}$$

$$= \frac{1}{C_\psi} \int_{\mathbb{R}^2 \times \mathbb{R} \times \mathbb{R} \times \mathbb{R}^+} f(a, s, b, \mathbf{t}) \left\{ \int_{\mathbb{R}^2} \psi_{a,s,b,\mathbf{t}}(\mathbf{x}) \, \overline{\psi_{a',s',b',\mathbf{t}'}}(\mathbf{x}) \, d\mathbf{x} \right\} \frac{da \, ds \, db \, d\mathbf{t}}{a^{2(1+\alpha)}}$$

$$= \frac{1}{C_\psi} \int_{\mathbb{R}^2 \times \mathbb{R} \times \mathbb{R} \times \mathbb{R}^+} f(a, s, b, \mathbf{t}) \left\langle \psi_{a,s,b,\mathbf{t}}, \, \psi_{a',s',b',\mathbf{t}'} \right\rangle_2 \frac{da \, ds \, db \, d\mathbf{t}}{a^{2(1+\alpha)}}$$

$$= f(a', s', b', \mathbf{t}'). \tag{3.379}$$

The expression (3.379) evidently establishes the converse part of the assertion. Therefore, the proof of Theorem 3.8.7 is complete. □

3.8.2 Applications of bendlet transform to partial differential equations

"How did Biot arrive at the partial differential equation? [the heat conduction equation]... Perhaps Laplace gave Biot the equation and left him to sink or swim for a few years in trying to derive it. That would have been merely an instance of the way great mathematicians since the very beginnings of mathematical research have effortlessly maintained their superiority over ordinary mortals."

-Clifford A. Truesdell

This subsection is aimed at describing a means for expressing the solutions of differential equations via the elements of a given family of analyzing functions, namely the bendlets. To facilitate the narrative, we shall take into consideration a couple of well-known partial differential equations; the Laplace equation, which is named after the French polymath Pierre-Simon Laplace, and the wave equation.

(A) Dirichlet problem for the three-dimensional Laplace equation in half-space

Consider the following boundary value problem:

$$\left.\begin{array}{l} \nabla^2 U := U_{xx} + U_{yy} + U_{zz} = 0, \quad -\infty < x, y < \infty, \quad z > 0 \\ U(x,y,0) = f(x,y), \quad -\infty < x, y < \infty \\ U(x,y,z) \to 0 \quad \text{as} \quad r = \sqrt{x^2 + y^2 + z^2} \to \infty \end{array}\right\}. \tag{3.380}$$

After applying the Fourier transform, the system of equations (3.380) transforms to a comparatively elegant form as

$$\left.\begin{array}{l} \dfrac{d^2 \widehat{U}}{dz^2} - \left(\omega_1^2 + \omega_2^2\right) \widehat{U} = 0, \quad z > 0 \\ \widehat{U}(\omega_1, \omega_2, 0) = \widehat{f}(\omega_1, \omega_2) \end{array}\right\}. \tag{3.381}$$

For $\mathbf{w} = (\omega_1, \omega_2)^T$, the solution of the transformed system of equations (3.381) is given by

$$\widehat{U}(\omega_1, \omega_2, z) = \widehat{f}(\omega_1, \omega_2) \exp\left\{-z\left(\omega_1^2 + \omega_2^2\right)^{1/2}\right\}$$
$$= \widehat{f}(\mathbf{w})\, \widehat{g}_z(\mathbf{w}), \tag{3.382}$$

where

$$\widehat{g}_z(\mathbf{w}) = \exp\left\{-z\left(\omega_1^2 + \omega_2^2\right)^{1/2}\right\}. \tag{3.383}$$

By virtue of the inverse Fourier transform, we can compute $g_z(x,y)$ to be

$$g_z(x,y) = \frac{z}{\left(x^2 + y^2 + z^2\right)^{3/2}}. \tag{3.384}$$

Invoking the inversion formula for the bendlet transform (3.366), we obtain

$$U(x,y,z) = \frac{1}{C_\psi} \int_{\mathbb{R}^2 \times \mathbb{R} \times \mathbb{R} \times \mathbb{R}^+} \mathbb{B}_\psi^{(\alpha)}\big[U(x,y,\cdot)\big](a,s,b,\mathbf{t})\, \psi_{a,s,b,\mathbf{t}}(x,y)\, \frac{da\,ds\,db\,d\mathbf{t}}{a^{2(1+\alpha)}}$$
$$= \frac{1}{C_\psi} \int_{\mathbb{R}^2 \times \mathbb{R} \times \mathbb{R} \times \mathbb{R}^+} H(a,s,b,\mathbf{t})\, \psi_{a,s,b,\mathbf{t}}(x,y)\, \frac{da\,ds\,db\,d\mathbf{t}}{a^{2(1+\alpha)}}, \tag{3.385}$$

where $H(a,s,b,\mathbf{t}) = \mathbb{B}_\psi^{(\alpha)}\big[U(x,y,\cdot)\big](a,s,b,\mathbf{t})$. In order to determine the coefficients $H(a,s,b,\mathbf{t})$ appearing in (3.385), we shall use the Parseval's formula as follows:

$$H(a,s,b,\mathbf{t}) = \Big\langle U,\, \psi_{a,s,b,\mathbf{t}}\Big\rangle_2$$
$$= \Big\langle \widehat{U},\, \widehat{\psi}_{a,s,b,\mathbf{t}}\Big\rangle_2$$
$$= a^{\frac{(1+\alpha)}{2}} \int_{\mathbb{R}^2} \widehat{U}(\omega_1, \omega_2, z)\, \overline{\widehat{\psi}\big(A_{a,\alpha}^T S_{(s,b)}^T \mathbf{w}\big)}\, e^{2\pi i \mathbf{t}^T \mathbf{w}}\, d\mathbf{w}. \tag{3.386}$$

Plugging (3.386) in (3.385), we get

$$U(x,y,z)$$
$$= \frac{1}{C_\psi} \int_{\mathbb{R}^2 \times \mathbb{R} \times \mathbb{R} \times \mathbb{R}^+} a^{\frac{(1+\alpha)}{2}} \left\{\int_{\mathbb{R}^2} \widehat{U}(\omega_1, \omega_2, z)\, \overline{\widehat{\psi}\big(A_{a,\alpha}^T S_{(s,b)}^T \mathbf{w}\big)}\, e^{2\pi i \mathbf{t}^T \mathbf{w}}\, d\mathbf{w}\right\} \psi_{a,s,b,\mathbf{t}}(x,y)\, \frac{da\,ds\,db\,d\mathbf{t}}{a^{2(1+\alpha)}}$$
$$= \frac{1}{C_\psi} \int_{\mathbb{R}^2 \times \mathbb{R}^2 \times \mathbb{R} \times \mathbb{R} \times \mathbb{R}^+} e^{2\pi i \mathbf{t}^T \mathbf{w}}\, \widehat{f}(\mathbf{w})\, \widehat{g}_z(\mathbf{w})\, \overline{\widehat{\psi}\big(A_{a,\alpha}^T S_{(s,b)}^T \mathbf{w}\big)} \psi_{a,s,b,\mathbf{t}}(x,y)\, \frac{da\,ds\,db\,d\mathbf{t}\,d\mathbf{w}}{a^{\frac{3(1+\alpha)}{2}}}. \tag{3.387}$$

Expression (3.387) is the desired integral representation of the solution of the Laplace equation (3.380) under the given boundary conditions.

(B) Two-dimensional homogeneous initial-value wave equation

Consider the two-dimensional homogeneous initial-value wave equation given by

$$\left. \begin{array}{ll} U_{tt} = c^2 \left(U_{xx} + U_{yy} \right), & -\infty < x, y < \infty, \quad t > 0 \\ U(x,y,0) = 0, \quad U_t(x,y,0) = f(x,y), & -\infty < x, y < \infty \end{array} \right\}, \tag{3.388}$$

where c is a constant. We assume that $U(x,y,t)$ and its first partial derivatives vanish at infinity. By virtue of the Fourier transform, the system of equations (3.388) transforms to

$$\left. \begin{array}{l} \dfrac{d^2 \widehat{U}}{dt^2} + c^2 \left| \mathbf{w} \right|^2 \widehat{U} = 0 \\[2mm] \widehat{U}(\omega_1, \omega_2, 0) = 0, \quad \left(\dfrac{d\widehat{U}}{dt} \right)_{t=0} = \widehat{f}(\mathbf{w}) \end{array} \right\}, \tag{3.389}$$

where $\mathbf{w} = (\omega_1, \omega_2)^T \in \mathbb{R}^2$. Moreover, the solution of the transformed system of equations (3.389) is given by

$$\begin{aligned} \widehat{U}(\omega_1, \omega_2, t) &= \widehat{f}(\omega_1, \omega_2) \, \frac{\sin \left(c\, t \left| \mathbf{w} \right| \right)}{c \left| \mathbf{w} \right|} \\[2mm] &= \widehat{f}(\mathbf{w}) \, \frac{\sin \left(c\, t \left| \mathbf{w} \right| \right)}{c \left| \mathbf{w} \right|} \\[2mm] &= \frac{1}{2ic} \, \frac{\widehat{f}(\mathbf{w})}{\left| \mathbf{w} \right|} \left\{ \exp \left(i c\, t \left| \mathbf{w} \right| \right) - \exp \left(-i c\, t \left| \mathbf{w} \right| \right) \right\} \\[2mm] &= \frac{\widehat{f}(\mathbf{w})}{2ic \left| \mathbf{w} \right|} \exp \left(i c\, t \left| \mathbf{w} \right| \right) - \frac{\widehat{f}(\mathbf{w})}{2ic \left| \mathbf{w} \right|} \exp \left(-i c\, t \left| \mathbf{w} \right| \right). \end{aligned} \tag{3.390}$$

Let $\mathscr{S}$ denotes the space of complex-valued solutions $U(x,y,t)$ of the wave equation given by (3.388), which are square integrable with respect to the spatial coordinate (x,y) when time t is fixed. Next, we decompose the space of solutions $\mathscr{S}$ into a direct sum of two subspaces $\mathscr{S}_+$ and $\mathscr{S}_-$ of positive and negative frequencies, respectively, as $\mathscr{S} = \mathscr{S}_+ \oplus \mathscr{S}_-$, where

$$\left. \begin{array}{l} \mathscr{S}_+ := \left\{ f_+ \in L^2 \left(\mathbb{R}^2 \times \mathbb{R}^+ \right) : \widehat{f}_+(\mathbf{w}, t) = \widehat{f}_+(\mathbf{w}, 0) \exp \left(-i c\, t \left| \mathbf{w} \right| \right) \right\} \\[2mm] \mathscr{S}_- := \left\{ f_- \in L^2 \left(\mathbb{R}^2 \times \mathbb{R}^+ \right) : \widehat{f}_-(\mathbf{w}, t) = \widehat{f}_-(\mathbf{w}, 0) \exp \left(+i c\, t \left| \mathbf{w} \right| \right) \right\} \end{array} \right\}. \tag{3.391}$$

Consequently, any solution $U(x,y,t)$ of the wave equation (3.388) can be expressed as a two-component sum as follows:

$$U(x,y,t) = U_+(x,y,t) + U_-(x,y,t). \tag{3.392}$$

Hence, we infer that the Fourier transform of any solution $U(x,y,t)$ is of the following form:

$$\widehat{U}(\omega_1, \omega_2, t) = \widehat{U}_+(\omega_1, \omega_2, 0) \exp \left(-i c\, t \left| \mathbf{w} \right| \right) + \widehat{U}_-(\omega_1, \omega_2, 0) \exp \left(i c\, t \left| \mathbf{w} \right| \right). \tag{3.393}$$

Next, for $f_+ \in \mathscr{S}_+$ and $f_- \in \mathscr{S}_-$, we define

$$\left. \begin{array}{l} C_{f_+} = \displaystyle\int_{\mathbb{R} \times \mathbb{R} \times \mathbb{R}^+} \left| \widehat{f}_+ \left(A_{a,\alpha}^T S_{(s,b)}^T \mathbf{w} \right) \right|^2 \dfrac{da\, ds\, db}{a^{(1+\alpha)}} \\[4mm] C_{f_-} = \displaystyle\int_{\mathbb{R} \times \mathbb{R} \times \mathbb{R}^+} \left| \widehat{f}_- \left(A_{a,\alpha}^T S_{(s,b)}^T \mathbf{w} \right) \right|^2 \dfrac{da\, ds\, db}{a^{(1+\alpha)}} \end{array} \right\}. \tag{3.394}$$

The inner product on the space $\mathscr{S}_\pm$ is given by

$$\left\langle U_\pm, V_\pm \right\rangle_{\mathscr{S}_\pm} = \int_{\mathbb{R}\times\mathbb{R}} U_\pm(x,y,t)\,\overline{U_\pm(x,y,t)}\,dx\,dy$$

$$= \int_{\mathbb{R}\times\mathbb{R}} \widehat{U}_\pm(\omega_1,\omega_2,0)\,\overline{\widehat{U}_\pm(\omega_1,\omega_2,0)}\,d\omega_1\,d\omega_2. \tag{3.395}$$

It is worth noticing that the inner product defined in (3.395) does not depend upon the time t. This is the main reason why we decompose the whole space $\mathscr{S}$ into a direct sum of the subspaces $\mathscr{S}_+$ and $\mathscr{S}_-$. If we use the standard inner product on $L^2(\mathbb{R}^2)$, then the exponents $\exp\left(ict\,|\mathbf{w}|\right)$ and $\exp\left(-ict\,|\mathbf{w}|\right)$ do not cancel, and the time dependence is not removed.

Finally, we seek a solution of the wave equation (3.388) bearing the following integral representation:

$$U(x,y,t) = \frac{1}{C_{\psi_+}} \int_{\mathbb{R}^2\times\mathbb{R}\times\mathbb{R}\times\mathbb{R}^+} H_+(a,s,b,\mathbf{t})\,\psi_{+a,s,b,\mathbf{t}}(x,y)\,\frac{da\,ds\,db\,d\mathbf{t}}{a^{2(1+\alpha)}}$$

$$+ \frac{1}{C_{\psi_-}} \int_{\mathbb{R}^2\times\mathbb{R}\times\mathbb{R}\times\mathbb{R}^+} H_-(a,s,b,\mathbf{t})\,\psi_{-a,s,b,\mathbf{t}}(x,y)\,\frac{da\,ds\,db\,d\mathbf{t}}{a^{2(1+\alpha)}}, \tag{3.396}$$

where $H_\pm(a,s,b,\mathbf{t}) = \mathbb{B}_{\psi_\pm}^{(\alpha)}\big[U_\pm(x,y,\cdot)\big](a,s,b,\mathbf{t})$. In order to determine the coefficients $H_\pm(a,s,b,\mathbf{t})$ appearing in (3.396), we shall invoke the Parseval's formula for the two-dimensional Fourier transform and proceed as

$$H_\pm(a,s,b,\mathbf{t}) = \left\langle U_\pm, \psi_{\pm a,s,b,\mathbf{t}} \right\rangle_{\mathscr{S}_\pm}$$

$$= \left\langle \widehat{U}_\pm, \widehat{\psi}_{\pm a,s,b,\mathbf{t}} \right\rangle_{\mathscr{S}_\pm}$$

$$= a^{\frac{(1+\alpha)}{2}} \int_{\mathbb{R}^2} \widehat{U}_\pm(\omega_1,\omega_2,0)\,\overline{\widehat{\psi}_\pm\big(A_{a,\alpha}^T S_{(s,b)}^T \mathbf{w}\big)}\,e^{2\pi i\mathbf{w}\cdot\mathbf{t}}\,d\mathbf{w}. \tag{3.397}$$

Implementing the expression (3.397) in (3.396) yields the following representation of the solution of wave equation:

$$U(x,y,t)$$

$$= \frac{1}{C_{\psi_+}} \int_{\mathbb{R}^2\times\mathbb{R}^2\times\mathbb{R}\times\mathbb{R}\times\mathbb{R}^+} \widehat{U}_+(\omega_1,\omega_2,0)\,\overline{\widehat{\psi}_+\big(A_{a,\alpha}^T S_{(s,b)}^T \mathbf{w}\big)}\,e^{2\pi it^T\mathbf{w}}\,\psi_{+a,s,b,\mathbf{t}}(x,y)\,\frac{da\,ds\,db\,d\mathbf{t}\,d\mathbf{w}}{a^{\frac{3(1+\alpha)}{2}}}$$

$$+ \frac{1}{C_{\psi_-}} \int_{\mathbb{R}^2\times\mathbb{R}^2\times\mathbb{R}\times\mathbb{R}\times\mathbb{R}^+} \widehat{U}_-(\omega_1,\omega_2,0)\,\overline{\widehat{\psi}_-\big(A_{a,\alpha}^T S_{(s,b)}^T \mathbf{w}\big)}\,e^{2\pi it^T\mathbf{w}}\,\psi_{-a,s,b,\mathbf{t}}(x,y)\,\frac{da\,ds\,db\,d\mathbf{t}\,d\mathbf{w}}{a^{\frac{3(1+\alpha)}{2}}}. \tag{3.398}$$

Moreover, upon comparing equations (3.390) and (3.393), we obtain

$$\left.\begin{aligned} \widehat{U}_+(\omega_1,\omega_2,0) &= -\frac{\widehat{f}(\mathbf{w})}{2ic\,|\mathbf{w}|} \\[2mm] \widehat{U}_-(\omega_1,\omega_2,0) &= \frac{\widehat{f}(\mathbf{w})}{2ic\,|\mathbf{w}|} \end{aligned}\right\}. \tag{3.399}$$

Consequently, the solution $U(x,y,t)$ corresponding to the wave equation (3.388) can be represented as

$U(x,y,t)$

$$= -\frac{1}{C_{\psi_+}} \int_{\mathbb{R}^2 \times \mathbb{R}^2 \times \mathbb{R} \times \mathbb{R} \times \mathbb{R}^+} e^{2\pi i t^T \mathbf{w}} \frac{\widehat{f}(\mathbf{w})}{2ic\,|\mathbf{w}|} \overline{\widehat{\psi}_+\left(A_{a,\alpha}^T S_{(s,b)}^T \mathbf{w}\right)} \psi_{+a,s,b,t}(x,y) \frac{da\,ds\,db\,dt\,d\mathbf{w}}{a^{3(1+\alpha)/2}}$$

$$+ \frac{1}{C_{\psi_-}} \int_{\mathbb{R}^2 \times \mathbb{R}^2 \times \mathbb{R} \times \mathbb{R} \times \mathbb{R}^+} e^{2\pi i t^T \mathbf{w}} \frac{\widehat{f}(\mathbf{w})}{2ic\,|\mathbf{w}|} \overline{\widehat{\psi}_-\left(A_{a,\alpha}^T S_{(s,b)}^T \mathbf{w}\right)} \psi_{-a,s,b,t}(x,y) \frac{da\,ds\,db\,dt\,d\mathbf{w}}{a^{3(1+\alpha)/2}}.$$

$$(3.400)$$

Expression (3.400) gives a superposition of the solutions of the wave equation as given by (3.388).

3.9   Exercises

Exercise 3.9.1. Show that the Gaussian function $f(t) = e^{-\sigma t^2}$, $\sigma > 0$ does not satisfy the admissibility condition (3.27).

Exercise 3.9.2. Verify that the Fourier transform of the Haar wavelet (3.31) is given by (3.32).

Exercise 3.9.3. For the Haar wavelet $\psi(t)$ defined in (3.31), discuss the orthogonality of the daughter wavelets $\psi_{a,b}(t)$, $a \in \mathbb{R}^+$ and $b \in \mathbb{R}$.

Exercise 3.9.4. Comment upon the vanishing moments of the Haar and Morlet wavelets given in (3.31) and (3.35), respectively.

Exercise 3.9.5. Verify the admissibility condition for the Mexican hat wavelet (3.45). Also, demonstrate that the wavelets

$$\psi^{(m)}(t) = \left(\frac{1}{i}\frac{d}{dt}\right)^m e^{-t^2/2} \qquad (3.401)$$

tend to be sensitive to increasingly sharper details as the order of the derivative m is increased.

Exercise 3.9.6. Compute the centre (mean) and radius (dispersion) of the daughter wavelets corresponding to the Haar (3.31), Morlet (3.35) and Mexican hat (3.45) wavelets. Also, find their respective duration-bandwidth products.

Exercise 3.9.7. Evaluate the wavelet transform of the function

$$f(t) = \begin{cases} t, & 0 \le t \le 1 \\ 2-t, & 1 \le t \le 2 \\ 0, & \text{otherwise,} \end{cases} \qquad (3.402)$$

with respect to the Haar wavelet (3.31).

Exercise 3.9.8. For the Gaussian function $f(t) = e^{-\sigma t^2}$, $\sigma > 0$, compute the wavelet transform with respect to the Morlet wavelet (3.35) with $\omega_0 = 7$.

Exercise 3.9.9. For $a \in \mathbb{R}^+$, $b \in \mathbb{R}$ and

$$\psi_{a,b}(t) = \begin{cases} 1, & b \leq t < b + \dfrac{a}{2} \\ -1, & b + \dfrac{a}{2} \leq t < b + a \\ 0, & \text{otherwise,} \end{cases} \tag{3.403}$$

show that the wavelet transform (3.47) can be expressed as

$$\mathscr{W}_\psi \big[f \big](a,b) = \frac{1}{\sqrt{a}} \int_b^{b+\frac{a}{2}} \left(f(t) - f\big(t + \tfrac{a}{2}\big) \right) dt. \tag{3.404}$$

Exercise 3.9.10. For any pair of functions $f, g \in L^2(\mathbb{R})$, compute the wavelet transform (3.47) of the convolution function $(f * g)(z)$ defined in (1.75).

Exercise 3.9.11. Compute the wavelet transform corresponding to the wave-train given by

$$f(t) = \frac{1}{\sqrt{2\pi}} \exp\left(-\frac{t^2}{2} \right) \cos(\omega_0 t), \quad \omega_0 \in \mathbb{R}, \tag{3.405}$$

with respect to the Haar wavelet (3.31).

Exercise 3.9.12. For the function $f(t) = \chi_{[-c,\, c]}(t)\, e^{i\omega_0 t}$, where $c, \omega_0 \in \mathbb{R}$, compute the corresponding wavelet transform with respect to the Mexican hat wavelet (3.45).

Exercise 3.9.13. With the aid of Theorem 3.2.7, obtain the convolution-based wavelets corresponding to the Haar wavelet (3.31) and the following pair of bounded integrable functions:

(i) $\phi(t) = \begin{cases} 1, & 0 \leq t < \dfrac{1}{2} \\ 0, & \text{otherwise,} \end{cases}$

(ii) $\phi(t) = \dfrac{1}{\sqrt{2\pi}}\, e^{-t^2}$.

Exercise 3.9.14. Find the Stockwell transform corresponding to the below given functions, with respect to the Gaussian window $g(t) = e^{-t^2/2}$:

(i) $f(t) = \chi_{[-c,\, c]}(t)$, $c > 0$,

(ii) $f(t) = e^{-\sigma|t|}$, $\sigma > 0$,

(iii) $f(t) = e^{i\omega_0 t}$, $\omega_0 \in \mathbb{R}$,

(iv) $f(t) = (1 - t^2)\, e^{-t^2}$,

(v) $f(t) = \begin{cases} e^{-\lambda t}, & t \geq 0,\ \lambda > 0 \\ 0, & \text{otherwise.} \end{cases}$

Exercise 3.9.15. For $a \in \mathbb{R}^+$, $\mathbf{b} \in \mathbb{R}^2$ and $\theta \in [0, 2\pi)$, show that the elementary operations of rigid translation $\mathcal{T}_{\mathbf{b}}$, planer dilation $\mathcal{D}_a$ and planer rotation R_θ satisfy the commutation rules (3.130).

Exercise 3.9.16. Compute the two-dimensional wavelet transform of the function

$$f(t_1, t_2) = \frac{1}{\pi^{1/4}} \exp\left\{-\frac{(t_1^2 + t_2^2)}{2}\right\}, \tag{3.406}$$

with respect to the truncated Morlet wavelet $\psi_M(t_1, t_2)$ given in (3.177).

Exercise 3.9.17. Compute the two-dimensional Fourier transform of the end-stopped wavelet

$$f(t_1, t_2) = \frac{1}{4} t_1 \exp\left\{-\frac{(t_1^2 + t_2^2) + k_0(k_0 - 2it_2)}{4}\right\}, \tag{3.407}$$

where k_0 is a real parameter chosen in accordance with the wave-vector $\mathbf{w}_0 = (0, k_0)$ in the two-dimensional Morlet wavelet. Also, discuss the spatial and spectral characteristics of the end-stopped wavelet (3.407).

Exercise 3.9.18. Compute the two-dimensional wavelet transform of the below given functions with respect to the end-stopped wavelet (3.407):

(i) $f(t_1, t_2) = \chi_{[-1, 1]}(t_1)\, \chi_{[-1, 1]}(t_2),$

(ii) $f(t_1, t_2) = \exp\left\{-\frac{(t_1^2 + t_2^2)}{2}\right\}.$

Exercise 3.9.19. For the function $f(t_1, t_2) = U(t_1)\, \delta(t_2)$, where

$$U(t_1) = \begin{cases} 1, & t_1 \geq 0 \\ 0, & \text{otherwise}, \end{cases} \tag{3.408}$$

compute the wavelet transform (3.139) with respect to the two-dimensional Morlet wavelet (3.174) with $\mathbf{w}_0 = (0, 1)$ and $\epsilon = 4$. Also, discuss the different partial energy densities of the corresponding two-dimensional wavelet transform.

Exercise 3.9.20. For $\sigma > 0$, compute the Radon transform (3.185) corresponding to the following functions:

(i) $f(x_1, x_2) = e^{-\sigma(x_1^2 + x_2^2)},$

(ii) $f(x_1, x_2) = t_1\, e^{-\sigma(x_1^2 + x_2^2)},$

(iii) $f(x_1, x_2) = t_2\, e^{-\sigma(x_1^2 + x_2^2)},$

(iv) $f(x_1, x_2) = (x_1^2 + x_2^2)\, e^{-\sigma(x_1^2 + x_2^2)}.$

Exercise 3.9.21. If $\nabla = \left(\dfrac{\partial}{\partial x_1}, \dfrac{\partial}{\partial x_2}\right)$ is the gradient operator, then find the Radon transform (3.185) of the bivariate function $\nabla f(x_1, x_2)$.

Exercise 3.9.22. Obtain the Radon transform (3.185) of the function

$$f(x_1, x_2) = H_{k_1}(x_1)\, H_{k_2}(x_2)\, e^{-(x_1^2 + x_2^2)}, \tag{3.409}$$

where $H_n(x)$ is the Hermite polynomial of degree n defined via the Rodrigues formula as

$$e^{-t^2} H_n(x) = (-1)^n \left(\frac{d}{dx}\right)^n e^{-x^2}. \tag{3.410}$$

Exercise 3.9.23. For any $f, g \in L^2(\mathbb{R}^2)$, find the Radon transform (3.185) of the convolution function $(f \circledast g)(\mathbf{z})$ defined in (3.226).

Exercise 3.9.24. If A is a 2×2, non-singular matrix, then compute the Radon transform (3.185) for the function $f(A\mathbf{x})$, where $f(\mathbf{x})$ is a given bivariate function.

Exercise 3.9.25. Comment upon the main differences between the ridgelet transform defined in (3.194) and the two-dimensional wavelet transform (3.139) with particular emphasis on the fact that for linear singularities, once the orientation of the ridgelet has been tuned, one can arbitrarily refine the orthogonal (singular) direction, leading to more efficient representations of two-dimensional signals.

Exercise 3.9.26. Given the function

$$W(r) = \begin{cases} 1, & |r| \leq \dfrac{1}{2} \\ \cos\left(\dfrac{(2|r| - 1)\pi}{2}\right), & \dfrac{1}{2} \leq |r| \leq 1 \\ 0, & \text{otherwise,} \end{cases} \tag{3.411}$$

supported on the interval $[-1, 1]$. Discuss the characteristics of the function

$$W'(r) = \chi_{[0,\,\infty]}(r) \sqrt{\left(W(r/2)\right)^2 - \left(W(r)\right)^2} \tag{3.412}$$

for acting as a radial window for the curvelet transform.

4

The Intertwining of Wavelet Transforms

"Wavelet theory is the result of a multidisciplinary effort that brought together mathematicians, physicists, and engineers.... This connection has created of flow of ideas that goes well beyond the construction of new bases or transforms."

-Stèphane G. Mallat

4.1 Introduction

The classical wavelet transform and its ramifications have proved to be of utmost significance in dealing with diverse problems in science and engineering, including some basic problems in digital signal processing, speech and image processing, pattern recognition, biomedical engineering, turbulence, geophysics, quantum physics, astronomy and so on. The mathematical underpinnings of the wavelet theory depend not only on the classical Fourier analysis, but also on the ideas from abstract harmonic analysis, including Von-Neumann algebras and some prominent group structures, such as the affine and shearlet groups. This unifying influence of the wavelet theory resulted in its recognition as an independent and board field of research in harmonic analysis.

On the other hand, the profound developments in the theory of Fourier transforms resulted in the advent of fractional Fourier transform, linear canonical transform and special affine Fourier transform. The prime advantage of these generalized Fourier transforms lies in the fact they are endowed with extra degrees of freedom which in turn enhances the analyzing capabilities. For instance, the fractional Fourier transform implements a sense of rotation in the time-frequency plane; the linear canonical transform performs twisting operations by independently rotating the time and frequency axes and the special affine Fourier transform shifts the twisted time-frequency objects by an offset vector. The visual description of the aforementioned facts is given in Figure 4.1. Owing to the flexibility inculcated into the time-frequency plane by the generalized Fourier transforms, they are best matched for the class of signals whose energy is not well concentrated in the usual Fourier domain. A typical class of such signals are the chirp-like signals, which are ubiquitous in nature as well as in man-made systems.

Based on the notion of generalized Fourier transforms, the theory of wavelets has also witnessed many significant lucubrations resulting in the birth of fractional wavelets, linear canonical wavelets and their modifications. Nevertheless, many other directional representation systems, such as, ridgelets, curvelets, ripplets and shearlets, have also been extended to the generalized Fourier domains. A detailed perspective on some of the aforementioned developments can be found in [133, 134, 192, 208, 255, 267, 271, 274, 286, 296, 306, 310]. In this chapter, we shall extensively study the theory of generalized wavelet transforms,

DOI: 10.1201/9781003175766-4

including the fractional wavelet transform, fractional Stockwell transform, linear canonical wavelet transform, linear canonical Stockwell transform, linear canonical ridgelet transform, linear canonical curvelet and ripplet transforms and the linear canonical shearlet transform.

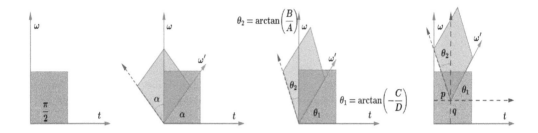

FIGURE 4.1: Visual description of the effect of generalized Fourier transforms on time-frequency objects.

4.2   The Fractional Wavelet Transform

"The more deeply we probe the fundamentals of physical behaviour, the more that it is very precisely controlled by mathematics."

-Roger Penrose

Although the wavelet transform performs exceptionally well while analyzing non-transient signals, however, since each wavelet component is actually a differently scaled bandpass filter in the Fourier domain, the efficiency of the wavelet transform is abysmally reduced while processing the signals whose energy is not well-concentrated in Fourier domain. For instance, many chirp-like signals arising in radar and other communication systems are not band-limited in usual Fourier domain but are band-limited either in the fractional Fourier domain or in other generalized Fourier domains. Therefore, it is quite appealing to carry the study of wavelet transforms from the classical Fourier domain to the much privileged fractional Fourier domain.

In this section, we shall adjunct the fractional Fourier transform to the well-known wavelet transform and study the fundamental properties of the novel integral transform coined as the "fractional wavelet transform." The idea behind this transform is to revamp the classical family of wavelets to fractional wavelets in the sense that each fractional wavelet component turns to be a differently scaled bandpass filter in the fractional Fourier domain. This can be achieved via several approaches; however, we shall focus on the one relying upon the convolution operations. The main reason for employing the convolution operation is to minimize the computational cost while accomplishing the aforementioned objective. This new approach generalizes the wavelet transform, with the generalization induced by the fractional Fourier transform. Besides the formulation of orthogonality relation, inversion formula and range characterization, our goal is to study the net time-fractional-frequency resolution and demonstrate the constant Q-property of the fractional wavelet transform.

4.2.1 Definition and basic properties

To begin with, it is imperative to mention that the classical wavelet transform (3.47) corresponding to any $f \in L^2(\mathbb{R})$ with respect to the mother wavelet $\psi \in L^2(\mathbb{R})$ can be expressed via the convolution operation $*$ defined in (1.75) as

$$
\begin{aligned}
\mathscr{W}_\psi\big[f\big](a,b) &= \frac{1}{\sqrt{a}} \int_{\mathbb{R}} f(t)\, \overline{\psi\left(\frac{t-b}{a}\right)}\, dt \\
&= \frac{1}{\sqrt{a}} \int_{\mathbb{R}} f(t)\, \overline{\breve{\psi}\left(\frac{b-t}{a}\right)}\, dt \\
&= \big(f * \mathcal{D}_a \breve{\psi}\big)(b), \quad a \in \mathbb{R}^+,\ b \in \mathbb{R},
\end{aligned}
\tag{4.1}
$$

where $\breve{\psi}(t) = \psi(-t)$ and $\mathcal{D}_a$ denotes the usual dilation operator defined as $\mathcal{D}_a \psi(t) = a^{-1/2} \psi(t/a)$. From expression (4.1), it is quite evident that the convolution operation $*$ can serve as an underpinning of the classical wavelet transform. Therefore, in order to extend the notion of the wavelet transform to the fractional Fourier domain, it suffices to replace the classical convolution $*$ appearing in (4.1) with the fractional convolution $\circledast_\alpha$ given by (1.141). Consequently, for $\alpha \neq n\pi$ with $n \in \mathbb{Z}$, we can construct a new family of daughter wavelets as under:

$$
\begin{aligned}
&\mathcal{F}_\psi^\alpha(a,b) \\
&:= \left\{ \psi_{a,b}^\alpha(t) = \frac{1}{\sqrt{a}}\, \psi\left(\frac{t-b}{a}\right) \exp\left\{ \frac{i(b^2 - t^2)\cot\alpha}{2} \right\};\ a \in \mathbb{R}^+, b \in \mathbb{R} \text{ and } \alpha \neq n\pi, n \in \mathbb{Z} \right\}.
\end{aligned}
\tag{4.2}
$$

The daughter wavelets $\psi_{a,b}^\alpha(t)$ appearing in (4.2) are governed by the fractional parameter α and are formally called as the "fractional wavelets." Moreover, it is quite interesting to note that for $\alpha = \pi/2$, the fractional wavelets boil down to the classical wavelets, which is completely in analogy to the fact that for $\alpha = \pi/2$, the fractional Fourier transform reduces to the classical Fourier transform.

Definition 4.2.1. For any $f \in L^2(\mathbb{R})$, the α-angle fractional wavelet transform with respect to the wavelet $\psi \in L^2(\mathbb{R})$ is denoted by $\mathscr{W}_\psi^\alpha[f]$ and is defined as

$$
\mathscr{W}_\psi^\alpha\big[f\big](a,b) = \big(f \circledast_\alpha \mathcal{D}_a \breve{\psi}\big)(b) = \int_{\mathbb{R}} f(t)\, \overline{\psi_{a,b}^\alpha(t)}\, dt,
\tag{4.3}
$$

where $\psi_{a,b}^\alpha(t)$ is given by (4.2).

The above definition allows us to make the following points regarding the fractional wavelet transform (4.3):

(i) For $\alpha = \pi/2$, Definition 4.2.1 reduces to the definition of the classical wavelet transform (3.47). This shows that the fractional wavelet transform (4.3) is indeed a generalization of the classical wavelet transform.

(ii) The fractional wavelet transform (4.3) can also be expressed via the classical wavelet transform in the following fashion:

$$
\begin{aligned}
\mathscr{W}_\psi^\alpha\big[f\big](a,b) &= \exp\left\{ -\frac{ib^2 \cot\alpha}{2} \right\} \frac{1}{\sqrt{a}} \int_{\mathbb{R}} \exp\left\{ \frac{it^2 \cot\alpha}{2} \right\} f(t)\, \overline{\psi\left(\frac{t-b}{a}\right)}\, dt \\
&= \exp\left\{ -\frac{ib^2 \cot\alpha}{2} \right\} \mathscr{W}_\psi\big[F_\alpha\big](a,b),
\end{aligned}
\tag{4.4}
$$

where

$$F_\alpha(t) = \exp\left\{\frac{it^2 \cot\alpha}{2}\right\} f(t). \tag{4.5}$$

From expression (4.4), we infer that the fractional wavelet transform corresponds to three fundamental actions; viz, multiplying the input signal $f(t)$ with the chirp $\exp\left\{i(t^2 \cot\alpha)/2\right\}$ to obtain the new signal $F_\alpha(t)$, computing the classical wavelet transform of the signal $F_\alpha(t)$ and finally, multiplying the output with another chirp $\exp\left\{-i(b^2 \cot\alpha)/2\right\}$.

(iii) From expression (4.4), it follows that the fractional wavelet transform (4.3) is practically reliable in the sense that the computational complexity of the fractional wavelet transform is completely determined by the computational complexity of the classical wavelet transform.

(iv) The function $\psi \in L^2(\mathbb{R})$ appearing in Definition 4.2.1 is to be considered as a wavelet in the sense that it satisfies the classical admissibility condition (3.27). That is, the fractional wavelet transform defined in (4.3) does not require any extra conditions to be imposed on the mother wavelet $\psi(t)$.

Next, we present an important lemma which allows us to study the spectral characteristics of the fractional wavelet transform (4.3).

Lemma 4.2.1. *For any $f \in L^2(\mathbb{R})$, the fractional wavelet transform $\mathscr{W}_\psi^\alpha[f](a, b)$ is expressible as follows:*

$$\mathscr{W}_\psi^\alpha[f](a, b) = \sqrt{2\pi a} \int_\mathbb{R} \mathscr{F}_\alpha[f](\omega) \, \overline{\mathscr{F}[\psi](a\omega \csc\alpha)} \, \overline{\mathcal{K}_\alpha(b, \omega)} \, d\omega, \tag{4.6}$$

where $\mathscr{F}_\alpha$ denotes the fractional Fourier transform, whereas $\mathscr{F}$ denotes the usual Fourier transform.

Proof. Using Theorem 1.3.6 associated with the fractional convolution $\circledast_\alpha$, we can compute the fractional Fourier transform of (4.3) as

$$\mathscr{F}_\alpha\left(\mathscr{W}_\psi^\alpha[f](a, b)\right)(\omega) = \sqrt{2\pi} \, \mathscr{F}_\alpha[f](\omega) \, \mathscr{F}\left[D_a \overline{\breve{\psi}}\right](\omega \csc\alpha), \tag{4.7}$$

where the fractional Fourier transform on L.H.S of (4.7) is computed with respect to the translation variable b. Observe that,

$$\begin{aligned}
\mathscr{F}\left[D_a \overline{\breve{\psi}}\right](\omega \csc\alpha) &= \frac{1}{\sqrt{2\pi}} \int_\mathbb{R} D_a \overline{\breve{\psi}}(t) \, e^{-i(\omega \csc\alpha)t} \, dt \\
&= \frac{1}{\sqrt{2\pi}} \int_\mathbb{R} \frac{1}{\sqrt{a}} \, \overline{\psi\left(-\frac{t}{a}\right)} \, e^{-i(\omega \csc\alpha)t} \, dt \\
&= \frac{\sqrt{a}}{\sqrt{2\pi}} \int_\mathbb{R} \overline{\psi(z)} \, e^{-i(a\omega \csc\alpha)z} \, dz \\
&= \sqrt{a} \, \overline{\mathscr{F}[\psi]}(a\omega \csc\alpha). \tag{4.8}
\end{aligned}$$

Plugging the relation (4.8) in (4.7), we obtain

$$\mathscr{F}_\alpha\left(\mathscr{W}_\psi^\alpha[f](a, b)\right)(\omega) = \sqrt{2\pi a} \, \mathscr{F}_\alpha[f](\omega) \, \overline{\mathscr{F}[\psi]}(a\omega \csc\alpha). \tag{4.9}$$

Finally, applying the inverse fractional Fourier transform on both sides of (4.9), we obtain the desired expression as

$$\mathscr{W}_\psi^\alpha \big[f\big](a,b) = \sqrt{2\pi a} \int_\mathbb{R} \mathscr{F}_\alpha\big[f\big](\omega)\, \overline{\mathscr{F}\big[\psi\big](a\omega \csc \alpha)}\, \overline{\mathcal{K}_\alpha(b,\omega)}\, d\omega.$$

This completes the proof of Lemma 4.2.1. □

From expression (4.6), it is quite evident that the fractional wavelet transform is equivalent to multiplying the fractional spectrum of the given signal by a fractionally scaled window and then applying the inverse fractional Fourier transform. Moreover, we infer that each fractional wavelet component is actually a differently scaled band-pass filter in the fractional domain. This distinguishing feature of the fractional wavelet transform can overcome the weakness of the traditional wavelet transform by enabling the processing of non-transient signals along non-orthogonal directions in the time-frequency plane.

In the following theorem, we study some elementary properties of the fractional wavelet transform (4.3) under certain alterations in the input signal.

Theorem 4.2.2. *Let $c_1, c_2 \in \mathbb{C}$, $k \in \mathbb{R}$, $\lambda > 0$ are scalars and ψ, ϕ be a given pair of wavelets. Then, for $f, g \in L^2(\mathbb{R})$, the fractional wavelet transform defined in (4.3) satisfies the following properties:*

(i) *Linearity:* $\mathscr{W}_\psi^\alpha \big[c_1 f + c_2 g\big](a,b) = c_1\, \mathscr{W}_\psi^\alpha \big[f\big](a,b) + c_2\, \mathscr{W}_\psi^\alpha \big[g\big](a,b),$

(ii) *Anti-linearity:* $\mathscr{W}_{c_1\psi+c_2\phi}^\alpha \big[f\big](a,b) = \overline{c_1}\, \mathscr{W}_\psi^\alpha \big[f\big](a,b) + \overline{c_2}\, \mathscr{W}_\phi^\alpha \big[f\big](a,b),$

(iii) *Translation:* $\mathscr{W}_\psi^\alpha \big[f(t-k)\big](a,b) = e^{i(k^2 - bk)\cot\alpha}\, \mathscr{W}_\psi^\alpha \big[e^{ikt\cot\alpha} f(t)\big](a, b-k),$

(iv) *Scaling:* $\mathscr{W}_\psi^\alpha \big[f(\lambda t)\big](a,b) = \dfrac{1}{\sqrt{\lambda}}\, \mathscr{W}_\psi^\beta \big[f\big](\lambda a, \lambda b),$ *where* $\beta = \operatorname{arccot}\left(\dfrac{\cot \alpha}{\lambda^2}\right),$

(v) *Reflection:* $\mathscr{W}_\psi^\alpha \big[f(-t)\big](a,b) = \mathscr{W}_{\check\psi}^\alpha \big[f\big](a,-b),$ *where* $\check\psi(t) = \psi(-t),$

(vi) *Symmetry:* $\mathscr{W}_\psi^\alpha \big[f\big](a,b) = e^{i(b^2 \cot\alpha)/2}\, \mathscr{W}_{\overline{f}}^\gamma \big[\Psi\big]\left(\dfrac{1}{a}, -\dfrac{b}{a}\right),$ $\Psi(t) = e^{iabt\cot\alpha}\, \overline{\psi(t)}$ *and* $\gamma = \operatorname{arccot}(a^2 \cot\alpha).$

Proof. For the reason of space, we completely omit the proof. □

Next, our motive is study the orthogonality and invertibility of the fractional wavelet transform defined in (4.3). Here, it is imperative to mention that the orthogonality relation and inversion formula for the fractional wavelet transform hold good without imposing any extra conditions on the mother wavelet $\psi(t)$.

Theorem 4.2.3. *If $\mathscr{W}_\psi^\alpha \big[f\big](a,b)$ and $\mathscr{W}_\psi^\alpha \big[g\big](a,b)$ are the fractional wavelet transforms of any pair of functions $f, g \in L^2(\mathbb{R})$ with respect to the wavelet ψ, then*

$$\int_\mathbb{R}\int_{\mathbb{R}^+} \mathscr{W}_\psi^\alpha \big[f\big](a,b)\, \overline{\mathscr{W}_\psi^\alpha \big[g\big](a,b)}\, \frac{da\,db}{a^2} = C_\psi \big\langle f, g \big\rangle_2, \tag{4.10}$$

where C_ψ is given by the usual admissibility condition (3.27).

Proof. Applying Lemma 4.2.1, we can express the fractional wavelet transforms of the given pair of square integrable functions f and g as follows:

$$\mathscr{W}_\psi^\alpha \left[f \right] (a,b) = \sqrt{2\pi a} \int_{\mathbb{R}} \mathscr{F}_\alpha \left[f \right] (\omega) \, \overline{\mathscr{F} \left[\psi \right] (a\omega \csc \alpha)} \, \overline{\mathcal{K}_\alpha (b, \omega)} \, d\omega \tag{4.11}$$

and

$$\mathscr{W}_\psi^\alpha \left[g \right] (a,b) = \sqrt{2\pi a} \int_{\mathbb{R}} \mathscr{F}_\alpha \left[g \right] (\eta) \, \overline{\mathscr{F} \left[\psi \right] (a\eta \csc \alpha)} \, \overline{\mathcal{K}_\alpha (b, \eta)} \, d\eta. \tag{4.12}$$

Consequently, we have

$$\int_{\mathbb{R}} \int_{\mathbb{R}^+} \mathscr{W}_\psi^\alpha \left[f \right] (a,b) \, \overline{\mathscr{W}_\psi^\alpha \left[g \right] (a,b)} \, \frac{da \, db}{a^2}$$

$$= \int_{\mathbb{R}} \int_{\mathbb{R}^+} \left\{ \sqrt{2\pi a} \int_{\mathbb{R}} \mathscr{F}_\alpha \left[f \right] (\omega) \, \overline{\mathscr{F} \left[\psi \right] (a\omega \csc \alpha)} \, \overline{\mathcal{K}_\alpha (b, \omega)} \, d\omega \right\}$$

$$\times \left\{ \sqrt{2\pi a} \int_{\mathbb{R}} \overline{\mathscr{F}_\alpha \left[g \right] (\eta)} \, \mathscr{F} \left[\psi \right] (a\eta \csc \alpha) \, \mathcal{K}_\alpha (b, \eta) \, d\eta \right\} \frac{da \, db}{a^2}$$

$$= (2\pi) \int_{\mathbb{R}} \int_{\mathbb{R}} \int_{\mathbb{R}^+} \mathscr{F}_\alpha \left[f \right] (\omega) \, \overline{\mathscr{F}_\alpha \left[g \right] (\eta)} \, \overline{\mathscr{F} \left[\psi \right] (a\omega \csc \alpha)} \, \mathscr{F} \left[\psi \right] (a\eta \csc \alpha)$$

$$\times \left\{ \int_{\mathbb{R}} \mathcal{K}_\alpha (b, \eta) \, \overline{\mathcal{K}_\alpha (b, \omega)} \, db \right\} \frac{da \, d\omega \, d\eta}{a}$$

$$= (2\pi) \int_{\mathbb{R}} \int_{\mathbb{R}} \int_{\mathbb{R}^+} \mathscr{F}_\alpha \left[f \right] (\omega) \, \overline{\mathscr{F}_\alpha \left[g \right] (\eta)} \, \overline{\mathscr{F} \left[\psi \right] (a\omega \csc \alpha)} \, \mathscr{F} \left[\psi \right] (a\eta \csc \alpha)$$

$$\times \, \delta(\omega - \eta) \frac{da \, d\omega \, d\eta}{a}$$

$$= (2\pi) \int_{\mathbb{R}} \int_{\mathbb{R}^+} \mathscr{F}_\alpha \left[f \right] (\omega) \, \overline{\mathscr{F}_\alpha \left[g \right] (\omega)} \, \left| \mathscr{F} \left[\psi \right] (a\omega \csc \alpha) \right|^2 \frac{da \, d\omega}{a}$$

$$= \int_{\mathbb{R}} \mathscr{F}_\alpha \left[f \right] (\omega) \, \overline{\mathscr{F}_\alpha \left[g \right] (\omega)} \left\{ (2\pi) \int_{\mathbb{R}^+} \left| \mathscr{F} \left[\psi \right] (a\omega \csc \alpha) \right|^2 \frac{da}{a} \right\} d\omega. \tag{4.13}$$

In order to simplify the inner integral on the R.H.S of (4.13), we substitute $a\omega \csc \alpha = \omega'$, so that

$$2\pi \int_{\mathbb{R}^+} \left| \mathscr{F} \left[\psi \right] (a\omega \csc \alpha) \right|^2 \frac{da}{a} = 2\pi \int_{\mathbb{R}} \frac{\left| \mathscr{F} \left[\psi \right] (\omega') \right|^2}{|\omega'|} \, d\omega' = C_\psi. \tag{4.14}$$

Using (4.14) in (4.13), we get

$$\int_{\mathbb{R}} \int_{\mathbb{R}^+} \mathscr{W}_\psi^\alpha \left[f \right] (a,b) \, \overline{\mathscr{W}_\psi^\alpha \left[g \right] (a,b)} \, \frac{da \, db}{a^2} = C_\psi \int_{\mathbb{R}} \mathscr{F}_\alpha \left[f \right] (\omega) \, \overline{\mathscr{F}_\alpha \left[g \right] (\omega)} \, d\omega$$

$$= C_\psi \left\langle \mathscr{F}_\alpha \left[f \right], \mathscr{F}_\alpha \left[g \right] \right\rangle_2$$

$$= C_\psi \left\langle f, g \right\rangle_2.$$

This completes the proof of Theorem 4.2.3 $\square$

Remark 4.2.1. For $f = g$, the orthogonality relation (4.10) boils down to the following compact form:

$$\int_{\mathbb{R}} \int_{\mathbb{R}^+} \left| \mathscr{W}_\psi^\alpha \left[f \right] (a,b) \right|^2 \frac{da \, db}{a^2} = C_\psi \left\| f \right\|_2^2. \tag{4.15}$$

The quantity $\left| \mathscr{W}_{\psi}^{\alpha}[f](a,b) \right|^2$ represents the energy density of the signal in the time-fractional-frequency plane and is formally called as the "fractional scalogram." From (4.15) it is evident that, except for the factor C_{ψ}, the net concentration of the fractional wavelet transform in the time-fractional-frequency plane is equal to the net concentration of the input signal in its natural domain. Therefore, choosing the wavelet ψ in such a way that $C_{\psi} = 1$, we infer that the total energy of the signal is preserved between the natural and transformed domains.

In the next theorem, we demonstrate that the fractional wavelet transform defined in (4.3) is invertible in the sense that any square integrable function can be reconstructed from the corresponding fractional wavelet transform.

Theorem 4.2.4. *If* $\mathscr{W}_{\psi}^{\alpha}[f](a,b)$ *is the fractional wavelet transform of an arbitrary function* $f \in L^2(\mathbb{R})$ *with respect to the wavelet* $\psi \in L^2(\mathbb{R})$, *then* f *can be reconstructed via the following formula:*

$$f(t) = \frac{1}{C_{\psi}} \int_{\mathbb{R}} \int_{\mathbb{R}^+} \mathscr{W}_{\psi}^{\alpha}[f](a,b)\, \psi_{a,b}^{\alpha}(t)\, \frac{da\,db}{a^2}, \quad a.e., \tag{4.16}$$

where $C_{\psi} \neq 0$ *is given by the usual admissibility condition* (3.27).

Proof. The proof is left as an exercise to the reader. □

In the sequel, we present a characterization of range of the fractional wavelet transform defined in (4.3). Moreover, we shall also show that the range constitutes a reproducing kernel Hilbert space whose kernel is the autocorrelation function of the given wavelet.

Theorem 4.2.5. *A function* $f \in L^2(\mathbb{R}^+ \times \mathbb{R})$ *is the fractional wavelet transform of a certain square integrable function on* $\mathbb{R}$ *with respect to the wavelet* ψ *if and only if it satisfies the following reproduction formula:*

$$f(a',b') = \frac{1}{C_{\psi}} \int_{\mathbb{R}} \int_{\mathbb{R}^+} f(a,b) \left\langle \psi_{a,b}^{\alpha}, \psi_{a',b'}^{\alpha} \right\rangle_2 \frac{da\,db}{a^2}, \tag{4.17}$$

where $C_{\psi} \neq 0$ *is given by the usual admissibility condition* (3.27).

Proof. Consider a function $f \in L^2(\mathbb{R}^+ \times \mathbb{R})$ belonging to the range of the fractional wavelet transform; that is, $f \in \mathscr{W}_{\psi}^{\alpha}(L^2(\mathbb{R}))$. Thus, we can find a function $h \in L^2(\mathbb{R})$, such that $\mathscr{W}_{\psi}^{\alpha}[h] = f$. Therefore, we have

$$f(a',b') = \mathscr{W}_{\psi}^{\alpha}[h](a',b')$$

$$= \int_{\mathbb{R}} h(t)\, \overline{\psi_{a',b'}^{\alpha}(t)}\, dt$$

$$= \int_{\mathbb{R}} \left\{ \frac{1}{C_{\psi}} \int_{\mathbb{R}} \int_{\mathbb{R}^+} \mathscr{W}_{\psi}^{\alpha}[h](a,b)\, \psi_{a,b}^{\alpha}(t)\, \frac{da\,db}{a^2} \right\} \overline{\psi_{a',b'}^{\alpha}(t)}\, dt$$

$$= \frac{1}{C_{\psi}} \int_{\mathbb{R}} \int_{\mathbb{R}^+} \mathscr{W}_{\psi}^{\alpha}[h](a,b) \left\{ \int_{\mathbb{R}} \psi_{a,b}^{\alpha}(t)\, \overline{\psi_{a',b'}^{\alpha}(t)}\, dt \right\} \frac{da\,db}{a^2}$$

$$= \frac{1}{C_{\psi}} \int_{\mathbb{R}} \int_{\mathbb{R}^+} \mathscr{W}_{\psi}^{\alpha}[h](a,b) \left\langle \psi_{a,b}^{\alpha}, \psi_{a',b'}^{\alpha} \right\rangle_2 \frac{da\,db}{a^2}$$

$$= \frac{1}{C_{\psi}} \int_{\mathbb{R}} \int_{\mathbb{R}^+} f(a,b) \left\langle \psi_{a,b}^{\alpha}, \psi_{a',b'}^{\alpha} \right\rangle_2 \frac{da\,db}{a^2}. \tag{4.18}$$

In order to show the converse implication, we assume that a function $f \in L^2(\mathbb{R}^+ \times \mathbb{R})$ satisfies (4.17). Then, it remains to show that $f \in \mathscr{W}_\psi^\alpha(L^2(\mathbb{R}))$; that is, we must fetch out a function $g \in L^2(\mathbb{R})$, such that $\mathscr{W}_\psi^\alpha[g] = f$. The obvious candidate for such a function is

$$g(t) = \frac{1}{C_\psi} \int_\mathbb{R} \int_{\mathbb{R}^+} f(a,b)\, \psi_{a,b}^\alpha(t)\, \frac{da\, db}{a^2}. \tag{4.19}$$

The function defined in (4.19) is indeed square integrable, because

$$
\begin{aligned}
\left\| g \right\|_2^2 &= \int_\mathbb{R} g(t)\overline{g(t)}\, dt \\
&= \int_\mathbb{R} \left\{ \frac{1}{C_\psi} \int_\mathbb{R} \int_{\mathbb{R}^+} f(a,b)\, \psi_{a,b}^\alpha(t)\, \frac{da\, db}{a^2} \right\} \left\{ \frac{1}{C_\psi} \int_\mathbb{R} \int_{\mathbb{R}^+} \overline{f(a',b')}\, \overline{\psi_{a',b'}^\alpha(t)}\, \frac{da'\, db'}{a'^2} \right\} dt \\
&= \frac{1}{C_\psi} \int_\mathbb{R} \int_{\mathbb{R}^+} \left\{ \frac{1}{C_\psi} \int_\mathbb{R} \int_{\mathbb{R}^+} f(a,b) \left\langle \psi_{a,b}^\alpha,\, \psi_{a',b'}^\alpha \right\rangle_2 \frac{da\, db}{a^2} \right\} \overline{f(a',b')}\, \frac{da'\, db'}{a'^2} \\
&= \frac{1}{C_\psi} \int_\mathbb{R} \int_{\mathbb{R}^+} f(a',b')\, \overline{f(a',b')}\, \frac{da'\, db'}{a'^2} \\
&= \frac{1}{C_\psi} \left\| f \right\|_2^2 < \infty. \tag{4.20}
\end{aligned}
$$

Finally, by virtue of the Fubini's theorem, we can again alter the order of integration to obtain the desired result as follows:

$$
\begin{aligned}
\mathscr{W}_\psi^\alpha\big[g\big](a',b') &= \int_\mathbb{R} g(t)\, \overline{\psi_{a',b'}^\alpha(t)}\, dt \\
&= \int_\mathbb{R} \left\{ \frac{1}{C_\psi} \int_\mathbb{R} \int_{\mathbb{R}^+} f(a,b)\, \psi_{a,b}^\alpha(t)\, \frac{da\, db}{a^2} \right\} \overline{\psi_{a',b'}^\alpha(t)}\, dt \\
&= \frac{1}{C_\psi} \int_\mathbb{R} \int_{\mathbb{R}^+} f(a,b) \left\langle \psi_{a,b}^\alpha,\, \psi_{a',b'}^\alpha \right\rangle_2 \frac{da\, db}{a^2} \\
&= f(a',b').
\end{aligned}
$$

This completes the proof of Theorem 4.2.5. $\qquad\qquad\qquad\qquad\qquad\qquad\qquad \Box$

As a consequence of Theorem 4.2.5, we could easily infer that the range of fractional wavelet transform is a reproducing kernel Hilbert space. Formally, we have the following corollary:

Corollary 4.2.6. *The fractional wavelet transform defined in (4.3) is a reproducing kernel Hilbert space embedded as a subspace in $L^2(\mathbb{R}^2)$ with kernel given by*

$$K_\psi^\alpha(a,b; a',b') = C_\psi^{-1} \left\langle \psi_{a,b}^\alpha,\, \psi_{a',b'}^\alpha \right\rangle_2. \tag{4.21}$$

4.2.2   Constant Q-property and time-fractional-frequency resolution

"Information is the resolution of uncertainty. As long as something can be relayed that resolves uncertainty, that is the fundamental nature of information."

-Claude E. Shannon

In this subsection, our aim is to demonstrate that the fractional wavelet transform defined in (4.3) is capable of providing a joint time-fractional-frequency resolution of finite energy signals. Indeed, we can observe from the Definition 4.2.1 and Lemma 4.2.1 that the localizing behaviour of the fractional wavelet transform is determined by the analyzing functions $\psi_{a,b}^{\alpha}(t)$; that is, if the analyzing functions are supported in the time domain or fractional-frequency domain, then the fractional wavelet transform (4.3) is accordingly supported in the respective domains. For the mathematical description of the preceding arguments, consider a wavelet $\psi(t)$ centred about the point E_{ψ} and having a dispersion of Δ_{ψ} in the time domain. Consequently, the centre (mean) and radius (dispersion about mean) of each of the daughter wavelets $\psi_{a,b}^{\alpha}(t)$ are given by

$$
E\left[\psi_{a,b}^{\alpha}(t)\right] = \frac{\displaystyle\int_{\mathbb{R}} t \left|\psi_{a,b}^{\alpha}(t)\right|^2 dt}{\displaystyle\int_{\mathbb{R}} \left|\psi_{a,b}^{\alpha}(t)\right|^2 dt} = \frac{\displaystyle\int_{\mathbb{R}} t \left|\psi_{a,b}(t)\right|^2 dt}{\displaystyle\int_{\mathbb{R}} \left|\psi_{a,b}(t)\right|^2 dt} = E\left[\psi_{a,b}(t)\right] = aE_{\psi} + b \tag{4.22}
$$

and

$$
\Delta\left[\psi_{a,b}^{\alpha}(t)\right] = \left(\frac{\displaystyle\int_{\mathbb{R}} (t - b - aE_{\psi}) \left|\psi_{a,b}^{\alpha}(t)\right|^2 dt}{\displaystyle\int_{\mathbb{R}} \left|\psi_{a,b}^{\alpha}(t)\right|^2 dt}\right)^{1/2} = \left(\frac{\displaystyle\int_{\mathbb{R}} (t - b - aE_{\psi}) \left|\psi_{a,b}(t)\right|^2 dt}{\displaystyle\int_{\mathbb{R}} \left|\psi_{a,b}(t)\right|^2 dt}\right)^{1/2}
$$

$$
= \Delta\left[\psi_{a,b}(t)\right] = a\Delta_{\psi}. \tag{4.23}
$$

From equations (4.22) and (4.23), we infer that the fractional wavelet transform of any function $f \in L^2(\mathbb{R})$ with respect to the analyzing wavelet $\psi \in L^2(\mathbb{R})$ provides the localized information in the time window

$$
\left[aE_{\psi} + b - a\Delta_{\psi}, \; aE_{\psi} + b + a\Delta_{\psi}\right] \tag{4.24}
$$

having a net width of $2a\Delta_{\psi}$, which is directly proportional to the scale a. Furthermore, from Lemma 4.2.1, it is quite interesting to note that unlike the classical wavelet transform, the spectral characteristics exhibited by the fractional wavelet transform (4.3) are localized by the fractional Fourier domain window $\mathscr{F}[\psi](a\omega \csc \alpha)$. Thus, if $E_{\mathscr{F}[\psi](a\omega \csc \alpha)}$ and $\Delta_{\mathscr{F}[\psi](a\omega \csc \alpha)}$ denote the centre and radius of the fractional domain window $\mathscr{F}[\psi](a\omega \csc \alpha)$, then we can easily deduce that

$$
E_{\mathscr{F}[\psi](a\omega \csc \alpha)} = \frac{E_{\mathscr{F}[\psi](\omega)}}{a \csc \alpha} \quad \text{and} \quad \Delta_{\mathscr{F}[\psi](a\omega \csc \alpha)} = \frac{\Delta_{\mathscr{F}[\psi](\omega)}}{a|\csc \alpha|}. \tag{4.25}
$$

Therefore, the Q-factor (Quality-factor) pertaining to the fractional wavelet transform defined in (4.3) is given by

$$
Q = \frac{\text{Radius of fractional-frequency window}}{\text{Centre of fractional-frequency window}} = \frac{\Delta_{\mathscr{F}[\psi](a\omega \csc \alpha)}}{E_{\mathscr{F}[\psi](a\omega \csc \alpha)}} = \pm \frac{\Delta_{\mathscr{F}[\psi](\omega)}}{E_{\mathscr{F}[\psi](\omega)}}, \tag{4.26}
$$

which is clearly independent of both the scaling parameter a as well as the fractional parameter α and is thus constant. Therefore, from (4.26), we conclude that the fractional wavelet transform (4.3) enjoys the constant Q-property. Moreover, from (4.25), it follows that the fractional wavelet transform $\mathscr{W}_{\psi}^{\alpha}[f](a,b)$ provides the spectral information of the function $f \in L^2(\mathbb{R})$ within the fractional-frequency window

$$
\left[\frac{E_{\mathscr{F}[\psi](\omega)}}{a \csc \alpha} - \frac{\Delta_{\mathscr{F}[\psi](\omega)}}{a|\csc \alpha|}, \; \frac{E_{\mathscr{F}[\psi](\omega)}}{a \csc \alpha} + \frac{\Delta_{\mathscr{F}[\psi](\omega)}}{a|\csc \alpha|}\right] \tag{4.27}
$$

having a net width of $2(a|\csc\alpha|)^{-1}\Delta_{\mathscr{F}[\psi](\omega)}$, which is inversely proportional to the fractional scale $(a\csc\alpha)$. Thus, the net time-fractional-frequency localization of the fractional wavelet transform defined in (4.3) is given by the window

$$\left[aE_\psi + b - a\Delta_\psi,\, aE_\psi + b + a\Delta_\psi\right] \times \left[\frac{E_{\mathscr{F}[\psi](\omega)}}{a\csc\alpha} - \frac{\Delta_{\mathscr{F}[\psi](\omega)}}{a|\csc\alpha|},\, \frac{E_{\mathscr{F}[\psi](\omega)}}{a\csc\alpha} + \frac{\Delta_{\mathscr{F}[\psi](\omega)}}{a|\csc\alpha|}\right],$$
$$(4.28)$$

occupying a net area of $4|\sin\alpha|\,\Delta_\psi\Delta_{\mathscr{F}[\psi](\omega)}$ in the time-fractional-frequency plane as shown in Figure 4.2. Besides, it is worth noticing that the area of the window (4.28) depends only upon the mother wavelet $\psi \in L^2(\mathbb{R})$ and the fractional parameter α, whereas it is completely independent of both the scaling parameter a and the translation parameter b. From the above discussion, it follows that the fractional parameter α can be invoked to optimize the concentration of the fractional wavelet transform in the time-fractional-frequency plane. As such, the fractional parameter α plays a key role in the applications of the fractional wavelet transform for the analysis of those signals whose energy is not well-concentrated in the Fourier domain, for instance chirp-like signals which are ubiquitous both in nature as well as in man-made systems.

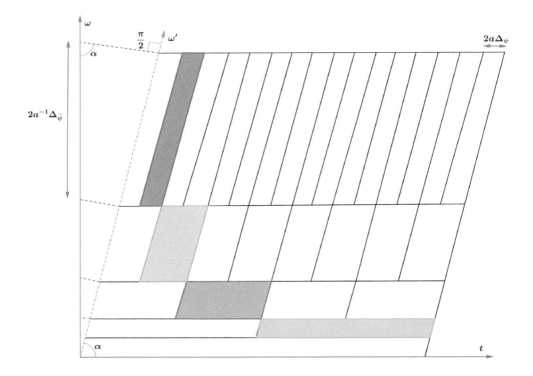

FIGURE 4.2: Tiling on the time-fractional-frequency plane due to fractional wavelet transform.

4.3 The Fractional Stockwell Transform

"Mathematical discoveries, small or great, are never born of spontaneous generation. They always presuppose a soil seeded with preliminary knowledge and well prepared by labour, both conscious and subconscious."

-Jules H. Poincaré

In this section, our goal is to formulate the fractional analogue of the Stockwell transform by appropriately adjoining the instincts of the fractional Fourier transform to the classical Stockwell transform. The prime advantage of the fractional Stockwell transform lies in the fact that it allows the signal analysis along non-orthogonal directions in the time-frequency plane, while providing the absolutely referenced phase information. Thus, the fractional Stockwell transform can provide richer signal representations of chirp-like signals arising in radar, sonar and other communication systems.

The formulation of the fractional Stockwell transform can be carried out via several approaches. For instance, the most trivial approach is to replace the Fourier kernel $(2\pi)^{-1/2}e^{-i\omega t}$ appearing in the definition of the classical Stockwell transform (3.79) with the fractional Fourier kernel $\mathcal{K}_\alpha(t,\omega)$ given by (1.128). Physically, the idea behind such a procedure is to segment a signal via a time-localized window of variable width then perform the fractional Fourier analysis for each segment. However, the major disadvantage of such a procedure is that it is computationally laborious. To circumvent such limitations, it is desirable to take advantage of the convolution structure exhibited by the classical Stockwell transform by replacing it with an appropriate fractional convolution. It is imperative to mention that, as of now, many variants of the fractional Stockwell transform have appeared in the literature, which rely upon different convolution-based approaches in the fractional Fourier domain [273, 274]. In the present section, we shall study the notion of fractional Stockwell transform by invoking the fractional convolution $\circledast_\alpha$ defined in (1.141). It is pertinent to mention that, the convolution operation $\circledast_\alpha$ lucidly inculcates the influence of the fractional parameter α into the novel Stockwell transform, while maintaining the elegance of the underlying window functions.

4.3.1 Definition and basic properties

To begin this subsection, we recall the elementary operators pertaining to the classical Stockwell transform (3.79). For $a \in \mathbb{R} \setminus \{0\}$ and $b \in \mathbb{R}$, the dilation $(\mathcal{D}_a)$, modulation $(\mathcal{M}_a)$ and translation $(\mathcal{T}_b)$ operators acting on $\psi \in L^2(\mathbb{R})$ are given by

$$(\mathcal{D}_a\psi)(t) = |a|\,\psi(at), \quad (\mathcal{M}_a\psi)(t) = e^{iat}\psi(t) \quad \text{and} \quad (\mathcal{T}_b\psi)(t) = \psi(t-b). \tag{4.29}$$

Then, the classical Stockwell transform $\mathcal{S}_\psi\big[f\big](a,b)$ of any $f \in L^2(\mathbb{R})$ with respect to the window function $\psi \in L^2(\mathbb{R})$ can be expressed via the classical convolution operation $*$ defined in (1.75) as

$$\mathcal{S}_\psi\big[f\big](a,b) = \frac{1}{\sqrt{2\pi}}\left((\mathcal{M}_{-a}f) * \left(\mathcal{D}_a\breve{\psi}\right)\right)(b), \qquad \breve{\psi}(t) = \psi(-t). \tag{4.30}$$

Therefore, in order to formulate the notion of fractional Stockwell transform, we ought to replace the convolution operation $*$ with the fractional convolution $\circledast_\alpha$ given by (1.141).

In doing so, we obtain a novel family of analyzing functions, governed by the fractional parameter α, as follows:

$$\mathcal{F}_\psi^\alpha(a,b) := \left\{ \psi_{a,b}^\alpha(t) = |a|\, \psi\big(a(t-b)\big) \exp\left\{ iat - \frac{i(t^2 - b^2)\cot\alpha}{2} \right\}; \right.$$

$$\left. a \in \mathbb{R} \setminus \{0\},\ b \in \mathbb{R}\ \text{ and }\ \alpha \neq n\pi,\ n \in \mathbb{Z} \right\}. \qquad (4.31)$$

Note that for $\alpha = \pi/2$, the fractional Stockwell windows $\psi_{a,b}^\alpha(t)$ reduce to the classical Stockwell windows. With the family of fractional Stockwell windows at hand, we have the following definition of the fractional Stockwell transform:

Definition 4.3.1. The α-angle fractional Stockwell transform of any $f \in L^2(\mathbb{R})$ with respect to the window function $\psi \in L^2(\mathbb{R})$ is denoted by $\mathcal{S}_\psi^\alpha[f]$ and is defined as

$$\mathcal{S}_\psi^\alpha\big[f\big](a,b) = \frac{1}{\sqrt{2\pi}} \left((\mathcal{M}_{-a}f) \circledast_\alpha \left(\mathscr{D}_a \breve{\psi}\right) \right)(b) = \frac{1}{\sqrt{2\pi}} \int_{\mathbb{R}} f(t)\, \overline{\psi_{a,b}^\alpha(t)}\, dt, \qquad (4.32)$$

where $\psi_{a,b}^\alpha(t)$ is given by (4.31).

Definition 4.3.1 allows us to make the following points regarding the fractional Stockwell transform (4.32):

(i) The fractional Stockwell transform defined in (4.32) is in consistency with the classical Stockwell transform in the sense that for $\alpha = \pi/2$, Definition 4.3.1 boils down to the definition of classical Stockwell transform.

(ii) The fractional Stockwell transform (4.32) can also be rewritten in the following fashion:

$$\mathcal{S}_\psi^\alpha\big[f\big](a,b) = \frac{|a|}{\sqrt{2\pi}} \int_{\mathbb{R}} f(t) \left\{ \overline{\psi\big(a(t-b)\big)} \exp\left\{ \frac{i(t^2 - b^2)\cot\alpha}{2} \right\} \right\} e^{-iat}\, dt, \qquad (4.33)$$

which implies that the fractional Stockwell transform (4.32) can be interpreted as an angular Stockwell transform using a chirp-modulated analysis window, with the angle of rotation determined by the fractional parameter α. Therefore, we conclude that the fractional Stockwell transform given by Definition 4.3.1 can be of substantial importance for optimizing the concentration of the Stockwell spectrum and thereby enhancing the efficiency of the Stockwell transform. Some of the areas where the fractional Stockwell transform (4.32) may be handy include, instantaneous frequency estimation, linear time-frequency filtering, adaptive linear time-frequency representations, sparse phase retrieval, source separation and non-linear system identification.

(iii) From expression (4.33), it readily follows that the fractional Stockwell transform (4.32) is easy to implement as the computational complexity is completely determined by the computational complexity of the classical Stockwell transform.

In the following lemma, our aim is to express the fractional Stockwell transform defined in (4.32) via the respective spectra of the input signal and the basic window function. Such an expression plays a key role in understanding the time-fractional-frequency resolution of the fractional Stockwell transform.

Lemma 4.3.1. *The fractional Stockwell transform of any function $f \in L^2(\mathbb{R})$ with respect to the window function $\psi \in L^2(\mathbb{R})$ can be expressed as*

$$\mathcal{S}_\psi^\alpha\big[f\big](a,b) = e^{-iab} \int_{\mathbb{R}} \mathscr{F}_\alpha\big[f\big](\omega)\, \overline{\mathscr{F}\big[e^{it}\,\psi(t)\big]\left(\frac{\omega \csc\alpha}{a}\right)}\, \mathcal{K}_\alpha(b,\omega)\, d\omega, \qquad (4.34)$$

where $\mathscr{F}_\alpha$ denotes the fractional Fourier transform, whereas $\mathscr{F}$ denotes the classical Fourier transform.

Proof. Using the convolution theorem for the fractional convolution (1.141), we can compute the fractional Fourier transform of (4.32) with respect to the translation variable b as follows:

$$\mathscr{F}_\alpha\left(S_\psi^\alpha\left[f\right](a,b)\right)(\omega) = \frac{1}{\sqrt{2\pi}}\mathscr{F}_\alpha\left[\left((M_{-a}f)\otimes_\alpha\left(\mathscr{D}_a\breve{\psi}\right)\right)\right](\omega)$$

$$= \mathscr{F}_\alpha\left[(M_{-a}f)\right](\omega)\,\mathscr{F}\left[\left(\mathscr{D}_a\breve{\psi}\right)\right](\omega\csc\alpha). \qquad (4.35)$$

Applying the definition of fractional Fourier transform (1.127), we can simplify the pre-factor in (4.35) as

$$\mathscr{F}_\alpha\left[(M_{-a}f)\right](\omega) = \sqrt{\frac{1-i\cot\alpha}{2\pi}}\int_{\mathbb{R}} e^{-iat}f(t)\exp\left\{\frac{i(t^2+\omega^2)\cot\alpha}{2}-i\omega t\csc\alpha\right\}dt$$

$$= \sqrt{\frac{1-i\cot\alpha}{2\pi}}\int_{\mathbb{R}} f(t)\exp\left\{\frac{i(t^2+\omega^2)\cot\alpha}{2}-i(\omega+a\sin\alpha)t\csc\alpha\right\}dt$$

$$= \sqrt{\frac{1-i\cot\alpha}{2\pi}}\exp\left\{-\frac{ia^2\sin\alpha\cos\alpha}{2}-ia\omega\cos\alpha\right\}$$

$$\times\int_{\mathbb{R}} f(t)\exp\left\{\frac{i(t^2+(\omega+a\sin\alpha)^2)}{2}\cot\alpha - i(\omega+a\sin\alpha)t\csc\alpha\right\}dt$$

$$= \exp\left\{-\frac{ia^2\sin\alpha\cos\alpha}{2}-ia\omega\cos\alpha\right\}\mathscr{F}_\alpha\left[f\right](\omega+a\sin\alpha). \qquad (4.36)$$

Furthermore, we observe that

$$\mathscr{F}\left[\left(\mathscr{D}_a\breve{\psi}\right)\right](\omega\csc\alpha) = \frac{|a|}{\sqrt{2\pi}}\int_{\mathbb{R}}\overline{\psi(-at)}\exp\left\{-it\,\omega\csc\alpha\right\}dt$$

$$= \frac{1}{\sqrt{2\pi}}\int_{\mathbb{R}}\overline{\psi(z)}\exp\left\{\frac{iz\omega\csc\alpha}{a}\right\}dz$$

$$= \overline{\mathscr{F}\left[\psi\right]\left(\frac{\omega\csc\alpha}{a}\right)}. \qquad (4.37)$$

Implementing (4.36) and (4.37) in (4.35) yields the following expression:

$$\mathscr{F}_\alpha\left(S_\psi^\alpha\left[f\right](a,b)\right)(\omega)$$

$$= \exp\left\{-\frac{ia^2\sin\alpha\cos\alpha}{2}-ia\omega\cos\alpha\right\}\mathscr{F}_\alpha\left[f\right](\omega+a\sin\alpha)\overline{\mathscr{F}\left[\psi\right]\left(\frac{\omega\csc\alpha}{a}\right)}. \qquad (4.38)$$

Finally, applying the inverse fractional Fourier transform (1.134) on both sides of the expression (4.38), we obtain the desired result as follows:

$$S_\psi^\alpha\left[f\right](a,b) = \sqrt{\frac{1+i\cot\alpha}{2\pi}}\int_{\mathbb{R}}\exp\left\{-\frac{ia^2\sin\alpha\cos\alpha}{2}-ia\omega\cos\alpha\right\}\mathscr{F}_\alpha\left[f\right](\omega+a\sin\alpha)$$

$$\times\overline{\mathscr{F}\left[\psi\right]\left(\frac{\omega\csc\alpha}{a}\right)}\exp\left\{-\frac{i(b^2+\omega^2)\cot\alpha}{2}+ib\omega\csc\alpha\right\}d\omega$$

$$= \sqrt{\frac{1+i\cot\alpha}{2\pi}}\int_{\mathbb{R}}\exp\left\{-\frac{ia^2\sin\alpha\cos\alpha}{2}-ia(\omega-a\sin\alpha)\cos\alpha\right\}\mathscr{F}_\alpha\left[f\right](\omega)$$

$$\times \overline{\mathscr{F}\big[\psi\big]\left(\frac{(\omega - a\sin\alpha)\csc\alpha}{a}\right)}$$

$$\times \exp\left\{-\frac{i(b^2 + (\omega - a\sin\alpha)^2)\cot\alpha}{2} + ib(\omega - a\sin\alpha)\csc\alpha\right\} d\omega$$

$$= \sqrt{\frac{1 + i\cot\alpha}{2\pi}} \int_{\mathbb{R}} \exp\left\{\frac{ia^2\sin\alpha\cos\alpha}{2} - ia\omega\cos\alpha\right\} \mathscr{F}_\alpha\big[f\big](\omega)$$

$$\times \overline{\mathscr{F}\big[\psi\big]\left(\frac{\omega\csc\alpha - a}{a}\right)}$$

$$\times \exp\left\{-\frac{i(b^2 + \omega^2 + a^2\sin^2\alpha - 2a\omega\sin\alpha)\cot\alpha}{2} + ib\omega\csc\alpha - iab\right\} d\omega$$

$$= \sqrt{\frac{1 + i\cot\alpha}{2\pi}} \int_{\mathbb{R}} e^{-iab}\,\mathscr{F}_\alpha\big[f\big](\omega)\,\overline{\mathscr{F}\big[e^{it}\psi(t)\big]\left(\frac{\omega\csc\alpha}{a}\right)}$$

$$\times \exp\left\{-\frac{i(b^2 + \omega^2)\cot\alpha}{2} + ib\omega\csc\alpha\right\} d\omega$$

$$= e^{-iab}\int_{\mathbb{R}} \mathscr{F}_\alpha\big[f\big](\omega)\,\overline{\mathscr{F}\big[e^{it}\,\psi(t)\big]\left(\frac{\omega\csc\alpha}{a}\right)}\,\overline{\mathcal{K}_\alpha(b,\omega)}\,d\omega.$$

This completes the proof of Lemma 4.3.1.   □

As a consequence of Lemma 4.3.1, we infer that the fractional Stockwell transform defined in (4.32) can be regarded as a unification of three processes, viz; multiplying the fractional spectrum of the input signal by a window function whose argument is fractionally scaled, inverting the resulting function via the inverse fractional Fourier transform and finally modulating the output. These features of the fractional Stockwell transform serve as the linchpin for efficient analysis of signals which reveal their characteristics beyond the classical Fourier domain.

In the next theorem, we present some elementary properties of the fractional Stockwell transform.

Theorem 4.3.2. *Let $c_1, c_2 \in \mathbb{C}$, $k \in \mathbb{R}$ and $\lambda \in \mathbb{R} \setminus \{0\}$, then for any $f, g \in L^2(\mathbb{R})$, the fractional Stockwell transform defined in (4.32) satisfies the following properties:*

(i) *Linearity:* $\mathcal{S}_\psi^\alpha\big[c_1 f + c_2 g\big](a, b) = c_1\,\mathcal{S}_\psi^\alpha\big[f\big](a, b) + c_2\,\mathcal{S}_\psi^\alpha\big[g\big](a, b),$

(ii) *Anti-linearity:* $\mathcal{S}_{c_1\psi + c_2\phi}^\alpha\big[f\big](a, b) = \overline{c_1}\,\mathcal{S}_\psi^\alpha\big[f\big](a, b) + \overline{c_2}\,\mathcal{S}_\phi^\alpha\big[f\big](a, b),$

(iii) *Translation:* $\mathcal{S}_\psi^\alpha\big[\mathcal{T}_k f\big](a, b) = e^{i(k^2 - bk)\cot\alpha - iak}\,\mathcal{S}_\psi^\alpha\big[\mathcal{M}_{k\cot\alpha} f\big](a, b - k),$

(iv) *Dilation:* $\mathcal{S}_\psi^\alpha\big[\mathcal{D}_\lambda f\big](a, b) = |\lambda|\,\mathcal{S}_\psi^\beta\big[f\big]\left(\frac{a}{\lambda}, \lambda b\right), \quad \beta = \operatorname{arccot}\left(\frac{\cot\alpha}{\lambda^2}\right),$

(v) *Translation in window:* $\mathcal{S}_{\mathcal{T}_k\psi}^\alpha\big[f\big](a, b) = \exp\left\{\frac{i(k^2 + 2abk)\cot\alpha}{2a^2}\right\}\mathcal{S}_\psi^\alpha\big[f\big](a, b + k/a),$

(vi) *Dilation in window:* $\mathcal{S}_{\mathcal{D}_\lambda\psi}^\alpha\big[f\big](a, b) = \mathcal{S}_\psi^\alpha\big[\mathcal{M}_{a(\lambda - 1)} f\big](a\lambda, b).$

Proof. The proofs of (i) and (ii) are easy to follow and are, therefore, omitted.

(iii) In case the input signal is translated by any $k \in \mathbb{R}$, then we observe that

$$
\mathcal{S}_\psi^\alpha \big[\mathcal{T}_k f \big](a,b) = \frac{|a|}{\sqrt{2\pi}} \int_{\mathbb{R}} f(t-k) \, \overline{\psi\big(a(t-b)\big)} \exp\left\{ \frac{i(t^2-b^2)\cot\alpha}{2} - iat \right\} dt
$$

$$
= \frac{|a|}{\sqrt{2\pi}} \int_{\mathbb{R}} f(z) \, \overline{\psi\big(a(z+k-b)\big)} \exp\left\{ \frac{i\big((z+k)^2-b^2\big)\cot\alpha}{2} - ia(z+k) \right\} dz
$$

$$
= \frac{|a|}{\sqrt{2\pi}} \int_{\mathbb{R}} f(z) \, \overline{\psi\big(a(z-(b-k))\big)} \exp\left\{ \frac{i\big(z^2-(b-k)^2\big)\cot\alpha}{2} - iaz \right\}
$$

$$
\times \exp\left\{ i\big(k^2 - bk + kz\big)\cot\alpha - iak \right\} dz
$$

$$
= \frac{1}{\sqrt{2\pi}} \int_{\mathbb{R}} e^{ikz\cot\alpha} f(z) \, \overline{\psi_{a,b-k}^\alpha(z)} \exp\left\{ i\big(k^2 - bk\big)\cot\alpha - iak \right\} dz
$$

$$
= e^{i(k^2-bk)\cot\alpha - iak} \, \mathcal{S}_\psi^\alpha \big[\mathcal{M}_{k\cot\alpha} f \big](a, b-k).
$$

(iv) Upon applying the dilation operator $\mathscr{D}_\lambda$, $\lambda \in \mathbb{R} \setminus \{0\}$ to the input signal, we have

$$
\mathcal{S}_\psi^\alpha \big[\mathscr{D}_\lambda f \big](a,b) = \frac{|a\lambda|}{\sqrt{2\pi}} \int_{\mathbb{R}} f(\lambda t) \, \overline{\psi\big(a(t-b)\big)} \exp\left\{ \frac{i(t^2-b^2)\cot\alpha}{2} - iat \right\} dt
$$

$$
= \frac{|a|}{\sqrt{2\pi}} \int_{\mathbb{R}} f(z) \, \overline{\psi\left(\frac{a(t-b)}{\lambda} \right)} \exp\left\{ \frac{i\big(z^2-(\lambda b)^2\big)\cot\alpha}{2\lambda^2} - i\left(\frac{a}{\lambda}\right)z \right\} dz
$$

$$
= |\lambda| \frac{1}{\sqrt{2\pi}} \int_{\mathbb{R}} f(z) \, \overline{\psi_{a/\lambda, \lambda b}^\beta(t)}
$$

$$
= \mathcal{S}_\psi^\beta \big[f \big] \left(\frac{a}{\lambda}, \lambda b \right), \quad \text{where} \quad \beta = \operatorname{arccot}\left(\frac{\cot\alpha}{\lambda^2} \right).
$$

(v) In case the window function $\psi(t)$ is translated by $k \in \mathbb{R}$, it follows that

$$
\mathcal{S}_{\mathcal{T}_k \psi}^\alpha \big[f \big](a,b) = \frac{1}{\sqrt{2\pi}} \int_{\mathbb{R}} f(t) \, \overline{\big(\mathcal{T}_k \psi\big)_{a,b}^\alpha(t)} \, dt
$$

$$
= \frac{|a|}{\sqrt{2\pi}} \int_{\mathbb{R}} f(t) \, \overline{\mathcal{T}_k \psi\big(a(t-b)\big)} \exp\left\{ \frac{i(t^2-b^2)\cot\alpha}{2} - iat \right\} dt
$$

$$
= \frac{|a|}{\sqrt{2\pi}} \int_{\mathbb{R}} f(t) \, \overline{\psi\big(a(t-b)-k\big)} \exp\left\{ \frac{i(t^2-b^2)\cot\alpha}{2} - iat \right\} dt
$$

$$
= \frac{|a|}{\sqrt{2\pi}} \int_{\mathbb{R}} f(t) \, \overline{\psi\left(a\left(t - \left(b + \frac{k}{a} \right) \right) \right)}
$$

$$
\times \exp\left\{ \frac{i\big(k^2 + 2abk\big)\cot\alpha}{2a^2} \right\} \exp\left\{ \frac{i\big(t^2-(b+k/a)^2\big)\cot\alpha}{2} - iat \right\} dt
$$

$$
= \exp\left\{ \frac{i\big(k^2 + 2abk\big)\cot\alpha}{2a^2} \right\} \left\{ \frac{1}{\sqrt{2\pi}} \int_{\mathbb{R}} f(t) \, \overline{\psi_{a,b+k/a}^\alpha(t)} \, dt \right\}
$$

$$
= \exp\left\{ \frac{i\big(k^2 + 2abk\big)\cot\alpha}{2a^2} \right\} \mathcal{S}_\psi^\alpha \big[f \big](a, b + k/a).
$$

(vi) Upon dilating the window function $\psi(t)$ via the operator $\mathscr{D}_\lambda$, $\lambda \in \mathbb{R} \setminus \{0\}$, we have

$$S^{\alpha}_{\mathcal{D}_{\lambda}\psi}\big[f\big](a,b) = \frac{1}{\sqrt{2\pi}} \int_{\mathbb{R}} f(t)\, \overline{\big(\mathcal{D}_{\lambda}\psi\big)^{\alpha}_{a,b}(t)}\, dt$$

$$= \frac{|a|}{\sqrt{2\pi}} \int_{\mathbb{R}} f(t)\, \overline{\mathcal{D}_{\lambda}\psi\big(a(t-b)\big)}\, \exp\left\{ \frac{i\big(t^2-b^2\big)\cot\alpha}{2} - iat \right\} dt$$

$$= \frac{|a\lambda|}{\sqrt{2\pi}} \int_{\mathbb{R}} f(t)\, \overline{\psi\big(a\lambda(t-b)\big)}\, \exp\left\{ \frac{i\big(t^2-b^2\big)\cot\alpha}{2} - iat \right\} dt$$

$$= \frac{|a\lambda|}{\sqrt{2\pi}} \int_{\mathbb{R}} e^{ia(\lambda-1)t} f(t)\, \overline{\psi\big(a\lambda(t-b)\big)}\, \exp\left\{ \frac{i\big(t^2-b^2\big)\cot\alpha}{2} - ia\lambda t \right\} dt$$

$$= \frac{1}{\sqrt{2\pi}} \int_{\mathbb{R}} \mathcal{M}_{a(\lambda-1)} f(t)\, \overline{\psi^{\alpha}_{a\lambda,b}(t)}\, dt$$

$$= S^{\alpha}_{\psi}\big[\mathcal{M}_{a(\lambda-1)}f\big](a\lambda,b).$$

This completes the proof of Theorem 4.3.2   □

The fractional Stockwell transform of a one-dimensional signal is a two-dimensional time-fractional-frequency joint representation. It is therefore desirable as well as fruitful that the transformation should be lossless and hence the resolution of identity must be satisfied. In pursuit of that, we shall demonstrate that indeed an orthogonality relation between the natural and transformed domains holds good. As a consequence of such a relationship, we obtain the energy preserving relation, which allows us to conclude that indeed the family of analyzing functions $\mathcal{F}^{\alpha}_{\psi}(a,b)$ defined in (4.31) generates a resolution of the identity.

Theorem 4.3.3. *If $S^{\alpha}_{\psi}\big[f\big](a,b)$ and $S^{\alpha}_{\psi}\big[g\big](a,b)$ are the fractional Stockwell transforms corresponding to a given pair of functions $f, g \in L^2(\mathbb{R})$, then*

$$\int_{\mathbb{R}} \int_{\mathbb{R}} S^{\alpha}_{\psi}\big[f\big](a,b)\, \overline{S^{\alpha}_{\psi}\big[g\big](a,b)}\, \frac{da\, db}{|a|} = C_{\psi}\big\langle f,\, g \big\rangle_2, \tag{4.39}$$

provided the window function $\psi \in L^2(\mathbb{R})$ satisfies the following condition:

$$C_{\psi} = \int_{\mathbb{R}} \frac{\big|\mathscr{F}[\psi](\omega-1)\big|^2}{|\omega|}\, d\omega < \infty. \tag{4.40}$$

Proof. Invoking Lemma 4.3.1, we can express the fractional Stockwell transform of the given pair of square integrable functions f and g as follows:

$$S^{\alpha}_{\psi}\big[f\big](a,b) = e^{-iab} \int_{\mathbb{R}} \mathscr{F}_{\alpha}\big[f\big](\omega)\, \overline{\mathscr{F}\big[e^{it}\,\psi(t)\big]\left(\frac{\omega\csc\alpha}{a}\right)}\, \mathcal{K}_{\alpha}(b,\omega)\, d\omega \tag{4.41}$$

and

$$S^{\alpha}_{\psi}\big[g\big](a,b) = e^{-iab} \int_{\mathbb{R}} \mathscr{F}_{\alpha}\big[g\big](\xi)\, \overline{\mathscr{F}\big[e^{it}\,\psi(t)\big]\left(\frac{\xi\csc\alpha}{a}\right)}\, \mathcal{K}_{\alpha}(b,\xi)\, d\xi. \tag{4.42}$$

Thus, we have

$$\int_{\mathbb{R}} \int_{\mathbb{R}} S^{\alpha}_{\psi}\big[f\big](a,b)\, \overline{S^{\alpha}_{\psi}\big[g\big](a,b)}\, \frac{da\, db}{|a|}$$

$$= \int_{\mathbb{R}} \int_{\mathbb{R}} \left\{ e^{-iab} \int_{\mathbb{R}} \mathscr{F}_{\alpha}\big[f\big](\omega)\, \overline{\mathscr{F}\big[e^{it}\,\psi(t)\big]\left(\frac{\omega\csc\alpha}{a}\right)}\, \mathcal{K}_{\alpha}(b,\omega)\, d\omega \right\}$$

$$\times \left\{ e^{iab} \int_{\mathbb{R}} \overline{\mathscr{F}_\alpha [g](\xi)} \, \mathscr{F} \left[e^{it} \, \psi(t) \right] \left(\frac{\xi \csc \alpha}{a} \right) \mathcal{K}_\alpha(b, \xi) \, d\xi \right\} \frac{da \, db}{|a|}$$

$$= \int_{\mathbb{R}} \int_{\mathbb{R}} \int_{\mathbb{R}} \mathscr{F}_\alpha [f](\omega) \, \overline{\mathscr{F}_\alpha [g](\xi)} \, \overline{\mathscr{F} \left[e^{it} \, \psi(t) \right] \left(\frac{\omega \csc \alpha}{a} \right)} \mathscr{F} \left[e^{it} \, \psi(t) \right] \left(\frac{\xi \csc \alpha}{a} \right)$$

$$\times \left\{ \int_{\mathbb{R}} \mathcal{K}_\alpha(b, \xi) \, \overline{\mathcal{K}_\alpha(b, \omega)} \, db \right\} \frac{da \, d\omega \, d\xi}{|a|}$$

$$= \int_{\mathbb{R}} \int_{\mathbb{R}} \int_{\mathbb{R}} \mathscr{F}_\alpha [f](\omega) \, \overline{\mathscr{F}_\alpha [g](\xi)} \, \overline{\mathscr{F} \left[e^{it} \, \psi(t) \right] \left(\frac{\omega \csc \alpha}{a} \right)} \mathscr{F} \left[e^{it} \, \psi(t) \right] \left(\frac{\xi \csc \alpha}{a} \right)$$

$$\times \, \delta(\omega - \xi) \frac{da \, d\omega \, d\xi}{|a|}$$

$$= \int_{\mathbb{R}} \int_{\mathbb{R}} \mathscr{F}_\alpha [f](\omega) \, \overline{\mathscr{F}_\alpha [g](\omega)} \left| \mathscr{F} \left[e^{it} \, \psi(t) \right] \left(\frac{\omega \csc \alpha}{a} \right) \right|^2 \frac{da \, d\omega}{|a|}$$

$$= \int_{\mathbb{R}} \mathscr{F}_\alpha [f](\omega) \, \overline{\mathscr{F}_\alpha [g](\omega)} \left\{ \int_{\mathbb{R}} \left| \mathscr{F} \left[e^{it} \, \psi(t) \right] \left(\frac{\omega \csc \alpha}{a} \right) \right|^2 \frac{da}{|a|} \right\} d\omega. \tag{4.43}$$

To facilitate the simplification of the second integral on the R.H.S of (4.43), we shall make use of the substitution $\omega' = a/\omega \csc \alpha$, so that

$$\int_{\mathbb{R}} \left| \mathscr{F} \left[e^{it} \, \psi(t) \right] \left(\frac{\omega \csc \alpha}{a} \right) \right|^2 \frac{da}{|a|} = \int_{\mathbb{R}} \left| \mathscr{F} \left[\psi(t) \right] \left(\frac{\omega \csc \alpha}{a} - 1 \right) \right|^2 \frac{da}{|a|}$$

$$= \int_{\mathbb{R}} \frac{\left| \mathscr{F} [\psi] (\omega' - 1) \right|^2}{|\omega'|} \, d\omega'$$

$$= C_\psi < \infty. \tag{4.44}$$

Plugging (4.44) into (4.43), we obtain the desired orthogonality relation as

$$\int_{\mathbb{R}} \int_{\mathbb{R}} \mathcal{S}_\psi^\alpha [f](a, b) \, \overline{\mathcal{S}_\psi^\alpha [g](a, b)} \, \frac{da \, db}{|a|} = C_\psi \int_{\mathbb{R}} \mathscr{F}_\alpha [f](\omega) \, \overline{\mathscr{F}_\alpha [g](\omega)} \, d\omega$$

$$= C_\psi \left\langle \mathscr{F}_\alpha [f], \, \mathscr{F}_\alpha [g] \right\rangle_2$$

$$= C_\psi \left\langle f, g \right\rangle_2.$$

This completes the proof of Theorem 4.3.3. $\qquad\qquad\qquad\qquad\qquad\square$

Remark 4.3.1. For $f = g$, the orthogonality relation (4.39) yields the following energy preserving relation:

$$\int_{\mathbb{R}} \int_{\mathbb{R}} \left| \mathcal{S}_\psi^\alpha [f](a, b) \right|^2 \frac{da \, db}{|a|} = C_\psi \left\| f \right\|_2^2, \tag{4.45}$$

which shows that the total energy of the signal in the transformed domain is a C_ψ-multiple of the total energy of the signal in the natural domain. Also, in case the window function ψ satisfies $C_\psi = 1$, the fractional Stockwell transform (4.32) sets up an isometry from $L^2(\mathbb{R})$ to $L^2(\mathbb{R}^2)$. Moreover, by virtue of (4.45), we conclude that the family of analyzing functions $F_\psi^\alpha(a, b)$ defined in (4.31) generates a resolution of the identity:

$$I = \frac{1}{C_\psi} \int_{\mathbb{R}} \int_{\mathbb{R}} \left\langle \psi_{a,b}^\alpha, \, \psi_{a,b}^\alpha \right\rangle_2 \frac{da \, db}{|a|}, \tag{4.46}$$

where I denotes the identity operator.

In the following theorem, we establish an inversion formula associated with the fractional Stockwell transform.

Theorem 4.3.4. *Any function $f \in L^2(\mathbb{R})$ can be recovered from the corresponding fractional Stockwell transform $\mathcal{S}_\psi^\alpha[f](a, b)$ via the following formula:*

$$f(t) = \frac{1}{\sqrt{2\pi}C_\psi} \int_\mathbb{R} \int_\mathbb{R} \mathcal{S}_\psi^\alpha[f](a, b)\, \psi_{a,b}^\alpha(t)\, \frac{da\, db}{|a|}, \quad a.e., \tag{4.47}$$

where $C_\psi \neq 0$ is given by (4.40).

Proof. For a given function $f \in L^2(\mathbb{R})$, we shall apply Theorem 4.3.3 with respect to an arbitrary function $g \in L^2(\mathbb{R})$. Then, we have

$$\begin{aligned}
\langle f, g \rangle_2 &= \frac{1}{C_\psi} \int_\mathbb{R} \int_\mathbb{R} \mathcal{S}_\psi^\alpha[f](a, b)\, \overline{\mathcal{S}_\psi^\alpha[g](a, b)}\, \frac{da\, db}{|a|} \\
&= \frac{1}{C_\psi} \int_\mathbb{R} \int_\mathbb{R} \mathcal{S}_\psi^\alpha[f](a, b) \left\{ \frac{1}{\sqrt{2\pi}} \int_\mathbb{R} \overline{g(t)}\, \psi_{a,b}^\alpha(t)\, dt \right\} \frac{da\, db}{|a|} \\
&= \frac{1}{\sqrt{2\pi}C_\psi} \int_\mathbb{R} \left\{ \int_\mathbb{R} \int_\mathbb{R} \mathcal{S}_\psi^\alpha[f](a, b)\, \psi_{a,b}^\alpha(t)\, \frac{da\, db}{|a|} \right\} \overline{g(t)}\, dt \\
&= \left\langle \frac{1}{\sqrt{2\pi}C_\psi} \int_\mathbb{R} \int_\mathbb{R} \mathcal{S}_\psi^\alpha[f](a, b)\, \psi_{a,b}^\alpha(t)\, \frac{da\, db}{|a|},\, g(t) \right\rangle_2. \tag{4.48}
\end{aligned}$$

Noting that the function $g \in L^2(\mathbb{R})$ is chosen arbitrarily, the expression (4.48) yields the following reconstruction formula:

$$f(t) = \frac{1}{\sqrt{2\pi}C_\psi} \int_\mathbb{R} \int_\mathbb{R} \mathcal{S}_\psi^\alpha[f](a, b)\, \psi_{a,b}^\alpha(t)\, \frac{da\, db}{|a|}, \quad a.e.$$

This completes the proof of Theorem 4.3.4. $\qquad\square$

Next, our aim is to study the pointwise convergence of the "weak sense" inversion formula (4.47) associated with the fractional Stockwell transform. To facilitate the narrative, we have the following lemma:

Lemma 4.3.5. *For $M, N \in \mathbb{R}^+$ and any $f \in L^2(\mathbb{R})$, define*

$$f_{M,N}(t) = \frac{1}{\sqrt{2\pi}C_\psi} \int_\mathbb{R} \int_{M<|a|<N} \mathcal{S}_\psi^\alpha[f](a, b)\, \psi_{a,b}^\alpha(t)\, \frac{da\, db}{|a|}. \tag{4.49}$$

Then, the function $f_{M,N}$ is uniformly continuous and the corresponding fractional Fourier transform is given by

$$\mathscr{F}_\alpha[f_{M,N}](\omega) = \frac{1}{C_\psi} \mathscr{F}_\alpha[f](\omega) \left\{ \int_{M<|a|<N} \left| \mathscr{F}\left[e^{iz}\psi(z)\right]\left(\frac{\omega \csc \alpha}{a}\right) \right|^2 \frac{da}{|a|} \right\}, \tag{4.50}$$

where $C_\psi \neq 0$ is given by (4.40).

Proof. Firstly, we shall demonstrate that $f_{M,N}$ is well defined. We proceed as

$$\begin{aligned}
\int_\mathbb{R} \int_{M<|a|<N} & \left| \mathcal{S}_\psi^\alpha[f](a, b)\, \psi_{a,b}^\alpha(t) \right| \frac{da\, db}{|a|} \\
&\leq \int_{M<|a|<N} \left\{ \int_\mathbb{R} \left| \mathcal{S}_\psi^\alpha[f](a, b) \right|^2 db \right\}^{1/2} \left\{ \int_\mathbb{R} \left| \psi_{a,b}^\alpha(t) \right|^2 db \right\}^{1/2} \frac{da}{|a|}
\end{aligned}$$

$$= \int_{M<|a|<N} \left\{ \int_{\mathbb{R}} \left| \mathcal{S}_\psi^\alpha \left[f \right] (a,b) \right|^2 db \right\}^{1/2} \left\| \psi \right\|_2 \frac{da}{|a|}$$

$$\leq \left\| \psi \right\|_2 \left\{ \int_{M<|a|<N} \frac{da}{|a|} \right\}^{1/2} \left\{ \int_{\mathbb{R}} \int_{M<|a|<N} \left| \mathcal{S}_\psi^\alpha \left[f \right] (a,b) \right|^2 \frac{da\,db}{|a|} \right\}^{1/2}$$

$$\leq \sqrt{C_\psi} \left\| f \right\|_2 \left\| \psi \right\|_2 \left\{ \int_{M<|a|<N} \frac{da}{|a|} \right\}^{1/2} < \infty. \tag{4.51}$$

Next, we claim $f_{M,N}$ is uniformly continuous on $\mathbb{R}$. For any $t, t' \in \mathbb{R}$, we have

$$\left| f_{M,N}(t) - f_{M,N}(t') \right|$$

$$= \frac{1}{\sqrt{2\pi C_\psi}} \left| \int_{\mathbb{R}} \int_{M<|a|<N} \mathcal{S}_\psi^\alpha \left[f \right] (a,b) \left(\psi_{a,b}^\alpha(t) - \psi_{a,b}^\alpha(t') \right) \frac{da\,db}{|a|} \right|$$

$$\leq \frac{1}{\sqrt{2\pi C_\psi}} \left\{ \int_{\mathbb{R}} \int_{M<|a|<N} \left| \mathcal{S}_\psi^\alpha \left[f \right] (a,b) \right|^2 \frac{da\,db}{|a|} \right\}^{1/2} \left\{ \int_{\mathbb{R}} \int_{M<|a|<N} \left| \psi_{a,b}^\alpha(t) - \psi_{a,b}^\alpha(t') \right|^2 \frac{da\,db}{|a|} \right\}^{1/2}$$

$$\leq \frac{1}{\sqrt{2\pi C_\psi}} \left\| f \right\|_2 \left\{ \int_{M<|a|<N} \left\| \psi(at - \cdot) - \psi(at' - \cdot) \right\|_2^2 \frac{da}{|a|} \right\}^{1/2}. \tag{4.52}$$

By virtue of inequality (4.52), it evidently follows that $\left| f_{M,N}(t) - f_{M,N}(t') \right| \to 0$, whenever $\left\| t - t' \right\|_2 \to 0$. Hence, we conclude that $f_{M,N}$ is uniformly continuous on $\mathbb{R}$.

Finally, in order to show that (4.50) holds good, we choose an arbitrary square integrable function g and proceed as follows:

$$\left\langle \mathscr{F}_\alpha \left[f_{M,N} \right], \mathscr{F}_\alpha \left[g \right] \right\rangle_2$$

$$= \left\langle f_{M,N}, g \right\rangle_2$$

$$= \int_{\mathbb{R}} \left\{ \frac{1}{\sqrt{2\pi C_\psi}} \int_{\mathbb{R}} \int_{M<|a|<N} \mathcal{S}_\psi^\alpha \left[f \right] (a,b) \, \psi_{a,b}^\alpha(t) \frac{da\,db}{|a|} \right\} \overline{g(t)}\, dt$$

$$= \frac{1}{\sqrt{2\pi C_\psi}} \int_{\mathbb{R}} \int_{M<|a|<N} \mathcal{S}_\psi^\alpha \left[f \right] (a,b) \left\{ \int_{\mathbb{R}} \psi_{a,b}^\alpha(t) \, \overline{g(t)}\, dt \right\} \frac{da\,db}{|a|}$$

$$= \frac{1}{C_\psi} \int_{\mathbb{R}} \int_{M<|a|<N} \mathcal{S}_\psi^\alpha \left[f \right] (a,b) \, \overline{\mathcal{S}_\psi^\alpha \left[g \right] (a,b)} \frac{da\,db}{|a|}$$

$$= \frac{1}{C_\psi} \int_{\mathbb{R}} \int_{M<|a|<N} \left\{ e^{-iab} \int_{\mathbb{R}} \mathscr{F}_\alpha \left[f \right] (\omega) \, \overline{\mathscr{F} \left[e^{iz} \, \psi(z) \right] \left(\frac{\omega \csc \alpha}{a} \right)} \, \overline{\mathcal{K}_\alpha(b,\omega)}\, d\omega \right\}$$

$$\times \left\{ e^{iab} \int_{\mathbb{R}} \overline{\mathscr{F}_\alpha \left[g \right] (\xi)} \, \mathscr{F} \left[e^{iz} \, \psi(z) \right] \left(\frac{\xi \csc \alpha}{a} \right) \mathcal{K}_\alpha(b,\xi)\, d\xi \right\} \frac{da\,db}{|a|}$$

$$= \frac{1}{C_\psi} \int_{\mathbb{R}} \int_{M<|a|<N} \mathscr{F}_\alpha \left[f \right] (\omega) \, \overline{\mathscr{F}_\alpha \left[g \right] (\omega)} \left| \mathscr{F} \left[e^{iz} \, \psi(z) \right] \left(\frac{\omega \csc \alpha}{a} \right) \right|^2 \frac{da\,d\omega}{|a|}$$

$$= \frac{1}{C_\psi} \int_{\mathbb{R}} \mathscr{F}_\alpha \left[f \right] (\omega) \left\{ \int_{M<|a|<N} \left| \mathscr{F} \left[e^{iz} \, \psi(z) \right] \left(\frac{\omega \csc \alpha}{a} \right) \right|^2 \frac{da}{|a|} \right\} \overline{\mathscr{F}_\alpha \left[g \right] (\omega)}\, d\omega. \tag{4.53}$$

Since the square integrable function g is chosen arbitrarily, therefore, by virtue of the fundamental properties of the inner product, the expression (4.53) yields the desired result:

$$\mathscr{F}_\alpha\Big[f_{M,N}\Big](\omega) = \frac{1}{C_\psi}\,\mathscr{F}_\alpha\big[f\big](\omega)\left\{\int_{M<|a|<N}\left|\mathscr{F}\big[e^{iz}\,\psi(z)\big]\left(\frac{\omega\csc\alpha}{a}\right)\right|^2\frac{da}{|a|}\right\}.$$

This completes the proof of Lemma 4.3.5. $\qquad\qquad\qquad\qquad\qquad\qquad\square$

Below, we present the formal convergence result regarding the inverse formula associated with the fractional Stockwell transform.

Theorem 4.3.6. *Given a window function $\psi \in L^2(\mathbb{R})$ satisfying $C_\psi \neq 0$, where C_ψ has its usual meaning and is given by (4.40). Then, for any $f \in L^2(\mathbb{R})$, we have*

$$\lim_{\substack{M\to0,\\N\to\infty}}\big\|f - f_{M,N}\big\|_2 = 0. \tag{4.54}$$

Moreover, in case $\mathscr{F}_\alpha[f] \in L^1(\mathbb{R})$, then we have

$$\lim_{\substack{M\to0,\\N\to\infty}}\big\|f - f_{M,N}\big\|_\infty = 0. \tag{4.55}$$

Proof. For the sake of brevity, we shall only prove (4.55). The assertion (4.54) can be verified on the similar lines. Using Lemma 4.3.5, we obtain

$$\begin{aligned}
\big\|f - f_{M,N}\big\|_\infty &\leq \big\|f - f_{M,N}\big\|_1\\[4pt]
&= \left\|\mathscr{F}_\alpha\big[f\big](\omega) - \mathscr{F}_\alpha\big[f_{M,N}\big](\omega)\right\|_1\\[4pt]
&= \left\|\mathscr{F}_\alpha\big[f\big](\omega) - \frac{1}{C_\psi}\,\mathscr{F}_\alpha\big[f\big](\omega)\left\{\int_{M<|a|<N}\left|\mathscr{F}\big[e^{iz}\,\psi(z)\big]\left(\frac{\omega\csc\alpha}{a}\right)\right|^2\frac{da}{|a|}\right\}\right\|_1\\[4pt]
&= \left\|\mathscr{F}_\alpha\big[f\big](\omega)\left(1 - \frac{1}{C_\psi}\left\{\int_{M<|a|<N}\left|\mathscr{F}\big[e^{iz}\,\psi(z)\big]\left(\frac{\omega\csc\alpha}{a}\right)\right|^2\frac{da}{|a|}\right\}\right)\right\|_1\\[4pt]
&= \int_{\mathbb{R}}\left|\mathscr{F}_\alpha\big[f\big](\omega)\right|\left|1 - \frac{1}{C_\psi}\left\{\int_{M<|a|<N}\left|\mathscr{F}\big[e^{iz}\,\psi(z)\big]\left(\frac{\omega\csc\alpha}{a}\right)\right|^2\frac{da}{|a|}\right\}\right|d\omega.
\end{aligned}$$
$$\tag{4.56}$$

Since the window function $\psi \in L^2(\mathbb{R})$ satisfies (4.40), therefore, it follows that

$$\int_{M<|a|<N}\left|\mathscr{F}\big[e^{iz}\,\psi(z)\big]\left(\frac{\omega\csc\alpha}{a}\right)\right|^2\frac{da}{|a|} \leq \int_{\mathbb{R}}\left|\mathscr{F}\big[e^{iz}\,\psi(z)\big]\left(\frac{\omega\csc\alpha}{a}\right)\right|^2\frac{da}{|a|}. \tag{4.57}$$

Upon making appropriate substitution, as in (4.44), on the R.H.S of (4.57), we get

$$\int_{M<|a|<N}\left|\mathscr{F}\big[e^{iz}\,\psi(z)\big]\left(\frac{\omega\csc\alpha}{a}\right)\right|^2\frac{da}{|a|} \leq C_\psi < \infty. \tag{4.58}$$

Consequently, we obtain

$$\lim_{\substack{M\to0,\\N\to\infty}}\left|1 - \frac{1}{C_\psi}\left\{\int_{M<|a|<N}\left|\mathscr{F}\big[e^{iz}\,\psi(z)\big]\left(\frac{\omega\csc\alpha}{a}\right)\right|^2\frac{da}{|a|}\right\}\right| = 0. \tag{4.59}$$

Finally, the conclusion follows by application of the dominated convergence theorem. This completes the proof of Theorem 4.3.6. $\qquad\qquad\qquad\qquad\qquad\square$

In the following theorem, we shall investigate on the range of the fractional Stockwell transform defined in (4.32). As a consequence of such a theorem, we can instantly infer that indeed the range of the fractional Stockwell transform constitutes a reproducing kernel Hilbert space.

Theorem 4.3.7. *A function $f \in L^2(\mathbb{R}^2)$ is the fractional Stockwell transform of a certain square integrable function with respect to the window function $\psi \in L^2(\mathbb{R})$ if and only if it satisfies the following reproduction formula:*

$$f(a', b') = \frac{1}{2\pi\, C_\psi} \int_\mathbb{R} \int_\mathbb{R} f(a, b) \left\langle \psi_{a,b}^\alpha, \psi_{a',b'}^\alpha \right\rangle_2 \frac{da\, db}{|a|}, \tag{4.60}$$

where $C_\psi \neq 0$ is given by (4.40).

Proof. Suppose that a function $f \in L^2(\mathbb{R}^2)$ is the fractional Stockwell transform of $h \in L^2(\mathbb{R})$; that is, $\mathcal{S}_\psi^\alpha[h] = f$. Then, we observe that

$$f(a', b') = \mathcal{S}_\psi^\alpha \Big[h \Big] (a', b')$$

$$= \frac{1}{\sqrt{2\pi}} \int_\mathbb{R} h(t)\, \overline{\psi_{a',b'}^\alpha}(t)\, dt$$

$$= \frac{1}{\sqrt{2\pi}} \int_\mathbb{R} \left\{ \frac{1}{\sqrt{2\pi}C_\psi} \int_\mathbb{R} \int_\mathbb{R} \mathcal{S}_\psi^\alpha \Big[h \Big] (a, b)\, \psi_{a,b}^\alpha(t) \frac{da\, db}{|a|} \right\} \overline{\psi_{a',b'}^\alpha}(t)\, dt$$

$$= \frac{1}{2\pi\, C_\psi} \int_\mathbb{R} \int_\mathbb{R} \mathcal{S}_\psi^\alpha \Big[h \Big] (a, b) \left\{ \int_\mathbb{R} \psi_{a,b}^\alpha(t)\, \overline{\psi_{a',b'}^\alpha}(t)\, dt \right\} \frac{da\, db}{|a|}$$

$$= \frac{1}{2\pi\, C_\psi} \int_\mathbb{R} \int_\mathbb{R} \mathcal{S}_\psi^\alpha \Big[h \Big] (a, b) \left\langle \psi_{a,b}^\alpha, \psi_{a',b'}^\alpha \right\rangle_2 \frac{da\, db}{|a|}$$

$$= \frac{1}{2\pi\, C_\psi} \int_\mathbb{R} \int_\mathbb{R} f(a, b) \left\langle \psi_{a,b}^\alpha, \psi_{a',b'}^\alpha \right\rangle_2 \frac{da\, db}{|a|}. \tag{4.61}$$

Conversely, we assume that a function $f \in L^2(\mathbb{R}^2)$ satisfies (4.60). Then, we have to show that $f \in \mathcal{S}_\psi^\alpha(L^2(\mathbb{R}))$; that is, there must exist a function $g \in L^2(\mathbb{R})$, such that $\mathcal{S}_\psi^\alpha[g] = f$. We claim that

$$g(t) = \frac{1}{\sqrt{2\pi}C_\psi} \int_\mathbb{R} \int_\mathbb{R} f(a, b)\, \psi_{a,b}^\alpha(t) \frac{da\, db}{|a|}. \tag{4.62}$$

It is quite easy to verify the square integrability of the function g defined in (4.62). Finally, we observe that

$$\mathcal{S}_\psi^\alpha \Big[g \Big] (a', b') = \frac{1}{\sqrt{2\pi}} \int_\mathbb{R} g(t)\, \overline{\psi_{a',b'}^\alpha}(t)\, dt$$

$$= \frac{1}{\sqrt{2\pi}} \int_\mathbb{R} \left\{ \frac{1}{\sqrt{2\pi}C_\psi} \int_\mathbb{R} \int_\mathbb{R} f(a, b)\, \psi_{a,b}^\alpha(t) \frac{da\, db}{|a|} \right\} \overline{\psi_{a',b'}^\alpha}(t)\, dt$$

$$= \frac{1}{2\pi\, C_\psi} \int_\mathbb{R} \int_\mathbb{R} f(a, b) \left\langle \psi_{a,b}^\alpha, \psi_{a',b'}^\alpha \right\rangle_2 \frac{da\, db}{|a|}$$

$$= f(a', b').$$

This completes the proof of Theorem 4.3.7. $\qquad\square$

Corollary 4.3.8. *The range of the fractional Stockwell transform defined in (4.32) is a reproducing kernel Hilbert space embedded as a subspace in $L^2(\mathbb{R}^2)$ with kernel given by*

$$K_\psi^\alpha(a, b; a', b') = \frac{1}{2\pi\, C_\psi} \left\langle \psi_{a,b}^\alpha, \psi_{a',b'}^\alpha \right\rangle_2. \tag{4.63}$$

4.4   The Linear Canonical Wavelet Transform

"The progress of mathematics is a collective enterprise. All of us are needed."

-Yves Meyer

This section deals with yet another ramification of the conventional wavelet transform, namely the linear canonical wavelet transform. Since the linear canonical transform is known for its state-of-the-art flexibility owing to the higher degrees of freedom, therefore, it is quite lucrative to inculcate the same to the classical wavelet transform. This is primarily achieved by constructing a new family of wavelets, known as the "linear canonical wavelets," via the fundamental convolution operations in the linear canonical domain. Due to the availability of several convolution structures in the linear canonical domain, different families of linear canonical wavelets can be constructed via the convolution operations. However, we shall remain confined to a specific convolution structure, which not only instils higher flexibility into the daughter wavelets, but also generalizes both the classical and fractional wavelets in an elegant fashion. The flexibility of the linear canonical wavelets can be ascertained from the fact that they are nicely adapted to the twisting operations in the linear canonical domain. Such distinguishing features of the linear canonical wavelets increase the efficiency and reliability of the linear canonical wavelet transform for the analysis of chirp-like signals, which is the most important class of signals in the contemporary communication systems. For automotive radar applications, such signals are usually called as the "linear frequency modulated waveforms."

4.4.1   Convolution-based linear canonical wavelet transform

"His [Meyer's] fundamental work in the theory of wavelets has transformed the world of signal processing and has led to a myriad of practical applications."

-Kenneth A. Ribet

In this subsection, our aim is to study the linear canonical wavelet transform by constructing a novel family of linear canonical wavelets via the fundamental convolution structure in the linear canonical domain. Recall that, for typographical convenience, the 2×2 real and unimodular matrix associated with the linear canonical transform is denoted as $M = (A, B; C, D)$. To facilitate the construction of linear canonical wavelets, we shall take into consideration the convolution operation $\otimes_M$ defined in (1.174). Thus, for a given unimodular matrix $M = (A, B; C, D)$ and any wavelet $\psi \in L^2(\mathbb{R})$, we define the family of linear canonical wavelets as under:

$$\mathcal{F}_\psi^M(a, b) := \left\{ \psi_{a,b}^M(t) = \frac{1}{\sqrt{a}} \, \psi\left(\frac{t-b}{a}\right) \exp\left\{\frac{iA(b^2 - t^2)}{2B}\right\}; \ a \in \mathbb{R}^+, \ b \in \mathbb{R} \right\}. \quad (4.64)$$

It is worth noticing that the daughter wavelets $\psi_{a,b}^M(t)$ defined in (4.64) are dictated by the parameters of the given unimodular matrix M. Moreover, for $M = (0, 1; -1, 0)$ the family of linear canonical wavelets $\mathcal{F}_\psi^M(a, b)$ boils down to the family of classical wavelets $\mathscr{F}_\psi(a, b)$ given by (3.42), whereas for $M = (\cos\alpha, \sin\alpha; -\sin\alpha, \cos\alpha)$, $\alpha \neq n\pi$, $n \in \mathbb{Z}$, it reduces to the family of fractional wavelets $\mathcal{F}_\psi^\alpha(a, b)$ defined in (4.2). Therefore, we infer that the linear canonical wavelets defined in (4.64) serve as a lucid generalization of both the classical and fractional wavelets.

Definition 4.4.1. Given a unimodular matrix $M = (A, B; C, D)$, the linear canonical wavelet transform of any $f \in L^2(\mathbb{R})$ with respect to the wavelet $\psi \in L^2(\mathbb{R})$ is denoted by $\mathscr{W}_\psi^M[f]$ and is defined as

$$\mathscr{W}_\psi^M[f](a, b) = \left(f \otimes_M \mathcal{D}_a \breve{\psi} \right)(b) = \int_{\mathbb{R}} f(t) \, \overline{\psi_{a,b}^M(t)} \, dt, \tag{4.65}$$

where $\mathcal{D}_a$ denotes the usual dilation operator with $\breve{\psi}(t) = \psi(-t)$ and the linear canonical wavelets $\psi_{a,b}^M(t)$ are given by (4.64).

The following important points regarding the linear canonical wavelet transform can be extracted from Definition 4.4.1:

(i) For $M = (0, 1; -1, 0)$, Definition 4.4.1 reduces to the definition of the classical wavelet transform (3.47).

(ii) For $M = (\cos\alpha, \sin\alpha; -\sin\alpha, \cos\alpha)$, $\alpha \neq n\pi$, $n \in \mathbb{Z}$, Definition 4.4.1 yields the fractional wavelet transform (4.3).

(iii) The linear canonical wavelet transform (4.65) can be expressed alternatively by appropriate chirp multiplications before and after the computation of the classical wavelet transform as

$$\mathscr{W}_\psi^M[f](a, b) = \exp\left\{ -\frac{iAb^2}{2B} \right\} \frac{1}{\sqrt{a}} \int_{\mathbb{R}} \exp\left\{ \frac{iAt^2}{2B} \right\} f(t) \, \overline{\psi\left(\frac{t - b}{a} \right)} \, dt$$

$$= \exp\left\{ -\frac{iAb^2}{2B} \right\} \mathscr{W}_\psi[F_M](a, b), \tag{4.66}$$

where

$$F_M(t) = \exp\left\{ \frac{iAt^2}{2B} \right\} f(t). \tag{4.67}$$

Therefore, we conclude that the computational cost of the of the linear canonical wavelet transform (4.65) is completely determined by the classical wavelet transform.

(iv) The time-chirping operation in the definition of linear canonical wavelets performs suitable adjustments to the analyzing elements and makes them adaptable to the twisting operations in the linear canonical domain.

(v) The function $\psi \in L^2(\mathbb{R})$ appearing in Definition 4.4.1 is to be considered as a wavelet in the sense that it satisfies the classical admissibility condition (3.27). That is, the linear canonical wavelet transform defined in (4.65) does not require any extra conditions to be imposed on the mother wavelet $\psi(t)$.

In the following proposition, we examine the behaviour of the linear canonical wavelets in the linear canonical domain.

Proposition 4.4.1. *For any wavelet $\psi \in L^2(\mathbb{R})$, we have*

$$\mathscr{L}_M\left[\psi_{a,b}^M\right](\omega) = \sqrt{2\pi a} \, \mathscr{F}[\psi]\left(\frac{a\omega}{B} \right) \mathcal{K}_M(b, \omega), \tag{4.68}$$

where $\mathscr{L}_M$ denotes the linear canonical transform with respect to the unimodular matrix $M = (A, B; C, D)$ and $\mathscr{F}$ denotes the usual Fourier transform.

Proof. Using the definition of linear canonical transform given in (1.164), we have

$$
\mathscr{L}_M\left[\psi_{a,b}^M\right](\omega) = \int_{\mathbb{R}} \psi_{a,b}^M(t)\, \mathcal{K}_M(t,\omega)\, dt
$$

$$
= \frac{1}{\sqrt{2\pi i B}} \int_{\mathbb{R}} \psi_{a,b}^M(t) \exp\left\{ \frac{i(At^2 - 2t\omega + D\omega^2)}{2B} \right\} dt
$$

$$
= \frac{1}{\sqrt{2\pi i B}} \int_{\mathbb{R}} \frac{1}{\sqrt{a}}\, \psi\left(\frac{t-b}{a}\right) \exp\left\{ \frac{iA(b^2 - t^2)}{2B} \right\} \exp\left\{ \frac{i(At^2 - 2t\omega + D\omega^2)}{2B} \right\} dt
$$

$$
= \frac{1}{\sqrt{2\pi i B}} \int_{\mathbb{R}} \frac{1}{\sqrt{a}}\, \psi\left(\frac{t-b}{a}\right) \exp\left\{ \frac{iAb^2}{2B} \right\} \exp\left\{ \frac{i(D\omega^2 - 2t\omega)}{2B} \right\} dt. \qquad (4.69)
$$

Making use of the substitution $(t-b)/a = z$ in (4.69), we obtain

$$
\mathscr{L}_M\left[\psi_{a,b}^M\right](\omega) = \sqrt{\frac{a}{2\pi i B}} \int_{\mathbb{R}} \psi(z) \exp\left\{ \frac{i(Ab^2 - 2b\omega + D\omega^2)}{2B} \right\} \exp\left\{ -\frac{ia\omega z}{B} \right\} dz
$$

$$
= \sqrt{2\pi a}\, \mathscr{F}\left[\psi\right]\left(\frac{a\omega}{B}\right) \mathcal{K}_M(b,\omega).
$$

This completes the proof of Proposition 4.4.1. $\qquad\qquad\qquad\qquad\qquad\qquad\square$

Remark 4.4.1. Since the classical daughter wavelets $\psi_{a,b}(t)$ act as band-pass filters in the Fourier domain, therefore, by virtue of Proposition 4.4.1 it is quite evident that each linear canonical wavelet $\psi_{a,b}^M(t)$ defined in (4.64) can be regraded as a differently scaled band-pass filter in the linear canonical domain.

In the following theorem, we shall demonstrate that a given wavelet can be convoluted with a bounded integrable function via the linear canonical convolution operation $\otimes_M$ defined in (1.174) to yield a new wavelet.

Theorem 4.4.2. *Given a real, unimodular matrix* $M = (A, B; C, D)$, *a wavelet* ψ *and a bounded integrable function* ϕ. *Then,* $\Psi_M(t) = e^{iAt^2/2B}(\phi \otimes_M \psi)(t)$ *is also a wavelet.*

Proof. Firstly, we show that the function $\Psi_M(t) = e^{iAt^2/2B}(\phi \otimes_M \psi)(t)$ is indeed square integrable. Using Definition 1.4.2, we have

$$
\int_{\mathbb{R}} \left|\Psi_M(t)\right|^2 dt = \int_{\mathbb{R}} \left|e^{iAt^2/2B}(\phi \otimes_M \psi)(t)\right|^2 dt
$$

$$
= \int_{\mathbb{R}} \left|(\phi \otimes_M \psi)(t)\right|^2 dt
$$

$$
= \int_{\mathbb{R}} \left| \int_{\mathbb{R}} \phi(x)\, \psi(t-x) \exp\left\{ \frac{iA(x^2 - t^2)}{2B} \right\} dx \right|^2 dt
$$

$$
\leq \int_{\mathbb{R}} \left(\int_{\mathbb{R}} |\phi(x)| |\psi(t-x)|\, dx \right)^2 dt
$$

$$
= \int_{\mathbb{R}} \left(\int_{\mathbb{R}} |\phi(x)|^{1/2} |\psi(t-x)| |\phi(x)|^{1/2}\, dx \right)^2 dt
$$

$$
\leq \int_{\mathbb{R}} \left\{ \left(\int_{\mathbb{R}} |\phi(x)| |\psi(t-x)|^2 dx \right) \left(\int_{\mathbb{R}} |\phi(x)|\, dx \right) \right\} dt
$$

$$
= \left\|\phi\right\|_1 \int_{\mathbb{R}} \int_{\mathbb{R}} |\phi(x)| |\psi(t-x)|^2 dt\, dx
$$

$$
= \left\|\phi\right\|_1^2 \left\|\psi\right\|_2^2 < \infty.
$$

Next, we show that the function $\Psi_M(t) = e^{iAt^2/2B}(\phi \otimes_M \psi)(t)$ satisfies the admissibility condition (3.27). To facilitate the intent, we invoke the fundamental relationship between the Fourier and linear canonical transforms. We proceed as follows:

$$\mathscr{F}\Big[\Psi_M\Big](\omega) = \mathscr{F}\Big[e^{iAt^2/2B}(\phi \otimes_M \psi)(t)\Big](\omega)$$

$$= \frac{1}{\sqrt{2\pi}} \exp\left\{-\frac{iD(B\omega)^2}{2B}\right\} \int_{\mathbb{R}} (\phi \otimes_M \psi)(t)$$

$$\times \exp\left\{\frac{i(At^2 - 2t(B\omega) + D(B\omega)^2)}{2B}\right\} dt$$

$$= \sqrt{iB} \exp\left\{-\frac{iD(B\omega)^2}{2B}\right\} \int_{\mathbb{R}} (\phi \otimes_M \psi)(t)\, \mathcal{K}_M(t, B\omega)\, dt$$

$$= \sqrt{iB} \exp\left\{-\frac{iDB\omega^2}{2}\right\} \mathscr{L}_M\Big[(\phi \otimes_M \psi)(t)\Big](B\omega)$$

$$= \sqrt{2\pi iB} \exp\left\{-\frac{iDB\omega^2}{2}\right\} \mathscr{L}_M\Big[\phi\Big](B\omega)\, \mathscr{F}\Big[\phi\Big](\omega). \tag{4.70}$$

Based on the relationship between the Fourier and linear canonical transforms obtained in (4.70), we have

$$\int_{\mathbb{R}} \frac{\big|\mathscr{F}[\Psi_M](\omega)\big|^2}{|\omega|}\, d\omega = 2\pi|B| \int_{\mathbb{R}} \frac{\big|\mathscr{L}_M[\phi](B\omega)\big|^2 \big|\mathscr{F}[\phi](\omega)\big|^2}{|\omega|}\, d\omega$$

$$\le |B| \sup\left(\big|\mathscr{L}_M[\phi](B\omega)\big|^2\right)\left\{2\pi \int_{\mathbb{R}} \frac{\big|\hat{\psi}(\omega)\big|^2}{|\omega|}\, d\omega\right\} < \infty.$$

Hence, we conclude that the function $\Psi_M(t) = e^{iAt^2/2B}(\phi \otimes_M \psi)(t)$ is a wavelet. □

Below, we present a lemma which deals with the frequency domain effect of the linear canonical wavelet transform defined in (4.65).

Lemma 4.4.3. *For any $f \in L^2(\mathbb{R})$, the linear canonical wavelet transform $\mathscr{W}_\psi^M[f](a, b)$ is expressible as follows:*

$$\mathscr{L}_M\left(\mathscr{W}_\psi^M\Big[f\Big](a,b)\right)(\omega) = \sqrt{2\pi a}\, \mathscr{L}_M\Big[f\Big](\omega)\, \overline{\mathscr{F}\Big[\psi\Big]\left(\frac{a\omega}{B}\right)}, \tag{4.71}$$

where the linear canonical transform on the L.H.S of (4.71) is computed with respect to the translation variable b.

Proof. In order to accomplish the proof, we shall employ the Parseval's formula (1.171) associated with the linear canonical transform and then invoke Proposition 4.4.1 as

$$\mathscr{W}_\psi^M\Big[f\Big](a,b) = \Big\langle f, \psi_{a,b}^M\Big\rangle_2$$

$$= \Big\langle \mathscr{L}_M\big[f\big], \mathscr{L}_M\big[\psi_{a,b}^M\big]\Big\rangle_2$$

$$= \int_{\mathbb{R}} \mathscr{L}_M\big[f\big](\omega)\, \overline{\mathscr{L}_M\big[\psi_{a,b}^M\big](\omega)}\, d\omega$$

$$= \sqrt{2\pi a} \int_{\mathbb{R}} \mathscr{L}_M\big[f\big](\omega)\, \overline{\mathscr{F}\big[\psi\big]\left(\frac{a\omega}{B}\right)}\, \overline{\mathcal{K}_M(b,\omega)}\, d\omega$$

$$= \sqrt{2\pi a}\, \mathscr{L}_M^{-1}\left(\mathscr{L}_M\big[f\big](\omega)\, \overline{\mathscr{F}\big[\psi\big]\left(\frac{a\omega}{B}\right)}\right)(b). \tag{4.72}$$

Applying linear canonical transform on both sides of (4.72), we obtain the desired expression:

$$\mathscr{L}_M\left(\mathscr{W}_\psi^M\big[f\big](a,b)\right)(\omega) = \sqrt{2\pi a}\,\mathscr{L}_M\big[f\big](\omega)\,\overline{\mathscr{F}[\psi]}\left(\frac{a\omega}{B}\right).$$

This completes the proof of Lemma 4.4.3.   □

In the following theorem, we assemble some basic properties of the convolution-based linear canonical wavelet transform.

Theorem 4.4.4. *For $k \in \mathbb{R}$ and $\lambda \in \mathbb{R}^+$, let $\mathcal{T}_k$ and $\mathcal{D}_\lambda$ be the usual translation and dilation operators pertaining to the wavelet transform. If $c_1, c_2 \in \mathbb{C}$ and $f, g \in L^2(\mathbb{R})$, then the linear canonical wavelet transform defined in (4.65) satisfies the following properties:*

(i) *Linearity:* $\mathscr{W}_\psi^M\big[c_1 f + c_2 g\big](a,b) = c_1\,\mathscr{W}_\psi^M\big[f\big](a,b) + c_2\,\mathscr{W}_\psi^M\big[g\big](a,b),$

(ii) *Anti-linearity:* $\mathscr{W}_{c_1\psi+c_2\phi}^M\big[f\big](a,b) = \overline{c_1}\,\mathscr{W}_\psi^M\big[f\big](a,b) + \overline{c_2}\,\mathscr{W}_\phi^M\big[f\big](a,b),$

(iii) *Translation:* $\mathscr{W}_\psi^M\big[(\mathcal{T}_k f)\big](a,b) = e^{iA(k^2-bk)/B}\,\mathscr{W}_\psi^M\big[e^{iAkt/B}f(t)\big](a,b-k),$

(iv) *Dilation:* $\mathscr{W}_\psi^M\big[(\mathcal{D}_\lambda f)\big](a,b) = \mathscr{W}_\psi^{M'}\big[f\big]\left(\frac{a}{\lambda},\frac{b}{\lambda}\right),$ *where* $M' = (A, B/\lambda^2; \lambda^2 C, D),$

(v) *Parity:* $\mathscr{W}_{(P\psi)}^M\big[(Pf)\big](a,b) = \mathscr{W}_\psi^M\big[f\big](a,-b),$ *where* $(Pf)(t) = f(-t),$

(vi) *Symmetry:* $\mathscr{W}_\psi^M\big[f\big](a,b) = e^{iAb^2/2B}\,\mathscr{W}_{\bar f}^{M''}\big[\Psi\big]\left(\frac{1}{a},-\frac{b}{a}\right),$ *where* $\Psi(t) = e^{iAabt/B}\overline{\psi(t)}$ *and* $M'' = (A, B/a^2; a^2 C, D),$

(vii) *Translation in wavelet:* $\mathscr{W}_{(\mathcal{T}_k\psi)}^M\big[f\big](a,b) = \exp\left\{\dfrac{iA(a^2 k^2 + 2abk)}{2B}\right\}\mathscr{W}_\psi^M\big[f\big](a,b+ak),$

(viii) *Dilation in wavelet:* $\mathscr{W}_{(\mathcal{D}_\lambda\psi)}^M\big[f\big](a,b) = \mathscr{W}_\psi^M\big[f\big](\lambda a,b).$

Proof. The proof of Theorem 4.4.4 follows directly by using Definition 4.4.1 and is omitted.   □

In the next theorem, we obtain an orthogonality relation for the linear canonical wavelet transform defined in (4.65) by making use of Lemma 4.4.3. It is worth mentioning that the orthogonality relation holds under the condition that the mother wavelet $\psi \in L^2(\mathbb{R})$ be admissible in the classical sense. In that case, we can also formulate an inversion formula guaranteeing the reconstruction of the input function $f \in L^2(\mathbb{R})$ from the corresponding linear canonical wavelet coefficients given by $\langle f, \psi_{a,b}^M\rangle_2$.

Theorem 4.4.5. *For any pair of functions $f, g \in L^2(\mathbb{R})$, the corresponding linear canonical wavelet transforms $\mathscr{W}_\psi^M\big[f\big](a,b)$ and $\mathscr{W}_\psi^M\big[g\big](a,b)$ satisfy the following orthogonality relation:*

$$\int_\mathbb{R}\int_{\mathbb{R}^+}\mathscr{W}_\psi^M\big[f\big](a,b)\,\overline{\mathscr{W}_\psi^M\big[g\big](a,b)}\,\frac{da\,db}{a^2} = C_\psi\,\big\langle f, g\big\rangle_2, \qquad (4.73)$$

where C_ψ is given by the classical admissibility condition (3.27).

Proof. Applying the Parseval's formula for the linear canonical transform to $\mathscr{W}_\psi^M[f](a,b)$ and $\mathscr{W}_\psi^M[g](a,b)$ with regard to the translation variable b, we can write

$$
\int_{\mathbb{R}} \int_{\mathbb{R}^+} \mathscr{W}_\psi^M[f](a,b) \overline{\mathscr{W}_\psi^M[g](a,b)} \frac{da\,db}{a^2}
$$

$$
= \int_{\mathbb{R}^+} \int_{\mathbb{R}} \mathscr{L}_M\left(\mathscr{W}_\psi^M[f](a,b)\right)(\omega) \overline{\mathscr{L}_M\left(\mathscr{W}_\psi^M[g](a,b)\right)(\omega)} \frac{d\omega\,da}{a^2}
$$

$$
= \int_{\mathbb{R}^+} \int_{\mathbb{R}} \left\{\sqrt{2\pi a}\,\mathscr{L}_M[f](\omega)\,\overline{\mathscr{F}[\psi]\left(\frac{a\omega}{B}\right)}\right\}\left\{\sqrt{2\pi a}\,\overline{\mathscr{L}_M[g](\omega)}\,\mathscr{F}[\psi]\left(\frac{a\omega}{B}\right)\right\} \frac{d\omega\,da}{a^2}
$$

$$
= \int_{\mathbb{R}} \mathscr{L}_M[f](\omega)\,\overline{\mathscr{L}_M[g](\omega)} \left\{2\pi \int_{\mathbb{R}^+} \left|\mathscr{F}[\psi]\left(\frac{a\omega}{B}\right)\right|^2 \frac{da}{a}\right\} d\omega. \tag{4.74}
$$

For an explicit expression of the inner integral on the R.H.S of (4.74), we make use of the substitution $a\omega/B = \omega'$, so that

$$
2\pi \int_{\mathbb{R}^+} \left|\mathscr{F}[\psi]\left(\frac{a\omega}{B}\right)\right|^2 \frac{da}{a} = 2\pi \int_{\mathbb{R}} \frac{\left|\mathscr{F}[\psi](\omega')\right|^2}{|\omega'|}\,d\omega' = C_\psi. \tag{4.75}
$$

Implementing relation (4.75) in (4.74) yields the following expression:

$$
\int_{\mathbb{R}} \int_{\mathbb{R}^+} \mathscr{W}_\psi^M[f](a,b) \overline{\mathscr{W}_\psi^M[g](a,b)} \frac{da\,db}{a^2} = C_\psi \int_{\mathbb{R}} \mathscr{L}_M[f](\omega)\,\overline{\mathscr{L}_M[g](\omega)}\,d\omega
$$

$$
= C_\psi \left\langle \mathscr{L}_M[f],\,\mathscr{L}_M[g]\right\rangle_2
$$

$$
= C_\psi \left\langle f,\,g\right\rangle_2.
$$

This completes the proof of Theorem 4.4.5 $\qquad\qquad\square$

Remark 4.4.2. For $f = g$, the orthogonality relation (4.73) yields the energy preserving relation:

$$
\int_{\mathbb{R}} \int_{\mathbb{R}^+} \left|\mathscr{W}_\psi^M[f](a,b)\right|^2 \frac{da\,db}{a^2} = C_\psi \left\|f\right\|_2^2. \tag{4.76}
$$

That is, the total energy of the transformed signal is a C_ψ-multiple of total energy of the input signal. In case $C_\psi = 1$, the linear canonical wavelet transform defined in (4.65) is an isometry from $L^2(\mathbb{R})$ to $L^2(\mathbb{R}^+ \times \mathbb{R})$.

Below, we shall formulate an inversion formula associated with the linear canonical wavelet transform.

Theorem 4.4.6. *Any function* $f \in L^2(\mathbb{R})$ *can be reconstructed from the corresponding linear canonical wavelet transform* $\mathscr{W}_\psi^M[f](a,b)$ *via the following reproducing formula:*

$$
f(t) = \frac{1}{C_\psi} \int_{\mathbb{R}} \int_{\mathbb{R}^+} \left\langle f,\,\psi_{a,b}^M\right\rangle_2 \psi_{a,b}^M(t) \frac{da\,db}{a^2}, \quad a.e., \tag{4.77}
$$

where $C_\psi \neq 0$ *is given by the usual admissibility condition (3.27).*

Proof. Using Theorem 4.4.5 with $g(t') = \delta(t - t')$ yields

$$
C_\psi \left\langle f,\,\delta(t - \cdot)\right\rangle_2 = \int_{\mathbb{R}} \int_{\mathbb{R}^+} \mathscr{W}_\psi^M[f](a,b) \overline{\mathscr{W}_\psi^M[\delta(t - \cdot)](a,b)} \frac{da\,db}{a^2}
$$

$$= \int_{\mathbb{R}} \int_{\mathbb{R}^+} \left\langle f, \psi_{a,b}^M \right\rangle_2 \left\{ \int_{\mathbb{R}} \delta(t - t')\, \psi_{a,b}^M(t')\, dt' \right\} \frac{da\, db}{a^2}$$

$$= \int_{\mathbb{R}} \int_{\mathbb{R}^+} \left\langle f, \psi_{a,b}^M \right\rangle_2 \psi_{a,b}^M(t)\, \frac{da\, db}{a^2}. \tag{4.78}$$

Note that,

$$\left\langle f, \delta(t - \cdot) \right\rangle_2 = \int_{\mathbb{R}} f(t')\, \delta(t - t')\, dt' = f(t). \tag{4.79}$$

Finally, using (4.79) in (4.78), we obtain the desired reproducing formula for the linear canonical wavelet transform:

$$f(t) = \frac{1}{C_\psi} \int_{\mathbb{R}} \int_{\mathbb{R}^+} \left\langle f, \psi_{a,b}^M \right\rangle_2 \psi_{a,b}^M(t)\, \frac{da\, db}{a^2}, \quad a.e.$$

This completes the proof of the Theorem 4.4.6. $\square$

We now present a characterization of range of the linear canonical wavelet transform defined in (4.65). Besides, we can easily deduce that the range of linear canonical wavelet transform constitutes a reproducing kernel Hilbert space whose kernel is the autocorrelation function of the given wavelet.

Theorem 4.4.7. *A function $f \in L^2(\mathbb{R}^+ \times \mathbb{R})$ is the linear canonical wavelet transform of a certain square integrable function on $\mathbb{R}$ with respect to the wavelet ψ if and only if it satisfies the following reproduction formula:*

$$f(a', b') = \frac{1}{C_\psi} \int_{\mathbb{R}} \int_{\mathbb{R}^+} f(a, b) \left\langle \psi_{a,b}^M, \psi_{a',b'}^M \right\rangle_2 \frac{da\, db}{a^2}, \tag{4.80}$$

where $C_\psi \neq 0$ is given by the usual admissibility condition (3.27).

Proof. The proof follows as a consequence of the inversion formula (4.77) and is omitted for the reason of space. $\square$

Corollary 4.4.8. *The range of the linear canonical wavelet transform defined in (4.65) is a reproducing kernel Hilbert space embedded as a subspace in $L^2(\mathbb{R}^2)$ with kernel given by*

$$K_\psi^M\big(a, b; a', b'\big) = C_\psi^{-1} \left\langle \psi_{a,b}^M, \psi_{a',b'}^M \right\rangle_2. \tag{4.81}$$

Moreover, the kernel $K_\psi^M\big(a, b; a', b'\big)$ is pointwise bounded:

$$\left| K_\psi^M\big(a, b; a', b'\big) \right| \leq C_\psi^{-1} \left\| \psi \right\|_2^2, \quad \forall\, (a, b), (a', b') \in \mathbb{R}^+ \times \mathbb{R}. \tag{4.82}$$

Remark 4.4.3. The reproducing kernel $K_\psi^M\big(a, b; a', b'\big)$ given by (4.81) measures the correlation of the two linear canonical wavelets $\psi_{a,b}^M(t)$ and $\psi_{a',b'}^M(t)$. According to (4.80), the linear canonical wavelet transform of a certain function $f \in L^2(\mathbb{R})$ about $(a', b') \in \mathbb{R}^+ \times \mathbb{R}$ can always be expressed via another $(a, b) \in \mathbb{R}^+ \times \mathbb{R}$ by virtue of the reproducing kernel $K_\psi^M\big(a, b; a', b'\big)$. This means that all the linear canonical wavelet coefficients $\left\langle f, \psi_{a,b}^M \right\rangle_2$ on the time-frequency plane are related to each other, and there always exists redundancy when the continuous linear canonical wavelet transform is used for signal reconstruction. In order to reduce the redundancy, it is desirable that the reproducing kernel satisfies the property that $K_\psi^M\big(a, b; a', b'\big) = \delta(a - a', b - b')$; however, it is quite difficult to find a set of orthonormal linear canonical wavelets satisfying the requirement when the translation and dilation parameters are both continuous. To circumvent such limitations, the translation and dilation parameters of the linear canonical wavelet transform are often discretized in practice and lay the foundation for the theory of discrete linear canonical wavelet transforms.

In analogy to the uncertainty principles governing the simultaneous localization of a function f and its Fourier transform, a different class of uncertainty principles comparing the localization of f (or its Fourier transform) with the localization of the corresponding Gabor or wavelet transform were studied by Wilczok [313]. Motivated by this fact, we shall also obtain an uncertainty inequality comparing the localization of the linear canonical transform of a function f with the linear canonical wavelet transform $\mathscr{W}_\psi^M[f](a,b)$, regarded as a function of the translation variable b. It is imperative to mention that, such an uncertainty inequality circumscribes the Heisenberg-type uncertainty inequalities for both the classical and fractional wavelet transforms.

Theorem 4.4.9. *If $\mathscr{W}_\psi^M[f](a,b)$ is the linear canonical wavelet transform of any non-trivial function $f \in L^2(\mathbb{R})$, with respect to the wavelet $\psi \in L^2(\mathbb{R})$, then the following uncertainty inequality holds:*

$$\left\{ \int_{\mathbb{R}} \int_{\mathbb{R}^+} b^2 \left| \mathscr{W}_\psi^M[f](a,b) \right|^2 \frac{da\, db}{a^2} \right\}^{1/2} \left\{ \int_{\mathbb{R}} \omega^2 \left| \mathscr{L}_M[f](\omega) \right|^2 d\omega \right\}^{1/2} \geq \frac{|B|\sqrt{C_\psi}}{2} \left\| f \right\|_2^2,$$
(4.83)

where C_ψ is given by the usual admissibility condition (3.27).

Proof. Recall that the Heisenberg's uncertainty principle pertaining to the linear canonical transform governs the simultaneous localization of a non-trivial function $f \in L^2(\mathbb{R})$ and the corresponding linear canonical transform $\mathscr{L}_M[f](\omega)$ via the following inequality:

$$\left\{ \int_{\mathbb{R}} t^2 |f(t)|^2 dt \right\}^{1/2} \left\{ \int_{\mathbb{R}} \omega^2 \left| \mathscr{L}_M[f](\omega) \right|^2 d\omega \right\}^{1/2} \geq \frac{|B|}{2} \left\{ \int_{\mathbb{R}} |f(t)|^2 dt \right\}.$$
(4.84)

We shall identify the linear canonical wavelet transform $\mathscr{W}_\psi^M[f](a,b)$ as a function of the translation variable b, so that the inequality (4.84) becomes

$$\left\{ \int_{\mathbb{R}} b^2 \left| \mathscr{W}_\psi^M[f](a,b) \right|^2 db \right\}^{1/2} \left\{ \int_{\mathbb{R}} \omega^2 \left| \mathscr{L}_M \left(\mathscr{W}_\psi^M[f](a,b) \right)(\omega) \right|^2 d\omega \right\}^{1/2}$$
$$\geq \frac{|B|}{2} \left\{ \int_{\mathbb{R}} \left| \mathscr{W}_\psi^M[f](a,b) \right|^2 db \right\}. \quad (4.85)$$

Applying Lemma 4.4.3, we can rewrite inequality (4.85) as

$$\left\{ \int_{\mathbb{R}} b^2 \left| \mathscr{W}_\psi^M[f](a,b) \right|^2 db \right\}^{1/2} \left\{ 2\pi a \int_{\mathbb{R}} \omega^2 \left| \mathscr{L}_M[f](\omega) \right|^2 \left| \mathscr{F}[\psi] \left(\frac{a\omega}{B} \right) \right|^2 d\omega \right\}^{1/2}$$
$$\geq \frac{|B|}{2} \left\{ \int_{\mathbb{R}} \left| \mathscr{W}_\psi^M[f](a,b) \right|^2 db \right\}. \quad (4.86)$$

Upon integrating (4.86) on both sides with respect to the measure da/a^2, we get

$$\int_{\mathbb{R}^+} \left\{ \int_{\mathbb{R}} b^2 \left| \mathscr{W}_\psi^M[f](a,b) \right|^2 db \right\}^{1/2} \left\{ 2\pi a \int_{\mathbb{R}} \omega^2 \left| \mathscr{L}_M[f](\omega) \right|^2 \left| \mathscr{F}[\psi] \left(\frac{a\omega}{B} \right) \right|^2 d\omega \right\}^{1/2} \frac{da}{a^2}$$
$$\geq \frac{|B|}{2} \left\{ \int_{\mathbb{R}^+} \int_{\mathbb{R}} \left| \mathscr{W}_\psi^M[f](a,b) \right|^2 \frac{da\, db}{a^2} \right\}. \quad (4.87)$$

Invoking the Cauchy-Schwarz inequality on the L.H.S of (4.87), followed by a change in the order of integration yields

$$\left\{ \int_{\mathbb{R}} \int_{\mathbb{R}+} b^2 \left| \mathscr{W}_{\psi}^M \big[f\big](a,b) \right|^2 \frac{da\,db}{a^2} \right\}^{1/2} \left\{ \int_{\mathbb{R}} \omega^2 \left| \mathscr{L}_M \big[f\big](\omega) \right|^2 \left(2\pi \int_{\mathbb{R}+} \left| \mathscr{F}\big[\psi\big] \left(\frac{a\omega}{B} \right) \right|^2 \frac{da}{a} \right) d\omega \right\}^{1/2}$$

$$\geq \frac{|B|}{2} \left\{ \int_{\mathbb{R}} \int_{\mathbb{R}+} \left| \mathscr{W}_{\psi}^M \big[f\big](a,b) \right|^2 \frac{da\,db}{a^2} \right\}$$

$$= \frac{|B|\,C_{\psi}}{2} \big\| f \big\|_2^2, \tag{4.88}$$

where the R.H.S of (4.88) is obtained by virtue of the energy preserving relation given by (4.76). Moreover, by making use of the substitution $a\omega/B = \omega'$, we can write

$$2\pi \int_{\mathbb{R}+} \left| \mathscr{F}\big[\psi\big] \left(\frac{a\omega}{B} \right) \right|^2 \frac{da}{a} = 2\pi \int_{\mathbb{R}} \frac{\left| \mathscr{F}\big[\psi\big](\omega') \right|^2}{|\omega'|} d\omega' = C_{\psi},$$

so that

$$\int_{\mathbb{R}} \omega^2 \left| \mathscr{L}_M \big[f\big](\omega) \right|^2 \left(2\pi \int_{\mathbb{R}+} \left| \mathscr{F}\big[\psi\big] \left(\frac{a\omega}{B} \right) \right|^2 \frac{da}{a} \right) d\omega = C_{\psi} \int_{\mathbb{R}} \omega^2 \left| \mathscr{L}_M \big[f\big](\omega) \right|^2 d\omega. \tag{4.89}$$

Using the estimate obtained in (4.89), we can simplify the inequality (4.88) to

$$\left\{ \int_{\mathbb{R}} \int_{\mathbb{R}+} b^2 \left| \mathscr{W}_{\psi}^M \big[f\big](a,b) \right|^2 \frac{da\,db}{a^2} \right\}^{1/2} \left\{ \int_{\mathbb{R}} \omega^2 \left| \mathscr{L}_M \big[f\big](\omega) \right|^2 d\omega \right\}^{1/2} \geq \frac{|B|\sqrt{C_{\psi}}}{2} \big\| f \big\|_2^2,$$

which is the desired Heisenberg-type uncertainty inequality associated with the linear canonical wavelet transform.   ☐

Remark 4.4.4. For $M = (0,1;-1,0)$, the uncertainty inequality obtained in (4.83) boils down to the Heisenberg-type uncertainty inequality for the classical wavelet transform. Moreover, for $M = (\cos\alpha, \sin\alpha; -\sin\alpha, \cos\alpha)$, $\alpha \neq n\pi$, $n \in \mathbb{Z}$, the inequality (4.83) reduces to the Heisenberg-type uncertainty inequality for the fractional wavelet transform.

4.4.2 Constant Q-property and resolution of the convolution-based linear canonical wavelet transform

"Only a few know, how much one must know, to know how little one knows."

-Werner K. Heisenberg

This subsection is intended to study the resolution of the linear canonical wavelet transform (4.65) in the time-linear-canonical-frequency plane . By virtue of Proposition 4.4.1, it is quite evident that each linear canonical wavelet component $\psi_{a,b}^M(t)$ acts as a differently scaled band-pass filter in the linear canonical domain. Thus, Definition 4.4.1 and Lemma 4.4.3 imply that the resolving power of the linear canonical wavelet transform is determined by the daughter wavelets $\psi_{a,b}^M(t)$; that is, if the daughter wavelets are supported in the time domain or linear canonical domain, then the linear canonical wavelet transform defined in (4.65) is accordingly supported in the respective domains. For an explicit description of the preceding arguments, consider a wavelet $\psi(t)$ centred about the point E_{ψ} and having a radius (dispersion) of Δ_{ψ} in the time domain. Consequently, the centre and radius of each of the daughter wavelets $\psi_{a,b}^M(t)$ are given by

$$E\Big[\psi_{a,b}^M(t)\Big] = \frac{\int_{\mathbb{R}} t \left| \psi_{a,b}^M(t) \right|^2 dt}{\int_{\mathbb{R}} \left| \psi_{a,b}^M(t) \right|^2 dt} = \frac{\int_{\mathbb{R}} t \left| \psi_{a,b}(t) \right|^2 dt}{\int_{\mathbb{R}} \left| \psi_{a,b}(t) \right|^2 dt} = E\Big[\psi_{a,b}(t)\Big] = aE_{\psi} + b \tag{4.90}$$

and

$$
\Delta\left[\psi_{a,b}^{M}(t)\right] = \left(\frac{\displaystyle\int_{\mathbb{R}} (t-b-aE_{\psi})\left|\psi_{a,b}^{M}(t)\right|^{2}dt}{\displaystyle\int_{\mathbb{R}}\left|\psi_{a,b}^{M}(t)\right|^{2}dt}\right)^{1/2} = \left(\frac{\displaystyle\int_{\mathbb{R}} (t-b-aE_{\psi})\left|\psi_{a,b}(t)\right|^{2}dt}{\displaystyle\int_{\mathbb{R}}\left|\psi_{a,b}(t)\right|^{2}dt}\right)^{1/2}
$$

$$
= \Delta\left[\psi_{a,b}(t)\right] = a\Delta_{\psi}. \tag{4.91}
$$

Therefore, by virtue of (4.90) and (4.91), we conclude that the linear canonical wavelet transform of any function $f \in L^{2}(\mathbb{R})$ with respect to the analyzing wavelet $\psi \in L^{2}(\mathbb{R})$ offers the localized information in the time window

$$
\left[aE_{\psi} + b - a\Delta_{\psi}, \; aE_{\psi} + b + a\Delta_{\psi}\right], \tag{4.92}
$$

having a net width of $2a\Delta_{\psi}$, which is directly proportional to the scale a. Moreover, Lemma 4.4.3 implies that the linear canonical spectrum of the linear canonical wavelet transform is characterized by a scalable window $\mathscr{F}[\psi](a\omega/B)$ dictated by the free parameter $B \in \mathbb{R}\backslash\{0\}$. Thus, if $E_{\mathscr{F}[\psi](a\omega/B)}$ and $\Delta_{\mathscr{F}[\psi](a\omega/B)}$ denote the centre and radius of the scaled window $\mathscr{F}[\psi](a\omega/B)$, capturing the linear canonical spectrum, then we can easily deduce that

$$
E_{\mathscr{F}[\psi](a\omega/B)} = B\left(\frac{E_{\mathscr{F}[\psi](\omega)}}{a}\right) \quad \text{and} \quad \Delta_{\mathscr{F}[\psi](a\omega/B)} = |B|\left(\frac{\Delta_{\mathscr{F}[\psi](\omega)}}{a}\right). \tag{4.93}
$$

Consequently, the Q-factor of the linear canonical wavelet transform (4.65) is obtained as

$$
Q = \frac{\text{Radius of linear-canonical-frequency window}}{\text{Centre of linear-canonical-frequency window}} = \frac{\Delta_{\mathscr{F}[\psi](a\omega/B)}}{E_{\mathscr{F}[\psi](a\omega/B)}} = \pm\frac{\Delta_{\mathscr{F}[\psi](\omega)}}{E_{\mathscr{F}[\psi](\omega)}}, \tag{4.94}
$$

which is clearly independent of both the scaling parameter a as well as the free parameter $B \in \mathbb{R}\backslash\{0\}$ and is thus constant. This is the constant Q-property of the linear canonical wavelet transform. Moreover, from (4.93), it follows that the linear canonical wavelet transform $\mathscr{W}_{\psi}^{M}[f](a,b)$ localizes the linear canonical spectral contents of a signal $f \in L^{2}(\mathbb{R})$ within the flexible window

$$
\left[B\left(\frac{E_{\mathscr{F}[\psi](\omega)}}{a}\right) - |B|\left(\frac{\Delta_{\mathscr{F}[\psi](\omega)}}{a}\right), \; B\left(\frac{E_{\mathscr{F}[\psi](\omega)}}{a}\right) + |B|\left(\frac{\Delta_{\mathscr{F}[\psi](\omega)}}{a}\right)\right], \tag{4.95}
$$

having a net width of $2|B|\,a^{-1}\Delta_{\mathscr{F}[\psi](\omega)}$. Thus, the net time-linear-canonical-frequency resolution of the linear canonical wavelet transform defined in (4.65) is given by the window

$$
\left[aE_{\psi} + b - a\Delta_{\psi}, \; aE_{\psi} + b + a\Delta_{\psi}\right]
$$

$$
\times \left[B\left(\frac{E_{\mathscr{F}[\psi](\omega)}}{a}\right) - |B|\left(\frac{\Delta_{\mathscr{F}[\psi](\omega)}}{a}\right), \; B\left(\frac{E_{\mathscr{F}[\psi](\omega)}}{a}\right) + |B|\left(\frac{\Delta_{\mathscr{F}[\psi](\omega)}}{a}\right)\right], \tag{4.96}
$$

occupying a net area of $4|B|\,\Delta_{\psi}\Delta_{\mathscr{F}[\psi](\omega)}$ in the time-linear-canonical-frequency plane as shown in Figure 4.3. Moreover, it is worth noticing that the area of the window (4.96) depends only upon the mother wavelet $\psi \in L^{2}(\mathbb{R})$ and is completely independent of both the scaling parameter a and the translation parameter b. However, note that the area of the window (4.96) is dictated by the free parameter $B \in \mathbb{R}\backslash\{0\}$ determined by the given unimodular matrix $M = (A, B; C, D)$. The free parameter $B \in \mathbb{R}\backslash\{0\}$ can be of substantial importance in optimizing the signal concentration of the linear canonical wavelet transform in the time-linear-canonical-frequency plane. Hence, we conclude that the linear canonical wavelet transform defined in (4.65) is of critical significance in signal analysis, especially in the processing of those signals whose energy is not well-concentrated in the Fourier domain, for instance chirp-like which often arise in modern communication systems.

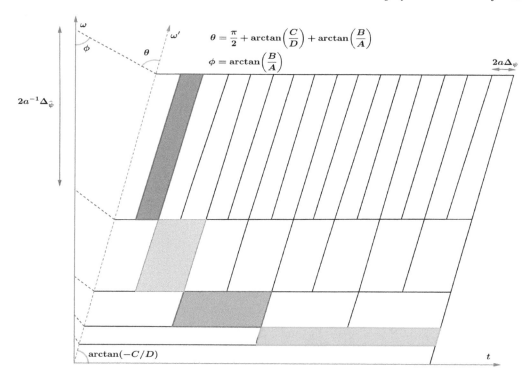

FIGURE 4.3: Tiling on the time-linear-canonical-frequency plane due to linear canonical wavelet transform.

4.4.3   Composition of linear canonical wavelet transforms

"When I have clarified and exhausted a subject, then I turn away from it, in order to go into darkness again."

-Carl F. Gauss

The main objective of this subsection is to study the composition of linear canonical wavelet transforms, which can be thought of as the successive application of two linear canonical wavelet transforms with respect to a distinct pair of wavelets. Besides, the formulation of the composition of linear canonical wavelet transforms, we shall also obtain the corresponding orthogonality relation and inversion formula.

As a consequence of Lemma 4.4.3, the linear canonical wavelet transform of any $f \in L^2(\mathbb{R})$ with respect to the wavelet $\psi \in L^2(\mathbb{R})$ can be expressed as

$$\mathscr{W}_\psi^M \big[f\big](a,b) = \sqrt{2\pi a} \int_\mathbb{R} \mathscr{L}_M \big[f\big](\omega) \, \overline{\mathscr{F}\big[\psi\big] \left(\frac{a\omega}{B}\right)} \, \overline{\mathcal{K}_M(b,\omega)} \, d\omega. \tag{4.97}$$

Moreover, for any other wavelet $\phi \in L^2(\mathbb{R})$ and $(a',b') \in \mathbb{R}^+ \times \mathbb{R}$, we can express the linear canonical transform of $\mathscr{W}_\phi^M \big[f\big](a',b')$, with respect to the translation variable b' as follows:

$$\mathscr{L}_M \left(\mathscr{W}_\phi^M \big[f\big](a',b')\right)(\omega) = \sqrt{2\pi a'} \, \mathscr{L}_M \big[f\big](\omega) \, \overline{\mathscr{F}\big[\phi\big] \left(\frac{a'\omega}{B}\right)}. \tag{4.98}$$

Based on (4.97) and (4.98), we define the composition of linear canonical wavelet transforms of any $f \in L^2(\mathbb{R})$ with respect to the wavelet pair (ψ, ϕ) in the following fashion:

$$CW_{(\psi,\phi)}^M [f](a, a', b) = \mathscr{W}_\psi^M \left(\mathscr{W}_\phi^M [f](a', b') \right)(a, b). \tag{4.99}$$

Equivalently, the composition of linear canonical wavelet transforms given by (4.99) can also be expressed as

$$CW_{(\psi,\phi)}^M [f](a, a', b) = \sqrt{2\pi a} \int_{\mathbb{R}} \mathscr{L}_M \left(\mathscr{W}_\phi^M [f](a', b') \right)(\omega) \overline{\mathscr{F}[\psi]\left(\frac{a\omega}{B}\right)} \overline{\mathcal{K}_M(b, \omega)} \, d\omega$$

$$= 2\pi \sqrt{a'a} \int_{\mathbb{R}} \mathscr{L}_M [f](\omega) \overline{\mathscr{F}[\psi]\left(\frac{a\omega}{B}\right)} \overline{\mathscr{F}[\phi]\left(\frac{a'\omega}{B}\right)} \overline{\mathcal{K}_M(b, \omega)} \, d\omega. \tag{4.100}$$

Having formulated the notion of composition of linear canonical wavelet transforms, we shall obtain an orthogonality relation and the inversion formula associated with composition of linear canonical wavelet transforms defined in (4.99).

Theorem 4.4.10. *If* $CW_{(\psi,\phi)}^M [f](a, a', b)$ *and* $CW_{(\psi,\phi)}^M [g](a, a', b)$ *are the compositions of wavelet transforms corresponding to a given pair of functions* $f, g \in L^2(\mathbb{R})$ *with respect to the wavelet pair* (ψ, ϕ), *then the following orthogonality relation holds:*

$$\int_{\mathbb{R}} \int_{\mathbb{R}^+} \int_{\mathbb{R}^+} CW_{(\psi,\phi)}^M [f](a, a', b) \overline{CW_{(\psi,\phi)}^M [g](a, a', b)} \frac{da \, da' \, db}{a^2 a'^2} = C_{(\psi,\phi)} \left\langle f, g \right\rangle_2, \tag{4.101}$$

where $C_{(\psi,\phi)}$ *is given by the cross admissibility condition of the wavelet pair* (ψ, ϕ):

$$C_{(\psi,\phi)} = (2\pi)^2 \int_{\mathbb{R}} \int_{\mathbb{R}} \frac{|\mathscr{F}[\psi](\omega)|^2 |\mathscr{F}[\phi](\omega')|^2}{|\omega\omega'|} \, d\omega \, d\omega' < \infty. \tag{4.102}$$

Proof. Using the definition of composition of linear canonical wavelet transforms given in (4.100), we have

$$\int_{\mathbb{R}} \int_{\mathbb{R}^+} \int_{\mathbb{R}^+} CW_{(\psi,\phi)}^M [f](a, a', b) \overline{CW_{(\psi,\phi)}^M [g](a, a', b)} \frac{da \, da' \, db}{a^2 a'^2}$$

$$= \int_{\mathbb{R}} \int_{\mathbb{R}^+} \int_{\mathbb{R}^+} \left\{ 2\pi\sqrt{a'a} \int_{\mathbb{R}} \mathscr{L}_M [f](\omega) \overline{\mathscr{F}[\psi]\left(\frac{a\omega}{B}\right)} \overline{\mathscr{F}[\phi]\left(\frac{a'\omega}{B}\right)} \overline{\mathcal{K}_M(b, \omega)} \, d\omega \right\}$$

$$\times \left\{ 2\pi\sqrt{a'a} \int_{\mathbb{R}} \overline{\mathscr{L}_M [g](\omega')} \mathscr{F}[\psi]\left(\frac{a\omega'}{B}\right) \mathscr{F}[\phi]\left(\frac{a'\omega'}{B}\right) \mathcal{K}_M(b, \omega') \, d\omega' \right\} \frac{da \, da' \, db}{a^2 a'^2}$$

$$= (2\pi)^2 \int_{\mathbb{R}} \int_{\mathbb{R}} \int_{\mathbb{R}^+} \int_{\mathbb{R}^+} \mathscr{L}_M [f](\omega) \overline{\mathscr{L}_M [g](\omega')} \mathscr{F}[\psi]\left(\frac{a\omega'}{B}\right) \overline{\mathscr{F}[\psi]\left(\frac{a\omega}{B}\right)}$$

$$\times \mathscr{F}[\phi]\left(\frac{a'\omega'}{B}\right) \overline{\mathscr{F}[\phi]\left(\frac{a'\omega'}{B}\right)} \left\{ \int_{\mathbb{R}} \mathcal{K}_M(b, \omega') \overline{\mathcal{K}_M(b, \omega)} \, db \right\} \frac{da \, da' \, d\omega \, d\omega'}{aa'}$$

$$= (2\pi)^2 \int_{\mathbb{R}} \int_{\mathbb{R}} \int_{\mathbb{R}^+} \int_{\mathbb{R}^+} \mathscr{L}_M [f](\omega) \overline{\mathscr{L}_M [g](\omega')} \mathscr{F}[\psi]\left(\frac{a\omega'}{B}\right) \overline{\mathscr{F}[\psi]\left(\frac{a\omega}{B}\right)}$$

$$\times \mathscr{F}[\phi]\left(\frac{a'\omega}{B}\right) \overline{\mathscr{F}[\phi]\left(\frac{a'\omega'}{B}\right)} \delta(\omega - \omega') \frac{da \, da' \, d\omega \, d\omega'}{aa'}$$

$$= (2\pi)^2 \int_{\mathbb{R}} \int_{\mathbb{R}^+} \int_{\mathbb{R}^+} \mathscr{L}_M [f](\omega) \overline{\mathscr{L}_M [g](\omega)} \left|\mathscr{F}[\psi]\left(\frac{a\omega}{B}\right)\right|^2 \left|\mathscr{F}[\phi]\left(\frac{a'\omega}{B}\right)\right|^2 \frac{da \, da' \, d\omega}{aa'}$$

$$= \int_{\mathbb{R}} \mathscr{L}_M\big[f\big](\omega)\, \overline{\mathscr{L}_M\big[g\big](\omega)} \left\{ (2\pi)^2 \int_{\mathbb{R}^+}\int_{\mathbb{R}^+} \left|\mathscr{F}\big[\psi\big]\left(\frac{a\omega}{B}\right)\right|^2 \left|\mathscr{F}\big[\phi\big]\left(\frac{a'\omega}{B}\right)\right|^2 \frac{da\,da'}{aa'} \right\} d\omega.$$

$$(4.103)$$

In order to simplify the inner integral on the R.H.S of (4.103), we shall make use of the substitution $a\omega/B = \xi$ and $a'\omega/B = \eta$, so that

$$(2\pi)^2 \int_{\mathbb{R}^+}\int_{\mathbb{R}^+} \left|\mathscr{F}\big[\psi\big]\left(\frac{a\omega}{B}\right)\right|^2 \left|\mathscr{F}\big[\phi\big]\left(\frac{a'\omega}{B}\right)\right|^2 \frac{da\,da'}{aa'}$$

$$= (2\pi)^2 \int_{\mathbb{R}}\int_{\mathbb{R}} \frac{\big|\mathscr{F}\big[\psi\big](\xi)\big|^2 \big|\mathscr{F}\big[\psi\big](\eta)\big|^2}{\xi\eta}\, d\xi\, d\eta$$

$$= C_{(\psi,\phi)}.$$

$$(4.104)$$

Plugging the estimate obtained in (4.104) into (4.103) and then invoking the Parseval's formula for the linear canonical transform, we get

$$\int_{\mathbb{R}}\int_{\mathbb{R}^+}\int_{\mathbb{R}^+} \mathcal{CW}^M_{(\psi,\phi)}\big[f\big](a,a',b)\, \overline{\mathcal{CW}^M_{(\psi,\phi)}\big[g\big](a,a',b)}\, \frac{da\,da'\,db}{a^2 a'^2} = C_{(\psi,\phi)} \big\langle f, g \big\rangle_2.$$

This completes the proof of Theorem 4.4.10. $\qquad\qquad\qquad\qquad\qquad\qquad\qquad\square$

Remark 4.4.5. For $f = g$, the orthogonality relation (4.101) yields the energy preserving relation for the composition of linear canonical wavelet transforms:

$$\int_{\mathbb{R}}\int_{\mathbb{R}^+}\int_{\mathbb{R}^+} \left|\mathcal{CW}^M_{(\psi,\phi)}\big[f\big](a,a',b)\right|^2 \frac{da\,da'\,db}{a^2 a'^2} = C_{(\psi,\phi)} \big\|f\big\|_2^2.$$

$$(4.105)$$

Moreover, the operator $\mathcal{CW}^M_{(\psi,\phi)}\big[f\big](a,a',b)$ is bounded, linear and for $C_{\psi,\phi} = 1$, it turns to be an isometry from $L^2(\mathbb{R})$ to $L^2(\mathbb{R}^+ \times \mathbb{R}^+ \times \mathbb{R})$.

In the next theorem, we formulate an inversion formula pertaining to the composition of linear canonical wavelet transforms (4.99).

Theorem 4.4.11. *Any function $f \in L^2(\mathbb{R})$ can be reconstructed from the composition of linear canonical wavelet transforms $\mathcal{CW}^M_{(\psi,\phi)}\big[f\big](a,a',b)$ via the following formula:*

$$f(t) = C^{-1}_{(\psi,\phi)} \int_{\mathbb{R}}\int_{\mathbb{R}^+}\int_{\mathbb{R}^+} \mathcal{CW}^M_{(\psi,\phi)}\big[f\big](a,a',b)\, \overline{\mathcal{CW}^M_{(\psi,\phi)}\big[\delta(\cdot - t)\big](a,a',b)}\, \frac{da\,da'\,db}{a^2 a'^2}, \quad (4.106)$$

where $C_{(\psi,\phi)} \neq 0$ is given by (4.102).

Proof. For the Dirac function $g_t(x) = \delta(t - x)$, $t \in \mathbb{R}$, we can write

$$g_t(x) = \int_{\mathbb{R}} \mathcal{K}_M(t,\omega)\, \overline{\mathcal{K}_M(x,\omega)}\, d\omega,$$

$$(4.107)$$

so that $\mathscr{L}_M\big[g_t\big](\omega) = \mathcal{K}_M(t,\omega)$. By virtue of the Parseval's formula for the linear canonical transform and the orthogonality relation (4.101), we have

$$f(t) = \int_{\mathbb{R}} \mathscr{L}_M\big[f\big](\omega)\, \overline{\mathcal{K}_M(t,\omega)}\, d\omega$$

$$= \int_{\mathbb{R}} \mathscr{L}_M\big[f\big](\omega)\, \overline{\mathscr{L}_M\big[g_t\big](\omega)}\, d\omega$$

$$= C_{(\psi,\phi)}^{-1} \int_{\mathbb{R}} \int_{\mathbb{R}^+} \int_{\mathbb{R}^+} \mathcal{CW}_{(\psi,\phi)}^{M}\Big[f\Big](a,a',b) \overline{\mathcal{CW}_{(\psi,\phi)}^{M}\Big[g_t\Big](a,a',b)} \, \frac{da \, da' \, db}{a^2 a'^2}. \qquad (4.108)$$

As a consequence of (4.108), we conclude that the reconstruction formula is given by

$$f(t) = C_{(\psi,\phi)}^{-1} \int_{\mathbb{R}} \int_{\mathbb{R}^+} \int_{\mathbb{R}^+} \mathcal{CW}_{(\psi,\phi)}^{M}\Big[f\Big](a,a',b) \overline{\mathcal{CW}_{(\psi,\phi)}^{M}\Big[\delta(\cdot - t)\Big](a,a',b)} \, \frac{da \, da' \, db}{a^2 a'^2}.$$

This completes the proof of Theorem 4.4.11. $\qquad\qquad\qquad\qquad\qquad\qquad\square$

4.4.4 Modified linear canonical wavelet transform

"I seem to have been only like a boy playing on the seashore, and diverting myself in now and then finding a smoother pebble or a prettier shell than ordinary, whilst the great ocean of truth lay all undiscovered before me."

-Isaac Newton

In this subsection, we shall present another variant of the linear canonical wavelet transform primarily defined in (4.65) by appropriately modifying the family of daughter wavelets (4.64) as follows:

$$\mathfrak{F}_{\psi}^{M}(a,b) := \left\{ \psi_{a,b}^{M}(t) = \frac{1}{\sqrt{a}} \psi\left(\frac{t-b}{a}\right) \exp\left\{ \frac{iA}{2B}\left(\left(\frac{b}{a}\right)^2 - t^2\right)\right\}; \ a \in \mathbb{R}^+, \, b \in \mathbb{R} \right\}. \qquad (4.109)$$

The use of scale-dependent complex amplitude $e^{iAb^2/2Ba^2}$ in lieu to $e^{iAb^2/2B}$ has the effect of eliminating the redundancy of the linear canonical wavelet transform when applied for signal decomposition and reconstruction. It is worth mentioning that such a modification in the family of daughter wavelets $\mathfrak{F}_{\psi}^{M}(a,b)$ does not affect on the partitioning of the time-frequency cells in the time-frequency plane. Moreover, the novel daughter wavelets $\psi_{a,b}^{M}(t)$, $a \in \mathbb{R}^+$, $b \in \mathbb{R}$ still act as band-pass filters in the linear canonical domain. For a detailed analysis regarding the significance of the modifications carried out in the family of analyzing functions (4.109), we refer the interest reader to [305].

Based on the modified linear canonical wavelets defined in (4.109), we have the following definition of the modified linear canonical wavelet transform:

Definition 4.4.2. Given a unimodular matrix $M = (A, B; C, D)$, the modified linear canonical wavelet transform of any $f \in L^2(\mathbb{R})$ with respect to the wavelet $\psi \in L^2(\mathbb{R})$ is denoted by $\mathcal{W}_{\psi}^{M}[f]$ and is defined as

$$\mathcal{W}_{\psi}^{M}\Big[f\Big](a,b) = \Big\langle f, \psi_{a,b}^{M}\Big\rangle_2 = \int_{\mathbb{R}} f(t) \overline{\psi_{a,b}^{M}(t)} \, dt, \qquad (4.110)$$

where $\psi_{a,b}^{M}(t)$ is given by (4.109).

Below, we mention some important points regarding the modified linear canonical wavelet transform given in Definition 4.4.2.

(i) For $M = (0, 1; -1, 0)$, Definition 4.4.2 reduces to the definition of the classical wavelet transform given in (3.47).

(ii) For $M = (\cos\alpha, \sin\alpha; -\sin\alpha, \cos\alpha)$, $\alpha \neq n\pi$, $n \in \mathbb{Z}$, Definition 4.4.2 yields a new variant of the fractional wavelet transform.

(iii) The modified linear canonical wavelet transform (4.110) can also be expressed in terms of the classical wavelet transform by appropriate chirp multiplications in the following fashion:

$$\mathcal{W}_\psi^M \big[f\big](a,b) = \exp\left\{-\frac{iA}{2B}\left(\frac{b}{a}\right)^2\right\} \int_{\mathbb{R}} \exp\left\{\frac{iAt^2}{2B}\right\} f(t)\, \overline{\psi\left(\frac{t-b}{a}\right)}\, dt$$

$$= \exp\left\{-\frac{iA}{2B}\left(\frac{b}{a}\right)^2\right\} \mathcal{W}_\psi \big[F_M\big](a,b), \qquad (4.111)$$

where

$$F_M(t) = \exp\left\{\frac{iAt^2}{2B}\right\} f(t). \qquad (4.112)$$

That is, the modified linear canonical wavelet transform defined in (4.110) has the same computational complexity as that of the classical wavelet transform.

(iv) Due to the extra degrees of freedom inherited from the conventional linear canonical transform, the modified linear canonical wavelet transform (4.110) enjoys both high concentrations and tunable resolutions when dealing with chirp signals.

(v) The generating function $\psi \in L^2(\mathbb{R})$ used for the construction of modified linear canonical wavelets in (4.109) is primarily a usual wavelet in the sense that it satisfies the classical admissibility condition (3.27). That is, the modified linear canonical wavelet transform defined in (4.110) does not require any extra conditions to be imposed on the mother wavelet $\psi(t)$.

As a consequence of the following proposition, we shall demonstrate that the modified linear canonical wavelets can be regarded as band-pass filters in the linear canonical domain.

Proposition 4.4.12. *The linear canonical transform of the modified linear canonical wavelets $\psi_{a,b}^M(t)$ defined in (4.109) can be expressed as*

$$\mathcal{L}_M\big[\psi_{a,b}^M\big](\omega) = \sqrt{\frac{a}{iB}}\, \exp\left\{\frac{i(A(b/a)^2 - 2b\omega + D\omega^2)}{2B}\right\} \mathscr{F}\big[\psi\big]\left(\frac{a\omega}{B}\right), \qquad (4.113)$$

where $\mathcal{L}_M$ denotes the linear canonical transform with respect to the unimodular matrix $M = (A, B; C, D)$ and $\mathscr{F}$ denotes the usual Fourier transform.

Proof. Using the definition of linear canonical transform given in (1.164), we have

$$\mathcal{L}_M\big[\psi_{a,b}^M\big](\omega) = \int_{\mathbb{R}} \psi_{a,b}^M(t)\, \mathcal{K}_M(t,\omega)\, dt$$

$$= \frac{1}{\sqrt{2\pi iB}} \int_{\mathbb{R}} \psi_{a,b}^M(t)\, \exp\left\{\frac{i(At^2 - 2t\omega + D\omega^2)}{2B}\right\} dt$$

$$= \frac{1}{\sqrt{2\pi iB}} \int_{\mathbb{R}} \frac{1}{\sqrt{a}}\, \psi\left(\frac{t-b}{a}\right) \exp\left\{\frac{iA}{2B}\left(\left(\frac{b}{a}\right)^2 - t^2\right)\right\}$$

$$\times \exp\left\{\frac{i(At^2 - 2t\omega + D\omega^2)}{2B}\right\} dt$$

$$= \frac{1}{\sqrt{2\pi iB}} \int_{\mathbb{R}} \frac{1}{\sqrt{a}}\, \psi\left(\frac{t-b}{a}\right) \exp\left\{\frac{iA}{2B}\left(\frac{b}{a}\right)^2\right\} \exp\left\{\frac{i(D\omega^2 - 2t\omega)}{2B}\right\} dt.$$

$$(4.114)$$

Substituting $(t - b)/a = z$ in (4.114) yields the following expression:

$$
\begin{aligned}
\mathscr{L}_M\left[\psi_{a,b}^M\right](\omega) &= \sqrt{\frac{a}{2\pi i B}} \int_{\mathbb{R}} \psi(z) \exp\left\{\frac{i\left(A(b/a)^2 - 2b\omega + D\omega^2\right)}{2B}\right\} \exp\left\{-\frac{ia\omega z}{B}\right\} dz \\
&= \sqrt{\frac{a}{iB}} \exp\left\{\frac{i\left(A(b/a)^2 - 2b\omega + D\omega^2\right)}{2B}\right\} \mathscr{F}[\psi]\left(\frac{a\omega}{B}\right).
\end{aligned}
$$

This completes the proof of Proposition 4.4.12. □

Remark 4.4.6. The linear canonical transform of the modified linear canonical wavelets can also be expressed as

$$
\begin{aligned}
\mathscr{L}_M\left[\psi_{a,b}^M\right](\omega) &= \sqrt{\frac{a}{iB}} \exp\left\{\frac{i\left(D\omega^2 - D(a\omega)^2\right)}{2B}\right\} \\
&\quad \times \exp\left\{\frac{i\left(A(b/a)^2 - 2(b/a)(a\omega) + D(a\omega)^2\right)}{2B}\right\} \mathscr{F}[\psi]\left(\frac{a\omega}{B}\right) \\
&= \sqrt{2\pi a}\, \exp\left\{\frac{i\left(D\omega^2 - D(a\omega)^2\right)}{2B}\right\} \mathscr{F}[\psi]\left(\frac{a\omega}{B}\right) \mathcal{K}_M\left(\frac{b}{a}, a\omega\right). \quad (4.115)
\end{aligned}
$$

Next, we establish an orthogonality relation for the modified linear canonical wavelet transform defined in (4.110).

Theorem 4.4.13. *If* $\mathcal{W}_\psi^M[f](a,b)$ *and* $\mathcal{W}_\psi^M[g](a,b)$ *are the modified linear canonical wavelet transforms of any pair of functions* $f, g \in L^2(\mathbb{R})$, *then we have*

$$
\int_{\mathbb{R}} \int_{\mathbb{R}^+} \mathcal{W}_\psi^M[f](a,b) \overline{\mathcal{W}_\psi^M[g](a,b)} \frac{da\, db}{a^2} = C_\psi \left\langle f, g \right\rangle_2, \quad (4.116)
$$

where C_ψ *is given by the classical admissibility condition* (3.27).

Proof. Invoking Proposition 4.4.12, we can express the modified linear canonical wavelet transform of the given pair of functions $f, g \in L^2(\mathbb{R})$ as

$$
\begin{aligned}
\mathcal{W}_\psi^M[f](a,b) &= \left\langle f, \psi_{a,b}^M \right\rangle_2 \\
&= \left\langle \mathscr{L}_M[f], \mathscr{L}_M[\psi_{a,b}^M] \right\rangle_2 \\
&= \sqrt{\frac{a}{-iB}} \int_{\mathbb{R}} \exp\left\{-\frac{i\left(A(b/a)^2 - 2b\omega + D\omega^2\right)}{2B}\right\} \mathscr{L}_M[f](\omega) \overline{\mathscr{F}[\psi]\left(\frac{a\omega}{B}\right)} d\omega
\end{aligned}
$$
$$(4.117)$$

and

$$
\mathcal{W}_\psi^M[g](a,b) = \sqrt{\frac{a}{-iB}} \int_{\mathbb{R}} \exp\left\{-\frac{i\left(A(b/a)^2 - 2b\omega' + D\omega'^2\right)}{2B}\right\} \mathscr{L}_M[g](\omega') \overline{\mathscr{F}[\psi]\left(\frac{a\omega'}{B}\right)} d\omega'.
$$
$$(4.118)$$

Therefore, we can write

$$
\begin{aligned}
&\int_{\mathbb{R}} \int_{\mathbb{R}^+} \mathcal{W}_\psi^M[f](a,b) \overline{\mathcal{W}_\psi^M[g](a,b)} \frac{da\, db}{a^2} \\
&= \int_{\mathbb{R}^+} \int_{\mathbb{R}} \left\{ \sqrt{\frac{a}{-iB}} \int_{\mathbb{R}} \exp\left\{-\frac{i\left(A(b/a)^2 - 2b\omega + D\omega^2\right)}{2B}\right\} \mathscr{L}_M[f](\omega) \overline{\mathscr{F}[\psi]\left(\frac{a\omega}{B}\right)} d\omega \right\}
\end{aligned}
$$

$$\times \left\{ \sqrt{\frac{a}{iB}} \int_{\mathbb{R}} \exp \left\{ \frac{i(A(b/a)^2 - 2b\omega' + D\omega'^2)}{2B} \right\} \overline{\mathscr{L}_M [g](\omega')} \, \mathscr{F} [\psi] \left(\frac{a\omega'}{B} \right) d\omega' \right\} \frac{da\, db}{a^2}$$

$$= \int_{\mathbb{R}^+} \int_{\mathbb{R}} \int_{\mathbb{R}} \exp \left\{ -\frac{iD(\omega^2 - \omega'^2)}{2B} \right\} \mathscr{L}_M [f](\omega) \overline{\mathscr{L}_M [g](\omega')} \, \overline{\mathscr{F} [\psi] \left(\frac{a\omega}{B} \right)} \, \mathscr{F} [\psi] \left(\frac{a\omega'}{B} \right)$$

$$\times 2\pi \left\{ \frac{1}{2\pi |B|} \int_{\mathbb{R}} \exp \left\{ \frac{ib(\omega - \omega')}{B} \right\} db \right\} \frac{da\, d\omega\, d\omega'}{a}$$

$$= 2\pi \int_{\mathbb{R}^+} \int_{\mathbb{R}} \int_{\mathbb{R}} \mathscr{L}_M [f](\omega) \overline{\mathscr{L}_M [g](\omega')} \, \overline{\mathscr{F} [\psi] \left(\frac{a\omega}{B} \right)} \, \mathscr{F} [\psi] \left(\frac{a\omega'}{B} \right)$$

$$\times \exp \left\{ -\frac{iD(\omega^2 - \omega'^2)}{2B} \right\} \delta(\omega - \omega') \frac{da\, d\omega\, d\omega'}{a}$$

$$= \int_{\mathbb{R}} \mathscr{L}_M [f](\omega) \overline{\mathscr{L}_M [g](\omega)} \left\{ 2\pi \int_{\mathbb{R}^+} \left| \mathscr{F} [\psi] \left(\frac{a\omega}{B} \right) \right|^2 \frac{da}{a} \right\} d\omega. \tag{4.119}$$

Solving the inner integral on the R.H.S of (4.119) by substituting $a\omega/B = \omega'$, we get

$$2\pi \int_{\mathbb{R}^+} \left| \mathscr{F} [\psi] \left(\frac{a\omega}{B} \right) \right|^2 \frac{da}{a} = 2\pi \int_{\mathbb{R}} \frac{|\mathscr{F} [\psi] (\omega')|^2}{|\omega'|} d\omega' = C_\psi. \tag{4.120}$$

Using (4.120) in (4.119) followed by application of the Parseval's formula for linear canonical transform, we obtain the desired orthogonality relation as

$$\int_{\mathbb{R}} \int_{\mathbb{R}^+} \mathcal{W}_\psi^M [f](a,b) \overline{\mathcal{W}_\psi^M [g](a,b)} \frac{da\, db}{a^2} = C_\psi \int_{\mathbb{R}} \mathscr{L}_M [f](\omega) \overline{\mathscr{L}_M [g](\omega)} \, d\omega$$

$$= C_\psi \left\langle \mathscr{L}_M [f], \mathscr{L}_M [g] \right\rangle_2$$

$$= C_\psi \left\langle f, g \right\rangle_2.$$

This completes the proof of Theorem 4.4.13. □

Remark 4.4.7. For $f = g$, the orthogonality relation (4.116) yields the energy preserving relation:

$$\int_{\mathbb{R}} \int_{\mathbb{R}^+} \left| \mathcal{W}_\psi^M [f](a,b) \right|^2 \frac{da\, db}{a^2} = C_\psi \left\| f \right\|_2^2. \tag{4.121}$$

The quantity $\left| \mathcal{W}_\psi^M [f](a,b) \right|^2$ represents the energy density of the signal in the time-linear-canonical-frequency plane and is formally called as the "linear canonical scalogram." Therefore, from (4.121), we infer that the total energy of the transformed signal is a C_ψ-multiple of the energy of the input signal. Moreover, choosing the mother wavelet ψ in such a way that $C_\psi = 1$, the modified linear canonical wavelet transform becomes an isometry from the space of signals $L^2(\mathbb{R})$ to the transformation space $L^2(\mathbb{R}^+ \times \mathbb{R})$.

Next, we present an inversion formula associated with the modified linear canonical wavelet transform (4.110).

Theorem 4.4.14. *Any function* $f \in L^2(\mathbb{R})$ *can be reconstructed from the modified linear canonical wavelet transform* $\mathcal{W}_\psi^M [f](a,b)$ *via the following reconstruction formula:*

$$f(t) = \frac{1}{C_\psi} \int_{\mathbb{R}} \int_{\mathbb{R}^+} \mathcal{W}_\psi^M [f](a,b) \, \psi_{a,b}^M(t) \frac{da\, db}{a^2}, \quad a.e., \tag{4.122}$$

with the factor $C_\psi \neq 0$ *given by the usual admissibility condition (3.27).*

Proof. The reconstruction formula (4.122) is a straightforward consequence of Theorem 4.4.13, so we omit the proof. □

We culminate this subsection with the following characterization of the modified linear canonical wavelet transform defined in (4.110):

Theorem 4.4.15. *Let* $(a, b), (a', b') \in \mathbb{R}^+ \times \mathbb{R}$. *Then,* $\mathcal{W}_\psi^M[f](a, b)$ *is the modified linear canonical wavelet transform of a certain signal* $f \in L^2(\mathbb{R})$ *if and only if it satisfies the following reproducing kernel equation:*

$$\mathcal{W}_\psi^M[f](a', b') = \int_\mathbb{R} \int_{\mathbb{R}^+} \mathcal{W}_\psi^M[f](a, b) \, \mathcal{K}_\psi^M(a, b; a', b') \, \frac{da \, db}{a^2}, \tag{4.123}$$

where $\mathcal{K}_\psi^M(a, b; a', b')$ *denotes the reproducing kernel and is given by*

$$\mathcal{K}_\psi^M(a, b; a', b') = C_\psi^{-1} \left\langle \psi_{a,b}^M, \psi_{a',b'}^M \right\rangle_2, \tag{4.124}$$

with $C_\psi \neq 0$ *given by the usual admissibility condition* (3.27).

Proof. The proof immediately follows by using the inversion formula (4.122) and noting that

$$\begin{aligned}
\mathcal{W}_\psi^M[f](a', b') &= \int_\mathbb{R} f(t) \, \overline{\psi_{a',b'}^M(t)} \, dt \\
&= \int_\mathbb{R} \left\{ \frac{1}{C_\psi} \int_\mathbb{R} \int_{\mathbb{R}^+} \mathcal{W}_\psi^M[f](a, b) \, \psi_{a,b}^M(t) \, \frac{da \, db}{a^2} \right\} \overline{\psi_{a',b'}^M(t)} \, dt \\
&= \frac{1}{C_\psi} \int_\mathbb{R} \int_{\mathbb{R}^+} \mathcal{W}_\psi^M[f](a, b) \left\{ \int_\mathbb{R} \psi_{a,b}^M(t) \, \overline{\psi_{a',b'}^M(t)} \, dt \right\} \frac{da \, db}{a^2} \\
&= \frac{1}{C_\psi} \int_\mathbb{R} \int_{\mathbb{R}^+} \mathcal{W}_\psi^M[f](a, b) \left\langle \psi_{a,b}^M, \psi_{a',b'}^M \right\rangle_2 \frac{da \, db}{a^2} \\
&= \int_\mathbb{R} \int_{\mathbb{R}^+} \mathcal{W}_\psi^M[f](a, b) \, \mathcal{K}_\psi^M(a, b; a', b') \, \frac{da \, db}{a^2}.
\end{aligned}$$

□

Remark 4.4.8. The reproducing kernel $\mathcal{K}_\psi^M(a, b; a', b')$ given by (4.124) measures the correlation of the two modified linear canonical wavelets $\psi_{a,b}^M(t)$ and $\psi_{a',b'}^M(t)$. According to (4.123), the modified linear canonical wavelet transform of a certain signal $f \in L^2(\mathbb{R})$ about $(a', b') \in \mathbb{R}^+ \times \mathbb{R}$ is always expressible via another $(a, b) \in \mathbb{R}^+ \times \mathbb{R}$ by virtue of the reproducing kernel $\mathcal{K}_\psi^M(a, b; a', b')$, satisfying

$$\begin{aligned}
\left| \mathcal{K}_\psi^M(a, b; a', b') \right| &= C_\psi^{-1} \left| \left\langle \psi_{a,b}^M, \psi_{a',b'}^M \right\rangle_2 \right| \\
&\leq C_\psi^{-1} \left\| \psi_{a,b}^M \right\|_2 \left\| \psi_{a',b'}^M \right\|_2 \\
&= C_\psi^{-1} \left\| \psi \right\|_2 \left\| \psi \right\|_2 \\
&= C_\psi^{-1} \left\| \psi \right\|_2^2, \tag{4.125}
\end{aligned}$$

for all $(a, b), (a', b') \in \mathbb{R}^+ \times \mathbb{R}$.

4.5   The Linear Canonical Stockwell Transform

"It is strange that only extraordinary men make the discoveries, which later appear so easy and simple."

-Georg C. Lichtenberg

This section deals with yet another ramification of the classical Stockwell transform, namely the linear canonical Stockwell transform. The notion of linear canonical Stockwell transform essentially relies on the idea of intertwining the merits of the well-known linear canonical transform with the Stockwell transform, facilitated by an appropriate convolution operation.

4.5.1   Definition and basic properties

In this subsection, we shall formally start the investigation on the linear canonical Stockwell transform by employing the linear canonical convolution operation $\otimes_M$ defined in (1.174), which nicely generalizes both the classical and fractional analogues of the Stockwell transform. The main advantage of the linear canonical Stockwell transform is that besides allowing the signal analysis along non-orthogonal directions, with absolutely referenced phase information in the time-frequency plane, each time-frequency cell is sheared by a certain degree determined by the free parameters of the 2×2, real, unimodular matrix $M = (A, B; C, D)$ involved in the definition of the linear canonical transform. Therefore, the linear canonical Stockwell transform can play a pivotal role in the analysis of signals which reveal their characteristics beyond the traditional Fourier domain. Such classes of signals frequently arise in radar, sonar and many other modern-day communication systems.

For a given real, unimodular matrix $M = (A, B; C, D)$ and a window function $\psi \in L^2(\mathbb{R})$, we define a novel family of analyzing functions as under:

$$\mathcal{F}_\psi^M(a,b) := \left\{ \psi_{a,b}^M(t) = |a|\, \psi\big(a(t-b)\big) \exp\left\{ iat - \frac{iA(t^2 - b^2)}{2B} \right\} \right\};\ a \in \mathbb{R} \setminus \{0\},\ b \in \mathbb{R} \right\}.$$

$$(4.126)$$

Note that the family of analyzing functions defined in (4.126) is governed the parameters of the given unimodular matrix M in such a way that for $M = (0, 1; -1, 0)$ it reduces to the classical version $\mathscr{F}_\psi(a,b)$ given by (3.78), whereas for $M = (\cos\alpha, \sin\alpha; -\sin\alpha, \cos\alpha)$, $\alpha \neq n\pi$, $n \in \mathbb{Z}$, it yields the fractional version $\mathcal{F}_\psi^\alpha(a,b)$ defined in (4.31). Hence, we conclude that the novel family of analyzing functions given by (4.126) circumscribes both the classical and fractional variants.

To justify the construction of the family of analyzing functions (4.126), for $a \in \mathbb{R} \setminus \{0\}$ and $b \in \mathbb{R}$, let $\mathscr{D}_a$, $\mathcal{M}_a$ and $\mathcal{T}_b$ denote the dilation, modulation and translation operators as defined in (4.29). Then, the continuous Stockwell transform $\mathcal{S}_\psi[f](a,b)$ of any $f \in L^2(\mathbb{R})$ with respect to the window function $\psi \in L^2(\mathbb{R})$ can be expressed via the classical convolution operation $*$ by virtue of the expression (4.30). Therefore, in order to formulate the notion of linear canonical Stockwell transform, we need to replace the classical convolution operation $*$ in (4.30) with the linear canonical convolution operation $\otimes_M$ defined in (1.174). In doing so, the family of analyzing functions involved in the classical Stockwell transform is modified to a novel family of analyzing functions, which is precisely the one defined in (4.126). More formally, we have the following definition:

Definition 4.5.1. Given a unimodular matrix $M = (A, B; C, D)$, the linear canonical Stockwell transform of any $f \in L^2(\mathbb{R})$ with respect to the window function $\psi \in L^2(\mathbb{R})$ is denoted by $\mathcal{S}_\psi^M[f]$ and is defined as

$$\mathcal{S}_\psi^M[f](a,b) = \frac{1}{\sqrt{2\pi}} \left((\mathcal{M}_{-a}f) \otimes_M \left(\mathcal{D}_a \breve{\psi} \right) \right)(b) = \frac{1}{\sqrt{2\pi}} \int_\mathbb{R} f(t)\, \overline{\psi_{a,b}^M(t)}\, dt, \qquad (4.127)$$

where $\breve{\psi}(t) = \psi(-t)$ and the analyzing functions $\psi_{a,b}^M(t)$ are given by (4.126).

Below, we mention some notable features of the linear canonical Stockwell transform given by Definition 4.5.1.

(i) For $M = (0, 1; -1, 0)$, Definition 4.5.1 boils down to the definition of the classical Stockwell transform given in (3.79).

(ii) For $M = (\cos\alpha, \sin\alpha; -\sin\alpha, \cos\alpha)$, $\alpha \neq n\pi$, $n \in \mathbb{Z}$, the linear canonical Stockwell transform given by Definition 4.5.1 reduces to the fractional Stockwell transform (4.32).

(iii) The linear canonical Stockwell transform (4.127) can also be rewritten in the following fashion:

$$\mathcal{S}_\psi^M[f](a,b) = \frac{|a|}{\sqrt{2\pi}} \int_\mathbb{R} f(t) \left\{ \overline{\psi(a(t-b))} \exp\left\{ \frac{iA(t^2 - b^2)}{2B} \right\} \right\} e^{-iat}\, dt. \qquad (4.128)$$

From expression (4.128), it follows that the linear canonical Stockwell transform is reliant upon a chirp-modulated window function, which is dictated by the parameters A and B appearing in the given unimodular matrix $M = (A, B; C, D)$. Therefore, by varying these parameters, one can achieve optimized concentrations of the Stockwell spectrum, especially for the class of signals whose energy is well-concentrated only beyond the Fourier domain.

(iv) The linear canonical Stockwell transform defined in (4.127) can be expressed via the classical Stockwell transform (3.79) as follows:

$$\mathcal{S}_\psi^M[f](a,b) = \frac{|a|}{\sqrt{2\pi}} \exp\left\{ -\frac{iAb^2}{2B} \right\} \int_\mathbb{R} \exp\left\{ \frac{iAt^2}{2B} \right\} f(t)\, \overline{\psi(a(t-b))}\, e^{-iat}\, dt$$

$$= \exp\left\{ -\frac{iAb^2}{2B} \right\} \mathcal{S}_\psi[F_M](a,b), \qquad (4.129)$$

where

$$F_M(t) = \exp\left\{ \frac{iAt^2}{2B} \right\} f(t). \qquad (4.130)$$

Therefore, by virtue of (4.129), it is quite evident that the linear canonical Stockwell transform (4.127) is easy to implement as the computational complexity is completely determined by that of the classical Stockwell transform (3.79).

In the following proposition, we shall compute the linear canonical transform of the analyzing functions given in (4.126).

Proposition 4.5.1. *For any window function* $\psi \in L^2(\mathbb{R})$, *the linear canonical transform of the daughter functions* $\psi_{a,b}^M(t)$ *defined in (4.126) satisfies:*

$$\mathcal{L}_M[\psi_{a,b}^M](\omega) = \sqrt{2\pi a}\, \mathcal{F}[\psi]\left(\frac{a\omega}{B} \right) \mathcal{K}_M(b, \omega), \qquad (4.131)$$

where $\mathcal{L}_M$ *denotes the linear canonical transform with respect to the unimodular matrix* $M = (A, B; C, D)$ *and* $\mathcal{F}$ *denotes the usual Fourier transform.*

Proof. Using the definition of linear canonical transform given in (1.164), we have

$$
\mathscr{L}_M\Big[\psi_{a,b}^M\Big](\omega) = \int_{\mathbb{R}} \psi_{a,b}^M(t)\, \mathcal{K}_M(t,\omega)
$$

$$
= \frac{1}{\sqrt{2\pi i B}} \int_{\mathbb{R}} \psi_{a,b}^M(t)\, \exp\left\{ \frac{i(At^2 - 2t\omega + D\omega^2)}{2B} \right\} dt
$$

$$
= \frac{1}{\sqrt{2\pi i B}} \int_{\mathbb{R}} |a|\, \psi\big(a(t-b)\big)\, \exp\left\{ iat - \frac{iA(t^2 - b^2)}{2B} \right\}
$$

$$
\times \exp\left\{ \frac{i(At^2 - 2t\omega + D\omega^2)}{2B} \right\} dt
$$

$$
= \frac{1}{\sqrt{2\pi i B}} \int_{\mathbb{R}} |a|\, \psi\big(a(t-b)\big)\, \exp\left\{ iat + \frac{iAb^2}{2B} \right\} \exp\left\{ \frac{i(D\omega^2 - 2t\omega)}{2B} \right\} dt.
$$

$$(4.132)$$

Making use of the substitution $a(t-b) = z$ in (4.132), we obtain

$$
\mathscr{L}_M\Big[\psi_{a,b}^M\Big](\omega) = \frac{e^{iab}}{\sqrt{2\pi i B}} \int_{\mathbb{R}} \psi(z)\, \exp\left\{ \frac{i(Ab^2 - 2b\omega + D\omega^2)}{2B} \right\} \exp\left\{ -\frac{iz(\omega - aB)}{aB} \right\} dz
$$

$$
= e^{iab} \int_{\mathbb{R}} \psi(z)\, \exp\left\{ -\frac{iz(\omega - aB)}{aB} \right\} \mathcal{K}_M(b,\omega)\, dz
$$

$$
= \sqrt{2\pi}\, e^{iab}\, \mathscr{F}\big[\psi\big]\left(\frac{\omega - aB}{aB} \right) \mathcal{K}_M(b,\omega).
$$

This completes the proof of Proposition 4.5.1. $\qquad\qquad\qquad\qquad\qquad\qquad\square$

Remark 4.5.1. By virtue of (4.131), it follows that each of the analyzing elements $\psi_{a,b}^M(t)$ act as band-pass filters in the linear canonical domain.

Below, we present an alternate form of the linear canonical Stockwell transform (4.127) in terms of the spectral representation of the input and windowing signals.

Lemma 4.5.2. *For any $f \in L^2(\mathbb{R})$, the linear canonical Stockwell transform $\mathcal{S}_\psi^M[f](a,b)$ has the following alternative integral representation:*

$$
\mathcal{S}_\psi^M\Big[f\Big](a,b) = e^{-iab} \int_{\mathbb{R}} \mathscr{L}_M\Big[f\Big](\omega)\, \overline{\mathscr{F}\big[\psi\big]\left(\frac{\omega - aB}{aB} \right)}\, \overline{\mathcal{K}_M(b,\omega)}\, d\omega. \qquad (4.133)
$$

Proof. The proof follows by virtue of the Parseval's formula (1.171) associated with the linear canonical transform together with the application of Proposition 4.5.1 as

$$
\mathcal{S}_\psi^M\Big[f\Big](a,b) = \frac{1}{\sqrt{2\pi}} \Big\langle f,\, \psi_{a,b}^M \Big\rangle_2
$$

$$
= \frac{1}{\sqrt{2\pi}} \Big\langle \mathscr{L}_M\big[f\big],\, \mathscr{L}_M\big[\psi_{a,b}^M\big] \Big\rangle_2
$$

$$
= \frac{1}{\sqrt{2\pi}} \int_{\mathbb{R}} \mathscr{L}_M\big[f\big](\omega)\, \overline{\mathscr{L}_M\big[\psi_{a,b}^M\big](\omega)}\, d\omega
$$

$$
= e^{-iab} \int_{\mathbb{R}} \mathscr{L}_M\big[f\big](\omega)\, \overline{\mathscr{F}\big[\psi\big]\left(\frac{\omega - aB}{aB} \right)}\, \overline{\mathcal{K}_M(b,\omega)}\, d\omega.
$$

This completes the proof of Lemma 4.5.2. $\qquad\qquad\qquad\qquad\qquad\qquad\qquad\square$

Remark 4.5.2. The expression (4.133) demonstrates that the localization of the linear canonical spectrum of the input signal via the linear canonical Stockwell transform (4.127) is governed by the real, non-zero parameter B appearing in the given unimodular matrix $M = (A, B; C, D)$.

Some of the elementary properties of the linear canonical Stockwell transform defined in (4.127) are assembled in the following theorem:

Theorem 4.5.3. *For $k \in \mathbb{R}$, $\lambda \in \mathbb{R} \setminus \{0\}$, let $\mathcal{T}_k$ and $\mathcal{D}_\lambda$ denote the usual translation and dilation operators associated with the Stockwell transform. Then, for any $f, g \in L^2(\mathbb{R})$ and $c_1, c_2 \in \mathbb{C}$, the linear canonical Stockwell transform defined in (4.127) satisfies the following properties:*

(i) *Linearity:* $\mathcal{S}_\psi^M \Big[c_1 f + c_2 g \Big] (a, b) = c_1 \, \mathcal{S}_\psi^M \Big[f \Big] (a, b) + c_2 \, \mathcal{S}_\psi^M \Big[g \Big] (a, b),$

(ii) *Anti-linearity:* $\mathcal{S}_{c_1 \psi + c_2 \phi}^M \Big[f \Big] (a, b) = \overline{c_1} \, \mathcal{S}_\psi^M \Big[f \Big] (a, b) + \overline{c_2} \, \mathcal{S}_\phi^M \Big[f \Big] (a, b),$

(iii) *Translation:* $\mathcal{S}_\psi^M \Big[\mathcal{T}_k f \Big] (a, b) = e^{iA(k^2 - bk)/B - iak} \, \mathcal{S}_\psi^M \Big[\mathcal{M}_{Ak/B} f \Big] (a, b - k),$

(iv) *Dilation:* $\mathcal{S}_\psi^M \Big[\mathcal{D}_\lambda f \Big] (a, b) = |\lambda| \, \mathcal{S}_\psi^{M'} \Big[f \Big] \Big(\frac{a}{\lambda}, \lambda b \Big),$ *where $M' = (A, \lambda^2 B; C/\lambda^2, D),$*

(v) *Translation in window:* $\mathcal{S}_{\mathcal{T}_k \psi}^M \Big[f \Big] (a, b) = \exp \left\{ \frac{iA(k^2 + 2abk)}{2Ba^2} \right\} \mathcal{S}_\psi^M \Big[f \Big] (a, b + k/a),$

(vi) *Dilation in window:* $\mathcal{S}_{\mathcal{D}_\lambda \psi}^M \Big[f \Big] (a, b) = \mathcal{S}_\psi^M \Big[\mathcal{M}_{a(\lambda - 1)} f \Big] (a\lambda, b).$

Proof. The proof of the theorem is easy to follow and is, therefore, omitted. $\qquad \square$

Next, our motive is to obtain an orthogonality relationship between a given pair of signals and the corresponding linear canonical Stockwell transforms. As a consequence of such a relationship, we can easily deduce the energy preserving relation associated with the linear canonical Stockwell transform. Besides, we shall also demonstrate that indeed the family of analyzing functions $\mathcal{F}_\psi^M (a, b)$ defined in (4.126) generates a resolution of the identity.

Theorem 4.5.4. *If $\mathcal{S}_\psi^M [f](a, b)$ and $\mathcal{S}_\psi^M [g](a, b)$ are the linear canonical Stockwell transforms corresponding to a given pair of functions $f, g \in L^2(\mathbb{R})$, then*

$$\int_\mathbb{R} \int_\mathbb{R} \mathcal{S}_\psi^M \Big[f \Big] (a, b) \, \overline{\mathcal{S}_\psi^M \Big[g \Big] (a, b)} \, \frac{da \, db}{|a|} = C_\psi \big\langle f, \, g \big\rangle_2, \qquad (4.134)$$

provided the window function $\psi \in L^2(\mathbb{R})$ satisfies the following condition:

$$C_\psi = \int_\mathbb{R} \frac{\big| \mathscr{F} [\psi] (\omega - 1) \big|^2}{|\omega|} \, d\omega < \infty. \qquad (4.135)$$

Proof. Using Lemma 4.5.2, we can express the linear canonical Stockwell transform of the given pair of square integrable functions f and g as

$$\mathcal{S}_\psi^M \Big[f \Big] (a, b) = e^{-iab} \int_\mathbb{R} \mathscr{L}_M \Big[f \Big] (\omega) \, \overline{\mathscr{F} \Big[\psi \Big] \Big(\frac{\omega - aB}{aB} \Big)} \, \mathcal{K}_M(b, \omega) \, d\omega \qquad (4.136)$$

and

$$\mathcal{S}_\psi^M\big[g\big](a,b) = e^{-iab}\int_{\mathbb{R}} \mathcal{L}_M\big[g\big](\omega')\, \overline{\mathscr{F}\big[\psi\big]\left(\frac{\omega'-aB}{aB}\right)}\, \overline{\mathcal{K}_M(b,\omega')}\, d\omega'. \qquad (4.137)$$

Thus, we have

$$\int_{\mathbb{R}}\int_{\mathbb{R}} \mathcal{S}_\psi^M\big[f\big](a,b)\, \overline{\mathcal{S}_\psi^M\big[g\big](a,b)}\, \frac{da\,db}{|a|}$$

$$= \int_{\mathbb{R}}\int_{\mathbb{R}} \left\{ e^{-iab}\int_{\mathbb{R}} \mathcal{L}_M\big[f\big](\omega)\, \overline{\mathscr{F}\big[\psi\big]\left(\frac{\omega-aB}{aB}\right)}\, \overline{\mathcal{K}_M(b,\omega)}\, d\omega \right\}$$

$$\times \left\{ e^{iab}\int_{\mathbb{R}} \overline{\mathcal{L}_M\big[g\big](\omega')}\, \mathscr{F}\big[\psi\big]\left(\frac{\omega'-aB}{aB}\right) \mathcal{K}_M(b,\omega')\, d\omega' \right\} \frac{da\,db}{|a|}$$

$$= \int_{\mathbb{R}}\int_{\mathbb{R}}\int_{\mathbb{R}} \mathcal{L}_M\big[f\big](\omega)\, \overline{\mathcal{L}_M\big[g\big](\omega')}\, \mathscr{F}\big[\psi\big]\left(\frac{\omega'-aB}{aB}\right) \overline{\mathscr{F}\big[\psi\big]\left(\frac{\omega-aB}{aB}\right)}$$

$$\times \left\{ \int_{\mathbb{R}} \mathcal{K}_M(b,\omega')\, \overline{\mathcal{K}_M(b,\omega)}\, db \right\} \frac{da\,d\omega\,d\omega'}{|a|}$$

$$= \int_{\mathbb{R}}\int_{\mathbb{R}}\int_{\mathbb{R}} \mathcal{L}_M\big[f\big](\omega)\, \overline{\mathcal{L}_M\big[g\big](\omega')}\, \mathscr{F}\big[\psi\big]\left(\frac{\omega'-aB}{aB}\right) \overline{\mathscr{F}\big[\psi\big]\left(\frac{\omega-aB}{aB}\right)}$$

$$\times\, \delta(\omega-\omega')\, \frac{da\,d\omega\,d\omega'}{|a|}$$

$$= \int_{\mathbb{R}}\int_{\mathbb{R}} \mathcal{L}_M\big[f\big](\omega)\, \overline{\mathcal{L}_M\big[g\big](\omega)}\, \left|\mathscr{F}\big[\psi\big]\left(\frac{\omega-aB}{aB}\right)\right|^2 \frac{da\,d\omega}{|a|}$$

$$= \int_{\mathbb{R}} \mathcal{L}_M\big[f\big](\omega)\, \overline{\mathcal{L}_M\big[g\big](\omega)}\, \left\{ \int_{\mathbb{R}} \left|\mathscr{F}\big[\psi\big]\left(\frac{\omega-aB}{aB}\right)\right|^2 \frac{da}{|a|} \right\} d\omega. \qquad (4.138)$$

In order to simplify the inner integral on the R.H.S of (4.138), we substitue $\eta = aB/\omega$, so that

$$\int_{\mathbb{R}} \left|\mathscr{F}\big[\psi\big]\left(\frac{\omega-aB}{aB}\right)\right|^2 \frac{da}{|a|} = \int_{\mathbb{R}} \frac{\left|\mathscr{F}\big[\psi\big](\eta-1)\right|^2}{|\eta|}\, d\eta = C_\psi < \infty. \qquad (4.139)$$

Making use of the estimate (4.139) in (4.138) yields the desired relationship as

$$\int_{\mathbb{R}}\int_{\mathbb{R}} \mathcal{S}_\psi^M\big[f\big](a,b)\, \overline{\mathcal{S}_\psi^M\big[g\big](a,b)}\, \frac{da\,db}{|a|} = C_\psi \int_{\mathbb{R}} \mathcal{L}_M\big[f\big](\omega)\, \overline{\mathcal{L}_M\big[g\big](\omega)}\, d\omega$$

$$= C_\psi \big\langle \mathcal{L}_M\big[f\big], \mathcal{L}_M\big[g\big] \big\rangle_2$$

$$= C_\psi \big\langle f, g \big\rangle_2.$$

This completes the proof of Theorem 4.5.4. $\qquad\qquad\qquad\qquad\qquad\qquad\qquad$ □

Remark 4.5.3. For $f = g$, the orthogonality relation (4.134) implies that the total energy of the transformed signal $\mathcal{S}_\psi^M\big[f\big](a,b)$ in the space $L^2(\mathbb{R}\setminus\{0\}\times\mathbb{R})$ is a C_ψ-multiple of the total energy of the input signal f in the space $L^2(\mathbb{R})$. More precisely, we can write

$$\int_{\mathbb{R}}\int_{\mathbb{R}} \left|\mathcal{S}_\psi^M\big[f\big](a,b)\right|^2 \frac{da\,db}{|a|} = C_\psi \left\|f\right\|_2^2. \qquad (4.140)$$

Thus, for $C_\psi = 1$, the linear canonical Stockwell transform defined in (4.127) turns to be an isometric map from $L^2(\mathbb{R})$ to $L^2(\mathbb{R}^2)$. Moreover, by virtue of (4.140), we can also infer that the family of analyzing functions $\mathcal{F}_\psi^M(a,b)$ defined in (4.126) generates a resolution of the identity:

$$I = \frac{1}{C_\psi} \int_\mathbb{R} \int_\mathbb{R} \left\langle \psi_{a,b}^M, \psi_{a,b}^M \right\rangle_2 \frac{da\,db}{|a|}, \tag{4.141}$$

with I being the identity operator.

The inversion formula pertaining to the linear canonical Stockwell transform defined in (4.127) is given in the following theorem:

Theorem 4.5.5. *If $\mathcal{S}_\psi^M[f](a,b)$ is the linear canonical Stockwell transform of any function $f \in L^2(\mathbb{R})$, with respect to the window function $\psi \in L^2(\mathbb{R})$, then the input function f can be recovered from the Stockwell coefficients $\langle f, \psi_{a,b}^M \rangle_2$ via the following integral formula:*

$$f(t) = \frac{1}{\sqrt{2\pi}C_\psi} \int_\mathbb{R} \int_\mathbb{R} \left\langle f, \psi_{a,b}^M \right\rangle_2 \psi_{a,b}^M(t) \frac{da\,db}{|a|}, \quad a.e., \tag{4.142}$$

with $C_\psi \neq 0$ being given by (4.135).

Proof. Denote $g_t(x) = \delta(t-x)$, $t \in \mathbb{R}$. Then, by virtue of the fundamental property of the linear canonical kernel given in Theorem 1.4.1, we have

$$g_t(x) = \int_\mathbb{R} \mathcal{K}_M(t,\omega) \overline{\mathcal{K}_M(x,\omega)}\,d\omega. \tag{4.143}$$

From (4.143), we infer that $\mathcal{L}_M[g_t](\omega) = \mathcal{K}_M(t,\omega)$. Therefore, by virtue of the Parseval's formula for the linear canonical transform and the orthogonality relation (4.134), we have

$$\begin{aligned} f(t) &= \int_\mathbb{R} \mathcal{L}_M[f](\omega) \overline{\mathcal{K}_M(t,\omega)}\,d\omega \\ &= \int_\mathbb{R} \mathcal{L}_M[f](\omega) \overline{\mathcal{L}_M[g_t](\omega)}\,d\omega \\ &= \frac{1}{C_\psi} \int_\mathbb{R} \int_\mathbb{R} \mathcal{S}_\psi^M[f](a,b) \overline{\mathcal{S}_\psi^M[g_t](a,b)} \frac{da\,db}{|a|} \\ &= \frac{1}{C_\psi} \int_\mathbb{R} \int_\mathbb{R} \left\langle f, \psi_{a,b}^M \right\rangle_2 \overline{\mathcal{S}_\psi^M[\delta(t-x)](a,b)} \frac{da\,db}{|a|}. \end{aligned} \tag{4.144}$$

Note that,

$$\mathcal{S}_\psi^M[\delta(t-x)](a,b) = \frac{1}{\sqrt{2\pi}} \int_\mathbb{R} \delta(t-x) \overline{\psi_{a,b}^M(x)}\,dx = \frac{1}{\sqrt{2\pi}} \overline{\psi_{a,b}^M(t)}. \tag{4.145}$$

Plugging (4.145) into equation (4.144), yields the desired reconstruction formula for the linear canonical Stockwell transform as

$$f(t) = \frac{1}{\sqrt{2\pi}C_\psi} \int_\mathbb{R} \int_\mathbb{R} \left\langle f, \psi_{a,b}^M \right\rangle_2 \psi_{a,b}^M(t) \frac{da\,db}{|a|}, \quad a.e.$$

This completes the proof of Theorem 4.5.5. $\qquad\square$

In the following theorem, we present a characterization of the range of the linear canonical Stockwell transform defined in (4.127).

Theorem 4.5.6. *A function $f \in L^2(\mathbb{R}^2)$ is the linear canonical Stockwell transform of a certain square integrable function with respect to the window function $\psi \in L^2(\mathbb{R})$ if and only if it satisfies the following reproduction formula:*

$$f(a',b') = \frac{1}{2\pi\,C_\psi} \int_{\mathbb{R}} \int_{\mathbb{R}} f(a,b) \left\langle \psi_{a,b}^M,\, \psi_{a',b'}^M \right\rangle_2 \frac{da\,db}{|a|}, \qquad (4.146)$$

where $C_\psi \neq 0$ is given by (4.135) and the functions $\psi_{a,b}^M(t)$ are as defined in (4.126).

Proof. With the inversion formula (4.142) in hindsight, it is quite easy to verify the arguments, so we omit the proof. □

Corollary 4.5.7. *The range of the linear canonical Stockwell transform defined in (4.127) is a reproducing kernel Hilbert space embedded as a subspace in $L^2(\mathbb{R}^2)$ with kernel given by*

$$K_\psi^M(a,b;a',b') = \frac{1}{2\pi\,C_\psi} \left\langle \psi_{a,b}^M,\, \psi_{a',b'}^M \right\rangle_2. \qquad (4.147)$$

Moreover, for all $(a,b),(a',b') \in \mathbb{R} \setminus \{0\} \times \mathbb{R}$, the kernel also satisfies

$$\left| K_\psi^M(a,b;a',b') \right| \leq (2\pi\,C_\psi)^{-1} \left\| \psi \right\|_2^2. \qquad (4.148)$$

4.6   The Linear Canonical Ridgelet Transform

"Nothing in life is to be feared, it is only to be understood. Now is the time to understand more, so that we may fear less."

-Marie S. Curie

The extension of the classical wavelet theory to the realm of fractional and linear canonical transforms has paved the way for further insights into the theory of directional representation systems, such as ridgelets, curvelets, ripplets, shearlets and so on. The reason for the study of generalized directional representation systems is to inculcate certain extra flexibility into the aforementioned directional waveforms. In pursuit of that, we intend to present a comprehensive analysis regarding such extended directional representation systems. In this section, we shall present the notion of linear canonical ridgelet transform and also explore the underlying mathematical theory.

4.6.1   Definition and basic properties

In order to facilitate the formulation of linear canonical ridgelet transform, it is imperative to recall that the classical ridgelet transform (3.194) corresponding to any $f \in L^2(\mathbb{R}^2)$ with respect to a one-dimensional wavelet $\psi \in L^2(\mathbb{R})$ can be expressed as the wavelet transform of the Radon transform (3.185). Mathematically,

$$R_\psi\!\left[f\right](a,b,\theta) = \mathscr{W}_\psi\!\left[\mathfrak{R}\!\left[f\right](t,\theta)\right](a,b), \quad a \in \mathbb{R}^+, b \in \mathbb{R} \text{ and } \theta \in [0,2\pi), \qquad (4.149)$$

where notations have their usual meanings and the wavelet transform on the R.H.S of (4.149) is computed with respect to $t \in \mathbb{R}$. With the expression (4.149) in hindsight, the linear canonical ridgelet transform can be formulated by computing the linear canonical wavelet transform (4.65) of the Radon transform $\mathfrak{R}[f](t,\theta)$, when regarded as a function of $t \in \mathbb{R}$. We observe that,

$$
\begin{aligned}
\mathscr{W}_\psi^M\Big[\mathfrak{R}[f](t,\theta)\Big](a,b) &= \frac{1}{\sqrt{a}}\int_{\mathbb{R}}\mathfrak{R}[f](t,\theta)\,\overline{\psi\left(\frac{t-b}{a}\right)}\exp\left\{\frac{iA(t^2-b^2)}{2B}\right\}dt \\
&= \frac{1}{\sqrt{a}}\int_{\mathbb{R}}\left\{\int_{\mathbb{R}^2}f(\mathbf{x})\,\delta(\Lambda_\theta\cdot\mathbf{x}-t)\,d\mathbf{x}\right\}\overline{\psi\left(\frac{t-b}{a}\right)}\exp\left\{\frac{iA(t^2-b^2)}{2B}\right\}dt \\
&= \frac{1}{\sqrt{a}}\int_{\mathbb{R}}\int_{\mathbb{R}^2}f(\mathbf{x})\,\delta(\Lambda_\theta\cdot\mathbf{x}-t)\,\overline{\psi\left(\frac{t-b}{a}\right)}\exp\left\{\frac{iA(t^2-b^2)}{2B}\right\}d\mathbf{x}\,dt \\
&= \frac{1}{\sqrt{a}}\int_{\mathbb{R}^2}f(\mathbf{x})\,\overline{\psi\left(\frac{\Lambda_\theta\cdot\mathbf{x}-b}{a}\right)}\exp\left\{\frac{iA(\Lambda_\theta\cdot\mathbf{x}-b)(\Lambda_\theta\cdot\mathbf{x}+b)}{2B}\right\}d\mathbf{x},
\end{aligned}
$$
$$(4.150)$$

where $M = (A, B; C, D)$ is a given 2×2, real and unimodular matrix. Based on the above obtained expression (4.150), we define a family of linear canonical ridgelets governed by the matrix M as

$$
\mathfrak{F}_\psi^M(a,b,\theta) := \left\{\psi_{a,b,\theta}^M(\mathbf{t}) = a^{-1/2}\,\psi\left(\frac{\Lambda_\theta\cdot\mathbf{t}-b}{a}\right)\exp\left\{-\frac{iA(\Lambda_\theta\cdot\mathbf{t}-b)(\Lambda_\theta\cdot\mathbf{t}+b)}{2B}\right\}\right\},
$$
$$(4.151)$$

where $\mathbf{t} = (t_1, t_2) \in \mathbb{R}^2$, $a \in \mathbb{R}^+$, $b \in \mathbb{R}$ and $\theta \in [0, 2\pi)$. From (4.151), it is quite evident that each linear canonical ridgelet behaves like a one-dimensional linear canonical wavelet in a specific direction.

Having constructed a novel family of ridgelets (4.151), coined as the "linear canonical ridgelets," we have the following definition of the linear canonical ridgelet transform:

Definition 4.6.1. Given a real, unimodular matrix $M = (A, B; C, D)$, the linear canonical ridgelet transform of any $f \in L^2(\mathbb{R}^2)$ is denoted by $R_\psi^M[f]$ and is defined as

$$
R_\psi^M[f](a,b,\theta) = \left\langle f, \psi_{a,b,\theta}^M\right\rangle_2 = \int_{\mathbb{R}^2}f(\mathbf{t})\,\overline{\psi_{a,b,\theta}^M(\mathbf{t})}\,d\mathbf{t},
$$
$$(4.152)$$

where $\psi_{a,b,\theta}^M(\mathbf{t})$ is given by (4.151).

Some important features of the linear canonical ridgelet transform (4.152) are given below:

(i) The classical ridgelet transform (3.194) is just a particular case of linear canonical ridgelet transform (4.152) and follows by plugging $M = (0, 1; -1, 0)$ in Definition 4.6.1.

(ii) For $M = (\cos\alpha, \sin\alpha; -\sin\alpha, \cos\alpha)$, $\alpha \neq n\pi$, $n \in \mathbb{Z}$, the linear canonical ridgelet transform given by Definition 4.6.1 yields a new variant of the ridgelet transform, namely the fractional ridgelet transform.

(iii) The linear canonical ridgelet transform (4.152) is endowed with higher degrees of freedom in lieu to the classical ridgelet transform. Nevertheless, it is worth emphasizing that no extra conditions are to be imposed on the given classical one-dimensional wavelet $\psi \in L^2(\mathbb{R})$.

(iv) The linear canonical ridgelet transform (4.152) can be regarded as an intertwining of the linear canonical wavelet transform (4.65) and the classical Radon transform (3.185).

(v) Since the computational complexity of the linear canonical wavelet transform is completely determined by that of the classical wavelet transform, therefore, from expression (4.150), we conclude that the computational complexity of the linear canonical ridgelet transform (4.152) is determined by the computational complexity of classical ridgelet transform.

Below, we shall study some fundamental properties of the linear canonical ridgelet transform defined in (4.152).

Theorem 4.6.1. *For any pair of functions $f, g \in L^2(\mathbb{R}^2)$ and the scalars $c_1, c_2 \in \mathbb{C}$, $\mathbf{k} \in \mathbb{R}^2$, $k \in \mathbb{R}$, $\lambda > 0$ and $\theta' \in [0, 2\pi)$, the linear canonical ridgelet transform (4.152) satisfies the following properties:*

(i) *Linearity:* $R_\psi^M \Big[c_1 f + c_2 g \Big](a, b, \theta) = c_1 \, R_\psi^M \Big[f \Big](a, b, \theta) + c_2 \, R_\psi^M \Big[g \Big](a, b, \theta),$

(ii) *Anti-linearity:* $R_{c_1 \psi + c_2 \phi}^M \Big[f \Big](a, b, \theta) = \overline{c_1} \, R_\psi^M \Big[f \Big](a, b, \theta) + \overline{c_2} \, R_\phi^M \Big[f \Big](a, b, \theta),$

(iii) *Translation:* $R_\psi^M \Big[\mathcal{T}_\mathbf{k} f \Big](a, b, \theta) = \exp \left\{ \dfrac{iA(\Lambda_\theta \cdot \mathbf{k} - b)(\Lambda_\theta \cdot \mathbf{k})}{B} \right\} R_\psi^M \Big[F_M \Big](a, b - \Lambda_\theta \cdot \mathbf{k}, \theta),$ *where* $F_M(\mathbf{t}) = \exp \left\{ \dfrac{iA(\Lambda_\theta \cdot \mathbf{t})(\Lambda_\theta \cdot \mathbf{k})}{B} \right\} f(\mathbf{t}),$

(iv) *Planer dilation:* $R_\psi^M \Big[\mathcal{D}_\lambda f \Big](a, b, \theta) = \dfrac{1}{\sqrt{\lambda}} R_\psi^N \Big[f \Big] \left(\dfrac{a}{\lambda}, \dfrac{b}{\lambda}, \theta \right),$ $N = (A, B/\lambda^2; \lambda^2 C, D),$

(v) *Parity:* $R_{P\psi}^M \Big[Pf \Big](a, b, \theta) = R_\psi^M \Big[f \Big](a, -b, \theta),$ *where* $Pf(\mathbf{t}) = f(-\mathbf{t}),$

(vi) *Rotation:* $R_\psi^M \Big[R_{\theta'} f \Big](a, b, \theta) = R_\psi^M \Big[f \Big](a, b, \theta - \theta'),$

(vii) *Translation in ridgelet:* $R_{\mathcal{T}_k \psi}^M \Big[f \Big](a, b, \theta) = \exp \left\{ \dfrac{iA(2abk + (ak)^2)}{2B} \right\} R_\psi^M \Big[f \Big](a, b + ak, \theta),$

(viii) *Dilation in ridgelet:* $R_{\mathcal{D}_\lambda \psi}^M \Big[f \Big](a, b, \theta) = R_\psi^M \Big[f \Big](a\lambda, b, \theta).$

Proof. It is straightforward to verify the assertions (i) and (ii), so we proceed to prove the rest.

(iii) Translating the input signal $f(\mathbf{t})$ with $\mathbf{k} \in \mathbb{R}^2$, we have

$$
R_\psi^M \Big[\mathcal{T}_\mathbf{k} f \Big](a, b, \theta) = \int_{\mathbb{R}^2} (\mathcal{T}_\mathbf{k} f)(\mathbf{t}) \, \overline{\psi_{a,b,\theta}^M(\mathbf{t})} \, d\mathbf{t}
$$

$$
= \frac{1}{\sqrt{a}} \int_{\mathbb{R}^2} f(\mathbf{t} - \mathbf{k}) \, \overline{\psi \left(\frac{\Lambda_\theta \cdot \mathbf{t} - b}{a} \right)} \exp \left\{ \frac{iA(\Lambda_\theta \cdot \mathbf{t} - b)(\Lambda_\theta \cdot \mathbf{t} + b)}{2B} \right\} d\mathbf{t}
$$

$$
= \frac{1}{\sqrt{a}} \int_{\mathbb{R}^2} f(\mathbf{z}) \, \overline{\psi \left(\frac{\Lambda_\theta \cdot (\mathbf{z} + \mathbf{k}) - b}{a} \right)}
$$

$$
\times \exp \left\{ \frac{iA(\Lambda_\theta \cdot (\mathbf{z} + \mathbf{k}) - b)(\Lambda_\theta \cdot (\mathbf{z} + \mathbf{k}) + b)}{2B} \right\} d\mathbf{z}
$$

$$
= \frac{1}{\sqrt{a}} \int_{\mathbb{R}^2} f(\mathbf{z}) \, \overline{\psi \left(\frac{\Lambda_\theta \cdot \mathbf{z} - (b - \Lambda_\theta \cdot \mathbf{k})}{a} \right)}
$$

$$\times \exp\left\{\frac{iA\big(\Lambda_\theta \cdot \mathbf{z} - (b - \Lambda_\theta \cdot \mathbf{k})\big)\big(\Lambda_\theta \cdot \mathbf{z} + (b + \Lambda_\theta \cdot \mathbf{k})\big)}{2B}\right\} d\mathbf{z}$$

$$= \frac{1}{\sqrt{a}} \int_{\mathbb{R}^2} f(\mathbf{z}) \, \overline{\psi\left(\frac{\Lambda_\theta \cdot \mathbf{z} - (b - \Lambda_\theta \cdot \mathbf{k})}{a}\right)}$$

$$\times \exp\left\{\frac{iA\big(\Lambda_\theta \cdot \mathbf{z} - (b - \Lambda_\theta \cdot \mathbf{k})\big)\big(\Lambda_\theta \cdot \mathbf{z} + (b - \Lambda_\theta \cdot \mathbf{k})\big)}{2B}\right\}$$

$$\times \exp\left\{\frac{iA\big(\Lambda_\theta \cdot \mathbf{z} - (b - \Lambda_\theta \cdot \mathbf{k})\big)\Lambda_\theta \cdot \mathbf{k}}{B}\right\} d\mathbf{z}$$

$$= \exp\left\{\frac{iA(\Lambda_\theta \cdot \mathbf{k} - b)(\Lambda_\theta \cdot \mathbf{k})}{B}\right\} \frac{1}{\sqrt{a}} \int_{\mathbb{R}^2} \exp\left\{\frac{iA(\Lambda_\theta \cdot \mathbf{z})(\Lambda_\theta \cdot \mathbf{k})}{B}\right\}$$

$$\times f(\mathbf{z}) \, \overline{\psi\left(\frac{\Lambda_\theta \cdot \mathbf{z} - (b - \Lambda_\theta \cdot \mathbf{k})}{a}\right)}$$

$$\times \exp\left\{\frac{iA\big(\Lambda_\theta \cdot \mathbf{z} - (b - \Lambda_\theta \cdot \mathbf{k})\big)\big(\Lambda_\theta \cdot \mathbf{z} + (b - \Lambda_\theta \cdot \mathbf{k})\big)}{2B}\right\} d\mathbf{z}$$

$$= \exp\left\{\frac{iA(\Lambda_\theta \cdot \mathbf{k} - b)(\Lambda_\theta \cdot \mathbf{k})}{B}\right\} R_\psi^M\big[F_M\big](a, b - \Lambda_\theta \cdot \mathbf{k}, \theta),$$

where $F_M(\mathbf{t}) = \exp\left\{\dfrac{iA(\Lambda_\theta \cdot \mathbf{t})(\Lambda_\theta \cdot \mathbf{k})}{B}\right\} f(\mathbf{t})$.

(iv) Under the application of planer dilation operator $\mathcal{D}_\lambda$, $\lambda \in \mathbb{R}^+$ to the input signal $f(\mathbf{t})$, we have

$$R_\psi^M\big[\mathcal{D}_\lambda f\big](a, b, \theta) = \int_{\mathbb{R}^2} (\mathcal{D}_\lambda f)(\mathbf{t}) \, \overline{\psi_{a,b,\theta}^M(\mathbf{t})} \, dt$$

$$= \frac{1}{\lambda\sqrt{a}} \int_{\mathbb{R}^2} f(\lambda^{-1}\mathbf{t}) \, \overline{\psi\left(\frac{\Lambda_\theta \cdot \mathbf{t} - b}{a}\right)} \exp\left\{\frac{iA(\Lambda_\theta \cdot \mathbf{t} - b)(\Lambda_\theta \cdot \mathbf{t} + b)}{2B}\right\} dt$$

$$= \frac{1}{\sqrt{a}} \int_{\mathbb{R}^2} f(\mathbf{z}) \, \overline{\psi\left(\frac{\Lambda_\theta \cdot (\lambda\mathbf{z}) - b}{a}\right)}$$

$$\times \exp\left\{\frac{iA\big(\Lambda_\theta \cdot (\lambda\mathbf{z}) - b\big)\big(\Lambda_\theta \cdot (\lambda\mathbf{z}) + b\big)}{2B}\right\} d\mathbf{z}$$

$$= \frac{1}{\sqrt{(a/\lambda)\lambda}} \int_{\mathbb{R}^2} f(\mathbf{z}) \, \overline{\psi\left(\frac{\Lambda_\theta \cdot \mathbf{z} - b/\lambda}{a/\lambda}\right)}$$

$$\times \exp\left\{\frac{iA(\Lambda_\theta \cdot \mathbf{z} - b/\lambda)(\Lambda_\theta \cdot \mathbf{z} + b/\lambda)}{2(B/\lambda^2)}\right\} d\mathbf{z}$$

$$= \frac{1}{\sqrt{\lambda}} R_\psi^N\big[f\big]\left(\frac{a}{\lambda}, \frac{b}{\lambda}, \theta\right), \quad N = (A, B/\lambda^2; \lambda^2 C, D).$$

(v) Upon the application of the parity operator P, we have

$$R_{P\psi}^M\big[Pf\big](a, b, \theta) = \int_{\mathbb{R}^2} (Pf)(\mathbf{t}) \, \overline{(P\psi)_{a,b,\theta}^M(\mathbf{t})} \, d\mathbf{t}$$

$$= \frac{1}{\sqrt{a}} \int_{\mathbb{R}^2} f(-\mathbf{t}) \, \overline{P\psi\left(\frac{\Lambda_\theta \cdot \mathbf{t} - b}{a}\right)} \exp\left\{\frac{iA(\Lambda_\theta \cdot \mathbf{t} - b)(\Lambda_\theta \cdot \mathbf{t} + b)}{2B}\right\} d\mathbf{t}$$

$$= \frac{1}{\sqrt{a}} \int_{\mathbb{R}^2} f(-\mathbf{t}) \, \overline{\psi\left(-\frac{\Lambda_\theta \cdot \mathbf{t} - b}{a}\right)} \exp\left\{\frac{iA(\Lambda_\theta \cdot \mathbf{t} - b)(\Lambda_\theta \cdot \mathbf{t} + b)}{2B}\right\} d\mathbf{t}$$

$$= \frac{1}{\sqrt{a}} \int_{\mathbb{R}^2} f(\mathbf{z}) \, \overline{\psi\left(-\frac{\Lambda_\theta \cdot (-\mathbf{z}) - b}{a}\right)}$$

$$\times \exp\left\{\frac{iA(\Lambda_\theta \cdot (-\mathbf{z}) - b)(\Lambda_\theta \cdot (-\mathbf{z}) + b)}{2B}\right\} d\mathbf{z}$$

$$= \frac{1}{\sqrt{a}} \int_{\mathbb{R}^2} f(\mathbf{z}) \, \overline{\psi\left(\frac{\Lambda_\theta \cdot \mathbf{z} + b}{a}\right)} \exp\left\{\frac{iA(\Lambda_\theta \cdot \mathbf{z} + b)(\Lambda_\theta \cdot \mathbf{z} - b)}{2B}\right\} d\mathbf{z}$$

$$= R_\psi^M\big[f\big](a, -b, \theta).$$

(vi) Upon applying the rotation operator $R_{\theta'}$, $\theta' \in [0, 2\pi)$ to the input signal $f(\mathbf{t})$, we have

$$R_\psi^M\big[R_{\theta'} f\big](a, b, \theta) = \int_{\mathbb{R}^2} f(R_{-\theta'} \mathbf{t}) \, \overline{\psi_{a,b,\theta}^M(\mathbf{t})} \, d\mathbf{t}$$

$$= \frac{1}{\sqrt{a}} \int_{\mathbb{R}^2} f(R_{-\theta'} \mathbf{t}) \, \overline{\psi\left(\frac{\Lambda_\theta \cdot \mathbf{t} - b}{a}\right)} \exp\left\{\frac{iA(\Lambda_\theta \cdot \mathbf{t} - b)(\Lambda_\theta \cdot \mathbf{t} + b)}{2B}\right\} d\mathbf{t}$$

$$= \frac{1}{\sqrt{a}} \int_{\mathbb{R}^2} f(\mathbf{z}) \, \overline{\psi\left(\frac{\Lambda_\theta \cdot (R_{\theta'} \mathbf{z}) - b}{a}\right)}$$

$$\times \exp\left\{\frac{iA(\Lambda_\theta \cdot (R_{\theta'} \mathbf{z}) - b)(\Lambda_\theta \cdot (R_{\theta'} \mathbf{z}) + b)}{2B}\right\} d\mathbf{z}$$

$$= \frac{1}{\sqrt{a}} \int_{\mathbb{R}^2} f(\mathbf{z}) \, \overline{\psi\left(\frac{\Lambda_{\theta-\theta'} \cdot \mathbf{z} - b}{a}\right)}$$

$$\times \exp\left\{\frac{iA(\Lambda_{\theta-\theta'} \cdot \mathbf{z} - b)(\Lambda_{\theta-\theta'} \cdot \mathbf{z} + b)}{2B}\right\} d\mathbf{z}$$

$$= R_\psi^M\big[f\big](a, b, \theta - \theta').$$

(vii) Under translation of the underlying wavelet ψ by $k \in \mathbb{R}$, the linear canonical ridgelet transform becomes

$$R_{\mathcal{T}_k \psi}^M\big[f\big](a, b, \theta) = \int_{\mathbb{R}^2} f(\mathbf{t}) \, \overline{(\mathcal{T}_k \psi)_{a,b,\theta}^M(\mathbf{t})} \, d\mathbf{t}$$

$$= \frac{1}{\sqrt{a}} \int_{\mathbb{R}^2} f(\mathbf{t}) \, \overline{\mathcal{T}_k \psi\left(\frac{\Lambda_\theta \cdot \mathbf{t} - b}{a}\right)} \exp\left\{\frac{iA(\Lambda_\theta \cdot \mathbf{t} - b)(\Lambda_\theta \cdot \mathbf{t} + b)}{2B}\right\} d\mathbf{t}$$

$$= \frac{1}{\sqrt{a}} \int_{\mathbb{R}^2} f(\mathbf{t}) \, \overline{\psi\left(\frac{\Lambda_\theta \cdot \mathbf{t} - b}{a} - k\right)} \exp\left\{\frac{iA(\Lambda_\theta \cdot \mathbf{t} - b)(\Lambda_\theta \cdot \mathbf{t} + b)}{2B}\right\} d\mathbf{t}$$

$$= \frac{1}{\sqrt{a}} \int_{\mathbb{R}^2} f(\mathbf{t}) \, \overline{\psi\left(\frac{\Lambda_\theta \cdot \mathbf{t} - (b + ak)}{a}\right)} \exp\left\{\frac{iA(2abk + (ak)^2)}{2B}\right\}$$

$$\times \exp\left\{\frac{iA(\Lambda_\theta \cdot \mathbf{t} - (b + ak))(\Lambda_\theta \cdot \mathbf{t} + (b + ak))}{2B}\right\} d\mathbf{t}$$

$$= \exp\left\{\frac{iA(2abk + (ak)^2)}{2B}\right\} R_\psi^M\big[f\big](a, b + ak, \theta).$$

(viii) Applying the usual dilation operator $\mathcal{D}_\lambda$, $\lambda \in \mathbb{R}^+$ to the basic wavelet ψ, we have

$$
\begin{aligned}
R^M_{\mathcal{D}_\lambda \psi}\big[f\big](a,b,\theta) &= \int_{\mathbb{R}^2} f(\mathbf{t})\, \overline{(\mathcal{D}_\lambda \psi)^M_{a,b,\theta}(\mathbf{t})}\, d\mathbf{t} \\
&= \frac{1}{\sqrt{a}} \int_{\mathbb{R}^2} f(\mathbf{t})\, \overline{\mathcal{D}_\lambda \psi\left(\frac{\Lambda_\theta \cdot \mathbf{t} - b}{a}\right)} \exp\left\{\frac{iA(\Lambda_\theta \cdot \mathbf{t} - b)(\Lambda_\theta \cdot \mathbf{t} + b)}{2B}\right\} d\mathbf{t} \\
&= \frac{1}{\sqrt{a\lambda}} \int_{\mathbb{R}^2} f(\mathbf{t})\, \overline{\psi\left(\frac{\Lambda_\theta \cdot \mathbf{t} - b}{a\lambda}\right)} \exp\left\{\frac{iA(\Lambda_\theta \cdot \mathbf{t} - b)(\Lambda_\theta \cdot \mathbf{t} + b)}{2B}\right\} d\mathbf{t} \\
&= R^M_\psi\big[f\big](a\lambda, b, \theta).
\end{aligned}
$$

This completes the proof of Theorem 4.6.1. $\qquad\qquad\qquad\qquad\qquad\qquad\Box$

The following proposition offers a lucid and elegant representation of the linear canonical ridgelet transform (4.152) in the linear canonical domain:

Proposition 4.6.2. *For any $f \in L^2(\mathbb{R}^2)$, the linear canonical ridgelet transform defined in (4.152) can be expressed via the linear canonical transform as*

$$
\mathscr{L}_M\left(R^M_\psi\big[f\big](a,b,\theta)\right)(\omega) = 2\pi \sqrt{\frac{a}{iB}}\, \exp\left\{\frac{iD\omega^2}{2B}\right\} \mathscr{F}\big[F_M\big]\left(\frac{\omega\Lambda_\theta}{B}\right) \overline{\mathscr{F}\big[\psi\big]\left(\frac{a\omega}{B}\right)}, \quad (4.153)
$$

where the linear canonical transform on the L.H.S of (4.153) is computed with respect to the translation variable b and

$$
F_M(\mathbf{t}) = f(\mathbf{t}) \exp\left\{\frac{iA(\Lambda_\theta \cdot \mathbf{t})^2}{2B}\right\}, \qquad \frac{\omega\Lambda_\theta}{B} = \left(\frac{\omega\cos\theta}{B}, \frac{\omega\sin\theta}{B}\right). \quad (4.154)
$$

Proof. Note that the linear canonical ridgelet transform defined in (4.152) can be regarded as the linear canonical wavelet transform of the Radon transform as illustrated in (4.150). Therefore, by virtue of Lemma 4.4.3, we can express the linear canonical transform of (4.152) as follows:

$$
\mathscr{L}_M\left(R^M_\psi\big[f\big](a,b,\theta)\right)(\omega) = \sqrt{2\pi a}\, \mathscr{L}_M\left(\mathfrak{R}\big[f\big](t,\theta)\right)(\omega) \overline{\mathscr{F}\big[\psi\big]\left(\frac{a\omega}{B}\right)}. \quad (4.155)
$$

To compute the linear canonical transform of $\mathfrak{R}\big[f\big](t,\theta)$, with respect to variable t, we proceed as

$$
\begin{aligned}
\mathscr{L}_M\left(\mathfrak{R}\big[f\big](t,\theta)\right)(\omega) &= \int_{\mathbb{R}} \mathfrak{R}\big[f\big](t,\theta)\, \mathcal{K}_M(t,\omega)\, dt \\
&= \frac{1}{\sqrt{2\pi iB}} \int_{\mathbb{R}} \mathfrak{R}\big[f\big](t,\theta) \exp\left\{\frac{i(At^2 - 2t\omega + D\omega^2)}{2B}\right\} dt \\
&= \frac{1}{\sqrt{2\pi iB}} \int_{\mathbb{R}} \left\{\int_{\mathbb{R}^2} f(\mathbf{x})\, \delta(\Lambda_\theta \cdot \mathbf{x} - t)\, d\mathbf{x}\right\} \\
&\qquad\qquad\qquad\qquad\qquad \times \exp\left\{\frac{i(At^2 - 2t\omega + D\omega^2)}{2B}\right\} dt \\
&= \frac{1}{\sqrt{2\pi iB}} \int_{\mathbb{R}^2} f(\mathbf{x}) \exp\left\{\frac{i(A(\Lambda_\theta \cdot \mathbf{x})^2 - 2(\Lambda_\theta \cdot \mathbf{x})\omega + D\omega^2)}{2B}\right\} d\mathbf{x} \\
&= \frac{1}{\sqrt{2\pi iB}} \int_{\mathbb{R}^2} f(\mathbf{x}) \exp\left\{\frac{i(A(\Lambda_\theta \cdot \mathbf{x})^2 - 2(\Lambda_\theta \cdot \mathbf{x})\omega + D\omega^2)}{2B}\right\} d\mathbf{x}
\end{aligned}
$$

$$= \frac{1}{\sqrt{2\pi iB}} \int_{\mathbb{R}^2} f(\mathbf{x}) \exp\left\{ \frac{iD\omega^2}{2B} \right\}$$

$$\times \exp\left\{ \frac{i(A(\Lambda_\theta \cdot \mathbf{x})^2 - 2(\Lambda_\theta \cdot \mathbf{x})\omega)}{2B} \right\} d\mathbf{x}$$

$$= \frac{1}{\sqrt{2\pi iB}} \exp\left\{ \frac{iD\omega^2}{2B} \right\} \int_{\mathbb{R}^2} f(\mathbf{x}) \exp\left\{ \frac{iA(\Lambda_\theta \cdot \mathbf{x})^2}{2B} \right\}$$

$$\times \exp\left\{ -\frac{i(\omega\Lambda_\theta) \cdot \mathbf{x}}{B} \right\} d\mathbf{x}$$

$$= \sqrt{\frac{2\pi}{iB}} \exp\left\{ \frac{iD\omega^2}{2B} \right\} \mathscr{F}\left[F_M\right]\left(\frac{\omega\Lambda_\theta}{B} \right), \tag{4.156}$$

where $\dfrac{\omega\Lambda_\theta}{B} = \left(\dfrac{\omega\cos\theta}{B}, \dfrac{\omega\sin\theta}{B} \right)$ and $F_M(\mathbf{t})$ is given by $F_M(\mathbf{t}) = f(\mathbf{t})\exp\left\{ \dfrac{iA(\Lambda_\theta \cdot \mathbf{t})^2}{2B} \right\}$.
Finally, using (4.156) in (4.155), we obtain the desired expression for the linear canonical transform of the linear canonical ridgelet transform as

$$\mathscr{L}_M\left(R_\psi^M[f](a,b,\theta) \right)(\omega) = 2\pi\sqrt{\frac{a}{iB}} \exp\left\{ \frac{iD\omega^2}{2B} \right\} \mathscr{F}\left[F_M\right]\left(\frac{\omega\Lambda_\theta}{B} \right) \overline{\mathscr{F}[\psi]\left(\frac{a\omega}{B} \right)}.$$

This completes the proof of Proposition 4.6.2   □

Having the Proposition 4.6.2 at hand, we are ready to obtain an orthogonality relation for the linear canonical ridgelet transform defined in (4.152). As a consequence of such a relation, we can infer that the linear canonical ridgelet transform (4.152) is indeed an isometry from $L^2(\mathbb{R}^2)$ to $L^2(\mathbb{R}^+ \times \mathbb{R} \times [0, 2\pi))$.

Theorem 4.6.3. *The linear canonical ridgelet transforms of any pair of functions $f, g \in L^2(\mathbb{R}^2)$ satisfy the following orthogonality relation:*

$$\int_0^{2\pi} \int_{\mathbb{R}} \int_{\mathbb{R}^+} R_\psi^M[f](a,b,\theta)\, \overline{R_\phi^M[g](a,b,\theta)}\, \frac{da\,db\,d\theta}{a^3} = C_{\psi,\phi}\left\langle f, g \right\rangle_2, \tag{4.157}$$

where $C_{\psi,\phi}$ is given by the usual cross admissibility condition (3.201).

Proof. Identifying the linear canonical ridgelet transform (4.152) as a function of the translation variable b and then applying the Parseval's formula for the linear canonical transform together with an implication of the Fubini's theorem, we have

$$\int_0^{2\pi} \int_{\mathbb{R}} \int_{\mathbb{R}^+} R_\psi^M[f](a,b,\theta)\, \overline{R_\phi^M[g](a,b,\theta)}\, \frac{da\,db\,d\theta}{a^3}$$

$$= \int_0^{2\pi} \int_{\mathbb{R}^+} \int_{\mathbb{R}} \mathscr{L}_M\left(R_\psi^M[f](a,b,\theta) \right)(\omega)\, \overline{\mathscr{L}_M\left(R_\phi^M[g](a,b,\theta) \right)(\omega)}\, \frac{d\omega\,da\,d\theta}{a^3}$$

$$= \int_0^{2\pi} \int_{\mathbb{R}^+} \int_{\mathbb{R}} \left\{ 2\pi\sqrt{\frac{a}{iB}} \exp\left\{ \frac{iD\omega^2}{2B} \right\} \mathscr{F}\left[F_M\right]\left(\frac{\omega\Lambda_\theta}{B} \right) \overline{\mathscr{F}[\psi]\left(\frac{a\omega}{B} \right)} \right\}$$

$$\times \left\{ 2\pi\sqrt{-\frac{a}{iB}} \exp\left\{ -\frac{iD\omega^2}{2B} \right\} \overline{\mathscr{F}\left[G_M\right]\left(\frac{\omega\Lambda_\theta}{B} \right)} \mathscr{F}[\phi]\left(\frac{a\omega}{B} \right) \right\} \frac{d\omega\,da\,d\theta}{a^3}$$

$$= \int_0^{2\pi} \int_{\mathbb{R}} \mathscr{F}\left[F_M\right]\left(\frac{\omega\Lambda_\theta}{B} \right) \overline{\mathscr{F}\left[G_M\right]\left(\frac{\omega\Lambda_\theta}{B} \right)}$$

$$\times \left\{ \frac{(2\pi)^2}{|B|} \int_{\mathbb{R}^+} \overline{\mathscr{F}[\psi]\left(\frac{a\omega}{B} \right)} \mathscr{F}[\phi]\left(\frac{a\omega}{B} \right) \frac{da}{a^2} \right\} d\omega\,d\theta, \tag{4.158}$$

where the functions $F_M(\mathbf{t})$ and $G_M(\mathbf{t})$ appearing on the R.H.S of (4.158) are given by

$$F_M(\mathbf{t}) = f(\mathbf{t}) \exp\left\{ \frac{iA(\Lambda_\theta \cdot \mathbf{t})^2}{2B} \right\} \quad \text{and} \quad G_M(\mathbf{t}) = g(\mathbf{t}) \exp\left\{ \frac{iA(\Lambda_\theta \cdot \mathbf{t})^2}{2B} \right\}.$$

For solving the inner integral on the R.H.S of (4.158), we shall make use of the substitution $a\omega/B = \omega'$, so that

$$\frac{(2\pi)^2}{|B|} \int_{\mathbb{R}^+} \overline{\mathscr{F}[\psi]\left(\frac{a\omega}{B}\right)} \mathscr{F}[\phi]\left(\frac{a\omega}{B}\right) \frac{da}{a^2} = \frac{|\omega|}{|B|^2}\left\{ (2\pi)^2 \int_{\mathbb{R}} \frac{\overline{\mathscr{F}[\psi](\omega')}\,\mathscr{F}[\phi](\omega')}{|\omega'|^2}\, d\omega' \right\}$$

$$= \frac{|\omega|}{|B|^2}\, C_{\psi,\phi}. \tag{4.159}$$

Using the estimate (4.159) in (4.158) and invoking the cross admissibility condition given by (3.201), we get

$$\int_0^{2\pi} \int_{\mathbb{R}} \int_{\mathbb{R}^+} R_\psi^M\big[f\big](a,b,\theta)\, \overline{R_\phi^M\big[g\big](a,b,\theta)}\, \frac{da\, db\, d\theta}{a^3}$$

$$= C_{\psi,\phi} \int_0^{2\pi} \int_{\mathbb{R}} \mathscr{F}\big[F_M\big]\left(\frac{\omega\Lambda_\theta}{B}\right) \overline{\mathscr{F}\big[G_M\big]\left(\frac{\omega\Lambda_\theta}{B}\right)} \frac{|\omega|}{|B|^2}\, d\omega\, d\theta$$

$$= C_{\psi,\phi} \int_{\mathbb{R}^2} \mathscr{F}\big[F_M\big](\mathbf{w})\, \overline{\mathscr{F}\big[G_M\big](\mathbf{w})}\, d(\mathbf{w}), \tag{4.160}$$

where the last expression is obtained via the substitution $\omega\Lambda_\theta/B = (\omega_1, \omega_2) = \mathbf{w}$ and the fact that the Jacobian of the transformation is ω/B^2. Finally, applying the Parseval's formula for the Fourier transform on the R.H.S of (4.160), we obtain the desired orthogonality relation as

$$\int_0^{2\pi} \int_{\mathbb{R}} \int_{\mathbb{R}^+} R_\psi^M\big[f\big](a,b,\theta)\, \overline{R_\phi^M\big[g\big](a,b,\theta)}\, \frac{da\, db\, d\theta}{a^3} = C_{\psi,\phi} \int_{\mathbb{R}^2} \mathscr{F}\big[F_M\big](\mathbf{w})\, \overline{\mathscr{F}\big[G_M\big](\mathbf{w})}\, d\mathbf{w}$$

$$= C_{\psi,\phi} \int_{\mathbb{R}^2} F_M(\mathbf{t})\, \overline{G_M(\mathbf{t})}\, d\mathbf{t}$$

$$= C_{\psi,\phi} \int_{\mathbb{R}^2} f(\mathbf{t})\, \overline{g(\mathbf{t})}\, d\mathbf{t}$$

$$= C_{\psi,\phi} \big\langle f, g \big\rangle_2.$$

This completes the proof of Theorem 4.6.3. $\qquad\qquad\square$

Remark 4.6.1. For $f = g$ and $\psi = \phi$, Theorem 4.6.3 boils down to the following expression:

$$\int_0^{2\pi} \int_{\mathbb{R}} \int_{\mathbb{R}^+} \big|R_\psi^M\big[f\big](a,b,\theta)\big|^2\, \frac{da\, db\, d\theta}{a^3} = C_\psi \|f\|_2^2, \tag{4.161}$$

where C_ψ is given by the admissibility condition (3.193). Evidently, if the wavelet ψ satisfies $C_\psi = 1$, then the linear canonical ridgelet transform (4.152) is indeed an isometry from $L^2(\mathbb{R}^2)$ to $L^2(\mathbb{R}^+ \times \mathbb{R} \times [0, 2\pi))$.

Following is the inversion formula associated with the linear canonical ridgelet transform defined in (4.152):

Theorem 4.6.4. *Any function $f \in L^2(\mathbb{R}^2)$ can be reconstructed from the corresponding linear canonical ridgelet transform $R_\psi^M[f](a, b, \theta)$ via the following formula:*

$$f(\mathbf{t}) = \frac{1}{C_{\psi,\phi}} \int_0^{2\pi} \int_\mathbb{R} \int_{\mathbb{R}^+} R_\psi^M[f](a, b, \theta) \, \phi_{a,b,\theta}^M(\mathbf{t}) \frac{da \, db \, d\theta}{a^3}, \quad a.e., \tag{4.162}$$

where $C_{\psi,\phi} \neq 0$ is given by the cross admissibility condition (3.201).

Proof. We shall accomplish the proof by virtue of Theorem 4.6.3. To facilitate the narrative, we choose $g(\mathbf{x}) = \delta(\mathbf{x} - \mathbf{t})$, $\mathbf{t} \in \mathbb{R}^2$ and note that

$$
\begin{aligned}
R_\phi^M[g](a, b, \theta) &= \frac{1}{\sqrt{a}} \int_{\mathbb{R}^2} \delta(\mathbf{x} - \mathbf{t}) \, \overline{\phi\left(\frac{\Lambda_\theta \cdot \mathbf{x} - b}{a}\right)} \exp\left\{\frac{iA(\Lambda_\theta \cdot \mathbf{x} - b)(\Lambda_\theta \cdot \mathbf{x} + b)}{2B}\right\} d\mathbf{x} \\
&= \frac{1}{\sqrt{a}} \, \overline{\phi\left(\frac{\Lambda_\theta \cdot \mathbf{x} - b}{a}\right)} \exp\left\{\frac{iA(\Lambda_\theta \cdot \mathbf{x} - b)(\Lambda_\theta \cdot \mathbf{x} + b)}{2B}\right\} \\
&= \overline{\phi_{a,b,\theta}^M(\mathbf{t})}. \tag{4.163}
\end{aligned}
$$

Besides, we observe that

$$\left\langle f, g \right\rangle_2 = \int_{\mathbb{R}^2} f(\mathbf{x}) \, \delta(\mathbf{x} - \mathbf{t}) \, d\mathbf{t} = f(\mathbf{t}). \tag{4.164}$$

Invoking the orthogonality relation (4.157), then using (4.163) and (4.164), we obtain the reconstruction formula for the linear canonical ridgelet transform as

$$f(\mathbf{t}) = \frac{1}{C_{\psi,\phi}} \int_0^{2\pi} \int_\mathbb{R} \int_{\mathbb{R}^+} R_\psi^M[f](a, b, \theta) \, \phi_{a,b,\theta}^M(\mathbf{t}) \frac{da \, db \, d\theta}{a^3}, \quad a.e.$$

This completes the proof of Theorem 4.6.4. $\square$

Towards the end of the subsection, we shall formulate the Heisenberg-type uncertainty inequality associated with the linear canonical ridgelet transform defined in (4.152).

Theorem 4.6.5. *If $R_\psi^M[f](a, b, \theta)$ is the linear canonical ridgelet transform of any non-trivial function $f \in L^2(\mathbb{R}^2)$, then the following inequality holds:*

$$\left\{\int_0^{2\pi} \int_\mathbb{R} \int_{\mathbb{R}^+} b^2 \left|R_\psi^M[f](a, b, \theta)\right|^2 \frac{da \, db \, d\theta}{a^3}\right\}^{1/2} \left\{\int_{\mathbb{R}^2} |\mathbf{w}|^2 \left|\mathscr{F}[F_M](\mathbf{w})\right|^2 d\mathbf{w}\right\}^{1/2} \geq \frac{\sqrt{C_\psi}}{2} \|f\|_2^2, \tag{4.165}$$

where C_ψ is given by the usual admissibility condition (3.193) and the function $F_M(\mathbf{t})$ is defined as

$$F_M(\mathbf{t}) = f(\mathbf{t}) \exp\left\{\frac{iA(\Lambda_\theta \cdot \mathbf{t})^2}{2B}\right\}.$$

Proof. For any non-trivial function $g \in L^2(\mathbb{R})$, the Heisenberg's uncertainty principle associated with the linear canonical transform reads:

$$\left\{\int_\mathbb{R} t^2 |g(t)|^2 dt\right\}^{1/2} \left\{\int_\mathbb{R} \omega^2 \left|\mathscr{L}_M[g](\omega)\right|^2 d\omega\right\}^{1/2} \geq \frac{|B|}{2} \left\{\int_\mathbb{R} |g(t)|^2 dt\right\}. \tag{4.166}$$

We shall identify the linear canonical ridgelet transform $R_\psi^M[f](a, b, \theta)$ as a function of the translation variable b and then invoke (4.166) to obtain

$$\left\{ \int_{\mathbb{R}} b^2 \left| R_\psi^M[f](a, b, \theta) \right|^2 db \right\}^{1/2} \left\{ \int_{\mathbb{R}} \omega^2 \left| \mathscr{L}_M \left(R_\psi^M[f](a, b, \theta) \right)(\omega) \right|^2 d\omega \right\}^{1/2}$$

$$\geq \frac{|B|}{2} \left\{ \int_{\mathbb{R}} \left| R_\psi^M[f](a, b, \theta) \right|^2 db \right\}. \quad (4.167)$$

Integrating the inequality (4.167) with respect to the measure $da\, d\theta/a^3$ yields

$$\int_0^{2\pi} \int_{\mathbb{R}+} \left\{ \int_{\mathbb{R}} b^2 \left| R_\psi^M[f](a, b, \theta) \right|^2 db \right\}^{1/2} \left\{ \int_{\mathbb{R}} \omega^2 \left| \mathscr{L}_M \left(R_\psi^M[f](a, b, \theta) \right)(\omega) \right|^2 d\omega \right\}^{1/2} \frac{da\, d\theta}{a^3}$$

$$\geq \frac{|B|}{2} \left\{ \int_0^{2\pi} \int_{\mathbb{R}+} \int_{\mathbb{R}} \left| R_\psi^M[f](a, b, \theta) \right|^2 \frac{db\, da\, d\theta}{a^3} \right\}. \quad (4.168)$$

Applying the Cauchy-Schwarz inequality, we can redraft inequality (4.168) as

$$\left\{ \int_0^{2\pi} \int_{\mathbb{R}} \int_{\mathbb{R}+} b^2 \left| R_\psi^M[f](a, b, \theta) \right|^2 \frac{da\, db\, d\theta}{a^3} \right\}^{1/2}$$

$$\times \left\{ \int_0^{2\pi} \int_{\mathbb{R}} \int_{\mathbb{R}+} \omega^2 \left| \mathscr{L}_M \left(R_\psi^M[f](a, b, \theta) \right)(\omega) \right|^2 \frac{da\, d\omega\, d\theta}{a^3} \right\}^{1/2} \geq C_\psi \frac{|B|}{2} \|f\|_2^2. \quad (4.169)$$

By virtue of Proposition 4.6.2, we can simplify the second set of integrals on the L.H.S of (4.169) as

$$\int_0^{2\pi} \int_{\mathbb{R}} \int_{\mathbb{R}+} \omega^2 \left| \mathscr{L}_M \left(R_\psi^M[f](a, b, \theta) \right)(\omega) \right|^2 \frac{da\, d\omega\, d\theta}{a^3}$$

$$= \frac{(2\pi)^2}{|B|} \int_0^{2\pi} \int_{\mathbb{R}} \int_{\mathbb{R}+} \omega^2 \left| \mathscr{F}[F_M] \left(\frac{\omega \Lambda_\theta}{B} \right) \right|^2 \left| \mathscr{F}[\psi] \left(\frac{a\omega}{B} \right) \right|^2 \frac{da\, d\omega\, d\theta}{a^2}$$

$$= \frac{1}{|B|} \int_0^{2\pi} \int_{\mathbb{R}} \omega^2 \left| \mathscr{F}[F_M] \left(\frac{\omega \Lambda_\theta}{B} \right) \right|^2 \left\{ (2\pi)^2 \int_{\mathbb{R}+} \left| \mathscr{F}[\psi] \left(\frac{a\omega}{B} \right) \right|^2 \frac{da}{a^2} \right\} d\omega\, d\theta, \quad (4.170)$$

where $F_M(\mathbf{t}) = f(\mathbf{t}) \exp \left\{ \dfrac{iA(\Lambda_\theta \cdot \mathbf{t})^2}{2B} \right\}$. The inner integral on the R.H.S of (4.170) can be solved by substituting $a\omega/B = \omega'$ and noting that

$$(2\pi)^2 \int_{\mathbb{R}+} \left| \mathscr{F}[\psi] \left(\frac{a\omega}{B} \right) \right|^2 \frac{da}{a^2} = \frac{|\omega|}{|B|} \int_{\mathbb{R}} \frac{\left| \mathscr{F}[\psi](\omega') \right|^2}{|\omega'|^2} d\omega' = \frac{|\omega|}{|B|} C_\psi, \quad (4.171)$$

where C_ψ is given by the admissibility condition (3.193). Using (4.171) in (4.170), we get

$$\int_0^{2\pi} \int_{\mathbb{R}} \int_{\mathbb{R}+} \omega^2 \left| \mathscr{L}_M \left(R_\psi^M[f](a, b, \theta) \right)(\omega) \right|^2 \frac{da\, d\omega\, d\theta}{a^3}$$

$$= C_\psi \int_0^{2\pi} \int_{\mathbb{R}} \omega^2 \left| \mathscr{F}[F_M] \left(\frac{\omega \Lambda_\theta}{B} \right) \right|^2 \frac{|\omega|}{|B|^2} d\omega\, d\theta$$

$$= B^2 C_\psi \int_{\mathbb{R}^2} |\mathbf{w}|^2 \left| \mathscr{F}[F_M](\mathbf{w}) \right|^2 d\mathbf{w}, \quad (4.172)$$

where the last expression is obtained via the substitution $\omega\Lambda_\theta/B = (\omega_1, \omega_2) = \mathbf{w}$ and the fact that the Jacobian of the transformation is ω/B^2. Finally, using (4.172) in (4.169), we obtain the desired uncertainty inequality for the linear canonical ridgelet transform as

$$\left\{\int_0^{2\pi}\int_{\mathbb{R}}\int_{\mathbb{R}+} b^2 \left|R_\psi^M\big[f\big](a,b,\theta)\right|^2 \frac{da\,db\,d\theta}{a^3}\right\}^{1/2} \left\{\int_{\mathbb{R}^2}|\mathbf{w}|^2 \left|\mathscr{F}\big[F_M\big](\mathbf{w})\right|^2 d\mathbf{w}\right\}^{1/2} \geq \frac{\sqrt{C_\psi}}{2}\left\|f\right\|_2^2.$$

This completes the proof of Theorem 4.6.5.   □

Remark 4.6.2. For $M = (0,1;-1,0)$ and $M = (\cos\alpha, \sin\alpha; -\sin\alpha, \cos\alpha)$, $\alpha \neq n\pi$, $n \in \mathbb{Z}$, the uncertainty inequality (4.165) boils down to the respective Heisenberg-type inequalities for the classical and fractional variants of the ridgelet transform.

4.7   The Linear Canonical Curvelet and Ripplet Transforms

"The tool which serves as intermediary between theory and practice, between thought and observation, is mathematics, it is mathematics which builds the linking bridges and gives the ever more reliable forms. From this it has come about that our entire contemporary culture, in as much as it is based on the intellectual penetration and the exploitation of nature, has its foundations in mathematics."

-David Hilbert

This section is exclusively devoted to extend the pre-eminent directional representation systems, namely curvelets and ripplets to the linear canonical domain. In fact, our goal is to answer the fundamental question of whether it is possible to increase the flexibility of curvelet and ripplet waveforms to optimize the concentration of curvelet and ripplet spectra? The answer to this question is affirmative and lies in intertwining the curvelet and ripplet transforms with the well-known linear canonical transform, which is best known for its flexibility and higher degrees of freedom in modelling physical phenomenon. To facilitate the intent, we subdivide the section into two subsections, respectively dealing with the notions of linear canonical curvelet and ripplet transforms.

In order to formulate the notion of linear canonical curvelet and ripplet transforms, it is imperative to present the notion of the two-dimensional linear canonical transform. In this regard, we have the following definition:

Definition 4.7.1. Given a real, unimodular matrix $M = (A, B; C, D)$, the two-dimensional linear canonical transform of any $f \in L^2(\mathbb{R}^2)$ is defined as

$$\mathcal{L}_M\big[f\big](\mathbf{w}) = \begin{cases} \displaystyle\int_{\mathbb{R}^2} f(\mathbf{t})\,\mathcal{K}_M(\mathbf{t},\mathbf{w})\,d\mathbf{t}, & B \neq 0 \\[2mm] \sqrt{D}\,\exp\left\{\dfrac{iCD|\mathbf{w}|^2}{2}\right\} f(D\mathbf{w}), & B = 0, \end{cases} \tag{4.173}$$

where $\mathcal{K}_M(\mathbf{t},\mathbf{w})$, with $\mathbf{t} = (t_1,t_2)^T$, $\mathbf{w} = (\omega_1,\omega_2)^T$, denotes the kernel of the two-dimensional linear canonical transform and is given by

$$\mathcal{K}_M(\mathbf{t},\mathbf{w}) = \frac{1}{2\pi B}\exp\left\{\frac{i(A|\mathbf{t}|^2 - 2\mathbf{t}^T\mathbf{w} + D|\mathbf{w}|^2)}{2B}\right\}, \quad B \neq 0. \tag{4.174}$$

It is pertinent to mention that for the case $B = 0$, the two-dimensional linear canonical transform (4.173) corresponds to a scaling transformation coupled with a chirp multiplication operation. Moreover, the case $B < 0$ is also of no particular interest to us. As such, we shall only focus our attention on the case $B > 0$. We also note that the phase-space transform (4.173) is lossless if and only if the matrix M is unimodular; that is, $AD - BC = 1$. The inversion formula corresponding to (4.173) is given by

$$f(\mathbf{t}) = \mathcal{L}_M^{-1}\left(\mathcal{L}_M\left[f\right](\mathbf{w})\right)(\mathbf{t}) := \int_{\mathbb{R}^2} \mathcal{L}_M\left[f\right](\mathbf{w}) \, \overline{\mathcal{K}_M(\mathbf{t}, \mathbf{w})} \, d\mathbf{w}. \qquad (4.175)$$

For any pair of functions $f, g \in L^2(\mathbb{R}^2)$, the Parseval's formula associated with (4.173) reads:

$$\left\langle f, g \right\rangle_2 = \left\langle \mathcal{L}_M\left[f\right], \mathcal{L}_M\left[g\right] \right\rangle_2. \qquad (4.176)$$

The two-dimensional linear canonical transform via polar coordinates is particulary interesting, because it allows a hassle-free construction of the linear canonical curvelet and ripplet waveforms. For $\omega_1 = r\cos\phi$, $\omega_2 = r\sin\phi$ and $t_1 = \rho\cos\eta$, $t_2 = \rho\sin\eta$, where $r, \rho \geq 0$ and $\phi, \eta \in [0, 2\pi)$, the polar linear canonical transform is given by

$$\mathcal{L}_M\left[f\right](r, \phi) = \frac{1}{2\pi B} \int_0^{2\pi} \int_0^\infty f(\rho, \eta) \, \exp\left\{ \frac{i(A\rho^2 + Dr^2 - 2\rho r \cos(\eta - \phi))}{2B} \right\} \rho \, d\rho \, d\eta. \qquad (4.177)$$

Also, the inversion formula corresponding to the integral transformation (4.177) is given by

$$f(\rho, \eta) = \frac{1}{2\pi B} \int_0^{2\pi} \int_0^\infty \mathcal{L}_M\left[f\right](r, \phi) \, \exp\left\{ -\frac{i(A\rho^2 + Dr^2 - 2\rho r \cos(\eta - \phi))}{2B} \right\} r \, dr \, d\phi. \qquad (4.178)$$

Remark 4.7.1. For $M = (0, 1; -1, 0)$ and $M = (\cos\alpha, \sin\alpha; -\sin\alpha, \cos\alpha)$, $\alpha \neq n\pi$, $n \in \mathbb{Z}$, the expressions (4.173) and (4.177) reduce to their respective counterparts for the two-dimensional Fourier and fractional Fourier transforms.

4.7.1 Linear canonical curvelet transform

"There cannot be a language more universal and more simple, more free from errors and obscurities...more worthy to express the invariable relations of all natural things [than mathematics]. [It interprets] all phenomena by the same language, as if to attest the unity and simplicity of the plan of the universe, and to make still more evident that unchangeable order which presides over all natural causes."

-Joseph Fourier

The classical curvelet transform aims to deal with certain interesting phenomena occurring along curved edges in higher dimensional signals. Unlike the wavelet transform, the curvelet transform provides time-frequency localization with a reasonable directionality and anisotropy by using angled polar wedges or angled trapezoid windows in frequency domain. The intrinsic multi-scale and anisotropic nature of curvelet waveforms have bestowed the curvelet transform a respectable status in the context of multi-scale analysis of multi-dimensional data. Besides many pleasant properties, the curvelet transform offers optimally sparse representations of objects which display curve-punctuated smoothness; that is, smoothness except for discontinuity along a general curve with bounded curvature. Keeping in view the merits of curvelet transform, we shall formulate the notion linear canonical curvelet transform-a novel integral transform endowed with higher degrees of freedom in lieu to the classical curvelet transform.

Consider the frequency plane $\mathbb{R}^2$ and let (r, ϕ), $r \geq 0$, $\phi \in [0, 2\pi)$ denote the polar coordinates of an arbitrary point $\mathbf{w} \in \mathbb{R}^2$. We choose a pair of window functions $W : (0, \infty) \to (0, \infty)$, called as "radial window," and $V : (-\infty, \infty) \to (0, \infty)$, called as "angular window," satisfying the following admissibility conditions:

$$B^2 \int_0^\infty |W(r)|^2 \, \frac{dr}{r} = 1, \qquad \text{supp}\,(W) \subseteq \left(\frac{1}{2}, 2\right) \qquad (4.179)$$

and

$$(2\pi)^2 \int_{-1}^1 |V(\phi)|^2 \, d\phi = 1, \qquad \text{supp}\,(V) \subseteq [-1, 1]. \qquad (4.180)$$

The window functions satisfying (4.179) and (4.180) are used to construct a novel family of complex-valued waveforms, known as "linear canonical curvelets," adopted to scale $a > 0$, location $\mathbf{b} \in \mathbb{R}^2$ and orientation $\theta \in [0, 2\pi)$ (or $(-\pi, \pi)$ according to convenience). For a fixed scale $a \in (0, a_0)$, where $a_0 < \pi^2$, consider a basic waveform $\Psi_a : \mathbb{R}^2 \to \mathbb{C}$ defined via the polar linear canonical transform (4.177) as

$$\mathcal{L}_M\left[\Psi_a\right](r, \phi) = a^{3/4} W\left(ar\right) V\left(\frac{\phi}{\sqrt{a}}\right). \qquad (4.181)$$

Applying the inverse transformation (4.178) on both sides of the expression (4.181), we have

$$\Psi_a(\rho, \eta) = \frac{a^{3/4}}{2\pi B} \int_0^{2\pi} \int_0^\infty W\left(ar\right) V\left(\frac{\phi}{\sqrt{a}}\right) \exp\left\{-\frac{i(A\rho^2 + Dr^2 - 2\rho r \cos(\eta - \phi))}{2B}\right\} r \, dr \, d\phi$$

$$= \frac{a^{3/4}}{2\pi B} \exp\left\{-\frac{iA\rho^2}{2B}\right\} \int_0^{2\pi} \int_0^\infty \exp\left\{-\frac{iDr^2}{2B}\right\} W\left(ar\right) V\left(\frac{\phi}{\sqrt{a}}\right)$$

$$\times \exp\left\{\frac{i\rho r \cos(\eta - \phi))}{B}\right\} r \, dr \, d\phi. \quad (4.182)$$

Upon simplifying (4.182), we obtain a novel basic waveform $\Psi_a^M(\mathbf{t})$ via the following expression:

$$\mathscr{F}\left[\Psi_a^M\right](r, \phi) = a^{3/4} B \, \exp\left\{-\frac{iDBr^2}{2}\right\} W\left(aBr\right) V\left(\frac{\phi}{\sqrt{a}}\right), \qquad (4.183)$$

where $\Psi_a^M(\mathbf{t}) = \exp\left\{\frac{iA|\mathbf{t}|^2}{2B}\right\} \Psi_a(\mathbf{t})$. Hence, the family of novel curvelets $\Psi_{a,\mathbf{b},\theta}^M(\mathbf{t})$ (or linear canonical curvelets) is obtained by translating the basic waveform $\Psi_a^M(\mathbf{t})$ with $\mathbf{b} \in \mathbb{R}^2$ and then inducing a rotation of $\theta \in [0, 2\pi)$ radians; that is,

$$\Psi_{a,\mathbf{b},\theta}^M(\mathbf{t}) = \Psi_a^M\left(R_\theta(\mathbf{t} - \mathbf{b})\right), \quad \mathbf{t} \in \mathbb{R}^2, \qquad (4.184)$$

where $R_\theta = \begin{pmatrix} \cos\theta & -\sin\theta \\ \sin\theta & \cos\theta \end{pmatrix}$ denotes the 2×2 rotation matrix effecting the planar rotation by θ radians.

Having formulated the notion of linear canonical curvelets $\Psi_{a,\mathbf{b},\theta}^M(\mathbf{t})$ defined in (4.184), we are ready to introduce the formal definition of the novel curvelet transform (or linear canonical curvelet transform).

Definition 4.7.2. Given a real, unimodular matrix $M = (A, B; C, D)$ with $B > 0$ and any square integrable function f on $\mathbb{R}^2$, the novel curvelet transform is denoted by $\Gamma_\Psi^M[f]$ and is defined as

$$\Gamma_\Psi^M[f](a, \mathbf{b}, \theta) = \left\langle f, \Psi_{a,\mathbf{b},\theta}^M \right\rangle_2 = \int_{\mathbb{R}^2} f(\mathbf{t}) \, \overline{\Psi_{a,\mathbf{b},\theta}^M(\mathbf{t})} \, d\mathbf{t}, \tag{4.185}$$

where $a \in (0, a_0)$ with $a_0 < \pi^2$, $\mathbf{b} \in \mathbb{R}^2$, $\theta \in [0, 2\pi)$ and $\Psi_{a,\mathbf{b},\theta}^M(\mathbf{t})$ is given by (4.184).

The following important points regarding the novel curvelet transform can be extracted from Definition 4.7.2:

(i) The ordinary curvelet transform (3.220) is just a special case of novel curvelet transform (4.185) and follows by plugging $M = (0, 1; -1, 0)$ in Definition 4.7.2.

(ii) For $M = (\cos\alpha, \sin\alpha; -\sin\alpha, \cos\alpha)$, $\alpha \neq n\pi$, $n \in \mathbb{Z}$, Definition 4.7.2 yields a new variant of the curvelet transform based on the fractional Fourier transform.

(iii) The distinguishing features of the newly constructed family of novel curvelet waveforms can be ascertained from expression (4.183), which demonstrates that apart from the chirp multipliers, the radial window is dictated by the parameter $B > 0$ appearing in the given unimodular matrix $M = (A, B; C, D)$. Such features of the novel curvelet waveforms can be of substantial importance in optimizing the concentration of the curvelet spectrum.

(iv) The computational complexity of the novel curvelet transform given in Definition 4.7.2 is completely determined by that of the ordinary curvelet transform.

Next, we present a proposition which interlinks the Fourier transform of the novel curvelet transform $\Gamma_\Psi^M[f](a, \mathbf{b}, \theta)$, as a function of the translation variable $\mathbf{b}$, with the respective Fourier transforms of the given function f and the basic waveform Ψ_a^M.

Proposition 4.7.1. *Given any $f \in L^2(\mathbb{R}^2)$, the novel curvelet transform $\Gamma_\Psi^M[f](a, \mathbf{b}, \theta)$ defined in (4.185) can be expressed as*

$$\mathscr{F}\left(\Gamma_\Psi^M[f](a, \mathbf{b}, \theta)\right)(\mathbf{w}) = 2\pi\, \mathscr{F}[f](\mathbf{w})\, \mathscr{F}\left[\Psi_a^M\right](R_\theta \mathbf{w}). \tag{4.186}$$

Proof. To accomplish the motive, we shall firstly compute the two-dimensional Fourier transform (1.288) of the novel curvelet waveforms $\Psi_{a,\mathbf{b},\theta}^M(\mathbf{t})$ defined in (4.184). We proceed as

$$\mathscr{F}\left[\Psi_{a,\mathbf{b},\theta}^M\right](\mathbf{w}) = \frac{1}{2\pi} \int_{\mathbb{R}^2} \Psi_{a,\mathbf{b},\theta}^M(\mathbf{t})\, e^{-i\mathbf{t}^T \mathbf{w}}\, d\mathbf{t}$$

$$= \frac{1}{2\pi} \int_{\mathbb{R}^2} \Psi_a^M\left(R_\theta(\mathbf{t} - \mathbf{b})\right) e^{-i\mathbf{t}^T \mathbf{w}}\, d\mathbf{t}$$

$$= \frac{1}{2\pi} \int_{\mathbb{R}^2} \Psi_a^M(\mathbf{z})\, e^{-i(\mathbf{b} + R_{-\theta}\mathbf{z})^T \mathbf{w}}\, d\mathbf{z}$$

$$= \frac{e^{-i\mathbf{b}^T \mathbf{w}}}{2\pi} \int_{\mathbb{R}^2} \Psi_a^M(\mathbf{z})\, e^{-i\mathbf{z}^T (R_\theta \mathbf{w})}\, d\mathbf{z}$$

$$= e^{-i\mathbf{b}^T \mathbf{w}}\, \mathscr{F}\left[\Psi_a^M\right](R_\theta \mathbf{w}). \tag{4.187}$$

Let (σ, μ), (ρ, η) and (r, ϕ) denote the polar coordinates of the variables $\mathbf{b}$, $\mathbf{t}$ and $\mathbf{w}$ respectively. Then, we can rewrite (4.187) as follows:

$$\mathscr{F}\left[\Psi_{a,\mathbf{b},\theta}^M\right](r, \phi) = e^{-ir\sigma\cos(\mu - \phi)}\, \mathscr{F}\left[\Psi_a^M\right](r, \phi - \theta)$$

$$= e^{-ir\sigma \cos(\mu-\phi)} a^{3/4} B \, \exp\left\{-\frac{iDBr^2}{2}\right\} W(aBr) V\left(\frac{\phi-\theta}{\sqrt{a}}\right). \quad (4.188)$$

Using Definition 4.7.2 and invoking the well-known Parseval's formula in polar coordinates, we have

$$\Gamma_\Psi^M\left[f\right](a,\mathbf{b},\theta) = \left\langle f, \, \Psi_{a,\mathbf{b},\theta}^M \right\rangle_2$$

$$= a^{3/4} B \int_0^{2\pi} \int_0^\infty e^{ir\sigma \cos(\mu-\phi)} \mathscr{F}\left[f\right](r,\phi)$$

$$\times \exp\left\{\frac{iDBr^2}{2}\right\} W(aBr) V\left(\frac{\phi-\theta}{\sqrt{a}}\right) r \, dr \, d\phi. \quad (4.189)$$

Next, translating the expression (4.189) into Cartesian coordinates yields the following:

$$\Gamma_\Psi^M\left[f\right](a,\mathbf{b},\theta) = \int_{\mathbb{R}^2} \mathscr{F}\left[f\right](\mathbf{w}) \overline{\mathscr{F}\left[\Psi_a^M\right](R_\theta \mathbf{w})} \, e^{i\mathbf{b}^T\mathbf{w}} \, d\mathbf{w}$$

$$= 2\pi \, \mathscr{F}^{-1}\left(\mathscr{F}\left[f\right](\mathbf{w}) \overline{\mathscr{F}\left[\Psi_a^M\right](R_\theta \mathbf{w})}\right)(\mathbf{b}). \quad (4.190)$$

Applying the Fourier transform on both sides of (4.190), we obtain the desired result

$$\mathscr{F}\left(\Gamma_\Psi^M\left[f\right](a,\mathbf{b},\theta)\right)(\mathbf{w}) = 2\pi \, \mathscr{F}\left[f\right](\mathbf{w}) \, \mathscr{F}\left[\Psi_a^M\right](R_\theta \mathbf{w}).$$

This completes the proof of Proposition 4.7.1. □

Next, our objective is to analyze the support and oscillatory behaviour of the novel curvelet transform by invoking Proposition 4.7.1. We shall demonstrate that the proposed transform enjoys a certain degree of freedom as the radial window is comparatively more flexible with the degree of flexibility governed by the matrix parameter $B > 0$. As such, the proposed transform is capable of optimizing the concentration of the curvelet spectrum.

Let (σ, μ), $\sigma \geq 0$, $\mu \in [0, 2\pi)$ be the polar coordinates of the translation variable $\mathbf{b}$. Then, as a consequence of Proposition 4.7.1, we can express the novel curvelet transform defined in (4.185) as follows:

$$\Gamma_\Psi^M\left[f\right](a,\mathbf{b},\theta) = \Gamma_\Psi^M\left[f\right](a,(\sigma,\mu),\theta)$$

$$= a^{3/4} B \int_0^{2\pi} \int_0^\infty e^{ir\sigma \cos(\mu-\phi)} \mathscr{F}\left[f\right](r,\phi)$$

$$\times \exp\left\{\frac{iDBr^2}{2}\right\} W(aBr) V\left(\frac{\phi-\theta}{\sqrt{a}}\right) r \, dr \, d\phi. \quad (4.191)$$

From (4.188), we observe that the support of the analyzing elements $\Psi_{a,\mathbf{b},\theta}^M(t)$ in the frequency domain is completely determined by the support of the radial window $W(aBr)$ and the angular window $V\left(\frac{\phi-\theta}{\sqrt{a}}\right)$. Moreover, we observe that

$$\left.\begin{array}{l} \operatorname{supp}\left(W(aBr)\right) \subseteq \left(\dfrac{1}{2aB}, \dfrac{2}{aB}\right), \\[3mm] \operatorname{supp}\left(V\left(\dfrac{\phi-\theta}{\sqrt{a}}\right)\right) \subseteq \left[-\sqrt{a}+\theta, \sqrt{a}+\theta\right] \end{array}\right\}. \quad (4.192)$$

Hence, we conclude that the support of the analyzing elements $\Psi^M_{a,\mathbf{b},\theta}(t)$ in the frequency domain depends upon the choice of the matrix parameter B and is completely independent of the translation parameter $\mathbf{b}$. Therefore, an appropriate matrix parameter B can be chosen to optimize the concentration of novel curvelet spectrum.

On the other hand, since the curvelet functions $\Psi^M_{a,\mathbf{b},\theta}(t)$ have compact support in the frequency domain, therefore, the well-known Heisenberg's uncertainty principle implies that the novel curvelet functions cannot have compact support in the time domain. We note that for large $|t|$, the decay of the novel curvelet functions $\Psi^M_{a,\mathbf{b},\theta}(\mathbf{t})$ depends upon the smoothness of the corresponding Fourier transform; the smoother the $\mathscr{F}\big[\Psi^M_{a,\mathbf{b},\theta}\big](\mathbf{w})$, the faster the decay. Also, by definition $\mathscr{F}\big[\Psi^M_a\big](\mathbf{w})$ is supported away from the vertical axis $\omega_1 = 0$ but near the horizontal axis $\omega_2 = 0$. Hence, for smaller values of $a < a_0$, the basic waveform $\Psi^M_a(\mathbf{t})$ is less oscillatory in t_2 direction and more oscillatory in t_1 direction.

Below, we shall present the formal reconstruction formula associated with the novel curvelet transform (4.185). We note that the said reconstruction formula is valid for high frequency signals. The analogue for low-frequency signals will be dealt afterwards. To facilitate the narrative, recall that for any pair functions $f, g \in L^2(\mathbb{R}^2)$, the two-dimensional convolution operation is denoted by $\circledast$ and is defined as

$$(f \circledast g)(\mathbf{z}) = \int_{\mathbb{R}^2} f(\mathbf{t})\, g(\mathbf{z} - \mathbf{t})\, d\mathbf{t}. \tag{4.193}$$

Upon applying the two-dimensional Fourier transform (1.288), we obtain the spectral representation of the convolution (4.193) as

$$\mathscr{F}\big[(f \circledast g)\big](\mathbf{w}) = 2\pi\, \mathscr{F}\big[f\big](\mathbf{w})\, \mathscr{F}\big[g\big](\mathbf{w}). \tag{4.194}$$

Theorem 4.7.2. *For any $f \in L^2(\mathbb{R}^2)$ satisfying $\mathscr{F}[f](\mathbf{w}) = 0$, $\forall\, |\mathbf{w}| < 2/a_0 B$, $a_0 < \pi^2$, the reconstruction formula for the novel curvelet transform $\Gamma^M_\Psi[f](a, \mathbf{b}, \theta)$ defined in (4.185) is given by*

$$f(\mathbf{t}) = \int_0^{2\pi} \int_{\mathbb{R}^2} \int_0^{a_0} \Gamma^M_\Psi\big[f\big](a, \mathbf{b}, \theta)\, \Psi^M_{a,\mathbf{b},\theta}(\mathbf{t})\, \frac{da\, d\mathbf{b}\, d\theta}{a^3}, \tag{4.195}$$

where the radial and angular windows W and V satisfy their respective admissibility conditions (4.179) and (4.180).

Proof. We note that the novel curvelet transform $\Gamma^M_\Psi[f](a, \mathbf{b}, \theta)$ defined in (4.192) can be expressed via the convolution operation $\circledast$ as follows:

$$\begin{aligned}
\Gamma^M_\Psi\big[f\big](a, \mathbf{b}, \theta) &= \int_{\mathbb{R}^2} f(\mathbf{t})\, \overline{\Psi^M_a\left(R_\theta(\mathbf{t} - \mathbf{b})\right)}\, d\mathbf{t} \\
&= \int_{\mathbb{R}^2} f(\mathbf{t})\, \overline{\Psi^M_{a,0,\theta}\left(-(\mathbf{b} - \mathbf{t})\right)}\, d\mathbf{t} \\
&= (f \circledast \tilde{\Psi}^M_{a,0,\theta})(\mathbf{b}), \qquad \tilde{\Psi}^M(\mathbf{t}) = \overline{\Psi^M(-\mathbf{t})}.
\end{aligned} \tag{4.196}$$

We now define a function

$$F^M_{a,\theta}(\mathbf{t}) = \int_{\mathbb{R}^2} \Gamma^M_\Psi\big[f\big](a, \mathbf{b}, \theta)\, \Psi^M_{a,\mathbf{b},\theta}(\mathbf{t})\, d\mathbf{b}. \tag{4.197}$$

Using the convolution representation (4.196), we can express (4.197) as follows:

$$F^M_{a,\theta}(\mathbf{t}) = \left(\big(f \circledast \tilde{\Psi}^M_{a,0,\theta} \big)(\mathbf{b}) \circledast \Psi^M_{a,0,\theta}(\mathbf{b}) \right)(\mathbf{t}). \tag{4.198}$$

Applying (4.194), we can compute the Fourier transform of the function $F_{a,\theta}(\mathbf{t})$ as

$$\mathscr{F}\left[F_{a,\theta}^{M}\right](\mathbf{w}) = 2\pi\, \mathscr{F}\left[\left(f \circledast \tilde{\Psi}_{a,0,\theta}^{M}\right)\right](\mathbf{w})\, \mathscr{F}\left[\Psi_{a,0,\theta}^{M}\right](\mathbf{w})$$

$$= (2\pi)^2 \mathscr{F}\left[f\right](\mathbf{w})\left|\mathscr{F}\left[\Psi_{a,0,\theta}^{M}\right](\mathbf{w})\right|^2. \tag{4.199}$$

Consequently, we have

$$\int_0^{a_0}\int_0^{2\pi} \mathscr{F}\left[F_{a,\theta}^{M}\right](\mathbf{w})\,\frac{d\theta\,da}{a^3} = (2\pi)^2 \mathscr{F}\left[f\right](\mathbf{w})\int_0^{a_0}\int_0^{2\pi}\left|\mathscr{F}\left[\Psi_{a,0,\theta}^{M}\right](\mathbf{w})\right|^2\frac{d\theta\,da}{a^3}. \tag{4.200}$$

Next, we shall evaluate the integral expression on the R.H.S of (4.200). To do so, we shall use the polar coordinates of $\mathbf{w}$ and invoke the admissibility conditions (4.179) and (4.180). For $r \geq 2/a_0 B$, $a_0 < \pi^2$, we have

$$(2\pi)^2 \int_0^{a_0}\int_0^{2\pi}\left|\mathscr{F}\left[\Psi_{a,0,\theta}^{M}\right](r,\phi)\right|^2\frac{d\theta\,da}{a^3}$$

$$= (2\pi B)^2 \int_0^{a_0}\int_0^{2\pi}\left|W(aBr)\right|^2\left|V\left(\frac{\phi-\theta}{\sqrt{a}}\right)\right|^2\frac{d\theta\,da}{a^{3/2}}$$

$$= B^2 \int_0^{a_0}\left|W(aBr)\right|^2\left\{(2\pi)^2\int_0^{2\pi}\left|V\left(\frac{\phi-\theta}{\sqrt{a}}\right)\right|^2 d\theta\right\}\frac{da}{a^{3/2}}$$

$$= B^2 \int_0^{a_0}\left|W(aBr)\right|^2\frac{da}{a}$$

$$= B^2 \int_0^{a_0 Br}\left|W(r')\right|^2\frac{dr'}{r'} = 1. \tag{4.201}$$

Implementing (4.201) in (4.200), we obtain

$$\mathscr{F}\left[f\right](\mathbf{w}) = \int_0^{a_0}\int_0^{2\pi}\mathscr{F}\left[F_{a,\theta}^{M}\right](\mathbf{w})\,\frac{d\theta\,da}{a^3}. \tag{4.202}$$

That is,

$$f(\mathbf{t}) = \int_0^{2\pi}\int_{\mathbb{R}^2}\int_0^{a_0}\Gamma_{\Psi}^{M}\left[f\right](a,\mathbf{b},\theta)\,\Psi_{a,\mathbf{b},\theta}^{M}(\mathbf{t})\,\frac{da\,d\mathbf{b}\,d\theta}{a^3}.$$

This completes the proof of Theorem 4.7.2.   □

Theorem 4.7.3. *For any $f \in L^2(\mathbb{R}^2)$ satisfying $\mathscr{F}\left[f\right](\mathbf{w}) = 0$, $\forall\,|\mathbf{w}| < 2/a_0 B$, $a_0 < \pi^2$, we have*

$$\left\|\Gamma_{\Psi}^{M}\left[f\right](a,\mathbf{b},\theta)\right\|_2^2 = \left\|f\right\|_2^2. \tag{4.203}$$

That is, the total energy of the signal is preserved in the process of transformation from the space $L^2(\mathbb{R}^2)$ to $L^2\big((0,a_0)\times\mathbb{R}^2\times[0,2\pi)\big)$, where $a_0 < \pi^2$.

Proof. Invoking the Parseval's formula for the two-dimensional Fourier transform together with (4.194) and (4.201), we have

$$\left\|\Gamma_{\Psi}^{M}\left[f\right](a,\mathbf{b},\theta)\right\|_2^2 = \int_0^{2\pi}\int_{\mathbb{R}^2}\int_0^{a_0}\left|\Gamma_{\Psi}^{M}\left[f\right](a,\mathbf{b},\theta)\right|^2\frac{da\,d\mathbf{b}\,d\theta}{a^3}$$

$$= \int_0^{2\pi} \int_{\mathbb{R}^2} \int_0^{a_0} \left| \left(f \circledast \tilde{\Psi}_{a,0,\theta}^M \right)(\mathbf{b}) \right|^2 \frac{da\, d\mathbf{b}\, d\theta}{a^3}$$

$$= \int_0^{2\pi} \int_{\mathbb{R}^2} \int_0^{a_0} \left| \mathscr{F} \left[\left(f \circledast \tilde{\Psi}_{a,0,\theta}^M \right) \right](\mathbf{w}) \right|^2 \frac{da\, d\mathbf{w}\, d\theta}{a^3}$$

$$= (2\pi)^2 \int_0^{2\pi} \int_{\mathbb{R}^2} \int_0^{a_0} \left| \mathscr{F} \left[f \right](\mathbf{w}) \right|^2 \left| \mathscr{F} \left[\Psi_{a,0,\theta}^M \right](\mathbf{w}) \right|^2 \frac{da\, d\mathbf{w}\, d\theta}{a^3}$$

$$= \int_{\mathbb{R}^2} \left| \mathscr{F} \left[f \right](\mathbf{w}) \right|^2 \left\{ (2\pi)^2 \int_0^{a_0} \int_0^{2\pi} \left| \mathscr{F} \left[\Psi_{a,0,\theta}^M \right](\mathbf{w}) \right|^2 \frac{d\theta\, da}{a^3} \right\} d\mathbf{w}$$

$$= \int_{\mathbb{R}^2} \left| \mathscr{F} \left[f \right](\mathbf{w}) \right|^2 d\mathbf{w}$$

$$= \left\| f \right\|_2^2,$$

which evidently completes the proof of Theorem 4.7.3. $\qquad\square$

Remark 4.7.2. From (4.203), we infer that the novel curvelet transform defined in (4.185) is an isometry from the space of signals in $L^2(\mathbb{R}^2)$ satisfying $\mathscr{F}[f](\mathbf{w}) = 0, \forall |\mathbf{w}| < 2/a_0 B$, $a_0 < \pi^2$ to the space of transforms $L^2\big((0, a_0) \times \mathbb{R}^2 \times [0, 2\pi)\big)$.

We note that the reconstruction formula (4.195) is concerned for those signals $f \in L^2(\mathbb{R}^2)$ satisfying $\mathscr{F}[f](\mathbf{w}) = 0, \forall |\mathbf{w}| < 2/a_0 B$, $a_0 < \pi^2$. In order to have a complete reconstruction formula, we need to take care of the other frequency components as well. To facilitate the narrative, we consider an arbitrary square integrable function f on $\mathbb{R}^2$ and define

$$(T_1 f)(\mathbf{t}) = \int_0^{2\pi} \int_{\mathbb{R}^2} \int_0^{a_0} \Gamma_\Psi^M \big[f \big](a, \mathbf{b}, \theta) \, \Psi_{a,\mathbf{b},\theta}^M(\mathbf{t}) \, \frac{da\, d\mathbf{b}\, d\theta}{a^3}$$

$$= \int_0^{2\pi} \int_0^{a_0} \left(\left(f \circledast \tilde{\Psi}_{a,0,\theta}^M \right)(\mathbf{b}) \circledast \Psi_{a,0,\theta}^M(\mathbf{b}) \right)(\mathbf{t}) \, \frac{da\, d\theta}{a^3} \qquad (4.204)$$

and

$$(T_0 f)(\mathbf{t}) = f(\mathbf{t}) - (T_1 f)(\mathbf{t}). \qquad (4.205)$$

Here, we note that

$$\mathscr{F} \big[(T_1 f) \big](\mathbf{w}) = (2\pi)^2 \int_0^{2\pi} \int_0^{a_0} \mathscr{F} \big[f \big](\mathbf{w}) \left| \mathscr{F} \left[\Psi_{a,0,\theta}^M \right](\mathbf{w}) \right|^2 \frac{da\, d\theta}{a^3}$$

$$= B^2 \, \mathscr{F} \big[f \big](\mathbf{w}) \int_0^{a_0 B |\mathbf{w}|} |W(a)|^2 \frac{da}{a}$$

$$= (2\pi)^2 \, \mathscr{F} \big[f \big](\mathbf{w}) \left(\mathscr{F} \big[\Omega^M \big](\mathbf{w}) \right)^2, \qquad (4.206)$$

where

$$\left(\mathscr{F} \big[\Omega^M \big](\mathbf{w}) \right)^2 = \frac{B^2}{(2\pi)^2} \int_0^{a_0 B |\mathbf{w}|} |W(a)|^2 \frac{da}{a}.$$

Furthermore, in view of the additivity of the Fourier transform, we have

$$\mathscr{F} \big[(T_0 f) \big](\mathbf{w}) = \mathscr{F} \big[f \big](\mathbf{w}) - \mathscr{F} \big[(T_1 f) \big](\mathbf{w})$$

$$= \mathscr{F}\big[f\big](\mathbf{w}) \left(1 - (2\pi)^2 \Big(\mathscr{F}\big[\Omega^M\big](\mathbf{w})\Big)^2\right)$$

$$= (2\pi)^2 \, \mathscr{F}\big[f\big](\mathbf{w}) \left(\frac{1}{(2\pi)^2} - \Big(\mathscr{F}\big[\Omega^M\big](\mathbf{w})\Big)^2\right)$$

$$= (2\pi)^2 \, \mathscr{F}\big[f\big](\mathbf{w}) \left(\mathscr{F}\big[\Phi^M\big](\mathbf{w})\right)^2, \tag{4.207}$$

where

$$\left(\mathscr{F}\big[\Phi^M\big](\mathbf{w})\right)^2 = \frac{1}{(2\pi)^2} - \Big(\mathscr{F}\big[\Omega^M\big](\mathbf{w})\Big)^2.$$

Thanks to the property (4.194) of the convolution operation $\circledast$, we infer from equations (4.206) and (4.207) that

$$(T_0 f)(\mathbf{t}) = \left(f \circledast \Phi^M \circledast \Phi^M\right)(\mathbf{t}) \quad \text{and} \quad (T_1 f)(\mathbf{t}) = \left(f \circledast \Omega^M \circledast \Omega^M\right)(\mathbf{t}). \tag{4.208}$$

Moreover, we observe that

$$(2\pi)^2 \left[\Big(\mathscr{F}\big[\Omega^M\big](\mathbf{w})\Big)^2 + \Big(\mathscr{F}\big[\Phi^M\big](\mathbf{w})\Big)^2\right] = 1. \tag{4.209}$$

Also,

$$\mathscr{F}\big[\Phi^M\big](\mathbf{w}) = 0, \quad |\mathbf{w}| > 2/a_0 B \quad \text{and} \quad \mathscr{F}\big[\Phi^M\big](\mathbf{w}) = 1/2\pi, \quad |\mathbf{w}| < 1/2a_0 B.$$

Finally, we define the father wavelet $\Phi_{\mathbf{b}}^M(\mathbf{t}) = \Phi^M(\mathbf{t} - \mathbf{b})$, so that

$$(T_0 f)(\mathbf{t}) = \int_{\mathbb{R}^2} \left\langle f, \Phi_{\mathbf{b}}^M \right\rangle_2 \Phi_{\mathbf{b}}^M(\mathbf{t}) \, d\mathbf{b}. \tag{4.210}$$

Consequently, (4.205) implies that

$$f(\mathbf{t}) = \int_{\mathbb{R}^2} \left\langle f, \Phi_{\mathbf{b}}^M \right\rangle_2 \Phi_{\mathbf{b}}^M(\mathbf{t}) \, d\mathbf{b} + \int_0^{2\pi} \int_{\mathbb{R}^2} \int_0^{a_0} \Gamma_\Psi^M\big[f\big](a, \mathbf{b}, \theta) \, \Psi_{a,\mathbf{b},\theta}^M(\mathbf{t}) \, \frac{da \, d\mathbf{b} \, d\theta}{a^3}.$$

Therefore, we conclude that the complete reconstruction formula for the novel curvelet transform (4.185) is composed of both curvelet waveforms and isotropic father wavelets. The above discussion can be summarized into the following theorem:

Theorem 4.7.4. *For any $f \in L^2(\mathbb{R}^2)$, the reproducing formula for the novel curvelet transform $\Gamma_\Psi^M\big[f\big](a, \mathbf{b}, \theta)$ defined in (4.185) is given by*

$$f(\mathbf{t}) = \int_{\mathbb{R}^2} \left\langle f, \Phi_{\mathbf{b}}^M \right\rangle_2 \Phi_{\mathbf{b}}^M(\mathbf{t}) \, d\mathbf{b} + \int_0^{2\pi} \int_{\mathbb{R}^2} \int_0^{a_o} \Gamma_\Psi^M\big[f\big](a, \mathbf{b}, \theta) \, \Psi_{a,\mathbf{b},\theta}^M(\mathbf{t}) \, \frac{da \, d\mathbf{b} \, d\theta}{a^3}, \tag{4.211}$$

where the radial and angular windows W and V satisfy their respective admissibility conditions (4.179) and (4.180).

Next, we shall formulate an uncertainty inequality comparing the localization of the novel curvelet transform defined in (4.185) with the localization of the Fourier transform of input signal. More precisely, we obtain a lower bound for optimal joint resolution of the transformed signal $\Gamma_\Psi^M\big[f\big](a, \mathbf{b}, \theta)$, regarded as a function of the translation variable $\mathbf{b}$ (integrating out the rest), and the Fourier transform of the input signal $\mathscr{F}\big[f\big](\mathbf{w})$.

Theorem 4.7.5. *If $\Gamma_\Psi^M[f](a, \mathbf{b}, \theta)$ is the novel curvelet transform of any non-trivial function $f \in L^2(\mathbb{R}^2)$ satisfying $\mathscr{F}[f](\mathbf{w}) = 0, \forall |\mathbf{w}| < 2/a_0 B, \; a_0 < \pi^2$, then the following uncertainty inequality holds:*

$$\left\{ \int_0^{2\pi} \int_{\mathbb{R}^2} \int_0^{a_0} |\mathbf{b}|^2 \left| \Gamma_\Psi^M[f](a, \mathbf{b}, \theta) \right|^2 \frac{da\,d\mathbf{b}\,d\theta}{a^3} \right\}^{1/2} \left\{ \int_{\mathbb{R}^2} |\mathbf{w}|^2 \left| \mathscr{F}[f](\mathbf{w}) \right|^2 d\mathbf{w} \right\}^{1/2} \geq \|f\|_2^2.$$

$$(4.212)$$

Proof. For the given non-trivial, square integrable function f on $\mathbb{R}^2$, the Heisenberg's uncertainty inequality (1.296) can be recast as

$$\left\{ \int_{\mathbb{R}^2} |\mathbf{t}|^2 |f(\mathbf{t})|^2 d\mathbf{t} \right\}^{1/2} \left\{ \int_{\mathbb{R}^2} |\mathbf{w}|^2 \left| \mathscr{F}[f](\mathbf{w}) \right|^2 d\mathbf{w} \right\}^{1/2} \geq \left\{ \int_{\mathbb{R}^2} |f(\mathbf{t})|^2 d\mathbf{t} \right\}. \quad (4.213)$$

Identifying $\Gamma_\Psi^M[f](a, \mathbf{b}, \theta)$ as a function of the planer translation variable $\mathbf{b}$ and then invoking (4.213) yields the following inequality:

$$\left\{ \int_{\mathbb{R}^2} |\mathbf{b}|^2 \left| \Gamma_\Psi^M[f](a, \mathbf{b}, \theta) \right|^2 d\mathbf{b} \right\}^{1/2} \left\{ \int_{\mathbb{R}^2} |\mathbf{w}|^2 \left| \mathscr{F}\left(\Gamma_\Psi^M[f](a, \mathbf{b}, \theta) \right)(\mathbf{w}) \right|^2 d\mathbf{w} \right\}^{1/2}$$

$$\geq \left\{ \int_{\mathbb{R}^2} \left| \Gamma_\Psi^M[f](a, \mathbf{b}, \theta) \right|^2 d\mathbf{b} \right\}. \quad (4.214)$$

Integrating (4.214) with respect to the measure $da\,d\theta/a^3$, we obtain

$$\int_0^{2\pi} \int_0^{a_0} \left\{ \int_{\mathbb{R}^2} |\mathbf{b}|^2 \left| \Gamma_\Psi^M[f](a, \mathbf{b}, \theta) \right|^2 d\mathbf{b} \right\}^{1/2}$$

$$\times \left\{ \int_{\mathbb{R}^2} |\mathbf{w}|^2 \left| \mathscr{F}\left(\Gamma_\Psi^M[f](a, \mathbf{b}, \theta) \right)(\mathbf{w}) \right|^2 d\mathbf{w} \right\}^{1/2} \frac{da\,d\theta}{a^3}$$

$$\geq \left\{ \int_0^{2\pi} \int_0^{a_0} \int_{\mathbb{R}^2} \left| \Gamma_\Psi^M[f](a, \mathbf{b}, \theta) \right|^2 \frac{d\mathbf{b}\,da\,d\theta}{a^3} \right\}. \quad (4.215)$$

Applying the Cauchy-Schwarz inequality followed by the implication of (4.203), the inequality (4.215) can be expressed as

$$\left\{ \int_0^{2\pi} \int_{\mathbb{R}^2} \int_0^{a_0} |\mathbf{b}|^2 \left| \Gamma_\Psi^M[f](a, \mathbf{b}, \theta) \right|^2 \frac{da\,d\mathbf{b}\,d\theta}{a^3} \right\}^{1/2}$$

$$\times \left\{ \int_0^{2\pi} \int_{\mathbb{R}^2} \int_0^{a_0} |\mathbf{w}|^2 \left| \mathscr{F}\left(\Gamma_\Psi^M[f](a, \mathbf{b}, \theta) \right)(\mathbf{w}) \right|^2 \frac{da\,d\mathbf{w}\,d\theta}{a^3} \right\}^{1/2} \geq \|f\|_2^2. \quad (4.216)$$

Making use of the relation (4.186), then invoking the polar coordinates (r, ϕ) for $\mathbf{w} \in \mathbb{R}^2$, together with the implication of (4.201), we observe that

$$\int_0^{2\pi} \int_{\mathbb{R}^2} \int_0^{a_0} |\mathbf{w}|^2 \left| \mathscr{F}\left(\Gamma_\Psi^M[f](a, \mathbf{b}, \theta) \right)(\mathbf{w}) \right|^2 \frac{da\,d\mathbf{w}\,d\theta}{a^3}$$

$$= (2\pi)^2 \int_0^{2\pi} \int_{\mathbb{R}^2} \int_0^{a_0} |\mathbf{w}|^2 \left| \mathscr{F}[f](\mathbf{w}) \right|^2 \left| \mathscr{F}[\Psi_a^M](R_\theta \mathbf{w}) \right|^2 \frac{da\,d\mathbf{w}\,d\theta}{a^3}$$

$$= (2\pi)^2 \int_{\mathbb{R}^2} |\mathbf{w}|^2 \left| \mathscr{F}[f](\mathbf{w}) \right|^2 \left\{ \int_0^{2\pi} \int_0^{a_0} \left| \mathscr{F}[\Psi_a^M](R_\theta \mathbf{w}) \right|^2 \frac{da\,d\theta}{a^3} \right\} d\mathbf{w}$$

$$= \int_{\mathbb{R}^2} |\mathbf{w}|^2 \left| \mathscr{F}\left[f\right](\mathbf{w}) \right|^2 \left\{ (2\pi)^2 \int_0^{2\pi} \int_0^{a_0} \left| \mathscr{F}\left[\Psi_{a,0,\theta}^M\right](\mathbf{w}) \right|^2 \frac{da\,d\theta}{a^3} \right\} d\mathbf{w}$$

$$= \int_0^{2\pi} \int_0^{\infty} r^2 \left| \mathscr{F}\left[f\right](r,\phi) \right|^2 \left\{ (2\pi)^2 \int_0^{2\pi} \int_0^{a_0} \left| \mathscr{F}\left[\Psi_{a,0,\theta}^M\right](r,\phi) \right|^2 \frac{da\,d\theta}{a^3} \right\} r\,dr\,d\phi$$

$$= \int_0^{2\pi} \int_0^{\infty} r^2 \left| \mathscr{F}\left[f\right](r,\phi) \right|^2 r\,dr\,d\phi$$

$$= \int_{\mathbb{R}^2} |\mathbf{w}|^2 \left| \mathscr{F}\left[f\right](\mathbf{w}) \right|^2 d\mathbf{w}. \tag{4.217}$$

Plugging (4.217) in (4.216), we obtain the desired Heisenberg-type uncertainty inequality:

$$\left\{ \int_0^{2\pi} \int_{\mathbb{R}^2} \int_0^{a_0} |\mathbf{b}|^2 \left| \Gamma_\Psi^M\left[f\right](a,\mathbf{b},\theta) \right|^2 \frac{da\,d\mathbf{b}\,d\theta}{a^3} \right\}^{1/2} \left\{ \int_{\mathbb{R}^2} |\mathbf{w}|^2 \left| \mathscr{F}\left[f\right](\mathbf{w}) \right|^2 d\mathbf{w} \right\}^{1/2} \geq \left\| f \right\|_2^2.$$

This completes the proof of Theorem 4.7.5.   □

We culminate this subsection with an example demonstrating the construction of novel curvelet waveforms via a pair of radial and angular windows satisfying the admissibility conditions (4.179) and (4.180).

Example 4.7.1. Given a 2×2 real, unimodular matrix $M = (A, B; C, D)$ with $B > 0$, we consider the following window functions:

$$W(r) = \begin{cases} \dfrac{1}{B} \cos\left[\dfrac{\pi}{2}\left(\nu(5 - 6r)\right)\right], & 2/3 \leq r \leq 5/6 \\[2mm] \dfrac{1}{B}, & 5/6 \leq r \leq 4/3 \\[2mm] \dfrac{1}{B} \cos\left[\dfrac{\pi}{2}\left(\nu(3r - 4)\right)\right], & 4/3 \leq r \leq 5/3 \\[2mm] 0, & \text{otherwise} \end{cases} \tag{4.218}$$

and

$$V(\phi) = \begin{cases} \dfrac{1}{2\pi} \cos\left[\dfrac{\pi}{2}\left(\nu(-3\phi - 1)\right)\right], & -2/3 \leq \phi \leq -1/3 \\[2mm] \dfrac{1}{2\pi}, & -1/3 \leq \phi \leq 1/3 \\[2mm] \dfrac{1}{2\pi} \cos\left[\dfrac{\pi}{2}\left(\nu(3\phi - 1)\right)\right], & 1/3 \leq \phi \leq 2/3 \\[2mm] 0, & \text{otherwise}, \end{cases} \tag{4.219}$$

where ν is a smooth function, such that

$$\nu(y) = \begin{cases} 0, & y \leq 0 \\ 1, & y \geq 1 \end{cases} \quad \text{and} \quad \nu(y) + \nu(1 - y) = 1, \quad y \in \mathbb{R}. \tag{4.220}$$

It is imperative to mention that the smoothness of the window functions (4.218) and (4.219) is dictated by the function ν; the smoother ν is, the smoother are the window functions and consequently faster is the decay of curvelets. Some explicit expressions for the function ν are mentioned in Example 3.6.1 dealing with the construction of classical curvelet waveforms.

Next, we show that the window functions (4.218) and (4.219) can be used to construct a pair of radial and angular windows satisfying the admissibility conditions given by (4.179) and (4.180). To facilitate the narrative, we compute the sum $\sum_{\ell=-\infty}^{\infty} |V(\phi - \ell)|^2$, where $\phi \in \mathbb{R}$. By definition, we have supp $(V) \subseteq [-2/3, 2/3]$. Moreover, for a fixed $\phi \in \mathbb{R}$, the aforementioned sum contains only two non-vanishing terms, and for $\phi \in [1/3, 2/3]$, we have

$$(2\pi)^2 \sum_{\ell=-\infty}^{\infty} |V(\phi - \ell)|^2 = (2\pi)^2 \Big(|V(\phi)|^2 + |V(\phi - 1)|^2 \Big)$$

$$= \cos^2\left[\frac{\pi}{2}\left(\nu(3\phi - 1)\right)\right] + \cos^2\left[\frac{\pi}{2}\left(\nu(-3\phi + 2)\right)\right]$$

$$= \cos^2\left[\frac{\pi}{2}\left(\nu(x)\right)\right] + \cos^2\left[\frac{\pi}{2}\left(\nu(1 - x)\right)\right]$$

$$= \cos^2\left[\frac{\pi}{2}\left(\nu(x)\right)\right] + \cos^2\left[\frac{\pi}{2}\left(1 - \nu(x)\right)\right]$$

$$= \cos^2\left[\frac{\pi}{2}\left(\nu(x)\right)\right] + \sin^2\left[\frac{\pi}{2}\left(\nu(x)\right)\right]$$

$$= 1. \tag{4.221}$$

Since, supp $(W) \subset [1/2, 2]$, therefore for $j \in \mathbb{Z}$, we have supp $(W(2^j r)) \subseteq [2^{-j-1}, 2^{1-j}]$. Consequently, the sum $\sum_{j\in\mathbb{Z}} |W(2^j r)|^2$ has only two non-vanishing terms corresponding to $r \in [1/2, 1]$, viz; $|W(r)|^2$ and $|W(2r)|^2$. Thus, for $r \in [1/2, 1]$, we have

$$B^2 \sum_{j=-\infty}^{\infty} |W(2^j r)|^2 = B^2 \Big(|W(r)|^2 + |W(2r)|^2 \Big)$$

$$= \begin{cases} 1, & 1/2 \le r \le 2/3 \\ \cos^2\left[\frac{\pi}{2}\left(\nu(6r - 4)\right)\right] + \cos^2\left[\frac{\pi}{2}\left(\nu(5 - 6r)\right)\right], & 2/3 \le r \le 5/6 \\ 1, & 5/6 \le r \le 1 \\ 0, & \text{otherwise.} \end{cases} \tag{4.222}$$

Note that

$$\cos^2\left[\frac{\pi}{2}\left(\nu(6r - 4)\right)\right] + \cos^2\left[\frac{\pi}{2}\left(\nu(5 - 6r)\right)\right] = 1. \tag{4.223}$$

Using the identity (4.223) in (4.222), it follows that

$$B^2 \sum_{j=-\infty}^{\infty} |W(2^j r)|^2 = 1. \tag{4.224}$$

Proceeding further, we choose a window function $W'(r)$ satisfying $\ln 2\, W'(r) = W(r)$. Then, we shall demonstrate that the window functions W' and V satisfy the admissibility conditions (4.179) and (4.180), respectively. Observe that,

$$1 = (2\pi)^2 \sum_{\ell=-\infty}^{\infty} |V(\phi - \ell)|^2 = (2\pi)^2 \int_0^1 \sum_{\ell=-\infty}^{\infty} |V(\phi - \ell)|^2 \, d\phi = (2\pi)^2 \int_{-\infty}^{\infty} |V(\phi)|^2 \, d\phi, \tag{4.225}$$

which validates the admissibility condition (4.180) for the window function $V(\phi)$ defined in (4.219). Finally, for $r \in (0, \infty)$, we take $r = 2^x$, $x \in (-\infty, \infty)$, so that

$$1 = B^2 \sum_{j=-\infty}^{\infty} |W(2^j r)|^2 = B^2 \sum_{j=-\infty}^{\infty} |W(2^{j+x})|^2 = B^2 \int_0^1 \sum_{j=-\infty}^{\infty} |W(2^{j+x})|^2 \, dx$$

$$= B^2 \sum_{j=-\infty}^{\infty} \frac{1}{\ln 2} \int_{2^j}^{2^{j+1}} |W(y)|^2 \frac{dy}{y}$$

$$= B^2 \sum_{j=-\infty}^{\infty} \int_{2^j}^{2^{j+1}} |W'(y)|^2 \frac{dy}{y}$$

$$= B^2 \int_0^{\infty} |W'(y)|^2 \frac{dy}{y}, \tag{4.226}$$

which verifies the admissibility condition (4.179) for the window function $W'(r)$. Hence, we conclude that the window functions W' and V can be used for a lucid construction of the novel curvelet waveforms.

4.7.2 Linear canonical ripplet transform

"There is geometry in the humming of the strings, there is music in the spacing of the spheres."

-Pythagoras

The ripplet transform is a refinement of the ordinary curvelet transform based on the modified curvelet waveforms, known as "ripplets," obeying a non-parabolic scaling law for resolving two-dimensional singularities. However, the ripplet transform offers flexibility only in the angular window of the underlying waveforms, whereas the radial window is unaltered. Therefore, one of the fundamental concerns is that, whether it is possible to induce flexibility in both the radial and angular windows of the ordinary ripplet transform? The answer to the query is affirmative and lies in revisiting the theory by revamping the notion of ripplets via the two-dimensional linear canonical transform given by Definition 4.7.1.

Let (r, ϕ), $r \geq 0$, $\phi \in [0, 2\pi)$ denote the polar co-ordinates of any $\mathbf{w} \in \mathbb{R}^2$. Then, for a pair of radial and angular window functions $W : (0, \infty) \to (0, \infty)$ and $V : (-\infty, \infty) \to (0, \infty)$, satisfying the admissibility conditions (4.179) and (4.180), consider a basic waveform $\Upsilon_a : \mathbb{R}^2 \to \mathbb{C}$ defined via the polar linear canonical transform (4.177) as

$$\mathcal{L}_M \left[\Upsilon_{a,0,0} \right] (r, \phi) = \frac{a^{(1+d)/2d}}{\sqrt{c}} W(ar) V \left(\frac{a^{1/d}\phi}{ca} \right), \quad c > 0, \, d > 1, \tag{4.227}$$

where $a \in (0, a_0)$ and $a_0 < \pi^2$. Applying the inverse linear canonical transform (4.178) on both sides of the expression (4.227), we obtain

$$\Upsilon_{a,0,0}(\rho, \eta) = \frac{a^{(1+d)/2d}}{2\pi B \sqrt{c}} \int_0^{2\pi} \int_0^{\infty} W(ar) V \left(\frac{a^{1/d}\phi}{ac} \right)$$

$$\times \exp \left\{ -\frac{i(A\rho^2 + Dr^2 - 2\rho r \cos(\eta - \phi))}{2B} \right\} r \, dr \, d\phi$$

$$= \frac{a^{(1+d)/2d}}{2\pi B \sqrt{c}} \exp \left\{ -\frac{iA\rho^2}{2B} \right\} \int_0^{2\pi} \int_0^{\infty} \exp \left\{ -\frac{iDr^2}{2B} \right\} W(ar) V \left(\frac{a^{1/d}\phi}{ac} \right)$$

$$\times \exp \left\{ \frac{i\rho r \cos(\eta - \phi))}{B} \right\} r \, dr \, d\phi. \tag{4.228}$$

Upon simplifying (4.228), we obtain a novel basic waveform $\Upsilon_{a,0,0}^M(\mathbf{t})$ via the following expression:

$$\mathscr{F}\left[\Upsilon_{a,0,0}^M\right](r,\phi) = \frac{a^{(1+d)/2d}B}{\sqrt{c}} \exp\left\{-\frac{iDBr^2}{2}\right\} W(aBr) V\left(\frac{a^{1/d}\phi}{ac}\right), \quad c>0, d>1,$$

$$(4.229)$$

where $\Upsilon_{a,0,0}^M(\mathbf{t}) = \exp\left\{\frac{iA|\mathbf{t}|^2}{2B}\right\} \Upsilon_{a,0,0}(\mathbf{t})$. Consequently, the family of novel complex-valued ripplet waveforms (or linear canonical ripplets) adopted to scale $a > 0$, location $\mathbf{b} \in \mathbb{R}^2$ and orientation $\theta \in [0, 2\pi)$ (or $(-\pi, \pi)$ according to convenience) is obtained by translating the basic waveform by $\mathbf{b} \in \mathbb{R}^2$ and then inducing a planer rotation of θ radians; that is,

$$\Upsilon_{a,\mathbf{b},\theta}^M(\mathbf{t}) = \Upsilon_a^M(R_\theta(\mathbf{t}-\mathbf{b})), \quad \mathbf{t} \in \mathbb{R}^2, \qquad (4.230)$$

where $R_\theta = \begin{pmatrix} \cos\theta & -\sin\theta \\ \sin\theta & \cos\theta \end{pmatrix}$ denotes the 2×2 rotation matrix effecting the planar rotation by θ radians.

Based on the newly constructed family of linear canonical ripplet waveforms given in (4.230), we have the following definition of the novel ripplet transform:

Definition 4.7.3. Given a real, unimodular matrix $M = (A, B; C, D)$ with $B > 0$. The novel ripplet transform (or linear canonical ripplet transform) of any $f \in L^2(\mathbb{R}^2)$ is denoted by $\Re_\Upsilon^M[f]$ and is defined as

$$\Re_\Upsilon^M[f](a,\mathbf{b},\theta) = \left\langle f, \Upsilon_{a,\mathbf{b},\theta}^M \right\rangle_2 = \int_{\mathbb{R}^2} f(\mathbf{t}) \overline{\Upsilon_{a,\mathbf{b},\theta}^M(\mathbf{t})} \, d\mathbf{t}, \qquad (4.231)$$

where $a \in (0, a_0)$ with $a_0 < \pi^2$, $\mathbf{b} \in \mathbb{R}^2$, $\theta \in [0, 2\pi)$ and $\Upsilon_{a,\mathbf{b},\theta}^M(\mathbf{t})$ is given by (4.230).

Definition 4.7.3 allows us to point out the following facts regarding the linear canonical ripplet transform:

(i) The ordinary ripplet transform (3.265) is just a special case of novel ripplet transform (4.231) and follows by choosing the unimodular matrix as $M = (0, 1; -1, 0)$.

(ii) For the unimodular matrix $M = (\cos\alpha, \sin\alpha; -\sin\alpha, \cos\alpha)$, $\alpha \neq n\pi$, $n \in \mathbb{Z}$, Definition 4.7.3 yields a new variant of the ripplet transform based on the fractional Fourier transform.

(iii) As a consequence of the expression (4.229), we infer that the novel ripplet transform (4.231) is based on a pair of highly flexible radial and angular windows. The matrix parameter $B > 0$ induces flexibility into the radial window, whereas the parameters $c > 0$ and $d > 1$ control the nature of angular window.

(iv) For $c = 1$ and $d = 2$, the linear canonical ripplet transform (4.231) boils down to the linear canonical curvelet transform (4.185).

(v) The computational complexity of the novel ripplet transform defined in (4.231) is completely determined by that of the ordinary ripplet transform.

Next, our main concern is to obtain a spectral representation of the novel ripplet transform $\Re_\Upsilon^M[f](a,\mathbf{b},\theta)$, when regarded as a function of the translation variable $\mathbf{b}$. In this regard, we have the following proposition:

Proposition 4.7.6. *If* $\Re_{\Upsilon}^{M}[f](a, \mathbf{b}, \theta)$ *is the novel ripplet transform of any* $f \in L^2(\mathbb{R}^2)$, *then we have*

$$\mathscr{F}\left(\Re_{\Upsilon}^{M}[f](a, \mathbf{b}, \theta)\right)(\mathbf{w}) = 2\pi \, \mathscr{F}[f](\mathbf{w}) \, \mathscr{F}[\Upsilon_a^{M}](R_\theta \mathbf{w}), \qquad (4.232)$$

where $\mathscr{F}$ *denotes the two-dimensional Fourier transform given by* (1.288).

Proof. To accomplish the proof, we need to compute the two-dimensional Fourier transform of the novel ripplet waveforms (4.230). Therefore, we have

$$\begin{aligned}
\mathscr{F}[\Upsilon_{a,\mathbf{b},\theta}^{M}](\mathbf{w}) &= \frac{1}{2\pi} \int_{\mathbb{R}^2} \Upsilon_{a,\mathbf{b},\theta}^{M}(\mathbf{t}) \, e^{-i\mathbf{t}^T \mathbf{w}} \, d\mathbf{t} \\
&= \frac{1}{2\pi} \int_{\mathbb{R}^2} \Upsilon_a^{M}(R_\theta(\mathbf{t} - \mathbf{b})) \, e^{-i\mathbf{t}^T \mathbf{w}} \, d\mathbf{t} \\
&= \frac{1}{2\pi} \int_{\mathbb{R}^2} \Upsilon_a^{M}(\mathbf{z}) \, e^{-i(\mathbf{b} + R_{-\theta}\mathbf{z})^T \mathbf{w}} \, d\mathbf{z} \\
&= \frac{e^{-i\mathbf{b}^T \mathbf{w}}}{2\pi} \int_{\mathbb{R}^2} \Upsilon_a^{M}(\mathbf{z}) \, e^{-i\mathbf{z}^T(R_\theta \mathbf{w})} \, d\mathbf{z} \\
&= e^{-i\mathbf{b}^T \mathbf{w}} \, \mathscr{F}[\Upsilon_a^{M}](R_\theta \mathbf{w}). \qquad (4.233)
\end{aligned}$$

Let (σ, μ), (ρ, η) and (r, ϕ) denote the polar coordinates of the variables $\mathbf{b}$, $\mathbf{t}$ and $\mathbf{w}$, respectively. Then, we can rewrite (4.233) as follows:

$$\begin{aligned}
\mathscr{F}[\Upsilon_{a,\mathbf{b},\theta}^{M}](r, \phi) &= e^{-ir\sigma \cos(\mu-\phi)} \mathscr{F}[\Upsilon_a^{M}](r, \phi - \theta) \\
&= e^{-ir\sigma \cos(\mu-\phi)} \frac{a^{(1+d)/2d} B}{\sqrt{c}} \, \exp\left\{-\frac{iDBr^2}{2}\right\} W(aBr) \, V\left(\frac{a^{1/d}(\phi - \theta)}{ac}\right).
\end{aligned}$$
$$(4.234)$$

Invoking the Parseval's formula for the polar Fourier transform, we get

$$\begin{aligned}
\Re_{\Upsilon}^{M}[f](a, \mathbf{b}, \theta) &= \langle f, \, \Upsilon_{a,\mathbf{b},\theta}^{M} \rangle_2 \\
&= \frac{a^{(1+d)/2d} B}{\sqrt{c}} B \int_0^{2\pi} \int_0^\infty e^{ir\sigma \cos(\mu-\phi)} \mathscr{F}[f](r, \phi) \\
&\qquad \times \exp\left\{\frac{iDBr^2}{2}\right\} W(aBr) \, V\left(\frac{a^{1/d}(\phi - \theta)}{ac}\right) r \, dr \, d\phi. \quad (4.235)
\end{aligned}$$

Upon reverting to the usual Cartesian coordinates, the expression (4.235) can be recast as

$$\begin{aligned}
\Re_{\Upsilon}^{M}[f](a, \mathbf{b}, \theta) &= \int_{\mathbb{R}^2} \mathscr{F}[f](\mathbf{w}) \, \overline{\mathscr{F}[\Upsilon_a^{M}](R_\theta \mathbf{w})} \, e^{i\mathbf{b}^T \mathbf{w}} \, d\mathbf{w} \\
&= 2\pi \, \mathscr{F}^{-1}\left(\mathscr{F}[f](\mathbf{w}) \, \overline{\mathscr{F}[\Upsilon_a^{M}](R_\theta \mathbf{w})}\right)(\mathbf{b}). \qquad (4.236)
\end{aligned}$$

Applying the 2D-Fourier transform on both sides of (4.236), we obtain the desired result as

$$\mathscr{F}\left(\Re_{\Upsilon}^{M}[f](a, \mathbf{b}, \theta)\right)(\mathbf{w}) = 2\pi \, \mathscr{F}[f](\mathbf{w}) \, \mathscr{F}[\Upsilon_a^{M}](R_\theta \mathbf{w}).$$

This completes the proof of Proposition 4.7.6.   □

Our next endeavour is to obtain a complete reconstruction formula associated with the linear canonical ripplet transform defined in (4.231). In this direction, we have the following theorem, which serves as the reconstruction formula for a specific class of signals:

Theorem 4.7.7. *For any $f \in L^2(\mathbb{R}^2)$ with $\mathscr{F}[f](\mathbf{w}) = 0$, $\forall \, |\mathbf{w}| < 2/a_0 B$, $a_0 < \pi^2$, the reconstruction formula for the linear canonical ripplet transform (4.231) is given by*

$$f(\mathbf{t}) = \int_0^{2\pi} \int_{\mathbb{R}^2} \int_0^{a_0} \Re_\Upsilon^M [f](a, \mathbf{b}, \theta) \, \Upsilon_{a,\mathbf{b},\theta}^M(\mathbf{t}) \, \frac{da \, d\mathbf{b} \, d\theta}{a^3}, \qquad (4.237)$$

where the radial and angular windows W and V satisfy the admissibility conditions given by (4.179) and (4.180), respectively.

Proof. Define the function

$$F_{a,\theta}^M(\mathbf{t}) = \int_{\mathbb{R}^2} \Re_\Upsilon^M [f](a, \mathbf{b}, \theta) \, \Upsilon_{a,\mathbf{b},\theta}^M(\mathbf{t}) \, d\mathbf{b}. \qquad (4.238)$$

Then, we observe that the expression (4.238) can be redrafted via the two-dimensional convolution operation $\circledast$ as follows:

$$F_{a,\theta}^M(\mathbf{t}) = \left((f \circledast \tilde{\Upsilon}_{a,0,\theta}^M)(\mathbf{b}) \circledast \Upsilon_{a,0,\theta}^M(\mathbf{b}) \right)(\mathbf{t}), \qquad \tilde{\Upsilon}^M(\mathbf{t}) = \overline{\Upsilon^M(-\mathbf{t})}. \qquad (4.239)$$

Consequently, the Fourier transform of the function $F_{a,\theta}^M(\mathbf{t})$ defined in (4.239) is given by

$$\mathscr{F}\left[F_{a,\theta}^M\right](\mathbf{w}) = (2\pi)^2 \mathscr{F}\left[f\right](\mathbf{w}) \left|\mathscr{F}\left[\Upsilon_{a,0,\theta}^M\right](\mathbf{w})\right|^2, \qquad (4.240)$$

which upon integration, with respect to the measure $d\theta \, da/a^3$, yields the following expression:

$$\int_0^{a_0} \int_0^{2\pi} \mathscr{F}\left[F_{a,\theta}^M\right](\mathbf{w}) \, \frac{d\theta \, da}{a^3} = (2\pi)^2 \mathscr{F}\left[f\right](\mathbf{w}) \int_0^{a_0} \int_0^{2\pi} \left|\mathscr{F}\left[\Upsilon_{a,0,\theta}^M\right](\mathbf{w})\right|^2 \frac{d\theta \, da}{a^3}. \qquad (4.241)$$

In order to simplify the set of integrals on the R.H.S of (4.241), we shall use the polar coordinates of $\mathbf{w}$ and the prescribed admissibility conditions (4.179) and (4.180). For $r \geq 2/a_0 B$ with $a_0 < \pi^2$, we have

$$(2\pi)^2 \int_0^{a_0} \int_0^{2\pi} \left|\mathscr{F}\left[\Upsilon_{a,0,\theta}^M\right](r, \phi)\right|^2 \frac{d\theta \, da}{a^3}$$

$$= (2\pi B)^2 \int_0^{a_0} \int_0^{2\pi} \left|W(aBr)\right|^2 \left|V\left(\frac{a^{1/d}(\phi - \theta)}{ac}\right)\right|^2 \frac{a^{(1+d)/d}}{c} \frac{d\theta \, da}{a^3}$$

$$= B^2 \int_0^{a_0} \left|W(aBr)\right|^2 \left\{(2\pi)^2 \int_0^{2\pi} \left|V\left(\frac{a^{1/d}(\phi - \theta)}{ac}\right)\right|^2 d\theta\right\} \frac{a^{(1+d)/d}}{c} \frac{da}{a^3}$$

$$= B^2 \int_0^{a_0} \left|W(aBr)\right|^2 \left(c \, a^{(d-1)/d}\right) \frac{a^{(1+d)/d}}{c} \frac{da}{a^3}$$

$$= B^2 \int_0^{a_0} \left|W(aBr)\right|^2 \frac{da}{a}$$

$$= B^2 \int_0^{a_0 Br} \left|W(r')\right|^2 \frac{dr'}{r'} = 1. \qquad (4.242)$$

Therefore, by virtue of (4.241), we obtain the desired reconstruction formula as

$$f(\mathbf{t}) = \int_0^{2\pi} \int_{\mathbb{R}^2} \int_0^{a_0} \Re_{\Upsilon}^M\left[f\right](a, \mathbf{b}, \theta)\, \Upsilon_{a,\mathbf{b},\theta}^M(\mathbf{t})\, \frac{da\, d\mathbf{b}\, d\theta}{a^3}.$$

This completes the proof of Theorem 4.7.7. □

Theorem 4.7.8. *Given any bivariate function $f \in L^2(\mathbb{R}^2)$ satisfying $\mathscr{F}\left[f\right](\mathbf{w}) = 0$, $\forall\, |\mathbf{w}| < 2/a_0 B$, $a_0 < \pi^2$, we have*

$$\left\|\Re_{\Upsilon}^M\left[f\right](a, \mathbf{b}, \theta)\right\|_2^2 = \left\|f\right\|_2^2, \tag{4.243}$$

with the radial and angular windows W and V satisfying the admissibility conditions given by (4.179) and (4.180), respectively.

Proof. By definition of the linear canonical ripplet transform, we have

$$\begin{aligned}
\left\|\Re_{\Upsilon}^M\left[f\right](a, \mathbf{b}, \theta)\right\|_2^2 &= \int_0^{2\pi} \int_{\mathbb{R}^2} \int_0^{a_0} \left|\Re_{\Upsilon}^M\left[f\right](a, \mathbf{b}, \theta)\right|^2 \frac{da\, d\mathbf{b}\, d\theta}{a^3} \\
&= \int_0^{2\pi} \int_{\mathbb{R}^2} \int_0^{a_0} \left|(f \circledast \tilde{\Upsilon}_{a,0,\theta}^M)(\mathbf{b})\right|^2 \frac{da\, d\mathbf{b}\, d\theta}{a^3} \\
&= \int_0^{2\pi} \int_{\mathbb{R}^2} \int_0^{a_0} \left|\mathscr{F}\left[(f \circledast \tilde{\Upsilon}_{a,0,\theta}^M)\right](\mathbf{w})\right|^2 \frac{da\, d\mathbf{w}\, d\theta}{a^3} \\
&= (2\pi)^2 \int_0^{2\pi} \int_{\mathbb{R}^2} \int_0^{a_0} \left|\mathscr{F}\left[f\right](\mathbf{w})\right|^2 \left|\mathscr{F}\left[\Upsilon_{a,0,\theta}^M\right](\mathbf{w})\right|^2 \frac{da\, d\mathbf{w}\, d\theta}{a^3} \\
&= \int_{\mathbb{R}^2} \left|\mathscr{F}\left[f\right](\mathbf{w})\right|^2 \left\{(2\pi)^2 \int_0^{a_0} \int_0^{2\pi} \left|\mathscr{F}\left[\Upsilon_{a,0,\theta}^M\right](\mathbf{w})\right|^2 \frac{d\theta\, da}{a^3}\right\} d\mathbf{w} \\
&= \int_{\mathbb{R}^2} \left|\mathscr{F}\left[f\right](\mathbf{w})\right|^2 d\mathbf{w} \\
&= \left\|f\right\|_2^2,
\end{aligned}$$

which evidently completes the proof of Theorem 4.7.8. □

Remark 4.7.3. By virtue of (4.243), it follows that the linear canonical ripplet transform defined in (4.231) is an isometry from the space of signals in $L^2(\mathbb{R}^2)$ satisfying $\mathscr{F}\left[f\right](\mathbf{w}) = 0$, $\forall\, |\mathbf{w}| < 2/a_0 B$, $a_0 < \pi^2$ to the space of transforms $L^2\left((0, a_0) \times \mathbb{R}^2 \times [0, 2\pi)\right)$.

In analogy to the linear canonical curvelet transform, the reconstruction formula obtained in Theorem 4.7.7 is valid for a specific class of signals $f \in L^2(\mathbb{R}^2)$ satisfying $\mathscr{F}\left[f\right](\mathbf{w}) = 0$, $\forall\, |\mathbf{w}| < 2/a_0 B$, $a_0 < \pi^2$. In order to have a complete reconstruction formula, we need to take care of the other frequency components as well. Therefore, proceeding in a manner similar to the case of curvelet transform, we can derive the following complete reconstruction formula for the linear canonical ripplet transform (4.231), which is composed of both ripplet waveforms and isotropic father wavelets:

Theorem 4.7.9. *If $\Re_{\Upsilon}^M\left[f\right](a, \mathbf{b}, \theta)$ is the novel ripplet transform of any $f \in L^2(\mathbb{R}^2)$, then there exists a band-limited, purely radial function $\Phi_{\mathbf{b}}^M \in L^2(\mathbb{R}^2)$ with $\Phi_{\mathbf{b}}^M(\mathbf{t}) = \Phi^M(\mathbf{t} - \mathbf{b})$, such that*

$$f(\mathbf{t}) = \int_{\mathbb{R}^2} \left\langle f, \Phi_{\mathbf{b}}^M \right\rangle_2 \Phi_{\mathbf{b}}^M(\mathbf{t})\, d\mathbf{b} + \int_0^{2\pi} \int_{\mathbb{R}^2} \int_0^{a_0} \Re_{\Upsilon}^M\left[f\right](a, \mathbf{b}, \theta)\, \Upsilon_{a,\mathbf{b},\theta}^M(\mathbf{t})\, \frac{da\, d\mathbf{b}\, d\theta}{a^3} \tag{4.244}$$

with the radial and angular windows W and V satisfying the admissibility conditions given by (4.179) and (4.180), respectively.

4.8   The Linear Canonical Shearlet Transform

"Equipped with his five senses, man explores the universe around him and calls the adventure Science."

-Edwin P. Hubble

Undoubtedly, the classical shearlet transform has emerged as one of the most effective frameworks for representing multidimensional data ranging over diverse areas of science and engineering, particularly in the field of signal and image processing, remote sensing, data compression and several other areas where the detection of geometrical features is the key problem. The shearlet transform is reliant up on an affine-like system of well-localized waveforms, known as "shearlets." These shearlet functions have the privilege of being adapted to various scales, locations and orientations and are generated by dilating and translating a single generating function, where the dilation matrix is the product of a parabolic scaling matrix and a shear matrix. Despite immense lucubrations in the context of shearlet transforms, the extension of continuous shearlet transform from the classical Fourier to linear canonical domain has not been investigated yet. With the advantages of linear canonical transform in hindsight, it is quite lucrative to revisit the classical shearlet transform by studying a new variant based on the linear canonical transform.

4.8.1   Definition and basic properties

In this subsection, our main aim is to introduce the notion of the linear canonical shearlet transform by constructing a novel family of shearlets based on the two-dimensional linear canonical transform. To begin with, we present some preliminaries regarding the two-dimensional linear canonical transform. For the sake of compliance to the classical shearlets, we shall slightly modify Definition 4.7.1 as follows:

Definition 4.8.1. Given a real, unimodular matrix $M = (A, B; C, D)$, $B \neq 0$, the two-dimensional linear canonical transform of any $f \in L^2(\mathbb{R}^2)$ is defined as

$$\mathcal{L}_M\big[f\big](\mathbf{w}) = \int_{\mathbb{R}^2} f(\mathbf{x})\, \mathcal{K}_M(\mathbf{x}, \mathbf{w})\, d\mathbf{x}, \qquad (4.245)$$

where $\mathcal{K}_M(\mathbf{x}, \mathbf{w})$, with $\mathbf{x} = (x_1, x_2)^T$, $\mathbf{w} = (\omega_1, \omega_2)^T$, denotes the kernel of the two-dimensional linear canonical transform and is given by

$$\mathcal{K}_M(\mathbf{x}, \mathbf{w}) = \frac{1}{B} \exp\left\{ \frac{i\pi(A|\mathbf{x}|^2 - 2\mathbf{x}^T\mathbf{w} + D|\mathbf{w}|^2)}{B} \right\}. \qquad (4.246)$$

The inversion formula associated with the two-dimensional linear canonical transform (4.245) asserts that

$$f(\mathbf{x}) = \mathcal{L}_M^{-1}\big(\mathcal{L}_M\big[f\big](\mathbf{w})\big)(\mathbf{x}) = \int_{\mathbb{R}^2} \mathcal{L}_M\big[f\big](\mathbf{w})\, \overline{\mathcal{K}_M(\mathbf{x}, \mathbf{w})}\, d\mathbf{w}. \qquad (4.247)$$

Also, for any pair of functions $f, g \in L^2(\mathbb{R}^2)$, the Parseval's formula corresponding to the two-dimensional linear canonical transform (4.245) reads:

$$\big\langle f, g \big\rangle_2 = \big\langle \mathcal{L}_M\big[f\big], \mathcal{L}_M\big[g\big] \big\rangle_2. \qquad (4.248)$$

We now define a novel convolution operation $\circledast_M$ associated with Definition 4.8.1, which plays a pivotal role in the formulation of the linear canonical shearlet transform.

Definition 4.8.2. For any pair of functions $f, g \in L^2(\mathbb{R}^2)$, the two-dimensional linear canonical convolution with respect to the real, unimodular matrix $M = (A, B; C, D)$, $B \neq 0$ is denoted by $\circledast_M$ and is defined as

$$(f \circledast_M g)(\mathbf{t}) = \int_{\mathbb{R}^2} f(\mathbf{x}) \, g(\mathbf{t} - \mathbf{x}) \, \exp\left\{ \frac{i\pi A(|\mathbf{x}|^2 - |\mathbf{t}|^2)}{B} \right\} d\mathbf{x}. \qquad (4.249)$$

It is imperative to mention that the convolution operation (4.249) in the spatial domain corresponds to simple multiplication in the spectral domain. More precisely, we have

$$\mathcal{L}_M\Big[(f \circledast_M g)(\mathbf{t})\Big](\mathbf{w}) = \mathcal{L}_M\Big[f\Big](\mathbf{w}) \, \mathscr{F}\Big[g\Big] \left(B^{-1}\mathbf{w}\right), \qquad (4.250)$$

where $\mathscr{F}$ denotes the two-dimensional Fourier transform (1.292). Since the classical shearlet transform is expressible via the classical convolution operation $\circledast$ as demonstrated in (3.306), therefore, in order to formulate the notion of linear canonical shearlet transform, we ought to replace the classical convolution operation with the linear canonical convolution operation $\circledast_M$ defined in (4.249). Based on the convolution operation $\circledast_M$ and a given shearlet function $\psi \in L^2(\mathbb{R}^2)$, we define a novel family of linear canonical shearlets as

$$\mathfrak{F}_\psi^M(a, s, \mathbf{t}) := \left\{ \psi_{a,s,\mathbf{t}}^M(\mathbf{x}) = |\det A_a|^{-1/2} \, \psi\big(A_a^{-1} S_s^{-1}(\mathbf{x} - \mathbf{t})\big) \exp\left\{ \frac{i\pi A(|\mathbf{t}|^2 - |\mathbf{x}|^2)}{B} \right\} \right\}, \qquad (4.251)$$

where $(a, s, \mathbf{t}) \in \mathbb{R}^+ \times \mathbb{R} \times \mathbb{R}^2$ with the parabolic scaling and shear matrices being given by

$$A_a = \begin{pmatrix} a & 0 \\ 0 & \sqrt{a} \end{pmatrix} \quad \text{and} \quad S_s = \begin{pmatrix} 1 & s \\ 0 & 1 \end{pmatrix}, \qquad (4.252)$$

respectively. As a consequence of (4.251), we infer that the novel shearlet functions $\psi_{a,s,\mathbf{t}}^M(\mathbf{x})$ are dictated by the parameters of the given unimodular matrix $M = (A, B; C, D)$. Moreover, for $M = (0, 1; -1, 0)$, the linear canonical shearlet family (4.251) reduces to the classical shearlet family (3.303), whereas for $M = (\cos\alpha, \sin\alpha; -\sin\alpha, \cos\alpha)$, $\alpha \neq n\pi$, $n \in \mathbb{Z}$, we obtain a fractional variant of the classical shearlet family. Thus, we conclude that the linear canonical shearlets given in (4.251) serve as an extreme generalization of the notion of classical shearlets.

Definition 4.8.3. Given a real, unimodular matrix $M = (A, B; C, D)$, with $B > 0$, the linear canonical shearlet transform of any $f \in L^2(\mathbb{R}^2)$ with respect to the shearlet $\psi \in L^2(\mathbb{R}^2)$ is denoted by $\mathcal{S}_\psi^M[f]$ and is defined as

$$\mathcal{S}_\psi^M\Big[f\Big](a, s, \mathbf{t}) = \left(f \circledast_M \breve{\psi}_{a,s,0}\right)(\mathbf{t}) = \int_{\mathbb{R}^2} f(\mathbf{x}) \, \overline{\psi_{a,s,\mathbf{t}}^M(\mathbf{x})} \, d\mathbf{x}, \qquad (4.253)$$

where $\psi_{a,s,\mathbf{t}}^M(\mathbf{x})$ is given by (4.251).

The following important points regarding the linear canonical shearlet transform can be extracted from Definition 4.8.3:

(i) For the unimodular matrix $M = (0, 1; -1, 0)$, Definition 4.8.3 boils down to the notion of classical shearlet transform (3.305).

(ii) Plugging $M = (\cos\alpha, \sin\alpha; -\sin\alpha, \cos\alpha)$, $\alpha \neq n\pi$, $n \in \mathbb{Z}$, Definition 4.8.3 yields a new variant of the shearlet transform, namely the fractional shearlet transform.

(iii) The linear canonical shearlet transform defined in (4.253) can be expressed via the classical shearlet transform as follows:

$$\mathcal{S}_\psi^M \big[f\big](a, s, \mathbf{t})$$
$$= |\det A_a|^{-1/2} \exp\left\{-\frac{i\pi A |\mathbf{t}|^2}{B}\right\} \int_{\mathbb{R}^2} \exp\left\{\frac{i\pi A |\mathbf{x}|^2}{B}\right\} f(\mathbf{x}) \, \overline{\psi\big(A_a^{-1} S_s^{-1}(\mathbf{x} - \mathbf{t})\big)} \, d\mathbf{x}$$
$$= \exp\left\{-\frac{i\pi A |\mathbf{t}|^2}{B}\right\} \mathcal{S}_\psi \big[F_M\big](a, s, \mathbf{t}), \tag{4.254}$$

where

$$F_M(\mathbf{x}) = \exp\left\{\frac{i\pi A |\mathbf{x}|^2}{B}\right\} f(\mathbf{x}). \tag{4.255}$$

From expression (4.254), we conclude that the computational cost of the of the linear canonical shearlet transform (4.253) is completely determined by the computational complexity of the classical shearlet transform.

(iv) The importance of linear canonical shearlets lies in the fact that, they are increasingly flexible with the degree of flexibility determined by the underlying unimodular matrix $M = (A, B; C, D)$, with $B > 0$. Moreover, the chirp multiplication operation used in the construction of linear canonical shearlets is also quite handy in achieving an elegant representation of the novel shearlets in the two-dimensional linear canonical domain.

(v) The function $\psi \in L^2(\mathbb{R}^2)$ appearing in Definition 4.8.3 is to be considered as a shearlet in the sense that it satisfies the classical admissibility condition (3.295). That is, the linear canonical shearlet transform defined in (4.253) is free from extraneous conditions to be imposed on the generating function $\psi(\mathbf{x})$.

In order to obtain the spectral representation of the linear canonical shearlet transform (4.253), we need to compute the two-dimensional linear canonical transform of the linear canonical shearlets given by (4.251). In this direction, we have the following proposition:

Proposition 4.8.1. *Given any shearlet* $\psi \in L^2(\mathbb{R}^2)$, *we have*

$$\mathcal{L}_M\big[\psi_{a,s,\mathbf{t}}^M\big](\mathbf{w}) = a^{3/4} \, \mathcal{K}_M(\mathbf{t}, \mathbf{w}) \, \mathscr{F}\big[\psi\big]\big(A_a^T S_s^T B^{-1} \mathbf{w}\big), \tag{4.256}$$

where $\mathcal{K}_M(\mathbf{t}, \mathbf{w})$ *is given by (4.246).*

Proof. By virtue of Definition 4.8.1, we observe that

$$\mathcal{L}_M\big[\psi_{a,s,\mathbf{t}}^M\big](\mathbf{w}) = \int_{\mathbb{R}} \psi_{a,s,\mathbf{t}}^M(\mathbf{x}) \, \mathcal{K}_M(\mathbf{x}, \mathbf{w}) \, d\mathbf{x}$$
$$= |\det A_a|^{-1/2} \int_{\mathbb{R}^2} \psi\big(A_a^{-1} S_s^{-1}(\mathbf{x} - \mathbf{t})\big) \exp\left\{\frac{i\pi A(|\mathbf{t}|^2 - |\mathbf{x}|^2)}{B}\right\} \mathcal{K}_M(\mathbf{x}, \mathbf{w}) \, d\mathbf{x}$$
$$= \frac{|\det A_a|^{-1/2}}{B} \int_{\mathbb{R}^2} \psi\big(A_a^{-1} S_s^{-1}(\mathbf{x} - \mathbf{t})\big) \exp\left\{\frac{i\pi A(|\mathbf{t}|^2 - |\mathbf{x}|^2)}{B}\right\}$$
$$\times \exp\left\{\frac{i\pi(A|\mathbf{x}|^2 - 2\mathbf{x}^T \mathbf{w} + D|\mathbf{w}|^2)}{B}\right\} d\mathbf{x}$$
$$= \frac{|\det A_a|^{-1/2}}{B} \int_{\mathbb{R}^2} \psi\big(A_a^{-1} S_s^{-1}(\mathbf{x} - \mathbf{t})\big) \exp\left\{\frac{i\pi A|\mathbf{t}|^2}{B}\right\}$$
$$\times \exp\left\{\frac{i\pi(D|\mathbf{w}|^2 - 2\mathbf{x}^T \mathbf{w})}{B}\right\} d\mathbf{x}. \tag{4.257}$$

Substituting $A_a^{-1} S_s^{-1}(\mathbf{x} - \mathbf{t}) = \mathbf{z}$ on the R.H.S of (4.257), so that $d\mathbf{x} = |\det A_a| \, d\mathbf{z}$, we get

$$
\mathcal{L}_M \left[\psi_{a,s,\mathbf{t}}^M \right](\mathbf{w}) = \frac{|\det A_a|^{1/2}}{B} \int_{\mathbb{R}^2} \psi(\mathbf{z}) \exp \left\{ \frac{i\pi A |\mathbf{t}|^2}{B} \right\}
$$

$$
\times \exp \left\{ \frac{i\pi (D|\mathbf{w}|^2 - 2(\mathbf{t} + S_s A_a \mathbf{z})^T \mathbf{w})}{B} \right\} d\mathbf{z}
$$

$$
= \frac{|\det A_a|^{1/2}}{B} \int_{\mathbb{R}^2} \psi(\mathbf{z}) \exp \left\{ -\frac{2\pi i \, \mathbf{z}^T (A_a^T S_s^T \mathbf{w})}{B} \right\}
$$

$$
\times \exp \left\{ \frac{i\pi (A|\mathbf{t}|^2 - 2\mathbf{t}^T \mathbf{w} + D|\mathbf{w}|^2)}{B} \right\} d\mathbf{z}
$$

$$
= a^{3/4} \, \mathcal{K}_M(\mathbf{t}, \mathbf{w}) \int_{\mathbb{R}^2} \psi(\mathbf{z}) \exp \left\{ -\frac{2\pi i \, \mathbf{z}^T (A_a^T S_s^T \mathbf{w})}{B} \right\} d\mathbf{z}
$$

$$
= a^{3/4} \, \mathcal{K}_M(\mathbf{t}, \mathbf{w}) \, \mathscr{F} \left[\psi \right] (A_a^T S_s^T B^{-1} \mathbf{w}).
$$

This completes the proof of Proposition 4.8.1. $\qquad\qquad\qquad\qquad\qquad\qquad\qquad\square$

Remark 4.8.1. From Proposition 4.8.1, we conclude that each shearlet component is a differently scaled and sheared band-pass filter in the linear canonical domain.

In the following lemma, we shall obtain the spectral representation of the linear canonical shearlet transform defined in (4.253).

Lemma 4.8.2. *For any $f \in L^2(\mathbb{R})$, the linear canonical shearlet transform $\mathcal{S}_\psi^M[f](a, s, \mathbf{t})$ satisfies:*

$$
\mathcal{L}_M \left(\mathcal{S}_\psi^M \left[f \right](a, s, \mathbf{t}) \right)(\mathbf{w}) = a^{3/4} \, \mathcal{L}_M \left[f \right](\mathbf{w}) \, \overline{\mathscr{F} \left[\psi \right] (A_a^T S_s^T B^{-1} \mathbf{w})}, \qquad (4.258)
$$

where the linear canonical transform on the L.H.S of (4.258) is computed with respect to the translation variable $\mathbf{t}$.

Proof. Invoking Proposition 4.8.1 together with the Parseval's formula (4.248), we have

$$
\mathcal{S}_\psi^M \left[f \right](a, s, \mathbf{t}) = \left\langle f, \psi_{a,s,\mathbf{t}}^M \right\rangle_2
$$

$$
= \left\langle \mathcal{L}_M \left[f \right], \mathcal{L}_M \left[\psi_{a,s,\mathbf{t}}^M \right] \right\rangle_2
$$

$$
= \int_{\mathbb{R}^2} \mathcal{L}_M \left[f \right](\mathbf{w}) \, \overline{\mathcal{L}_M \left[\psi_{a,s,\mathbf{t}}^M \right](\mathbf{w})} \, d\mathbf{w}
$$

$$
= a^{3/4} \int_{\mathbb{R}^2} \mathcal{L}_M \left[f \right](\mathbf{w}) \, \overline{\mathscr{F} \left[\psi \right] (A_a^T S_s^T B^{-1} \mathbf{w})} \, \overline{\mathcal{K}_M(\mathbf{t}, \mathbf{w})} \, d\mathbf{w}
$$

$$
= a^{3/4} \, \mathcal{L}_M^{-1} \left(\mathcal{L}_M \left[f \right](\mathbf{w}) \, \overline{\mathscr{F} \left[\psi \right] (A_a^T S_s^T B^{-1} \mathbf{w})} \right)(\mathbf{t}). \qquad (4.259)
$$

Applying the two-dimensional linear canonical transform on both sides of (4.259), we obtain the desired expression as

$$
\mathcal{L}_M \left(\mathcal{S}_\psi^M \left[f \right](a, s, \mathbf{t}) \right)(\mathbf{w}) = a^{3/4} \, \mathcal{L}_M \left[f \right](\mathbf{w}) \, \overline{\mathscr{F} \left[\psi \right] (A_a^T S_s^T B^{-1} \mathbf{w})}.
$$

This completes the proof of Lemma 4.8.2. $\qquad\qquad\qquad\qquad\qquad\qquad\qquad\qquad\square$

Some basic properties of the linear canonical shearlet transform (4.253) are assembled in the next theorem. Prior to that, we recall that for $\mathbf{k} \in \mathbb{R}^2$, the planer translation operator $\mathcal{T}_{\mathbf{k}}$ acting on any $f \in L^2(\mathbb{R}^2)$ is defined by $\mathcal{T}_{\mathbf{k}} f(\mathbf{x}) = f(\mathbf{x} - \mathbf{k})$.

Theorem 4.8.3. *Let ψ and ϕ are two given shearlets and $c_1, c_2 \in \mathbb{C}$, $\mathbf{k} \in \mathbb{R}^2$, $\lambda > 0$ are scalar, with $\mathcal{T}_{\mathbf{k}}$ as the planer translation operator. Then for any pair of functions $f, g \in L^2(\mathbb{R}^2)$, the linear canonical shearlet transform (4.253) satisfies the following properties:*

(i) *Linearity:* $\mathcal{S}_\psi^M \big[c_1 f + c_2 g \big](a, s, t) = c_1 \mathcal{S}_\psi^M \big[f \big](a, s, t) + c_2 \mathcal{S}_\psi \big[g \big](a, s, t),$

(ii) *Anti-linearity:* $\mathcal{S}_{c_1 \psi + c_2 \phi}^M \big[f \big](a, s, t) = \overline{c_1} \, \mathcal{S}_\psi^M \big[f \big](a, s, t) + \overline{c_2} \, \mathcal{S}_\phi^M \big[f \big](a, s, t),$

(iii) *Translation:* $\mathcal{S}_\psi^M \big[\mathcal{T}_{\mathbf{k}} f \big](a, s, t) = \exp \left\{ -\dfrac{2\pi i A(\mathbf{t} - \mathbf{k})^T \mathbf{k}}{B} \right\} \mathcal{S}_\psi^M \big[F_M \big](a, s, \mathbf{t} - \mathbf{k}),$

where $F_M(\mathbf{x}) = \exp \left\{ \dfrac{2\pi i A(\mathbf{x}^T \mathbf{k})}{B} \right\} f(\mathbf{x}),$

(iv) *Translation in shearlet:*
$$\mathcal{S}_{\mathcal{T}_{\mathbf{k}} \psi}^M \big[f \big](a, s, t) = \exp \left\{ \dfrac{i\pi A(|S_s A_a \mathbf{k}|^2 + 2\mathbf{t}^T (S_s A_a \mathbf{k}))}{B} \right\} \mathcal{S}_\psi^M \big[f \big](a, s, \mathbf{t} + S_s A_a \mathbf{k}),$$

(v) *Scaling:* $\mathcal{S}_\psi^M \big[f(\lambda \mathbf{x}) \big](a, s, t) = \dfrac{1}{\lambda} \mathcal{S}_{\tilde{\psi}}^{M'} \big[f \big](a, s, \lambda t),$ where $\tilde{\psi}(\mathbf{x}) = \psi \left(\dfrac{\mathbf{x}}{\lambda} \right)$, $M' = (A, \lambda^2 B; C/\lambda^2, D),$

(vi) *Reflection:* $\mathcal{S}_\psi^M \big[f(-\mathbf{x}) \big](a, s, t) = \mathcal{S}_{\check{\psi}}^M \big[f \big](a, s, -t),$ where $\check{\psi}(\mathbf{x}) = \psi(-\mathbf{x}).$

Proof. The proofs of (i) and (ii) are straightforward and are omitted. We shall proceed to prove the rest of the properties of linear canonical shearlet transform.

(iii) In order to evaluate the linear canonical shearlet transform corresponding to the translated input function $\mathcal{T}_{\mathbf{k}} f(\mathbf{x})$, we proceed as

$$\mathcal{S}_\psi^M \big[\mathcal{T}_{\mathbf{k}} f \big](a, s, t) = \int_{\mathbb{R}^2} f(\mathbf{x} - \mathbf{k}) \, \overline{\psi_{a,s,t}^M(\mathbf{x})} \, d\mathbf{x}$$

$$= |\det A_a|^{-1/2} \int_{\mathbb{R}^2} f(\mathbf{x} - \mathbf{k}) \, \overline{\psi\big(A_a^{-1} S_s^{-1}(\mathbf{x} - \mathbf{t})\big)}$$

$$\times \exp \left\{ \dfrac{i\pi A(|\mathbf{x}|^2 - |\mathbf{t}|^2)}{B} \right\} d\mathbf{x}$$

$$= |\det A_a|^{-1/2} \int_{\mathbb{R}^2} f(\mathbf{z}) \, \overline{\psi\big(A_a^{-1} S_s^{-1}(\mathbf{z} - (\mathbf{t} - \mathbf{k}))\big)}$$

$$\times \exp \left\{ \dfrac{i\pi A(|\mathbf{z}|^2 - |\mathbf{t} - \mathbf{k}|^2)}{B} \right\} \exp \left\{ \dfrac{2\pi i A(\mathbf{z}^T \mathbf{k} + \mathbf{k}^T \mathbf{k} - \mathbf{t}^T \mathbf{k})}{B} \right\} d\mathbf{z}$$

$$= \exp \left\{ -\dfrac{2\pi i A(\mathbf{t} - \mathbf{k})^T \mathbf{k}}{B} \right\} \int_{\mathbb{R}^2} \exp \left\{ \dfrac{2\pi i A(\mathbf{z}^T \mathbf{k})}{B} \right\} f(\mathbf{z}) \, \overline{\psi_{a,s,\mathbf{t}-\mathbf{k}}^M(\mathbf{z})} \, d\mathbf{z}$$

$$= \exp \left\{ -\dfrac{2\pi i A(\mathbf{t} - \mathbf{k})^T \mathbf{k}}{B} \right\} \mathcal{S}_\psi^M \big[F_M \big](a, s, \mathbf{t} - \mathbf{k}),$$

where $F_M(\mathbf{x}) = \exp \left\{ \dfrac{2\pi i A(\mathbf{x}^T \mathbf{k})}{B} \right\} f(\mathbf{x}).$

(iv) If the translation operator $\mathcal{T}_{\mathbf{k}}$ is applied to the generating shearlet $\psi \in L^2(\mathbb{R}^2)$, then

$$\mathcal{S}_{\mathcal{T}_{\mathbf{k}}\psi}\big[f\big](a,s,\mathbf{t}) = \int_{\mathbb{R}^2} f(\mathbf{x})\,\overline{(\mathcal{T}_{\mathbf{k}}\psi)_{a,s,\mathbf{t}}^{M}(\mathbf{x})}\,d\mathbf{x}$$

$$= |\det A_a|^{-1/2} \int_{\mathbb{R}^2} f(\mathbf{x})\,\overline{(\mathcal{T}_{\mathbf{k}}\psi)\big(A_a^{-1}S_s^{-1}(\mathbf{x}-\mathbf{t})\big)}\,d\mathbf{x}$$

$$= |\det A_a|^{-1/2} \int_{\mathbb{R}^2} f(\mathbf{x})\,\overline{\psi\big(A_a^{-1}S_s^{-1}(\mathbf{x}-\mathbf{t})-\mathbf{k}\big)}$$

$$\times \exp\left\{\frac{i\pi A(|\mathbf{x}|^2 - |\mathbf{t}|^2)}{B}\right\}$$

$$= |\det A_a|^{-1/2} \int_{\mathbb{R}^2} f(\mathbf{x})\,\overline{\psi\big(A_a^{-1}S_s^{-1}(\mathbf{x}-(\mathbf{t}+S_s A_a \mathbf{k}))\big)}$$

$$\times \exp\left\{\frac{i\pi A(|\mathbf{x}|^2 - |\mathbf{t}|^2)}{B}\right\}d\mathbf{x}$$

$$= |\det A_a|^{-1/2} \int_{\mathbb{R}^2} f(\mathbf{x})\,\overline{\psi\big(A_a^{-1}S_s^{-1}(\mathbf{x}-(\mathbf{t}+S_s A_a \mathbf{k}))\big)}$$

$$\times \exp\left\{\frac{i\pi A(|\mathbf{x}|^2 - |\mathbf{t}+S_s A_a \mathbf{k}|^2)}{B}\right\}$$

$$\times \exp\left\{\frac{i\pi A\big(|S_s A_a \mathbf{k}|^2 + 2\mathbf{t}^T(S_s A_a \mathbf{k})\big)}{B}\right\}d\mathbf{x}$$

$$= \exp\left\{\frac{i\pi A\big(|S_s A_a \mathbf{k}|^2 + 2\mathbf{t}^T(S_s A_a \mathbf{k})\big)}{B}\right\}\int_{\mathbb{R}^2} f(\mathbf{x})\,\overline{\psi_{a,s,\mathbf{t}+S_s A_a \mathbf{k}}^{M}(\mathbf{z})}\,d\mathbf{z}$$

$$= \exp\left\{\frac{i\pi A\big(|S_s A_a \mathbf{k}|^2 + 2\mathbf{t}^T(S_s A_a \mathbf{k})\big)}{B}\right\}\mathcal{S}_{\psi}^{M}\big[f\big](a,s,\mathbf{t}+S_s A_a \mathbf{k}).$$

(v) Upon scaling the input signal isotropically by $\lambda \in \mathbb{R}^+$, we have

$$\mathcal{S}_{\psi}^{M}\big[f(\lambda\mathbf{x})\big](a,s,\mathbf{t}) = \int_{\mathbb{R}^2} f(\lambda\mathbf{x})\,\overline{\psi_{a,s,\mathbf{t}}^{M}(\mathbf{x})}\,d\mathbf{x}$$

$$= |\det A_a|^{-1/2} \int_{\mathbb{R}^2} f(\lambda\mathbf{x})\,\overline{\psi\big(A_a^{-1}S_s^{-1}(\mathbf{x}-\mathbf{t})\big)}$$

$$\times \exp\left\{\frac{i\pi A(|\mathbf{x}|^2 - |\mathbf{t}|^2)}{B}\right\}d\mathbf{x}$$

$$= |\det A_a|^{-1/2} \int_{\mathbb{R}^2} f(\mathbf{z})\,\overline{\tilde{\psi}\big(A_a^{-1}S_s^{-1}(\mathbf{z}-\lambda\mathbf{t})\big)}$$

$$\times \exp\left\{\frac{i\pi A(|\mathbf{z}|^2 - |\lambda\mathbf{t}|^2)}{\lambda^2 B}\right\}\frac{d\mathbf{z}}{\lambda}$$

$$= \frac{1}{\lambda}\int_{\mathbb{R}^2} f(\mathbf{z})\,\overline{\psi_{a,s,\lambda\mathbf{t}}^{M'}(\mathbf{x})}\,d\mathbf{z}$$

$$= \frac{1}{\lambda}\mathcal{S}_{\tilde{\psi}}^{M'}\big[f\big](a,s,\lambda\mathbf{t}),$$

where $M' = (A, \lambda^2 B; C/\lambda^2, D)$ and $\tilde{\psi}(\mathbf{x}) = \psi\left(\dfrac{\mathbf{x}}{\lambda}\right)$.

(vi) Finally, under the reflection of input signal, we have

$$\mathcal{S}_{\psi}^{M}\big[f(-\mathbf{x})\big](a,s,\mathbf{t}) = \int_{\mathbb{R}^2} f(-\mathbf{x})\,\overline{\psi_{a,s,\mathbf{t}}^{M}(\mathbf{x})}\,d\mathbf{x}$$

$$= |\det A_a|^{-1/2} \int_{\mathbb{R}^2} f(-\mathbf{x}) \, \overline{\psi\big(A_a^{-1} S_s^{-1}(\mathbf{x} - \mathbf{t})\big)}$$

$$\times \exp\left\{ \frac{i\pi A(|\mathbf{x}|^2 - |\mathbf{t}|^2)}{B} \right\} d\mathbf{x}$$

$$= |\det A_a|^{-1/2} \int_{\mathbb{R}^2} f(\mathbf{z}) \, \overline{\psi\big(A_a^{-1} S_s^{-1}(-\mathbf{z} - \mathbf{t})\big)}$$

$$\times \exp\left\{ \frac{i\pi A(|\mathbf{z}|^2 - |\mathbf{t}|^2)}{B} \right\} d\mathbf{z}$$

$$= |\det A_a|^{-1/2} \int_{\mathbb{R}^2} f(\mathbf{z}) \, \overline{\psi\big(-A_a^{-1} S_s^{-1}(\mathbf{z} + \mathbf{t})\big)}$$

$$\times \exp\left\{ \frac{i\pi A(|\mathbf{z}|^2 - |\mathbf{t}|^2)}{B} \right\} d\mathbf{z}$$

$$= \int_{\mathbb{R}^2} f(\mathbf{z}) \, \overline{\check{\psi}^M_{a,s,\mathbf{t}}(\mathbf{z})}$$

$$= \mathcal{S}^M_{\check{\psi}}\big[f\big](a, s, -\mathbf{t}), \quad \text{where} \quad \check{\psi}(\mathbf{x}) = \psi(-\mathbf{x}).$$

This completes the proof of Theorem 4.8.3. $\qquad\square$

Next, our aim is to obtain an orthogonality relation between a given pair of signals and the corresponding linear canonical shearlet transforms as defined in (4.253). As a consequence of such a relationship, we can infer that the linear canonical shearlet transform is an isometry from the space of two-dimensional finite energy signals $L^2(\mathbb{R}^2)$ to the space of transformations $L^2(\mathbb{R}^+ \times \mathbb{R} \times \mathbb{R}^2)$.

Theorem 4.8.4. *If $\mathcal{S}^M_\psi[f](a, s, \mathbf{t})$ and $\mathcal{S}^M_\psi[g](a, s, \mathbf{t})$ are the linear canonical shearlet transforms of a given pair of functions $f, g \in L^2(\mathbb{R}^2)$, then the following orthogonality relation holds:*

$$\int_{\mathbb{R}^2} \int_{\mathbb{R}} \int_{\mathbb{R}^+} \mathcal{S}^M_\psi[f](a, s, \mathbf{t}) \, \overline{\mathcal{S}^M_\psi[g](a, s, \mathbf{t})} \, \frac{da \, ds \, d\mathbf{t}}{a^3} = C_\psi \big\langle f, \, g \big\rangle_2, \qquad (4.260)$$

provided $C_\psi^+ = C_\psi^- = C_\psi$, where the entities C_ψ^+ and C_ψ^- have their usual meanings and are given by (3.297).

Proof. Identifying $\mathcal{S}^M_\psi[f](a, s, \mathbf{t})$ and $\mathcal{S}^M_\psi[g](a, s, \mathbf{t})$ as functions of the translation variable $\mathbf{t}$, then applying the Parseval's formula for the two-dimensional linear canonical transform (4.248) together with Lemma 4.8.2, we get

$$\int_{\mathbb{R}^2} \int_{\mathbb{R}} \int_{\mathbb{R}^+} \mathcal{S}^M_\psi[f](a, s, \mathbf{t}) \, \overline{\mathcal{S}^M_\psi[g](a, s, \mathbf{t})} \, \frac{da \, ds \, d\mathbf{t}}{a^3}$$

$$= \int_{\mathbb{R}^2} \int_{\mathbb{R}} \int_{\mathbb{R}^+} \mathcal{L}_M\big(\mathcal{S}^M_\psi[f](a, s, \mathbf{t})\big)(\mathbf{w}) \, \overline{\mathcal{L}_M\big(\mathcal{S}^M_\psi[f](a, s, \mathbf{t})\big)(\mathbf{w})} \, \frac{da \, ds \, d\mathbf{w}}{a^3}$$

$$= \int_{\mathbb{R}^2} \int_{\mathbb{R}} \int_{\mathbb{R}^+} \left\{ a^{3/4} \, \mathcal{L}_M[f](\mathbf{w}) \, \overline{\mathscr{F}[\psi]\big(A_a^T S_s^T B^{-1} \mathbf{w}\big)} \right\}$$

$$\times \left\{ a^{3/4} \, \overline{\mathcal{L}_M[g](\mathbf{w})} \, \mathscr{F}[\psi]\big(A_a^T S_s^T B^{-1} \mathbf{w}\big) \right\} \frac{da \, ds \, d\mathbf{w}}{a^3}$$

$$= \int_{\mathbb{R}^2} \int_{\mathbb{R}} \int_{\mathbb{R}^+} \mathcal{L}_M[f](\mathbf{w}) \, \overline{\mathcal{L}_M[g](\mathbf{w})} \, \Big| \mathscr{F}[\psi]\big(A_a^T S_s^T B^{-1} \mathbf{w}\big) \Big|^2 \, \frac{da \, ds \, d\mathbf{w}}{a^{3/2}}. \qquad (4.261)$$

In order to simplify the R.H.S of (4.261), we use $\mathbf{w} = (\omega_1, \omega_2)^T$, so that $A_a^T S_s^T B^{-1} \mathbf{w} = (aB^{-1}\omega_1, s\sqrt{a}B^{-1}\omega_1 + \sqrt{a}B^{-1}\omega_2)$ and then proceed as

$$\int_{\mathbb{R}^2} \int_{\mathbb{R}} \int_{\mathbb{R}^+} \mathcal{L}_M\big[f\big](\mathbf{w}) \overline{\mathcal{L}_M\big[g\big](\mathbf{w})} \left|\mathscr{F}\big[\psi\big]\left(A_a^T S_s^T B^{-1}\mathbf{w}\right)\right|^2 \frac{da\,ds\,d\mathbf{w}}{a^{3/2}}$$

$$= \int_{\mathbb{R}^+} \int_{\mathbb{R}} \int_{\mathbb{R}} \int_{\mathbb{R}} \mathcal{L}_M\big[f\big](\omega_1, \omega_2) \overline{\mathcal{L}_M\big[g\big](\omega_1, \omega_2)}$$

$$\times \left|\mathscr{F}\big[\psi\big]\left(aB^{-1}\omega_1, s\sqrt{a}B^{-1}\omega_1 + \sqrt{a}B^{-1}\omega_2\right)\right|^2 \frac{ds\,d\omega_1\,d\omega_2\,da}{a^{3/2}}$$

$$= \int_{\mathbb{R}} \int_0^\infty \int_0^\infty \int_{\mathbb{R}} \mathcal{L}_M\big[f\big](\omega_1, \omega_2) \overline{\mathcal{L}_M\big[g\big](\omega_1, \omega_2)} \left|\mathscr{F}\big[\psi\big]\left(aB^{-1}\omega_1, \xi_2\right)\right|^2 \frac{d\xi_2\,da\,d\omega_1\,d\omega_2}{a^2(|B|^{-1}\omega_1)}$$

$$- \int_{\mathbb{R}} \int_{-\infty}^0 \int_0^\infty \int_{\mathbb{R}} \mathcal{L}_M\big[f\big](\omega_1, \omega_2) \overline{\mathcal{L}_M\big[g\big](\omega_1, \omega_2)} \left|\mathscr{F}\big[\psi\big]\left(aB^{-1}\omega_1, \xi_2\right)\right|^2 \frac{d\xi_2\,da\,d\omega_1\,d\omega_2}{a^2(|B|^{-1}\omega_1)}$$

$$= \int_{\mathbb{R}} \int_0^\infty \mathcal{L}_M\big[f\big](\omega_1, \omega_2) \overline{\mathcal{L}_M\big[g\big](\omega_1, \omega_2)} \left\{ \int_0^\infty \int_{\mathbb{R}} \frac{\left|\mathscr{F}\big[\psi\big](\xi_1, \xi_2)\right|^2}{\xi_1^2} d\xi_2\,d\xi_1 \right\} d\omega_1\,d\omega_2$$

$$+ \int_{\mathbb{R}} \int_{-\infty}^0 \mathcal{L}_M\big[f\big](\omega_1, \omega_2) \overline{\mathcal{L}_M\big[g\big](\omega_1, \omega_2)} \left\{ \int_{-\infty}^0 \int_{\mathbb{R}} \frac{\left|\mathscr{F}\big[\psi\big](\xi_1, \xi_2)\right|^2}{\xi_1^2} d\xi_2\,d\xi_1 \right\} d\omega_1\,d\omega_2$$

$$= C_\psi^+ \int_{\mathbb{R}} \int_0^\infty \mathcal{L}_M\big[f\big](\omega_1, \omega_2) \overline{\mathcal{L}_M\big[g\big](\omega_1, \omega_2)} \, d\omega_1\,d\omega_2$$

$$+ C_\psi^- \int_{\mathbb{R}} \int_{-\infty}^0 \mathcal{L}_M\big[f\big](\omega_1, \omega_2) \overline{\mathcal{L}_M\big[g\big](\omega_1, \omega_2)} \, d\omega_1\,d\omega_2. \tag{4.262}$$

In case the shearlet $\psi \in L^2(\mathbb{R}^2)$ is such that $C_\psi^+ = C_\psi^- = C_\psi$, then the above expression (4.262) boils down to

$$\int_{\mathbb{R}^2} \int_{\mathbb{R}} \int_{\mathbb{R}^+} \mathcal{L}_M\big[f\big](\mathbf{w}) \overline{\mathcal{L}_M\big[g\big](\mathbf{w})} \left|\mathscr{F}\big[\psi\big]\left(A_a^T S_s^T B^{-1}\mathbf{w}\right)\right|^2 \frac{da\,ds\,d\mathbf{w}}{a^3}$$

$$= C_\psi \left\{ \int_{\mathbb{R}} \int_0^\infty \mathcal{L}_M\big[f\big](\omega_1, \omega_2) \overline{\mathcal{L}_M\big[g\big](\omega_1, \omega_2)} \, d\omega_1\,d\omega_2 \right.$$

$$\left. + \int_{\mathbb{R}} \int_{-\infty}^0 \mathcal{L}_M\big[f\big](\omega_1, \omega_2) \overline{\mathcal{L}_M\big[g\big](\omega_1, \omega_2)} \, d\omega_1\,d\omega_2 \right\}$$

$$= C_\psi \int_{\mathbb{R}} \int_{\mathbb{R}} \mathcal{L}_M\big[f\big](\omega_1, \omega_2) \overline{\mathcal{L}_M\big[g\big](\omega_1, \omega_2)} \, d\omega_1\,d\omega_2$$

$$= C_\psi \int_{\mathbb{R}^2} \mathcal{L}_M\big[f\big](\mathbf{w}) \overline{\mathcal{L}_M\big[g\big](\mathbf{w})} \, d\mathbf{w}$$

$$= C_\psi \left\langle \mathcal{L}_M\big[f\big](\mathbf{w}), \mathcal{L}_M\big[g\big](\mathbf{w}) \right\rangle_2$$

$$= C_\psi \left\langle f, g \right\rangle_2. \tag{4.263}$$

Finally, implementing (4.263) in (4.261), we obtain the desired orthogonality relation as

$$\int_{\mathbb{R}^2} \int_{\mathbb{R}} \int_{\mathbb{R}^+} \mathcal{S}_\psi^M\big[f\big](a, s, \mathbf{t}) \overline{\mathcal{S}_\psi^M\big[g\big](a, s, \mathbf{t})} \frac{da\,ds\,d\mathbf{t}}{a^3} = C_\psi \left\langle f, g \right\rangle_2.$$

This completes the proof of Theorem 4.8.4. $\qquad\qquad\qquad\qquad\qquad\qquad\square$

Remark 4.8.2. For $f = g$, Theorem 4.8.4 yields the following energy preserving relation for the linear canonical shearlet transform (4.253):

$$\int_{\mathbb{R}^2} \int_{\mathbb{R}} \int_{\mathbb{R}^+} \left| \mathcal{S}_\psi^M [f](a, s, \mathbf{t}) \right|^2 \frac{da\, ds\, d\mathbf{t}}{a^3} = C_\psi \left\| f \right\|_2^2. \tag{4.264}$$

From expression (4.264), we infer that the net concentration of the linear canonical shearlet transform $\mathcal{S}_\psi^M [f](a, s, \mathbf{t})$ in the space $L^2(\mathbb{R}^+ \times \mathbb{R} \times \mathbb{R}^2)$ is a C_ψ-multiple of total energy of the input signal f in the natural domain $L^2(\mathbb{R}^2)$. Therefore, choosing the shearlet ψ such that $C_\psi = 1$, the linear canonical shearlet transform defined in (4.253) constitutes an isometry from the space of two-dimensional finite energy signals $L^2(\mathbb{R}^2)$ to the space of transforms $L^2(\mathbb{R}^+ \times \mathbb{R} \times \mathbb{R}^2)$.

In the following theorem, we obtain the reconstruction formula for the linear canonical shearlet transform defined in (4.253).

Theorem 4.8.5. *Any function $f \in L^2(\mathbb{R}^2)$ can be retrieved from the corresponding linear canonical shearlet transform $\mathcal{S}_\psi^M [f](a, s, \mathbf{t})$ via the following formula:*

$$f(\mathbf{x}) = C_\psi^{-1} \int_{\mathbb{R}^2} \int_{\mathbb{R}} \int_{\mathbb{R}^+} \mathcal{S}_\psi^M [f](a, s, \mathbf{t})\, \psi_{a,s,\mathbf{t}}^M(\mathbf{x}) \frac{da\, ds\, d\mathbf{t}}{a^3}, \quad a.e., \tag{4.265}$$

where $0 \neq C_\psi$ has its usual meaning.

Proof. By virtue of the orthogonality relation (4.260) with $g(\mathbf{x}') = \delta(\mathbf{x} - \mathbf{x}')$, we obtain

$$C_\psi \left\langle f, \delta(\mathbf{x} - \mathbf{x}') \right\rangle_2 = \int_{\mathbb{R}^2} \int_{\mathbb{R}} \int_{\mathbb{R}^+} \mathcal{S}_\psi^M [f](a, s, \mathbf{t}) \overline{\mathcal{S}_\psi^M \left[\delta(\mathbf{x} - \cdot) \right](a, s, \mathbf{t})} \frac{da\, ds\, d\mathbf{t}}{a^3}. \tag{4.266}$$

On the other hand, we can compute the linear canonical shearlet transform of the function $g(\mathbf{x}') = \delta(\mathbf{x} - \mathbf{x}')$ as follows:

$$\mathcal{S}_\psi^M \left[\delta(\mathbf{x} - \cdot) \right](a, s, \mathbf{t}) = \int_{\mathbb{R}^2} \delta(\mathbf{x} - \mathbf{x}') \overline{\psi_{a,s,\mathbf{t}}^M(\mathbf{x}')}\, d\mathbf{x}' = \overline{\psi_{a,s,\mathbf{t}}^M(\mathbf{x})}. \tag{4.267}$$

Moreover, the L.H.S of (4.266) can be be simplified as

$$C_\psi \left\langle f, \delta(\mathbf{x} - \mathbf{x}') \right\rangle_2 = C_\psi \int_{\mathbb{R}^2} f(\mathbf{x}')\, \delta(\mathbf{x} - \mathbf{x}')\, d\mathbf{x}' = C_\psi\, f(\mathbf{x}). \tag{4.268}$$

Finally, using (4.267) and (4.268) in (4.266), we obtain the desired reconstruction formula:

$$f(\mathbf{x}) = C_\psi^{-1} \int_{\mathbb{R}^2} \int_{\mathbb{R}} \int_{\mathbb{R}^+} \mathcal{S}_\psi^M [f](a, s, \mathbf{t})\, \psi_{a,s,\mathbf{t}}^M(\mathbf{x}) \frac{da\, ds\, d\mathbf{t}}{a^3}, \quad a.e.$$

This completes the proof of Theorem 4.8.5. $\qquad\square$

Below, we shall present a characterization for the linear canonical shearlet transform defined in (4.253).

Theorem 4.8.6. *Let $(a, s, \mathbf{t}), (a', s', \mathbf{t}') \in \mathbb{R}^+ \times \mathbb{R} \times \mathbb{R}^2$. Then, $\mathcal{S}_\psi^M [f](a, s, \mathbf{t})$ is the linear canonical shearlet transform of a certain signal $f \in L^2(\mathbb{R}^2)$ if and only if it satisfies the following reproducing kernel equation:*

$$\mathcal{S}_\psi [f](a', s', \mathbf{t}') = \int_{\mathbb{R}^2} \int_{\mathbb{R}} \int_{\mathbb{R}^+} \mathcal{S}_\psi^M [f](a, s, \mathbf{t})\, \mathfrak{K}^M(a, s, \mathbf{t}; a', s', \mathbf{t}') \frac{da\, ds\, d\mathbf{t}}{a^3}, \tag{4.269}$$

where $\mathfrak{K}^M(a,s,\mathbf{t};a',s',\mathbf{t}')$ denotes the reproducing kernel and is given by

$$\mathfrak{K}^M(a,s,\mathbf{t};a',s',\mathbf{t}') = C_\psi^{-1} \left\langle \psi_{a,s,\mathbf{t}}^M, \psi_{a',s',\mathbf{t}'}^M \right\rangle_2, \tag{4.270}$$

and $0 \neq C_\psi$ has its usual meaning.

Proof. In order to prove the theorem, it suffices to to observe that

$$
\begin{aligned}
\mathcal{S}_\psi\big[f\big](a',s',\mathbf{t}') &= \int_{\mathbb{R}^2} f(\mathbf{x})\,\overline{\psi_{a',s',\mathbf{t}'}^M(\mathbf{x})}\,d\mathbf{x} \\
&= \int_{\mathbb{R}^2} \left\{ C_\psi^{-1} \int_{\mathbb{R}^2} \int_{\mathbb{R}} \int_{\mathbb{R}^+} \mathcal{S}_\psi^M\big[f\big](a,s,\mathbf{t})\,\psi_{a,s,\mathbf{t}}^M(\mathbf{x})\,\frac{da\,ds\,d\mathbf{t}}{a^3} \right\} \overline{\psi_{a',s',\mathbf{t}'}^M(\mathbf{x})}\,d\mathbf{x} \\
&= C_\psi^{-1} \int_{\mathbb{R}^2} \int_{\mathbb{R}} \int_{\mathbb{R}^+} \mathcal{S}_\psi^M\big[f\big](a,s,\mathbf{t}) \left\{ \int_{\mathbb{R}^2} \psi_{a,s,\mathbf{t}}^M(\mathbf{x})\,\overline{\psi_{a',s',\mathbf{t}'}^M(\mathbf{x})}\,d\mathbf{x} \right\} \frac{da\,ds\,d\mathbf{t}}{a^3} \\
&= C_\psi^{-1} \int_{\mathbb{R}^2} \int_{\mathbb{R}} \int_{\mathbb{R}^+} \mathcal{S}_\psi^M\big[f\big](a,s,\mathbf{t}) \left\langle \psi_{a,s,\mathbf{t}}^M, \psi_{a',s',\mathbf{t}'}^M \right\rangle_2 \frac{da\,ds\,d\mathbf{t}}{a^3} \\
&= \int_{\mathbb{R}^2} \int_{\mathbb{R}} \int_{\mathbb{R}^+} \mathcal{S}_\psi^M\big[f\big](a,s,\mathbf{t})\,\mathfrak{K}^M(a,s,\mathbf{t};a',s',\mathbf{t}')\,\frac{da\,ds\,d\mathbf{t}}{a^3}.
\end{aligned}
$$

$\square$

Remark 4.8.3. The reproducing kernel $\mathfrak{K}^M(a,s,\mathbf{t};a',s',\mathbf{t}')$ given by (4.270) measures the correlation of a given pair of linear canonical shearlets $\psi_{a,s,\mathbf{t}}^M(\mathbf{x})$ and $\psi_{a',s',\mathbf{t}'}^M(\mathbf{x})$. As a consequence of (4.269), it follows that the linear canonical shearlet transform of a certain signal $f \in L^2(\mathbb{R}^2)$ about $(a',s',\mathbf{t}') \in \mathbb{R}^+ \times \mathbb{R} \times \mathbb{R}^2$ is always expressible via another $(a,s,\mathbf{t}) \in \mathbb{R}^+ \times \mathbb{R} \times \mathbb{R}^2$ by virtue of the reproducing kernel $\mathfrak{K}^M(a,s,\mathbf{t};a',s',\mathbf{t}')$, satisfying

$$\left| \mathfrak{K}^M(a,s,\mathbf{t};a',s',\mathbf{t}') \right| \leq C_\psi^{-1} \big\|\psi\big\|_2^2, \quad \forall\,(a,s,\mathbf{t}),\,(a',s',\mathbf{t}') \in \mathbb{R}^+ \times \mathbb{R} \times \mathbb{R}^2. \tag{4.271}$$

4.8.2 Uncertainty principle for the linear canonical shearlet transform

> "All analysts spend half their time hunting through the literature for inequalities which they want to use and cannot prove."

> -Godfrey H. Hardy

This subsection marks closure of the ongoing discourse on linear canonical shearlet transform. Here, we focus on formulating a Heisenberg-type uncertainty inequality for the linear canonical shearlet transform defined in (4.253). Such an inequality sets a lower bound upon the simultaneous resolution of the transformed signal $\mathcal{S}_\psi^M\big[f\big](a,s,\mathbf{t})$, regarded as a function of the translation variable $\mathbf{t}$, and the two-dimensional linear canonical transform of the input signal $f \in L^2(\mathbb{R}^2)$. To facilitate the motive, we note that the variances of a given square integrable function $f(\mathbf{x})$ and the corresponding two-dimensional linear canonical transform $\mathcal{L}_M\big[f\big](\mathbf{w})$ are defined as

$$\Delta_{f(\mathbf{x})}^2 = \frac{\int_{\mathbb{R}^2} |\mathbf{x} - \mathbf{x}_0|^2 \big|f(\mathbf{x})\big|^2 d\mathbf{x}}{\int_{\mathbb{R}^2} \big|f(\mathbf{x})\big|^2 d\mathbf{x}} \quad \text{and} \quad \Delta_{\mathcal{L}_M[f](\mathbf{w})}^2 = \frac{\int_{\mathbb{R}^2} |\mathbf{w} - \mathbf{w}_0|^2 \big|\mathcal{L}_M\big[f\big](\mathbf{w})\big|^2 d\mathbf{w}}{\int_{\mathbb{R}^2} \big|\mathcal{L}_M\big[f\big](\mathbf{w})\big|^2 d\mathbf{w}},$$

$$\tag{4.272}$$

where the first order moments $\mathbf{x}_0$ and $\mathbf{w}_0$ are, respectively, given by

$$\mathbf{x}_0 = \frac{\int_{\mathbb{R}^2} |\mathbf{x}|^2 |f(\mathbf{x})|^2 d\mathbf{x}}{\int_{\mathbb{R}^2} |f(\mathbf{x})|^2 d\mathbf{x}} \quad \text{and} \quad \mathbf{w}_0 = \frac{\int_{\mathbb{R}^2} |\mathbf{w}|^2 |\mathcal{L}_M[f](\mathbf{w})|^2 d\mathbf{w}}{\int_{\mathbb{R}^2} |\mathcal{L}_M[f](\mathbf{w})|^2 d\mathbf{w}}. \tag{4.273}$$

In the following proposition, we obtain the Heisenberg's uncertainty inequality for the two-dimensional linear canonical transform defined in (4.245). Such an inequality serves as the pedestal for obtaining the desired Heisenberg-type uncertainty inequality for the linear canonical shearlet transform (4.253).

Proposition 4.8.7. *For any non-trivial function $f \in L^2(\mathbb{R}^2)$, the spreads in the spatial and linear canonical domains satisfy the following inequality:*

$$\Delta^2_{f(\mathbf{x})} \cdot \Delta^2_{\mathcal{L}_M[f](\mathbf{w})} \geq |B|^2, \tag{4.274}$$

with equality if and only if the function f has the following form:

$$f(\mathbf{x}) = K \exp\left\{\left(\frac{i\pi\gamma A - B}{\gamma B}\right)|\mathbf{x}|^2\right\}, \quad \gamma \in \mathbb{R}^+, \, K \in \mathbb{C}. \tag{4.275}$$

Proof. Define the function

$$F(\mathbf{w}) = \int_{\mathbb{R}^2} f(\mathbf{x}) \exp\left\{\frac{i\pi(A|\mathbf{x}|^2 - 2\mathbf{x}^T\mathbf{w})}{B}\right\} d\mathbf{x}. \tag{4.276}$$

Then, by using Definition 4.8.1, we observe that the function (4.276) can be expressed as

$$F(\mathbf{w}) = B \exp\left\{-\frac{i\pi D|\mathbf{w}|^2}{B}\right\} \mathcal{L}_M[f](\mathbf{w}), \tag{4.277}$$

so that

$$\Delta^2_{\mathcal{L}_M[f](\mathbf{w})} = \frac{\int_{\mathbb{R}^2} |\mathbf{w} - \mathbf{w}_0|^2 |(\mathbf{w})|^2 d\mathbf{w}}{\int_{\mathbb{R}^2} |F(\mathbf{w})|^2 d\mathbf{w}} = \Delta^2_{F(\mathbf{w})}. \tag{4.278}$$

For $\sigma = B^{-1}\mathbf{w}$, it is worth noticing that the function $F(\sigma)$ is the two-dimensional Fourier transform of the function $G(\mathbf{x})$ defined by

$$G(\mathbf{x}) = f(\mathbf{x}) \exp\left\{\frac{i\pi A|\mathbf{x}|^2}{B}\right\}. \tag{4.279}$$

Therefore, using the Heisenberg's uncertainty principle for the two-dimensional Fourier transform, we get

$$\Delta^2_{G(\mathbf{x})} \cdot \Delta^2_{F(\sigma)} \geq 1. \tag{4.280}$$

Moreover, we observe that

$$\Delta^2_{F(\sigma)} = \frac{\Delta^2_{F(\mathbf{w})}}{|B|^2} = \frac{\Delta^2_{\mathcal{L}_M[f](\mathbf{w})}}{|B|^2} \tag{4.281}$$

and

$$\Delta^2_{G(\mathbf{x})} = \frac{\int_{\mathbb{R}^2} |\mathbf{x} - \mathbf{x}_0|^2 \left| f(\mathbf{x}) \exp\left\{ \frac{i\pi A |\mathbf{x}|^2}{B} \right\} \right|^2 d\mathbf{x}}{\int_{\mathbb{R}^2} \left| f(\mathbf{x}) \exp\left\{ \frac{i\pi A |\mathbf{x}|^2}{B} \right\} \right|^2 d\mathbf{x}} = \frac{\int_{\mathbb{R}^2} |\mathbf{x} - \mathbf{x}_0|^2 |f(\mathbf{x})|^2 d\mathbf{x}}{\int_{\mathbb{R}^2} |f(\mathbf{x})|^2 d\mathbf{x}} = \Delta^2_{f(\mathbf{x})}.$$

$$\text{(4.282)}$$

By virtue of the relations (4.281) and (4.282), we obtain the desired Heisenberg's uncertainty inequality for the two-dimensional linear canonical transform as

$$\Delta^2_{f(\mathbf{x})} \cdot \Delta^2_{\mathcal{L}_M[f](\mathbf{w})} \geq |B|^2. \qquad (4.283)$$

Furthermore, since inequality (4.280) is best possible provided the function G is a generalized Guassian function; that is, $G(\mathbf{x}) = K e^{-|\mathbf{x}|^2/\gamma}$, $\gamma \in \mathbb{R}^+$, $K \in \mathbb{C}$, therefore, we conclude that equality holds in (4.283) if and only if

$$f(\mathbf{x}) = K \exp\left\{ \left(\frac{i\pi \gamma A - B}{\gamma B} \right) |\mathbf{x}|^2 \right\}.$$

This completes the proof of Proposition 4.8.7.   □

Remark 4.8.4. In case the first order moments are chosen as $\mathbf{x}_0 = \mathbf{0}$ and $\mathbf{w}_0 = \mathbf{0}$, then the Heisenberg's uncertainty inequality (4.274) can be recast as

$$\left\{ \int_{\mathbb{R}^2} |\mathbf{x}|^2 |f(\mathbf{x})|^2 d\mathbf{x} \right\}^{1/2} \left\{ \int_{\mathbb{R}^2} |\mathbf{w}|^2 |\mathcal{L}_M[f](\mathbf{w})|^2 d\mathbf{w} \right\}^{1/2} \geq |B| \left\{ \int_{\mathbb{R}^2} |f(\mathbf{x})|^2 d\mathbf{x} \right\}. \quad (4.284)$$

The aforementioned choice for the first order moments is reasonable in the sense that any function $f \in L^2(\mathbb{R}^2)$ can be translated and modulated appropriately so that the first order moments satisfy $\mathbf{x}_0 = \mathbf{0}$ and $\mathbf{w}_0 = \mathbf{0}$, without affecting the respective variances.

Following is the Heisenberg-type uncertainty inequality associated with the linear canonical shearlet transform defined in (4.253):

Theorem 4.8.8. *If $\mathcal{S}^M_\psi[f](a, s, \mathbf{t})$ is the linear canonical shearlet transform corresponding to any non-trivial function $f \in L^2(\mathbb{R}^2)$, with respect to the shearlet $\psi \in L^2(\mathbb{R}^2)$ satisfying $C^+_\psi = C^-_\psi = C_\psi$, then the following uncertainty inequality holds:*

$$\left\{ \int_{\mathbb{R}^2} \int_{\mathbb{R}} \int_{\mathbb{R}^+} |\mathbf{t}|^2 \left| \mathcal{S}^M_\psi[f](a, s, \mathbf{t}) \right|^2 \frac{da\, ds\, d\mathbf{t}}{a^3} \right\}^{1/2} \left\{ \int_{\mathbb{R}^2} |\mathbf{w}|^2 \left| \mathcal{L}_M[f](\mathbf{w}) \right|^2 d\mathbf{w} \right\}^{1/2}$$

$$\geq |B| \sqrt{C_\psi} \left\| f \right\|^2_2, \quad (4.285)$$

where the entities C^+_ψ and C^-_ψ have their usual meanings and are given by (3.297).

Proof. Identifying $\mathcal{S}^M_\psi[f](a, s, \mathbf{t})$ as a function of the translation variable $\mathbf{t}$ and then invoking (4.284), we get

$$\left\{ \int_{\mathbb{R}^2} |\mathbf{t}|^2 \left| \mathcal{S}^M_\psi[f](a, s, \mathbf{t}) \right|^2 d\mathbf{t} \right\}^{1/2} \left\{ \int_{\mathbb{R}^2} |\mathbf{w}|^2 \left| \mathcal{L}_M\left(\mathcal{S}^M_\psi[f](a, s, \mathbf{t}) \right)(\mathbf{w}) \right|^2 d\mathbf{w} \right\}^{1/2}$$

$$\geq |B| \left\{ \int_{\mathbb{R}^2} \left| \mathcal{S}^M_\psi[f](a, s, \mathbf{t}) \right|^2 d\mathbf{t} \right\}. \quad (4.286)$$

Invoking Lemma 4.8.2, we can express (4.286) in the following fashion:

$$\left\{ \int_{\mathbb{R}^2} |\mathbf{t}|^2 \left| \mathcal{S}_\psi^M \big[f \big] (a, s, \mathbf{t}) \right|^2 d\mathbf{t} \right\}^{1/2}$$

$$\times \left\{ \int_{\mathbb{R}^2} a^{3/2} |\mathbf{w}|^2 \left| \mathcal{L}_M \big[f \big] (\mathbf{w}) \right|^2 \left| \mathscr{F} \big[\psi \big] \big(A_a^T S_s^T B^{-1} \mathbf{w} \big) \right|^2 d\mathbf{w} \right\}^{1/2}$$

$$\geq |B| \left\{ \int_{\mathbb{R}^2} \left| \mathcal{S}_\psi^M \big[f \big] (a, s, \mathbf{t}) \right|^2 d\mathbf{t} \right\}. \quad (4.287)$$

Integrating (4.287) on both sides with respect to the measure $da\, ds/a^3$ followed by the application of Cauchy-Schwarz inequality, we get

$$\left\{ \int_{\mathbb{R}^2} \int_{\mathbb{R}} \int_{\mathbb{R}^+} |\mathbf{t}|^2 \left| \mathcal{S}_\psi^M \big[f \big] (a, s, \mathbf{t}) \right|^2 \frac{da\, ds\, d\mathbf{t}}{a^3} \right\}^{1/2}$$

$$\times \left\{ \int_{\mathbb{R}^2} \int_{\mathbb{R}} \int_{\mathbb{R}^+} |\mathbf{w}|^2 \left| \mathcal{L}_M \big[f \big] (\mathbf{w}) \right|^2 \left| \mathscr{F} \big[\psi \big] \big(A_a^T S_s^T B^{-1} \mathbf{w} \big) \right|^2 \frac{da\, ds\, d\mathbf{w}}{a^{3/2}} \right\}^{1/2}$$

$$\geq |B| \left\{ \int_{\mathbb{R}^2} \int_{\mathbb{R}} \int_{\mathbb{R}^+} \left| \mathcal{S}_\psi^M \big[f \big] (a, s, \mathbf{t}) \right|^2 \frac{da\, ds\, d\mathbf{t}}{a^3} \right\}$$

$$= |B|\, C_\psi \left\| f \right\|_2^2. \quad (4.288)$$

Next, we shall simplify the second set of integrals on the L.H.S of (4.288). For $\mathbf{w} = (\omega_1, \omega_2)^T$, we note that $A_a^T S_s^T B^{-1} \mathbf{w} = (aB^{-1}\omega_1,\, s\sqrt{a}B^{-1}\omega_1 + \sqrt{a}B^{-1}\omega_2)$. Then, we have

$$\int_{\mathbb{R}^2} \int_{\mathbb{R}} \int_{\mathbb{R}^+} |\mathbf{w}|^2 \left| \mathcal{L}_M \big[f \big] (\mathbf{w}) \right|^2 \left| \mathscr{F} \big[\psi \big] \big(A_a^T S_s^T B^{-1} \mathbf{w} \big) \right|^2 \frac{da\, ds\, d\mathbf{w}}{a^{3/2}}$$

$$= \int_{\mathbb{R}^+} \int_{\mathbb{R}} \int_{\mathbb{R}} \int_{\mathbb{R}} (\omega_1^2 + \omega_2^2) \left| \mathcal{L}_M \big[f \big] (\omega_1, \omega_2) \right|^2$$

$$\times \left| \mathscr{F} \big[\psi \big] \big(aB^{-1}\omega_1,\, s\sqrt{a}B^{-1}\omega_1 + \sqrt{a}B^{-1}\omega_2 \big) \right|^2 \frac{ds\, d\omega_1\, d\omega_2\, da}{a^{3/2}}$$

$$= \int_{\mathbb{R}} \int_0^\infty \int_0^\infty \int_{\mathbb{R}} (\omega_1^2 + \omega_2^2) \left| \mathcal{L}_M \big[f \big] (\omega_1, \omega_2) \right|^2 \left| \mathscr{F} \big[\psi \big] \big(aB^{-1}\omega_1, \xi_2 \big) \right|^2 \frac{d\xi_2\, da\, d\omega_1\, d\omega_2}{a^2(|B|^{-1}\omega_1)}$$

$$- \int_{\mathbb{R}} \int_{-\infty}^0 \int_0^\infty \int_{\mathbb{R}} (\omega_1^2 + \omega_2^2) \left| \mathcal{L}_M \big[f \big] (\omega_1, \omega_2) \right|^2 \left| \mathscr{F} \big[\psi \big] \big(aB^{-1}\omega_1, \xi_2 \big) \right|^2 \frac{d\xi_2\, da\, d\omega_1\, d\omega_2}{a^2(|B|^{-1}\omega_1)}$$

$$= \int_{\mathbb{R}} \int_0^\infty (\omega_1^2 + \omega_2^2) \left| \mathcal{L}_M \big[f \big] (\omega_1, \omega_2) \right|^2 \left\{ \int_0^\infty \int_{\mathbb{R}} \frac{\left| \mathscr{F} \big[\psi \big] (\xi_1, \xi_2) \right|^2}{\xi_1^2} d\xi_2\, d\xi_1 \right\} d\omega_1\, d\omega_2$$

$$+ \int_{\mathbb{R}} \int_{-\infty}^0 (\omega_1^2 + \omega_2^2) \left| \mathcal{L}_M \big[f \big] (\omega_1, \omega_2) \right|^2 \left\{ \int_{-\infty}^0 \int_{\mathbb{R}} \frac{\left| \mathscr{F} \big[\psi \big] (\xi_1, \xi_2) \right|^2}{\xi_1^2} d\xi_2\, d\xi_1 \right\} d\omega_1\, d\omega_2$$

$$= C_\psi^+ \int_{\mathbb{R}} \int_0^\infty (\omega_1^2 + \omega_2^2) \left| \mathcal{L}_M \big[f \big] (\omega_1, \omega_2) \right|^2 d\omega_1\, d\omega_2$$

$$+ C_\psi^- \int_{\mathbb{R}} \int_{-\infty}^0 (\omega_1^2 + \omega_2^2) \left| \mathcal{L}_M \big[f \big] (\omega_1, \omega_2) \right|^2 d\omega_1\, d\omega_2. \quad (4.289)$$

By using the given hypothesis, we have $C_\psi^+ = C_\psi^- = C_\psi$, therefore the relation obtained in (4.289) yields

$$\int_{\mathbb{R}^2} \int_{\mathbb{R}} \int_{\mathbb{R}^+} |\mathbf{w}|^2 \left| \mathcal{L}_M \big[f\big](\mathbf{w}) \right|^2 \left| \mathscr{F}\big[\psi\big](A_a^T S_s^T B^{-1} \mathbf{w}) \right|^2 \frac{da\,ds\,d\mathbf{w}}{a^{3/2}}$$

$$= C_\psi \int_{\mathbb{R}^2} |\mathbf{w}|^2 \left| \mathcal{L}_M\big[f\big](\mathbf{w}) \right|^2 d\mathbf{w}. \quad (4.290)$$

Finally, using (4.290) in (4.288), we obtain the desired Heisenberg-type uncertainty inequality for the linear canonical shearlet transform as

$$\left\{ \int_{\mathbb{R}^2} \int_{\mathbb{R}} \int_{\mathbb{R}^+} |\mathbf{t}|^2 \left| \mathcal{S}_\psi^M\big[f\big](a,s,\mathbf{t}) \right|^2 \frac{da\,ds\,dt}{a^3} \right\}^{1/2} \left\{ \int_{\mathbb{R}^2} |\mathbf{w}|^2 \left| \mathcal{L}_M\big[f\big](\mathbf{w}) \right|^2 d\mathbf{w} \right\}^{1/2}$$

$$\geq |B|\, \sqrt{C_\psi}\, \big\| f \big\|_2^2.$$

This completes the proof of Theorem 4.8.8.   □

Remark 4.8.5. For $M = (0,1;-1,0)$ and $M = (\cos\alpha, \sin\alpha; -\sin\alpha, \cos\alpha)$, $\alpha \neq n\pi$, $n \in \mathbb{Z}$, the uncertainty inequality (4.285) boils down to the respective Heisenberg-type inequalities for the classical and fractional variants of the shearlet transform.

4.9   Exercises

Exercise 4.9.1. Obtain the fractional Fourier transform of the Haar wavelet (3.31) and discuss the influence of the fractional parameter α.

Exercise 4.9.2. If $\psi(t)$ is the Mexican Hat wavelet (3.45), then for $\alpha \neq n\pi$, $n \in \mathbb{Z}$ discuss the family of fractional daughter wavelets

$$\psi_{a,b}^\alpha(t) = \frac{1}{\sqrt{a}}\, \psi\left(\frac{t-b}{a}\right) \exp\left\{ \frac{i(b^2 - t^2)\cot\alpha}{2} \right\}, \quad a \in \mathbb{R}^+, b \in \mathbb{R}. \quad (4.291)$$

Also, discuss the variations of the fractional parameter α on the daughter wavelets $\psi_{a,b}^\alpha(t)$.

Exercise 4.9.3. Compute the fractional wavelet transform of the unit-step

$$U(t) := \begin{cases} 1, & t \geq 0 \\ 0, & \text{otherwise,} \end{cases} \quad (4.292)$$

with respect to the Mexican Hat wavelet (3.45).

Exercise 4.9.4. Use Theorem 4.2.2 to compute the fractional wavelet transform of the translated and dilated versions of the Gaussian function $f(t) = e^{-t^2/2}$ with respect to the Haar wavelet (3.31).

Exercise 4.9.5. For the Mexican Hat wavelet (3.45), find the Q-factor and the total area occupied by the time-fractional-frequency window (4.28) at $\alpha = \pi/4, \pi/6, \pi/8$ and $\pi/10$.

Exercise 4.9.6. Compute the fractional Stockwell transform of the following functions, with respect to the Gaussian window $g(t) = e^{-t^2/2}$:

(i) $f(t) = \delta(t - t_0)$, $t_0 \in \mathbb{R}$,

(ii) $f(t) = \chi_{[-1,\,1]}(t)$,

(iii) $f(t) = t\,U(t)$,

(iv) $f(t) = t^2$,

(v) $f(t) = e^{-t}\,U(t)$,

where $U(t)$ is the usual unit-step given by (4.292).

Exercise 4.9.7. Use Proposition 4.4.1 to evaluate the linear canonical transform of the family of daughter wavelets (4.64) obtained from the Haar wavelet (3.31) with the following choices of the unimodular matrix $M = (A, B; C, D)$:

(i) $M = (1, 2; 1/2, 2)$,

(ii) $M = (1, 1/2; 2, 2)$,

(iii) $M = (3, 1; 5, 2)$,

(iv) $M = (3, 5; 1, 2)$,

(v) $M = (1/4, 1/6; -3, 2)$,

(vi) $M = (1/5, 1/10; 0, 5)$.

Also, in each case find the Q-factor and the net time-linear-canonical-frequency resolution of the window (4.96).

Exercise 4.9.8. Using Definition 4.4.1, compute the linear canonical wavelet transform of the Gaussian function $f(t) = e^{-t^2/2}$, with respect to the Morlet wavelet (3.35).

Exercise 4.9.9. Consider the function

$$\phi(t) = \begin{cases} 1, & 0 \le t \ge 1 \\ 0, & \text{otherwise,} \end{cases} \tag{4.293}$$

then use Theorem 4.4.2 to obtain a new wavelet by convoluting (4.293) with the Haar wavelet (3.31).

Exercise 4.9.10. Compute the linear canonical ridgelet transform of the function $f(\mathbf{t}) = K\, e^{-\sigma|\mathbf{t}|^2}$, $\sigma > 0$, $K \in \mathbb{C}$, with respect to the Haar wavelet (3.31).

Exercise 4.9.11. For $B > 0$, consider the function

$$G(r) = \begin{cases} \dfrac{1}{B}, & |r| \le \dfrac{1}{2} \\ \dfrac{1}{B} \cos\left(\dfrac{(2|r| - 1)\pi}{2}\right), & \dfrac{1}{2} \le |r| \le 1 \\ 0, & \text{otherwise.} \end{cases} \tag{4.294}$$

Then, discuss the characteristics of the function

$$H(r) = \chi_{[0,\,\infty]}(r)\, \sqrt{\left(G(r/2)\right)^2 - \left(G(r)\right)^2} \tag{4.295}$$

for acting as a radial window for the linear canonical curvelet transform.

Exercise 4.9.12. If $\psi(t)$ is the classical shearlet discussed in Example 3.7.1, then use Proposition 4.8.1 to obtain the linear canonical spectral representation of the novel analyzing functions $\psi_{a,s,\mathbf{t}}^M(\mathbf{x})$ given in (4.251).

5

The Wavelet Transforms and Kith

"Universe is a "grand book" written in the language of mathematics."

-Galileo Galilei

5.1 Introduction

This chapter is entirely devoted for a detailed study of several new classes of wavelet transforms based on certain well-known orthogonal polynomials and special functions [321]. Among the class of orthogonal polynomials, we shall be focussed on the Laguerre and Legendre polynomials and formulate the associated wavelet transforms by employing the tools of Laguerre and Legendre transforms. This is followed by the formulation of another couple of wavelet transforms by using the fundamental notions of the Bessel and Dunkl transforms, which are based on the well-know class of special functions, namely the Bessel and Dunkl functions. Nevertheless, complementary to these developments, we shall formulate yet another hybrid wavelet transform in the guise of Mehler-Fock wavelet transform, which relies on a special class of the Legendre functions. The chapter is concluded with the exploration of an interesting interface between the classical wavelet transform and the Hartley transform.

5.2 The Laguerre Wavelet Transform

"The mathematical studies done by Edmond Nicolas Laguerre in the 19-th century laid the foundation for contemporary optical communications."

-Mario Martinelli and Paolo Martinelli

In this section, we shall present the notion of Laguerre wavelet transform obtained via the class of generalized Laguerre polynomials which are orthogonal over $[0, \infty)$ under a suitable weighting factor. To facilitate the intent, we recapitulate the notions of classical Laguerre polynomials, generalized Laguerre polynomials and the associated Laguerre transform. In the sequel, we investigate upon the fundamental aspects of the Laguerre wavelet transform.

5.2.1 Laguerre polynomials and transform

The Laguerre polynomials are named after the French mathematician Edmond Nicolas Laguerre who was working on the class of orthogonal polynomials. These polynomials often

DOI: 10.1201/9781003175766-5

arise in diverse aspects of quantum mechanics, for instance in the radial part of the solution of Schrödinger equation for a one electron atom, in the static Wigner functions of oscillator systems and so on. The classical Laguerre polynomials are defined as the polynomial solutions of the Laguerre's differential equation:

$$tf''(t) + (1 - t)f'(t) + nf(t) = 0, \tag{5.1}$$

which is a second order linear differential equation and has non-singular solutions only if n is a non-negative integer. In case n is not a non-negative integer, the corresponding solutions of the differential equation (5.1) are referred as the "Laguerre functions." The Laguerre polynomials are denoted by $L_n(t)$ and are defined as

$$\left.\begin{array}{lll}
n = 0; & L_n(t) = 1 \\[4pt]
n = 1; & L_n(t) = -t + 1 \\[4pt]
n = 2; & L_n(t) = \dfrac{1}{2}\left(t^2 - 4t + 2\right) \\[8pt]
n = 3; & L_n(t) = \dfrac{1}{6}\left(-t^3 + 9t^2 - 18t + 6\right) \\[8pt]
n = 4; & L_n(t) = \dfrac{1}{24}\left(t^4 - 16t^3 + 72t^2 - 96t + 24\right)
\end{array}\right\} \tag{5.2}$$

and, in general,

$$L_n(t) = \frac{1}{n!}\left\{(-t)^n + n^2(-t)^{n-1} + \cdots + n(n!)(-t) + n!\right\}. \tag{5.3}$$

These polynomials can also be defined recursively as

$$L_0(t) = 1, \; L_1(t) = 1 - t, \quad L_k(t) = \frac{(2k + 1 - t)L_k(t) - kL_{k-1}(t)}{k + 1}, \quad \text{for} \quad k \geq 1. \tag{5.4}$$

Moreover, the Rodrigues formula for the Laguerre polynomial $L_n(t)$ is given by

$$L_n(t) = \frac{e^t}{n!}\frac{d^n}{dt^n}\left(e^{-t}t^n\right) = \frac{1}{n!}\left(\frac{d}{dt} - 1\right)^n t^n, \tag{5.5}$$

whereas the closed form is given by

$$L_n(t) = \sum_{k=0}^{n}\binom{n}{k}\frac{(-1)^k}{k!}t^k. \tag{5.6}$$

It is pertinent to mention that the Laguerre polynomials are orthogonal over the interval $[0, \infty)$ with respect to the inner product

$$\left\langle f, g \right\rangle = \int_0^\infty f(t)\, g(t)\, e^{-t} dt. \tag{5.7}$$

With gentle modifications to the Laguerre's differential equation (5.1), the generalized Laguerre polynomials are defined as the polynomial solutions of the second order linear differential equation:

$$tf''(t) + (\alpha + 1 - t)f'(t) + nf(t) = 0, \quad \alpha \in \mathbb{R}. \tag{5.8}$$

The generalized Laguerre polynomials are denoted by $L_n^{(\alpha)}(t)$ and are defined recursively as follows:

$$L_0^{(\alpha)}(t) = 1, \quad L_1^{(\alpha)}(t) = 1 + \alpha - t \quad \text{and}$$

$$L_k^{(\alpha)}(t) = \frac{(2k+1+\alpha-t)L_k^{(\alpha)}(t) - (k+\alpha)L_{k-1}^{(\alpha)}(t)}{k+1}, \quad \text{for} \quad k \geq 1. \quad (5.9)$$

Based on the recurrence relation (5.9), the first few generalized Laguerre polynomials are given below:

$$\left.\begin{array}{ll} n = 0; & L_n^{(\alpha)}(t) = 1 \\[2mm] n = 1; & L_n^{(\alpha)}(t) = -t + (\alpha+1) \\[2mm] n = 2; & L_n^{(\alpha)}(t) = \dfrac{t^2}{2} - (\alpha+2)t + \dfrac{(\alpha+1)(\alpha+2)}{2} \\[2mm] n = 3; & L_n^{(\alpha)}(t) = -\dfrac{t^3}{6} + \dfrac{(\alpha+3)t^2}{2} - \dfrac{(\alpha+2)(\alpha+3)t}{2} + \dfrac{(\alpha+1)(\alpha+2)(\alpha+3)t}{6} \end{array}\right\}. \quad (5.10)$$

The Rodrigues formula for the generalized Laguerre polynomials is given by

$$L_n^{(\alpha)}(t) = \frac{t^{-\alpha}e^t}{n!}\frac{d^n}{dt^n}\left(e^{-t}t^{n+\alpha}\right) = \frac{t^{-\alpha}}{n!}\left(\frac{d}{dt}-1\right)^n t^{n+\alpha}, \quad n = 0,1,2,\ldots, \quad (5.11)$$

whereas the closed form is given by

$$L_n^{(\alpha)}(t) = \sum_{k=0}^{n} \binom{n+\alpha}{k+\alpha} \frac{(-1)^k}{k!} t^k. \quad (5.12)$$

The generalized Laguerre polynomials are orthogonal over $[0,\infty)$ with respect to the measure having weight function $t^\alpha e^{-t}$ and the orthogonality relation reads:

$$\int_0^\infty L_n^{(\alpha)}(t)\,L_m^{(\alpha)}(t)\,d\Omega(t) = \frac{\delta_{n,m}}{\rho(n)}, \quad (5.13)$$

where $d\Omega(t) = t^\alpha e^{-t}dt$ and $\rho(n) = \dfrac{n!}{\Gamma(n+\alpha+1)}$.

Note that the classical Laguerre polynomials are just a special case of the generalized Laguerre polynomials for the case $\alpha = 0$. That is,

$$L_n(t) = L_n^{(0)}(t), \quad n = 0,1,2,\ldots.$$

Having presented a sound overview of the Laguerre polynomials, our motive is to study the Laguerre transform. The Laguerre transform is reliant upon a special class of measurable functions $P_n^{(\alpha)}(t)$, $\alpha > -1$, defined as

$$P_n^{(\alpha)}(t) = \rho(n)\Gamma(\alpha+1)L_n^{(\alpha)}(t), \quad t \in [0,\infty), \quad (5.14)$$

where $\rho(n)$ has the usual meaning and $L_n^{(\alpha)}(t)$ is the generalized Laguerre polynomial of degree n and order $\alpha > -1$. Prior to the formal definition of the Laguerre transform, we denote $L_\Lambda^2[0,\infty)$ as the space of measurable functions f satisfying

$$\left\|f\right\|_\Lambda = \left\{\int_0^\infty |f(t)|^2 d\Lambda(t)\right\}^{1/2} < \infty, \quad (5.15)$$

where

$$d\Lambda(t) = \frac{d\Omega(t)}{\Gamma(\alpha+1)} = \frac{1}{\Gamma(\alpha+1)}t^\alpha e^{-t}dt. \quad (5.16)$$

The norm in (5.15) is induced by the inner product

$$\left\langle f, g \right\rangle_\Lambda = \int_0^\infty f(t)\, \overline{g(t)}\, d\Lambda(t), \quad \forall\, f, g \in L_\Lambda^2[0, \infty). \tag{5.17}$$

Definition 5.2.1. For any $f \in L_\Lambda^2[0, \infty)$, the Laguerre transform is denoted by $\mathscr{L}[f]$ and is defined as

$$\mathscr{L}[f](n) = \int_0^\infty f(t)\, P_n^{(\alpha)}(t)\, d\Lambda(t). \tag{5.18}$$

The inverse Laguerre transform corresponding to Definition 5.2.1 is given as follows:

$$f(t) = \mathscr{L}^{-1}\left(\mathscr{L}[f](n)\right)(t) := \sum_{n=0}^\infty \mathscr{L}[f](n)\, P_n^{(\alpha)}(t)\, \sigma(n), \quad \sigma(n) = \frac{1}{\rho(n)\Gamma(\alpha+1)}. \tag{5.19}$$

Furthermore, the Parseval's formula associated with the Laguerre transform (5.18) reads:

$$\sum_{n=0}^\infty \sigma(n)\mathscr{L}[f](n)\, \mathscr{L}[g](n) = \left\langle f, g \right\rangle_\Lambda. \tag{5.20}$$

In particular, for $f = g$, we have

$$\sum_{n=0}^\infty \sigma(n)\, \left|\mathscr{L}[f](n)\right|^2 = \left\|f\right\|_\Lambda^2. \tag{5.21}$$

5.2.2   Laguerre wavelet transform: Definition and basic properties

In this subsection, our aim is to execute the main idea regarding the formulation of the Laguerre wavelet transform. The Laguerre wavelet transform is based upon the fundamental notions of the Laguerre translation and convolution operations. In order to define the notion of Laguerre translation and convolution operations, we need the basic function $\Delta_\alpha(t, y, z)$ defined as

$$\Delta_\alpha(t, y, z) = \sum_{n=0}^\infty P_n^{(\alpha)}(t)\, P_n^{(\alpha)}(y)\, P_n^{(\alpha)}(z)\sigma(n), \tag{5.22}$$

which is symmetric in all the three variables t, y, z. Therefore, as a consequence of the transformation (5.18) together with (5.19), we have

$$\int_0^\infty \Delta_\alpha(t, y, z)\, P_n^{(\alpha)}(z)\, d\Lambda(z) = P_n^{(\alpha)}(t)\, P_n^{(\alpha)}(y). \tag{5.23}$$

Setting $n = 0$ in (5.23), we obtain

$$\int_0^\infty \Delta_\alpha(t, y, z)\, d\Lambda(z) = 1. \tag{5.24}$$

We note that the basic function $\Delta_\alpha(t, y, z)$, defined in (5.22), is to be interpreted as the Laguerre analogue of the translates of the Dirac-delta function on the real line.

Definition 5.2.2. For any $f \in L_\Lambda^2[0, \infty)$, the Laguerre translation operator is denoted by $\mathscr{T}_y f$ and is defined as

$$\mathscr{T}_y f(t) = f(t, y) = \int_0^\infty f(z)\, \Delta_\alpha(t, y, z)\, d\Lambda(z), \quad t > 0,\ y < \infty. \tag{5.25}$$

The Laguerre translation operator (5.25) is linear, continuous in $L^2_\Lambda[0,\infty)$ and satisfies the norm inequality $\|\mathscr{T}_y f\|_\Lambda \leq \|f\|_\Lambda$.

Definition 5.2.3. For any pair of functions $f, g \in L^2_\Lambda[0,\infty)$, the Laguerre convolution is denoted by $\circledast_{\mathscr{L}}$ and is defined as

$$(f \circledast_{\mathscr{L}} g)(t) = \int_0^\infty f(y)\,\mathscr{T}_t g(y)\, d\Lambda(y)$$
$$= \int_0^\infty \int_0^\infty f(y)\, g(z)\, \Delta_\alpha(t, y, z)\, d\Lambda(z)\, d\Lambda(y). \qquad (5.26)$$

Using the fact that $f, g \in L^2_\Lambda[0,\infty)$, it can be easily demonstrated that the Laguerre convolution (5.26) satisfies the following norm inequality:

$$\operatorname*{ess.\ sup}_{t \in [0,\infty)} \left|(f \circledast_{\mathscr{L}} g)(t)\right| \leq \|f\|_\Lambda \|g\|_\Lambda. \qquad (5.27)$$

Next, we shall employ the Laguerre translation operator (5.25) to construct a new family of wavelets, known as the "Laguerre wavelets." For $a > 0$ and $b \geq 0$, let $\mathcal{D}_a$ denotes the usual dilation operator pertaining to the wavelet transform and $\mathscr{T}_b$ be the Laguerre translation operator defined in (5.25). Then, the Laguerre wavelets are defined by the combined action of the operators $\mathcal{D}_a$ and $\mathscr{T}_b$ on $\psi \in L^2_\Lambda[0,\infty)$ as

$$\psi^{(\alpha)}_{a,b}(t) = \mathscr{T}_b \mathcal{D}_a \psi(t) = \frac{1}{\sqrt{a}} \int_0^\infty \psi\left(\frac{z}{a}\right) \Delta_\alpha(t, b, z)\, d\Lambda(z), \qquad (5.28)$$

where the convergence of the integral is guaranteed by the norm inequality satisfied by the Laguerre translation (5.25). Based on the family of Laguerre wavelets (5.28), we have the formal definition of the Laguerre wavelet transform.

Definition 5.2.4. The Laguerre wavelet transform of any $f \in L^2_\Lambda[0,\infty)$ with respect to the Laguerre wavelet $\psi \in L^2_\Lambda[0,\infty)$ is denoted by $\mathcal{LW}_\psi[f]$ and is defined as

$$\mathcal{LW}_\psi[f](a, b) = \left\langle f, \psi^{(\alpha)}_{a,b} \right\rangle_\Lambda = \frac{1}{\sqrt{a}} \int_0^\infty \int_0^\infty f(t)\, \overline{\psi\left(\frac{z}{a}\right)} \Delta_\alpha(t, b, z)\, d\Lambda(z)\, d\Lambda(t). \qquad (5.29)$$

Note that the Laguerre wavelet transform (5.29) can also be expressed via the Laguerre convolution (5.26) as follows:

$$\mathcal{LW}_\psi[f](a, b) = (f \circledast_{\mathscr{L}} \mathcal{D}_a \overline{\psi})(b). \qquad (5.30)$$

Therefore, the convergence of the integral (5.29) is validated by the norm inequality (5.27) satisfied by the Laguerre convolution. In fact, as a consequence of Hölder's inequality, we have

$$\left|\mathcal{LW}_\psi[f](a, b)\right| \leq \frac{1}{\sqrt{a}} \left\{ \int_0^\infty \int_0^\infty |f(t)|^2 \Delta(t, b, z)\, d\Lambda(z)\, d\Lambda(t) \right\}^{1/2}$$
$$\times \left\{ \int_0^\infty \int_0^\infty \left|\psi\left(\frac{z}{a}\right)\right|^2 \Delta(t, b, z)\, d\Lambda(z)\, d\Lambda(t) \right\}^{1/2}$$
$$\leq \frac{1}{\sqrt{a}} \left\{ \int_0^\infty |f(t)|^2 \left(\int_0^\infty \Delta(t, b, z)\, d\Lambda(z) \right) d\Lambda(t) \right\}^{1/2}$$
$$\times \left\{ \int_0^\infty \left|\psi\left(\frac{z}{a}\right)\right|^2 \left(\int_0^\infty \Delta(t, b, z)\, d\Lambda(t) \right) d\Lambda(z) \right\}^{1/2}$$

$$\leq \left\{ \int_0^\infty |f(t)|^2 d\Lambda(t) \right\}^{1/2} \left\{ \int_0^\infty \frac{1}{a} \left| \psi \left(\frac{z}{a} \right) \right|^2 d\Lambda(z) \right\}^{1/2}$$

$$\leq \left\| f \right\|_\Lambda \left\| \mathcal{D}_a \psi \right\|_\Lambda. \tag{5.31}$$

In pursuit of the Moyal's principle and reconstruction formula associated with the Laguerre wavelet transfom defined in (5.29), we have the following lemma:

Lemma 5.2.1. *For any $f \in L_\Lambda^2[0, \infty)$, the Lagurre wavelet transform (5.29) and the Laguerre transform (5.18) admit the following relationship:*

$$\mathscr{L}\left(\mathcal{L}\mathscr{W}_\psi \left[f \right] (a, b) \right)(n) = \mathscr{L}\left[f \right](n) \, \overline{\mathscr{L}\left[\psi \right](a, n)}, \tag{5.32}$$

where

$$\mathscr{L}\left[\psi \right](a, n) = \frac{1}{\sqrt{a}} \int_0^\infty \psi \left(\frac{z}{a} \right) P_n^{(\alpha)}(z) \, d\Lambda(z). \tag{5.33}$$

Proof. Using Definition 5.2.4, we have

$$\mathcal{L}\mathscr{W}_\psi \left[f \right](a, b) = \frac{1}{\sqrt{a}} \int_0^\infty \int_0^\infty f(t) \, \overline{\psi \left(\frac{z}{a} \right)} \Delta(t, b, z) \, d\Lambda(z) \, d\Lambda(t)$$

$$= \frac{1}{\sqrt{a}} \int_0^\infty \int_0^\infty f(t) \, \overline{\psi \left(\frac{z}{a} \right)} \left\{ \sum_{n=0}^\infty P_n^{(\alpha)}(t) \, P_n^{(\alpha)}(b) \, P_n^{(\alpha)}(z) \sigma(n) \right\} d\Lambda(z) \, d\Lambda(t)$$

$$= \sum_{n=0}^\infty P_n^{(\alpha)}(b) \, \sigma(n) \left\{ \int_0^\infty f(t) \, P_n^{(\alpha)}(t) \, d\Lambda(t) \right\} \left\{ \frac{1}{\sqrt{a}} \int_0^\infty \overline{\psi \left(\frac{z}{a} \right)} P_n^{(\alpha)}(z) \, d\Lambda(z) \right\}$$

$$= \sum_{n=0}^\infty \mathscr{L}\left[f \right](n) \, \overline{\mathscr{L}\left[\psi \right](a, n)} \, P_n^{(\alpha)}(b) \, \sigma(n)$$

$$= \mathscr{L}^{-1} \left(\mathscr{L}\left[f \right](n) \, \overline{\mathscr{L}\left[\psi \right](a, n)} \right)(b). \tag{5.34}$$

Applying the Laguerre transform on both sides of (5.34) with respect to b, we obtain

$$\mathscr{L}\left(\mathcal{L}\mathscr{W}_\psi \left[f \right] (a, b) \right)(n) = \mathscr{L}\left[f \right](n) \, \overline{\mathscr{L}\left[\psi \right](a, n)}.$$

This completes the proof of Lemma 5.2.1. □

Following is the Moyal's principle associated with the Laguerre wavelet transform defined in (5.29):

Theorem 5.2.2. *If $\mathcal{L}\mathscr{W}_\psi \left[f \right](a, b)$ and $\mathcal{L}\mathscr{W}_\psi \left[g \right](a, b)$ are the Laguerre wavelet transforms of any pair of functions $f, g \in L_\Lambda^2[0, \infty)$, then*

$$\int_0^\infty \int_0^\infty \mathcal{L}\mathscr{W}_\psi \left[f \right](a, b) \, \overline{\mathcal{L}\mathscr{W}_\psi \left[g \right](a, b)} \, \mathcal{K}(a) \, d\Lambda(a) \, d\Lambda(b) = \sum_{n=0}^\infty \sigma(n) \, \mathscr{L}\left[f \right](n) \, \overline{\mathscr{L}\left[g \right](n)} \, C_\psi(n), \tag{5.35}$$

where $\mathcal{K}(a)$ is the weight function and

$$C_\psi(n) = \int_0^\infty \left| \mathscr{L}\left[\psi \right](a, n) \right|^2 \mathcal{K}(a) \, d\Lambda(a). \tag{5.36}$$

Proof. Using Lemma 5.2.1, we have

$$\int_0^\infty \int_0^\infty \mathcal{LW}_\psi\big[f\big](a,b) \overline{\mathcal{LW}_\psi\big[g\big](a,b)} \, \mathcal{K}(a) \, d\Lambda(a) \, d\Lambda(b)$$

$$= \int_0^\infty \int_0^\infty \left(\sum_{n=0}^\infty \mathscr{L}\big[f\big](n) \overline{\mathscr{L}\big[\psi\big](a,n)} \, P_n^{(\alpha)}(b) \, \sigma(n) \right)$$

$$\times \left(\sum_{m=0}^\infty \overline{\mathscr{L}\big[g\big](m)} \, \mathscr{L}\big[\psi\big](a,m) \, P_m^{(\alpha)}(b) \, \sigma(m) \right) \mathcal{K}(a) \, d\Lambda(a) \, d\Lambda(b)$$

$$= \int_0^\infty \int_0^\infty \sum_{n=0}^\infty \sum_{m=0}^\infty \mathscr{L}\big[f\big](n) \overline{\mathscr{L}\big[g\big](m)} \, \overline{\mathscr{L}\big[\psi\big](a,n)} \, \mathscr{L}\big[\psi\big](a,m)$$

$$\times \left(P_n^{(\alpha)}(b) \, P_m^{(\alpha)}(b) \, \sigma(n) \, \sigma(m) \right) \mathcal{K}(a) \, d\Lambda(a) \, d\Lambda(b)$$

$$= \int_0^\infty \sum_{n=0}^\infty \sum_{m=0}^\infty \mathscr{L}\big[f\big](n) \overline{\mathscr{L}\big[g\big](m)} \, \overline{\mathscr{L}\big[\psi\big](a,n)} \, \mathscr{L}\big[\psi\big](a,m)$$

$$\times \left(\int_0^\infty P_n^{(\alpha)}(b) \, P_m^{(\alpha)}(b) \, \sigma(n) \, \sigma(m) \, d\Lambda(b) \right) \mathcal{K}(a) \, d\Lambda(a). \qquad (5.37)$$

Invoking the orthogonality of generalized Laguerre polynomials (5.13), we have

$$\int_0^\infty P_n^{(\alpha)}(b) \, P_m^{(\alpha)}(b) \, \sigma(n) \, \sigma(m) \, d\Lambda(b)$$

$$= \int_0^\infty \rho(n) \Gamma(\alpha+1) L_n^{(\alpha)}(b) \, \rho(m) \, \Gamma(\alpha+1) L_m^{(\alpha)}(b) \, \sigma(n) \, \sigma(m) \, d\Lambda(b)$$

$$= \int_0^\infty L_n^{(\alpha)}(b) \, L_m^{(\alpha)}(b) \, \frac{d\Omega(b)}{\Gamma(\alpha+1)}$$

$$= \frac{\delta_{n,m}}{\rho(n) \, \Gamma(\alpha+1)}. \qquad (5.38)$$

Implementing (5.38) in (5.37), we get

$$\int_0^\infty \int_0^\infty \mathcal{LW}_\psi\big[f\big](a,b) \overline{\mathcal{LW}_\psi\big[g\big](a,b)} \, \mathcal{K}(a) \, d\Lambda(a) \, d\Lambda(b)$$

$$= \int_0^\infty \sum_{n=0}^\infty \sigma(n) \mathscr{L}\big[f\big](n) \overline{\mathscr{L}\big[g\big](n)} \left| \mathscr{L}\big[\psi\big](a,n) \right|^2 \mathcal{K}(a) \, d\Lambda(a)$$

$$= \sum_{n=0}^\infty \sigma(n) \, \mathscr{L}\big[f\big](n) \overline{\mathscr{L}\big[g\big](n)} \, C_\psi(n).$$

This completes the proof of Theorem 5.2.2. □

Remark 5.2.1. For $f = g$, Theorem 5.2.2 boils down to the following compact form:

$$\int_0^\infty \int_0^\infty \left| \mathcal{LW}_\psi\big[f\big](a,b) \right|^2 \mathcal{K}(a) \, d\Lambda(a) \, d\Lambda(b) = \sum_{n=0}^\infty \sigma(n) \left| \mathscr{L}\big[f\big](n) \right|^2 C_\psi(n). \qquad (5.39)$$

Finally, we shall formulate the inversion formula associated with the Laguerre wavelet transform defined in (5.29).

Theorem 5.2.3. *If $\mathcal{LW}_\psi[f](a,b)$ is the Laguerre wavelet transform of any $f \in L^2_\Lambda[0,\infty)$, then the following inversion formula holds:*

$$f(t) = \int_0^\infty \int_0^\infty \mathcal{LW}_\psi[f](a,b)\, \Phi_{a,b}^{(\alpha)}(t)\, \mathcal{K}(a)\, d\Lambda(a)\, d\Lambda(b), \tag{5.40}$$

where $\Phi_{a,b}^{(\alpha)}(t)$ is defined as

$$\mathscr{L}\left[\Phi_{a,b}^{(\alpha)}\right](n) = \frac{\mathscr{L}\left[\psi_{a,b}^{(\alpha)}\right](n)}{C_\psi(n)} \tag{5.41}$$

and $C_\psi(n) > 0$ is given by (5.36).

Proof. As a consequence of Lemma 5.2.1, we have

$$\int_0^\infty \mathcal{LW}_\psi[f](a,b)\, P_n^{(\alpha)}(b)\, d\Lambda(b) = \mathscr{L}[f](n)\, \overline{\mathscr{L}[\psi](a,n)}. \tag{5.42}$$

Multiplying (5.42) on both sides by the factor $\mathscr{L}[\psi](a,n)\,\mathcal{K}(a)$ and then integrating with respect to the measure $d\Lambda(a)$, we obtain

$$\int_0^\infty \int_0^\infty \mathcal{LW}_\psi[f](a,b)\, P_n^{(\alpha)}(b)\, \mathscr{L}[\psi](a,n)\, \mathcal{K}(a) d\Lambda(b)\, d\Lambda(a)$$

$$= \int_0^\infty \mathscr{L}[f](n)\, \big|\mathscr{L}[\psi](a,n)\big|^2 d\Lambda(a)$$

$$= C_\psi(n)\, \mathscr{L}[f](n). \tag{5.43}$$

Also, we note that

$$\psi_{a,b}^{(\alpha)}(t) = \frac{1}{\sqrt{a}} \int_0^\infty \psi\left(\frac{z}{a}\right)\left(\sum_{n=0}^\infty P_n^{(\alpha)}(t)\, P_n^{(\alpha)}(b)\, P_n^{(\alpha)}(z)\sigma(n)\right) d\Lambda(z)$$

$$= \sum_{n=0}^\infty P_n^{(\alpha)}(t)\, P_n^{(\alpha)}(b)\, \sigma(n)\left\{\frac{1}{\sqrt{a}}\int_0^\infty \psi\left(\frac{z}{a}\right) P_n^{(\alpha)}(z)\, d\Lambda(z)\right\}$$

$$= \sum_{n=0}^\infty \mathscr{L}[\psi](a,n)\, P_n^{(\alpha)}(b)\, P_n^{(\alpha)}(t)\, \sigma(n)$$

$$= \mathscr{L}^{-1}\Big(\mathscr{L}[\psi](a,n)\, P_n^{(\alpha)}(b)\Big)(t). \tag{5.44}$$

Applying Laguerre transform on both sides of (5.44), we obtain

$$\mathcal{L}\left[\psi_{a,b}^{(\alpha)}\right](n) = \mathscr{L}[\psi](a,n)\, P_n^{(\alpha)}(b). \tag{5.45}$$

Using equation (5.45) in (5.43), we get

$$\mathscr{L}[f](n) = \int_0^\infty \int_0^\infty \mathcal{LW}_\psi[f](a,b)\left(\frac{\mathcal{L}\left[\psi_{a,b}^{(\alpha)}\right](n)}{C_\psi(n)}\right)\mathcal{K}(a)\, d\Lambda(a)\, d\Lambda(b)$$

$$= \int_0^\infty \int_0^\infty \mathcal{LW}_\psi[f](a,b)\, \mathcal{L}\left[\Phi_{a,b}^{(\alpha)}\right](n)\, \mathcal{K}(a)\, d\Lambda(a)\, d\Lambda(b). \tag{5.46}$$

Finally, applying the inverse Laguerre transform (5.19) on both sides of (5.46), we obtain

$$f(t) = \sum_{n=0}^{\infty} \sigma(n) \, P_n^{(\alpha)}(t) \left\{ \int_0^{\infty} \int_0^{\infty} \mathcal{LW}_\psi\big[f\big](a,b) \, \mathcal{L}\left[\Phi_{a,b}^{(\alpha)}\right](n) \, \mathcal{K}(a) \, d\Lambda(a) \, d\Lambda(b) \right\}$$

$$= \int_0^{\infty} \int_0^{\infty} \mathcal{LW}_\psi\big[f\big](a,b) \left(\sum_{n=0}^{\infty} \mathcal{L}\left[\Phi_{a,b}^{(\alpha)}\right](n) \, P_n^{(\alpha)}(t) \, \sigma(n) \right) \mathcal{K}(a) \, d\Lambda(a) \, d\Lambda(b)$$

$$= \int_0^{\infty} \int_0^{\infty} \mathcal{LW}_\psi\big[f\big](a,b) \, \Phi_{a,b}^{(\alpha)}(t) \, \mathcal{K}(a) \, d\Lambda(a) \, d\Lambda(b),$$

which is the desired inversion formula. This completes the proof of Theorem 5.2.3.   □

5.3   The Legendre Wavelet Transform

"All the truths of mathematics are linked to each other, and all means of discovering them are equally admissible."

-Adrien-Marie Legendre

In this section, our concern is to study another hybrid wavelet transform obtained from another class of orthogonal polynomials known as the "Legendre polynomials." These polynomials give birth to an intriguing transform known as the "Legendre transform" which serves as the pedestal for the Legendre wavelet transform. Primarily, we present a heuristic overview of the Legendre polynomials, Legendre transform and their fundamental properties. Subsequently, we shall deal with the study of Legendre wavelet transform and investigate upon the fundamental properties.

5.3.1   Legendre polynomials and transform

The Legendre polynomials are named after the French mathematician Adrien-Marie Legendre $(1752 - 1833)$ who discovered them in 1782. These polynomials constitute a class of complete orthogonal polynomials and have found numerous applications in mathematical and physical sciences. The Legendre polynomials are defined as the polynomial solutions of the Lagendre's differential equation:

$$(1 - t^2) f''(t) - 2t f'(t) + n(n+1) f(t) = 0, \tag{5.47}$$

which is a linear, second order differential equation with regular points at $t = \pm 1$. These polynomials are denoted by $P_n(t)$ and can be defined recursively as follows:

$$P_0(t) = 1, \quad P_1(t) = t, \quad P_{k+1}(t) = \frac{(2k+1)t P_k(t) - k P_{k-1}(t)}{k+1}, \quad \text{for} \quad k \geq 1. \tag{5.48}$$

It is pertinent to mention that the differential equation (5.47) also admits non-polynomial solutions called as the "Legendre functions of second kind" which are denoted by $Q_n(t)$. However, such solutions are of no particular interest to us and are, as such, omitted from the discourse.

The first few Legendre polynomials obtained from the recurrence relation (5.48) are given below:

$$
\left.
\begin{aligned}
n = 0; \quad & P_n(t) = 1 \\
n = 1; \quad & P_n(t) = t \\
n = 2; \quad & P_n(t) = \frac{1}{2}\left(3t^2 - 1\right) \\
n = 3; \quad & P_n(t) = \frac{1}{2}\left(5t^3 - 3t\right) \\
n = 4; \quad & P_n(t) = \frac{1}{8}\left(35t^4 - 30t^2 + 3\right) \\
n = 5; \quad & P_n(t) = \frac{1}{8}\left(63t^5 - 7015t\right)
\end{aligned}
\right\}. \tag{5.49}
$$

The Legendre polynomials can also be written in a compact form via the Rodrigues formula as

$$
P_n(t) = \frac{1}{2^n n!}\frac{d^n}{dt^n}\left(t^2 - 1\right)^n, \quad n \in \mathbb{N}_0,\ t \in [-1,1]. \tag{5.50}
$$

It is imperative to mention that, the Rodrigues formula (5.50) enables us to derive the following explicit representations of the Legendre polynomials:

$$
\left.
\begin{aligned}
P_n(t) &= \frac{1}{2^n}\sum_{k=0}^{n}\binom{n}{k}^2 (t-1)^{n-k}(t+1)^k \\
P_n(t) &= \sum_{k=0}^{n}\binom{n}{k}\binom{n+k}{k}\left\{\left(\frac{t-1}{2}\right)^k\right\} \\
P_n(t) &= \frac{1}{2^n}\sum_{k=0}^{[n/2]}(-1)^k\binom{n}{k}\binom{2n-2k}{n}t^{n-2k} \\
P_n(t) &= 2^n\sum_{k=0}^{n}\binom{n}{k}\binom{\frac{n+k+1}{2}}{n}t^k
\end{aligned}
\right\}. \tag{5.51}
$$

The Legendre polynomials are orthogonal over the interval $[-1,1]$ and the orthogonality relation reads:

$$
\int_{-1}^{1} P_n(t)\,P_m(t)\,dx = \frac{2}{2n+1}\,\delta_{n,m}. \tag{5.52}
$$

Some fundamental properties of the Legendre polynomials are summarized below:

(i) $\left|P_n(t)\right| \le P_n(1) = 1$,

(ii) $(1-t^2)P_n''(t) - 2tP_n'(t) + n(n+1)f(t) = 0$,

(iii) $P_n'(1) = \dfrac{n(n+1)}{2}$.

Next, our aim is to study the notion of Legendre transform abreast of its fundamental properties. Prior to that, we denote $L^2[-1,1]$ as the space of measurable functions f defined on $[-1,1]$ and satisfying

$$
\left\|f\right\|_{L^2[-1,1]} = \left\{\frac{1}{2}\int_{-1}^{1}|f(t)|^2 d(t)\right\}^{1/2} < \infty, \tag{5.53}
$$

where the norm in (5.53) is induced by the inner product

$$\left\langle f, g \right\rangle_{L^2[-1,1]} = \frac{1}{2} \int_{-1}^{1} f(t)\,\overline{g(t)}\,d(t), \quad \forall\, f, g \in L^2[-1,1]. \tag{5.54}$$

Definition 5.3.1. For any $f \in L^2[-1,1]$, the Legendre transform is denoted by $\mathbb{L}\left[f\right]$ and is defined as

$$\mathbb{L}\left[f\right](n) = \frac{1}{2} \int_{-1}^{1} f(t)\, P_n(t)\, dt. \tag{5.55}$$

The Legendre transform (5.55) is invertible and the inversion formula associated with Definition 5.3.1 is given by

$$f(t) = \mathbb{L}^{-1}\left(\mathbb{L}\left[f\right](n)\right)(t) := \sum_{n=0}^{\infty} (2n+1)\,\mathbb{L}\left[f\right](n)\, P_n(t). \tag{5.56}$$

Moreover, for any pair of functions $f, g \in L^2[-1,1]$, the Parseval's formula associated with the Legendre transform reads:

$$\frac{1}{2} \int_{-1}^{1} f(t)\,\overline{g(t)}\,d(t) = \sum_{n=0}^{\infty} (2n+1)\,\mathbb{L}\left[f\right](n)\,\overline{\mathbb{L}\left[g\right](n)}. \tag{5.57}$$

Some fundamental properties of the Legendre transform (5.55) are assembled below:

(i) $\left|\mathbb{L}\left[f\right](n)\right| \leq \left\|f\right\|_{L^2[-1,1]}$,

(ii) $\mathbb{L}\left[f+g\right](n) = \mathbb{L}\left[f\right](n) + \mathbb{L}\left[g\right](n)$,

(iii) $\mathbb{L}\left[cf\right](n) = c\,\mathbb{L}\left[f\right](n)$, where c is a scalar,

(iv) For any $n, m \in \mathbb{N}_0$, we have

$$\mathbb{L}\left[P_m\right](n) = \begin{cases} \dfrac{1}{2n+1}, & n = m \\ 0, & \text{otherwise.} \end{cases} \tag{5.58}$$

5.3.2 Legendre wavelet transform: Definition and basic properties

In this subsection, we shall formally initiate the study of Legendre wavelet transform by making an effective use of the Legendre transform. The Legendre wavelet transform relies on the fundamental notions of the Legendre translation and convolution operations obtained via the basic function $\Delta(t, y, z)$ defined as

$$\Delta(t, y, z) = \frac{\pi}{2} \sum_{n=0}^{\infty} (2n+1) P_n(t)\, P_n(y)\, P_n(z). \tag{5.59}$$

Evidently, the function $\Delta(t, y, z)$ is symmetric in all the three variables t, y, z and satisfies

$$\frac{1}{\pi} \int_{-1}^{1} \Delta(t, y, z)\, P_n(z)\, dz = P_n(t)\, P_n(y), \tag{5.60}$$

so that, for $n = 0$,

$$\int_{-1}^{1} \Delta(t, y, z)\, dz = \pi. \tag{5.61}$$

Definition 5.3.2. For any $y \in [-1, 1]$, the Legendre translation operator acting on $f \in L^2[-1, 1]$ is denoted by $\mathfrak{T}_y f$ and is defined as

$$\mathfrak{T}_y f(t) = f(t, y) = \frac{1}{\pi} \int_{-1}^{1} f(z) \, \Delta(t, y, z) \, dz. \tag{5.62}$$

The Legendre translation operator $y \to \mathfrak{T}_y f$ is linear from $L^2[-1, 1]$ into itself. Moreover, as a consequence of Hölder's inequality and (5.61), it can be easily verified that the translation operator (5.62) satisfies the norm inequality $\|\mathfrak{T}_y f\|_{L^2[-1,1]} \leq \|f\|_{L^2[-1,1]}$.

Based on the Legendre translation (5.62), following is the definition of Legendre convolution operator:

Definition 5.3.3. For any pair of functions $f, g \in L^2[-1, 1]$, the Legendre convolution is denoted by $\circledast_{\mathbb{L}}$ and is defined as

$$(f \circledast_{\mathbb{L}} g)(t) = \frac{1}{2} \int_{-1}^{1} f(y) \, \mathfrak{T}_t g(y) \, dy$$

$$= \frac{1}{2\pi} \int_{-1}^{1} \int_{-1}^{1} f(y) \, g(z) \, \Delta(t, y, z) \, dz \, dy. \tag{5.63}$$

The convolution theorem associated with the Legendre convolution operation defined in (5.63) states that

$$\mathbb{L}\Big[(f \circledast_{\mathbb{L}} g)\Big](n) = \mathbb{L}\Big[f\Big](n) \, \mathbb{L}\Big[g\Big](n). \tag{5.64}$$

Having presented a healthy overview of the prerequisites, we are now in a position to formulate the notion of Legendre wavelet transform by constructing a new family of wavelets known as the "Legendre wavelets." For $0 < a \leq 1$ and $-1 \leq b \leq 1$, let $\mathcal{D}_a$ and $\mathfrak{T}_b$ be the usual dilation and Legendre translation operators acting on $\psi \in L^2[-1, 1]$. Then, the Legendre wavelets are defined by the joint action of the operators $\mathcal{D}_a$ and $\mathfrak{T}_b$ on ψ as

$$\psi_{a,b}(t) = \mathfrak{T}_b \mathcal{D}_a \psi(t) = \frac{1}{\pi \sqrt{a}} \int_{-1}^{1} \psi\left(\frac{z}{a}\right) \Delta(t, b, z) \, dz. \tag{5.65}$$

Based upon the family of Legendre wavelets (5.65), we have the definition of the Legendre wavelet transform.

Definition 5.3.4. For any $f \in L^2[-1, 1]$, the continuous Legendre wavelet transform with respect to the Legendre wavelet ψ is denoted by $\mathbb{L}\mathscr{W}_\psi[f]$ and is defined as

$$\mathbb{L}\mathscr{W}_\psi\Big[f\Big](a, b) = \Big\langle f, \psi_{a,b} \Big\rangle_{L^2[-1,1]} = \frac{1}{2\pi\sqrt{a}} \int_{-1}^{1} \int_{-1}^{1} f(t) \, \overline{\psi\left(\frac{z}{a}\right)} \Delta(t, b, z) \, dz \, dt. \tag{5.66}$$

In the following lemma, we obtain a relationship between the Legendre transform (5.55) and the Legendre wavelet transform (5.66). Such a relationship shall be invoked in formulating the Moyal's principle associated with the Legendre wavelet transform (5.66).

Lemma 5.3.1. *For any $f \in L^2[-1, 1]$, the Legendre wavelet transform (5.66) and the Laguerre transform (5.55) admit the following relationship:*

$$\mathbb{L}\Big(\mathbb{L}\mathscr{W}_\psi\Big[f\Big](a, b)\Big)(n) = \mathbb{L}\Big[f\Big](n) \, \overline{\mathbb{L}\Big[\psi_a\Big](n)}, \tag{5.67}$$

where

$$\mathbb{L}\Big[\psi_a\Big](n) = \frac{1}{2\sqrt{a}} \int_{-1}^{1} \psi\left(\frac{z}{a}\right) P_n(z) \, dz. \tag{5.68}$$

Proof. Using Definition 5.3.4 together with (5.56), we get

$$\mathbb{L}\mathscr{W}_\psi\big[f\big](a,b) = \frac{1}{2\pi\sqrt{a}} \sum_{n=0}^{\infty} \frac{(2n+1)\pi}{2} \int_{-1}^{1}\int_{-1}^{1} f(t)\, \overline{\psi\left(\frac{z}{a}\right)} P_n(t)\, P_n(b)\, P_n(z)\, dz\, dt$$

$$= \sum_{n=0}^{\infty}(2n+1)\left(\frac{1}{2}\int_{-1}^{1} f(t)P_n(t)\, dt\right)\left(\frac{1}{2\sqrt{a}}\int_{-1}^{1} \overline{\psi\left(\frac{z}{a}\right)} P_n(z)\, dz\right) P_n(b)$$

$$= \sum_{n=0}^{\infty}(2n+1)\left(\frac{1}{2}\int_{-1}^{1} f(t)P_n(t)\, dt\right)\left(\frac{1}{2}\int_{-1}^{1} \overline{\mathcal{D}_a\psi(z)}\, P_n(z)\, dz\right)$$

$$= \sum_{n=0}^{\infty}(2n+1)\,\mathbb{L}\big[f\big](n)\,\overline{\mathbb{L}\big[\psi_a\big](n)}\, P_n(b)$$

$$= \mathbb{L}^{-1}\left(\mathbb{L}\big[f\big](n)\,\overline{\mathbb{L}\big[\psi_a\big](n)}\right)(b). \tag{5.69}$$

Applying the Legendre transform on both sides on (5.69) with respect to b, we obtain

$$\mathbb{L}\left(\mathbb{L}\mathscr{W}_\psi\big[f\big](a,b)\right)(n) = \mathbb{L}\big[f\big](n)\,\overline{\mathbb{L}\big[\psi_a\big](n)}.$$

This completes the proof of Lemma 5.3.1. □

Before concluding the subsection, we aim to study the Moyal's principle and obtain the inversion formula associated with the Legendre wavelet transform. Following is the Moyal's principle pertaining to the Legendre wavelet transform (5.66):

Theorem 5.3.2. *If ψ, ϕ are any two Legendre wavelets, then for $f, g \in L^2[-1,1]$, we have*

$$\int_{-1}^{1}\int_{0}^{1} \mathbb{L}\mathscr{W}_\psi\big[f\big](a,b)\,\overline{\mathbb{L}\mathscr{W}_\phi\big[g\big](a,b)}\,\mathcal{K}(a)\, da\, db = \sum_{n=0}^{\infty} 2(2n+1)\,\mathbb{L}\big[f\big](n)\,\overline{\mathbb{L}\big[g\big](n)}\, C_{\psi,\phi}(n), \tag{5.70}$$

where $\mathcal{K}(a)$ is the weight function and

$$C_{\psi,\phi}(n) = \int_{0}^{1} \mathbb{L}\big[\phi_a\big](n)\,\overline{\mathbb{L}\big[\psi_a\big](n)}\,\mathcal{K}(a)\, da < \infty. \tag{5.71}$$

Proof. As a consequence of Lemma 5.3.1, we have

$$\int_{-1}^{1}\int_{0}^{1} \mathbb{L}\mathscr{W}_\psi\big[f\big](a,b)\,\overline{\mathbb{L}\mathscr{W}_\phi\big[g\big](a,b)}\,\mathcal{K}(a)\, da\, db$$

$$= \int_{-1}^{1}\int_{0}^{1} \left\{\sum_{n=0}^{\infty}(2n+1)\,\mathbb{L}\big[f\big](n)\,\overline{\mathbb{L}\big[\psi_a\big](n)}\, P_n(b)\right\}$$

$$\times \left\{\sum_{m=0}^{\infty}(2m+1)\overline{\mathbb{L}\big[g\big](m)}\,\mathbb{L}\big[\phi_a\big](m)\, P_m(b)\right\}\mathcal{K}(a)\, da\, db$$

$$= \sum_{n=0}^{\infty}\sum_{m=0}^{\infty}(2n+1)(2m+1)$$

$$\times \left\{\int_{-1}^{1}\int_{0}^{1} \mathbb{L}\big[f\big](n)\,\overline{\mathbb{L}\big[g\big](m)}\,\mathbb{L}\big[\phi_a\big](m)\,\overline{\mathbb{L}\big[\psi_a\big](n)}\, P_n(b)\, P_m(b)\,\mathcal{K}(a)\, da\, db\right\}$$

$$= \sum_{n=0}^{\infty} \sum_{m=0}^{\infty} (2n+1)(2m+1)$$

$$\times \left\{ \int_0^1 \mathbb{L}\Big[f\Big](n)\, \overline{\mathbb{L}\Big[g\Big]}(m)\, \mathbb{L}\Big[\phi_a\Big](m)\, \overline{\mathbb{L}\Big[\psi_a\Big]}(n) \left(\int_{-1}^1 P_n(b)\, P_m(b)\, db \right) \mathcal{K}(a)\, da \right\}. \tag{5.72}$$

Using the orthogonality of Legendre polynomials (5.52), we can express (5.72) as follows:

$$\int_{-1}^1 \int_0^1 \mathbb{L}\mathscr{W}_\psi\Big[f\Big](a,b)\, \overline{\mathbb{L}\mathscr{W}_\phi\Big[g\Big]}(a,b)\, \mathcal{K}(a)\, da\, db$$

$$= \sum_{n=0}^{\infty} \sum_{m=0}^{\infty} (2n+1)(2m+1)\, \mathbb{L}\Big[f\Big](n)\, \overline{\mathbb{L}\Big[g\Big]}(m)$$

$$\times \left\{ \int_0^1 \overline{\mathbb{L}\Big[\psi_a\Big]}(n)\, \mathbb{L}\Big[\phi_a\Big](m)\, \mathcal{K}(a)\, da \right\} \frac{2}{2n+1}\, \delta_{n,m}$$

$$= \sum_{n=0}^{\infty} 2(2n+1)\, \mathbb{L}\Big[f\Big](n)\, \overline{\mathbb{L}\Big[g\Big]}(n) \left\{ \int_0^1 \overline{\mathbb{L}\Big[\psi_a\Big]}(n)\, \mathbb{L}\Big[\phi_a\Big](n)\, \mathcal{K}(a)\, da \right\}$$

$$= \sum_{n=0}^{\infty} 2(2n+1)\, \mathbb{L}\Big[f\Big](n)\, \overline{\mathbb{L}\Big[g\Big]}(n)\, C_{\psi,\phi}(n).$$

This completes the proof of Theorem 5.3.2. □

Remark 5.3.1. For $\psi = \phi$ and $f = g$, the relation (5.70) reduces to the following compact form:

$$\int_{-1}^1 \int_0^1 \Big| \mathbb{L}\mathscr{W}_\psi\Big[f\Big](a,b)\Big|^2 da\, db = \sum_{n=0}^{\infty} 2(2n+1)\, \Big|\mathbb{L}\Big[f\Big](n)\Big|^2 C_\psi(n), \tag{5.73}$$

where

$$C_\psi(n) = \int_0^1 \Big|\mathbb{L}\Big[\psi_a\Big](n)\Big|^2 \mathcal{K}(a)\, da < \infty. \tag{5.74}$$

Next, we have the inversion formula associated with the Legendre wavelet transform defined in (5.66).

Theorem 5.3.3. *If $\mathbb{L}\mathscr{W}_\psi\Big[f\Big](a,b)$ is the Legendre wavelet transform of any $f \in L^2[-1,1]$ with respect to the Legendre wavelet ψ, then f can be reconstructed as*

$$f(t) = \int_{-1}^1 \int_0^1 \mathbb{L}\mathscr{W}_\psi\Big[f\Big](a,b)\, \Phi_{a,b}(t)\, \mathcal{K}(a)\, da\, db, \tag{5.75}$$

where $\Phi_{a,b}(t)$ is defined as

$$\mathbb{L}\Big[\Phi_{a,b}\Big](n) = \frac{\mathbb{L}\Big[\psi_{a,b}\Big](n)}{2C_\psi(n)} \tag{5.76}$$

and $C_\psi(n) > 0$ is given by (5.74).

Proof. By virtue of Lemma 5.3.1, we have

$$\frac{1}{2} \int_{-1}^1 \mathbb{L}\mathscr{W}_\psi\Big[f\Big](a,b)\, P_n(b)\, db = \mathbb{L}\Big[f\Big](n)\, \overline{\mathbb{L}\Big[\psi_a\Big]}(n). \tag{5.77}$$

Multiplying (5.77) on both sides by $\mathbb{L}[\psi_a](n)\,\mathcal{K}(a)$ and then integrating with respect to the measure da, we get

$$\frac{1}{2}\int_{-1}^{1}\int_{0}^{1}\mathbb{L}\mathscr{W}_{\psi}\big[f\big](a,b)\,\mathbb{L}\big[\psi_a\big](n)\,P_n(b)\,\mathcal{K}(a)\,da\,db$$

$$=\int_{0}^{\infty}\mathbb{L}\big[f\big](n)\,\Big|\mathbb{L}\big[\psi_a\big](n)\Big|^2\,da$$

$$=C_{\psi}(n)\,\mathbb{L}\big[f\big](n). \tag{5.78}$$

Also, note that

$$\psi_{a,b}(t)=\frac{1}{\pi\sqrt{a}}\int_{-1}^{1}\psi\left(\frac{z}{a}\right)\Delta(t,b,z)\,dz$$

$$=\frac{1}{\pi\sqrt{a}}\int_{-1}^{1}\psi\left(\frac{\pi}{2}\sum_{n=0}^{\infty}(2n+1)P_n(t)\,P_n(b)\,P_n(z)\right)dz$$

$$=\sum_{n=0}^{\infty}(2n+1)P_n(t)\,P_n(b)\left\{\frac{1}{2\sqrt{a}}\int_{-1}^{1}\psi\left(\frac{z}{a}\right)P_n(z)\,dz\right\}$$

$$=\sum_{n=0}^{\infty}(2n+1)P_n(t)\,P_n(b)\,\mathbb{L}\big[\psi_a\big](n)$$

$$=\mathbb{L}^{-1}\Big(\mathbb{L}\big[\psi_a\big](n)\,P_n(b)\Big)(t). \tag{5.79}$$

Applying Legendre transform on both sides of (5.79) yields

$$\mathbb{L}\big[\psi_{a,b}\big](n)=\mathbb{L}\big[\psi_a\big](n)\,P_n(b). \tag{5.80}$$

Implementing (5.80) in (5.78), we get

$$\mathbb{L}\big[f\big](n)=\int_{-1}^{1}\int_{0}^{1}\mathbb{L}\mathscr{W}_{\psi}\big[f\big](a,b)\left(\frac{\mathbb{L}\big[\psi_{a,b}\big](n)}{2C_{\psi}(n)}\right)\mathcal{K}(a)\,da\,db$$

$$=\int_{-1}^{1}\int_{0}^{1}\mathbb{L}\mathscr{W}_{\psi}\big[f\big](a,b)\,\mathbb{L}\big[\Phi_{a,b}\big](n)\,\mathcal{K}(a)\,da\,db. \tag{5.81}$$

Finally, the desired inversion formula is obtained by applying the inverse Legendre transform (5.56) on both sides of (5.81) as

$$f(t)=\sum_{n=0}^{\infty}(2n+1)\,P_n(t)\left\{\int_{-1}^{1}\int_{0}^{1}\mathbb{L}\mathscr{W}_{\psi}\big[f\big](a,b)\,\mathbb{L}\big[\Phi_{a,b}\big](n)\,\mathcal{K}(a)\,da\,db\right\}$$

$$=\int_{-1}^{1}\int_{0}^{1}\mathbb{L}\mathscr{W}_{\psi}\big[f\big](a,b)\left(\sum_{n=0}^{\infty}(2n+1)\,\mathbb{L}\big[\Phi_{a,b}\big](n)\,P_n(t)\right)\mathcal{K}(a)\,da\,db$$

$$=\int_{-1}^{1}\int_{0}^{1}\mathbb{L}\mathscr{W}_{\psi}\big[f\big](a,b)\,\Phi_{a,b}(t)\,\mathcal{K}(a)\,da\,db.$$

This completes the proof of Theorem 5.3.3. $\qquad\qquad\qquad\qquad\qquad\qquad\square$

5.4   The Bessel Wavelet Transform

"Bessel was also an outstanding mathematician whose name became generally known through a special class of functions that have become an indispensable tool in applied mathematics, physics and engineering. The interest in these functions ... arose in the treatment of the problem of the perturbation in the planetary system."

-Walter E. Fricke

In this section, we shall formulate the Bessel wavelet transform based on the prominent class of special functions known as the "Bessel functions." The Bessel functions are an interesting class of functions, which are orthogonal over $[0, \infty)$ under an appropriate weighting factor and give rise to an important self-inversive integral transform, known as the "Hankel transform." To accomplish the objective, we present a rigorous introduction circumscribing the Bessel functions and the associated Hankel transform. Subsequently, we formulate the notion of Bessel wavelet transform by virtue of the fundamental concepts of Bessel dilation and the Hankel translation and convolution operations.

5.4.1   Bessel functions and Hankel transform

The Bessel functions were first defined by the Swiss mathematician Daniel Bernoulli $(1700-1782)$ and were subsequently generalized by the German mathematician Friedrich Bessel $(1784-1846)$. These functions are particularly important for solving diverse problems of wave propagation and static potentials. The Bessel functions are the canonical solutions of the Bessel's differential equation:

$$t^2 f''(t) + t f'(t) + (t^2 - \alpha^2) f(t) = 0. \tag{5.82}$$

The parameter α involved in (5.82) is arbitrary; however, the most important cases are when α is an integer or half integer. When α is an integer, the Bessel functions are also called as "cylindrical harmonics," because they appear in the solution of Laplace's equation in cylindrical coordinates. On the other hand, when α is a half-integer, the Bessel functions are precisely called as "spherical Bessel functions," because they are obtained while solving the Helmholtz equation in spherical coordinates. Although, α and $-\alpha$ produce the same differential equation, it is conventional to define different Bessel functions for these two values in such a way that the Bessel functions are mostly smooth functions of α.

It is pertinent to mention that the solutions of the Bessel's differential equation (5.82) are often referred as the "Bessel functions of first kind" and are denoted as $J_\alpha(t)$. For integer or positive α, the Bessel functions of first kind are finite at origin $t = 0$, while for negative, non-integer α, these functions diverge as t approaches to zero. The series expansion of the Bessel functions of first kind about $t = 0$ is given below:

$$J_\alpha(t) = \sum_{n=0}^{\infty} \frac{(-1)^n}{n! \Gamma(n + \alpha + 1)} \left(\frac{t}{2}\right)^{2n+\alpha}, \tag{5.83}$$

where $\Gamma(\cdot)$ is the well-known gamma function. The Bessel function of first kind is an entire function when α is an integer, otherwise it is a multi-valued function with singularity at $t = 0$. For non-integer α, the Bessel functions $J_\alpha(t)$ and $J_{-\alpha}(t)$ are linearly independent, thus they constitute two solutions of the Bessel's differential equation (5.82). On the other hand, if α is an integer, say $\alpha = m$, we have $J_{-m}(t) = (-1)^m J_m(t)$; that is, the solutions are no longer linearly independent. The second linearly independent solution in the former case is then found to be the Bessel function of second kind and is occasionally denoted by $Y_\alpha(t)$ or sometimes as $N_\alpha(t)$. Moreover, for non-integer α, the functions $J_\alpha(t)$ and $Y_\alpha(t)$ are related as

$$Y_\alpha(t) = \frac{J_\alpha(t)\cos(\alpha\pi) - J_{-\alpha}(t)}{\sin(\alpha\pi)}. \tag{5.84}$$

In the case of integer order n, the function is defined by taking limit as a non-integer α tends to n; that is, $Y_n(t) = \lim_{\alpha \to n} Y_\alpha(t)$.

The Bessel functions form an orthogonal basis with respect to the weighting factor r and the orthogonality relation is given by

$$\int_0^\infty J_\alpha(rt) J_\alpha(rz) r\, dr = \frac{\delta(t-z)}{t}, \quad t, z > 0. \tag{5.85}$$

The two linearly independent solutions of the Bessel's differential equation (5.82) are the building blocks of yet another important class of functions, known as the "Hankel functions of first and second kind," which are denoted by $H_\alpha^{(1)}(t)$ and $H_\alpha^{(2)}(t)$, respectively, and are defined as

$$\left. \begin{aligned} H_\alpha^{(1)}(t) &= J_\alpha(t) + i\, Y_\alpha(t) \\ H_\alpha^{(2)}(t) &= J_\alpha(t) - i\, Y_\alpha(t) \end{aligned} \right\}. \tag{5.86}$$

The linear combinations (5.86) are also known as the "Bessel functions of third kind." These functions are also expressible as

$$H_\alpha^{(1)}(t) = \frac{J_{-\alpha}(t) - e^{-\alpha\pi i} J_\alpha(t)}{i\sin(\alpha\pi)} \tag{5.87}$$

and

$$H_\alpha^{(2)}(t) = \frac{J_{-\alpha}(t) - e^{\alpha\pi i} J_\alpha(t)}{-i\sin(\alpha\pi)}. \tag{5.88}$$

If α is an integer, then the limit has to be calculated. Nevertheless, the following relationship holds whether α is an integer or not:

$$\left. \begin{aligned} H_{-\alpha}^{(1)}(t) &= e^{\alpha\pi i} H_\alpha^{(1)}(t) \\ H_{-\alpha}^{(2)}(t) &= e^{-\alpha\pi i} H_\alpha^{(2)}(t) \end{aligned} \right\}. \tag{5.89}$$

Particularly, when $\alpha = m + \frac{1}{2}$, where m is a non-negative integer, the expressions given in (5.89) imply that

$$\left. \begin{aligned} J_{-(m+\frac{1}{2})}(t) &= (-1)^{m+1} Y_{(m+\frac{1}{2})}(t) \\ Y_{-(m+\frac{1}{2})}(t) &= (-1)^m\ \ J_{(m+\frac{1}{2})}(t) \end{aligned} \right\}. \tag{5.90}$$

After a sound overview of the Bessel functions, our aim is to study the Hankel transform and recall some of its fundamental properties. The Hankel transform was introduced by the German mathematician Hermannd Hankel and is sometimes also referred as the "Fourier-Bessel transform." This transform expresses a given function as the weighted sum of an infinite number of the Bessel functions of first kind. Prior to recalling the formal definition of the Hankel transform, we note that although the Hankel transform is often defined for $L^1[0, \infty)$, however, the integral can be taken as the limit when the upper limit approaches to infinity; that is, as an improper integral, so that the Hankel transform can be extended for the space of square integrable functions $L^2[0, \infty)$.

Definition 5.4.1. For any $f \in L^2[0, \infty)$, the Hankel transform is denoted by $\mathcal{H}_\alpha[f]$ and is defined as

$$\mathcal{H}_\alpha\Big[f\Big](\omega) = \int_0^\infty f(t) J_\alpha(t\omega) \sqrt{t\omega}\, dt, \quad \omega \in (0, \infty), \ \alpha \geq -1/2, \tag{5.91}$$

where J_α is the Bessel function of first kind of order α.

The Hankel transform (5.91) is its own inverse in the sense that the original function $f \in L^2[0, \infty)$ can be reconstructed from the corresponding Hankel transform as

$$f(t) = \mathcal{H}_\alpha^{-1}\Big(\mathcal{H}_\alpha\Big[f\Big](\omega)\Big)(t) = \mathcal{H}_\alpha\Big[f\Big](\omega) = \int_0^\infty \mathcal{H}_\alpha\Big[f\Big](\omega) J_\alpha(t\omega) \sqrt{t\omega}\, d\omega. \tag{5.92}$$

In the following proposition, we obtain the Parseval's formula associated with the Hankel transform defined in (5.91).

Proposition 5.4.1. *For any pair of square integrable functions* $f, g \in L^2[0, \infty)$, *we have*

$$\int_0^\infty \mathcal{H}_\alpha\Big[f\Big](\omega) \overline{\mathcal{H}_\alpha\Big[g\Big](\omega)}\, d\omega = \int_0^\infty f(t)\, \overline{g(t)}\, dt. \tag{5.93}$$

Proof. Using Definition 5.4.1, we have

$$\int_0^\infty \mathcal{H}_\alpha\Big[f\Big](\omega) \overline{\mathcal{H}_\alpha\Big[g\Big](\omega)}\, d\omega = \int_0^\infty \left\{ \int_0^\infty f(t) J_\alpha(t\omega) \sqrt{t\omega}\, dt \right\} \left\{ \int_0^\infty \overline{g(z)} J_\alpha(z\omega) \sqrt{z\omega}\, dz \right\} d\omega$$

$$= \int_0^\infty \int_0^\infty f(t)\, \overline{g(z)} \sqrt{tz} \left\{ \int_0^\infty J_\alpha(t\omega)\, J_\alpha(z\omega)\, \omega\, d\omega \right\} dt\, dz. \tag{5.94}$$

As a consequence of orthogonality of the Bessel functions (5.85), we can express (5.94) as follows:

$$\int_0^\infty \mathcal{H}_\alpha\Big[f\Big](\omega) \overline{\mathcal{H}_\alpha\Big[g\Big](\omega)}\, d\omega = \int_0^\infty \int_0^\infty f(t)\, \overline{g(z)} \sqrt{tz}\, \frac{\delta(t-z)}{t}\, dt\, dz$$

$$= \int_0^\infty f(t)\, \overline{g(t)}\, dt.$$

This completes the proof of Proposition 5.4.1. $\qquad\qquad\qquad\qquad\qquad\qquad\square$

5.4.2   Bessel wavelet transform: Definition and basic properties

In this subsection, our endeavour is to formulate the notion of the Bessel wavelet transform. To facilitate the narrative, we present the fundamental notions of the Hankel translation and convolution operators on $L^2[0, \infty)$.

Definition 5.4.2. Given any $f \in L^2[0, \infty)$ and $y \geq 0$, the Hankel translation operator is denoted by T_y and is defined as

$$\mathsf{T}_y f(t) = f(t, y) = \int_0^\infty f(z)\, \Delta_\alpha(t, y, z)\, dz, \tag{5.95}$$

where

$$\Delta_\alpha(t, y, z) = \int_0^\infty \omega^{-\alpha-1/2} \sqrt{t\omega}\, J_\alpha(t\omega)\, \sqrt{y\omega}\, J_\alpha(y\omega)\, \sqrt{z\omega}\, J_\alpha(z\omega)\, d\omega, \qquad (5.96)$$

provided the above integral exists.

The Hankel translation operation (5.95) gives rise to another important operation on $L^2[0, \infty)$, known as the "Hankel convolution." Below, we present the notion of such a convolution operation.

Definition 5.4.3. For any pair of functions $f, g \in L^2[0, \infty)$, the Hankel convolution operator is denoted by $\circledast_\mathcal{H}$ and is defined as

$$
\begin{aligned}
(f \circledast_\mathcal{H} g)(t) &= \int_0^\infty f(x)\, \mathsf{T}_t g(x)\, dx \\
&= \int_0^\infty \int_0^\infty f(x)\, g(z)\, \Delta_\alpha(x, t, z)\, dz\, dx.
\end{aligned}
\qquad (5.97)$$

The convolution theorem associated with the Hankel convolution operation defined in (5.97) states that

$$\mathcal{H}_\alpha\Big[(f \circledast_\mathcal{H} g)\Big](\omega) = \omega^{-\alpha-1/2}\, \mathcal{H}_\alpha\Big[f\Big](\omega)\, \mathcal{H}_\alpha\Big[g\Big](\omega). \qquad (5.98)$$

In order to construct a new family of Bessel wavelets and formulate the notion of the Bessel wavelet transform, we need the Bessel dilation operator. For $a > 0$, let $\mathbb{D}_a$ denotes the Bessel dilation operator acting on $\psi \in L^2[0, \infty)$ as

$$\mathbb{D}_a \psi(t) = a^{\alpha-1/2} \psi\left(\frac{t}{a}\right). \qquad (5.99)$$

Then, the Bessel wavelets are defined by the combined action of the Hankel translation operator (5.95) and Bessel dilation operator (5.99) on $\psi \in L^2[0, \infty)$ as

$$
\begin{aligned}
\psi_{a,b}^{(\alpha)}(t) &= \mathbb{D}_a \mathsf{T}_b \psi(t) \\
&= \mathbb{D}_a \psi(t, b) \\
&= a^{\alpha-1/2} \psi\left(\frac{b}{a}, \frac{t}{a}\right) \\
&= a^{\alpha-1/2} \int_0^\infty \psi(z)\, \Delta_\alpha\left(\frac{b}{a}, \frac{t}{a}, z\right) dz, \quad a > 0, b \geq 0.
\end{aligned}
\qquad (5.100)$$

The function $\psi(t)$ is known as the "Bessel mother wavelet," whereas the functions $\psi_{a,b}^{(\alpha)}(t)$ are known as the "Bessel daughter wavelets." Based upon (5.100), we have the formal definition of the Bessel wavelet transform.

Definition 5.4.4. The Bessel wavelet transform of any $f \in L^2[0, \infty)$ with respect to the Bessel wavelet ψ is denoted by $\mathcal{BW}_\psi^{(\alpha)}[f]$ and is defined as

$$\mathcal{BW}_\psi^{(\alpha)}[f](a, b) = \left\langle f, \psi_{a,b}^{(\alpha)} \right\rangle_{L^2[0,\infty)} = a^{\alpha-1/2} \int_0^\infty \int_0^\infty f(t)\, \overline{\psi(z)}\, \Delta_\alpha\left(\frac{b}{a}, \frac{t}{a}, z\right) dz\, dt,$$

$$(5.101)$$

where $a > 0$, $b \geq 0$ and $\alpha \geq -1/2$.

After the formulation of the Bessel wavelet transform, our main motive is to study Moyal's principle corresponding to Definition 5.4.4. To do so, we need to obtain a relationship between the Bessel wavelet transform (5.101) and the Hankel transform (5.91).

Lemma 5.4.2. *For any $f \in L^2[0, \infty)$, the Bessel wavelet transform (5.101) can be expressed as*

$$\mathcal{BW}_\psi^{(\alpha)}\big[f\big](a, b) = \int_0^\infty \omega^{-\alpha - 1/2} \sqrt{b\omega} \; \mathcal{H}_\alpha\big[f\big](\omega) \overline{\mathcal{H}_\alpha\big[\psi\big](a\omega)} \, J_\alpha(b\omega) \, d\omega. \qquad (5.102)$$

Proof. Using Definition 5.4.4, we have

$$\mathcal{BW}_\psi^{(\alpha)}\big[f\big](a, b) = a^{\alpha - 1/2} \int_0^\infty \int_0^\infty f(t) \, \overline{\psi(z)}$$

$$\times \left\{ \int_0^\infty \omega^{-\alpha - 1/2} \sqrt{\frac{b\omega}{a}} \, J_\alpha\left(\frac{b\omega}{a}\right) \sqrt{\frac{t\omega}{a}} \, J_\alpha\left(\frac{t\omega}{a}\right) \sqrt{z\omega} \, J_\alpha(z\omega) \, d\omega \right\} dz \, dt$$

$$= a^{\alpha - 1/2} \int_0^\infty \int_0^\infty \int_0^\infty \omega^{-\alpha - 1/2} \sqrt{\frac{b\omega}{a}} \, J_\alpha\left(\frac{b\omega}{a}\right) \left\{ f(t) \, J_\alpha\left(\frac{t\omega}{a}\right) \sqrt{\frac{t\omega}{a}} \right\}$$

$$\times \left\{ \overline{\psi(z)} \, J_\alpha(z\omega) \sqrt{z\omega} \right\} d\omega \, dz \, dt$$

$$= a^{\alpha - 1/2} \int_0^\infty \omega^{-\alpha - 1/2} \sqrt{\frac{b\omega}{a}} \, J_\alpha\left(\frac{b\omega}{a}\right) \left\{ \int_0^\infty f(t) \, J_\alpha\left(\frac{t\omega}{a}\right) \sqrt{\frac{t\omega}{a}} \, dt \right\}$$

$$\times \left\{ \int_0^\infty \overline{\psi(z)} \, J_\alpha(z\omega) \sqrt{z\omega} \, dz \right\} d\omega$$

$$= a^{\alpha - 1/2} \int_0^\infty r^{-\alpha - 1/2} \sqrt{\frac{br}{a}} \, J_\alpha\left(\frac{br}{a}\right) \mathcal{H}_\alpha\big[f\big]\left(\frac{\omega}{a}\right) \overline{\mathcal{H}_\alpha[\psi](\omega)} \, d\omega. \qquad (5.103)$$

Making the change of variables $\omega/a \to \omega$ in (5.103), we obtain

$$\mathcal{BW}_\psi^{(\alpha)}\big[f\big](a, b) = a^{\alpha + 1/2} \int_0^\infty (a\omega)^{-\alpha - 1/2} \sqrt{b\omega} \, J_\alpha(b\omega) \mathcal{H}_\alpha\big[f\big](\omega) \overline{\mathcal{H}_\alpha\big[\psi\big](a\omega)} \, d\omega$$

$$= \int_0^\infty \omega^{-\alpha - 1/2} \sqrt{b\omega} \; \mathcal{H}_\alpha\big[f\big](\omega) \overline{\mathcal{H}_\alpha\big[\psi\big](a\omega)} \, J_\alpha(b\omega) \, d\omega.$$

This completes the proof of Lemma 5.4.2.   □

Following is the Moyal's principle associated with the Bessel wavelet transform defined via the integral transformation (5.101):

Theorem 5.4.3. *If $\mathcal{BW}_\psi^{(\alpha)}\big[f\big](a, b)$ and $\mathcal{BW}_\psi^{(\alpha)}\big[g\big](a, b)$ are the Bessel wavelet transforms of any pair of functions $f, g \in L^2[0, \infty)$, then we have*

$$\int_0^\infty \int_0^\infty \mathcal{BW}_\psi^{(\alpha)}\big[f\big](a, b) \, \overline{\mathcal{BW}_\psi^{(\alpha)}\big[g\big](a, b)} \, \frac{da \, db}{a^2} = C_\psi^{(\alpha)} \int_0^\infty \omega^{-2\alpha - 1} \, \mathcal{H}_\alpha\big[f\big](\omega) \, \overline{\mathcal{H}_\alpha\big[g\big](\omega)} \, d\omega, \qquad (5.104)$$

where

$$C_\psi^{(\alpha)} = \int_0^\infty \left| \mathcal{H}_\alpha\big[\psi\big](a\omega) \right|^2 \frac{da}{a^2} < \infty, \quad a.e. \quad \omega \in (0, \infty). \qquad (5.105)$$

Proof. Using Lemma 5.4.2, we can express the Bessel wavelet transform (5.101) as

$$\int_0^\infty \int_0^\infty B\mathscr{W}_\psi^{(\alpha)}\big[f\big](a,b)\,\overline{B\mathscr{W}_\psi^{(\alpha)}\big[g\big](a,b)}\,\frac{da\,db}{a^2}$$

$$= \int_0^\infty \int_0^\infty \left\{\int_0^\infty \omega^{-\alpha-1/2}\sqrt{b\omega}\,\mathcal{H}_\alpha\big[f\big](\omega)\,\overline{\mathcal{H}_\alpha\big[\psi\big](a\omega)}\,J_\alpha(b\omega)\,d\omega\right\}$$

$$\times \left\{\int_0^\infty \eta^{-\alpha-1/2}\sqrt{b\eta}\,\overline{\mathcal{H}_\alpha\big[g\big](\eta)}\,\mathcal{H}_\alpha\big[\psi\big](a\eta)\,J_\alpha(b\eta)\,d\eta\right\}\frac{da\,db}{a^2}$$

$$= \int_0^\infty \int_0^\infty \int_0^\infty \mathcal{H}_\alpha\big[f\big](\omega)\,\overline{\mathcal{H}_\alpha\big[g\big](\eta)}\,\overline{\mathcal{H}_\alpha\big[\psi\big](a\omega)}\,\mathcal{H}_\alpha\big[\psi\big](a\eta)\,(\omega\eta)^{-\alpha}$$

$$\times \left\{\int_0^\infty J_\alpha(b\omega)\,J_\alpha(b\eta)\,b\,db\right\}\frac{d\omega\,d\eta\,da}{a^2}. \qquad (5.106)$$

In view of the orthogonality of Bessel functions given by (5.85), we can further simplify the expression (5.106) as

$$\int_0^\infty \int_0^\infty B\mathscr{W}_\psi^{(\alpha)}\big[f\big](a,b)\,\overline{B\mathscr{W}_\psi^{(\alpha)}\big[g\big](a,b)}\,\frac{da\,db}{a^2}$$

$$= \int_0^\infty \int_0^\infty \int_0^\infty \mathcal{H}_\alpha\big[f\big](\omega)\,\overline{\mathcal{H}_\alpha\big[g\big](\eta)}\,\overline{\mathcal{H}_\alpha\big[\psi\big](a\omega)}\,\mathcal{H}_\alpha\big[\psi\big](a\eta)\,(\omega\eta)^{-\alpha}\frac{\delta(\omega-\eta)}{\omega}\frac{d\omega\,d\eta\,da}{a^2}$$

$$= \int_0^\infty \int_0^\infty \mathcal{H}_\alpha\big[f\big](\omega)\,\overline{\mathcal{H}_\alpha\big[g\big](\omega)}\,\big|\mathcal{H}_\alpha\big[\psi\big](a\omega)\big|^2\,\omega^{-2\alpha-1}\frac{d\omega\,da}{a^2}$$

$$= \int_0^\infty \omega^{-2\alpha-1}\mathcal{H}_\alpha\big[f\big](\omega)\,\overline{\mathcal{H}_\alpha\big[g\big](\omega)}\left\{\int_0^\infty \big|\mathcal{H}_\alpha\big[\psi\big](a\omega)\big|^2\frac{da}{a^2}\right\}d\omega$$

$$= C_\psi^{(\alpha)}\int_0^\infty \omega^{-2\alpha-1}\mathcal{H}_\alpha\big[f\big](\omega)\,\overline{\mathcal{H}_\alpha\big[g\big](\omega)}\,d\omega.$$

This completes the proof of Theorem 5.4.3. $\qquad\square$

Corollary 5.4.4. *For* $\alpha = -1/2$, *the Bessel wavelet transform* (5.101) *satisfies the relation:*

$$\int_0^\infty \int_0^\infty B\mathscr{W}_\psi^{(\alpha)}\big[f\big](a,b)\,\overline{B\mathscr{W}_\psi^{(\alpha)}\big[g\big](a,b)}\,\frac{da\,db}{a^2} = C_\psi^{(\alpha)}\big\langle f,g\big\rangle_{L^2[0,\infty)}. \qquad (5.107)$$

Finally, based on the orthogonality relation (5.107), we shall obtain a reconstruction formula associated with the Bessel wavelet transform for the case $\alpha = -1/2$.

Theorem 5.4.5. *Any function* $f \in L^2[0,\infty)$ *can be reconstructed from the corresponding Bessel wavelet transform* $B\mathscr{W}_\psi^{(\alpha)}\big[f\big](a,b)$, *with* $\alpha = -1/2$, *via the formula:*

$$f(t) = \frac{1}{C_\psi^{(\alpha)}}\int_0^\infty \int_0^\infty \int_0^\infty B\mathscr{W}_\psi^{(\alpha)}\big[f\big](a,b)\,\psi(z)\,\Delta_\alpha\left(\frac{b}{a},\frac{t}{a},z\right)\frac{dz\,da\,db}{a^3}, \qquad (5.108)$$

where $C_\psi^{(\alpha)} > 0$ *is given by* (5.105) *and the equality in* (5.108) *holds almost everywhere.*

Proof. By virtue of the orthogonality relation (5.107), we have

$$C_\psi^{(\alpha)}\int_0^\infty f(t)\,\overline{g(t)}\,dt$$

$$= \int_0^\infty \int_0^\infty B\mathscr{W}_\psi^{(\alpha)}\big[f\big](a,b)\,\overline{B\mathscr{W}_\psi^{(\alpha)}\big[g\big](a,b)}\,\frac{da\,db}{a^2}$$

$$= \int_0^\infty \int_0^\infty \mathcal{BW}_\psi^{(\alpha)} \big[f\big](a,b) \left\{ a^{\alpha - 1/2} \int_0^\infty \int_0^\infty \overline{g(t)}\, \psi(z)\, \Delta_\alpha \left(\frac{b}{a}, \frac{t}{a}, z \right) dz\, dt \right\} \frac{da\, db}{a^2}$$

$$= \int_0^\infty \left\{ \int_0^\infty \int_0^\infty \int_0^\infty \mathcal{BW}_\psi^{(\alpha)} \big[f\big](a,b)\, \psi(z)\, \Delta_\alpha \left(\frac{b}{a}, \frac{t}{a}, z \right) \frac{dz\, da\, db}{a^3} \right\} \overline{g(t)}\, dt. \quad (5.109)$$

As a consequence of (5.109), we conclude that

$$f(t) = \frac{1}{C_\psi^{(\alpha)}} \int_0^\infty \int_0^\infty \int_0^\infty \mathcal{BW}_\psi^{(\alpha)} \big[f\big](a,b)\, \psi(z)\, \Delta_\alpha \left(\frac{b}{a}, \frac{t}{a}, z \right) \frac{dz\, da\, db}{a^3}, \quad a.e.,$$

which is the desired reconstruction formula. This completes the proof of Theorem 5.4.5. $\quad\square$

5.5   The Dunkl Wavelet Transform

"The concept of wavelets has its origins in many fields, and part of the accomplishment of Daubechies is finding those places where the concept arose and showing how all the approaches relate to one another. The use of wavelets as an analytical tool is like Fourier analysis-simple and yet very powerful. In fact, wavelets are an extension of Fourier analysis to the case of localization in both frequency and space."

-American Mathematical Society (1994).

In this section, we shall formulate another hybrid wavelet transform namely the Dunkl wavelet transform obtained via the Dunkl differential-difference operator, which is one of the notable operators broadly employed in harmonic analysis and quantum mechanics. The Dunkl operator gives rise to the Dunkl kernel which constitutes the Dunkl transform and serves as the main tool for the construction of the Dunkl wavelet transform. In a first, we shall present a gentle overview of the preliminaries, including the Dunkl transform and then proceed to the formal aspects of the Dunkl wavelet transform.

5.5.1   Dunkl transform

The Dunkl operators were introduced and studied by Charles Dunkl (1941 − −) in 1989, in connection with an extension of the classical theory of spherical harmonics. Apart from their mathematical relevance, these operators are also applied in quantum mechanics in the study of one-dimensional harmonic oscillators governed by Wigner's commutation rules. The Dunkl operator with parameter $\alpha + 1/2$ is defined as a differential-difference operator:

$$\Lambda_\alpha f(t) = f'(t) + \left(\alpha + \frac{1}{2} \right) \left[\frac{f(t) - f(-t)}{t} \right], \quad \alpha > -1/2. \quad (5.110)$$

Based on the Dunkl operator Λ_α defined in (5.110), we consider the following initial value problem:

$$\Lambda_\alpha f(t) = \omega f(t), \quad f(0) = 1, \quad t \in \mathbb{R}, \ \omega \in \mathbb{C}. \quad (5.111)$$

The initial value problem (5.111) admits a unique solution called as the "Dunkl kernel" and is given by

$$E_\alpha(\omega t) = \dot{J}_\alpha(i\omega t) + \frac{\omega t}{2(\alpha + 1)}\, \dot{J}_{\alpha+1}(i\omega t), \quad t \in \mathbb{R}, \quad (5.112)$$

where $\dot{J}_\alpha$ is the normalized Bessel function of first kind of order α and is defined by

$$\dot{J}_\alpha(z) = \frac{\Gamma(\alpha+1)}{(z/2)^\alpha} J_\alpha(z) = \sum_{n=0}^{\infty} \frac{(-1)^n}{n!\,\Gamma(n+\alpha+1)} \left(\frac{z}{2}\right)^{2n}, \quad z \in \mathbb{C}. \tag{5.113}$$

Also, for $t \in \mathbb{R}$ and $\omega \in \mathbb{C}$, the Dunkl kernel yields the following integral representation:

$$E_\alpha(-i\,\omega t) = \frac{\Gamma(\alpha+1)}{\sqrt{\pi}\,\Gamma(\alpha+1/2)} \int_{-1}^{1} (1-x^2)^{\alpha-1/2}\,(1-x)\,e^{i\,\omega xt}\,dx. \tag{5.114}$$

Moreover, we note that the Dunkl kernel (5.112) gives rise to an integral transform known as the "Dunkl transform." Before proceeding to the notion of the Dunkl transform, it is needful to denote $L_\alpha^2\left(\mathbb{R}, d\Omega_\alpha\right)$ as the space of measurable functions f on $\mathbb{R}$, satisfying

$$\left\|f\right\|_{2,\alpha} = \left\{\int_{\mathbb{R}} |f(t)|^2 d\Omega_\alpha(t)\right\}^{1/2} < \infty, \tag{5.115}$$

where $d\Omega_\alpha(t)$ is the weighted Lebesgue measure on $\mathbb{R}$ and is given by

$$d\Omega_\alpha(t) = \left(2^{\alpha+1}\Gamma(\alpha+1)\right)^{-1} |t|^{2\alpha+1} dt. \tag{5.116}$$

Prior to the formal definition of the Dunkl transform, it is pertinent to mention that the Dunkl transform is often defined for integrable functions with respect to the measure $d\Omega_\alpha(t)$ and can be extended to square integrable functions with respect to the same measure by a Parseval-type relation.

Definition 5.5.1. For any $f \in L_\alpha^2\left(\mathbb{R}, d\Omega_\alpha\right)$, the Dunkl transform is denoted by $\mathscr{D}_\alpha[f]$ and is defined as

$$\mathscr{D}_\alpha\big[f\big](\omega) = \int_{\mathbb{R}} f(t)\,E_\alpha(-i\,\omega t)\,d\Omega_\alpha(t), \quad \omega \in \mathbb{R} \tag{5.117}$$

where $E_\alpha(-i\,\omega t)$ is given by (5.114).

The inversion formula corresponding to the Dunkl transform (5.117) is given by

$$f(t) = \mathscr{D}_\alpha^{-1}\Big(\mathscr{D}_\alpha\big[f\big](\omega)\Big)(t) := \int_{\mathbb{R}} \mathscr{D}_\alpha[f](\omega)\,E_\alpha(i\,\omega t)\,d\Omega_\alpha(\omega), \quad \omega \in \mathbb{R} \tag{5.118}$$

and the Parseval's formula asserts that

$$\int_{\mathbb{R}} \mathscr{D}_\alpha\big[f\big](\omega)\,\overline{\mathscr{D}_\alpha\big[g\big](\omega)}\,d\Omega_\alpha(\omega) = \int_{\mathbb{R}} f(t)\,\overline{g(t)}\,d\Omega_\alpha(t). \tag{5.119}$$

5.5.2 Dunkl wavelet transform: Definition and basic properties

To facilitate the formulation of the Dunkl wavelet transform, we shall present the fundamental notions of the Dunkl translation and convolution operators. Consider, the basic function

$$W_\alpha(t,y,z) = \big(1 - \sigma(t,y,z) + \sigma(z,t,y) + \sigma(z,y,t)\big)\Delta_\alpha(t,y,z), \tag{5.120}$$

where

$$\sigma(t,y,z) = \begin{cases} \dfrac{t^2 + y^2 + z^2}{2ty}, & t,y \neq 0 \\ 0, & \text{otherwise,} \end{cases} \tag{5.121}$$

and $\Delta_\alpha(t, y, z)$ is the Bessel kernel given by

$$\Delta(t, y, z) = \begin{cases} C_\alpha \left[\dfrac{\left(\left((|t| + |y|)^2 - z^2\right)\left(z^2 - (|t| - |y|)^2\right) \right)^{\alpha - 1/2}}{|tyz|^{2\alpha}} \right], & |z| \in A_{t,y} \\ 0, & \text{otherwise} \end{cases} \tag{5.122}$$

with

$$C_\alpha = \frac{\left(\Gamma(\alpha + 1)\right)^2}{2^{\alpha - 1}\sqrt{\pi}\,\Gamma(\alpha + 1/2)} \tag{5.123}$$

and

$$A_{t,y} = \left(\big| |t| - |y| \big|, |t| + |y| \right). \tag{5.124}$$

Note that the basic function $W_\alpha(t, y, z)$ satisfies the following inequality:

$$\int_{\mathbb{R}} \big| W_\alpha(t, y, z) \big| \, d\Omega_\alpha(z) \le 4. \tag{5.125}$$

By virtue of the basic function $W_\alpha(t, y, z)$ defined in (5.120), we now present the formal definitions of the Dunkl translation and convolution operators together with the respective fundamental properties.

Definition 5.5.2. For any $f \in L_\alpha^2(\mathbb{R}, d\Omega_\alpha)$, the Dunkl translation operator is denoted by $\mathcal{T}_y$ and is defined as

$$\mathcal{T}_y f(t) = f(t, y) = \int_{\mathbb{R}} f(z)\, W_\alpha(t, y, z)\, d\Omega_\alpha(z), \tag{5.126}$$

where $W_\alpha(t, y, z)$ is given by (5.120).

Some fundamental properties of the Dunkl translation operator given in (5.126) are listed below:

(i) $\left\| \mathcal{T}_y f \right\|_{2,\alpha} \le 4 \left\| f \right\|_{2,\alpha}$,

(ii) $\mathscr{D}_\alpha \left[\mathcal{T}_y f \right](\omega) = E_\alpha(i\,\omega t)\, \mathscr{D}_\alpha \left[f \right](\omega).$

Definition 5.5.3. Given any pair of functions $f, g \in L_\alpha^2(\mathbb{R}, d\Omega_\alpha)$, the Dunkl convolution operator is denoted by $\circledast_\mathscr{D}$ and is defined as

$$
\begin{aligned}
(f \circledast_\mathscr{D} g)(t) &= \int_{\mathbb{R}} f(x)\, \mathcal{T}_t g(-x)\, d\Omega_\alpha(x) \\
&= \int_{\mathbb{R}} \int_{\mathbb{R}} f(x)\, g(z)\, W_\alpha(-x, t, z)\, d\Omega_\alpha(z)\, d\Omega_\alpha(x).
\end{aligned} \tag{5.127}
$$

Alongside the notion of Dunkl convolution is the corresponding convolution theorem, which describes the behaviour of the convolution operation (5.127) under the Dunkl transform (5.117). The convolution theorem reads:

$$\mathscr{D}_\alpha \left[(f \circledast_\mathscr{D} g) \right](\omega) = \mathscr{D}_\alpha \left[f \right](\omega)\, \mathscr{D}_\alpha \left[g \right](\omega). \tag{5.128}$$

Now, we are in a position to construct a new family of wavelets called as "Dunkl wavelets." For $a \in \mathbb{R}^+$ and $b \in \mathbb{R}$, let $\mathcal{D}_a$ denotes the usual dilation operator and $\mathcal{T}_b$ be the Dunkl translation operator as defined in (5.126). Then, the Dunkl wavelets are defined by the combined action of these translation and dilation operators on $\psi \in L^2_\alpha(\mathbb{R}, d\Omega_\alpha)$ as

$$\psi^{(\alpha)}_{a,b}(t) = \mathcal{T}_b \mathcal{D}_a \psi(t) = \frac{1}{\sqrt{a}} \int_\mathbb{R} \psi\left(\frac{z}{a}\right) W_\alpha(b,t,z) \, d\Omega_\alpha(z), \qquad (5.129)$$

where the convergence of the integral is guaranteed by the norm inequality satisfied by the Dunkl translation operator.

Based on the family of Dunkl wavelets (5.129), following is the definition of the continuous Dunkl wavelet transform:

Definition 5.5.4. For any $f \in L^2_\alpha(\mathbb{R}, d\Omega_\alpha)$, the continuous Dunkl wavelet transform is denoted by $\mathcal{DW}^{(\alpha)}_\psi[f]$ and is defined as

$$\mathcal{DW}^{(\alpha)}_\psi[f](a,b) = \left\langle f, \psi^{(\alpha)}_{a,b} \right\rangle_{2,\alpha} = \frac{1}{\sqrt{a}} \int_\mathbb{R} \int_\mathbb{R} f(t)\, \overline{\psi\left(\frac{z}{a}\right)} W_\alpha(b,t,z) \, d\Omega_\alpha(z) \, d\Omega_\alpha(t). \quad (5.130)$$

From Definition 5.5.4, it is quite evident that the continuous Dunkl wavelet transform can be expressed via the Dunkl convolution (5.127) as

$$\mathcal{DW}^{(\alpha)}_\psi[f](a,b) = \left(f(t) \circledast_\mathcal{D} \mathcal{D}_a \overline{\psi}(-t) \right)(b). \qquad (5.131)$$

Proposition 5.5.1. *If* $\mathcal{DW}^{(\alpha)}_\psi[f](a,b)$ *is the continuous Dunkl wavelet transform of any* $f \in L^2_\alpha(\mathbb{R}, d\Omega_\alpha)$, *then we have*

$$\left| \mathcal{DW}^{(\alpha)}_\psi[f](a,b) \right| \le 4 \left\| f \right\|_{2,\alpha} \left\| \mathcal{D}_a \psi \right\|_{2,\alpha}, \qquad (5.132)$$

where

$$\mathcal{D}_a \psi(t) = \frac{1}{\sqrt{a}} \psi\left(\frac{t}{a}\right). \qquad (5.133)$$

Proof. Invoking Definition 5.5.4 and using Hölders inequality, we obtain

$$\left| \mathcal{DW}^{(\alpha)}_\psi[f](a,b) \right| \le \left\{ \int_\mathbb{R} \int_\mathbb{R} |f(t)|^2 \, W_\alpha(b,t,z) \, d\Omega_\alpha(z) \, d\Omega_\alpha(t) \right\}^{1/2}$$

$$\times \left\{ \frac{1}{a} \int_\mathbb{R} \int_\mathbb{R} \left| \psi\left(\frac{z}{a}\right) \right|^2 W_\alpha(b,t,z) \, d\Omega_\alpha(z) \, d\Omega_\alpha(z) \right\}^{1/2}$$

$$= \left\{ \int_\mathbb{R} |f(t)|^2 \left(\int_\mathbb{R} W_\alpha(b,t,z) \, d\Omega_\alpha(z) \right) d\Omega_\alpha(t) \right\}^{1/2}$$

$$\times \left\{ \int_\mathbb{R} |\mathcal{D}_a \psi(z)|^2 \left(\int_\mathbb{R} W_\alpha(b,t,z) \, d\Omega_\alpha(t) \right) d\Omega_\alpha(z) \right\}^{1/2}. \qquad (5.134)$$

Implementing (5.125) into (5.134), we get

$$\left| \mathcal{DW}^{(\alpha)}_\psi[f](a,b) \right| \le 4 \left\{ \int_\mathbb{R} |f(t)|^2 d\Omega_\alpha(t) \right\}^{1/2} \left\{ \int_\mathbb{R} |\mathcal{D}_a \psi(z)|^2 d\Omega_\alpha(z) \right\}^{1/2}$$

$$= 4 \left\| f \right\|_{2,\alpha} \left\| \mathcal{D}_a \psi \right\|_{2,\alpha}.$$

This completes the proof of Proposition 5.5.1. □

In the following lemma, we shall express the Dunkl wavelet transform (5.130) in terms of the Dunkl transform (5.117) by virtue of (5.131).

Lemma 5.5.2. *For any $f \in L^2_\alpha(\mathbb{R}, d\Omega_\alpha)$, the Dunkl wavelet transform $\mathcal{DW}^{(\alpha)}_\psi[f](a, b)$ can be expressed as*

$$\mathcal{DW}^{(\alpha)}_\psi[f](a, b) = a^{2\alpha+1/2} \mathcal{D}^{-1}_\alpha \left(\mathcal{D}_\alpha[f](\omega) \, \overline{\mathcal{D}_\alpha[\psi](a\omega)} \right)(b). \tag{5.135}$$

Proof. As a consequence of (5.128) and (5.131), we have

$$\mathcal{D}_\alpha \left(\mathcal{DW}^{(\alpha)}_\psi[f](a, b) \right)(\omega) = \mathcal{D}_\alpha \left[\left(f(t) \circledast_D D_a \overline{\psi}(-t) \right)(b) \right](\omega)$$
$$= \mathcal{D}_\alpha[f(t)](\omega) \, \mathcal{D}_\alpha[D_a \overline{\psi}(-t)](\omega). \tag{5.136}$$

Also, we observe that

$$\mathcal{D}_\alpha[D_a \overline{\psi}(-t)](\omega) = \frac{1}{\sqrt{a}} \int_{\mathbb{R}} \overline{\psi\left(-\frac{t}{a}\right)} E_\alpha(-i\,\omega t) \, d\Omega_\alpha(t). \tag{5.137}$$

Making use of the substitution $-t/a = t'$, we can rewrite (5.137) as

$$\mathcal{D}_\alpha[D_a \overline{\psi}(-t)](\omega) = \frac{a^{2\alpha+1}}{\sqrt{a}} \int_{\mathbb{R}} \overline{\psi(t')} E_\alpha(i\,a\omega t') \, d\Omega_\alpha(t')$$
$$= a^{2\alpha+1/2} \, \overline{\mathcal{D}_\alpha[\psi](a\omega)}. \tag{5.138}$$

Plugging (5.138) into (5.136), we obtain

$$\mathcal{D}_\alpha \left(\mathcal{DW}^{(\alpha)}_\psi[f](a, b) \right)(\omega) = a^{2\alpha+1/2} \mathcal{D}_\alpha[f(t)](\omega) \, \overline{\mathcal{D}_\alpha[\psi](a\omega)}. \tag{5.139}$$

Applying the inverse Dunkl transform (5.118) on both sides of (5.139) yields

$$\mathcal{DW}^{(\alpha)}_\psi[f](a, b) = a^{2\alpha+1/2} \int_{\mathbb{R}} \mathcal{D}_\alpha[f(t)](\omega) \, \overline{\mathcal{D}_\alpha[\psi](a\omega)} \, E_\alpha(i\,\omega b) \, d\Omega_\alpha(\omega)$$
$$= a^{2\alpha+1/2} \mathcal{D}^{-1}_\alpha \left(\mathcal{D}_\alpha[f](\omega) \, \overline{\mathcal{D}_\alpha[\psi](a\omega)} \right)(b).$$

This completes the proof of Lemma 5.5.2.   □

In the next theorem, we take an account of the Moyal's principle associated with the continuous Dunkl wavelet transform (5.130).

Theorem 5.5.3. *If $\mathcal{DW}^{(\alpha)}_\psi[f](a, b)$ and $\mathcal{DW}^{(\alpha)}_\psi[g](a, b)$ are the continuous Dunkl wavelet transforms of a given pair of functions $f, g \in L^2_\alpha(\mathbb{R}, d\Omega_\alpha)$, then we have*

$$\int_{\mathbb{R}} \int_{\mathbb{R}+} \mathcal{DW}^{(\alpha)}_\psi[f](a, b) \, \overline{\mathcal{DW}^{(\alpha)}_\psi[g](a, b)} \, \mathcal{K}(a) \, d\Omega_\alpha(a) \, d\Omega_\alpha(b) = C^{(\alpha)}_\psi \int_{\mathbb{R}} f(t) \, \overline{g(t)} \, d\Omega_\alpha(t), \tag{5.140}$$

where $\mathcal{K}(a)$ is the weight function and $C^{(\alpha)}_\psi$ is given by

$$C^{(\alpha)}_\psi = \int_{\mathbb{R}} a^{4\alpha+1} \left| \mathcal{D}_\alpha[\psi](a\omega) \right|^2 \mathcal{K}(a) \, d\Omega_\alpha(a) < \infty, \quad a.e. \quad \omega \in \mathbb{R}. \tag{5.141}$$

Proof. As a consequence of Lemma 5.5.2, we have

$$\int_{\mathbb{R}} \int_{\mathbb{R}+} \mathcal{DW}_\psi^{(\alpha)} \big[f\big](a,b) \, \overline{\mathcal{DW}_\psi^{(\alpha)} \big[g\big]}(a,b) \, \mathcal{K}(a) \, d\Omega_\alpha(a) \, d\Omega_\alpha(b)$$

$$= \int_{\mathbb{R}+} a^{4\alpha+1} \left\{ \int_{\mathbb{R}} \mathscr{D}_\alpha^{-1} \Big(\mathscr{D}_\alpha \big[f\big](\omega) \, \overline{\mathscr{D}_\alpha \big[\psi\big]}(a\omega) \Big)(b) \right.$$

$$\left. \times \overline{\mathscr{D}_\alpha^{-1} \Big(\mathscr{D}_\alpha \big[g\big](\omega) \, \overline{\mathscr{D}_\alpha \big[\psi\big]}(a\omega) \Big)}(b) \, d\Omega_\alpha(b) \right\} \mathcal{K}(a) \, d\Omega_\alpha(a). \qquad (5.142)$$

Applying the Parseval's formula for the Dunkl transform (5.119) to the inner integral in (5.142), we get

$$\int_{\mathbb{R}} \int_{\mathbb{R}+} \mathcal{DW}_\psi^{(\alpha)} \big[f\big](a,b) \, \overline{\mathcal{DW}_\psi^{(\alpha)} \big[g\big]}(a,b) \, \mathcal{K}(a) \, d\Omega_\alpha(a) \, d\Omega_\alpha(b)$$

$$= \int_{\mathbb{R}+} a^{4\alpha+1} \left\{ \int_{\mathbb{R}} \mathscr{D}_\alpha \big[f\big](\omega) \, \overline{\mathscr{D}_\alpha \big[\psi\big]}(a\omega) \, \overline{\mathscr{D}_\alpha \big[g\big]}(\omega) \, \mathscr{D}_\alpha \big[\psi\big](a\omega) \, d\Omega_\alpha(\omega) \right\} \mathcal{K}(a) \, d\Omega_\alpha(a)$$

$$= \int_{\mathbb{R}} \mathscr{D}_\alpha \big[f\big](\omega) \, \overline{\mathscr{D}_\alpha \big[g\big]}(\omega) \left\{ \int_{\mathbb{R}+} a^{4\alpha+1} \Big| \mathscr{D}_\alpha \big[\psi\big](a\omega) \Big|^2 \mathcal{K}(a) \, d\Omega_\alpha(a) \right\} d\Omega_\alpha(\omega)$$

$$= C_\psi^{(\alpha)} \int_{\mathbb{R}} \mathscr{D}_\alpha \big[f\big](\omega) \, \overline{\mathscr{D}_\alpha \big[g\big]}(\omega) \, d\Omega_\alpha(\omega)$$

$$= C_\psi^{(\alpha)} \int_{\mathbb{R}} f(t) \, \overline{g(t)} \, d\Omega_\alpha(t).$$

This completes the proof of Theorem 5.5.3. □

Remark 5.5.1. For $f = g$, Theorem 5.5.3 yields the energy preserving relation for the continuous Dunkl wavelet transform:

$$\int_{\mathbb{R}} \int_{\mathbb{R}+} \Big| \mathcal{DW}_\psi^{(\alpha)} \big[f\big](a,b) \Big|^2 \mathcal{K}(a) \, d\Omega_\alpha(a) \, d\Omega_\alpha(b) = C_\psi^{(\alpha)} \big\|f\big\|_{2,\alpha}^2. \qquad (5.143)$$

Finally, our aim is to obtain an inversion formula associated with the continuous Dunkl wavelet transform (5.130).

Theorem 5.5.4. *Any function $f \in L_\alpha^2(\mathbb{R}, d\Omega_\alpha)$ can be reconstructed from the corresponding continuous Dunkl wavelet transform $\mathcal{DW}_\psi^{(\alpha)} \big[f\big](a,b)$ via the formula:*

$$f(t) = \frac{1}{C_\psi^{(\alpha)}} \int_{\mathbb{R}} \int_{\mathbb{R}} \int_{\mathbb{R}+} \mathcal{DW}_\psi^{(\alpha)} \big[f\big](a,b) \, D_a\psi(z) \, W_\alpha(b,t,z) \, \mathcal{K}(a) \, d\Omega_\alpha(a) \, d\Omega_\alpha(z) \, d\Omega_\alpha(b),$$

$$(5.144)$$

where $C_\psi^{(\alpha)} > 0$ is given by (5.141) and the equality in (5.144) holds in the almost everywhere sense.

Proof. By virtue of Theorem 5.5.3, it follows that

$$C_\psi^{(\alpha)} \int_{\mathbb{R}} f(t) \, \overline{g(t)} \, d\Omega_\alpha(t)$$

$$= \int_{\mathbb{R}} \int_{\mathbb{R}+} \mathcal{DW}_\psi^{(\alpha)} \big[f\big](a,b) \, \overline{\mathcal{DW}_\psi^{(\alpha)} \big[g\big]}(a,b) \, \mathcal{K}(a) \, d\Omega_\alpha(a) \, d\Omega_\alpha(b)$$

$$= \int_{\mathbb{R}} \int_{\mathbb{R}^+} \mathcal{DW}_\psi^{(\alpha)} \Big[f \Big] (a,b) \left\{ \frac{1}{\sqrt{a}} \int_{\mathbb{R}} \int_{\mathbb{R}} \overline{g(t)} \, \psi\left(\frac{z}{a}\right) W_\alpha(b,t,z) \, d\Omega_\alpha(z) \, d\Omega_\alpha(t) \right\}$$
$$\times \, \mathcal{K}(a) \, d\Omega_\alpha(a) \, d\Omega_\alpha(b)$$

$$= \int_{\mathbb{R}} \left\{ \int_{\mathbb{R}} \int_{\mathbb{R}} \int_{\mathbb{R}^+} \mathcal{DW}_\psi^{(\alpha)} \Big[f \Big] (a,b) \, \mathcal{D}_a\psi(z) \, W_\alpha(b,t,z) \, \mathcal{K}(a) \, d\Omega_\alpha(a) \, d\Omega_\alpha(z) \, d\Omega_\alpha(b) \right\}$$
$$\times \, \overline{g(t)} \, d\Omega_\alpha(t). \qquad (5.145)$$

As a consequence of (5.145), we conclude that

$$f(t) = \frac{1}{C_\psi^{(\alpha)}} \int_{\mathbb{R}} \int_{\mathbb{R}} \int_{\mathbb{R}^+} \mathcal{DW}_\psi^{(\alpha)} \Big[f \Big] (a,b) \, \mathcal{D}_a\psi(z) \, W_\alpha(b,t,z) \, \mathcal{K}(a) \, d\Omega_\alpha(a) \, d\Omega_\alpha(z) \, d\Omega_\alpha(b), \quad a.e.,$$

which is the desired reconstruction formula. This completes the proof of Theorem 5.5.4. $\square$

5.6 The Mehler-Fock Wavelet Transform

"The true laboratory is the mind, where behind illusions we uncover the laws of truth."

-Jagadish C. Bose

This section is concerned with the study of the Mehler-Fock wavelet transform which is an off-shoot of the wavelet and Mehler-Fock transforms. The Mehler-Fock transform is based on an important class of functions known as the "cone functions," which are the zero order Legendre functions of first kind. Primarily, we ought to present the preliminaries encapsulating the Mehler-Fock transform, adjoint Mehler-Fock transform and the basic notions of Mehler-Fock translation and convolution operations. Subsequently, we shall examine the Mehler-Fock wavelet transform abreast of its fundamental properties.

5.6.1 Mehler-Fock transform

The Mehler-Fock transform was introduced by the Germen mathematician Gustav F. Mehler $(1835-1895)$ in 1881 while he was working on the transmission of electromagnetical waves in cone-domains. Later on, the transform was revisited by the Soviet physicist Vladimir A. Fock $(1898-1974)$, who did foundational work on quantum mechanics and quantum electrodynamics, and gained a deeper insight into this transformation. The Mehler-Fock transform is an integral transform based on the zero order Legendre functions of first kind, which constitute a particular class of solutions of the linear, second order Legendre differential equation:

$$\left(1 - t^2\right) f''(t) - 2t f'(t) + \left[\lambda(\lambda + 1) - \frac{\mu}{1 - t^2} \right] f(t) = 0, \quad \lambda \in \mathbb{C}, \ \mu \in \mathbb{R}. \qquad (5.146)$$

With regard to the parameters λ and μ appearing in (5.146), the two linearly independent solutions of (5.146) are denoted by P_λ^μ and Q_λ^μ. The solution P_λ^μ is called as the "Legendre function of first kind," whereas Q_λ^μ is called as the "Legendre function of second kind." An additional qualifier "associated" is assigned to the solutions in case the order of the Legendre functions μ is non-zero. Some particular cases of the solutions corresponding to (5.146) are often discussed separately, for instance, when $\lambda = n$ is an integer and $\mu = 0$, the solutions correspond to a special class of polynomials, known as the "Legendre polynomials."

It is pertinent to mention that the second order linear differential equation (5.146) has three regular singular points, viz; 1, -1 and ∞, therefore, like all such equations, it can be converted into a hypergeometric differential equation by a change of variable, and the solutions P_λ^μ and Q_λ^μ can be expressed using the hypergeometric function $_2F_1(\cdot)$ as

$$P_\lambda^\mu(t) = \frac{1}{\Gamma(1-\mu)} \left(\frac{1+t}{1-t}\right)^{\mu/2} {}_2F_1\left(-\lambda, 1+\lambda; 1-\mu; \frac{1-t}{2}\right), \quad \text{for } |1-t| < 2 \quad (5.147)$$

and

$$Q_\lambda^\mu(t) = \frac{\sqrt{\pi}\,\Gamma(\lambda+\mu+1)}{2^{\lambda+1}\Gamma(\lambda+3/2)} \frac{e^{i\mu\pi} (t^2-1)^{\mu/2}}{t^{\lambda+\mu+1}}$$

$$\times {}_2F_1\left(\frac{\lambda+\mu+1}{2}, \frac{\lambda+\mu+2}{2}; \lambda+\frac{3}{2}; \frac{1}{t^2}\right), \quad \text{for } |t| > 1, \quad (5.148)$$

where $\Gamma(\cdot)$ is the gamma function. Moreover, for $\mu = 0$, the expressions (5.147) and (5.148) reduce to their respective counterparts for the zero order Legendre functions of first and second kind; that is,

$$P_\lambda(t) = {}_2F_1\left(-\lambda, 1+\lambda; 1; \frac{1-t}{2}\right), \quad \text{for } |1-t| < 2 \quad (5.149)$$

and

$$Q_\lambda(t) = \frac{\sqrt{\pi}\,\Gamma(\lambda+1)}{(2t)^{\lambda+1}\Gamma(\lambda+3/2)} {}_2F_1\left(\frac{\lambda+1}{2}, \frac{\lambda+2}{2}; \lambda+\frac{3}{2}; \frac{1}{t^2}\right), \quad \text{for } |t| > 1. \quad (5.150)$$

Prior to recalling the formal definition of the Mehler-Fock transform, we denote $L^2[1,\infty)$ as the space of measurable functions satisfying

$$\left\| f \right\|_{L^2[1,\infty)} = \left\{ \int_1^\infty |f(t)|\, dt \right\}^{1/2} < \infty. \quad (5.151)$$

Definition 5.6.1. For any $f \in L^2[1,\infty)$, the Mehler-Fock transform is denoted by $\mathscr{M}[f]$ and is defined as

$$\mathscr{M}[f](\omega) = \int_1^\infty f(t)\, P_{i\omega-1/2}(t)\, dt, \quad \omega > 0, \quad (5.152)$$

where $P_{i\omega-1/2}(t)$ is called as the "cone function" and is the zero order Legendre function of first kind.

The inversion and Parseval's formulae associated with the Mehler-Fock transform (5.152) are given by

$$f(t) = \mathscr{M}^{-1}\Big(\mathscr{M}[f](\omega)\Big)(t) = \int_0^\infty \omega \tanh(\pi\omega)\, \mathscr{M}[f](\omega)\, P_{i\omega-1/2}(t)\, d\omega \quad (5.153)$$

and

$$\int_0^\infty \omega \tanh(\pi\omega)\, \mathscr{M}[f](\omega)\, \overline{\mathscr{M}[g](\omega)}\, d\omega = \int_1^\infty f(t)\, \overline{g(t)}\, dt, \quad (5.154)$$

respectively. Also, the cone-function $P_{i\omega-1/2}(t)$ involved in (5.152) satisfies many pleasant properties. One of the distinguishing features of the cone function is that it satisfies the product formula or the linearization formula given by

$$P_{i\omega-1/2}(t)\, P_{i\omega-1/2}(y) = \int_1^\infty P_{i\omega-1/2}(z)\, \Delta(t,y,z)\, dz, \quad t,y,z \in [1,\infty), \qquad (5.155)$$

where the kernel $\Delta(t,y,z)$ is defined as

$$\Delta(t,y,z) = \begin{cases} \pi^{-1}\left(2tyz - 1 - t^2 - y^2 - z^2\right)^{-1/2}, & 2tyz - 1 - t^2 - y^2 - z^2 > 0 \\ 0, & \text{otherwise.} \end{cases} \qquad (5.156)$$

Evidently, the kernel $\Delta(t,y,z)$ is non-negative, symmetrical and also satisfies the integral equation:

$$\int_1^\infty \Delta(t,y,z)\, dz = 1. \qquad (5.157)$$

Besides, some other useful properties of the cone function $P_{i\omega-1/2}(t)$ are listed as follows:

(i) $\left|P_{i\omega-1/2}(t)\right| \le k\, P_{-1/2}(t), \quad k > 0,$

(ii) $P_{-1/2}(t) \approx 1 \quad \text{as} \quad t \to 1,$

(iii) $P_{-1/2}(t) = \dfrac{\sqrt{2}\,\ln(t)}{\pi\sqrt{t}} \quad \text{as} \quad t \to \infty.$

Moreover, as a consequence of (5.152), (5.153) and (5.155), we can obtain an integral representation of the kernel $\Delta(t,y,z)$ as follows:

$$\Delta(t,y,z) = \int_0^\infty \omega \tanh(\pi\omega)\, P_{i\omega-1/2}(z) \left\{ \int_1^\infty \Delta(t,y,z)\, P_{i\omega-1/2}(z)\, dz \right\} d\omega$$

$$= \int_0^\infty \omega \tanh(\pi\omega)\, P_{i\omega-1/2}(t)\, P_{i\omega-1/2}(y)\, P_{i\omega-1/2}(z)\, d\omega. \qquad (5.158)$$

Intimately intertwined with the Mehler-Fock transform (5.152) is the adjoint Mehler-Fock transform denoted by $\mathscr{M}^*$ and defined as

$$\mathscr{M}^*\big[f\big](t) = \int_0^\infty f(\omega)\, P_{i\omega-1/2}(t)\, d\omega. \qquad (5.159)$$

The inversion formula corresponding to the adjoint Mehler-Fock transform (5.159) is given by

$$f(\omega) = \mathscr{M}^{*-1}\left(\mathscr{M}^*\big[f\big](t)\right)(\omega) := \omega \tanh(\pi\omega) \int_1^\infty \mathscr{M}^*\big[f\big](t)\, P_{i\omega-1/2}(t)\, dt, \qquad (5.160)$$

whereas the Parseval's formula reads:

$$\int_1^\infty \mathscr{M}^*\big[f\big](t)\, \overline{\mathscr{M}^*\big[g\big](t)}\, dt = \int_0^\infty \frac{1}{\omega \tanh(\pi\omega)}\, f(\omega)\, \overline{g(\omega)}\, d\omega. \qquad (5.161)$$

5.6.2   Mehler-Fock wavelet transform: Definition and basic properties

In this subsection, we formally study the Mehler-Fock wavelet transform by virtue of the Mehler-Fock translation and convolution operations defined below:

Definition 5.6.2. For $y \in [1, \infty)$, the Mehler-Fock translation operator acting on $f \in L^2[1, \infty)$ is denoted by $\mathbb{T}_y f$ and is defined as

$$\mathbb{T}_y f(t) = \int_1^\infty f(z)\, \Delta(t, y, z)\, dz, \quad \omega > 0, \tag{5.162}$$

where $\Delta(t, y, z)$ is given by (5.156).

Note that the translation operator (5.162) is a positive, linear and continuous operator from $L^2[1, \infty)$ into itself. Also, we have

$$\mathscr{M}\Big[\mathbb{T}_y f\Big](\omega) = P_{i\omega - 1/2}(y)\, \mathscr{M}\Big[f\Big](\omega). \tag{5.163}$$

Definition 5.6.3. For any pair of functions $f, g \in L^2[1, \infty)$, the Mehler-Fock convolution is denoted by $\circledast_{\mathscr{M}}$ and is defined as

$$\begin{aligned}
(f \circledast_{\mathscr{M}} g)(t) &= \int_1^\infty f(y)\, \mathbb{T}_t g(y)\, dy \\
&= \int_1^\infty \int_1^\infty f(y)\, g(z)\, \Delta(t, y, z)\, dz\, dy,
\end{aligned} \tag{5.164}$$

where $\Delta(t, y, z)$ is given by (5.156).

The convolution theorem associated with the Mehler-Fock convolution operation given in (5.164) asserts that

$$\mathscr{M}\Big[(f \circledast_{\mathscr{M}} g)\Big](\omega) = \mathscr{M}\Big[f\Big](\omega)\, \mathscr{M}\Big[g\Big](\omega). \tag{5.165}$$

In the sequel, we construct a novel family of wavelets, known as "Mehler-Fock wavelets" by applying appropriate dilation and translation operators on the generating function $\psi \in L^2[1, \infty)$. For $a > 0$ and $b \in [1, \infty)$, let $\mathcal{D}_a$ and $\mathbb{T}_b$ denote the usual dilation and the Mehler-Fock translation operators acting on $\psi \in L^2[1, \infty)$. Then, the Mehler-Fock wavelets are defined by the joint action of these translation and dilation operators as

$$\psi_{a,b}(t) = \mathbb{T}_b \mathcal{D}_a \psi(t) = \frac{1}{\sqrt{a}} \int_1^\infty \psi\left(\frac{z}{a}\right) \Delta(t, b, z)\, dz, \tag{5.166}$$

where $\Delta(t, y, z)$ is given by (5.156).

Based upon the novel family of wavelets (5.166), we have the following definition of the Mehler-Fock wavelet transform:

Definition 5.6.4. For any $f \in L^2[1, \infty)$, the Mehler-Fock wavelet transform with respect to the wavelet ψ is denoted by $\mathcal{MW}_\psi[f]$ and is defined as

$$\mathcal{MW}_\psi\big[f\big](a, b) = \Big\langle f, \psi_{a,b} \Big\rangle_{L^2[1,\infty)} = \frac{1}{\sqrt{a}} \int_1^\infty \int_1^\infty f(t)\, \overline{\psi\left(\frac{z}{a}\right)} \Delta(t, b, z)\, dz\, dt, \tag{5.167}$$

where $\Delta(t, y, z)$ is given by (5.156).

From Definition 5.6.4, it is worth noticing that the Mehler-Fock wavelet transform can also be expressed via the convolution operation (5.164) as

$$\mathcal{MW}_\psi\big[f\big](a,b) = \big(f \circledast_\mathcal{M} \overline{\mathcal{D}_a\psi}\,\big)(b). \tag{5.168}$$

Moreover, as a consequence of Hölder's inequality and (5.157), we observe that the Mehler-Fock wavelet transform is indeed a bounded operator, since

$$\left|\mathcal{MW}_\psi\big[f\big](a,b)\right| \leq \left\{\int_1^\infty \int_1^\infty |f(t)|^2 \Delta(t,b,z)\, dz\, dt\right\}^{1/2}$$

$$\times \left\{\int_1^\infty \int_1^\infty |\mathcal{D}_a\psi(z)|^2 \Delta(t,b,z)\, dz\, dt\right\}^{1/2}$$

$$\leq \left\{\int_1^\infty |f(t)|^2 \left(\int_1^\infty \Delta(t,b,z)\, dz\right) dt\right\}^{1/2}$$

$$\times \left\{\int_1^\infty |\mathcal{D}_a\psi(z)|^2 \left(\int_1^\infty \Delta(t,b,z)\, dt\right) dz\right\}^{1/2}$$

$$\leq \left\{\int_1^\infty |f(t)|^2 dt\right\}^{1/2} \left\{\int_1^\infty |\mathcal{D}_a\psi(z)|^2 dz\right\}^{1/2}$$

$$\leq \big\|f\big\|_{L^2[1,\infty)} \big\|\mathcal{D}_a\psi(z)\big\|_{L^2[1,\infty)}. \tag{5.169}$$

Next, our main aim is to investigate upon the Moyal's principle and the inversion formula corresponding to the continuous Mehler-Fock wavelet transform given in Definition 5.6.4. In this direction, we have the following lemma:

Lemma 5.6.1. *If $\mathcal{MW}_\psi\big[f\big](a,b)$ is the Mehler-Fock wavelet transform of any $f \in L^2[1,\infty)$, then we have*

$$\mathcal{M}\Big(\mathcal{MW}_\psi\big[f\big](a,b)\Big)(\omega) = \mathcal{M}\big[f\big](\omega)\, \mathcal{M}\big[\overline{\mathcal{D}_a\psi}\big](\omega), \tag{5.170}$$

where

$$\mathcal{M}\big[\overline{\mathcal{D}_a\psi}\big](\omega) = \frac{1}{\sqrt{a}}\int_1^\infty \overline{\psi\left(\frac{t}{a}\right)}\, P_{i\omega-1/2}(t)\, dt, \quad \omega > 0. \tag{5.171}$$

Proof. Using Definition 5.6.4 together with (5.158), we obtain

$$\mathcal{MW}_\psi\big[f\big](a,b)$$

$$= \frac{1}{\sqrt{a}}\int_1^\infty \int_1^\infty f(t)\, \overline{\psi\left(\frac{z}{a}\right)}\, \Delta(t,b,z)\, dz\, dt$$

$$= \frac{1}{\sqrt{a}}\int_1^\infty \int_1^\infty f(t)\, \overline{\psi\left(\frac{z}{a}\right)} \left\{\int_0^\infty \omega \tanh(\pi\omega)\, P_{i\omega-1/2}(t)\, P_{i\omega-1/2}(b)\, P_{i\omega-1/2}(z)\, d\omega\right\} dz\, dt. \tag{5.172}$$

Upon switching the order of integration in (5.172), we get

$$\mathcal{MW}_\psi\big[f\big](a,b) = \int_0^\infty \omega \tanh(\pi\omega)\, P_{i\omega-1/2}(b) \left\{\int_1^\infty f(t)\, P_{i\omega-1/2}(t)\, dt\right\}$$

$$\times \left\{\frac{1}{\sqrt{a}}\int_1^\infty \overline{\psi\left(\frac{z}{a}\right)}\, P_{i\omega-1/2}(z)\, dz\right\} d\omega$$

$$= \int_0^\infty \tanh(\pi\omega)\, P_{i\omega-1/2}(b)\, \mathscr{M}\Big[f\Big](\omega)\, \mathscr{M}\Big[\overline{\mathcal{D}_a\psi}\Big](\omega)\, d\omega$$

$$= \mathscr{M}^{-1}\Big(\mathscr{M}\Big[f\Big](\omega)\, \mathscr{M}\Big[\overline{\mathcal{D}_a\psi}\Big](\omega)\Big)(b). \tag{5.173}$$

Applying the inverse Mehler-Fock transform (5.153), we obtain the desired result as

$$\mathscr{M}\Big(M\mathscr{W}_\psi\Big[f\Big](a,b)\Big)(\omega) = \mathscr{M}\Big[f\Big](\omega)\, \mathscr{M}\Big[\overline{\mathcal{D}_a\psi}\Big](\omega).$$

This completes the proof of Lemma 5.6.1. □

Remark 5.6.1. The expression (5.170) could also be obtained directly by using the convolution form of the Mehler-Fock wavelet transform (5.168) together with (5.165) as

$$\mathscr{M}\Big(M\mathscr{W}_\psi\Big[f\Big](a,b)\Big)(\omega) = \mathscr{M}\Big[(f\circledast_\mathscr{M}\overline{\mathcal{D}_a\psi})(b)\Big](\omega)$$

$$= \mathscr{M}\Big[f\Big](\omega)\, \mathscr{M}\Big[\overline{\mathcal{D}_a\psi}\Big](\omega). \tag{5.174}$$

Following is the Moyal's principle for the continuous Mehler-Fock wavelet transform defined in (5.167):

Theorem 5.6.2. *If* $M\mathscr{W}_\psi[f](a,b)$ *and* $M\mathscr{W}_\phi[g](a,b)$ *are the respective Mehler-Fock wavelet transforms of a given pair of functions* $f, g \in L^2[1,\infty)$, *then the following orthogonality relation holds:*

$$\int_1^\infty \int_0^\infty M\mathscr{W}_\psi\Big[f\Big](a,b)\, \overline{M\mathscr{W}_\phi\Big[g\Big](a,b)}\, \frac{da\, db}{a^2} = C_{\psi,\phi} \int_1^\infty f(t)\, \overline{g(t)}\, dt, \tag{5.175}$$

where

$$C_{\psi,\phi} = \int_0^\infty \mathscr{M}\Big[\overline{\mathcal{D}_a\psi}\Big](\omega)\, \overline{\mathscr{M}\Big[\overline{\mathcal{D}_a\phi}\Big](\omega)}\, \frac{da}{a^2} < \infty, \quad a.e. \quad \omega > 0. \tag{5.176}$$

Proof. Invoking the relationship between the Mehler-Fock transform and the Mehler-Fock wavelet transform obtained in (5.170), we get

$$\int_1^\infty \int_0^\infty M\mathscr{W}_\psi\Big[f\Big](a,b)\, \overline{M\mathscr{W}_\phi\Big[g\Big](a,b)}\, \frac{da\, db}{a^2}$$

$$= \int_1^\infty \int_0^\infty \mathscr{M}^{-1}\Big(\mathscr{M}\Big[f\Big](\omega)\, \mathscr{M}\Big[\overline{\mathcal{D}_a\psi}\Big](\omega)\Big)(b)\, \overline{\mathscr{M}^{-1}\Big(\mathscr{M}\Big[g\Big](\omega)\, \mathscr{M}\Big[\overline{\mathcal{D}_a\phi}\Big](\omega)\Big)(b)}\, \frac{da\, db}{a^2}$$

$$= \int_1^\infty \int_0^\infty \left\{\int_0^\infty \omega \tanh(\pi\omega)\, P_{i\omega-1/2}(b)\, \mathscr{M}\Big[f\Big](\omega)\, \mathscr{M}\Big[\overline{\mathcal{D}_a\psi}\Big](\omega)\, d\omega\right\}$$

$$\times \left\{\int_0^\infty \overline{\omega \tanh(\pi\omega)\, P_{i\omega-1/2}(b)\, \mathscr{M}\Big[g\Big](\omega)\, \mathscr{M}\Big[\overline{\mathcal{D}_a\phi}\Big](\omega)}\, d\omega\right\}\, \frac{da\, db}{a^2}. \tag{5.177}$$

Using (5.159), we can express (5.177) alternatively as

$$\int_1^\infty \int_0^\infty M\mathscr{W}_\psi\Big[f\Big](a,b)\, \overline{M\mathscr{W}_\phi\Big[g\Big](a,b)}\, \frac{da\, db}{a^2}$$

$$= \int_1^\infty \int_0^\infty \left\{\mathscr{M}^*\Big(\omega \tanh(\pi\omega)\, \mathscr{M}\Big[f\Big](\omega)\, \mathscr{M}\Big[\overline{\mathcal{D}_a\psi}\Big](\omega)\Big)(b)\right\}$$

$$\times \left\{\overline{\mathscr{M}^*\Big(\omega \tanh(\pi\omega)\, \mathscr{M}\Big[g\Big](\omega)\, \mathscr{M}\Big[\overline{\mathcal{D}_a\phi}\Big](\omega)\Big)(b)}\right\}\, \frac{da\, db}{a^2}. \tag{5.178}$$

By virtue of the Parseval's formula for the adjoint Mehler-Fock transform (5.161), we can rewrite (5.177) as

$$
\int_1^\infty \int_0^\infty \mathcal{MW}_\psi\big[f\big](a,b)\overline{\mathcal{MW}_\phi\big[g\big](a,b)}\frac{da\,db}{a^2}
$$

$$
= \int_0^\infty \int_0^\infty \omega \tanh(\pi\omega)\Big(\mathcal{M}\big[f\big](\omega)\,\mathcal{M}\big[\overline{\mathcal{D}_a\psi}\big](\omega)\Big)\Big(\overline{\mathcal{M}\big[g\big](\omega)\,\mathcal{M}\big[\overline{\mathcal{D}_a\phi}\big](\omega)}\Big)\frac{da\,d\omega}{a^2}
$$

$$
= \int_0^\infty \omega \tanh(\pi\omega)\,\mathcal{M}\big[f\big](\omega)\overline{\mathcal{M}\big[g\big](\omega)}\left\{\int_0^\infty \mathcal{M}\big[\overline{\mathcal{D}_a\psi}\big](\omega)\overline{\mathcal{M}\big[\overline{\mathcal{D}_a\phi}\big](\omega)}\frac{da}{a^2}\right\}d\omega
$$

$$
= C_{\psi,\phi}\int_0^\infty \omega \tanh(\pi\omega)\,\mathcal{M}\big[f\big](\omega)\overline{\mathcal{M}\big[g\big](\omega)}\,d\omega. \tag{5.179}
$$

Finally, using the Parseval's formula for the Mehler-Fock transform (5.154), we obtain

$$
\int_1^\infty \int_0^\infty \mathcal{MW}_\psi\big[f\big](a,b)\overline{\mathcal{MW}_\phi\big[g\big](a,b)}\frac{da\,db}{a^2} = C_{\psi,\phi}\int_1^\infty f(t)\,\overline{g(t)}\,dt.
$$

This completes the proof of Theorem 5.6.2.   □

Remark 5.6.2. For $\psi = \phi$, Theorem 5.6.2 yields the following orthogonality relation:

$$
\int_1^\infty \int_0^\infty \mathcal{MW}_\psi\big[f\big](a,b)\,\overline{\mathcal{MW}_\psi\big[g\big](a,b)}\frac{da\,db}{a^2} = C_\psi\int_1^\infty f(t)\,\overline{g(t)}\,dt, \tag{5.180}
$$

where

$$
C_\psi = \int_0^\infty \Big|\mathcal{M}\big[\overline{\mathcal{D}_a\psi}\big](\omega)\Big|^2\frac{da}{a^2} < \infty, \quad a.e. \quad \omega > 0. \tag{5.181}
$$

Moreover, in case $f = g$ and $\psi = \phi$, Theorem 5.6.2 yields the energy preserving relation for the Mehler-Fock wavelet transform:

$$
\int_1^\infty \int_0^\infty \Big|\mathcal{MW}_\psi\big[f\big](a,b)\Big|^2\frac{da\,db}{a^2} = C_\psi\big\|f\big\|_{L^2[1,\infty)}. \tag{5.182}
$$

We conclude this subsection with the formulation of a reconstruction formula corresponding to the Mehler-Fock wavelet transform defined in (5.167).

Theorem 5.6.3. *If $\mathcal{MW}_\psi\big[f\big](a,b)$ is the Mehler-Fock wavelet transform of $f \in L^2[1,\infty)$ with respect to the Mehler-Fock wavelet ψ, then f can be recovered via another wavelet ϕ as*

$$
f(t) = \frac{1}{C_{\psi,\phi}}\int_1^\infty \int_1^\infty \int_0^\infty \mathcal{MW}_\psi\big[f\big](a,b)\,\mathcal{D}_a\phi(z)\Delta(t,b,z)\frac{da\,dz\,db}{a^2}, \quad a.e., \tag{5.183}
$$

where $C_{\psi,\phi} > 0$ is given by (5.176).

Proof. Using the orthogonality relation (5.175), the proof follows immediately.   □

Remark 5.6.3. By using the integral representation of the kernel $\Delta(t,b,z)$ given in (5.158), we can express the inversion formula (5.183) in terms of the cone functions $P_{i\omega-1/2}(\cdot)$ as

$$
f(t) = \frac{1}{C_{\psi,\phi}}\int_1^\infty \int_1^\infty \int_0^\infty \int_0^\infty \mathcal{MW}_\psi\big[f\big](a,b)\,\mathcal{D}_a\phi(z)\,\omega \tanh(\pi\omega)
$$

$$
\times P_{i\omega-1/2}(t)\,P_{i\omega-1/2}(b)\,P_{i\omega-1/2}(z)\frac{da\,d\omega\,dz\,db}{a^2}. \tag{5.184}
$$

5.7 The Interface of Wavelet and Hartley Transforms

"The number of arguments is unimportant unless some of them are correct."

-Ralph L. Hartley

It is a fact that the wavelet transform has achieved humongous success for offering simultaneous information about the temporal and spatial characteristics of a non-stationary signal. For instance, neither a piece of music nor the changing colour of a patch of sky over the course of sunset are best described by a function of time or frequency alone. In fact, the essence of such events lies in the often subtle dependence of their spectral properties on time. Despite being called as a "transform," the wavelet transform is not limited for providing spectral analysis, although appropriate choice of the wavelet can cause it to function that way. The implementation of the wavelet transform can be computationally expensive, involving repeated convolutions of the signal to be analyzed with an appropriately scaled version of the wavelet. Such cumbersome approaches could be efficiently tackled by the Hartley-based wavelet transform algorithms which offer greater efficiency and simplicity than the conventional Fourier-based techniques. In this chapter, we shall take a tour of the wavelet transform from the perspective of the Hartley transform, proposed as an alternative to the Fourier transform by Ralph L. Hartley $(1888 - 1970)$ in 1942 [50, 300]. The prime feature of such a wavelet transform lies in its computational supremacy and the fact that the transform is purely real and is, therefore, easier to display, understand and process.

5.7.1 Hartley transform

Although, the Hartley transform dates back to 1940s, however, only recently it has gained widespread acceptance. The most promising feature of the Hartley transform is the characteristic of producing a real output from a real input. Nevertheless, the availability of fast algorithms in the realm of Hartley transform makes it competitive with the well-known fast Fourier transform. Another pivotal feature of the Hartley transform is that it makes no distinction between the forward and backward transforms. Therefore, the Hartley transform is quite befitting for diverse aspects in signal and image processing and serves as a wonderful tool to rely upon. Prior to recalling the formal definition of the Hartley transform, we note that $L^2(\mathbb{R}, \mathbb{R})$ denotes the space of real-valued, square integrable functions defined on $\mathbb{R}$.

Definition 5.7.1. For any $f \in L^2(\mathbb{R}, \mathbb{R})$, the Hartley transform is denoted by $\mathscr{H}[f]$ and is defined as

$$\mathscr{H}[f](\omega) = \frac{1}{\sqrt{2\pi}} \int_{\mathbb{R}} f(t) \operatorname{cas}(\omega t) \, dt, \qquad (5.185)$$

where $\operatorname{cas}\theta = \cos\theta + \sin\theta$.

It is pertinent to mention that the fast Hartley transform (FHT) algorithm requires exactly half as many adds, multiplies, fetches, stores as an identically programmed complex fast Fourier transform (FFT) algorithm. Real FFT algorithms can at best match the FHT algorithm for speed, but at the cost of requiring two separate algorithms for the forward and backward transforms.

Next, our aim is to examine the Hartley transform of the convolution of two functions. Prior to that, we note that the even and odd parts of the Hartley trasform (5.185) are denoted by $\left\{\mathcal{H}\left[f\right](\omega)\right\}_e$ and $\left\{\mathcal{H}\left[f\right](\omega)\right\}_o$, respectively, and are defined as

$$\left\{\mathcal{H}\left[f\right](\omega)\right\}_e = \frac{\mathcal{H}\left[f\right](\omega) + \mathcal{H}\left[f\right](-\omega)}{2} = \int_{\mathbb{R}} f(t)\cos(\omega t)\, dt \qquad (5.186)$$

and

$$\left\{\mathcal{H}\left[f\right](\omega)\right\}_o = \frac{\mathcal{H}\left[f\right](\omega) - \mathcal{H}\left[f\right](-\omega)}{2} = \int_{\mathbb{R}} f(t)\sin(\omega t)\, dt. \qquad (5.187)$$

In lieu to the nature of convolution in the classical Fourier domain, the Hartley transform of a convolved input satisfies:

$$\mathcal{H}\left[(f * g)\right](\omega) = \mathcal{H}\left[f\right](\omega)\left\{\mathcal{H}\left[g\right](\omega)\right\}_e + \mathcal{H}\left[f\right](-\omega)\left\{\mathcal{H}\left[g\right](\omega)\right\}_o. \qquad (5.188)$$

The relationship (5.188) indicates that the convolution in the spatial domain leads to a chopping of the Hartley spectra into even and odd parts.

In the following theorem, we assemble some useful properties of the Hartley trasform defined in (5.185).

Theorem 5.7.1. *For the scalars* $k, \omega_0 \in \mathbb{R}$, $\lambda \in \mathbb{R} \setminus \{0\}$ *and any function* $f \in L^2(\mathbb{R}, \mathbb{R})$, *the Hartley transform (5.185) satisfies the following properties:*

(i) *Translation:* $\mathcal{H}\left[f(t - k)\right](\omega) = \cos(\omega k)\,\mathcal{H}\left[f\right](\omega) + \sin(\omega k)\,\mathcal{H}\left[f\right](-\omega)$,

(ii) *Scaling:* $\mathcal{H}\left[f(\lambda t)\right](\omega) = \dfrac{1}{|\lambda|}\,\mathcal{H}\left[f(t)\right]\left(\dfrac{\omega}{\lambda}\right)$,

(iii) *Parity:* $\mathcal{H}\left[f(-t)\right](\omega) = \mathcal{H}\left[f\right](-\omega)$,

(iv) *Modulation:* $\mathcal{H}\left[\cos(\omega_0 t)f(t)\right](\omega) = \dfrac{1}{2}\left\{\mathcal{H}\left[f\right](\omega - \omega_0) + \mathcal{H}\left[f\right](\omega + \omega_0)\right\}$.

Proof. We shall only prove (i) and (iv), the rest of the properties being quite straightforward.

(i) For any $k \in \mathbb{R}$, we have

$$\begin{aligned}
\mathcal{H}\left[f(t - k)\right](\omega) &= \frac{1}{\sqrt{2\pi}}\int_{\mathbb{R}} f(t - k)\,\mathrm{cas}(\omega t)\, dt \\
&= \frac{1}{\sqrt{2\pi}}\int_{\mathbb{R}} f(z)\Big(\cos(\omega k + \omega z) + \sin(\omega k + \omega z)\Big)\, dt \\
&= \frac{1}{\sqrt{2\pi}}\int_{\mathbb{R}} f(z)\Big(\cos(\omega k)\cos(\omega z) - \sin(\omega k)\sin(\omega z) \\
&\qquad\qquad + \sin(\omega k)\cos(\omega z) + \cos(\omega k)\sin(\omega z)\Big)\, dz \\
&= \frac{1}{\sqrt{2\pi}}\int_{\mathbb{R}} f(z)\cos(\omega k)\big(\cos(\omega z) + \sin(\omega z)\big)\, dz \\
&\qquad\qquad + \int_{\mathbb{R}} f(z)\sin(\omega k)\big(\cos(\omega z) - \sin(\omega z)\big)\, dz \\
&= \cos(\omega k)\,\mathcal{H}\left[f\right](\omega) + \sin(\omega k)\,\mathcal{H}\left[f\right](-\omega).
\end{aligned}$$

(iv) For any $\omega_0 \in \mathbb{R}$, we have

$$\mathscr{H}\Big[\cos(\omega_0 t)\, f(t)\Big](\omega) = \frac{1}{\sqrt{2\pi}} \int_{\mathbb{R}} \cos(\omega_0 t)\, f(t)\, \mathrm{cas}(\omega t)\, dt$$

$$= \frac{1}{\sqrt{2\pi}} \int_{\mathbb{R}} f(t)\Big(\cos(\omega_0 t)\, \cos(\omega t) + \cos(\omega_0 t)\, \sin(\omega t)\Big)\, dt$$

$$= \frac{1}{2}\Big\{\mathscr{H}\big[f\big](\omega - \omega_0) + \mathscr{H}\big[f\big](\omega + \omega_0)\Big\}.$$

This completes the proof of Theorem 5.7.1.   □

In order to have an intuition of the nuances, we shall obtain a relationship between the Fourier and Hartley transforms. Given any $f \in L^2(\mathbb{R}, \mathbb{R})$, the classical Fourier transform (1.7) can be expressed as

$$\mathscr{F}\big[f\big](\omega) = \frac{1}{\sqrt{2\pi}} \int_{\mathbb{R}} f(t)\, e^{-i\omega t}\, dt$$

$$= \frac{1}{\sqrt{2\pi}} \left\{ \int_{\mathbb{R}} f(t)\, \cos(\omega t)\, dt - i \int_{\mathbb{R}} f(t)\, \sin(\omega t)\, dt \right\}$$

$$= \Big\{\mathscr{H}\big[f\big](\omega)\Big\}_e - i\Big\{\mathscr{H}\big[f\big](\omega)\Big\}_o. \tag{5.189}$$

From relation (5.189), we infer that the real and imaginary parts of the Fourier transform are related to even and odd parts of the Hartley transform as

$$\mathrm{Real}\Big\{\mathscr{F}\big[f\big](\omega)\Big\} = \Big\{\mathscr{H}\big[f\big](\omega)\Big\}_e, \quad \mathrm{Imaginary}\Big\{\mathscr{F}\big[f\big](\omega)\Big\} = -\Big\{\mathscr{H}\big[f\big](\omega)\Big\}_o. \tag{5.190}$$

Equivalently, the Hartley transform can be obtained as the difference of the real and imaginary parts of the Fourier transform; that is,

$$\mathscr{H}\big[f\big](\omega) = \mathrm{Real}\Big\{\mathscr{F}\big[f\big](\omega)\Big\} - \mathrm{Imaginary}\Big\{\mathscr{F}\big[f\big](\omega)\Big\}. \tag{5.191}$$

Example 5.7.1. Consider the unilateral exponential function

$$f(t) = \begin{cases} e^{-\lambda t}, & t \geq 0 \\ 0, & \text{otherwise.} \end{cases} \tag{5.192}$$

The Fourier transform corresponding to (5.192) is obtained in (1.27) and reads:

$$\mathscr{F}\big[f\big](\omega) = \frac{1}{\sqrt{2\pi}} \left[\frac{\lambda}{\lambda^2 + \omega^2} + \frac{-i\omega}{\lambda^2 + \omega^2} \right]. \tag{5.193}$$

Next, we shall compute the Hartley transform (5.185) corresponding to (5.192) as

$$\mathscr{H}\big[f\big](\omega) = \frac{1}{\sqrt{2\pi}} \int_{\mathbb{R}} e^{-\lambda t}\, \mathrm{cas}(\omega t)\, dt$$

$$= \frac{1}{\sqrt{2\pi}} \left\{ \int_{\mathbb{R}} e^{-\lambda t} \cos(\omega t)\, dt + \int_{\mathbb{R}} e^{-\lambda t} \sin(\omega t)\, dt \right\}$$

$$= \frac{1}{\sqrt{2\pi}} \left\{ \int_{\mathbb{R}} e^{-\lambda t} \left(\frac{e^{i\omega t} + e^{-i\omega t}}{2} \right) dt + \int_{\mathbb{R}} e^{-\lambda t} \left(\frac{e^{i\omega t} - e^{-i\omega t}}{2i} \right) dt \right\}$$

$$= \frac{1}{\sqrt{2\pi}} \left\{ \frac{1}{2} \int_0^\infty e^{(i\omega - \lambda)t}\, dt + \frac{1}{2} \int_0^\infty e^{(-i\omega - \lambda)t}\, dt \right\}$$

$$+ \frac{1}{2i} \int_0^\infty e^{(i\omega - \lambda)t}\, dt - \frac{1}{2i} \int_0^\infty e^{(-i\omega - \lambda)t}\, dt \Bigg\}$$

$$= \frac{1}{\sqrt{2\pi}} \left\{ \frac{1}{2}\left(\frac{1}{\lambda - i\omega} + \frac{1}{\lambda + i\omega} \right) + \frac{1}{2i}\left(\frac{1}{\lambda - i\omega} + \frac{1}{\lambda + i\omega} \right) \right\}$$

$$= \frac{1}{\sqrt{2\pi}} \left\{ \frac{\lambda}{\lambda^2 + \omega^2} + \frac{\omega}{\lambda^2 + \omega^2} \right\}$$

$$= \text{Real}\Big\{ \mathscr{F}\big[f\big](\omega) \Big\} - \text{Imaginary}\Big\{ \mathscr{F}\big[f\big](\omega) \Big\}, \tag{5.194}$$

which validates (5.191).

5.7.2 Hartley-based wavelet transform

In this subsection, we shall demonstrate that the Hartley transform can be effectively employed as an alternate approach for obtaining the wavelet transform. The strategy is to make use of the fact that the Hartley transform makes no distinction between a forward and inverse transformation. Here, it is pertinent to mention that the translation and scaling properties of the Hartley transform obtained in Theorem 5.7.1 play a key role in expressing the wavelet transform via the Hartley transform.

Theorem 5.7.2. *For $a \in \mathbb{R}^+$, $b \in \mathbb{R}$, let $\mathscr{W}_\psi[f](a,b)$ be the wavelet transform of an any $f \in L^2(\mathbb{R}, \mathbb{R})$ with respect to the wavelet $\psi \in L^2(\mathbb{R}, \mathbb{R})$. Then, $\mathscr{W}_\psi[f](a,b)$ can be expressed via the Hartley transform (5.185) as follows:*

$$\mathscr{W}_\psi\big[f\big](a,b) = \sqrt{a} \int_\mathbb{R} \mathscr{H}\big[f\big](\omega)\Big\{ \cos(b\omega)\mathscr{H}\big[\psi\big](a\omega) + \sin(b\omega)\mathscr{H}\big[\psi\big](-a\omega) \Big\}\, d\omega. \tag{5.195}$$

Proof. Using the formal definition of wavelet transform given in (3.47), we have

$$\mathscr{W}_\psi\big[f\big](a,b) = \frac{1}{\sqrt{a}} \int_\mathbb{R} f(t)\, \psi\left(\frac{t-b}{a} \right) dt$$

$$= \frac{1}{\sqrt{a}} \int_\mathbb{R} \left\{ \int_\mathbb{R} \mathscr{H}\big[f\big](\omega)\, \text{cas}(\omega t)\, d\omega \right\} \left\{ \int_\mathbb{R} \mathscr{H}\left[\psi\left(\frac{t-b}{a} \right) \right](\eta)\, \text{cas}(\eta t)\, d\eta \right\} dt. \tag{5.196}$$

Now,

$$\mathscr{H}\left[\psi\left(\frac{t-b}{a} \right) \right](\eta)\, d\eta = \frac{1}{\sqrt{2\pi}} \int_\mathbb{R} \psi\left(\frac{t-b}{a} \right) \text{cas}(\eta t)\, dt$$

$$= \frac{a}{\sqrt{2\pi}} \int_\mathbb{R} \psi(z)\, \text{cas}\left(a\eta z + \eta b \right) dz$$

$$= a\Big\{ \cos(b\eta)\, \mathscr{H}\big[\psi\big](a\eta) + \sin(b\eta)\, \mathscr{H}\big[\psi\big](-a\eta) \Big\}. \tag{5.197}$$

Plugging (5.197) in (5.196), we get

$$\mathscr{W}_\psi\big[f\big](a,b) = \frac{1}{\sqrt{a}} \int_\mathbb{R} f(t)\, \psi\left(\frac{t-b}{a} \right) dt$$

$$= \sqrt{a} \int_\mathbb{R} \left\{ \int_\mathbb{R} \mathscr{H}\big[f\big](\omega)\, \text{cas}(\omega t)\, d\omega \right\}$$

$$\times \left\{ \int_\mathbb{R} \Big(\cos(b\eta)\, \mathscr{H}\big[\psi\big](a\eta) + \sin(b\eta)\, \mathscr{H}\big[\psi\big](-a\eta) \Big) \text{cas}(\eta t)\, d\eta \right\} dt. \tag{5.198}$$

Switching the order of integration and grouping together all the terms containing the variable t, we obtain

$$\mathscr{W}_\psi\Big[f\Big](a,b) = \sqrt{a} \int_{\mathbb{R}} \int_{\mathbb{R}} \mathscr{H}\Big[f\Big](\omega) \Big\{ \Big(\cos(b\eta)\, \mathscr{H}\Big[\psi\Big](a\eta) + \sin(b\eta)\, \mathscr{H}\Big[\psi\Big](-a\eta) \Big) \Big\}$$
$$\times \Big\{ \int_{\mathbb{R}} \mathrm{cas}(\omega t)\, \mathrm{cas}(\eta t)\, dt \Big\}\, d\omega\, d\eta$$
$$= \sqrt{a} \int_{\mathbb{R}} \int_{\mathbb{R}} \mathscr{H}\Big[f\Big](\omega) \Big\{ \Big(\cos(b\eta)\, \mathscr{H}\Big[\psi\Big](a\eta) + \sin(b\eta)\, \mathscr{H}\Big[\psi\Big](-a\eta) \Big) \Big\}$$
$$\times \delta(\omega - \eta)\, d\omega\, d\eta. \qquad (5.199)$$

Performing the integration of (5.199) over the variable η, we get

$$\mathscr{W}_\psi\Big[f\Big](a,b) = \sqrt{a} \int_{\mathbb{R}} \mathscr{H}\Big[f\Big](\omega) \Big\{ \cos(b\omega)\mathscr{H}\Big[\psi\Big](a\omega) + \sin(b\omega)\mathscr{H}\Big[\psi\Big](-a\omega) \Big\}\, d\omega.$$

This completes the proof of Theorem 5.7.2. $\qquad\qquad\qquad\square$

Next, our aim is to establish the Moyal's principle and inversion formula associated with the wavelet transform given by (5.195). Prior to that, we have the following admissibility condition:

Definition 5.7.2. A function $\psi \in L^2(\mathbb{R}, \mathbb{R})$ is said to be "admissible" if

$$C_\psi = \int_{\mathbb{R}+} \left[\left(\Big\{ \mathscr{H}\Big[\psi\Big](\omega) \Big\}_e \right)^2 + \left(\Big\{ \mathscr{H}\Big[\psi\Big](\omega) \Big\}_o \right)^2 \right] \frac{d\omega}{\omega} < \infty. \qquad (5.200)$$

Following is the Moyal's principle associated with the wavelet transform given by the expression (5.195):

Theorem 5.7.3. *If $\mathscr{W}_\psi[f](a,b)$ and $\mathscr{W}_\psi[g](a,b)$ are the wavelet transforms of any pair of functions $f, g \in L^2(\mathbb{R}, \mathbb{R})$ as defined in (5.195), then we have*

$$\int_{\mathbb{R}} \int_{\mathbb{R}+} \mathscr{W}_\psi\Big[f\Big](a,b)\, \mathscr{W}_\psi\Big[g\Big](a,b)\, \frac{da\, db}{a^2} = C_\psi \int_{\mathbb{R}} f(t)\, g(t)\, dt, \qquad (5.201)$$

where C_ψ is given by (5.200).

Proof. For the given pair of square integrable functions f and g, we have

$$\int_{\mathbb{R}} \int_{\mathbb{R}+} \mathscr{W}_\psi\Big[f\Big](a,b)\, \mathscr{W}_\psi\Big[g\Big](a,b)\, \frac{da\, db}{a^2}$$
$$= \int_{\mathbb{R}} \int_{\mathbb{R}+} \Big\{ \int_{\mathbb{R}} \mathscr{H}\Big[f\Big](\omega) \Big[\cos(b\omega)\mathscr{H}\Big[\psi\Big](a\omega) + \sin(b\omega)\mathscr{H}\Big[\psi\Big](-a\omega) \Big]\, d\omega \Big\}$$
$$\Big\{ \int_{\mathbb{R}} \mathscr{H}\Big[g\Big](\eta) \Big[\cos(b\eta)\mathscr{H}\Big[\psi\Big](a\eta) + \sin(b\eta)\mathscr{H}\Big[\psi\Big](-a\eta) \Big]\, d\eta \Big\}\, \frac{da\, db}{a}. \qquad (5.202)$$

Upon decomposing the Hartley transform of the wavelet ψ, appearing in (5.202), into even and odd parts, we obtain

$$\cos(b\omega)\mathscr{H}\Big[\psi\Big](a\omega) + \sin(b\omega)\mathscr{H}\Big[\psi\Big](-a\omega)$$
$$= \cos(b\omega)\Big[\Big\{ \mathscr{H}\Big[\psi\Big](a\omega) \Big\}_e + \Big\{ \mathscr{H}\Big[\psi\Big](a\omega) \Big\}_o \Big]$$

$$+ \sin(b\omega)\left[\left\{\mathscr{H}\left[\psi\right](-a\omega)\right\}_e + \left\{\mathscr{H}\left[\psi\right](-a\omega)\right\}_o\right]$$

$$= \cos(b\omega)\left[\left\{\mathscr{H}\left[\psi\right](a\omega)\right\}_e + \left\{\mathscr{H}\left[\psi\right](a\omega)\right\}_o\right]$$

$$+ \sin(b\omega)\left[\left\{\mathscr{H}\left[\psi\right](a\omega)\right\}_e - \left\{\mathscr{H}\left[\psi\right](a\omega)\right\}_o\right]$$

$$= \mathrm{cas}(b\omega)\left\{\mathscr{H}\left[\psi\right](a\omega)\right\}_e + \mathrm{cas}(-b\omega)\left\{\mathscr{H}\left[\psi\right](a\omega)\right\}_o. \tag{5.203}$$

Analogous to (5.203), we obtain

$$\cos(b\eta)\mathscr{H}\left[\psi\right](a\eta) + \sin(b\eta)\mathscr{H}\left[\psi\right](-a\eta)$$

$$= \mathrm{cas}(b\eta)\left\{\mathscr{H}\left[\psi\right](a\eta)\right\}_e + \mathrm{cas}(-b\eta)\left\{\mathscr{H}\left[\psi\right](a\eta)\right\}_o. \tag{5.204}$$

Implementing (5.203) and (5.204) in (5.202), we get

$$\int_{\mathbb{R}}\int_{\mathbb{R}^+} \mathscr{W}_\psi\left[f\right](a,b)\,\mathscr{W}_\psi\left[g\right](a,b)\,\frac{da\,db}{a^2}$$

$$= \int_{\mathbb{R}}\int_{\mathbb{R}^+}\int_{\mathbb{R}}\int_{\mathbb{R}} \mathscr{H}\left[f\right](\omega)\,\mathscr{H}\left[g\right](\omega)\left[\mathrm{cas}(b\omega)\,\mathrm{cas}(b\eta)\left\{\mathscr{H}\left[\psi\right](a\omega)\right\}_e\left\{\mathscr{H}\left[\psi\right](a\eta)\right\}_e\right.$$

$$+ \mathrm{cas}(b\omega)\,\mathrm{cas}(-b\eta)\left\{\mathscr{H}\left[\psi\right](a\omega)\right\}_e\left\{\mathscr{H}\left[\psi\right](a\eta)\right\}_o$$

$$+ \mathrm{cas}(-b\omega)\,\mathrm{cas}(b\eta)\left\{\mathscr{H}\left[\psi\right](a\omega)\right\}_o\left\{\mathscr{H}\left[\psi\right](a\eta)\right\}_e$$

$$\left. + \mathrm{cas}(-b\omega)\,\mathrm{cas}(-b\eta)\left\{\mathscr{H}\left[\psi\right](a\omega)\right\}_o\left\{\mathscr{H}\left[\psi\right](a\eta)\right\}_o\right]\frac{d\eta\,d\omega\,da\,db}{a}. \tag{5.205}$$

As a consequence of (5.205), we obtain

$$\int_{\mathbb{R}}\int_{\mathbb{R}^+} \mathscr{W}_\psi\left[f\right](a,b)\,\mathscr{W}_\psi\left[g\right](a,b)\,\frac{da\,db}{a^2}$$

$$= \int_{\mathbb{R}}\int_{\mathbb{R}^+}\int_{\mathbb{R}} \mathscr{H}\left[f\right](\omega)\,\mathscr{H}\left[g\right](\omega)\left\{\mathscr{H}\left[\psi\right](a\omega)\right\}_e\left\{\mathscr{H}\left[\psi\right](a\eta)\right\}_e$$

$$\times \left\{\int_{\mathbb{R}} \mathrm{cas}(b\omega)\,\mathrm{cas}(b\eta)\,db\right\}\frac{d\eta\,d\omega\,da}{a}$$

$$+ \int_{\mathbb{R}}\int_{\mathbb{R}^+}\int_{\mathbb{R}} \mathscr{H}\left[f\right](\omega)\,\mathscr{H}\left[g\right](\omega)\left\{\mathscr{H}\left[\psi\right](a\omega)\right\}_e\left\{\mathscr{H}\left[\psi\right](a\eta)\right\}_o$$

$$\times \left\{\int_{\mathbb{R}} \mathrm{cas}(b\omega)\,\mathrm{cas}(-b\eta)\,db\right\}\frac{d\eta\,d\omega\,da}{a}$$

$$+ \int_{\mathbb{R}}\int_{\mathbb{R}^+}\int_{\mathbb{R}} \mathscr{H}\left[f\right](\omega)\,\mathscr{H}\left[g\right](\omega)\left\{\mathscr{H}\left[\psi\right](a\omega)\right\}_o\left\{\mathscr{H}\left[\psi\right](a\eta)\right\}_e$$

$$\times \left\{\int_{\mathbb{R}} \mathrm{cas}(-b\omega)\,\mathrm{cas}(b\eta)\,db\right\}\frac{d\eta\,d\omega\,da}{a}$$

$$+ \int_{\mathbb{R}}\int_{\mathbb{R}^+}\int_{\mathbb{R}} \mathscr{H}\left[f\right](\omega)\,\mathscr{H}\left[g\right](\omega)\left\{\mathscr{H}\left[\psi\right](a\omega)\right\}_o\left\{\mathscr{H}\left[\psi\right](a\eta)\right\}_o$$

$$\times \left\{\int_{\mathbb{R}} \mathrm{cas}(-b\omega)\,\mathrm{cas}(-b\eta)\,db\right\}\frac{d\eta\,d\omega\,da}{a}. \tag{5.206}$$

Upon switching the order of integration, the expression (5.206) boils down to

$$\int_{\mathbb{R}} \int_{\mathbb{R}+} \mathscr{W}_{\psi}\big[f\big](a,b)\, \mathscr{W}_{\psi}\big[g\big](a,b)\, \frac{da\,db}{a^2}$$

$$= \int_{\mathbb{R}} \int_{\mathbb{R}+} \mathscr{H}\big[f\big](\omega)\, \mathscr{H}\big[g\big](\omega) \left(\left\{ \mathscr{H}\big[\psi\big](a\omega) \right\}_e \right)^2 \frac{da\,d\omega}{a}$$

$$+ \int_{\mathbb{R}} \int_{\mathbb{R}+} \mathscr{H}\big[f\big](\omega)\, \mathscr{H}\big[g\big](\omega) \left(\left\{ \mathscr{H}\big[\psi\big](a\omega) \right\}_o \right)^2 \frac{da\,d\omega}{a}$$

$$= \int_{\mathbb{R}} \int_{\mathbb{R}+} \mathscr{H}\big[f\big](\omega)\, \mathscr{H}\big[g\big](\omega) \left[\left(\left\{ \mathscr{H}\big[\psi\big](a\omega) \right\}_e \right)^2 + \left(\left\{ \mathscr{H}\big[\psi\big](a\omega) \right\}_o \right)^2 \right] \frac{da\,d\omega}{a}$$

$$= \int_{\mathbb{R}} \mathscr{H}\big[f\big](\omega)\, \mathscr{H}\big[g\big](\omega) \left\{ \int_{\mathbb{R}+} \left[\left(\left\{ \mathscr{H}\big[\psi\big](a\omega) \right\}_e \right)^2 + \left(\left\{ \mathscr{H}\big[\psi\big](a\omega) \right\}_o \right)^2 \right] \frac{da}{a} \right\} d\omega.$$

$$(5.207)$$

Making use of the substitution $a\omega = \xi$ into the inner integral in (5.207), we get

$$\int_{\mathbb{R}} \int_{\mathbb{R}+} \mathscr{W}_{\psi}\big[f\big](a,b)\, \mathscr{W}_{\psi}\big[g\big](a,b)\, \frac{da\,db}{a^2}$$

$$= \int_{\mathbb{R}} \mathscr{H}\big[f\big](\omega)\, \mathscr{H}\big[g\big](\omega) \left\{ \int_{\mathbb{R}+} \left[\left(\left\{ \mathscr{H}\big[\psi\big](\xi) \right\}_e \right)^2 + \left(\left\{ \mathscr{H}\big[\psi\big](\xi) \right\}_o \right)^2 \right] \frac{d\xi}{\xi} \right\} d\omega$$

$$= C_{\psi} \int_{\mathbb{R}} \mathscr{H}\big[f\big](\omega)\, \mathscr{H}\big[g\big](\omega)\, d\omega$$

$$= C_{\psi} \int_{\mathbb{R}} f(t)\, g(t)\, dt.$$

This completes the proof of Theorem 5.7.3. $\qquad\square$

Remark 5.7.1. For $f = g$, the assertion in Theorem 5.7.3 reduces to a more compact form:

$$\int_{\mathbb{R}} \int_{\mathbb{R}+} \left(\mathscr{W}_{\psi}\big[f\big](a,b) \right)^2 \frac{da\,db}{a^2} = C_{\psi} \int_{\mathbb{R}} \left(f(t) \right)^2 dt. \qquad (5.208)$$

Finally, we formulate the inversion formula associated with the wavelet transform given by the expression (5.195).

Theorem 5.7.4. *Any function $f \in L^2(\mathbb{R}, \mathbb{R})$ can be reconstructed from the wavelet transform $\mathscr{W}_{\psi}\big[f\big](a,b)$ defined in (5.195) as*

$$f(t) = \frac{1}{C_{\psi}} \int_{\mathbb{R}} \int_{\mathbb{R}+} \mathscr{W}_{\psi}\big[f\big](a,b)\, \mathscr{W}_{\psi}\big[\delta(\cdot - t)\big](a,b)\, \frac{da\,db}{a^2}, \quad a.e., \qquad (5.209)$$

where $C_{\psi} > 0$ is given by (5.200).

Proof. Consider the function $g(x) = \delta(t - x)$, $t \in \mathbb{R}$, then we observe that

$$\delta(t - x) = \int_{\mathbb{R}} \mathrm{cas}(\omega t)\, \mathrm{cas}(\omega x)\, d\omega. \qquad (5.210)$$

Using the definition of inverse Hartley transform, we obtain

$$g(x) = \delta(t - x) = \int_{\mathbb{R}} \mathscr{H}\big[g\big](\omega)\, \mathrm{cas}(\omega x) d\omega. \qquad (5.211)$$

By virtue of (5.210) and (5.211), we get $\mathcal{H}\big[g\big](\omega) = \mathrm{cas}(\omega t)$. Therefore, as a consequence of Theorem 5.7.3, we obtain

$$f(t) = \int_{\mathbb{R}} \mathcal{H}\big[f\big](\omega)\,\mathrm{cas}(\omega t)\,d\omega$$

$$= \int_{\mathbb{R}} \mathcal{H}\big[f\big](\omega)\,\mathcal{H}\big[g\big](\omega)\,d\omega$$

$$= \frac{1}{C_\psi} \int_{\mathbb{R}} \int_{\mathbb{R}+} \mathscr{W}_\psi\big[f\big](a,b)\,\mathscr{W}_\psi\big[g\big](a,b)\,\frac{da\,db}{a^2}$$

$$= \frac{1}{C_\psi} \int_{\mathbb{R}} \int_{\mathbb{R}+} \mathscr{W}_\psi\big[f\big](a,b)\,\mathscr{W}_\psi\big[\delta(\cdot - t)\big](a,b)\,\frac{da\,db}{a^2}, \quad a.e.,$$

which is the desired reconstruction formula. This completes the proof of Theorem 5.7.4. □

5.8   Exercises

Exercise 5.8.1. Show that the Laguerre transform of the function $f(t) = t^{\beta-\alpha-1}$, $\beta > 0, \alpha > -1$ is given by

$$\mathscr{L}\big[f\big](n) = \rho(n)\frac{\Gamma(\alpha-\beta+n+1)\,\Gamma(\beta)}{n!\,\Gamma(\alpha-\beta+1)}. \tag{5.212}$$

Exercise 5.8.2. Use (5.22) to show that the Laguerre wavelet transform of $f(t) = t^{\beta-\alpha-1}$, $\beta > 0, \alpha > -1$ with respect to the Haar wavelet (3.31) is given by

$$\mathscr{L}\mathscr{W}_\psi\big[f\big](a,b)$$

$$= \frac{\Gamma(\beta)}{\Gamma(\alpha+1)\,\Gamma(\alpha-\beta+1)} \left\{ \sum_n \frac{\Gamma(\alpha-\beta+n+1)\,\Gamma(\beta)}{n!} \Big(\Psi_1(n,\alpha) - \Psi_2(n,\alpha)\Big) L_n^{(\alpha)}(b) \right\}, \tag{5.213}$$

where

$$\Psi_1(n,\alpha) = \int_0^{a/2} P_n^{(\alpha)}(t)\,d\Lambda(t) \quad \text{and} \quad \Psi_2(n,\alpha) = \int_{a/2}^a P_n^{(\alpha)}(t)\,d\Lambda(t). \tag{5.214}$$

Exercise 5.8.3. Compute the Legendre transform of the function $f(t) = \chi_{[-1,\,1]}(t)$, where χ denotes the usual characteristic function.

Exercise 5.8.4. Compute the Legendre translation $\mathfrak{T}_{1/2}f(t)$ corresponding to the function

$$f(t) = \begin{cases} t, & 0 \le t \le 1 \\ 0, & \text{otherwise.} \end{cases} \tag{5.215}$$

Exercise 5.8.5. Compute the Legendre convolution of the following pairs of functions:

(i) $f(t) = \chi_{[0,\,1]}(t)$ and $g(t) = \chi_{[-1,\,0]}(t)$,

(ii) $f(t) = \begin{cases} t, & 0 \le t \le 1 \\ 0, & \text{otherwise} \end{cases}$ and $g(t) = \begin{cases} t, & -1 \le t \le 0 \\ 0, & \text{otherwise.} \end{cases}$

Exercise 5.8.6. Compute the Legendre wavelet transform of the function $f(t) = t$, with respect to the Haar wavelet (3.31).

Exercise 5.8.7. Discuss the family of Bessel daughter wavelets $\mathbb{D}_a \mathsf{T}_b \psi(t)$, where ψ is the Haar wavelet defined in (3.31).

Exercise 5.8.8. Compute the Bessel wavelet transform of the function $f(t) = t^{-3} \sin(\omega_0 t)$, $\omega_0 > 0$, with respect to the Haar wavelet (3.31).

Exercise 5.8.9. Obtain an expression for the Bessel wavelet transform of the function $f(t) = t^{-1/2}$, with respect to an arbitrary wavelet $\psi(t)$.

Exercise 5.8.10. Compute the Dunkl transform of the family of daughter wavelets $\mathsf{T}_b \mathcal{D}_a \psi(t)$ given by (5.129), where $\psi(t)$ is the Haar wavelet (3.31).

Exercise 5.8.11. Evaluate the Mehler-Fock transform of the function

$$f(t) = \frac{1}{\pi^2 (t-1)} \ln \left(\frac{t+1}{2} \right). \tag{5.216}$$

Also, using (5.158) obtain an expression for the Mehler-Fock wavelet transform of the function (5.216) with respect to an arbitrary wavelet $\psi(t)$.

Exercise 5.8.12. Compute the Hartley transforms for the following signals:

(i) *Unit-step:* $f(t) = U(t) := \begin{cases} 1, & t \geq 0 \\ 0, & \text{otherwise,} \end{cases}$

(ii) *Unit-ramp:* $f(t) = t\, U(t)$,

(iii) *Exponential:* $f(t) = \beta\, e^{-\alpha t} U(t)$, $\alpha > 0$, $\beta \in \mathbb{R}$,

(iv) *Double exponential:* $f(t) = e^{-\alpha |t|}$, $\alpha > 0$,

(v) *One-sided exponential:* $f(t) = (\beta - \alpha)^{-1} \left(e^{-\alpha t} - e^{-\beta t} \right)$, $\alpha > 0$, $\beta \in \mathbb{R}$, where $\alpha \neq \beta$,

(vi) *Complex sinusoid:* $f(t) = e^{i \omega_0 t}$, $\omega_0 \in \mathbb{R}$,

(vii) *Periodic wave:* $f(t) = \sum_n \alpha_n\, e^{2\pi i \omega_0 t}$, $\alpha_n, \omega_0 \in \mathbb{R}$,

(viii) *Sine wave:* $f(t) = \sin(\omega_0 t)$, $\omega_0 \in \mathbb{R}$,

(ix) *Cosine wave:* $f(t) = \cos(\omega_0 t)$, $\omega_0 \in \mathbb{R}$,

(x) *Damped sine:* $f(t) = e^{-\alpha t} \sin(\omega_0 t)\, U(t)$, $\alpha > 0$, $\omega_0 \in \mathbb{R}$,

(xi) *Damped cosine:* $f(t) = e^{-\alpha t} \cos(\omega_0 t)\, U(t)$, $\alpha > 0$, $\omega_0 \in \mathbb{R}$,

(xii) *Impulse:* $f(t) = K\, \delta(t)$, $K \in \mathbb{R}$,

(xiii) *Impulse train:* $f(t) = \sum_n \delta(t - nT)$, where T is the period.

Exercise 5.8.13. Compute the Hartley transform of the Haar wavelet $\psi(t)$ defined in (3.31). Also use Theorem 5.7.2 to obtain the wavelet transform of the following functions:

(i) $f(t) = U(t)$,

(ii) $f(t) = t\, U(t)$,

(iii) $f(t) = e^{-t^2/2}$,

where $U(t)$ is the unit-step.

Bibliography

The following bibliography is not, by any means, a complete one for the subject. For the most part, it consists of books and research papers to which reference is made in the text. Many other selected books and papers related to material of the subject have been included, so that they may serve to stimulate new interest in future study and research.

[1] Abe, S., and Sheridan, J.T. (1994a). Optical operations on wave functions as the Abelian subgroups of the special affine Fourier transformation, Optics Lett. 19, 1801-1803.

[2] Abe, S., and Sheridan, J.T. (1994b). Generalization of the fractional Fourier transformation to an arbitrary linear lossless transformation: An operator approach, J. Phys. A: Math. Gen. 27, 4179-4187.

[3] Abramowitz, M., and Stegun, I.A. (1972). Handbook of Mathematical Functions, Dover Publications Inc., New York.

[4] Abry, P., Clausel, M., Jaffard, S., Roux, S.G., and Vedel, B. (2015). The hyperbolic wavelet transform: An efficient tool for multifractal analysis of anisotropic fields, Rev. Mat. Iberoam. 31, 313-348.

[5] Addison, P.S. (2005). Wavelet transforms and the ECG: A review, Physiol. Meas. 26, 155-199.

[6] Addison, P.S. (2010). The Illustrated Wavelet Transform Handbook: Introductory Theory and Applications in Science, Engineering, Medicine and Finance, Second Edition, CRC Press.

[7] Akansua, A.N., Serdijnc, W.A., and Selesnick, I.W. (2010). Emerging applications of wavelets: A review, Phys. Communi. 3, 1-18.

[8] Akay, O., and Boudreaux-Bartels, G.F. (2001). Fractional convolution and correlation via operator methods and application to detection of linear FM signals, IEEE Trans. Signal Process. 49, 979-993.

[9] Akujuobi, C.M. (2022). Wavelets and Wavelet Transform Systems and Their Applications, Springer Nature, Switzerland AG.

[10] Aldroubi, A., and Unser, M. (1996). Wavelets in Medicine and Biology, CRC Press, Boca Raton.

[11] Ali, S.T., Antoine, J.P., and Gazeau, J.P. (1991). Square integrability of group representations on homogenous space-I. Reproducing triples and frames, Ann. Inst. H. Poincaré, 55, 829-855.

[12] Ali, S.T., Antoine, J.P., and Gazeau, J.P. (1991). Square integrability of group representations on homogeneous spaces-II. Coherent and quasi-coherent states. The case of Poincaré group, Ann. Inst. H. Poincaré, 55, 857-890.

457

[13] Ali, S.T., Antoine, J.P., and Gazeau, J.P. (2017). Coherent States, Wavelets and Their Generalizations, Second Edition, Springer, New York.

[14] Allen, R.L., and Mills, D.W. (2004). Signal Analysis: Time, Frequency, Scale and Structure, IEEE Press, U.S.A.

[15] Almeida, L.B. (1994). The fractional Fourier transform and time-frequency representations, IEEE Trans. Signal Process. 42, 3084-3091.

[16] Almeida, L.B. (1997). Product and convolution theorems for the fractional Fourier transform, IEEE Signal Process. Lett. 4, 15-17.

[17] Amiri, Z., Bagherzadeh, H., Harati, A., and Kamyabi-Gol, R.A. (2018). Study of shearlet transform using block matrix dilation, J. Appl. Math. Comput. 56, 665-689.

[18] Angrisani, L., and D'Arco, M. (2002). A measurement method based on a modified version of the chirplet transform for instantaneous frequency estimation, IEEE Trans. Instrument. Measure. 51(4), 704-711.

[19] Antoine, J.P., Murenzi, R., and Vandergheynst, P. (1999). Directional wavelets revisited: Cauchy wavelets and symmetry detection in patterns, Appl. Comput. Harmon. Anal. 6(3), 314-345.

[20] Antoine, J.P., Murenzi, R., Vandergheynst, P., and Ali, S.T. (2004). Two-Dimensional Wavelets and Their Relatives, Cambridge University Press, Cambridge, U.K.

[21] Antoine, J.P., Bogdanova, I., and Vandergheynst, P. (2008). The continuous wavelet transform on conic sections, Int. J. Wavelets, Multiresol. Informat. Process. 6(2), 137-156.

[22] Antoine, J.P., and Murenzi, R. (1996). Two-dimensional directional wavelets and the scale-angle representation, Signal Process. 52(3), 259-281.

[23] Aslaksen, E.W., and Klauder, J.R. (1968). Unitary representations of the affine group, J. Math. Phys. 9, 206-211.

[24] Aslaksen, E.W., and Klauder, J.R. (1969). Continuous representation theory using the affine group, J. Math. Phys. 10, 2267-2275.

[25] Baccar, C. (2016). Uncertainty principles for the continuous Hankel wavelet transform, Integ. Transf. Spec. Funct. 27(6), 413-429.

[26] Battle, G. (1987). A block spin construction of ondelettes. Part I: Lemariè functions, Commun. Math. Phys., 110, 601-615.

[27] Battle, G. (1997). Heisenberg inequalities for wavelet states, Appl. Comput. Harmon. Anal. 4, 119-146.

[28] Bahri, M., and Ashino, R. (2016). Some properties of windowed linear canonical transform and its logarithmic uncertainty principle, Int. J. Wavelets, Multiresol. Informat. Process. 14(3), 1650015.

[29] Bahri, M., and Ashino, R. (2017). Logarithmic uncertainty principle, convolution theorem related to continuous fractional wavelet transform and its properties on a generalized Sobolev space, Int. J. Wavelets Multiresolut. Inf. Process. 15(5), 1750050.

[30] Bahri, M., Amir, A.K., and Ashino, R. (2021). Linear canonical wavelet transform: Properties and inequalities, Int. J. Wavelets, Multiresol. Informat. Process. 19(6), 2150027.

[31] Bahri, M., Shah, F.A., and Tantary, A.Y. (2020). Uncertainty principles for the continuous shearlet transforms in arbitrary space dimensions, Integ. Transf. Special Funct. 31(7), 538-555.

[32] Bartolucci, F., Mari, F.D., Vito, E.D., and Odone, F. (2019). The Radon transform intertwines wavelets and shearlets, Appl. Comput. Harmon. Anal. 47(3), 822-847.

[33] Battisti, U., and Riba, L. (2016). Window-dependent bases for efficient representations of the Stockwell transform, Appl. Comput. Harmon. Anal. 40, 292-320.

[34] Beckner, W. (1995). Pitt's inequality and the uncertainty principle, Proc. Amer. Math. Soc. 123, 1897-1905.

[35] Bell, W.W. (1968). Special Functions for Scientists and Engineers, Dover Publications, Van Nostrand.

[36] Benedicks, M. (1985). On Fourier transforms of functions supported on sets of finite Lebesgue measure, J. Math. Anal. Appl. 106, 180-183.

[37] Bernardo, L.M. (1996). ABCD matrix formalism of fractional Fourier optics, Opt. Eng. 35(3), 732-740.

[38] Beylkin, G., Coifman, R., and Rokhlin, V. (1991). Fast wavelet transforms and numerical algorithms I, Comm. Pure Appl. Math. 44(2), 141-183.

[39] Beylkin, G., Coifman, R., and Rokhlin, V. (1992). Wavelets in numerical analysis, in: Wavelets and Their Applications, Jones and Bartlett, Boston, 181-210.

[40] Bhandari, A., and Zayed, A.I. (2018). Convolution and product theorems for the special affine Fourier transform, in Eds: Nashed, Li, Frontiers in Orthogonal Polynomials and q-Series, World Scientific, 119-137.

[41] Bhandari, A., and Zayed, A.I. (2019). Shift-invariant and sampling spaces associated with the special affine Fourier transform, Appl. Comput. Harmon. Anal. 47, 30-52.

[42] Bhatnagar, G., Wu, Q.M.J., and Raman, B. (2012). A new fractional random wavelet transform for fingerprint security, IEEE Trans. Syst. Man, Cybern. Part A: Systems and Humans 42, 262-275.

[43] Bhatnagar, G., and Wu, Q.M.J. (2013). A new logo watermarking based on redundant fractional wavelet transform, Math. Comput. Modell. 58, 204-218.

[44] Blackman, R.B., and Tukey, J.W. (1958). The Measurement of Power Spectra, Dover Publications, New York.

[45] Blatter, C.(1998). Wavelets: A Primer, A. K. Peters, Ltd., Massachusetts.

[46] Bogdanova, I., Vandergheynst, P., and Gazeau, J.P. (2007). Continuous wavelet transform on the hyperboloid, Appl. Comput. Harmon. Anal. 23, 285-306.

[47] Boggiatto, P., Fernandez, C., and Galbis, A. (2009). A group representation related to the Stockwell transform, Indiana Univ. Math. J. 58(5), 2277-2304.

[48] Boggess, A., and Narcowich, F. J. (2001). A First Course in Wavelets with Fourier Analysis, Prentice Hall, Up Saddle River, New Jersey.

[49] Bowman, F.(2003). Introduction to Bessel Functions, Dover Publications Inc., New York.

[50] Bracewell, R.N. (1986). The Hartley Transform, Oxford University Press, New York.

[51] Bracewell, R.N. (2000). The Fourier Transform and Its Applications, Third Edition, McGraw-Hill, Boston.

[52] Brackx, F.F., and Sommen, F.C. (2002). Clifford–Bessel wavelets in Euclidean space, Math. Meth. Appl. Sci. 25, 1479-1491.

[53] Bremaud, P. (2002). Mathematical Principles of Signal Processing, Fourier and Wavelet Analysis, Springer, New York.

[54] Bultheel, A., and Martinez-Sulbaran, H.(2007). Recent developments in the theory of the fractional Fourier and linear canonical transforms, Bull. Belg. Math. Soc. Simon Stevin. 13, 971-1005.

[55] Bultan, A. (1999). A four-parameter atomic decomposition of chirplets, IEEE Trans. Signal Process. 47(3), 731-745.

[56] Burrus, C.S., Gopinath, R.A., and Guo, H. (1998). Introduction to Wavelets and Wavelet Transforms: A Primer, Prentice Hall, New Jersey.

[57] Butz, T. (2006). Fourier Transformation for Pedestrians, Springer, Berlin.

[58] Cai, L.Z. (2000). Special affine Fourier transformation in frequency-domain, Optics Commun. 185, 271-276.

[59] Candés E.J., and Donoho, D.L. (1999). Ridgelets: A key to higher-dimensional intermittency?, Phil. Trans. R. Soc. London A. 357, 2495-2509.

[60] Candés E.J., and Donoho, D.L. (2002). Recovering edges in ill-posed inverse problems: Optimality of curvelet frames, Ann. Statist. 30, 784-842.

[61] Candés E.J., and Demanet, L. (2004). New tight frames of curvelets and optimal representations of objects with piecewise-C^2 singularities, Comm. Pure Appl. Math. 57 , 219-266.

[62] Candés E.J., and Donoho, D.L. (2005). Continuous curvelet transform-I. Resolution of the wavefront set, Appl. Comput. Harmon. Anal. 19, 162-197.

[63] Candés E.J., and Donoho, D.L. (2005). Continuous curvelet transform-II. Discretization and frames, Appl. Comput. Harmon. Anal. 19, 198-222.

[64] Candés E.J., and Demanet, L. (2005). The curvelet representation of wave propagators is optimally sparse, Comm. Pure Appl. Math. 58, 1472-1528.

[65] Candés E.J., Demanet, L., Donoho, D.L., and Ying, L. (2006). Fast discrete curvelet transforms, Multiscale Model. Simul. 5(3), 861-899.

[66] Cariolaro, G., Erseghe, T., Kraniauskas, P., and Laurenti, N. (1998). A unified framework for the fractional Fourier transform, IEEE Trans. Signal Process. 46(12), 3206-3219.

[67] Castro, L.P., Haque, M.R., Murshed, M.M., Saitoh, S., and Tuan, N.M. (2014). Quadratic Fourier transforms, Ann. Funct. Anal. 5, 10-23.

[68] Castro, L.P., Minh, L.T., and Tuan, N.M. (2018). New convolutions for quadratic-phase Fourier integral operators and their applications, Mediterr. J. Math. 15, 1-17.

[69] Catana, V. (2017). Abelian and Tauberian results for the one-dimensional modified Stockwell transforms, Appl. Anal. 96(6), 1047-1057.

[70] Chatterjee, P. (2015). Wavelet Analysis in Civil Engineering, CRC Press, Boca Raton, New York.

[71] Chui, C.K. (l992). An Introduction to Wavelets, Academic Press, New York.

[72] Chui, C.K., and Wang, J.Z. (1991). A cardinal spline approach to wavelets, Proc. Amer. Math. Soc. 113, 785-793.

[73] Chui, C.K., and Wang, J.Z. (1992). On compactly supported spline wavelets and a duality principle, Trans. Amer. Math. Soc. 330, 903-915.

[74] Cnops, J. (2001). The wavelet transform in Clifford analysis, Comput. Meth. Funct. Theo. 12, 353-374.

[75] Cohen, A. (1995). Wavelets and Multi-scale Signal Processing, Chapman and Hall, London.

[76] Cohen, L. (1989). Time-frequency distributions: A review, Proc. IEEE, 77, 941-981.

[77] Collins, S.A. (1970). Lens-system diffraction integral written in terms of matrix optics, J. Optical Soc. Amer. A, 60, 1772-1780.

[78] Colonna, F., Easley, G., Guo, K., and Labate, D. (2010). Radon transform inversion using the shearlet representation, Appl. Comput. Harmon. Anal. 29(2), 232-250.

[79] Cowling, M.G., and Price, J.F. (1994). Bandwidth verses time concentration: The Heisenberg-Pauli-Weyl inequality, SIAM J. Math. Anal. 15, 151-165.

[80] Dahlke, S., and Maass, P. (1995). The affine uncertainty principle in one and two dimensions, Comput. Math. Appl. 30, 293-305.

[81] Dahlke, S., Kutyniok, G., Maass, P., Sagiv, C., and Stark, H.G. (2008). The uncertainty principle associated with the continuous shearlet transform, Int. J. Wavelets, Multiresol. Informat. Process. 6, 157-181.

[82] Dahlke, S., Steidl, G., and Teschke, G. (2010). The continuous shearlet transform in arbitrary space dimensions, J. Fourier Anal. Appl. 16, 340-364.

[83] Dai, H., Zheng, Z., and Wang, W. (2017). A new fractional wavelet transform. Commun. Nonlinear Sci. Numer. Simulat. 44, 19-36.

[84] Daubechies, I. (1988a). Time-frequency localization operators: A geometric phase space approach, IEEE Trans. Inform. Theory. 34, 605-612.

[85] Daubechies, I. (1988b). Orthogonal bases of compactly supported wavelets, Commun. Pure Appl. Math. 41, 909-996.

[86] Daubechies, I. (1990). The wavelet transform, time-frequency localization and signal analysis, IEEE Trans. Inform. Theory 36, 961-1005.

[87] Daubechies, I. (1992). Ten Lectures on Wavelets, NSF-CBMS Regional Conference Series in Applied Math. 61, SIAM, Philadelphia.

[88] Daubechies, I. (1996). Where do wavelets come from? Proc. IEEE, 84(4), 510-513.

[89] Daubechies, I. (2021). Wavelets at your service, Art. Pract. Math. 48-57.

[90] Daubechies, I., Grossmann, A., and Meyer, Y. (1986), Painless non-orthogonal expansion, J. Math. Phys. 27, 1271-1283.

[91] Deans, S.R. (1983). The Radon Transform and Some of Its Applications, John Wiley & Sons, New York.

[92] Debnath, L. (1998b). Brief introduction to history of wavelets, Int. J. Math. Edu. Sci. Tech. 29, 677-688.

[93] Debnath, L., and Bhatta, D. (2015). Integral Transforms and Their Applications, Third Edition, Chapman and Hall, CRC Press, Boca Raton, Florida.

[94] Debnath, L., and Shah, F.A. (2015). Wavelet Transforms and Their Applications, Second Edition, Birkhäuser, Boston.

[95] Debnath, L., and Shah, F.A. (2017). Lecture Notes on Wavelet Transforms, Birkhäuser, New York.

[96] Deng, B., Tao, R., and Wang, Y. (2006). Convolution theorems for the linear canonical transform and their applications, Sci China, Ser F: Inform Sci. 49, 592-603.

[97] Do, M.N., and Vetterli, M. (2003). Contourlets, in: G. V. Welland (Ed.), Beyond Wavelets, Academic Press, New York.

[98] Do, M.N., and Vetterli, M. (2003). The finite ridgelet transform for image representation, IEEE Trans. Image Process. 12, 16-28.

[99] Do, M.N., and Vetterli, M. (2005). The contourlet transform: An efficient directional multiresolution image representation, IEEE Trans. Image Proc. 14, 2091-2106.

[100] Du, J., Wong, M.W., and Zhu, H. (2007). Continuous and discrete inversion formulas for the Stockwell transform, Integr. Transf. Spec. Funct. 18, 537-543.

[101] Duval-Poo, M.A., Odone, F., and Vito, E.D. (2015). Edges and corners with shearlets, IEEE Trans. Image Process. 24(11), 3768-3780.

[102] Erseghe, T., Kraniauskas, P., and Carioraro, G. (1999). Unified fractional Fourier transform and sampling theorem, IEEE Trans. Signal Process. 47(12), 3419-3423.

[103] Fadili, J., and Starck, J.L. (2009). Curvelets and ridgelets, in: Encyclopedia of Complexity and Systems Science, Springer, 1718-1738.

[104] Farashahi, A.G. (2020). Generalized wavelet transforms over finite fields, Linear Multilinear Algeb. 68(8), 1585-1604.

[105] Fashandi, M., Kamyabi-Gol, R.A., Niknam, A., and Pourabdollah, M.A. (2003). Continuous wavelet transform on a special homogeneous space, J. Math. Phys. 44, 42-60.

[106] Fashandi, M. (2018). Quaternionic continuous wavelet transform on a quaternionic Hilbert space, RACSAM 112, 1049-1057.

[107] Feichtinger, H.G., and Weisz, F. (2006). Inversion formulas for the short-time Fourier transform, J. Geom. Anal. 16, 507-521.

[108] Feng, Q., and Li, B.Z. (2016). Convolution and correlation theorems for the two-dimensional linear canonical transform and its applications, IET Signal Process. 10(2), 125-132.

[109] Ferreira, M. (2008). Spherical wavelet transform, Adv. Appl. Clifford Algebra. 18, 611-619.

[110] Ferreira, M. (2009). Spherical continuous wavelet transforms arising from sections of the Lorentz group, Appl. Comput. Harmon. Anal. 26, 212-229.

[111] Folland, G.B., and Sitaram, A. (1997). The uncertainty principle: A mathematical survey, J. Fourier Anal. Appl. 3, 207-238.

[112] Fourier, J.B. (1822). Thèorie analytique de la chaleur. Paris: Chez Firmin Didot, Père et Fils.

[113] Führ, H. (1998). Continuous wavelet transforms with abelian dilation groups, J. Math. Phys. 39(8), 3974-3986.

[114] Führ, H. (2005). Abstract Harmonic Analysis of Continuous Wavelet Transforms, Springer, Berlin.

[115] Fukuda, N., Kinoshita, T., and Yoshino, K. (2017). Wavelet transforms on Gelfand-Shilov spaces and concrete examples, J. Inequalit. Appl. 119, 1-24.

[116] Gabor, D. (1946), Theory of communications, J. Inst. Electr. Eng. London 93, 429-457.

[117] Geckinli, N.C., and Yavuz, D. (1978). Some novel windows and a concise tutorial comparison of window families, IEEE Trans. Acoust. Speech Signal Process. 26, 501-507.

[118] Gibson, P.C., Lamoureux, M.P. and Margrave G.F. (2006). Letter to the editor: Stockwell and wavelet transforms, J Fourier Anal. Appl. 12, 713-721.

[119] Ghobber, S. (2017). Concentration operators in the Dunkl wavelet theory, Mediterr. J. Math. 14, 41.

[120] Giv, H.H. (2013). Directional short-time Fourier transform, J. Math. Anal. Appl. 399(1), 100-107.

[121] Goel, N., and Singh, K. (2013). Analysis of Dirichlet, generalized Hamming and triangular window functions in the linear canonical transform domain, Signal Image Video Process. 7(5), 911-923.

[122] Goel, N., and Singh, K. (2016). Convolution and correlation theorems for the offset fractional Fourier transform and its application, AEU-Int. J. Elect. Commun. 72(2), 138-150.

[123] Gomes, J., and Velho, L. (2015). From Fourier Analysis to Wavelets, Springer, New York.

[124] Grafakos, L., and Sansing, C. (2008). Gabor frames and directional time-frequency analysis, Appl. Comput. Harmon. Anal. 25, 47-67.

[125] Grochenig, K. (2001). Foundations of Time Frequency Analysis, Birkhäuser, Boston.

[126] Grossmann, A. and Morlet, J. (1984). Decomposition of Hardy functions into square integrable wavelets of constant shape, SIAM J. Math. Anal. 15, 723-736.

[127] Guo, K., Labate, D., Lim, W.Q., Weiss, G., and Wilson, E. (2004). Wavelets with composite dilations, Electron. Res. Announc. Amer. Math. Soc. 10(9), 78-87.

[128] Guo, K., and Labate, D. (2007). Optimally sparse multidimensional representation using shearlets, SIAM J. Math. Anal. 39, 298-318.

[129] Guo, K., and Labate, D. (2008). Representation of Fourier integral operators using shearlets, J. Fourier Anal. Appl. 14, 327-371.

[130] Guo, K., and Labate, D. (2009). Characterization and analysis of edges using the continuous shearlet transform, SIAM J. Imaging. Sci. 2, 959-986.

[131] Guo, K., and Labate, D. (2012). Characterization of piecewise-smooth surfaces using the 3D continuous shearlet transform, J. Fourier Anal. Appl. 18, 488-516.

[132] Guo Q., Molahajloo S., and Wong M.W. (2010). Phases of modified Stockwell transforms and instantaneous frequencies, J. Math. Phys. 51, 052101.

[133] Guo, Y., Li, B.Z., and Yang, L.D. (2021). Novel fractional wavelet transform: Principles, MRA and application, Digit. Signal Process. 110, 102937.

[134] Guo, Y., and Li, B.Z. (2018). The linear canonical wavelet transform on some function spaces, Int. J. Wavelets Multiresolut. Inf. Process. 16(01), 1850010.

[135] Haar, A. (1910). Zur Theorie der orthogonalen funktionen-systeme, Math. Ann. 69, 331-371.

[136] Harris, F.J. (1978). On the use of windows for harmonic analysis with the discrete Fourier transform, IEEE Proc. 66, 51-83.

[137] Havin, V., and Jöricke, B. (1994). The Uncertainty Principle in Harmonic Analysis, Springer, Berlin.

[138] He, J. (2002). Wavelet transforms associated to square integrable group representations on $L^2(\mathbb{R}, \mathbb{H}, dz)$, Appl. Anal. 81, 495-512.

[139] He, J., and Yu, B. (2004). Continuous wavelet transforms on the space $L^2(\mathbb{R}, \mathbb{H}, dx)$, Appl. Math. Lett. 17, 11-121.

[140] Healy, J.J., Kutay, M.A., Ozaktas , H.M., and Sheridan, J.T. (2016). Linear Canonical Transforms, Springer, New York.

[141] Heil, C.E., and Walnut, D.F. (1989). Continuous and discrete wavelet transforms, SIAM Rev. 31, 628-666.

[142] Heisenberg, W. (1948a). Zur statistischen theori der turbulenz, Z. Phys. 124, 628-657.

[143] Helgason, S. (1999). The Radon Transform, Second Edition, Birkhäuser, Boston.

[144] Hennelly, B.M., and Sheridan, J.T. (2005). Fast numerical algorithm for the linear canonical transform, J. Opt. Soc. Amer. A. 22, 928-937.

[145] Hernandez, E., and Weiss, G. (1996). A First Course on Wavelets, CRC Press, Boca Raton.

[146] Hirschman, I., and Widder, D.V. (2005). The Convolution Transform, Dover Publications, New York.

[147] Hogan, J.A., and Lakey, J.D. (2005). Time-frequency and Time-scale Methods, Birkhäuser, Boston.

[148] Holschneider, M. (1995). Wavelets: An Analysis Tool, Oxford University Press, Oxford.

[149] Holschneider, M. (1996). Continuous wavelet transforms on the sphere, J. Math. Phys. 37, 4156-4165.

[150] Holschneider, M. (1991). Inverse Radon transforms through inverse wavelet transforms, Inverse Problems 7, 853-861.

[151] Howell, K.B. (2001). Principles of Fourier Analysis, Chapman & Hall/CRC, Boca Raton.

[152] Hramov, A.E., Koronovskii, A.A., Makarov, V.A., Pavlov, A.N., and Sitnikova, E. (2015). Wavelets in Neuroscience, Springer, New York.

[153] Huang, Y., and Suter, B. (1998). The fractional wave packet transform, Multidimens. Syst. Signal Process. 9, 399-402.

[154] Hubbard, B. (1996). The World According to Wavelets, A. K. Peters., Wellesley, Massachusetts.

[155] Huo, H. (2019). Uncertainty principles for the offset linear canonical transform, Circuits, Syst. Signal Process. 38, 395-406.

[156] Hutníková., M, and Misková, A. (2015). Continuous Stockwell transforms: Coherent states and localization operators, J. Math. Phys. 256, 073504.

[157] James, D.F.V., and Agarwal, G.S. (1996). The generalized Fresnel transform and its application to optics, Opt. Commun. 126, 207-212.

[158] Jiang, S., and Jiang, Z. (2018). Inversion formula for shearlet transform in arbitrary space dimensions, Numer. Funct. Anal. Optimiz. 37(11),1438-1463.

[159] Juliet, S., Rajsingh, E.B., and Ezra, K. (2016). A novel medical image compression using Ripplet transform, J. Real-Time Image Proc. 11, 401-412.

[160] Kaiser, G. (2010). A Friendly Guide to Wavelets, Birkhäuser, New York.

[161] Kaiser, J.F., and Schafer, R.W. (1980). On the use of the I_0-sinh window for spectrum analysis, IEEE Trans. Acoust. Speech Signal Process. 28, 105-107.

[162] Kalisa, C., and Torrésani, B. (1993). N-dimensional affine Weyl-Heisenberg wavelets, Ann. Inst. H. Poincaré 59, 201-236.

[163] Kawazoe, T. (1996). Wavelet transform associated to an induced representation of $SL(n+2, \mathbb{R})$, Ann. Henri Poincare 65(1), 1-13.

[164] Koc, A., Ozaktas, H.M., Candan, C., and Kutay, M.A. (2008). Digital computation of linear canonical transforms, IEEE Trans. Signal Process. 56(6), 2383-2394.

[165] Korenev, B.G. (2002). Bessel Functions and Their Applications, CRC Press, New York.

[166] Kou, K.I., and Xu, R.H. (2012). Windowed linear canonical transform and its applications, Signal Process. 92, 179-188.

[167] Kumar, D., Kumar, S., and Singh, B. (2016). Cone-adapted continuous shearlet transform and reconstruction formula, J. Non-linear Sci. Appl. 9, 262-269.

[168] Kumar S., Singh, K., and Saxena, R. (2011). Analysis of Dirichlet and generalized Hamming window functions in the fractional Fourier transform domains, Signal Process. 91(3), 600-606.

[169] Kunche, P., and Manikanthababu, N. (2020). Fractional Fourier Transform Techniques for Speech Enhancement, Springer.

[170] Kutyniok, G., and Labate, D. (2009). Resolution of the wave front set using continuous shearlets, Trans. Amer. Math. Soc. 361(5), 2719-2754.

[171] Kutyniok, G., and Labate, D. (2012). Shearlets: Multiscale Analysis for Multivariate Data. Birkhäuser, New York.

[172] Labate, D., Li, W.Q, Kutyniok, G., and Weiss, G. (2005). Sparse multidimensional representation using shearlets, SPIE Proc. SPIE 5914, 254-262.

[173] Lakshmanan, M.K., and Nikookar, H. (2006). A Review of wavelets for digital wireless communication, Wireless Pers. Communicat. 37, 387-420.

[174] Lemariè, P.G. (1988). Ondelettes a localisation exponentielle, J. Math. Pures Appl. 67, 227-236.

[175] Lepik, U., and Hein, H. (2014). Haar Wavelets, Springer, London.

[176] Lessig, C., Petersen, P., and Schäfer, M. (2019). Bendlets: A second-order shearlet transform with bend elements, Appl. Comput. Harmon. Anal. 46(2), 384-399.

[177] Li, J., and Wong, M.W. (2007). Polar wavelet transforms and localization operators, Integr. Equ. Oper. Theory 58, 99-110.

[178] Li, J., and Wong, M.W. (2012). Localization operators for curvelet transforms, J. Pseudo-Differ. Operat. Appl. 3, 121-143.

[179] Li, J., and Wong, M.W. (2014). Localization operators for ridgelet transforms, Math. Model. Nat. Phenom. 9(5), 194-203.

[180] Li, Y.M., and Wei, D. (2015). The wave packet transform associated with the linear canonical transform, Optik 126(21), 3168-3172.

[181] Lian, P. (2021). Uncertainty principles in linear canonical domains, Integ. Transf. Special Funct. 32, 67-77.

[182] Lim, W.Q. (2010). The discrete shearlet transform: A new directional transform and compactly supported shearlet frames, IEEE Trans. Image Process. 19(5), 1166-1180.

[183] Linglan, B., and Jianxun, H. (2002). Generalized wavelets and inversion of the radon transform on the Laguerre hypergroup, Approx. Theory Appl. 18(4), 48-62.

[184] Liu, H.P., and Peng, L.Z. (1997). Admissible wavelets associated with the Heisenberg group, Pacific J. Math. 180, 101-121.

[185] Liu, B., and Sun, W. (2008). Homogeneous approximation property for continuous wavelet transforms, J. Approx. Theory 155, 111-124.

[186] Liu, B., and Sun, W. (2009). Homogeneous approximation property for multivariate continuous wavelet transforms, Numer. Funct. Anal. Optimizat. 30(7-8), 784-798.

[187] Liu, B., and Sun, W., (2009). Inversion of the wavelet transform using Riemannian sums, Appl. Comput. Harmon. Anal. 27, 289-302.

[188] Lizhong, P., and Ruiqin, M. (2004). Wavelets associated with Hankel transform and their Weyl transforms, Science in China Ser. A Math. 47(3), 393-400.

[189] Louis, A.K., Maab, P., and Rieder, A. (1997). Wavelets: Theory and Applications, John Wiley and Sons, New York.

[190] Lu, Y., and Do, M.N. (2007). Multidimensional directional filter banks and surfacelets, IEEE Trans. Image Process. 16(4), 918-931.

[191] Ma, J., and Plonka, G. (2010). The curvelet transform, IEEE Signal Process. Magazine. 118-133.

[192] Mahato, K. (2018). The product of generalized wavelet transform involving fractional Hankel-type transform on some function spaces, Int. J. Wavelets, Multiresol. Informat. Process. 17, 1950002.

[193] Mallat, S. (1989a). Multiresolution approximations and wavelet orthonormal basis of $L^2(\mathbb{R})$, Trans. Amer. Math. Soc. 315, 69-88.

[194] Mallat, S. (1989b). A theory for multiresolution signal decomposition: The wavelet representation, IEEE Trans. Patt. Recog. Mach. Intelt. 11, 678-693.

[195] Mallat, S. (1989c). Multi-frequency channel decompositions of images and wavelet models, IEEE Trans. Acoust. Signal Speech Process. 37, 2091-2110.

[196] Mallat, S. (2009). A Wavelet Tour of Signal Processing, Third Edition, Academic Press, New York.

[197] Mann, S., and Haykin, S. (1992). Chirplets and warblets: Novel time-frequency methods, Electron. Lett. 28(2), 114-116.

[198] Mann, S., and Haykin, S. (1995). The chirplet transform: Physical considerations, IEEE Trans. Signal Process. 43(11), 2745-2761.

[199] Marzouqia, H.A., and Regibb, G.A. (2017). Curvelet transform with learning-based tiling, Signal Process. Image Commun. 53, 24-39.

[200] Mecklenbräuker, W., and Hlawatsch, F. (1997). The Wigner Distribution, Elsevier, Amsterdam.

[201] Mehra, M. (2018). Wavelets Theory and Its Applications, Springer, Singapore.

[202] Mejjaoli, H. (2017). Dunkl two-wavelet theory and localization operators, J. Pseudo-Differ. Oper. Appl. 8, 349-387.

[203] Mejjaoli, H. (2020). (k, a)-generalized wavelet transform and applications, J. Pseudo-Differ. Oper. Appl. 11, 55-92.

[204] Mejjaoli, H., and Sraieb, N. (2008). Uncertainty principles for the continuous Dunkl Gabor transform and the Dunkl continuous wavelet transform, Mediterr. J. Math. 5, 443-466.

[205] Mejjaoli, H., and Sraieb, N. (2021). Localization operators associated with the q-Bessel wavelet transform and applications, Int. J. Wavelets, Multiresol. Informat. Process. 19(4), 2050094.

[206] Mendlovic, D., and Ozaktas, H.M. (1993a). Fractional Fourier transforms and their optical implementation-I, J. Opt. Sco. AM. A 10(10), 1875-1881.

[207] Mendlovic, D., and Ozaktas, H.M. (1993b). Fractional Fourier transforms and their optical implementation-II, J. Opt. Sco. AM. A 10(12), 2522-2531.

[208] Mendlovic, D., Zalevsky, Z., Mas, D., García, J., and Ferreira, C. (1997). Fractional wavelet transform, Appl. Opt. 36, 4801-4806.

[209] Meyer, Y. (1993a). Wavelets: Algorithms and Applications, Society for Industrial and Applied Mathematics (SIAM), Philadelphia.

[210] Meyer, Y. (1993b), Wavelets and Operators, Cambridge University Press, Cambridge.

[211] Molahajloo, S., and Wong, M.W. (2009). Square-integrable group representations and localization operators for modified Stockwell transforms, Rend. Semin. Mat. Univ. Politec. Torino 67, 215-227.

[212] Moshinsky, M., and Quesne, C. (1971a). Linear canonical transformations and their unitary representations, J. Math. Phys. 12(8), 1772-1780.

[213] Moshinsky, M., and Quesne, C. (1971b). Linear canonical transformations and matrix elements, J. Math. Phys. 12(8), 1780-1783.

[214] Namias, V. (1980). The fractional order Fourier transform and its application to quantum mechanics, J. Inst. Math. Appl. 25, 241-265.

[215] Nazarathy, M., and Shamir, J. (1982). First-order optics–a canonical operator representation: Lossless systems, J. Opt. Soc. Am. 72, 356-364.

[216] Nickolas, P. (2017). Wavelets: A Student Guide, Cambridge University Press, Cambridge, U.K.

[217] Olson, T. (2017). Applied Fourier Analysis, Birkhäuser, New York.

[218] Oppenheim, A.V., and Schafer, R. W. (1975). Digital Signal Processing, Prentice-Hall, Inc., New Jersey.

[219] Osgood, B.G. (2019). Lectures on the Fourier Transform and Its Applications, American Mathematical Society, Providence, Rhode Island USA.

[220] Ozaktas, H.M., Barshan, B., and Mendlovic, D. (1994). Convolution and filtering in fractional Fourier domains, Optical Rev. 1, 15-16.

[221] Ozaktas, H.M., Zalevsky, Z., and Kutay, M.A. (2000). The Fractional Fourier Transform with Applications in Optics and Signal Processing, Wiley, New York.

[222] Palma, C., and Bagini, V. (1997). Extension of the Fresnel transform to ABCD systems, J. Opt. Soc. Amer. A 14, 1774-1779.

[223] Pandey, J.N., Jha, N.K, and Singh, O.P. (2016). The continuous wavelet transform in n-dimensions, Int. J. Wavelets Multiresolut. Inf. Process. 14, 1650037.

[224] Papoulis, A. (1973). Minimum-bias windows for high-resolution spectral estimates, IEEE Trans. Informat. Theory IT-19, 9-12.

[225] Papoulis, A. (1977). Signal Analysis, McGraw-Hill, New York.

[226] Pathak, R.S. (2009). The Wavelet Transform, Atlantis Press, Amsterdam.

[227] Pathak, R.S., and Dixit, M.M. (2003). Continuous and discrete Bessel wavelet transforms, J. Comput. Appl. Math. 160, 241-250.

[228] Pathak, R.S., and Pandey, C.P. (2009). Laguerre wavelet transforms, Integ. Transf. Spec. Funct. 20(7), 505-518.

[229] Patra, B. (2018). An Introduction to Integral Transforms, CRC Press, Boca Raton.

[230] Perel, M.V., and Sidorenko, M.S. (2009). Wavelet-based integral representation for solutions of the wave equation, J. Phys. A: Math. Theor. 42, 375211.

[231] Pinsky, M. (2001). Introduction to Fourier Analysis and Wavelets, Brooks-Cole, USA.

[232] Poularikas, A.D. (2010). Handbook of Transforms and Applications, Third Edition, CRC Press, Boca Raton.

[233] Prabhu, K.M. (2013). Window Functions and Their Applications in Signal Processing, CRC Press, Boca Raton.

[234] Prasad, L., and Iyengar, S.S. (1997). Wavelet Analysis with Applications to Image Processing, CRC Press, Boca Raton.

[235] Prasad, A., Manna, S., Mahato, A., and Singh, V.K. (2014). The generalized continuous wavelet transform associated with the fractional Fourier transform, J. Comput. Appl. Math. 259, 660-671.

[236] Prasad, A., and Kumar, P. (2016). Composition of the continuous fractional wavelet transforms, Nat. Acad. Sci. Lett. 39(2), 115-120.

[237] Prasad, A., and Mandal, U.K. (2017). Wavelet transforms associated with the Kontorovich-Lebedev transform, Int. J. Wavelets Multiresolut. Inf. Process. 15(2), 1750011.

[238] Prasad, A., and Singh, R. (2017). Bessel wavelet transform on the spaces with exponential growth, Filomat 31(8), 2459-2466.

[239] Prasad, A., and Mahato, A. (2018). Canonical Hankel wavelet transformation and Calderon's reproducing formula, Filomat 32(8), 2735-2743.

[240] Prasad, A., and Verma, S.K. (2018). Continuous wavelet transform associated with zero order Mehler-Fock transform and its composition, Int. J. Wavelets Multiresolut. Inf. Process. 16(6), 1850050.

[241] Prasad, A., and Sharma, P.B. (2020). The quadratic-phase Fourier wavelet transform, Math. Methods Appl. Sci. 43(4), 1953-1969.

[242] Prasad, A., and Singh, R. (2020). Abelian theorems for the Bessel wavelet transform, J. Anal. 28, 179-190.

[243] Resnikoff, H.L., and Wells, R. O. (1998). Wavelet Analysis: The Scalable Structure of Information, Springer, New York.

[244] Riba, L., and Wong, M.W. (2013). Continuous inversion formulas for multi-dimensional Stockwell transforms, Math. Model. Nat. Phenom. 8, 215-229.

[245] Riba, L., and Wong, M.W. (2015). Continuous inversion formulas for multi-dimensional modified Stockwell transforms, Integ. Transf. Spec. Funct. 26, 9-19.

[246] Rieder, A. (1991). The wavelet transform on Sobolev spaces and its approximation properties, Numer. Math. 58, 875-894.

[247] Saitoh, S. (1983). Hilbert spaces induced by Hilbert space valued functions, Proc. Amer. Math. Soc. 89, 74-78.

[248] Saitoh, S. (1983). The Weierstrass transform and an isometry in the heat equation, Applicable Analysis 16, 1-6.

[249] Saitoh, S. (1997). Integral Transforms, Reproducing Kernels and their Applications, Pitman Research Notes in Mathematics Series 369, Addison Wesley Longman, London.

[250] Sejdić, E., Djurović, I., and Stanković, L. (2011), Fractional Fourier transform as a signal processing tool: An overview of recent developments, Signal Process. 91, 1351-1369.

[251] Shah, F.A., and Debnath, L. (2018). Fractional wavelet frames in $L^2(\mathbb{R})$, Fract. Calculus Appl. Anal. 21(2), 399-422.

[252] Shah, F.A., Ahmad, O., and Jorgensen, P.E. (2018). Fractional wave packet systems in $L^2(\mathbb{R})$, J. Math. Phys. 59, 1-20.

[253] Shah, F.A., and Tantary, A.Y. (2018). Polar wavelet transform and the associated uncertainty principles. Int. J. Theor. Phys. 57(6), 1774-1786.

[254] Shah, F.A., and Tantary, A.Y. (2018). Quaternionic shearlet transform, Optik. 175, 115-125.

[255] Shah, F.A., and Tantary, A.Y. (2020). Linear canonical Stockwell transform, J. Math. Anal. Appl. 443, 1-28.

[256] Shah, F.A. and Tantary, A.Y. (2021). Lattice-based multi-channel sampling theorem for linear canonical transform, Digit. Signal Process. 117, 103168.

[257] Shah, F.A. and Tantary, A.Y. (2021). Non-isotropic angular Stockwell transform and the associated uncertainty principles, Appl Anal. 100(4), 835-859.

[258] Shah, F.A., Tantary, A.Y., and Zayed, A.I. (2021). A convolution-based special affine wavelet transform, Integ. Trans. Special Funct. 32(10), 780-800.

[259] Shah, F.A., Teali, A.A., and Tantary, A.Y. (2021). Special affine wavelet transform and the corresponding Poisson summation formula, Int. J. Wavelets Multiresolut. Inf. Process. 19(3), 2050086.

[260] Shah, F.A., Lone, W.Z., and Tantary, A.Y. (2021). Short-time quadratic-phase Fourier transform, Optik 245, 167689.

[261] Shah, F.A., Nisar, K.S., Lone, W.Z., and Tantary, A.Y. (2021). Uncertainty principles for the quadratic-phase Fourier transforms, Math. Methods Appl. Sci. 44, 10416-10431.

[262] Shah, F.A., and Lone, W.Z. (2022). Quadratic-phase wavelet transform with applications to generalized differential equations, Math. Methods Appl. Sci. 45(3), 1153-1175.

[263] Shah, F.A., Qadri, H.L., and Lone, W.Z. (2022). Non-separable windowed linear canonical transform, Optik 251, 168192.

[264] Shah, F.A., and Tantary, A.Y. (2022). Multidimensional linear canonical transform with applications to sampling and multiplicative filtering, Multidimens. Sys. Signal Process. 33, 621-650.

[265] Shah, F.A., and Tantary, A.Y. (2022). Sampling and multiplicative filtering associated with the quadratic-phase Fourier transforms, Signal Image Video Process. (to appear).

[266] Shannon, C.E. (1948). A mathematical theory of communication, Bell System Tech. J. 27, 623-656.

[267] Shi, J., Zhang, N., and Liu, X. (2012). A novel fractional wavelet transform and its applications, Sci. China Inf. Sci. 55, 1270-1279.

[268] Shi, J., Sha, X.J., Song, X., and Naitong, Z. (2014). Generalized convolution theorem associated with fractional Fourier transform, Wirel. Commun. Mob. Comput. 14, 1340-1351.

[269] Shi, J., Liu, X., and Zhang, N. (2014). Generalized convolution and product theorems associated with linear canonical transform, Signal Image Video Process. 8, 967-974.

[270] Shi, J., ·Liu, X., and Zhang, N. (2015). Multiresolution analysis and orthogonal wavelets associated with fractional wavelet transform. Signal Image Video Process. 9, 211-220.

[271] Shi, J., Liu, X., Sha, X., Zhang, Q., and Naitong, Z. (2017). A sampling theorem for fractional wavelet transform with error estimates, IEEE Trans. Signal Process. 65(18), 4797-4811.

[272] Shi, J., Liu, X., Xiang, W., Han, M., and Zhang, Q. (2020). Novel fractional wavelet packet transform: Theory, implementation, and applications, IEEE Trans. Signal Process. 68, 4041-4054.

[273] Shi, J., Zheng, J., Liu, X., Xiang, W., and Zhang, Q. (2020). Novel short-time fractional Fourier transform: Theory, implementation, and applications, IEEE Trans. Signal Process. 68, 3280-3295.

[274] Srivastava, H.M., Shah, F.A., and Tantary, A.Y. (2020). A family of convolution-based generalized Stockwell transforms, J. Pseudo-Differ. Oper. Appl. 11, 1505-1536.

[275] Srivastava, H.M., Shah, F.A., Garg, T.K., Lone, W.Z., and Qadri, H.L. (2021). Non-separable linear canonical wavelet transform, Symmetry 13, 2182.

[276] Srivastava, H.M., Singh, A., Rawat, A., and Singh, S. (2021). A family of Mexican hat wavelet transforms associated with an isometry in the heat equation, Math. Methods Appl. Sci. 44, 11340-11349.

[277] Starck, J.L., Candés E.J., and Donoho, D.L. (2002). The curvelet transform for image denoising, IEEE Trans. Image Process. 11, 670-684.

[278] Stern, A. (2006). Sampling of linear canonical transformed signals, Signal Process. 86, 1421-1425.

[279] Stern, A. (2007). Sampling of compact signals in offset linear canonical transform domains, Signal Image Video Process. 1, 359-367.

[280] Stern, A. (2008). Uncertainty principles in linear canonical transform domains and some of their implications in optics, J. Optical Soc. Amer. A 25(3), 647-652.

[281] Stockwell, R.G., Mansinha L., and Lowe, R.P. (1996). Localization of the complex sectrum: The S transform, IEEE Trans. Signal Process. 44, 998-1001.

[282] Stockwell, R.G. (2007). A basis for efficient representation of the S-transform, Digit. Signal Process. 17, 371-393.

[283] Strang, G., and Nguyen, T. (1996). Wavelets and Filter Banks, Wellesley-Cambridge Press, Wellesley, Massachusetts.

[284] Su, Y. (2016). Heisenberg type uncertainty principle for continuous shearlet transform, J. Nonlinear Sci. Appl. 9, 778-786.

[285] Sweldens, W. (1996). Wavelets: What next? Proc. IEEE 84(4), 680-685.

[286] Tantary A.Y., and Shah F.A. (2020). An intertwining of curvelet and linear canonical transforms, J. Math. 8814998.

[287] Tao, R., Deng, B., and Wang, Y. (2009). Fractional Fourier Transform and Its Applications, Tsinghua University Press, Beijing.

[288] Tao, R., Li, X.M., Li, Y.L., and Wang, Y. (2009). Time-delay estimation of chirp signals in the fractional Fourier domain, IEEE Trans. Signal Process. 57(7), 2852-2855.

[289] Tao, R., Li, Y. L., and Wang, Y. (2010). Short-time fractional Fourier transform and its applications, IEEE Trans. Signal Process. 58, 2568-2580.

[290] Torrésani, B. (1991). Wavelets associated with representations of the affine Weyl-Heisenberg group, J. Math. Phys. 32, 1273-1279.

[291] Torres, R., Pellat-Finet, P., and Torres, Y. (2010). Fractional convolution, fractional correlation and their translation invariance properties, Signal Process. 90(6), 1976-1984.

[292] Unser, M. (2000). Sampling-50 years after Shannon, Proc. IEEE 88(4), 569-587.

[293] Unser, M., and Aldroubi, A. (1996). A review of wavelets in biomedical applications, Proceed. IEEE 84(4), 626-638.

[294] Upadhyay, S.K., and Tripathi, A. (2015). Calderon's reproducing formula for Watson wavelet transform, Indian J. Pure Appl. Math. 46(3), 269-277.

[295] Upadhyay, S.K., Singh, R., and Tripathi, A. (2017). The relation between Bessel wavelet convolution product and Hankel convolution product involving Hankel transform, Int. J. Wavelets Multiresolut. Inf. Process. 15, 1750030.

[296] Upadhyay, S.K., and Khatterwani, K. (2019). Fractional wavelet transform through heat equation, J. Therm. Stress 42(11), 1386-1414.

[297] Vaidyanathan, P.P. (1993). Multirate Systems and Filter Banks, Prentice-Hall, Englewood Cliffs, New Jersey.

[298] Ventosa, S., Simon, C., Schimmel, M., Dañobeitia, J.J., and Mánuel, A. (2008). The *S*-transform from a wavelet point of view, IEEE Trans. Signal Process. 56(7), 2771-2780.

[299] Vera, D. (2013). Triebel-Lizorkin spaces and shearlets on the cone in $\mathbb{R}^2$, Appl. Comput. Harmon. Anal. 35, 130-150.

[300] Villasenor, J. (1990). Wavelet transform via the Hartley transform, Twenty-Fourth Asilomar Conference on Signals, Systems and Computers, 98-102.

[301] Vyas, A., Yu, S., and Paik, J. (2018). Multiscale Transforms with Application to Image Processing, Springer, Singapore.

[302] Walker, J.S. (1999). A Primer on Wavelets and Their Scientific Applications, Chapman & Hall/CRC Press LLC, Boca Raton, Florida.

[303] Walnut, D.F. (2002). An Introduction to Wavelet Analysis, Birkhäuser, Boston.

[304] Walter, G.G. (1994). Wavelets and Other Orthogonal Systems with Applications, CRC Press, Boca Raton.

[305] Wang, J., Wang, Y., Wang, W., and Ren, S. (2018). Discrete linear canonical wavelet transform and its applications, EURASIP J. Adv. Signal Process. 29, 1-18.

[306] Wang, J., Ding, Y., Ren, S., and Wang, W. (2019). Sampling and reconstruction of multi-band signals in multiresolution subspaces associated with the fractional wavelet transform, IEEE Signal Process. Lett. 26, 174-178.

[307] Wang, R. (2012). Introduction to Orthogonal Transforms, Cambridge University Press, Cambridge, U.K.

[308] Welland, G. (2003). Beyond Wavelets, Academic Press, New York.

[309] Wei, D., Ran, Q., and Li, Y. (2012). A convolution and correlation theorem for the linear canonical transform and its application, Circuits Syst. Signal Process. 31(1), 301-312.

[310] Wei, D., and Li, Y.M. (2014). Generalized wavelet transform based on the convolution operator in the linear canonical transform domain, Optik 125, 4491-4496.

[311] Wei, D., and Zhang, Y. (2021). Fractional Stockwell transform: Theory and applications, Digit. Signal Process. 115, 103090.

[312] Wiener, N. (1956). I Am a Mathematician, MIT Press, Cambridge.

[313] Wilczok, E. (2000). New uncertainty principles for the continuous Gabor transform and the continuous wavelet transform, Doc. Math. 5, 201-226.

[314] Willett, R., and Nowak, K. (2003). Platelets: A multi-scale approach for recovering edges and surfaces in photon-limited medical imaging, IEEE Trans. Med. Imaging 22(3), 332-350.

[315] Wojtaszczyk, P. (1997). A Mathematical Introduction to Wavelets, Cambridge University Press, Cambridge.

[316] Wong, M.W. (2002). Wavelet Transforms and Localization Operators, Birkhäuser, New York.

[317] Xiang, Q., and Qin, K. (2014). Convolution, correlation, and sampling theorems for the offset linear canonical transform, Signal Image Video Process. 8(3), 433-442.

[318] Xu, J., Yang, L., and Wu, D. (2010). Ripplet-a new transform for image processing, J. Vis. Commun. Image Represent. 21(7), 627-639.

[319] Xu, J., and Wu, D. (2012). Ripplet transform type II transform for feature extraction, IET Image Process. 6(4), 374-385.

[320] Xu, T.Z., and Li, B.Z. (2013). Linear Canonical Transform and Its Applications, Science Press, Beijing, China.

[321] Yang, X.J. (2022). Theory and Applications of Special Functions for Scientists and Engineers, Springer, Singapore.

[322] Yin, M., and He, J. (2014). Radial wavelet and Radon transform on the Heisenberg group, Appl. Anal. 93, 1-13.

[323] Zayed, A.I. (1993). Advances in Shannon's Sampling Theory, CRC Press, Boca Raton, Florida.

[324] Zayed, A.I. (1996). Handbook of Function and Generalized Function Transformations, CRC Press, New York.

[325] Zayed, A.I. (1996). On the relationship between the Fourier and fractional Fourier transforms, IEEE Signal Process. Lett. 3(12), 310-311.

[326] Zayed, A.I. (2018). Two-dimensional fractional Fourier transform and some of its properties, Integ. Transf. Special Funct. 29(7), 553-570.

[327] Zayed, A.I. (2018). Sampling of signals bandlimited to a disc in the linear canonical transform domain, IEEE Signal Process. Lett. 25(12), 1765-1769.

[328] Zayed, A.I. (2019). A new perspective on the two-dimensional fractional Fourier transform and its relationship with the Wigner distribution, J. Fourier Anal. Appl. 25(2), 460-487.

[329] Zayed, A.I., and Garcíía, A.G. (1999). New sampling formulae for the fractional Fourier transform, Signal Process. 77, 111-114.

[330] Zayed, A.I., and Roopkumar, R. (2020). Multidimensional fractional Fourier transform and generalized fractional convolution, Integ. Transf. Special Funct. 31(2), 152-165.

[331] Zhang, Z.C. (2015). Sampling theorem for the short-time linear canonical transform and its applications, Optik 113, 138-146.

[332] Zhang, Z.C. (2016). New convolution and product theorem for the linear canonical transform and its applications, Optik 127, 4894-4902.

[333] Zhang, Z.C. (2019). Linear canonical transform's differentiation properties and their application in solving generalized differential equations, Optik 188, 287-293.

[334] Zhao, J., Tao, R., Li, Y., and Wang, Y. (2009). Uncertainty principles for linear canonical transform, IEEE Trans. Signal Proces. 57(7), 2856-2858.

[335] Zhi, X., Wei, D., and Zhang, W. (2016). A generalized convolution theorem for the special affine Fourier transform and its application to filtering, Optik 127(5), 2613-2616.

Index

Milton Keynes UK
Ingram Content Group UK Ltd.
UKHW020845141024
449569UK00003B/76

9 781032 007960